Diversity, Ecology and Evolution of Odonata

Diversity, Ecology and Evolution of Odonata

Editors

M. Olalla Lorenzo-Carballa
Ricardo Koroiva

MDPI

Basel • Beijing • Wuhan • Barcelona • Belgrade • Novi Sad • Cluj • Manchester

Editors
M. Olalla Lorenzo-Carballa
CRETUS, Department of Zoology, Genetics and Physical Anthropology
Universidade de Santiago de Compostela
Santiago de Compostela
Spain

Ricardo Koroiva
Biological Science Institute,
Federal University of Pará
Belém
Brazil

Editorial Office
MDPI
St. Alban-Anlage 66
4052 Basel, Switzerland

This is a reprint of articles from the Special Issue published online in the open access journal *Diversity* (ISSN 1424-2818) (available at: https://www.mdpi.com/journal/diversity/special_issues/diversity_odonata).

For citation purposes, cite each article independently as indicated on the article page online and as indicated below:

Lastname, A.A.; Lastname, B.B. Article Title. *Journal Name* **Year**, *Volume Number*, Page Range.

ISBN 978-3-7258-0709-3 (Hbk)
ISBN 978-3-7258-0710-9 (PDF)
doi.org/10.3390/books978-3-7258-0710-9

Cover image courtesy of Ricardo Koroiva

Contents

diversity

Editorial

A Special Issue on the Diversity, Ecology and Evolution of Dragonflies and Damselflies (Insecta: Odonata)

M. Olalla Lorenzo-Carballa [1,*] and Ricardo Koroiva [2]

1 CRETUS, Department of Zoology, Genetics and Physical Anthropology, Universidade de Santiago de Compostela, 15782 Santiago de Compostela, Spain
2 Biological Science Institute, Federal University of Pará, Belém 66075-110, Pará, Brazil; ricardo.koroiva@gmail.com
* Correspondence: m.o.lorenzo.carballa@gmail.com

1. Introduction

The Odonata is an order of insects commonly known as dragonflies and damselflies, with a worldwide distribution except in Antarctica. The group includes more than 6300 species, which are divided into three suborders [1]: the Zygoptera (damselflies), which have a slender body and hold their wings closed over the back when at rest; the Anisoptera (dragonflies), which have a thicker body and hold their wings perpendicular to the body when at rest; and the Anisozygoptera, a smaller suborder that includes four species that have intermediate characteristics between dragonflies and damselflies (Figure 1).

Figure 1. Specimens from three suborders: (**A**) Anisoptera, (**B**) Anisozygoptera, and (**C**) Zygoptera. Reproduced with permission from Adolfo Cordero-Rivera and Diogo Silva Vilela.

Members of the order Odonata play an important ecological role as predators (see [2]). They consume large volumes of prey such as mosquitoes, flies, and other small insects, thus contributing to the regulation of insect populations; they also constitute important prey for larger animals such as birds, fish, and amphibians. In addition, Odonata occupy an important evolutionary position, being one of the oldest groups of flying insects, with fossils dating back more than 300 million years [3]. Due to their sensitivity to environmental changes, Odonata are used as bioindicators for freshwater quality assessment, wetland

Citation: Lorenzo-Carballa, M.O.; Koroiva, R. A Special Issue on the Diversity, Ecology and Evolution of Dragonflies and Damselflies (Insecta: Odonata). *Diversity* **2024**, *16*, 117. https://doi.org/10.3390/d16020117

Received: 7 February 2024
Accepted: 9 February 2024
Published: 12 February 2024

protection monitoring, and environmental impact assessment. Finally, the beauty and diversity of Odonata species also make them attractive to the general public, which has led to an increased interest in both their natural history and conservation [4].

In this Special Issue of *Diversity*, we invited fellow odonatologists to share their research findings on some of the most important topics related to the knowledge of this insect order: Diversity, Ecology, and Evolution. The twenty-one articles published in this issue address all of these topics to some extent.

2. Diversity

This section of the Special Issue opens with the article "Diversity of Palaearctic Dragonflies and Damselflies (Odonata)" [5], in which the authors present the results of a study in which they created distribution models for 402 dragonfly and damselfly species in the Palaearctic region, the largest biogeographical realm area, using more than 1.2 million distribution data. The study revealed a clear pattern of longitudinally declining diversity, with lentic species dominating in colder and drier areas. The article highlights the importance of understanding diversity patterns and threats to dragonflies in the Palearctic region for the conservation of these important insect species.

The second article, "Diversity of Nearctic Dragonflies and Damselflies (Odonata)" [6], presents a study on the diversity of dragonflies and damselflies in the Nearctic, a biogeographic region that includes Canada, the United States, and Mexico. The study used species distribution modeling and found greater species richness in the eastern part of the region and high levels of endemism in the southeastern United States, likely due to glacial refugia. The study emphasizes the importance of understanding the aquatic life cycle and requirements of dragonflies and damselflies to understand the distribution patterns of their diversity in the Nearctic.

The article "Taxonomic revision of eastern part of Western Palaearctic *Cordulegaster* using molecular phylogeny and morphology, with the description of two new species (Odonata: Anisoptera: Cordulegastridae)" [7] combines molecular and morphological techniques to clarify the taxonomy of the genus *Cordulegaster* Leach in Brewster, 1815. This study confirms the existence of the two traditional groups (*boltonii* and *bidentata*) and describes little-known or new taxa with their phenotypic variation. The molecular analyses support three known and one new species in the *boltonii* group, and a complex of four closely related species in the *bidentata* group, with one additional new species described for this group. The study also provides an identification key for all western Palaearctic *Cordulegaster*. Its importance for dragonfly research lies in its contribution to the taxonomic knowledge of this genus in the Eastern part of the Western Palaearctic region.

The article "Molecular and morphological analyses support different taxonomic units for Asian and Australo-Pacific forms of *Ischnura aurora* (Odonata, Coenagrionidae)" [8] presents the results of a study on the taxonomic status of *Ischnura aurora* (Brauer, 1865), a dragonfly species that has been the subject of taxonomic debate for many years. Using morphological and DNA sequencing analyses, this study concludes that the Australo-Pacific and Asian forms of *I. aurora* should be separated into two distinct species: *I. aurora* and *I. rubilio*, respectively. Furthermore, the study highlights the need to review all available material of the different subspecies of *I. aurora* and emphasizes the importance of the careful examination of DNA sequence data and voucher specimens in taxonomic studies to avoid specimen misidentifications and/or the amplification of non-orthologous copies of mitochondrial gene markers.

Finally, the article "DNA barcoding and new records of odonates (Insecta: Odonata) from Paraíba state, Brazil" [9] reports the results of a study of Odonata species in the Brazilian state of Paraíba and the creation of a DNA barcoding database for 70% of the species found in this state. The results show that the use of the COI fragment at the regional level can help in the identification and delimitation of species, and that morphological characteristics can be used to further confirm such identifications. The establishment of this DNA barcode library is an important milestone for the taxonomy and conservation of the biodiversity of Neotropical odonate species.

3. Ecology

The Ecology section opens with the article "Climate change is driving shifts in dragonfly species richness across Europe via differential dynamics of taxonomic and biogeographic groups" [10], which discusses the importance for conservation of understanding how changes in species richness correlate with changes in the range of different taxonomic and biogeographic groups. This study found that large-scale changes in dragonfly species richness are the result of several divergent dynamics that differ for taxonomic suborders and biogeographic groups, with thermal releases during climate-driven range expansion shifting local species richness across Europe. Dragonflies are proving to be important indicators of environmental status and conservation needs, highlighting the relevance of the topic to the Odonata field.

The article "Niche breadth predicts geographical range size and northern range shift in European dragonfly species (Odonata)" [11] discusses the relationship between niche breadth, range size, and range shifts in European dragonflies over a 22-year period. The study found that stream species with narrower niches and smaller ranges are more vulnerable to habitat loss and climate change, while species living in temporary water bodies are more resilient to climate change than species living in permanent water bodies. The results suggest that ongoing climate change and changes in land use will mostly affect species with narrow habitat requirements, leading to biotic homogenization in which specialists are displaced and replaced by generalists. The results of the study have important implications for understanding the effects of climate change and land use on dragonfly populations and can be used to better inform conservation measures.

The article "Population density and abundance of the northernmost population of *Cordulegaster heros* (Anisoptera: Cordulegastridae) in Europe (Czech Republic) with notes on its biogeographical range" [12], focuses on a long term study focused on characterizing habitat preferences and population trends of the species *Cordulegaster heros*, a Balkan endemic species. The authors found that current habitat conditions ensure the persistence of this species' populations in its northernmost distribution area, which are therefore evaluated as viable and stable. They stress the need to monitor the response of these populations to major interventions in the streams' catchment áreas, and to continue monitoring the trends in population abundance.

In the article "Detection and monitoring of riverine dragonfly of community interest (Insecta: Odonata): proposal for a standardised protocol based on exuviae collection" [13], the authors propose a protocol for monitoring riverine dragonfly species, based on the sampling of exuviae. To assess the suitability of the protocol, they carried out extensive samplings focusing on the species *Oxygastra curtisii*, *Macromia splendens*, and *Gomphus graslinii*; which are currently listed in the European Habitats Directive. Establishing robust sampling methodologies to quantify species' populations is important to ensure that decision-makers' judgments are well-founded when informing conservation decisions; which highlights the relevance of this study.

"Similar response of a range expanding dragonfly to low- and high-elevation predators" [14] describes a common garden experiment with the dragonfly species *Sympetrum striolatum* to test whether they can cope with new biotic interactions in expanding ranges. The study found that the dragonflies responded similarly to low- and high-elevation predators in terms of growth and feeding, but they responded more significantly to the familiar predator in terms of morphology. These results suggest that species that expand their range can successfully colonize new areas at higher elevations because they respond similarly to dominant predators at high elevations as they do to familiar predators at low elevations.

The study "First record of microsporidia infection in the damselfly *Ischnura elegans* larvae: temperature and predator cue effects on the host's life history" [15] reports on the first microsporidia infection in laboratory-reared damselfly *Ischnura elegans* larvae from adult females collected in the field in Poland. The study found that higher rearing temperatures and predation cues from alien signal crayfish increased the number of infected larvae leading to distorted wing development and death before hatching. The results suggest that ecologically challenging conditions can increase the risk of parasitism and emphasize the importance of considering the effects of microsporidian infection on dragonflies, which are often used as model organisms in eco-evolutionary studies.

The article "Evaluating potential distribution and niche divergence among populations of the world's largest living damselfly, *Megaloprepus caerulatus* (Drury, 1782)" [16] reports on a study of ecological divergence between populations of *Megaloprepus caerulatus* in Mexico, Costa Rica, and Panama. The authors used Ecological Niche Modeling (ENM) to compare potential distribution ranges and found evidence of strong ecological divergence between the Corcovado and Barro Colorado populations. The study lays the stage for further research on the factors driving niche divergence, and the effects of anthropogenic land use change and climate change on the distribution and conservation of this emblematic Neotropical species.

Two articles in this section dealt with land use for pasture and cocoa cultivation in the Brazilian state of Bahia. The first one, entitled "Land uses for pasture and cacao cultivation modify the Odonata assemblages in Atlantic Forest areas" [17], investigates the effects of changes in land use on Odonata assemblages in the Atlantic Forest of Brazil. The study found that changes such the conversion of original forest areas to pasture significantly alter the richness and composition of Odonata assemblages. The study shows the importance of conserving riparian vegetation and implementing sustainable land use practices to protect aquatic ecosystems and biodiversity. The second article, "Dragonflies (Odonata) in cocoa growing areas in the Atlantic Forest: taxonomic diversity and relationships with environmental and spatial variables" [18], investigated the impact of cocoa cultivation on Odonata assemblages and determined the relationship between different life stages of Odonata and local and spatial environmental variables. The study found that agroforestry cabruca areas where cocoa is grown in the shade of native trees harbor a variety of dragonfly species, including forest specialists, and that local and spatial environmental characteristics are important factors in structuring these assemblages. The results emphasize the importance of this agroforestry system for the conservation of dragonfly species in areas of the Atlantic Forest.

In "Odonata from Iberá wetland system (Corrientes, Argentina), are regional biogeographic schemes useful to assess Odonata biodiversity and its conservation?" [19], the authors analyzed the distribution patterns of Odonata species in the wetlands of the Iberá depression in Argentina to determine whether this region functions as an ecological and functional unit. They found that the Iberá Depression is not a functional unit and that Odonata respond to specific physical features of the wetlands rather than to biogeographical or ecoregional schemes. This study is important for dragonfly research because it emphasizes the need to understand the specific environmental factors that influence their distribution and the importance of local conservation measures for these species.

4. Evolution

The section on Evolution opens with the article "Genetic diversity and structure of *Anax imperator* Leach, 1815 populations (Odonata: Aeshnidae) in ponds at regional and European scales" [20], which examines the genetic diversity and structure of *Anax imperator* populations in Europe using microsatellite markers. The study found high gene flow at both regional and European scales and no pattern of isolation by distance, indicating historical or recent movements of individuals. The results highlight the potential role of the English Channel as a barrier to gene flow and the need for further studies to investigate the relationship between individuals and major wind currents. Overall, the study provides insights into the gene flow and dispersal patterns of *A. imperator* that may aid in the conservation of the species and the management of fragmented habitats.

The article "Evolution and biogeographic history of rubyspot damselflies (Hetaerininae: Calopterygidae: Odonata)" [21] presents the results of a study on the biogeography, ecology, and color evolution of Neotropical Hetaerininae damselflies. The study concludes that the genus *Hetaerina* is paraphyletic and that a reclassification of the genera within the Hetaerininae is needed. The study also provides evidence for a gradual dispersal of the Hetaerininae from North to South America that began in the Oligocene and ended in the Pliocene, and suggests that the expansion of the Isthmus of Panama during the Oligocene contributed to their dispersal. The relevance of the topic lies in its contribution to our understanding of the relationship between morphology, biogeography, and habitat in a charismatic group of damselflies.

The study "Testing the effect of Sampling effort on inferring phylogeographic history in *Psolodesmus mandarinus* (Calopterygidae, Odonata)" [22] discusses the effect of sampling design on phylogeographic inference and its implications for the study of spatial genetic structure and evolutionary units. The authors demonstrate the importance of a comprehensive sampling design for understanding the phylogeographic history of a dwarf dragonfly endemic to Taiwan, *Psolodesmus mandarinus*, and point out the potential bias in inferring the effects of isolation by a physical barrier. The study highlights the need to use careful spatial sampling strategies in future phylogeographic studies and to test the effects of sampling on the resulting inferences, which is crucial for progress in the field of dragonfly genetics and evolution.

The article "The quality of sequence data affects biodiversity and conservation perspectives in the Neotropical Damselfly *Megaloprepus caerulatus*" [23] discusses the importance of high-quality raw sequence data for species delimitation and discovery in odonate research, using the Neotropical damselfly genus *Megaloprepus* as an example. The study compares two sets of sequence markers used in previous research and identifies unresolved features and internal gaps as reasons for the different results in species delimitation and population genetic relationships. The article emphasizes the importance of accurate species delimitation for conservation management, especially for sensitive species such as those within the genus *Megaloprepus*.

In "Meiotic analysis of Gomphidae species sheds light on the large X chromosome of the family (Anisoptera, Odonata)" [24], a hypothesis about the original diploid number and sex-determining systems in the dragonfly family Gomphidae is proposed. The study analyzes the meiosis and heterochromatin characteristics of three species of Gomphidae from Argentina and suggests that the diploid number of the family was 23 and the large size of the sex chromosome is due to an increase in heterochromatin rather than structural rearrangements. The results of the study and the proposed hypothesis offer new insights into the evolutionary history and sex-determining mechanisms in Odonata.

5. Review Article

The Special Issue closes with a bibliometric analysis of Odonata research at a global scale, entitled: "A bibliometric analysis of the global research in Odonata: trends and gaps" [25]. The authors present the results of a study on the patterns of research on dragonflies and damselflies over the last ten years based on a bibliometric analysis of publications. They show that the number of publications on Odonata has increased, with ecology, taxonomy, and behavior being the main topics of the studies published in this time period. However, there are still some research gaps, especially in basic biology, biogeography, and knowledge about the larval stage of Odonata. This review also highlights the importance of increasing research efforts in neglected areas to better understand species' responses to various factors and to expand the necessary background information for other types of studies.

Author Contributions: Conceptualization, M.O.L.-C. and R.K.; data curation, M.O.L.-C. and R.K.; writing—original draft preparation, M.O.L.-C. and R.K.; writing—review and editing, M.O.L.-C. and R.K.; project administration, M.O.L.-C. and R.K. All authors have read and agreed to the published version of the manuscript.

Funding: R.K. is partially financed by National Council for Scientific and Technological Development (CNPq/Universal- 405218/2023-0) and the Coordenação de Aperfeiçoamento de Pessoal de Nível Superior—Brasil (CAPES)—Finance Code 001.

Institutional Review Board Statement: Not applicable.

Data Availability Statement: Not applicable.

Conflicts of Interest: The authors declare no conflicts of interest.

References

1. Paulson, D.; Schorr, M.; Deliry, C. World Odonata List. Available online: https://www.pugetsound.edu/puget-sound-museum-natural-history/biodiversity-resources/insects/dragonflies/world-odonata-list (accessed on 5 February 2024).
2. May, M. Odonata: Who They Are and What They Have Done for Us Lately: Classification and Ecosystem Services of Dragonflies. *Insects* **2019**, *10*, 62. [CrossRef] [PubMed]
3. Kohli, M.; Letsch, H.; Greve, C.; Béthoux, O.; Deregnaucourt, I.; Liu, S.; Zhou, X.; Donath, A.; Mayer, C.; Podsiadlowski, L.; et al. Evolutionary History and Divergence Times of Odonata (Dragonflies and Damselflies) Revealed through Transcriptomics. *iScience* **2021**, *24*, 103324. [CrossRef] [PubMed]
4. Abbott, J.C. *Dragonflies of Texas: A Field Guide*; University of Texas Press: Austin, TX, USA, 2015.
5. Kalkman, V.J.; Boudot, J.-P.; Futahashi, R.; Abbott, J.C.; Bota-Sierra, C.A.; Guralnick, R.; Bybee, S.M.; Ware, J.; Belitz, M.W. Diversity of Palaearctic Dragonflies and Damselflies (Odonata). *Diversity* **2022**, *14*, 966. [CrossRef]
6. Abbott, J.C.; Bota-Sierra, C.A.; Guralnick, R.; Kalkman, V.; González-Soriano, E.; Novelo-Gutiérrez, R.; Bybee, S.; Ware, J.; Belitz, M.W. Diversity of Nearctic Dragonflies and Damselflies (Odonata). *Diversity* **2022**, *14*, 575. [CrossRef]
7. Schneider, T.; Vierstraete, A.; Müller, O.; van Pelt, G.J.; Caspers, M.; Ikemeyer, D.; Snegovaya, N.; Dumont, H.J. Taxonomic Revision of Eastern Part of Western Palaearctic Cordulegaster Using Molecular Phylogeny and Morphology, with the Description of Two New Species (Odonata: Anisoptera: Cordulegastridae). *Diversity* **2021**, *13*, 667. [CrossRef]
8. Lorenzo-Carballa, M.O.; Sanmartín-Villar, I.; Cordero-Rivera, A. Molecular and Morphological Analyses Support Different Taxonomic Units for Asian and Australo-Pacific Forms of Ischnura Aurora (Odonata, Coenagrionidae). *Diversity* **2022**, *14*, 606. [CrossRef]
9. Koroiva, R.; Gomes, V.G.N.; Vilela, D.S. DNA Barcoding and New Records of Odonates (Insecta: Odonata) from Paraíba State, Brazil. *Diversity* **2022**, *14*, 203. [CrossRef]
10. Olsen, K.; Svenning, J.-C.; Balslev, H. Climate Change Is Driving Shifts in Dragonfly Species Richness across Europe via Differential Dynamics of Taxonomic and Biogeographic Groups. *Diversity* **2022**, *14*, 1066. [CrossRef]
11. Olsen, K.; Svenning, J.-C.; Balslev, H. Niche Breadth Predicts Geographical Range Size and Northern Range Shift in European Dragonfly Species (Odonata). *Diversity* **2022**, *14*, 719. [CrossRef]
12. Holuša, O.; Holušová, K. Population Density and Abundance of the Northernmost Population of *Cordulegaster heros* (Anisoptera: Cordulegastridae) in Europe (Czech Republic) with Notes on Its Biogeographical Range. *Diversity* **2022**, *14*, 854. [CrossRef]
13. Arguel, L.; Denis, A.S.; Danflous, S.; Gouix, N.; Santoul, F.; Buisson, L.; Pelozuelo, L. Detection and Monitoring of Riverine Dragonfly of Community Interest (Insecta: Odonata): Proposal for a Standardised Protocol Based on Exuviae Collection. *Diversity* **2022**, *14*, 728. [CrossRef]
14. Zebsa, R.; Mahdjoub, H.; Khelifa, R. Similar Response of a Range Expanding Dragonfly to Low- and High-Elevation Predators. *Diversity* **2022**, *14*, 302. [CrossRef]
15. Antoł, A.; Labecka, A.M.; Larsson, J.I.R.; Sniegula, S. First Record of Microsporidia Infection in the Damselfly Ischnura Elegans Larvae: Temperature and Predator Cue Effects on the Host's Life History. *Diversity* **2022**, *14*, 428. [CrossRef]
16. Encarnación-Luévano, A.; Escoto-Moreno, J.A.; Villalobos-Jiménez, G. Evaluating Potential Distribution and Niche Divergence among Populations of the World's Largest Living Damselfly, Megaloprepus Caerulatus (Drury, 1782). *Diversity* **2022**, *14*, 84. [CrossRef]
17. Santos, L.R.; Rodrigues, M.E. Land Uses for Pasture and Cacao Cultivation Modify the Odonata Assemblages in Atlantic Forest Areas. *Diversity* **2022**, *14*, 672. [CrossRef]
18. Santos, L.R.; Rodrigues, M.E. Dragonflies (Odonata) in Cocoa Growing Areas in the Atlantic Forest: Taxonomic Diversity and Relationships with Environmental and Spatial Variables. *Diversity* **2022**, *14*, 919. [CrossRef]
19. del Palacio, A.; Lozano, F.; Ramos, L.S.; de las Mercedes Navarro, M.; Muzón, J. Odonata from Iberá Wetland System (Corrientes, Argentina) Are Regional Biogeographic Schemes Useful to Assess Odonata Biodiversity and Its Conservation? *Diversity* **2022**, *14*, 842. [CrossRef]
20. Minot, M.; Husté, A. Genetic Diversity and Structure of Anax Imperator Leach, 1815 Populations (Odonata: Aeshnidae) in Ponds at Regional and European Scales. *Diversity* **2022**, *14*, 68. [CrossRef]
21. Standring, S.; Sánchez-Herrera, M.; Guillermo-Ferreira, R.; Ware, J.L.; Vega-Sánchez, Y.M.; Clement, R.; Drury, J.P.; Grether, G.F.; González-Rodríguez, A.; Mendoza-Cuenca, L.; et al. Evolution and Biogeographic History of Rubyspot Damselflies (Hetaerininae: Calopterygidae: Odonata). *Diversity* **2022**, *14*, 757. [CrossRef]

22. Wang, L.-J.; Chou, Y.-W.; Huang, J.-P. Testing the Effect of Sampling Effort on Inferring Phylogeographic History in Psolodesmus Mandarinus (Calopterygidae, Odonata). *Diversity* **2022**, *14*, 809. [CrossRef]
23. Feindt, W.; Hadrys, H. The Quality of Sequence Data Affects Biodiversity and Conservation Perspectives in the Neotropical Damselfly Megaloprepus Caerulatus. *Diversity* **2022**, *14*, 1056. [CrossRef]
24. Mola, L.M.; Rebagliati, P.J.; Fourastié, M.F.; Agopian, S.S. Meiotic Analysis of Gomphidae Species Sheds Light on the Large X Chromosome of the Family (Anisoptera, Odonata). *Diversity* **2022**, *14*, 874. [CrossRef]
25. Oliveira-Junior, J.M.B.; Rocha, T.S.; Vinagre, S.F.; Miranda-Filho, J.C.; Mendoza-Penagos, C.C.; Dias-Silva, K.; Juen, L.; Calvão, L.B. A Bibliometric Analysis of the Global Research in Odonata: Trends and Gaps. *Diversity* **2022**, *14*, 1074. [CrossRef]

diversity

MDPI

Article

Diversity of Palaearctic Dragonflies and Damselflies (Odonata)

Vincent J. Kalkman [1,*], Jean-Pierre Boudot [2], Ryo Futahashi [3], John C. Abbott [4], Cornelio A. Bota-Sierra [4,5], Robert Guralnick [6], Seth M. Bybee [7], Jessica Ware [8] and Michael W. Belitz [6]

1 Naturalis Biodiversity Center, 2300 RA Leiden, The Netherlands
2 IUCN Odonata Specialist Group, 54710 Ludres, France
3 Bioproduction Research Institute, National Institute of Advanced Industrial Science and Technology (AIST), Central 6, Tsukuba 305-8566, Japan
4 Alabama Museum of Natural History & UA Museums Department of Research and Collections, The University of Alabama, Tuscaloosa, AL 35487, USA
5 Grupo de Entomología Universidad de Antioquia (GEUA), Universidad de Antioquia, Medellín 50010, Colombia
6 Florida Museum of Natural History, University of Florida, Gainesville, FL 32611, USA
7 Department of Biology and Monte L. Bean Museum, Brigham Young University, Provo, UT 84602, USA
8 Division of Invertebrate Zoology, American Museum of Natural History, New York, NY 10024, USA
* Correspondence: vincent.kalkman@naturalis.nl

Abstract: More than 1.2 million distribution records were used to create species distribution models for 402 Palaearctic species of dragonflies and damselflies. On the basis of these diversity maps of total, lentic and lotic diversity for the whole of the Palaearctic (excluding China and the Himalayan region) are presented. These maps show a clear pattern of decreasing diversity longitudinally, with species numbers dropping in the eastern half of Europe and remaining low throughout a large part of Russia, then increasing again towards Russia's Far East and Korea. There are clear differences in diversity patterns of lentic and lotic species, with lentic species being dominant in colder and more arid areas. Areas with a high diversity of species assessed as threatened on the IUCN red list are largely restricted to the Mediterranean, Southwest Asia, and Japan, with clear hotspots found in the Levant and the southern half of Japan. The diversity at species, generic, and family level is higher in the south of Japan than in areas at a similar latitude in the western Mediterranean. This is likely to be the result of the more humid climate of Japan resulting in a higher diversity of freshwater habitats and the stronger impact of the glacial periods in the Western Palaearctic in combination with the Sahara, preventing tropical African lineages dispersing northwards.

Keywords: zygoptera; anisoptera; species diversity; distribution; biodiversity and conservation; biogeographical patterns

Citation: Kalkman, V.J.; Boudot, J.-P.; Futahashi, R.; Abbott, J.C.; Bota-Sierra, C.A.; Guralnick, R.; Bybee, S.M.; Ware, J.; Belitz, M.W. Diversity of Palaearctic Dragonflies and Damselflies (Odonata). *Diversity* **2022**, *14*, 966. https://doi.org/10.3390/d14110966

Academic Editors: M. Olalla Lorenzo-Carballa and Ricardo Koroiva

Received: 2 October 2022
Accepted: 29 October 2022
Published: 11 November 2022

Publisher's Note: MDPI stays neutral with regard to jurisdictional claims in published maps and institutional affiliations.

1. Introduction

At the start of this century, a first effort was made to depict the global patterns of diversity of dragonflies and damselflies (Odonata) [1]. This paper shows a map of Europe and the world, indicating the estimated diversity per grid cell of 250 by 250 km^2. At the time when this study was undertaken, only a few countries had a database with distribution records and rarely were maps showing the diversity available. Now, twenty years later, the availability of distribution data has shown a strong increase with databases currently available for Africa, Europe, Australia, North America, and parts of Asia [2–5], and it seems likely that within a decade, such databases will span the global range of Odonata. However, this will not mean that distribution patterns of all species will be known, let alone understood, as these databases simply contain an overview of available records and, for many species, distribution maps will often illustrate a lack of field work for some areas.

A map of the diversity of dragonflies and damselflies for Europe based on distribution databases published in 2018 [6] shows that the diversity map of Europe as published by [1]

is fairly accurate, and data published for some other continents suggest that the same is true for their global maps. Those maps show a pattern which is found in many other animal groups, with the highest diversity found in tropical regions of Asia and America and slightly lower diversity in Africa, the latter probably due to long dry periods in the last million years in combination with the, at present, more seasonal and irregular rainfall in this continent [7].

Looking at temperate regions, a clear difference exists between the more species-rich Nearctic and the comparatively species-poor Palaearctic; it is apparent that large parts of the Palaearctic are among the least diverse regions for dragonflies and damselflies. Within the Palaearctic, the diversity patterns shown on the map of [1] are very crude, and diversity is depicted as being largely identical throughout much of the region, with a lower diversity in the far north and the arid region of Central Asia and a higher diversity in Japan. Diversity maps built from richer sources of updated distribution data will likely show more complex patterns at a finer resolution, better reflecting the historical and contemporary factors determining diversity. The key contemporary factors for odonates on the continental scale are temperature and precipitation [2]. Although temperature generally increases towards the south and decreases with altitude, temperature zones do not run fully longitudinally but show a clear trend, with temperature in the east lower than those at comparable longitudes in the west of the Palaearctic. Regarding aridity, in addition to the desert areas in the Middle East, Iran, parts of Central Asia, and the Gobi, there is also a large, more arid area found spanning most of the centre of the Palaearctic running from the east of the Ukraine through Kazakhstan where it narrowly connects with the arid region of Mongolia and surrounding regions.

The key historical factor shaping the odonate fauna of the Palaearctic is the periods of glaciation during which the northern parts of the Palaearctic were uninhabitable for all but the hardiest of odonates, while in the south, higher diversity was limited to a small number of refugia. The most recent glacial period ended only approximately 11,700 years ago (end of the Younger Dryas). Although ice sheets did not reach as far south and east of Eurasia during the last glacial maximum (21,000 years BP) as during previous glacial periods, large parts of the Palaearctic were, nonetheless, uninhabitable for dragonflies and damselflies during this period, meaning that in most of the Palaearctic, the odonate fauna is composed of species which arrived only in the past ~10,000 years. The impact of the glacial periods varied regionally in the Palaearctic, with the ice sheets reaching farther south in the west, extending to Berlin and Moscow, than to the east, where it hardly reached the Novaya Zemlya and the Severnaya Zemlya archipelagos.

In the east, lowered sea level resulted in a large expanse of land known as Beringia, running from eastern Siberia to Alaska, connecting Eurasia to America. During and at the end of the last glacial period, Beringia (like a large part of mid-latitude Europe), had a sufficiently mild climate with a grassland steppe vegetation (the famous Mammoth Steppe), due to which it served both as a refuge and a land bridge allowing faunal exchange between the Palaearctic and the Nearctic up to ca. 11,000 years BP when it was recovered by the ocean [8,9]. Regional differences in the impact of the glacial period are furthermore caused by geographical barriers with the largely east–west running mountain chains of the Pyrenees, the Alps, the Caucasus, and the Himalayan regions preventing species from retracting southwards during glacial periods. These east–west ranges also likely prevented species' northward expansions after the last glacial period. Additional barriers in the Western Palaearctic are formed by the Mediterranean Sea and, at present, by the belt of desert running from the Sahara to the Middle East, Central Asia, and the Gobi.

The current paper aims to provide improved diversity maps of the dragonflies and damselflies of the Palaearctic, making use of the large amounts of georeferenced distribution data which became available in the last two decades and best practice species-distribution modelling approaches. On the basis of these maps, we will address the following questions:

- Are there differences in diversity patterns shown by lotic and lentic species?
- Are there areas with relatively high endemism?

- Are there areas with a relatively high percentage of globally threatened species, and do these match with Odonate endemism and richness hotspots?

2. Materials and Methods

The methods used are largely identical to the paper on the diversity of Nearctic Dragonflies and Damselflies [2], for which reason the description of the methods is largely identical as well.

2.1. Definition of Palaearctic Realm

We follow the definition of [10] for the Palaearctic realm. The available data for the Palaearctic part of China, India, Nepal, and Bhutan does not reflect the true diversity in these regions, for which reason these areas have been excluded from our analyses and are shown on our maps in a uniform grey colour. In this paper, we refer to the complete Palaearctic (thus including the Palaearctic parts of China, India, Nepal, and Bhutan) as the Palaearctic realm, and our study area (the Palaearctic with exclusion of China, India, Nepal, and Bhutan) is referred to as Palaearctic.

2.2. Species Occurrence Data

For the Western Palaearctic, distribution data have been brought together, resulting in atlases for the Mediterranean and North Africa [11], Europe [3], and West and Central Asia [12]. Distribution data for Japan were derived from a database constructed by the National Biodiversity Center of Japan [13] to which data used for the maps presented in the field guide of the Japanese odonates were added [14]. For the intervening areas of South Korea, North Korea, Mongolia, and Russia, a database was created by J.-P.B.containing most of the published records from that area. In total, 1,292,642 data points (a species on a location) were available for a total 402 species. This includes all species found in the Palaearctic region, with the exception of those which are in the Palaearctic region found only in India, Nepal, Bhutan, or China.

While we included only species occurring within the Palaearctic realm, occurrences for those species with ranges outside the realm were included in our downstream modelling steps. Once the initial occurrence data for Palaearctic species were assembled, we ran the occurrence records through a cleaning pipeline in the R package *CoordinateCleaner* [15] that flagged records (1) with equal latitude and longitude coordinates, (2) within a 1000 m radius around the geographic centroids of political countries and provinces, and (3) with either zero longitude or latitude. Maps displaying both unflagged and flagged occurrence records were generated for each species for expert review. During this step, expert review (by V.J.K., J.-P.B. and R.F.) decided which occurrence records were removed from the database, generating a final dataset of curated occurrence records to be used for distribution modelling.

2.3. Functional Traits and Conservation Status

All species were categorised as being either lentic- or lotic-dependent on the basis of the literature and expert knowledge. The following questions were used to classify each species: Can the species survive without a lotic environment? Those species for which the answer was "no" were labelled as "lotic obligate", and when the answer was yes, they were labelled as lentic. In addition, information on the IUCN conservation category of the 402 species in the Palaearctic region was downloaded from the IUCN portal (www.iucnredlist.org, accessed on 2 March 2022). Threatened species included species with red list categories classified by the IUCN as either near-threatened, vulnerable, endangered, or critically endangered. To test whether the type of aquatic habitat (lotic or lentic) used by odonates has an effect on the overall range size, a Wilcoxon test was performed using R software (R core team), with, as the dependent variable, the range size predicted for every species measured by the total number of pixels in which the species was predicted to occur (see below). We expected lentic species to have larger range sizes, as had been hypothesised, since lentic bodies of water are likely more ephemeral, thus favouring species with more effective dispersal abilities; as a result, lentic species would have larger ranges [16,17].

2.4. Species Distribution Modelling

We built a species distribution pipeline in R to predict the distribution of all 402 species found in the Palaearctic. This pipeline was strategically designed to efficiently model the distributions of hundreds of species, while including multiple steps that customise the process for each species.

First, we defined the accessible area, which was the geographic area where the distribution model was both fit and projected, by generating a buffered alpha hull around the accepted occurrence records. The alpha hull was calculated using the getDynamicAlphaHull function from the R package *rangeBuilder* [18], where we set the fraction of occurrences that can fall outside of the polygon to zero, an initial alpha value of 20, and an allowed maximum of three disjunct polygons. We then buffered the alpha hull by the larger value of either 75 km or the 80th percentile distance between an occurrence record and the nearest occurrence records to ensure the accessible area included areas that are accessible to a species through time [19]. These hulls were vetted for quality by expert curators (V.J.K., J.-P.B. and R.F.).

Next, we spatially thinned the occurrence records to remove potential spatial biases, where certain areas had more records than other areas, which likely reflected differences in human sampling effort more than changes in relative abundance across a landscape. Spatial thinning of occurrence records has been demonstrated to improve species distribution models using low-structure data sources [20]. We calculated the area of each accessible area in square metres using the area function in the R package raster [21] and retained all data points if a species' accessible area was less than 100,000 km^2. If a species had an accessible area >= 100,000 km^2 and < 250,000 km^2, we retained only one occurrence record per 25 km grid; if accessible area was >= 250,000 km^2 and < 1,000,000 km^2, one record per 50 km grid was retained; if accessible area was >= 1,000,000 km^2 and < 2,500,000 km^2, one record per 100 km grid was retained; and if accessible area was >= 2,500,000 km^2, one record per 200 km grid was retained. Even with thinning, there were still issues with data biases, requiring further efforts to tune model outputs, as discussed below.

After generating species-specific accessible areas and spatially thinning the occurrence records, we fit an initial Maxent model [22] using default settings in the dismo package in R [23]. Maxent uses a machine learning algorithm to fit relationships between species occurrence records and background samples to environmental predictors [24]. Our initial model included 13 of the 19 bioclimatic variables provided by WorldClim (Table 1; [25]). These initial 13 variables were chosen to reduce multicollinearity in our initial model, while still including a number of bioclimatic variables which we expect to be important to the ecological niche of Odonata. Initial bioclimatic variables had a spatial resolution of 30 s (~900 m at the equator) and were aggregated fivefold to the coarser resolution of approximately 4.5 km at the equator. To further avoid potentially problematic multicollinearity in our models, we calculated the variance inflation factors (VIF) of our initial model with all 13 bioclimatic variables [26]. If any predictor variable had a VIF greater than 5, we removed the variable with the lowest permutation contribution to the model. We then fit a new Maxent model with default settings and repeated this step until no variables were retained in the model with a VIF greater than 5.

Table 1. Description of predictor variables included in our SDM modelling framework and the mean permutation contribution of each variable averaged across all of our top models.

Bioclimatic Variable	Description	Mean Permutation Contribution
Bio 2	Mean diurnal range	15.8
Bio 4	Temperature seasonality	12.0
Bio 1	Annual mean temperature	11.6
Bio 5	Max. temperature of warmest month	10.7
Bio 15	Precipitation seasonality	9.2
Bio 8	Mean temperature of wettest quarter	7.2
Bio 9	Mean temperature of driest quarter	6.9
Bio 13	Precipitation of wettest month	5.7
Bio 12	Annual precipitation	5.3
Bio 6	Min. temperature of coldest month	5.1
Bio 14	Precipitation of driest month	4.5
Bio 17	Precipitation of driest quarter	3.2
Bio 16	Precipitation of wettest quarter	2.8

Using the species-specific predictor variables determined by following the above process, we next used the R package *ENMeval* [27] to quantitatively evaluate a suite of Maxent models with different tuning parameters in an effort to optimise model complexity and prevent overfitting. We fit models individually for each species, using every combination of tuning parameters with regularisation multipliers of 0.5, 1, 2, 3, and 4 and feature classes of "linear", "linear + quadratic", "hinge", "linear + quadratic + hinge", "linear + quadratic + hinge + product", and "linear + quadratic + hinge + product + threshold". Block partitioning of five random partitions was used to separate occurrence and background localities into training and testing bins. The model with the lowest AICc value was selected as the top model if it had training and validation AUC values greater than 0.7. In the rare cases where training or validation AUC were less than 0.7, the top model was selected as the model with the highest validation AUC. To select a threshold value to transform our predicted Maxent model into a binary (presence/absence) surface, we reclassified our predicted Maxent model surface into a binary surface on the basis of five different thresholding values. These values were the 0th, 1st, 2.5th, 5th, and 10th percentiles of the predicted SDM on a ClogLog scale. Given these five binary surfaces, we calculated the sensitivity (percentage of actual presences predicted) and specificity (percentage of actual pseudo-absences predicted) for reclassified surfaces, where pseudo-absences were randomly generated within the accessible area and the number of pseudo-absences matched the number of spatially thinned occurrence records. An adapted true skill statistic (Equation (1)) was calculated to find a thresholding value that balances type I and type II errors, although specificity was given one-third the weight of sensitivity given the presence-only nature of our occurrence records. The percentile value that led to the highest true skill statistic was selected as our final thresholding value and used to generate the predicted presence/absence distribution.

$$TSS = (Sensitivity + 1/3 * Specificity) - 1 \quad (1)$$

The top Maxent models and binary surfaces were mapped for each species and underwent expert evaluation by authors V.J.K., J.-P.B., and R.F. Species with predicted distributions that did not pass expert evaluation were rerun after making custom changes to the modelling framework to improve predicted distributions. These custom changes included altering accessible areas, decreasing the number of background points for species with small accessible areas or few sample points, and altering the thresholding value.

Altering threshold values was undertaken when there was clear evidence of over- or under-commission in model results. All final models required final curatorial approval.

2.5. Calculating Richness and Endemism

Predicted distributions were stacked for all species across their entire ranges, including areas outside of the Palaearctic realm. While [28] suggest using continuous values for stacking ENMs, we custom-tuned models during thresholding to avoid overfitting as described below, and thus opted for stacking the thresholded outputs directly. Species richness (SR) and corrected weighted endemism (CWE) were calculated for each grid cell. Species richness is defined here as the number of species per cell. Weighted endemism (Equation (2)) uses a moving window analysis including the central cell and the eight neighbouring cells to sum for each taxon t in the set of taxa T in the neighbourhood: the number of cells in the neighbourhood containing taxon t (the local range, r_t) divided by its range (R_t, the number of cells in which it is found). CWE is the quotient of weighted endemism (WE) divided by richness (Equation (3)); [29].

$$WE = \sum_t \in T \frac{r_t}{R_t} \quad (2)$$

$$CWE = WE/Richness \quad (3)$$

Since our endemism calculation involved range-weighting, ranges (R_t) should ideally be generated for an entire species range [30]. Here, we compensated for missing data from part of a species' range by determining a coarse estimate of overall range size using country-level range maps [31]. If a species occurred in a country outside of the Palaearctic, the range size was calculated as the sum of the areal extent of both the thresholded SDMs and the total country level area outside the Palaearctic region. For species found only in the Palaearctic, range size was simply the sum of the areal extent of the thresholded SDM.

Finally, we generated bivariate maps to visualise SR and CWE for lentic/lotic species on a single map. Bivariate categories were calculated by determining cells with less than the 33 percentile, between the 33 and 66 percentiles, and greater than the 66 percentile of species richness given a certain trait.

3. Results

Each species had a custom combination of bioclimatic variables that best predicted the species distribution given our occurrence records and had variance inflation factors less than five. Across all 402 species, the variables that had the highest permutation importance were the mean diurnal range, temperature seasonality, annual mean temperature, maximum temperature of the warmest month, and precipitation seasonality (Table 1).

3.1. Richness and Corrected Weighted Endemism (CWE)

Figure 1A shows the patterns in total diversity based on data of the 402 Palaearctic species of dragonflies and damselflies. The map shows clear differences between regions with Europe, parts of the Middle East, and Japan and adjacent mainland Asia having the highest diversity compared with colder areas in the north and desert areas, such as the Sahara, parts of the Middle East, and the Gobi, which were less species-rich. The Corrected Weighted Endemism map (Figure 1B) is strikingly different from the map showing overall diversity, with hotspots being clearly centred in the Mediterranean, areas in Iran and in Japan, Korea, and Russia's Far East.

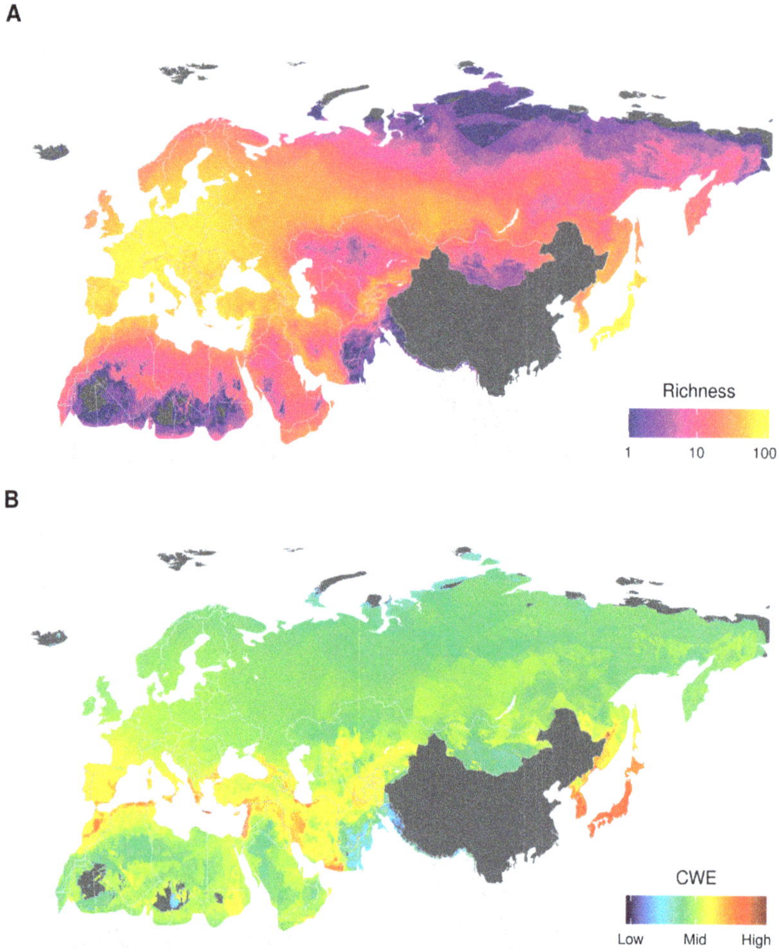

Figure 1. (**A**) Distribution of Odonata richness and (**B**) corrected weighted endemism (CWE) of Palaearctic odonates. Grey shading indicates the parts of the Palearctic for which our data are insufficient to make predictions.

3.2. Richness and Endemism by Aquatic Habitats

Figure 2 shows that diversity patterns are clearly different between species of standing (lentic, 244 species) and running water (lotic, 158 species). Low diversity areas tend to be dominated by lentic species. This is true for both northern areas, which have a lower diversity due to the lower temperatures, as well as for the arid regions (Sahara, deserts of the Middle East, Central Asia, and the Gobi). Figure 2C,D, respectively, show the corrected weighted endemism for lentic and lotic species. The areas with a high CWE are more pronounced for those of lotic environments than for those of lentic environments. This is due to lotic species being more regionally concentrated and having smaller ranges than those of lentic environments (Figure 3, W = 24,254, $p = 1.805 \times 10^{-7}$).

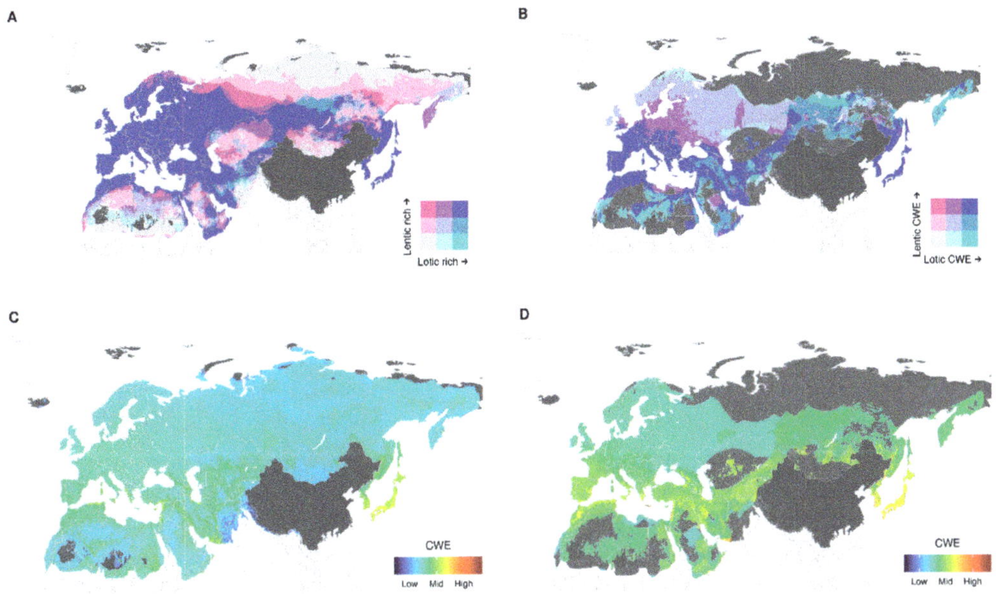

Figure 2. Species richness and corrected weighted endemism for aquatic habitats used by odonate species in the Palaearctic. (**A**) Bivariate plot showing distribution of richness for lotic-dependent and lentic species; (**B**) bivariate plot showing distribution of corrected weighted endemism (CWE) for lotic-dependent and lentic species; (**C**) corrected weighted endemism (CWE) for lentic species; and (**D**) corrected weighted endemism (CWE) for lotic-dependent species.

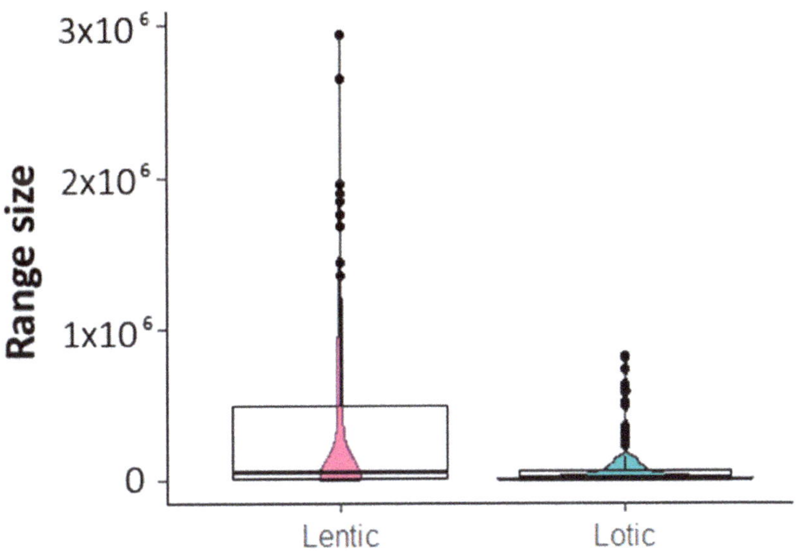

Figure 3. Range size (defined by the number of cells with predicted presence) for lentic- and lotic-dependent Odonata species.

3.3. Richness of Species According to IUCN Red List Category

Of the 402 species, 45 species have not yet been officially assessed for the IUCN Red List. Eleven of the 357 assessed species are Data Deficient, and 293 are of Least Concern.

The distribution of the remaining 53 species is shown in Figure 4 (Critically Endangered, 2 species; Endangered, 20 species; Vulnerable, 12 species; Near Threatened, 19 species). The threatened species are clearly concentrated in three areas: (a) the western Mediterranean, (b) Iran, Turkey, the Southern Caucasus, and the Levant, and (c) Japan and Korean Peninsula (Figure 4). Of the 45 species not assessed, only 8 are likely to be in one of these categories and all of these are from Japan or adjacent mainland so that their inclusion would not alter the general pattern.

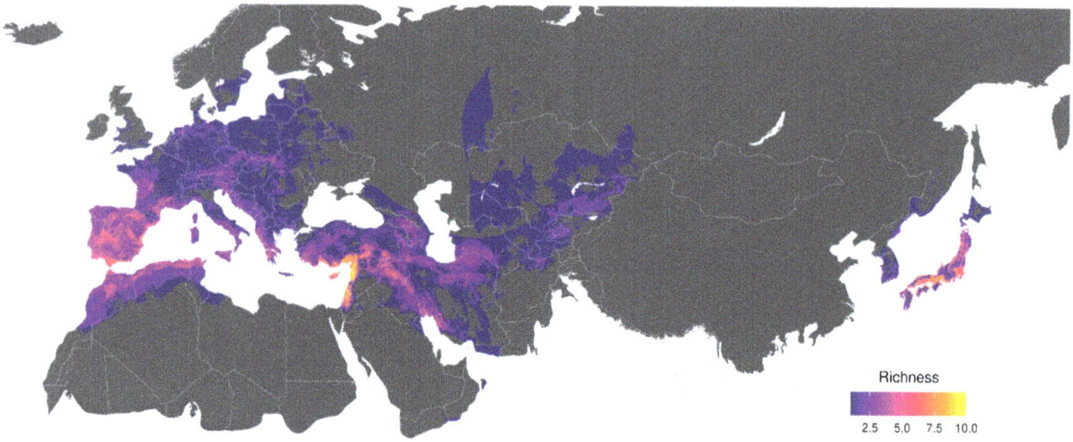

Figure 4. Distribution of the species richness for the IUCN Red List category Critically Endangered (2), Endangered (19), Vulnerable (12), or Near Threatened (16).

3.4. *Sampling Effort*

The distribution data is highly unevenly distributed across the Palaearctic with high densities of records available for Europe (although with strong regional differences) and Japan, while lower amounts of records available for North Africa, the Middle East, and West and Central Asia (Figure 5). Russia, Kazakhstan, and Mongolia are, on average, very poorly explored. For large areas of the latter three countries, which together cover almost 40 percent of the Palaearctic realm, none to a very few records are available and only small parts of these countries have been well explored (for instance, the southern Ural [32,33]. The methods used to make the SDMs partially compensate for this geographical imbalance, such that our maps reflect, to a large extent, actual richness patterns rather than sampling bias. However, it is also clear that sampling gaps are still an issue, and increasing field work is likely to show some of the low diversity areas, such as Kazakhstan, to be more diverse than the current map shows.

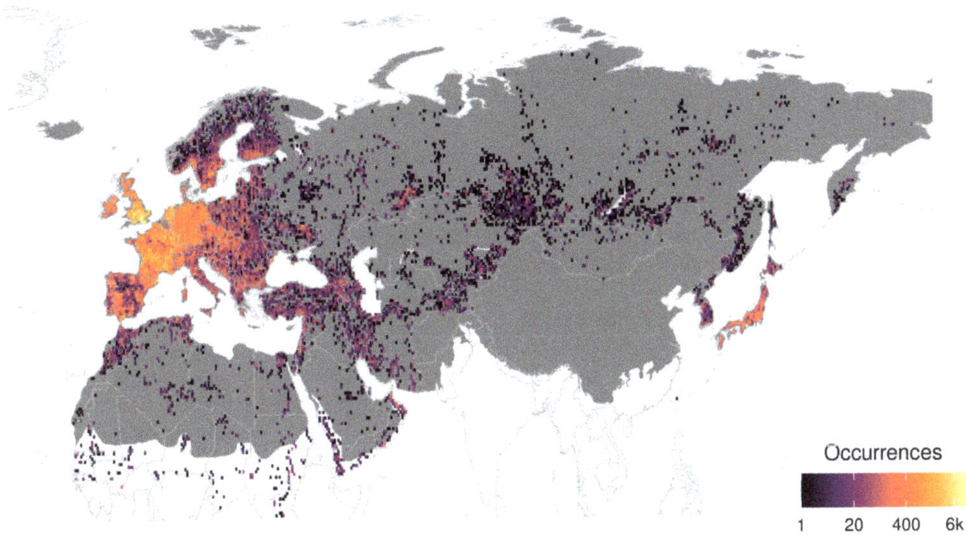

Figure 5. Odonata sampling effort in the Palaearctic.

4. Discussion

4.1. General Diversity Patterns

The general diversity pattern shown in Figure 1 is correlated largely to temperature and precipitation. Higher temperature towards the south generally results in an increase in diversity, whereas lower temperature at higher altitudes results in a decrease in diversity. The former can be seen as a general pattern throughout the whole of the Palaearctic, while the latter can be observed in the lower diversity found in, for instance, the Alps, the highlands of Afghanistan, and parts of western Mongolia. This pattern of higher diversity in warmer areas is offset by low precipitation, which results in decreased diversity, examples of which can be seen in the Middle East and the central deserts of Iran. Figure 1 also shows a strong east–west (longitudinal) pattern, with areas in the central two-thirds of Palearctic being less diverse than areas to the east and west. For Europe, such a pattern was already described by [1,34] and it seems to be governed by the oceanic climate in the west of Europe transcending towards a continental climate farther east. The warm summer of the continental climate allows some species to occur farther north than in areas with an oceanic climate, but for many others, the stronger and longer winters in these areas is limiting their distribution. This can, for instance, be seen when comparing the fauna between the Netherlands and areas at a similar latitude to the south of Moscow. In the latter, many species common in west or central Europe are already absent (for example, *Chalcolestes viridis*, *Pyrrhosoma nymphula*, *Ceriagrion tenellum*, *Erythromma lindenii*, and *Gomphus pulchellus*). A similar pattern, although less obvious due to the absence of data from China, can be seen in the east of the Palaearctic. The relatively sharp contrast between the north and the south of the Caucasus is caused by this mountain range preventing cold air from penetrating farther south, resulting in mild winters to the south of the mountains. In Central Asia, a sharp contrast in diversity is visible between the arid lowlands and the mountains of the Kopet Dagh in northeast Iran and the mountains to the east of Tajikistan and Kyrgyzstan, caused by the higher precipitation in these mountain ranges. The diversity at species, generic, and family level is higher in the south of Japan than in areas at a similar latitude in the western Mediterranean. This is likely to be the result of the more humid climate of Japan, resulting in a higher diversity of freshwater habitats and the stronger impact of the glacial periods in the Western Palaearctic in combination with the Sahara, preventing tropical African lineages dispersing northwards.

In the western two-thirds of the Palaearctic, the borders with the Afrotropical and Oriental regions are formed by clear geographical barriers with the Sahara and the desert of the Arabian Peninsula, forming the demarcation with the Afrotropical region and the Himalaya, forming a well-marked boundary with the Oriental region. In the east, such a clear geographical barrier is absent, and it is, therefore, expected that the Palaearctic odonate fauna more gradually merges into the Oriental fauna, with isolated Palaearctic 'islands' expected to occur at higher elevations in southwest China. We currently lack the data to study this in detail, but the patterns shown in Korea and Japan give some insight. In both countries, a large portion of the fauna consists of species whose range is largely limited to the Palaearctic, often belonging to genera that themselves are also largely restricted to the Palearctic (for instance, *Coenagrion*, *Aeshna*, *Somatochlora*, and *Sympetrum*). In addition, however, there is a substantial number of genera whose distribution is centred on the Oriental region. In both countries, the north has a higher percentage of distinctly Palaearctic species, which is easiest to observe when comparing the fauna of Hokkaido, Japan's northernmost Island, with that of Kyushu, the southernmost of Japan's large islands. The indigenous Odonata fauna of Hokkaido (seven species) consist completely of species with a Palaearctic distribution largely from genera centred on the Palaearctic. In contrast, approximately 10 percent of the species found on the southern Kyushu are species restricted largely to the Oriental region. Despite this distinct difference between these two islands, there is no obvious demarcation line, and areas dominated by Palaearctic species just gradually merge into areas dominated by Oriental species. Further efforts examining phylogenetic beta diversity may provide a more resolved view of the regionalization of fauna and the historical forces that may have shaped such regions.

4.2. Are There Differences in Diversity Patterns Shown by Lotic and Lentic Species?

Figure 2 shows that diversity patterns of lotic and lentic species are not identical and show clear regional differences. The north of the Paleartic is dominated largely by lentic species (Figure 2A). This northern area dominated by lentic species reaches farther south in the eastern two-thirds of the Palaeartic, which might be correlated with the temperature in the east being lower than at comparable longitudes in the west. Arid regions, such as the Sahara, Middle East, parts of Kazakhstan and Central Asia, and the Gobi Desert, are also dominated by lentic species. Whereas the dominance of lentic species in arid regions is likely to be caused by the general scarcity of lotic habitats, the dominance of lentic species in the north seems to be caused by lotic habitats simply being too cold for most parts of the year. The CWE for lentic species shows a fairly uniform pattern, while that of the lotic species shows clear regions with higher diversity mostly found in the Mediterranean, southwest and central Asia and Japan and the Korean Peninsula. The differences between CWE shown by lentic and lotic species are the result of lotic species having smaller ranges (Figure 3), which matches the results found for the Nearctic [2] and supports the hypothesis that lotic species are more specialised insects [16,17] and, therefore, more susceptible to harsh or changing environmental conditions.

4.3. Are There Areas with Relatively High Endemism?

Figure 1B very clearly shows that there are marked regions with high endemism, with CWE being especially high in Japan and the Korean Peninsula. The Palaearctic part of Japan is home to approximately 130 species of which no less than 30 are endemic to the islands. Most of these are fairly widespread on the islands but, nonetheless, contribute to the high CWE of the islands. The high number of endemics is easily explained by the isolation of the main Japanese islands, but, surprisingly, the Korean Peninsula shows a CWE matching that of Japan. This suggests that the warm south of the peninsula is isolated from the rest of mainland Asia by the colder climate in the north of the peninsula.

The areas with high CWE in the Western Palaearctic are concentrated on the Mediterranean and the mountain regions of Turkey and Iran. All these areas match with known areas of endemism for odonates [34–37]. The Palaearctic part of the Arabian Peninsula lacks distinct areas of high CWE, with all endemics with small ranges being found largely in areas in Oman and Yemen that fall outside the Palaearctic, as redefined by [10]. Most of the north and the central part of the Palaearctic have a very low CWE, with most species having large ranges. In Central Asia, a slightly higher CWE can be noticed, resulting from the presence of a few species largely restricted to the mountains of Tian Shan and the Pamir Mountains [12].

4.4. Are There Areas with a Relatively High Percentage of Globally Threatened Species?

Figure 4 shows three regions with a high number of species listed as either Critically Endangered, Endangered, Vulnerable, or Near Threatened on the IUCN Red List: (a) the western Mediterranean, (b) Iran, Turkey, the Southern Caucasus, and the Levant, and (c) Japan and the Korean Peninsula. Not surprisingly, these areas are also the areas with the highest density of species, with a small range being endemic to the Palaearctic, resulting in the map of threatened species having a high congruence with the map showing the corrected weighted endemism (Figure 1B). Nearly all species found outside these areas have large ranges, often with a part of the range found in areas with relatively low human impact. For this reason, the northern two-thirds of the Palaearctic hardly has any species threatened on a global scale, although many of these species are likely declining at a regional scale [34,35]. Within the three regions with a high number of threatened species, two well-defined hot spots are visible: the Levant and the southern half of Japan. Each of these two have both a high number of species with a small range and are strongly impacted by human activities. In the case of the Levant, freshwater habitats in the coastal region are impacted by increased intake of water for consumption and agriculture, the construction of hydroelectric dams, gravel mining, and wastewater pollution [12]. Climate change is expected to have an additional deleterious effect as the area is predicted to become both hotter and drier. In Japan, insecticides and the impact of alien species have in recent years become important factors in the decline of species [38].

5. Further Research

This paper provides a description for the patterns of diversity of dragonflies and damselflies found in the Palaearctic region and an overview of possible explanations for these patterns. The data on which the maps presented in this paper are based, together with the increasingly available molecular data, provide a valuable source of data for further studies focused on describing and understanding diversity patterns of dragonflies and damselflies in the Palaearctic. The possibilities for such studies are further increased by the availability of a similar set of data for the Nearctic region [2]. In order to understand the historical and contemporary factors determining the patterns of diversity in more detail, the following studies are deemed the most relevant:

Limits between Oriental and Palaearctic regions. In the western two-thirds of the Palaearctic, the border with the Afrotropical and the Oriental regions is formed by well-defined geographical barriers such as the Sahara and the Himalayas. In the east, no such clear barriers exist between the Palaearctic and Oriental regions. In a recent review of freshwater bioregions in China, Huang et al. [39] placed the border between the Palaearctic and the Oriental region on the line between the Qin Mountains of southern Shaanxi to the mouth of the Yangtze River. Expanding our database with data from China will allow us to test whether a clear demarcation line between the two faunas really exists in China or whether that is a large area where the fauna gradually turns over. Furthermore, such data can be used to establish which mountain ranges in southwest China have a clearly Palearctic fauna.

Establishing the timing and direction of exchange between the Palaearctic and the Nearctic. More than 40 percent of the 402 Odonata species occurring in the Palearctic belong to genera also occurring in the Nearctic. It is likely that many of these genera crossed from the Nearctic to the Palaearctic or vice versa. It is unknown whether these dispersal events were predominantly in one direction or to what extent these dispersal events were succeeded by rapid speciation. An analysis of haplotypes of five species shared between the Nearctic and Palaearctic indicated that dispersal though Beringia went both ways, with the data indicating that the populations of the continents have been separated for more than 400,000 years [40]. Molecular data and associated bio- and phylogeographic studies leveraging distribution models presented here and in [2] can help determine which parts of diversity originated from dispersal events and subsequent speciation between both parts of the Holarctic.

Locating and dating glacial refugia. One potential next step now that distribution models are available for all species is to backcast distributions to the Last Glacial Maximum. Backcasts have often been used to locate pleistocene refugial area, and these have often been confirmed with molecular techniques [41]. However, most studies have focused on only a few species, and here the possibility to simultaneously backcast all palearctic dragonflies might provide a novel means to establish fauna-wide refugia locations. For example, our results show a centre of endemism concentrated in the Mediterranean region especially in Morocco and Tunisia in North Africa, which may indicate an area of long-term climate stability that could also serve as a refugium [42,43]. Combining such approaches with time-calibrated trees and population genetic data, provides a means to understanding the timing of formation of refugia and recolonization dynamics from those locales.

Modelling the expected impact of climate change. Climate change is already having a substantial impact on the distribution of species, but at present, evidence for this is restricted largely to the best-investigated regions of the Palaearctic. Studies on shifts in distribution and changes in phenology in the Palaearctic are almost exclusively from north and western Europe (e.g., [17,44–47] and Japan [48,49]. The data at hand would allow us to determine climatological envelopes for species for different climatological scenarios and different time periods. Developing these scenarios is of importance for conservation planning but would also allow us to determine whether climate change will allow species to break through their biogeographical boundaries, which would, for instance, happen when climate change facilitates an eastwards jump of Western Palaearctic species and vice versa.

Supplementary Materials: The following supporting information can be downloaded at: https://www.mdpi.com/article/10.3390/d14110966/s1, Supplementary Table S1. List of the Odonata species recorded in the Palaearctic (as defined in this paper), with their IUCN conservation status, their terrestrial and aquatic habitat. En: endangered, LC: least concern, NA: not assessed, NT: near-threatened, Vu: vulnerable.

Author Contributions: Conceptualization, V.J.K., J.C.A., C.A.B.-S., R.G. and S.M.B.; methodology, J.C.A., C.A.B.-S., R.G., V.J.K. and S.M.B.; software, R.G. and M.W.B.; validation, V.J.K., J.-P.B. and R.F.; formal analysis, V.J.K., J.C.A., C.A.B.-S., R.G. and M.W.B.; data curation, J.-P.B., R.F. and V.J.K.; writing—original draft preparation, V.J.K., J.-P.B., R.F., J.C.A., C.A.B.-S., R.G., S.M.B., J.W. and M.W.B.; writing—review and editing, V.J.K., J.-P.B., R.F., J.C.A., C.A.B.-S., R.G., S.M.B., J.W. and M.W.B.; project administration, J.C.A., R.G., S.M.B. and J.W.; funding acquisition, J.C.A., R.G., S.M.B. and J.W. Contribution to databases: J.-P.B. and R.F. All authors have read and agreed to the published version of the manuscript.

Funding: This work was supported by NSF Awards (Abbott: 2002489; Ware: 2002473; Bybee: 2002432; Guralnick: 2002457), Genealogy of Odonata (GEODE): Dispersal and color as drivers of 300 million years of global dragonfly evolution.

Institutional Review Board Statement: Not applicable.

Informed Consent Statement: Not applicable.

Data Availability Statement: Data are provided below and in Supplementary Table S1. The composite richness and endemism values for each cell (released as CSV); the composite richness and endemism values for each cell broken out by trait (lentic, lotic; released as 2 CSVs) are available from a Zenodo repository (https://doi.org/10.5281/zenodo.7195870; accessed on 13 October 2022). Species-specific SDMs showing the predicted areas of presence for each species released as a single zip file are available at (https://osf.io/szqkw/) DOI 10.17605/OSF.IO/SZQKW (accessed on 15 September 2022).

Conflicts of Interest: The authors declare no conflict of interest.

References

1. Wasscher, M.T.; Bos, F.G. The European dragonflies: Notes on the checklist and on species diversity. *Odonatologica* **2000**, *29*, 31–43.
2. Abbott, J.C.; Bota-Sierra, C.A.; Guralnick, R.; Kalkman, V.; González-Soriano, E.; Novelo-Gutiérrez, R.; Bybee, S.; Ware, J.; Belitz, M.W. Diversity of Nearctic Dragonflies and Damselflies (Odonata). *Diversity* **2022**, *14*, 575. [CrossRef]
3. Boudot, J.-P.; Kalkman, V.J. *Atlas of the European Dragonflies and Damselflies*; KNNV-Publishing: Utrecht, The Netherlands, 2015.
4. Clausnitzer, V.; Dijkstra, K.-D.B.; Koch, R.; Boudot, J.-P.; Darwall, W.R.T.; Kipping, J.; Samraoui, B.; Samways, M.J.; Simaika, J.P.; Suhling, F. Focus on African Freshwaters: Hotspots of dragonfly diversity and conservation concern. *Front. Ecol. Environ.* **2012**, *10*, 129–134. [CrossRef]
5. Endersby, I. *The Distribution of Australian Dragonflies*; Busybird Publishing: Montmorency, VIC, Australia, 2021.
6. Kalkman, V.J.; Boudot, J.P.; Bernard, R.; De Knijf, G.; Suhling, F.; Termaat, T. Diversity and conservation of European dragonflies and damselflies (Odonata). *Hydrobiologia* **2018**, *811*, 269–282. [CrossRef]
7. Darwall, W.R.T.; Smith, K.G.; Allen, D.J.; Holland, R.A.; Harrison, I.J.; Brooks, E.G.E. (Eds.) *The Diversity of Life in African Freshwaters: Under Water, under Threat. An Analysis of the Status and Distribution of Freshwater Species throughout Mainland Africa*; IUCN: Cambridge, UK; Gland, Switzerland, 2011.
8. Combe, F.J.; Sikes, D.S.; Tkach, V.V.; Hope, A.G. Origins and diversity of the Bering Sea Island fauna: Shifting linkages across the northern continents. *Biodivers. Conserv.* **2021**, *30*, 1205–1232. [CrossRef]
9. Wen, J.; Nie, Z.-L.; Ickert-Bond, S.M. Intercontinental disjunctions between eastern Asia and western North America in vascular plants highlight the biogeographic importance of the Bering land bridge from late Cretaceous to Neogene. *J. Syst. Evol.* **2016**, *54*, 469–490. [CrossRef]
10. Olson, D.M.; Dinerstein, E.; Wikramanayake, E.D.; Burgess, N.D.; Powell, G.V.N.; Underwood, E.C.; D'amico, J.; Itoua, I.; Strand, H.E.; Morrison, J.C.; et al. Terrestrial Ecoregions of the World: A New Map of Life on Earth. *BioScience* **2001**, *51*, 933–938. [CrossRef]
11. Boudot, J.-P.; Kalkman, V.J.; Azpilicueta Amorin, M.; Bogdanović, T.; Cordero Rivera, A.; Degabriele, G.; Dommanget, J.-L.; Ferreira, S.; Garrigós, B.; Jović, M.; et al. Atlas of the Odonata of the Mediterranean and North Africa. *Libellula Suppl.* **2009**, *9*, 1–256.
12. Boudot, J.P.; Borisov, S.; De Knijf, G.; Grunsven, R.H.A.; van Schröter, A.; Kalkman, V.J. Atlas of the dragonflies and damselflies of West and Central Asia. *Brachytron Suppl.* **2021**, *22*, 3–248.
13. National Biodiversity Center of Japan, Ministry of the Environment, Japan, Tokyo. Data from: The National Survey on the Natural Environment Report of the Distributional Survey of Japanese Animals (Dragonflies). 2002. Available online: http://www.biodic.go.jp/reports/4-09/h000.html (accessed on 21 January 2021).
14. Ozono, A.; Kawashima, I.; Futahashi, R. *Dragonflies of Japan, Revised ed.*; Bun-ichi Sogo Shuppan Co.: Tokyo, Japan, 2021.
15. Zizka, A.; Silvestro, D.; Andermann, T.; Azevedo, J.; Duarte Ritter, C.; Edler, D.; Farooq, H.; Herdean, A.; Ariza, M.; Scharn, R.; et al. CoordinateCleaner: Standardized Cleaning of Occurrence Records from Biological Collection Databases. *Methods Ecol. Evol.* **2019**, *10*, 744–751. [CrossRef]
16. Hof, C.; Brändle, M.; Brandl, R. Lentic Odonates Have Larger and More Northern Ranges than Lotic Species. *J. Biogeogr.* **2006**, *33*, 63–70. [CrossRef]
17. Olsen, K.; Svenning, J.-C.; Balslev, H. Niche Breadth Predicts Geographical Range Size and Northern Range Shift in European Dragonfly Species (Odonata). *Diversity* **2022**, *14*, 719. [CrossRef]
18. Davis Rabosky, A.R.; Cox, C.L.; Rabosky, D.L.; Title, P.O.; Holmes, I.A.; Feldman, A.; McGuire, J.A. Coral Snakes Predict the Evolution of Mimicry across New World Snakes. *Nat. Commun.* **2016**, *7*, 11484. [CrossRef] [PubMed]
19. Barve, N.; Barve, V.; Jiménez-Valverde, A.; Lira-Noriega, A.; Maher, S.P.; Peterson, A.T.; Soberón, J.; Villalobos, F. The Crucial Role of the Accessible Area in Ecological Niche Modeling and Species Distribution Modeling. *Ecol. Model.* **2011**, *222*, 1810–1819. [CrossRef]
20. Steen, V.A.; Elphick, C.S.; Tingley, M.W. An Evaluation of Stringent Filtering to Improve Species Distribution Models from Citizen Science Data. *Divers. Distrib.* **2019**, *25*, 1857–1869. [CrossRef]
21. Hijmans, R.J. Package 'Raster'—Geographic Data Analysis and Modeling; CRAN Repos. 2019. Available online: https://rspatial.org/raster (accessed on 10 May 2022).
22. Phillips, S.J.; Anderson, R.P.; Dudík, M.; Schapire, R.E.; Blair, M.E. Opening the Black Box: An Open-Source Release of Maxent. *Ecography* **2017**, *40*, 887–893. [CrossRef]
23. Hijmans, R.J.; Phillips, S.; Leathwick, J.; Elith, J. Dismo: Species Distribution Modeling; R Package Version 1.1-4; Cran. 2017. Available online: https://cran.r-project.org/web/packages/dismo/dismo.pdf (accessed on 10 May 2022).
24. Merow, C.; Smith, M.J.; Silander, J.A. A Practical Guide to MaxEnt for Modeling Species' Distributions: What It Does, and Why Inputs and Settings Matter. *Ecography* **2013**, *36*, 1058–1069. [CrossRef]

25. Fick, S.E.; Hijmans, R.J. WorldClim 2: New 1-Km Spatial Resolution Climate Surfaces for Global Land Areas. *Int. J. Climatol.* **2017**, *37*, 4302–4315. [CrossRef]
26. Naimi, B.; Hamm, N.A.S.; Groen, T.A.; Skidmore, A.K.; Toxopeus, A.G. Where Is Positional Uncertainty a Problem for Species Distribution Modelling? *Ecography* **2014**, *37*, 191–203. [CrossRef]
27. Kass, J.M.; Muscarella, R.; Galante, P.J.; Bohl, C.L.; Pinilla-Buitrago, G.E.; Boria, R.A.; Soley-Guardia, M.; Anderson, R.P. ENMeval 2.0: Redesigned for Customizable and Reproducible Modeling of Species' Niches and Distributions. *Methods Ecol. Evol.* **2021**, *12*, 1602–1608. [CrossRef]
28. Calabrese, J.M.; Certain, G.; Kraan, C.; Dormann, C.F. Stacking Species Distribution Models and Adjusting Bias by Linking Them to Macroecological Models. *Glob. Ecol. Biogeogr.* **2014**, *23*, 99–112. [CrossRef]
29. González-Orozco, C.E.; Laffan, S.W.; Miller, J.T. Spatial Distribution of Species Richness and Endemism of the Genus Acacia in Australia. *Aust. J. Bot.* **2011**, *59*, 601–609. [CrossRef]
30. Earl, C.; Belitz, M.W.; Laffan, S.W.; Barve, V.; Barve, N.; Soltis, D.E.; Allen, J.M.; Soltis, P.S.; Mishler, B.D.; Kawahara, A.Y.; et al. Spatial phylogenetics of butterflies in relation to environmental drivers and angiosperm diversity across North America. *iScience* **2021**, *24*, 102239. [CrossRef] [PubMed]
31. Sandall, E.L.; Pinkert, S.; Jetz, W. Country-level checklists and occurrences for the world's Odonata (dragonflies and damselflies). *J. Biogeogr.* **2022**, *49*, 1586–1598. [CrossRef]
32. Yanybaeva, V.A.; Dumont, H.J.; Haritonov, A.Y.; Popova, O.N. The Odonata of the South Ural, Russia, with special reference to *Ischnura aralensis* Haritonov, 1979. *Odonatologica* **2006**, *35*, 167–185.
33. Haritonov, A.Y.; Eremina, E.E. The dragonflies (Odonata) of South Ural: The value of regional faunistic research. *Euroasian Entomol. J.* **2010**, *9*, 263–273.
34. Kalkman, V.J.; Boudot, J.-P.; Bernard, R.; Conze, K.-J.; De Knijf, G.; Dyatlova, E.; Ferreira, S.; Jović, M.; Ott, J.; Riservato, E.; et al. *European Red List of Dragonflies*; IUCN: Gland, Switzerland; Cambridge, UK; Office for Official Publications of the European Communities: Luxembourg, 2010.
35. Riservato, E.; Boudot, J.-P.; Ferreira, S.; Jović, M.; Kalkman, V.J.; Schneider, W.; Samraoui, B.; Cuttelod, A. *The Status and Distribution of Dragonflies of the Mediterranean Basin*; IUCN Red List of Threatened Species, Regional Assessments Series; IUCN: Gland, Switzerland; Malaga, Spain, 2009.
36. Samraoui, B.; Boudot, J.-P.; Ferreira, S.; Riservato, E.; Jović, M.; Kalkman, V.J.; Schneider, W. The Status and Distribution of Dragonflies (Odonata). In *The Status and Distribution of Freshwater Biodiversity in Northern Africa*; IUCN Red List of Threatened Species, Regional Assessments Series; Garcia, N., Cuttelod, A., Abdul Malak, D., Eds.; IUCN: Gland, Switzerland; Malaga, Spain, 2010; Chapter 5; pp. 51–70.
37. Kosterin, O.; Malikova, E.I.; Haritonov, A.Y. Critical species of Odonata in the Asian part of the former USSR and the Republic of Mongolia. *Int. J. Odonatol.* **2004**, *7*, 341–370. [CrossRef]
38. Karube, H. Conservation of endangered dragonflies. *Nat. Insects* **2020**, *55*, 2–3.
39. Huang, C.; Ebach, M.; Ahyong, S.T. Bioregionalisation of the freshwater zoogeographical areas of mainland China. *Zootaxa* **2020**, *4742*, 271–298. [CrossRef]
40. Kohli, M.; Djernæs, M.; Sanchez Herrera, M.; Sahlen, G.; Pilgrim, E.; Simonsen, T.J.; Olsen, K.; Ware, J. Comparative phylogeography uncovers evolutionary past of Holarctic dragonflies. *PeerJ* **2021**, *9*, e11338. [CrossRef]
41. Waltari, E.; Hijmans, R.J.; Peterson, A.T.; Nyári, A.S.; Perkins, S.L.; Guralnick, R.P. Locating Pleistocene Refugia: Comparing Phylogeographic and Ecological Niche Model Predictions. *PLoS ONE* **2007**, e563. [CrossRef] [PubMed]
42. Husemann, M.; Schmitt, T.; Zachos, F.E.; Ulrich, W.; Habel, J.C. Palaearctic biogeography revisited: Evidence for the existence of a North African refugium for Western Palaearctic biota. *J Biogeogr.* **2014**, *41*, 81–94. [CrossRef]
43. Harrison, S.; Noss, R. Endemism hotspots are linked to stable climatic refugia. *Ann Bot.* **2017**, *119*, 207–214. [CrossRef]
44. De Knijf, G.; Flenker, U.; Vanappelghem, C.; Manci, C.O.; Kalkman, V.J.; Demolder, H. The status of two boreo-alpine species, *Somatochlora alpestris* and *S. arctica*, in Romania and their vulnerability to the impact of climate change (Odonata: Corduliidae). *Int. J. Odonatol.* **2011**, *14*, 111–126. [CrossRef]
45. Dingemanse, N.J.; Kalkman, V.J. Changing temperature regimes have advanced the phenology of Odonata in the Netherlands. *Ecol. Entomol.* **2008**, *33*, 394–402. [CrossRef]
46. Hickling, R.; Roy, D.B.; Hill, J.K.; Thomas, C.D. A northward shift of range margins in British Odonata. *Glob. Change Biol.* **2005**, *11*, 502–506. [CrossRef]
47. Hassall, C.; Thompson, D.J.; French, G.C.; Harvey, I.F. Historical changes in the phenology of British Odonata are related to climate. *Glob. Change Biol.* **2007**, *13*, 933–941. [CrossRef]
48. Doi, H. Delayed phenological timing of dragonfly emergence in Japan over five decades. *Biol. Lett.* **2008**, *4*, 388–391. [CrossRef]
49. Aoki, T. Northward expansion of *Ictinogomphus pertinax* (Selys, 1854) in Japan. *Nat. Insects* **2021**, *56*, 12–17.

diversity

Article

Diversity of Nearctic Dragonflies and Damselflies (Odonata)

John C. Abbott [1,*], Cornelio A. Bota-Sierra [1,2], Robert Guralnick [3], Vincent Kalkman [4], Enrique González-Soriano [5], Rodolfo Novelo-Gutiérrez [6], Seth Bybee [7], Jessica Ware [8] and Michael W. Belitz [3]

1 Alabama Museum of Natural History & UA Museums Department of Research and Collections, The University of Alabama, Tuscaloosa, AL 35487, USA; cabotasierra@ua.edu
2 Grupo de Entomología Universidad de Antioquia (GEUA), Universidad de Antioquia, Medellin 50010, Colombia
3 Florida Museum of Natural History, University of Florida, Gainesville, FL 32611, USA; robgur@gmail.com (R.G.); michaelbelitz06@gmail.com (M.W.B.)
4 Naturalis Biodiversity Center, 2333 CR Leiden, The Netherlands; vincent.kalkman@naturalis.nl
5 Departamento de Zoología, Instituto de Biología, Universidad Nacional Autónoma de México, Apartado Postal 70-153, Mexico City 04510, Mexico; esoriano@ib.unam.mx
6 Red de Biodiversidad y Sistemática, Instituto de Ecología AC, Xalapa 91073, Mexico; rodolfo.novelo@inecol.mx
7 Department of Biology and Monte L. Bean Museum, Brigham Young University, Provo, UT 84602, USA; seth.bybee@gmail.com
8 American Museum of Natural History, Division of Invertebrate Zoology, New York, NY 10024, USA; jware@amnh.org
* Correspondence: jabbott1@ua.edu

Abstract: Rarely have studies assessed Odonata diversity for the entire Nearctic realm by including Canada, the United States, and Mexico. For the first time, we explored Odonata diversity in this region according to a definition of natural community assemblages and generated species distribution models (SDMs). Species occurrence data were assembled by reviewing databases of specimens held by significant Odonata repositories and through an extensive search of literature references. Species were categorized as forest-dependent or non-forest-dependent, as lentic or lotic-dependent, and according to conservation status. Predicted distributions were stacked for all species across their entire ranges, including areas outside of the Nearctic. Species richness and corrected weighted endemism (CWE) were then calculated for each grid cell. We found a pattern of greater species richness in the eastern portion of the Nearctic, which can be explained by the higher aquatic habitat diversity at micro and macroscales east of the Rocky Mountains, promoting niche partitioning and specialization. In the Nearctic region, the southeastern US has the highest number of endemic species of dragonflies and damselflies; this degree of endemism is likely due to glacial refuges providing a foundation for the evolution of a rich and unique biota.

Keywords: biogeography; North America; glaciation; species occurrence

Citation: Abbott, J.C.; Bota-Sierra, C.A.; Guralnick, R.; Kalkman, V.; González-Soriano, E.; Novelo-Gutiérrez, R.; Bybee, S.; Ware, J.; Belitz, M.W. Diversity of Nearctic Dragonflies and Damselflies (Odonata). *Diversity* **2022**, *14*, 575. https://doi.org/10.3390/d14070575

Academic Editors: M. Olalla Lorenzo-Carballa and Ricardo Koroiva

Received: 19 May 2022
Accepted: 12 July 2022
Published: 18 July 2022

1. Introduction

Dragonflies and damselflies (Odonata) are amongst the most recognizable insects. Their study in the Nearctic dates back to the 18th century (e.g., [1], but see [2] for a review). Moreover, their cultural significance stretches further back in time, serving a role in the traditions of multiple Native American cultures [3,4]. Knowledge about Nearctic Odonata is most complete for the United States and Canada, with significant efforts to close gaps in our understanding of conservation, taxonomy, ecology, physiology, and evolutionary biology (e.g., [5–10]). In northern Mexico some regional assessments and studies in key areas have been published (e.g., [11–16]). Nevertheless, knowledge gaps of Odonate distributions across the Nearctic portions of Mexico make it difficult to address questions about richness and endemism within the region.

The full Nearctic realm is defined by a distinct assemblage of natural communities in the North American continent, whose northern boundaries are Greenland west to Alaska and whose southern boundaries are three mountain ranges of northern and central Mexico: Sierra Madre Occidental, Sierra Madre Oriental, and Eje Neovolcánico Transversal [17]. These mountainous regions are considered part of the Mexican transition zone where the Neotropical and Nearctic biotas converge, making them biodiversity hotspots [18,19].

The biodiversity of Odonata has not been studied on the basis of a Nearctic definition according to natural community assemblages. Rather, most studies loosely defined the Nearctic realm politically, as north of the Mexican border, a fauna that currently includes 471 species [20]. Rarely have studies assessed Odonata diversity for the entire Nearctic realm by including Canada, the United States, and Mexico [2,21]. Usually, the US/Canada fauna [22,23] and Mexican faunas [24,25] have been treated separately. To date, no complete assessment for the entire Nearctic has been published.

In this study, we showcase production of species distribution models (SDMs) for the Odonate diversity of the entire Nearctic, utilizing a best practices approach with strong attention to curation and expert assessment at every step during production. This effort led to predicted distributions at relatively fine grain for 509 species that occur north of the Mexican mountain ranges to the Arctic pole. Input occurrence data used for modeling were scrupulously assembled from the published literature, museum databases, and the citizen science repository Odonata Central [26] to create these models, which were then used to create maps of richness and endemism.

We used these maps to address a set of questions about the structure of Odonate spatial diversity. First, we used the data in the IUCN red list [27] to map the distribution of the threatened and endangered species occurring in the Nearctic. We also mapped the aquatic habitat (lotic vs. lentic) used by the immature stages and the terrestrial habitat (forest vs. non-forest obligates) used by adults. Lastly, we mapped sampling efforts to identify potential gaps in our knowledge. These maps provide key means to assess hotspots of diversity and endemism, which are often poorly understood in insects, with importance in fields such as conservation and ecology. We explicitly address the following questions empirically: Are there differences in diversity patterns shown by forest and non-forest species? Are there differences in diversity patterns shown by lotic and lentic species? Are there areas with relatively high endemism, and do these match areas of endemism shown for other insect or animal groups? Are there areas with a relatively high percentage of globally threatened species, and how do these match with Odonate hotspots? Can areas be defined with a strong mismatch between predicted diversity and recorded diversity, and where is sampling least strong, so that information gaps can be closed?

2. Materials and Methods

2.1. Definition of Nearctic Realm

We followed the strict definition of Olson et al. [28] for the Nearctic realm, which does not include the three mountain ranges of northern and central Mexico, leaving this transition zone, along with several Neotropical Odonates, out of the study (Figure 1).

2.2. Species Occurrence Data

Species occurrence data were assembled by reviewing databases at the following entomological collections in Mexico and the USA: Alabama Museum of Natural History (ALMNH), Colección Nacional de Insectos Instituto de Biología Universidad Autónoma de México (CNIN/IBUNAM), Florida State Collection of Arthropods (FSCA), Instituto de Ecología Colección Entomológica, Xalapa (IEXA), Natalia von Ellenrieder (NvE), and Rosser W. Garrison (RWG). Furthermore, an extensive search of literature references was performed using the following keywords in English and Spanish: Mexico, USA, Canada, Odonata, dragonflies, damselflies, libélulas, and Norte América. The main references consulted were Behrstock et al. [29], Calvert [30,31], Cuevas–Yañez et al. [12], González-Soriano et al. [32], González-Soriano & Novelo-Gutiérrez [33], Escoto-Moreno et al. [13,34], Ortega-Salas and

González-Soriano [15], and Upson et al. [11]. Every locality lacking geographic coordinates was georeferenced using Google Earth [35] by searching for the location presented. The resulting coordinates were chosen when no more accurate information was available. The data were vetted and curated by examination of distribution maps by J.C.A., C.B.S., E.G.S., and R.N.G. Additionally, expert vetted records stored at Odonata Central [26], a public database for Odonate citizen science (www.odonatacentral.org, accessed on 10 May 2022), were also compiled for the Nearctic. All data used are compiled in Supplementary Material Table S1.

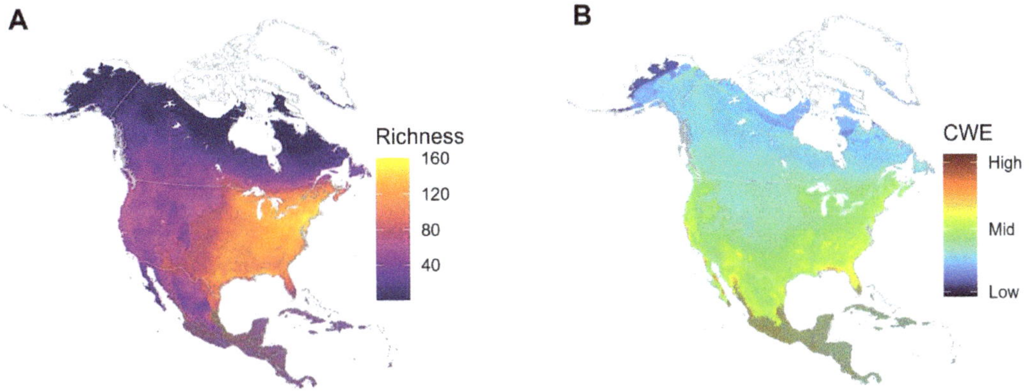

Figure 1. (**A**) Distribution of Odonata richness and (**B**) corrected weighted endemism (CWE) of Nearctic Odonates. Gray shading represents the Neotropical realm.

Using these occurrence records, we generated a list of all species of Odonata with non-vagrant records in the Nearctic. The Nearctic realm was defined using the WWF Biome 2 definition from a shapefile provided by The Nature Conservancy [28]. In total, there were 509 species that were determined to be residents of the Nearctic. While we only included species from the Nearctic, occurrences for those species with ranges outside the realm were included in our downstream modeling steps. Full species ranges are particularly critical for appropriately determining endemism and conservation status [36].

Once the initial occurrence data for Nearctic species were assembled, we ran the occurrence records through a cleaning pipeline in the R package *CoordinateCleaner* [37] that flagged records (1) with equal latitude and longitude coordinates, (2) within a 1000 m radius around the geographic centroids of political countries and provinces, and (3) with either zero longitude or latitude. Maps displaying both unflagged and flagged occurrence records were generated for each species in our Nearctic species list for expert review. During this step, expert review (J.C.A., C.B.S., E.G.S., and R.N.G.) was used to determine which occurrence records should be removed from the database because they were determined to be incorrect, generating a final dataset of curated occurrence records to be used for distribution modeling.

2.3. Functional Traits and Conservation Status

We subdivided our species list on the basis of two functional traits and conservation status as determined by the International Union for the Conservation of Nature (IUCN). Species were categorized as forest-dependent or non-forest dependent, as lentic or lotic-dependent, and according to conservation status. Information on habitat use, aquatic habitat by immature stages and terrestrial habitat by adults, were collected on the basis of the literature and expert knowledge. The following questions were used to classify each species: Can the species survive without forests? Can the species survive without a lotic environment? Those where the answer was "no" were labeled "forest obligate" or "lotic obligate", whereas, when the answer was yes, they were labeled as non-forest or lentic. We also included the IUCN conservation category of the 509 species in the Nearctic realm. All but one species were assessed, and the data were downloaded from the IUCN portal [27]. Threatened species included species with red list categories classified by the IUCN as either near-threatened, vulnerable, endangered, or critically endangered. After subdividing the species, we stacked the distributions of all species within each category and calculated the richness and CWE (corrected weighted endemism) values for each grid cell as explained below. While Calabrese et al. [38] suggested using continuous values for stacking ENMs (ecological niche models), we custom-tuned models during thresholding to avoid overfitting as described below and, thus, opted for stacking the thresholded outputs. We generated bivariate maps to visualize species richness and CWE for forest/non-forest species and lentic/lotic species on a single map. Bivariate categories were calculated by determining cells lower than the 33rd percentile, between the 33rd and 66th percentile, and greater than the 66th percentile of species richness given a certain trait.

To test if terrestrial or aquatic habitats used by Odonates has an effect on the overall range size, two Wilcoxon tests were performed using R software version 4.1.2 (R Core Team; Vienna, Austria) [39]. The first test used terrestrial habitat categorized as forest- and non-forest-dependent species as the predictor variable. The second test used aquatic habitat categorized as lotic or lentic species. The dependent variable for both tests was the range size predicted for every species measured by the total number of pixels in which the species was predicted to occur. We expected lentic species to have larger range sizes, as hypothesized by [40]. Similarly, forest species are expected to have smaller ranges than non-forest-dependent ones since they are highly specialized [41–43] in patchy distributed habitats.

2.4. Species Distribution Modeling

We built a species distribution pipeline in R to predict the distribution of all 509 species found in the Nearctic. This pipeline was strategically designed to efficiently model the distributions of hundreds of species, while including multiple steps that customize the process for each species.

First, we defined the accessible area, which was the geographic area where the distribution model was both fit and projected by generating a buffered alpha hull around the accepted occurrence records. The alpha hull was calculated using the getDynamicAlphaHullfunction from the R package *rangeBuilder* [44], where we set the fraction of occurrences that can fall outside of the polygon to be zero, with an initial alpha value of 20 and an allowed maximum of three disjunct polygons. We then buffered the alpha hull by the larger value of either 75 km or the 80th percentile distance between an occurrence record and the nearest occurrence records to ensure that the accessible area included areas that were accessible to a species through time [45]. These hulls were vetted for quality by expert curators (J.C.A., C.B.S., E.G.S., and R.N.G.).

Next, we spatially thinned occurrence records to remove potential spatial biases, where certain areas had more records than other areas, likely reflecting differences in human sampling effort more than changes in relative abundance across a landscape. Spatially thinning occurrence records were demonstrated to improve species distribution models using low-structure data sources [46]. We calculated the area of each accessible area in square meters using the area function in the R package raster [47] and retained all data points if a species' accessible area was less than 100,000 km^2. If a species had an accessible area $\geq$100,000 km^2 and <250,000 km^2, we only retained one occurrence record per 25 km grid; if accessible area was $\geq$250,000 km^2 and <1,000,000 km^2, one record per 50 km grid was retained; if accessible area was $\geq$1,000,000 km^2 and <2,500,000 km^2, one record per 100 km grid was retained; if accessible area was $\geq$2,500,000 km^2, one record per 200 km grid was retained. Even with thinning, there were still issues with data biases, requiring further efforts to tune model outputs, as discussed below.

After generating species-specific accessible areas and spatially thinning occurrence records, we fit an initial Maxent model [48] using default settings in the dismo package in R [49]. Maxent uses a machine learning algorithm to fit relationships between species occurrence records and background samples to environmental predictors [50]. Our initial model included 13 of the 19 bioclimatic variables provided by WorldClim (Table 1) [51]. These initial 13 variables were chosen to reduce multicollinearity in our initial model, while still including a number of bioclimatic variables we expect to be important to the ecological niche of Odonata. Initial bioclimatic variables had a spatial resolution of 30 s (~900 m at the equator) and were aggregated fivefold to the coarser resolution of approximately 4.5 km at the equator. Bioclimatic variables and occurrence records were reprojected to Lambert azimuthal equal area projection before analysis. To further avoid potentially problematic multicollinearity in our models, we calculated the variance inflation factors (VIF) of our initial model with all 13 bioclimatic variables [52]. If any predictor variable had a VIF >5, we removed the variable with the lowest permutation contribution to the model. We then fit a new Maxent model with default settings and repeated this step until no variables were retained in the model with a VIF greater than 5.

Table 1. Description of predictor variables included in our SDM modeling framework, and the mean permutation contribution of each variable averaged across all our top models.

Bioclimatic Variable	Description	Mean Permutation Contribution
Bio 8	Mean temperature of wettest quarter	14.4
Bio 2	Mean diurnal range	13.7
Bio 1	Annual mean temperature	11.1
Bio 4	Temperature seasonality	10.3
Bio 15	Precipitation seasonality	10.2
Bio 9	Mean temperature of driest quarter	9.6
Bio 5	Max temperature of warmest month	6.7
Bio 13	Precipitation of wettest month	5.4
Bio 14	Precipitation of driest month	5.0
Bio 12	Annual precipitation	3.7
Bio 6	Min temperature of coldest month	3.45
Bio 16	Precipitation of wettest quarter	3.2
Bio 17	Precipitation of driest quarter	3.1

Using the species-specific predictor variables determined by following the above process, we next used the R package *ENMeval* [53] to quantitatively evaluate a suite of Maxent models with different tuning parameters in an effort to optimize model complexity and prevent overfitting. We fit models for every combination of tuning parameters with regularization multipliers of 0.5, 1, 2, 3, and 4 and feature classes of "linear", "linear + quadratic", "hinge", "linear + quadratic + hinge", "linear + quadratic + hinge + product", and "linear + quadratic + hinge + product + threshold". Block partitioning of five random partitions was used to partition occurrence and background localities into training and testing bins. The model with the lowest AICc value was selected as our top model if it had training and validation AUC values greater than 0.7. In the rare cases where training or validation AUCs were less than 0.7, our top model was selected as the model with the highest validation AUC. To select a threshold value to transform our predicted Maxent model into a binary (presence/absence) surface, we reclassified our predicted Maxent model surface into a binary surface based on five different thresholding values. These values were the zeroth, first, 2.5th, fifth, and 10th percentiles of the predicted SDM on a ClogLog scale. Given these five binary surfaces, we calculated the sensitivity (percentage of actual presences predicted) and specificity (percentage of actual pseudo-absences predicted) for reclassified surfaces, where pseudo-absences were randomly generated within the accessible area and the number of pseudo-absences matched the number of spatially thinned occurrence records. An adapted true skill statistic (Equation (1)) was calculated to find a thresholding value that balances type I and type II errors, although specificity was given one-third the weight of sensitivity given the presence-only nature of our occurrence records. The percentile value that led to the highest true skill statistic was selected as our final thresholding value and used to generate the predicted presence/absence distribution.

$$TSS = \left(Sensitivity + \frac{1}{3} \times Sensitivity \right) - 1. \quad (1)$$

The top Maxent models and binary surfaces were mapped for each species and underwent expert evaluation by J.C.A. and C.B.S. Species with predicted distributions that did not pass expert evaluation were rerun after making custom changes to the modeling framework to improve predicted distributions. These custom changes included altering accessible areas, decreasing the number of background points for species with small accessible areas or few sample points, and altering the thresholding value. Altering threshold values was undertaken when there was clear evidence of over- or under-commission in model results.

2.5. Calculating Richness and Endemism

Predicted distributions were stacked for all species across their entire ranges, including areas outside of the Nearctic realm. Species richness and corrected weighted endemism (*CWE*) were calculated for each grid cell. Species richness here is defined as the number of species per cell. Weighted endemism (Equation (2)) uses a moving window analysis including the central cell and the eight neighboring cells to sum for each taxon t in the set of taxa T in the neighborhood, and the number of cells in the neighborhood containing taxon t (the local range, r_t) divided by its range (R_t, the number of cells in which it is found). *CWE* is the quotient of weighted endemism (*WE*) divided by richness (Equation (3) [54]).

$$WE = \Sigma_t \in T \frac{r_t}{R_t}. \quad (2)$$

$$CWE = \frac{WE}{Richness}. \quad (3)$$

3. Results

We found that each species had a custom combination of bioclimatic variables that best predicted the species distribution given our occurrence records and had variance inflation factors <5 (Table 1). Across all 509 species, the variables that had the highest permutation importance were the mean temperature of wettest quarter, mean diurnal range, annual mean temperature, temperature seasonality, and precipitation seasonality (Table 1).

3.1. Richness and Corrected Weighted Endemism (CWE)

Among the 509 species recorded in the Nearctic, 77 reach the transition zone within the central Mexico mountains ranges, 119 species also occur in the Neotropics beyond the Mexican transition zone, eight are shared with the Palearctic region, and one species, *Pantala flavescens*, occurs in the tropical and subtropical areas on all continents except Antarctica. *Crocothemis servilia* is an invasive species which is widespread in Asia. In sum, a total of 303 species are fully unique to the Nearctic (Figure 1A, Supplementary Table S1).

Species richness increases in the eastern region. The deserts in northern Mexico and the Rocky Mountains are notable with low species richness. The greatest number of endemics occurs along the southeastern coastal regions and the Mexican transition zone (Figure 1B).

3.2. Richness and Endemism by Terrestrial and Aquatic Habitats

There are a total of 116 forest-dependent species (22.7%, see Supplementary Table S1). Most forest-dependent species belong to the families Gomphidae (35.4% of its species), Coenagrionidae (17.5%), and Corduliidae (37.3%), along with all the species in the families

Platystictidae and Cordulegastridae. They are distributed throughout the Nearctic, but the highest diversity is found east of the 95° W meridian (Figure 2A). Forest-dependent species are also common in the southern part of the Nearctic along the Mexican transition zone and the southern forests of Canada (Figure 2A). There are 393 non-forest species, most of them in the families Libellulidae (95% of its species), Coenagrionidae (82.5%), Gomphidae (64.6%), and Aeshnidae (77.1%), with the families Lestidae and Macromiidae categorized as non-forest-dependent. CWE is higher for forest species in the northeast when compared to the non-forest species and higher for non-forest species in the west compared to forest species (Figure 2B). CWE for forest-dependent species is highest in the southeast (Figure 2C), whereas, for non-forest dependent species, it is highest in the southeast (central Florida) and California and Baja California in the west (Figure 2D).

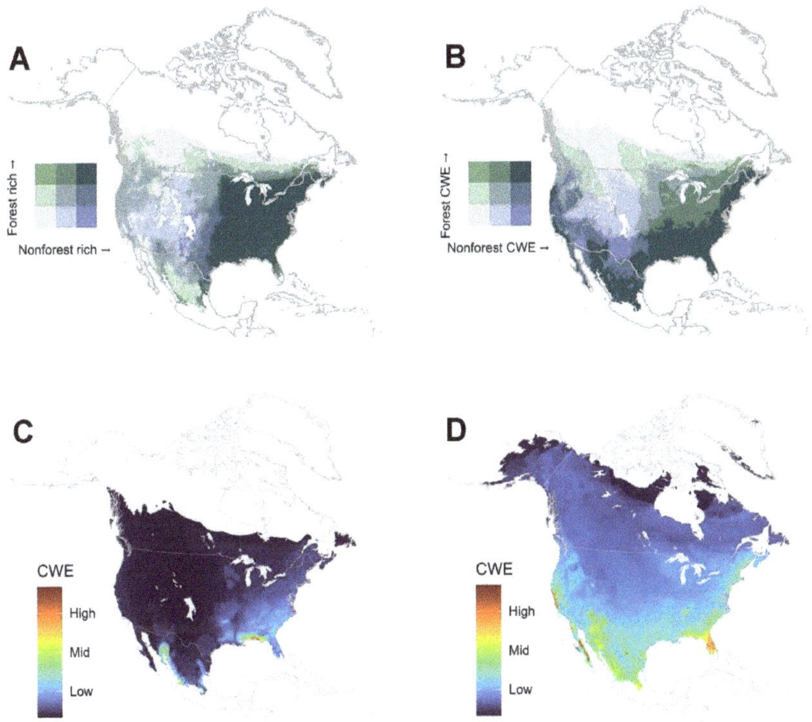

Figure 2. Species richness and corrected weighted endemism for terrestrial habitats used by Odonate species in the Nearctic realm. (**A**) Bivariate plot showing distribution of richness for forest-dependent and non-forest-dependent species; (**B**) bivariate plot showing distribution of corrected weighted endemism (CWE) for forest-dependent and non-forest-dependent species; (**C**) corrected weighted endemism (CWE) for forest-dependent species; (**D**) corrected weighted endemism (CWE) for non-forest-dependent species.

There are 221 strictly lotic species in the Nearctic (43.4%, see Supplementary Table S1), most of them in the families Gomphidae (82.3% of its species) and Coenagrionidae (46.8%), with species in the families Calopterygidae, Macromiidae, and Platystictidae being strictly lotic. There are proportionately more lotic species than lentic in the Mexican plateau near the transition zone and in southern Canada (Figure 3A). There are 288 lentic species, most of them in the families Libellulidae (88.3% of its species), Coenagrionidae (53.2%), and Corduliidae (52.9%); the families Aeshnidae, Lestidae, and Petaluridae do not depend on lotic habitats in the Nearctic. There is an equal proportion of relative lotic CWE to lentic CWE in much of the eastern, southern, and western areas of the Nearctic (Figure 3B).

CWE is highest for lentic species in the southeast (central Florida) and California and Baja California in the west (Figure 3C), whereas, for lotic-dependent species, it is highest in the southeast and Mexican transition zone (Figure 3D).

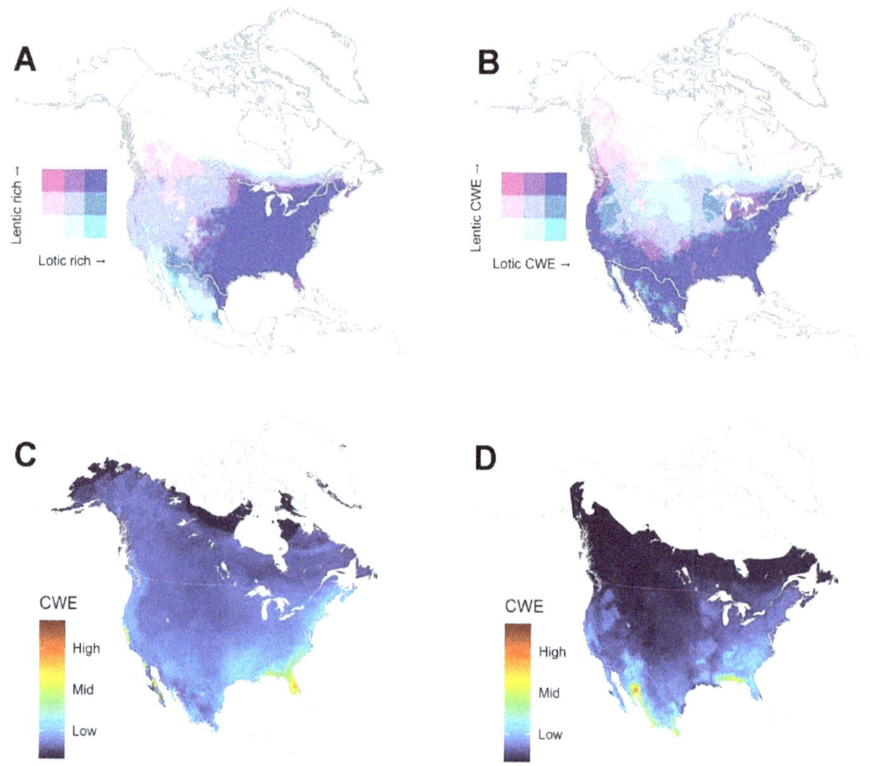

Figure 3. Species richness and corrected weighted endemism for aquatic habitats used by Odonate species in the Nearctic realm. (**A**) Bivariate plot showing distribution of richness for lotic-dependent and lentic species; (**B**) bivariate plot showing distribution of corrected weighted endemism (CWE) for lotic-dependent and lentic species; (**C**) corrected weighted endemism (CWE) for lentic-dependent species; (**D**) corrected weighted endemism (CWE) for lotic-dependent species.

We found that forest-dependent species, on average, have smaller ranges than non-forest-dependent species (W = 14,122, p= 1.567 $\times$ 10^{-9}, Figure 4A). Similarly, we found that lotic species, on average, have a smaller range than lentic species (W = 44,150, p = 2.347 $\times$ 10^{-15}, Figure 4B).

Most families contained both forest and non-forest dependent species, as well as lentic and lotic-dependent species (Figure 5). The eight largest families (Aeshnidae, Calopterygidae, Coenagrionidae, Corduliidae, Gomphidae, Lestidae, Libellulidae, and Macromiidae) had species categorized as exclusively non-forest-dependent or had most of their species characterized as non-forest-dependent.

The Petaluridae and Lestidae were characterized as uniformly requiring lentic environments. The largest family, the Coenagrionidae, contained about half lentic-, half lotic-dependent species. Within the next two largest families, there were predominately lotic-dependent species (Gomphidae) or lentic-dependent species (Libellulidae). Further examination of shifts between lotic and lentic environments will require a more complete phylogeny for North American taxa.

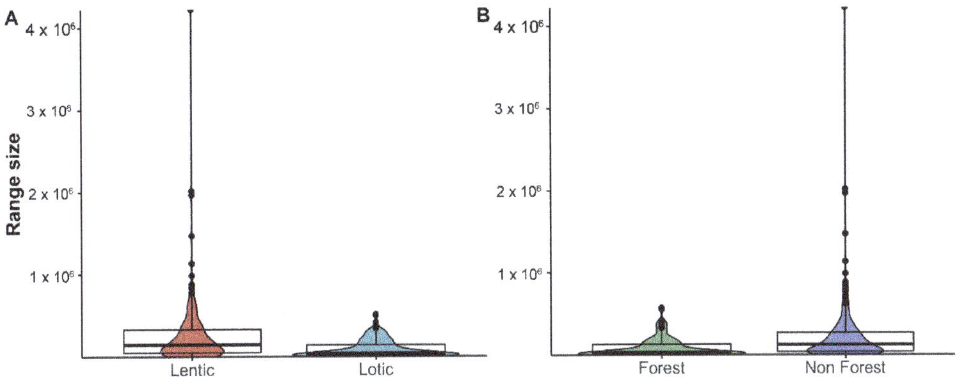

Figure 4. Range size (defined by the number of cells with predicted presence) for (**A**) lentic- and lotic-dependent Odonata species and (**B**) forest- and non-forest-dependent species.

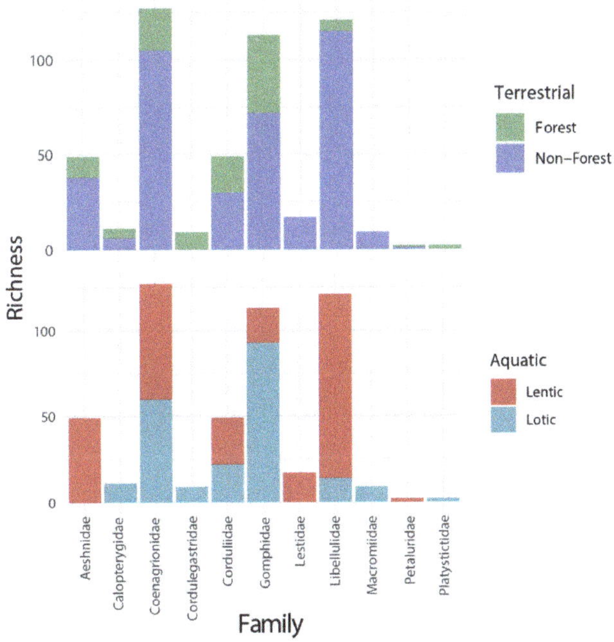

Figure 5. Species richness by family for forest- and non-forest-dependent species (above) and lentic- and lotic-dependent species (below).

3.3. Richness of Species According to IUCN Red List Category

Of the species found in the Nearctic, 10 are listed as data-deficient, six as vulnerable, seven as near-threatened, and three as endangered, while the majority, 482 species, are listed as least concern (Supplementary Table S1; Figure 6). Only *Hetaerina calverti*, a recently described species, is not yet assessed. The southeast and Mexican transition zone contain the majority of the endangered and data-deficient species (Figure 7).

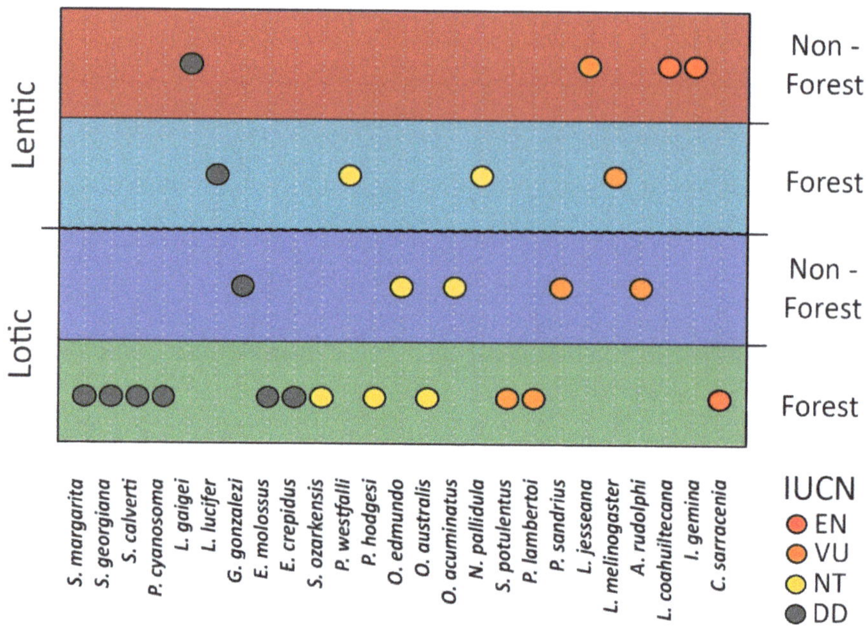

Figure 6. Terrestrial and aquatic habitats of the 24 species categorized as EN (endangered), VU (vulnerable), NT (near-threatened), and DD (data-deficient), according to the IUCN red list.

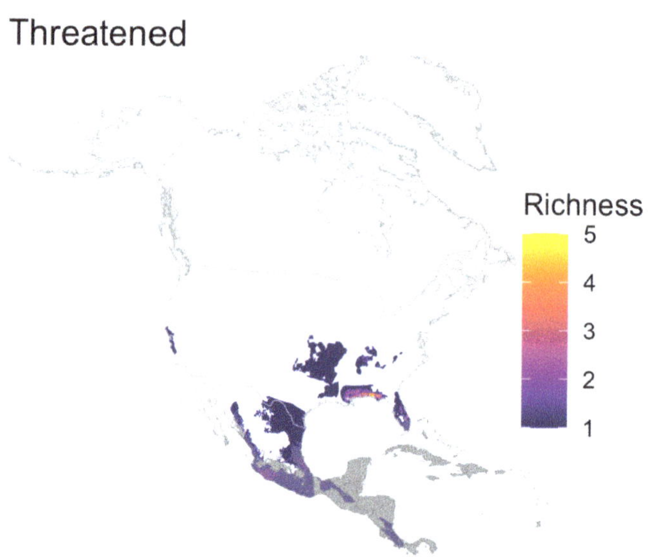

Figure 7. Distribution of the species richness for the IUCN Red List category of least concern.

3.4. Sampling Effort

Figure 8 shows the sampling effort and concentration across the Nearctic, where a high sampling effort can generally be observed except in Mexico, some places in the Rocky Mountains, parts of Canada, and Alaska. Some of the areas with the lowest sampling effort may indeed be areas where dragonflies are effectively absent; however, closing gaps via direct reporting of absences is still needed.

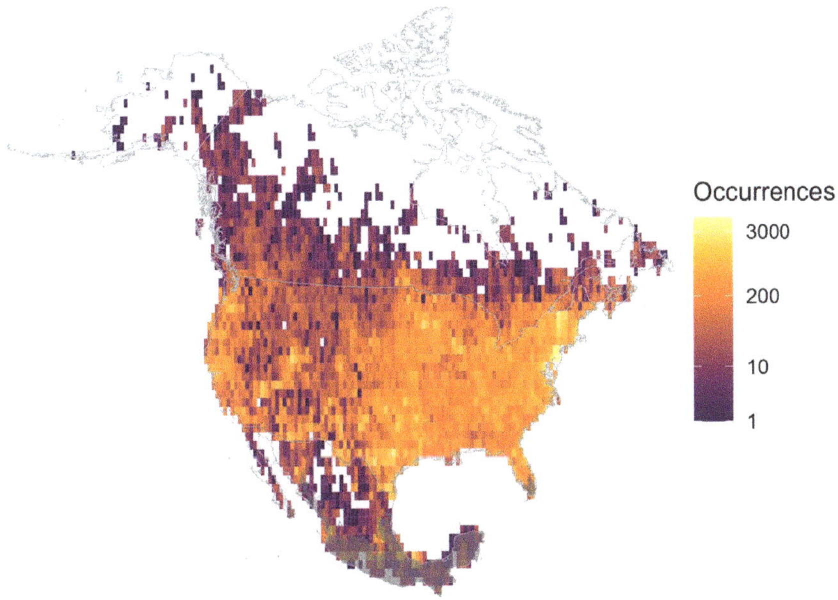

Figure 8. Odonata sampling effort in the Nearctic realm.

4. Discussion

4.1. General Diversity Patterns

The pattern of greater species richness in the eastern portion of the Nearctic (Figure 1A) can be explained by the higher aquatic habitat diversity at micro and macroscales east of the Rocky Mountains, promoting niche partitioning and specialization [55]. Much of the eastern portion of the Nearctic fauna may have benefitted from refuges for aquatic faunas being created by Pleistocene glaciation events, while extensive portions of western faunas were extirpated by glaciation [56–58]. This idea has not been directly tested, however, and deserves greater attention. Greater species richness in the east has been previously reported for Nearctic Odonata [20,56]. Higher species richness is also seen east of the Rocky Mountains within the Nearctic for other freshwater groups, e.g., rotifers, bivalves, amphipods, crayfish, fish, and turtles [59].

The southeastern US is home to the highest number of endemic dragonflies and damselflies in the Nearctic, likely due to glacial refuges, which provided a foundation for the evolution of a rich and unique biota in this area. The southeastern Nearctic is recognized as a hotspot of endemicity in groups such as turtles, fish, bivalves, gastropods, crayfish, and amphipods [57,58,60–63].

The greater topographical changes in the west have resulted in less aquatic habitat diversity than in the east [64,65], which is probably one of the main reasons for the lower species richness in these areas. Nonetheless, some species are unique to western Nearctic areas (Figure 1B). Similar distributions have been recorded in other aquatic insects, including stoneflies (Plecoptera) and caddisflies (Trichoptera), in which two clearly distinct components within the Nearctic fauna are observed: one in the east and one in the west [66–68]. The pattern is the opposite for butterflies and bumblebees, which do not depend on aquatic habitats and have the highest species richness in the western Nearctic [36,69], a mountainous region where plant diversity is also the highest [70].

We recognize that there is bias when using opportunistic naturalist occurrences (Odonata Central), as well as collection records from museums [71,72]. Most of the records are close to urbanized areas and roads, whereas some regions such as northern Mexico and northern Canada are clearly undersampled compared to the other Nearctic areas. Efforts toward closing these gaps should be conducted soon, especially in northern Mexico where new species have been found in recent years [73,74]. Nevertheless, we expect that the large-scale biodiversity pattern will not change significantly.

Compared to the Palearctic region (404 species), Odonate richness is similar, albeit with the whole of Europe supporting fewer than 140 species [75]. A number of freshwater groups show higher overall diversity in the Palearctic compared to the Nearctic, e.g., Ephemeroptera [76], Plecoptera [67], rotifers [55], gastropods [63], and amphipods [58]. The relative paucity, however, of freshwater vertebrate groups such as fish and turtles in the Palearctic, as compared to the Nearctic, is likely, in part, a result of the Pleistocene glaciations since there were more refugees in the Nearctic [57,61].

The composition of the Nearctic Odonate fauna is strongly influenced by Neotropical species. There is overlap in the Mexican transition zone with 23.4% of Nearctic species occurring in both the Nearctic and Neotropical realms. Outside of the Mexican transition zone, the shared species are mainly distributed along the coastal areas and the southeastern portion of the Nearctic.

Some species (e.g., *Lestes dryas, Aeshna juncea, Aeshna subarctica, Somatochlora sahlbergi*, and *Libellula quadrimaculata*) probably historically dispersed through the Bering Strait and are found in both the Palearctic and the Nearctic. They are mostly widely distributed within these areas or represent the northernmost-occurring Odonate species (e.g., *Somatochlora sahlbergi*), found north of the treeline. *Pantala flavescens* is probably the only species capable of consistent transoceanic dispersions without the aid of humans. Its amazing gliding abilities, together with a very fast larval cycle [77] and its adaptation to become dormant enabling it to survive drought conditions [78], make it the most widespread Odonata species in the world [79].

Outside of *P. flavescens*, probably only two species have crossed the Atlantic Ocean unaided by humans on more than a single occasion: *Anax junius*, native to the Nearctic which has not successfully reproduced in the Palearctic, and *Anax ephippiger*, native to Africa and southwest Asia, but with migrations spanning across Europe. Within the Neotropics, it has been found in the Lesser Antilles and French Guiana where it has successfully reproduced; it has, thus far, not been documented in the Nearctic.

In addition, *Crocothemis servilia*, a widespread Asian species, was first recorded in Florida in 1977 [80] and probably was carried as a nymph in the roots of aquarium plants (Buczynski and Bielak-Bielecki (2012) [81]); it is known as invasive in different countries. *Ischnura hastata* is one of the smallest species on the American continent. It is widespread in the Pacific and Caribbean Islands, and it occurs in both the Palearctic and the Neotropical realms. In the Palearctic, a population occurs on the Azores Islands within the Atlantic Ocean and is only composed of females where parthenogenetic reproduction occurs. It is thought that a gravid female was likely carried by wind to the islands and managed to survive the transatlantic flight [82,83].

4.2. Are There Differences in Diversity Patterns Shown by Forest and Non-Forest Species?

We classified 22.6% (115 species) as not being able to establish a population in the absence of forested habitat. The proportion of forest-dependent and non-forest-dependent species tends to be equal throughout much of the eastern Nearctic. This is explained by the greater diversity of Odonates throughout the east and this region supporting one of the largest areas of hardwood forests in the Nearctic. Heading westward, there is a greater proportion of non-forest-dependent species, many of which are wide-ranging with broad habitat requirements. Forest-dependent species are generally absent from the Great Plains and western Nearctic except for the Pacific Northwest, home to the only temperate rainforest in the west.

Northern Mexico is also home to a higher proportion of forest-dependent species which are largely Neotropical in origin. Northern Mexico is largely arid (except the Cuatro Cienagas area, which stands out for its high biodiversity [15]) and is known for a low species richness. However, forested areas within the zone run along the river margins and support a larger number of habitat restricted specialists. The area is also under-sampled compared to the rest of the Nearctic.

We found that non-forest-dependent species generally have larger range sizes than forest-dependent species, similar to the pattern reported in the tropical Andes, where non-forest species occupy larger elevation ranges compared to forest-dependent species [84]. This is likely due to forest-dependent species being more specialized [41–43] than open-area species which can reproduce in a wide variety of broadly available habitats.

4.3. Are There Differences in Diversity Patterns Shown by Lotic and Lentic Species?

We classified 43.4% (221 species) as unable to have lasting populations in the absence of a lotic habitat. The proportion of lotic-dependent species tends to be higher than the lentic species in the arid southwest and mountain areas in northern Mexico and some areas in the west. Most of the lentic environments in these regions are ephemeral [85], which does not favor the establishment of most Odonate species needing permanent water bodies to complete their life cycle [86]. There is an increase in the proportion of lentic species in Florida and in the Pacific northwest, likely due to the predominance of wetlands in these regions.

We found that lentic species generally have larger range sizes than lotic-dependent ones (Figure 5). Moreover, most of the species that are shared between the Nearctic and other realms (e.g., Neotropical and Palearctic) are lentic. These observations align with Hof et al. (2006) who analyzed latitudinal ranges for Odonates occurring in Europe and North America and hypothesized that lentic bodies of water are likely more ephemeral, thus favoring species with more effective dispersal abilities; as a result, lentic species would have larger ranges.

In the east, where the highest number of species occurs, the proportion of lentic and lotic species is similar, but there are still some areas where lentic endemicity is higher. Most of the species endemicity is in the southern area of the Nearctic toward coastlines, where the proportion of lotic and lentic endemic species is similar throughout most of the area, aligning with the glacial refugia explanation. Lotic CWE is proportionally higher in the mountainous regions of northern Mexico and the Rockies, where the overall lotic species richness is also higher than the lentic species richness.

4.4. Are There Areas with Relatively High Endemism?

The general pattern of CWE reveals hotspots in the southeast and west coast extending southward into Baja California and northern Mexico, extending northward into the southwestern United States (Figure 1B). The areas of highest CWE for both forest-dependent and non-forest-dependent species is in the extreme southeastern portion of the Nearctic (Figure 2C,D) along the surrounding areas of the Gulf Coast of Mexico, likely a result of refugia that resulted from the Pleistocene glaciation [56]. Non-forest-dependent CWE is highest in central Florida or the southeasternmost portion of the Nearctic and a result of species "spillover" from Caribbean Islands and northern expansion of the tropical Mexican fauna. Other areas of high CWE for non-forest-dependent species are found along the coastline of California and southward into Baja California. The former is due to low species richness (only 51 species known along the central coastline) and a relatively high proportion (10 species) with a restricted range in that area. Range-restricted species in Baja California are remnant northern extensions of the Neotropical fauna.

CWE for lentic- and lotic-dependent species shows a similar pattern to forest- and non-forest-dependent species with hotspots occurring in the southeast and northern Mexico (Figure 3C,D). This is expected, because most species that have limited ranges are associated with lotic environments which are, in turn, generally found in forested areas, and, as discussed previously, in the Nearctic, these refugial habitats are found in the southeast. The Cuatro Ciénegas Basin in north central Mexico is a particular hotspot for lotic species (Figure 3D). It is a protected nature reserve supporting inflowing rivers and streams, as well as pools providing unique aquatic habitats for Odonates.

4.5. Are There Areas with a Relatively High Percentage of Globally Threatened Species?

The areas of highest globally threatened species (Figure 7) correspond to areas of high CWE, which is no surprise given that range size is a key criterion used for the assessment of the Odonate species in the IUCN red list, and that data on the population dynamics are available for very few species [87]. As expected, most of the endangered species are found principally in the southeast. Within the Nearctic, there are three species ranked by the IUCN Red List as endangered (*Cordulegaster sarracenia, Ischnura gemina,* and *Libellula coahuiltecana*); two of them inhabit lentic environments in non-forested areas and one is found in lotic forested habitats. Six additional species listed as vulnerable (*Argia rudolphi, Leptobasis melinogaster, Libellula jesseana, Phanogomphus sandrius, Progomphus lambertoi,* and *Stylurus potulensis*), and seven additional species are assessed as near-threatened (*Nehalennia pallidula, Ophiogomphus acuminatus, Ophiogomphus australis, Ophiogomphus edmundo, Phanogomphus hodgesi, Phanogomphus westfalli,* and *Somatochlora ozarkensis*). Of these 16 species, half belong to the family Gomphidae; eight are forest-dependent and eight are lotic-dependent. Eleven species are found in the southeast, and, of the remaining three, one is restricted to the central west coast of California and two are restricted to the northern Mexican area. Exploration of less sampled areas in northern Mexico especially may lead to discovery of range-restricted species given the existing evidence of endemism and relative paucity of sampling compared to most other regions. This area certainly deserves more collection focus and has already yielded the discovery of new species [73,74,88].

5. Conclusions

Evaluating the aquatic life cycle and requirements of dragonflies and damselflies along with the recent geological history is key to understanding their diversity distribution patterns in the Nearctic. As aquatic insects, Pleistocene glaciations likely strongly constrain Odonate distributions to refugia mainly found in the southeast. These refugia also likely served as dispersal corridors for Neotropical species through the Caribbean Islands and Central America. These long-term areas of ample aquatic habitat, including today, serve as a major driver for Odonate diversity. Nevertheless, further work utilizing phylogenetic tools can provide an even sharper view of historical forces shaping current Nearctic diversity.

Terrestrial habitats are also important for Odonate distribution, with forest specialized species occupying smaller geographic ranges compared with non-specialists. Nevertheless, this pattern is different from that observed in strictly terrestrial insects such as butterflies and bumblebees, which are tied to plants. In these groups the diversity increases along with plant diversity in mountainous areas. Odonates do not follow this pattern, probably because they are hunters, and their specialization in forest habitats depends more on the structure of the vegetation (e.g., open-understory and thick-branched habitats), rather than on the diversity of the plant species.

Supplementary Materials: The following supporting information can be downloaded at https://www.mdpi.com/article/10.3390/d14070575/s1: Supplementary Table S1. List of the Odonata species recorded in the Nearctic, with their IUCN conservation status, their terrestrial and aquatic habitat, and the biogeographic realms where they occur. En: endangered, LC: least concern, NA: not assessed, NT: near-threatened, Vu: vulnerable. The table is also available at (https://doi.org/10.5281/zenodo.6544045 (accessed on 13 July 2022).

Author Contributions: Conceptualization, J.C.A., C.A.B.-S., R.G., V.K. and S.B.; methodology, J.C.A., C.A.B.-S., R.G., V.K. and S.B.; software, R.G. and M.W.B.; validation, J.C.A., C.A.B.-S., E.G.-S. and R.N.-G.; formal analysis, J.C.A., C.A.B.-S., R.G. and M.W.B.; data curation, J.C.A., C.A.B.-S., E.G.-S. and R.N.-G.; writing—original draft preparation, J.C.A., C.A.B.-S., R.G., V.K., E.G.-S., R.N.-G. and M.W.B.; writing—review and editing, J.C.A., C.A.B.-S., R.G., V.K., E.G.-S., R.N.-G., S.B., J.W. and M.W.B.; project administration, J.C.A., R.G., S.B. and J.W.; funding acquisition, J.C.A., R.G., S.B. and J.W. All authors contributed critically to the drafts and gave final approval for publication. All authors have read and agreed to the published version of the manuscript.

Funding: This work was supported by NSF Awards (Abbott: 2002489; Ware: 2002473; Bybee: 2002432; Guaralnick: 2002457), Genealogy of Odonata (GEODE): Dispersal and color as drivers of 300 million years of global dragonfly evolution.

Institutional Review Board Statement: Not applicable.

Informed Consent Statement: Not applicable.

Data Availability Statement: Data are provided in the paper and Supplementary Table S1. The composite richness and endemism values for each cell (released as GeoTIFF), the composite richness and endemism values for each cell broken out by trait (lentic, lotic, forest, and non-forest; released as 4 CSVs), and the species-specific SDMs showing the predicted areas of presence for each species (released as seven zip files) are available from a Zenodo repository (https://doi.org/10.5281/zenodo.6544045 (accessed on 13 July 2022)).

Acknowledgments: We are grateful to Rosser Garrison and Natalia von Ellenrieder for sharing their distributional database for Odonates. Bill Mauffray shared the database for specimens in the FSCA.

Conflicts of Interest: The authors declare no conflict of interest.

References

1. Hagen, H.A. Synopsis of the Odonata of America. *Proc. Boston Soc. Nat. Hist.* **1875**, *18*, 20–96.
2. Needham, J.G.; Westfall, M.J.; May, M. *Dragonflies of North America: The Odonata (Anisoptera) Fauna of Canada, the Continental United States, Northern Mexico and the Greater Antilles*; Scientific Publishers: Singapore, 2014.
3. Davis, J. *Dragon Flies: The Fascinating World of Odonata in Art, Culture and Religion*; 2011.
4. Hillerman, T. *The Boy Who Made Dragonfly*, 2nd ed.; University of New Mexico Press: Albuquerque, NM, USA, 1986.
5. Fincke, O.M. Sperm Competition in the Damselfly Enallagma Hageni Walsh (Odonata: Coenagrionidae): Benefits of Multiple Mating to Males and Females. *Behav. Ecol. Sociobiol.* **1984**, *14*, 235–240. [CrossRef]
6. Garrison, R. A Synopsis of the Genus Argia of the United States with Keys and Descriptions of New Species, *Argia sabino, A. leonorae*, and *A. pima* (Odonata: Coenagrionidae). *Trans. Am. Entomol. Soc.* **1994**, *120*, 287–368.
7. May, M. Simultaneous Control of Head and Thoracic Temperature by the Green Darner Dragonfly *Anax junius* (Odonata: Aeshnidae). *J. Exp. Biol.* **1995**, *198*, 2373–2384. [CrossRef] [PubMed]
8. Ware, J.; May, M.; Kjer, K. Phylogeny of the Higher Libelluloidea (Anisoptera: Odonata): An Exploration of the Most Speciose Superfamily of Dragonflies. *Mol. Phylogenet. Evol.* **2007**, *45*, 289–310. [CrossRef] [PubMed]
9. Dunkle, S.W. *Dragonflies through Binoculars: A Field Guide to Dragonflies of North America*; Oxford University Press: New York, NY, USA, 2000.
10. Tennessen, K.J. *Dragonfly Nymphs of North America: An Identification Guide*; Springer: Berlin, Germany, 2019.
11. Upson, S.; Danforth, D.; Gonzalez-Soriano, E.; Behrstock, R.A.; Bailowitz, R.A. A Preliminary Checklist of the Odonata of Sonora, Mexico. *Bull. Am. Odonatol.* **2007**, *10*, 23–51.
12. Cuevas-Yañez, K. Los Odonatos (Insecta: Odonata) de La Hidroeléctrica de Patla (El Pozo) y del Río Tecpatlán, Zihuateutla, Puebla, México. *Dugesiana* **2007**, *14*, 83–91.
13. Escoto-Moreno, J.A.; Novelo-Gutiérrez, R.; Sigala-Rodríguez, J.; Escoto-Rocha, J.; Carrillo-Lara, D.E.; Reynoso-Velasco, D. First Records of Odonata from Zacatecas State, Mexico. *Not. Odonatol.* **2015**, *8*, 151–155.
14. Novelo-Gutiérrez, R.; Gonzalez-Soriano, E. Odonata de La Reserva de La Biosfera La Michilia, Durango, Mexico. Parte II. Nayades. *Folia Entomol. Mex.* **1991**, *164*, 107–164.
15. Ortega-Salas, H.; Gonzalez-Soriano, E. Odonata of the Cuatro Ciénegas Basin. In *Animal Diversity and Biogeography of the Cuatro Ciénegas Basin*; Álvarez, F., Ojeda, M., Eds.; Springer: Berlin, Germany, 2019; pp. 117–128. ISBN 978-3-030-11261-5.
16. Cuevas–Yañez, K.; Espinosa–Rivera, J.C.; Martínez–Falcón, A.P.; Córdoba–Aguilar, A. Are All Mexican Odonate Species Documented? An Assessment of Species Richness. *Syst. Biodivers.* **2017**, *15*, 253–258. [CrossRef]
17. Escalante, T.; Morrone, J.J. Toward a Biogeographic Regionalization of the Nearctic Region: Area Nomenclature and Digital Map. *Zootaxa* **2021**, *5027*, 351–375. [CrossRef] [PubMed]
18. Halffter, G.; Morrone, J.J. An Analytical Review of Halffter's Mexican Transition Zone, and Its Relevance for Evolutionary Biogeography, Ecology and Biogeographical Regionalization. *Zootaxa* **2017**, *4226*, 1–46. [CrossRef] [PubMed]
19. Morrone, J.J. *The Mexican Transition Zone*; Springer: Cham, Switzerland, 2020; ISBN 9783030479169.
20. Paulson, D.R.; Dunkle, S.W. A Checklist of North American Odonata: Including English Name, Etymology, Type Locality, and Distribution. 2012 Edition. *Occas. Pap. Slater Museum Nat. Hist.* **2021**, *56*, 1–86.
21. Bailowitz, R.; Danforth, D.; Upson, S. *A Field Guide to the Dragonflies & Damselflies of Arizona and Sonora*; Nova Granada Publications: Tucson, AZ, USA, 2015.
22. Paulson, D.R. *Dragonflies and Damselflies of the West*; Princeton University Press: Princeton, NJ, USA, 2009.
23. Paulson, D.R. *Dragonflies and Damselflies of the East.*; Princeton University Press: Princeton, NJ, USA, 2011.

24. González-Soriano, E.; Novelo-Gutiérrez, R. Odonata. In *Biodiversidad, Taxonomia y Biogeografia de Artropodos de Mexico: Hacia una Sintesis de Su Conocimiento*; Llorente-Bousquets, J., Garcia-Aldrete, A.N., González-Soriano, E., Eds.; Facultad de Ciencias UNAM México City, Mexico, 1996; pp. 147–167.
25. González-Soriano, E.; Novelo-Gutiérrez, R. Odonata of Mexico revisited. In *Odonata: Biology of Dragonflies*; Tyagi, B.K., Ed. Scientific Publishers: Jodhpur, India, 2007; pp. 105–136.
26. Abbott, J.C. Odonata Central. Available online: http://www.odonatacentral.org (accessed on 3 February 2022).
27. IUCN. The IUCN Red List of Threatened Species. Available online: https://www.iucnredlist.org (accessed on 2 March 2022).
28. Olson, D.M.; Dinerstein, E.; Wikramanayake, E.D.; Burgess, N.D.; Powell, G.V.N.; Underwood, E.C.; D'amico, J.A.; Itoua, I.; Strand, H.E.; Morrison, J.C.; et al. Terrestrial Ecoregions of the World: A New Map of Life on Earth. *Bioscience* **2001**, *51*, 933 [CrossRef]
29. Behrstock, R.A.; Danforth, D.; Upson, S. A List of the Odonata of Chihuahua State, Mexico, Including New State Records and the First Mexican Record of *Argia Alberta*, Kennedy, 1918. *Bull. Am. Odonatol.* **2007**, *10*, 52–63.
30. Calvert, P.P. Odonata. In *Biologia Centrali Americana: Insecta Neuroptera*; R. Porter & Dulau Co.: London, UK, 1892–1908; pp. 17–420.
31. Calvert, P.P. Contributions to a Knowledge of the Odonata of the Neotropical Region, Exclusive of Mexico and Central America. *Ann. Carnegie Museum* **1909**, *6*, 73–280.
32. González-Soriano, E.; Noguera, F.A.; Zaragoza-Caballero, S.; Ramírez-García, E. Odonata de un Bosque Tropical Caducifolio: Sierra de San Javier, Sonora, México. *Rev. Mex. Biodivers.* **2009**, *80*, 341–348. [CrossRef]
33. Gonzalez-Soriano, E.; Novelo-Gutiérrez, R. Odonata de La Reserva de La Biosfera de La Michilia, Durango, Mexico. Parte I. Imagos. *Folia Entomol. Mex.* **1991**, *105*, 67–105.
34. Escoto-Moreno, J.A.; Márquez, J.; Novelo-Gutiérrez, R. Los Odonatos (Insecta: Odonata) Del Estado de Hidalgo, México: Situación Actual y Perspectivas. *Rev. Mex. Biodivers.* **2014**, *85*, 1043–1053. [CrossRef]
35. Google Earth Pro 2021.
36. Earl, C.; Belitz, M.W.; Laffan, S.W.; Barve, V.; Barve, N.; Soltis, D.E.; Allen, J.M.; Soltis, P.S.; Mishler, B.D.; Kawahara, A.Y.; et al. Spatial Phylogenetics of Butterflies in Relation to Environmental Drivers and Angiosperm Diversity across North America. *iScience* **2021**, *24*, 102239. [CrossRef]
37. Zizka, A.; Silvestro, D.; Andermann, T.; Azevedo, J.; Duarte Ritter, C.; Edler, D.; Farooq, H.; Herdean, A.; Ariza, M.; Scharn, R.; et al. CoordinateCleaner: Standardized Cleaning of Occurrence Records from Biological Collection Databases. *Methods Ecol. Evol.* **2019**, *10*, 744–751. [CrossRef]
38. Calabrese, J.M.; Certain, G.; Kraan, C.; Dormann, C.F. Stacking Species Distribution Models and Adjusting Bias by Linking Them to Macroecological Models. *Glob. Ecol. Biogeogr.* **2014**, *23*, 99–112. [CrossRef]
39. R Core Team. *R: A Language and Environment for Statistical Computing*; R Foundation for Statistical Computing: Vienna, Austria, 2019.
40. Hof, C.; Brändle, M.; Brandl, R. Lentic Odonates Have Larger and More Northern Ranges than Lotic Species. *J. Biogeogr.* **2006**, *33*, 63–70. [CrossRef]
41. Corbet, P.S. Forests as habitats for dragonflies (Odonata). In *Forests and Dragonflies*; Cordero-Rivera, A., Ed.; Pensoft: Sofia, Bulgaria; Moscow, Russia, 2006; pp. 13–35.
42. Sahlén, G. Specialists vs. generalists among dragonflies—The importance of forest environments in the formation of diverse species pools. In *Forests and Dragonflies*; Cordero-Rivera, A., Ed.; Pensoft: Sofia, Bulgaria; Moscow, Russia, 2006; pp. 153–179.
43. Bota-Sierra, C.A.; García-Robledo, C.; Escobar, F.; Novelo-Gutiérrez, R.; Londoño, G.A. Environment, Taxonomy and Morphology Constrain Insect Thermal Physiology along Tropical Mountains. *Funct. Ecol.* **2022**. early view. [CrossRef]
44. Davis Rabosky, A.R.; Cox, C.L.; Rabosky, D.L.; Title, P.O.; Holmes, I.A.; Feldman, A.; McGuire, J.A. Coral Snakes Predict the Evolution of Mimicry across New World Snakes. *Nat. Commun.* **2016**, *7*, 11484. [CrossRef]
45. Barve, N.; Barve, V.; Jiménez-Valverde, A.; Lira-Noriega, A.; Maher, S.P.; Peterson, A.T.; Soberón, J.; Villalobos, F. The Crucial Role of the Accessible Area in Ecological Niche Modeling and Species Distribution Modeling. *Ecol. Modell.* **2011**, *222*, 1810–1819. [CrossRef]
46. Steen, V.A.; Elphick, C.S.; Tingley, M.W. An Evaluation of Stringent Filtering to Improve Species Distribution Models from Citizen Science Data. *Divers. Distrib.* **2019**, *25*, 1857–1869. [CrossRef]
47. Hijmans, R.J. *Package 'Raster'—Geographic Data Analysis and Modeling*; CRAN Repos, 2019. Available online: https://rspatial.org/raster (accessed on 10 May 2022).
48. Phillips, S.J.; Anderson, R.P.; Dudík, M.; Schapire, R.E.; Blair, M.E. Opening the Black Box: An Open-Source Release of Maxent. *Ecography* **2017**, *40*, 887–893. [CrossRef]
49. Hijmans, R.J.; Phillips, S.; Leathwick, J.; Elith, J. Dismo: Species Distribution Modeling; R Package Version 1.1-4; Cran. 2017. Available online: https://cran.r-project.org/web/packages/dismo/dismo.pdf (accessed on 10 May 2022).
50. Merow, C.; Smith, M.J.; Silander, J.A. A Practical Guide to MaxEnt for Modeling Species' Distributions: What It Does, and Why Inputs and Settings Matter. *Ecography* **2013**, *36*, 1058–1069. [CrossRef]
51. Fick, S.E.; Hijmans, R.J. WorldClim 2: New 1-Km Spatial Resolution Climate Surfaces for Global Land Areas. *Int. J. Climatol.* **2017**, *37*, 4302–4315. [CrossRef]
52. Naimi, B.; Hamm, N.A.S.; Groen, T.A.; Skidmore, A.K.; Toxopeus, A.G. Where Is Positional Uncertainty a Problem for Species Distribution Modelling? *Ecography* **2014**, *37*, 191–203. [CrossRef]

53. Kass, J.M.; Muscarella, R.; Galante, P.J.; Bohl, C.L.; Pinilla-Buitrago, G.E.; Boria, R.A.; Soley-Guardia, M.; Anderson, R.P. ENMeval 2.0: Redesigned for Customizable and Reproducible Modeling of Species' Niches and Distributions. *Methods Ecol. Evol.* **2021**, *12*, 1602–1608. [CrossRef]
54. González-Orozco, C.E.; Laffan, S.W.; Miller, J.T. Spatial Distribution of Species Richness and Endemism of the Genus Acacia in Australia. *Aust. J. Bot.* **2011**, *59*, 601–609. [CrossRef]
55. Segers, H. Global Diversity of Rotifers (Rotifera) in Freshwater. *Hydrobiologia* **2008**, *595*, 49–59. [CrossRef]
56. Kalkman, V.J.; Clausnitzer, V.; Dijkstra, K.D.B.; Orr, A.G.; Paulson, D.R.; Van Tol, J. Global Diversity of Dragonflies (Odonata) in Freshwater. *Hydrobiologia* **2008**, *595*, 351–363. [CrossRef]
57. Lévêque, C.; Oberdorff, T.; Paugy, D.; Stiassny, M.L.J.; Tedesco, P.A. Global Diversity of Fish (Pisces) in Freshwater. *Hydrobiologia* **2008**, *595*, 545–567. [CrossRef]
58. Väinölä, R.; Witt, J.D.S.; Grabowski, M.; Bradbury, J.H.; Jazdzewski, K.; Sket, B. Global Diversity of Amphipods (Amphipoda; Crustacea) in Freshwater. *Hydrobiologia* **2008**, *595*, 241–255. [CrossRef]
59. Balian, E.V.; Segers, H.; Lévèque, C.; Martens, K. The Freshwater Animal Diversity Assessment: An Overview of the Results. *Hydrobiologia* **2008**, *595*, 627–637. [CrossRef]
60. Bogan, A.E. Global Diversity of Freshwater Mussels (Mollusca, Bivalvia) in Freshwater. *Hydrobiologia* **2008**, *595*, 139–147. [CrossRef]
61. Bour, R. Global Diversity of Turtles (Chelonii; Reptilia) in Freshwater. *Hydrobiologia* **2008**, *595*, 593–598. [CrossRef]
62. Crandall, K.A.; Buhay, J.E. Global Diversity of Crayfish (Astacidae, Cambaridae, and Parastacidae—Decapoda) in Freshwater. *Hydrobiologia* **2008**, *595*, 295–301. [CrossRef]
63. Strong, E.E.; Gargominy, O.; Ponder, W.F.; Bouchet, P. Global Diversity of Gastropods (Gastropoda; Mollusca) in Freshwater. *Hydrobiologia* **2008**, *595*, 149–166. [CrossRef]
64. Batzer, D.P.; Baldwin, A.H. *Wetland Habitats of North America: Ecology and Conservation Concerns*; University of California Press: Berkley, CA, USA, 2012.
65. Rey, J.R.; Walton, W.E.; Wolfe, R.J.; Roxanne, C.; O'Connell, S.M.; Berg, J.; Sakolsky-Hoopes, G.E.; Laderman, A.D. North American Wetlands and Mosquito Control. *Int. J. Environ. Res. Public Health* **2012**, *9*, 4537–4605. [CrossRef]
66. Stewart, K.W.; Stark, B.P. *Nymphs of North American Stonefly Genera (Plecoptera)*; The Caddis Press: Columbus, OH, USA, 2002.
67. Fochetti, R.; Tierno De Figueroa, J.M. Global Diversity of Stoneflies (Plecoptera; Insecta) in Freshwater. *Hydrobiologia* **2008**, *595*, 365–377. [CrossRef]
68. De Moor, F.C.; Ivanov, V.D. Global Diversity of Caddisflies (Trichoptera: Insecta) in Freshwater. *Hydrobiologia* **2008**, *595*, 393–407. [CrossRef]
69. Williams, P.H. Mapping Variations in the Strength and Breadth of Biogeographic Transition Zones Using Species Turnover. *Proc. R. Soc. B Biol. Sci.* **1996**, *263*, 579–588. [CrossRef]
70. Mishler, B.D.; Guralnick, R.; Soltis, P.S.; Smith, S.A.; Soltis, D.E.; Barve, N.; Allen, J.M.; Laffan, S.W. Spatial Phylogenetics of the North American Flora. *J. Syst. Evol.* **2020**, *58*, 393–405. [CrossRef]
71. Collins, S.D.; Abbott, J.C.; McIntyre, N.E. Quantifying the Degree of Bias from Using County-Scale Data in Species Distribution Modeling: Can Increasing Sample Size or Using County-Averaged Environmental Data Reduce Distributional Overprediction? *Ecol. Evol.* **2017**, *7*, 6012–6022. [CrossRef] [PubMed]
72. Botella, C.; Joly, A.; Monestiez, P.; Bonnet, P.; Munoz, F. Bias in Presence-Only Niche Models Related to Sampling Effort and Species Niches: Lessons for Background Point Selection. *PLoS ONE* **2020**, *15*, e0232078. [CrossRef]
73. Ortega-Salas, H.; González-Soriano, E. A New Species of Libellula Linnaeus, 1758, from the Cuatro Ciénegas Basin, Coahuila, México (Anisoptera: Libellulidae). *Zootaxa* **2015**, *4028*, 589–594. [CrossRef]
74. Vega-Sánchez, Y.M.; Mendoza-Cuenca, L.F.; González-Rodríguez, A. Hetaerina Calverti (Odonata: Zygoptera: Calopterygidae) sp. Nov., a New Cryptic Species of the American Rubyspot Complex. *Zootaxa* **2020**, *4766*, 485–497. [CrossRef] [PubMed]
75. Kalkman, V.J.; Boudot, J.-P.; Futahashi, R.; Abbott, J.C.; Bota-Sierra, C.A.; Guralnick, R.; Bybee, S.M.; Ware, J.L.; Belitz, M.W. Diversity of Palaearctic Dragonflies and Damselflies (Odonata). *Diversity* **2022**. in preparation.
76. Barber-James, H.M.; Gattolliat, J.L.; Sartori, M.; Hubbard, M.D. Global Diversity of Mayflies (Ephemeroptera, Insecta) in Freshwater. *Hydrobiologia* **2008**, *595*, 339–350. [CrossRef]
77. Suhling, F.; Schenk, K.; Padeffke, T.; Martens, A. A Field Study of Larval Development in a Dragonfly Assemblage in African Desert Ponds (Odonata). *Hydrobiologia* **2004**, *528*, 75–85. [CrossRef]
78. Van Damme, K.; Dumont, H.J. A Drought-Resistant Larva of *Pantala flavescens* (Fabricius, 1798) (Odonata: Libellulidae) in the Lençóis Maranhenses, NE-Brazil. *Int. J. Odonatol.* **1999**, *2*, 69–76. [CrossRef]
79. Ware, J.; Kohli, M.K.; Mendoza, C.M.; Troast, D.; Jinguji, H.; Hobson, K.A.; Sahlén, G.; Anderson, R.C.; Suhling, F. Evidence for Widespread Gene Flow and Migration in the Globe Skimmer Dragonfly Pantala Flavescens. *Int. J. Odonatol.* **2022**, *25*, 43–55. [CrossRef]
80. Paulson, D.R. An Asiatic Dragonfly, *Crocothemis servilia* (Drury), Established in Florida (Anisoptera: Libellulidae). *Not. Odonatol.* **1978**, *1*, 9–10.
81. Buczyński, P.; Bielak-Bielecki, P. *Crocothemis servilia* (Drury, 1773) (Odonata: Libellulidae) Introduced with Aquarium Plants to Lublin (Poland). *Ann. UMCS Biol.* **2012**, *67*, 21–26. [CrossRef]

82. Cordero Rivera, A.; Lorenzo Carballa, M.O.; Utzeri, C.; Vieira, V. Parthenogenetic *Ischnura hastata* (Say), Widespread in the Azores (Zygoptera: Coenagrionidae). *Odonatologica* **2005**, *34*, 1–9.
83. Lorenzo-Carballa, M.O.; Beatty, C.D.; Utzeri, C.; Vieira, V.; Cordero-Rivera, A. Parthenogenetic Ischnura Hastata Revisited: Present Status and Notes on Population Ecology and Behaviour (Odonata: Coenagrionidae). *Int. J. Odonatol.* **2009**, *12*, 395–411. [CrossRef]
84. Bota-Sierra, C.A.; Flórez-V, C.; Escobar, F.; Sandoval-H, J.; Novelo-Gutiérrez, R.; Londoño, G.A.; Cordero-Rivera, A. The Importance of Tropical Mountain Forests for the Conservation of Dragonfly Biodiversity: A Case from the Colombian Western Andes. *Int. J. Odonatol.* **2021**, *24*, 233–247. [CrossRef]
85. Bogan, M.T.; Noriega-Felix, N.; Vidal-Aguilar, S.L.; Findley, L.T.; Lytle, D.A.; Gutiérrez-Ruacho, O.G.; Alvarado-Castro, J.A.; Varela-Romero, A. Biogeography and Conservation of Aquatic Fauna in Spring-Fed Tropical Canyons of the Southern Sonoran Desert, Mexico. *Biodivers. Conserv.* **2014**, *23*, 2705–2748. [CrossRef]
86. Corbet, P.S. *Dragonflies: Behaviour and Ecology of Odonata*; Harley Books: Colchester, UK, 1999.
87. Tognelli, M.F.; Lasso, C.A.; Bota-Sierra, C.A.; Jiménez-Segura, L.F.; Cox, N.A. *Estado de Conservación y Distribución de la Biodiversidad de Agua Dulce en Los Andes Tropicales*, 1st ed.; Tognelli, M.F., Lasso, C.A., Bota-Sierra, C.A., Jiménez-Segura, L.F., Cox, N.A., Eds.; IUCN: Gland, Switzerland; Cambridge, UK; Arlington, TX, USA, 2016; ISBN 9782831717913.
88. González-Soriano, E. *Leptobasis melinogaster* spec. nov., a New Species from Mexico (Zygoptera: Coenagrionidae). *Odonatologica* **2002**, *31*, 181–185.

diversity

MDPI

Article

Taxonomic Revision of Eastern Part of Western Palaearctic *Cordulegaster* Using Molecular Phylogeny and Morphology, with the Description of Two New Species (Odonata: Anisoptera: Cordulegastridae) †

Thomas Schneider [1,*,‡], Andy Vierstraete [2], Ole Müller [3], Gert Jan van Pelt [4], Max Caspers [4], Dietmar Ikemeyer [5], Nataly Snegovaya [6] and Henri J. Dumont [2]

1 Independent Researcher, Arnold-Knoblauch-Ring 76, 14109 Wannsee, Germany
2 Department of Biology, University of Gent, 9000 Gent, Belgium; Andy.Vierstraete@UGent.be (A.V.); Henri.Dumont@UGent.be (H.J.D.)
3 Independent Researcher, Birkenweg 6d, 15306 Libbenichen, Germany; mueller.ole@gmail.com
4 Naturalis Biodiversity Center, P.O. Box 9517, 2300 RA Leiden, The Netherlands; gpelt@its.jnj.com (G.J.v.P.); max.caspers@naturalis.nl (M.C.)
5 Independent Researcher, Billerbecker Str. 6, 48329 Havixbeck, Germany; DKJIkemeyer@t-online.de
6 Zoological Institute NAS of Azerbaijan, Proezd 1128, Kvartal 504, Baku AZ 1073, Azerbaijan; snegovaya@yahoo.com
* Correspondence: thomas.rs@gmx.de
† urn:lsid:zoobank.org:pub:F0BC496E-587F-4008-91CD-C549DDD1BF4F.
‡ Volunteer at the Museum für Naturkunde Berlin, 10115 Berlin, Germany.

Citation: Schneider, T.; Vierstraete, A.; Müller, O.; van Pelt, G.J.; Caspers, M.; Ikemeyer, D.; Snegovaya, N.; Dumont, H.J. Taxonomic Revision of Eastern Part of Western Palaearctic *Cordulegaster* Using Molecular Phylogeny and Morphology, with the Description of Two New Species (Odonata: Anisoptera: Cordulegastridae). *Diversity* **2021**, *13*, 667. https://doi.org/10.3390/d13120667

Academic Editors: M. Olalla Lorenzo-Carballa, Ricardo Koroiva and Michael Wink

Received: 14 November 2021
Accepted: 7 December 2021
Published: 15 December 2021

Abstract: Taxonomy of the genus *Cordulegaster* Leach in Brewster, 1815 in the Eastern part of the Western Palaearctic is poorly resolved. A two-step approach was applied: sequences of mitochondrial and nuclear DNA fragments were used to sort specimens; poorly known or new taxa with their phenotypic variation were described. The existence of two traditional groups (*boltonii*- and *bidentata*-group) was confirmed. *Cordulegaster coronata* Morton, 1916, however, belongs to a different group. Molecular-analysis supported three known and one new species (*C. heros* Theischinger, 1979, *C. picta* Selys, 1854, *C. vanbrinkae* Lohmann, 1993, and *C. kalkmani* sp. nov.) in the *boltonii*-group. In the *bidentata*-group, all specimens from West-Turkey belonged to *C. insignis* Schneider, 1845, all specimens further east to a complex of four closely related species, which we name *charpentieri*-complex (*C. amasina* Morton, 1916, stat. rev., *C. mzymtae* Bartenev, 1929 *C. charpentieri* (Kolenati, 1846), stat. rev. and *C. cilicia* sp. nov.). The following taxa: *C. insignis nobilis* Morton, 1916, syn. nov., *C. nachitschevanica* Skvortsov and Snegovaya, 2015, syn. nov. *C. plagionyx* Skvortsov and Snegovaya, 2015, syn. nov. and the Caucasian subspecies *C. insignis lagodechica* Bartenev, 1930, syn. nov., were synonymized with *C. charpentieri*. Finally, we provide a key for all Western Palaearctic *Cordulegaster*.

Keywords: Caucasus-Anatolian-Iranian-region; hybridization; identification key; new species; new status; new synonym; variation

1. Introduction

1.1. General Remarks on the Genus Cordulegaster

Cordulegastridae is a family, containing three genera: *Cordulegaster* Leach in Brewester, 1815, *Anotogaster* Selys, 1854 and *Neallogaster* Cowley, 1934. Cordulegastridae is a rather small family with 51 species [1]. They are a morphologically conserved family of robust, black-and-yellow dragonflies. The genus *Cordulegaster* occurs in North Africa, Europe, and South-West Asia as far east as the Himalayan region of India and Nepal with species reported also from China and Vietnam [2,3]. In addition, there is a group of *Cordulegaster* species in North America, with one species extending to Central America [4].

1.2. Threats to Western Palaearctic Cordulegaster

Cordulegaster are reophiles, and mainly restricted to clean running springs and brooks. These habitats are under particular threat in the south-eastern part of the Western Palaearctic due to water crisis [5,6]. Therefore, some of the European *Cordulegaster* species are currently classified as endangered, like *C. helladica*, or even critically endangered, like *C. helladica kastalia* [5,7,8]. Other Western Palaearctic members of the genus are listed as data deficient, for example *C. vanbrinkae* [9]. A better taxonomic understanding of these taxa is therefore also of interest from the view of conservation aspects.

1.3. History of the Taxonomy of Western Palaearctic Cordulegaster

After the actual understanding, Western Palaearctic *Cordulegaster* are separated in two groups the *boltonii*- and the *bidentata*-group [10,11]. The main difference is that the males of the *boltonii*-group have one tooth on the upper appendages, while the males of the *bidentata*-group have two. This latter point was name giving for the *bidentata*-group. More differences between the two groups are listed and depicted in the Field Guide to the Dragonflies of Britain and Europe [11]. This view is further supported by a study on larvae, and recently by a molecular genetic analysis [12,13].

Species of *Cordulegaster* are remarkably similar in the structure of male appendices and female valvular scale, while the pattern of yellow markings on abdomen, thorax, and occipital triangle may vary, even within a single taxon [14,15]. This has resulted in confusion regarding the identification, distribution and infraspecific division of the species of the Western Palaearctic region. A comprehensive revision is urgently needed, to complement a recent study, which left the Eastern part of the Western Palaearctic region with about 10 taxa unresolved [13]. Recently, two additional taxa from the Caucasus were described, adding even more complexity to the taxonomy of the genus [16].

The research history of this genus in the Western Palaearctic includes some of the earliest works on dragonflies. *Cordulegaster insignis* was described based on a female from Kellemisch (Anatolia, Turkey) [17]. The location of Kellemisch is absent from recent maps and was searched for by Erich Schmidt, who localized it in the Western Mediterranean region of Turkey [18]. *C. insignis* was re-investigated and two males and two females from Syria were assigned to this species [19,20]. Kolenati described a further taxon from the Caucasus, *Aeshna charpentieri* Kolenati, 1846 [21]. The later taxon has caused much confusion since then. Unfortunately, in the region of the Caucasus, where Kolenati collected the type male, two members of this genus occur, one of the *bidentata*-group (*C. charpentieri* (Kolenati 1846)) and one of the *boltonii*-group (*C. picta* Selys, 1854). Selys confused *Aeshna charpentieri* with his *C. picta* and this confusion was ongoing until it culminated in the designation of a neotype of *C. charpentieri* by Waterston which belongs in fact to *C. picta* [22–24]. The complex history of confusion is described in detail by Dumont. In this paper *Aeshna charpentieri* is synonymized with a member of the *bidentata*-group *C. insignis* [25].

Morton proposed the taxa *C. boltonii algirica* Morton, 1916, *C. princeps* Morton, 1916, *C. insignis amasina* Morton, 1916, *C. insignis nobilis* Morton, 1916, and *C. coronata* Morton, 1916 on the basis of colour markings and with no clear structural characters [26]. Fraser made a further revision of the genus, did not change much to Morton's conclusions, but provided new figures for many taxa, including those from the east of the Western Palaearctic [2]. Fraser treated *C. coronata* as a subspecies of *C. insignis*. Bartenev described *C. mzymtae* and *C. insignis lagodechica* from the Caucasus, and *C. magnifica* from an unknown locality [27,28]. Dumont synonymized *C. i. lagodechica* with *C. insignis*, Waterston described *C. trinacriae* and Theischinger recognized *C. heros* as a further species in the *boltonii*-group [24,25,29]. Lohmann's revision of the genus involved new genera, species and subspecies, including *C. helladica* and *C. vanbrinkae* [30]. He assigned some members of the *bidentata*-group like *C. helladica*, *C. insignis charpentieri*, *C. insignis amasina* and *C. insignis* in a new genus *Sonjagaster*, which, however, has not been followed so far by others, and we synonymize here *Sonjagaster* with *Cordulegaster*. Moreover, he raised several subspecies of *C. insignis* (e.g., *C. i. amasina*, *C. i. charpentieri*, and *C. i. montandoni* [31]) to species level without giving details or plausible reasons for this change. In the Eastern part of the Western Palaearctic, another species, *C. coronata*, is found, occurring from North-East Iran to Kyrgyzstan. This taxon was described by Morton as a separate species, later treated as a subspecies of *C. insignis* [2,26].

Neallogaster schmidti was reported once from North-East Afghanistan near the border to Pakistan [32,33], which is just outside of the Western Palearctic according to most authors and therefore not treated here.

Thus, the exact taxonomic status of most *Cordulegaster* of the Eastern part of the Western Palaearctic region is unsettled, and the relations between the members of this genus in this region, as well as their exact distribution limits, are not known.

1.4. Geological and Climatic Events Influencing Diversification of Western Palaearctic Cordulegaster

The diversification of the biota in the Western Palaearctic has been shaped by geological and climatic events of the last millions of years. We assume that these events were also important for the evolutionary history of the genus *Cordulegaster*. One dramatic episode was the Messinian Salinity Crisis, about 6 million years ago, leading to the evaporation of the Mediterranean Sea. It connected many Mediterranean Islands and North Africa with Europe and allowed for their colonization from the continent [34]. As the Mediterranean Basin refilled, the *Cordulegaster* species of North Africa and Mediterranean Islands may have started to evolve in an allopatric way. A second major driver of *Cordulegaster* diversification may have been the Pleistocene Ice Ages. According to the calculation by Solano *C. trinacriae* and *C. boltonii* separated in the early Pleistocene about 1.32 million years ago [35]. Thus, the European but also the whole Western Palaearctic biota may be a result of the postglacial expansion, which has structured the current distribution concepts of species [36,37]. The last Glacial Maximum shaped the phylogeny and phytogeography of most of the existing taxa. Hewitt emphasized the importance of postglacial colonization events in animals and plants following a similar pattern of expansion from refugia [36,38,39]. These events seem to be particularly important for the understanding of species development of Western Palaearctic *Cordulegaster*, as already recently discussed for the Apennine Peninsula [35]. Following those concepts, the Western Palaearctic *Cordulegaster* may have become restricted to regions like the Mediterranean, Black Sea and Caspian Sea shores. These refugia may have kept populations isolated for a few millions of years and allowed them to evolve into species [35–37]. During interglacials, one of which we live in today, species may have expanded and hybridised in secondary contact zones [35–37]. Hot spots of such events in the case of *Cordulegaster* may be the Aegean region, Anatolia and the Caucasus region, where three glacial refuge population (Mediterranean, Black Sea and Caspian Sea) meet [40,41]. A final factor for species diversification may be aridity and the uplift of the Anatolian Diagonal [42].

The availability of molecular tools has recently revolutionised our species concept. It revealed that the definition of a species as an entity that is reproductively isolated is relative. Many species indeed hybridise under natural conditions, and reproductive isolation is only partial.

The complexity of the systematics of the genus *Cordulegaster* indicates that any revision solely based on morphology and colour patterns is bound to fail. A combination of molecular methods for reconstructing phylogenies with traditional morphological techniques is needed. We provide a revision of the Eastern part of the Western Palaearctic *Cordulegaster* by molecular genetic approach followed by phenotype description.

2. Material and Methods

2.1. Material

A total of 156 *Cordulegaster* specimens of the Eastern part of the Western Palaearctic were investigated (Table 1, Figures 1 and 2, Supplementary Material Figure S1). In the molecular genetic analysis sequences from 72 specimens from Eastern Europe, Turkey, the Caucasus countries, Iran and Kyrgyzstan were included. From these 72 specimens,

59 specimens and 107 genetic data are new; genetic data from 13 specimens (*C. heros* *C. picta*, and *C. insignis*) were already published [13]. New and published sequences of *Cordulegaster* from the Western part of the Western Palaearctic and American *Cordulegaster* were used for the COI overview tree (Figure S2). Many species investigated in this study were collected by Gert Jan van Pelt (G.J.v.P.) and are deposited in the Naturalis Biodiversity Center, Leiden (RMNH). The types and most specimens for re-description are deposited in RMNH (Table 1, Figure S1). Additional material was more recently collected by some of the authors (Thomas Schneider, T.S.; Nataly Snegovaya, N.S.; Henri J. Dumont, H.D.) and others listed in Table 1 and the acknowledgement. For the descriptive part material from RMNH and the private collection of one of the authors (T.S.) was used (detailed information Table 1, Figure S1).

Figure 1. Map showing the origin of investigated specimens of the *C. boltonii*-group.

2.2. General Methods

Our molecular analysis included Genbank sequences of New World and Asian species (Figure S2). Already published sequences of Western Palaearctic *Cordulegaster* are included [13,35,43]. From Central Turkey and east of this region, no sequences of *Cordulegaster* from the Western Palaearctic have been published so far. Thus, only the sequences published with this manuscript are available. Andy Vierstrate (A.V.) did the molecular phylogenetic work. He received only legs, with exact geographic data, however, without knowing the assumed taxa.

Table 1. *Cordulegaster* specimens investigated are listed here with group name, species names, notes, geo-coordinates, collectors, reference, and gene bank numbers. For further details see Figure S1.

Group	Species Identified	Former Taxon/Notes	Latitude	Longitude	Country	Collector/Reference	Genbank COI	Genbank ITS
bidentata-group	*amasina*		39.5629	34.1894	Turkey	leg. G.J. van Pelt	MK779843	MK861474
bidentata-group	*amasina*		40.9765	34.1894	Turkey	leg. G.J. van Pelt	MK779844	MK861470
bidentata-group	*amasina*	**male redescription**	40.5915	32.6597	Turkey	leg. G.J. van Pelt	MK779845	MK861469
bidentata-group	*amasina*		39.5629	34.1894	Turkey	leg. G.J. van Pelt	no data	MK861473
bidentata-group	*bidentata*		50.8101	14.7763	Czech Republic	leg. M. Waldhauser	MK779814	no data
bidentata-group	*bidentata*	morphology only			France	leg. T. Schneider		
bidentata-group	*bidentata*	morphology only			France	leg. T. Schneider		
bidentata-group	*bidentata*	morphology only			France	leg. T. Schneider		
bidentata-group	*bidentata*	morphology only			France	leg. T. Schneider		
bidentata-group	*bidentata*	morphology only			France	leg. T. Schneider		
bidentata-group	*bidentata*	morphology only			France	leg. T. Schneider		
bidentata-group	*bidentata*	morphology only			France	leg. T. Schneider		
bidentata-group	*bidentata*	morphology only			Germany	leg. T. Schneider		
bidentata-group	*bidentata*	morpholgy only	39.4386	23.0464	Greece	leg. T. Schneider		
bidentata-group	*bidentata*	was *C. b. sicilica*, morphology only			Italy	leg. T. Schneider		
bidentata-group	*bidentata*	was *C. b. sicilica*, morphology only			Italy	leg. T. Schneider		
bidentata-group	*bidentata*	was *C. b. sicilica*, morphology only			Italy	leg. T. Schneider		
bidentata-group	*charpentieri*		38.9487	46.1958	Armenia	leg. V. Ananian	MK779820	MK861491
bidentata-group	*charpentieri*		38.9131	46.4604	Armenia	leg. V. Ananian	MK779821	MK861463
bidentata-group	*charpentieri*		38.8956	46.1875	Armenia	leg. V. Ananian	no data	MK86149
bidentata-group	*charpentieri*		38.8956	46.1875	Armenia	leg. V. Ananian	no data	MK861467
bidentata-group	*charpentieri*		38.9487	46.1958	Armenia	leg. V. Ananian	no data	MK861460

Table 1. *Cont.*

Group	Species Identified	Former Taxon/Notes	Latitude	Longitude	Country	Collector/Reference	Genbank COI	Genbank ITS
bidentata-group	*charpentieri*		39.7038	45.3468	Armenia	leg. V. Ananian	no data	MK861462
bidentata-group	*charpentieri*	was *nachitschevanica*	39.1103	45.9144	Azerbaijan	leg. N. Snegovaya	MK779823	MK861463
bidentata-group	*charpentieri*	was *plagionyx*	41.6761	46.4931	Azerbaijan	leg. N. Snegovaya	MK779828	MK861466
bidentata-group	*charpentieri*	male, was *nachitschevanica*, morphology only	39.1103	45.9144	Azerbaijan	leg. N. Snegovaya		
bidentata-group	*charpentieri*	female, was *plagionyx*, morphology only	41.6761	46.4931	Azerbaijan	leg. N. Snegovaya		
bidentata-group	*charpentieri*	male, was *plagionyx*, morphology only	41.6761	46.4931	Azerbaijan	leg. N. Snegovaya		
bidentata-group	*charpentieri*	*charpentieri lagodechica*, *larva*	41.8404	46.2846	Georgia	leg. M. Lemke	MK779824	MK861465
bidentata-group	*charpentieri*		41.7458	44.7426	Georgia	leg. N. Snegovaya	MK779825	no data
bidentata-group	*charpentieri*	*charpentieri lagodechica*, *larva*	41.8404	46.2846	Georgia	leg. M. Lemke	MK779826	MK861477
bidentata-group	*charpentieri*	was *nobilis*	29.4311	53.0842	Iran	leg. T. Schneider	MN623223	MN612638
bidentata-group	*charpentieri*	was *nobilis*, male	33.4597	49.9472	Iran	leg. T. Schneider	MK779819	MK861457
bidentata-group	*charpentieri*	was *nobilis*	35.8926	52.0125	Iran	leg. M. Waldhauser	MK779829	MK861459
bidentata-group	*charpentieri*	was *nobilis*, female **paratype**	30.3339	52.1564	Iran	leg. T. Schneider	MK779830	MK861496
bidentata-group	*charpentieri*	was *nobilis*	29.0022	57.5869	Iran	leg. T. Schneider	MK779831	MK861471
bidentata-group	*charpentieri*	was *nobilis*; morphology only	29.0022	57.5869	Iran	leg. T. Schneider		
bidentata-group	*charpentieri*	was *nobilis*; morphology only	29.0022	57.5869	Iran	leg. T. Schneider		
bidentata-group	*charpentieri*	was *nobilis*; morphology only	29.0022	57.5869	Iran	leg. T. Schneider		

Table 1. *Cont.*

Group	Species Identified	Former Taxon/Notes	Latitude	Longitude	Country	Collector/Reference	Genbank COI	Genbank ITS
bidentata-group	*charpentieri*	was *nobilis*, male **paratype**, morphology only	30.3339	52.1564	Iran	leg. T. Schneider		
bidentata-group	*charpentieri*	was *nobilis*, female, morphology only	30.3339	52.1564	Iran	leg. T. Schneider		
bidentata-group	*charpentieri*	was *nobilis*, female, morphology only	30.3339	52.1564	Iran	leg. T. Schneider		
bidentata-group	*charpentieri*	was *nobilis*, female, morphology only	30.3339	52.1564	Iran	leg. T. Schneider		
bidentata-group	*charpentieri*	was *nobilis*, male, morphology only	30.3339	52.1564	Iran	leg. T. Schneider		
bidentata-group	*charpentieri*	was *nobilis*, male, morphology only	30.3339	52.1564	Iran	leg. T. Schneider		
bidentata-group	*charpentieri*	was *nobilis*, male, morphology only	30.3339	52.1564	Iran	leg. T. Schneider		
bidentata-group	*charpentieri*	was *nobilis*, **paratype** male, morphology only	30.3339	52.1564	Iran	leg. T. Schneider		
bidentata-group	*charpentieri*	was *nobilis*, **paratype** male, morphology only	30.3339	52.1564	Iran	leg. T. Schneider		
bidentata-group	*charpentieri*		40.5616	42.3471	Turkey	leg. G.J. van Pelt	MK779822	MK861458
bidentata-group	*charpentieri*	**male redescription, neotype**	38.3834	42.7723	Turkey	leg. G.J. van Pelt	MK779827	MK861476
bidentata-group	*cilicia* sp. nov.	was *insignis*, **paratype**, morphology only	34.2511	36.0111	Lebanon	leg. G.A. Mavromoustakis		
bidentata-group	*cilicia* sp. nov.	was *insignis*, **paratype**, morphology only	34.2511	36.0111	Lebanon	leg. G.A. Mavromoustakis		

Table 1. *Cont.*

Group	Species Identified	Former Taxon/Notes	Latitude	Longitude	Country	Collector/Reference	Genbank COI	Genbank ITS
bidentata-group	*cilicia* sp. nov.	was *insignis*, morphology only	34.2511	36.0111	Lebanon	leg. G.A. Mavromoustakis		
bidentata-group	*cilicia* sp. nov.	was *insignis*, female **paratype**, morphology only	34.4578	35.8414	Lebanon	leg. G.A. Mavromoustakis		
bidentata-group	*cilicia* sp. nov.	was *insignis*, morphology only	34.4578	35.8414	Lebanon	leg. G.A. Mavromoustakis		
bidentata-group	*cilicia* sp. nov.	was *insignis*, morphology only	34.4578	35.8414	Lebanon	leg. G.A. Mavromoustakis		
bidentata-group	*cilicia* sp. nov.	was *charpentieri*, **paratype**	40.6114	41.6286	Turkey	leg. G.J. van Pelt	MK779832	MK861494
bidentata-group	*cilicia* sp. nov.	was *insignis*, **paratype**, morphology only	38.7205	36.3950	Turkey	leg. G.J. van Pelt	MK779833	MK861475
bidentata-group	*cilicia* sp. nov.	was *insignis*, **paratype**, morphology only	38.0164	35.0311	Turkey	leg. G.J. van Pelt	MK779834	no data
bidentata-group	*cilicia* sp. nov.	was *insignis*, male **holotype**	38.0614	36.4722	Turkey	leg. G.J. van Pelt	MK779835	MK861498
bidentata-group	*cilicia* sp. nov.	was *insignis*, male **paratype**	37.9698	34.6767	Turkey	leg. G.J. van Pelt	MK779836	MK861468
bidentata-group	*cilicia* sp. nov.	was *insignis*	37.6512	34.5328	Turkey	leg. G.J. van Pelt	MK779837	MK861461
bidentata-group	*cilicia* sp. nov.	was *insignis*	37.9698	34.6766	Turkey	leg. G.J. van Pelt	MK779839	MK861472
bidentata-group	*cilicia* sp. nov.	was *insignis*	40.3334	42.5905	Turkey	leg. G.J. van Pelt	MK779840	MK861495
bidentata-group	*cilicia* sp. nov./*insignis*	hybrid: *cilcia* × *insignis*	37.9293	32.5000	Turkey	leg. G.J. van Pelt	MK779838	MK861497
coronata-group	*coronata*	was sometimes treated as ssp. of *insignis*	34.9569	60.1672	Iran	leg. T. Schneider	MK779857	MK861452
coronata-group	*coronata*	was sometimes treated as ssp. of *insignis*	34.9569	60.1672	Iran	leg. T. Schneider		

Table 1. *Cont.*

Group	Species Identified	Former Taxon/Notes	Latitude	Longitude	Country	Collector/Reference	Genbank COI	Genbank ITS
coronata-group	*coronata*	was sometimes treated as ssp. of *insignis*	34.9569	60.1672	Iran	leg. T. Schneider		
coronata-group	*coronata*	was sometimes treated as ssp. of *insignis*	34.9569	60.1672	Iran	leg. T. Schneider		
coronata-group	*coronata*	was sometimes treated as ssp. of *insignis*	34.9569	60.1672	Iran	leg. T. Schneider		
coronata-group	*coronata*	was sometimes treated as ssp. of *insignis*	34.9569	60.1672	Iran	leg. T. Schneider		
coronata-group	*coronata*	was sometimes treated as ssp. of *insignis*	34.9569	60.1672	Iran	leg. T. Schneider		
coronata-group	*coronata*	was sometimes treated as ssp. of *insignis*	34.9569	60.1672	Iran	leg. T. Schneider		
coronata-group	*coronata*	was sometimes treated as ssp. of *insignis*	41.2044	74.7661	Kyrgyzstan	leg. H. Dumont	MK779858	MK861453
bidentata-group	*helladica*	morphology only	38.0353	22.1098	Greece	leg. T. Schneider		
bidentata-group	*helladica*	morphology only	38.0353	22.1098	Greece	leg. T. Schneider		
bidentata-group	*helladica*	morphology only	38.0353	22.1098	Greece	leg. T. Schneider		
bidentata-group	*helladica*	morphology only	38.0353	22.1098	Greece	leg. T. Schneider		
bidentata-group	*helladica*	morphology only	38.0353	22.1098	Greece	leg. T. Schneider		
bidentata-group	*helladica*	morphology only	38.0353	22.1098	Greece	leg. T. Schneider		
bidentata-group	*helladica*	morphology only	38.0353	22.1098	Greece	leg. T. Schneider		
bidentata-group	*helladica*	morphology only	38.0353	22.1098	Greece	leg. T. Schneider		
bidentata-group	*helladica*	morphology only	38.0353	22.1098	Greece	leg. T. Schneider		
bidentata-group	*helladica*	morphology only	38.0353	22.1098	Greece	leg. T. Schneider		
bidentata-group	*helladica*	morphology only	38.0353	22.1098	Greece	leg. T. Schneider		
bidentata-group	*helladica*	morphology only	38.0353	22.1098	Greece	leg. T. Schneider		
bidentata-group	*helladica*	morphology only	38.4887	22.5048	Greece	leg. T. Schneider		

Table 1. *Cont.*

Group	Species Identified	Former Taxon/Notes	Latitude	Longitude	Country	Collector/Reference	Genbank COI	Genbank ITS
bidentata-group	*helladica*	morphology only	38.4887	22.5048	Greece	leg. T. Schneider		
boltonii-group	*heros*	morphology only	48.2132	16.3847	Austria	leg. T. Schneider		
boltonii-group	*heros*	morphology only	48.2132	16.3847	Austria	leg. T. Schneider		
boltonii-group	*heros*		42.4929	27.4733	Bulgaria	leg. G.J. van Pelt	MK779810	no data
boltonii-group	*heros*		42.7339	25.4858	Bulgaria	leg. G.J. van Pelt	MK779810	no data
boltonii-group	*heros*		42.5937	23.4102	Bulgaria	leg. G.J. van Pelt	MK779812	MK861451
boltonii-group	*heros*	morphology only	38.0353	22.1098	Greece	leg. T. Schneider		
boltonii-group	*heros*	morphology only	38.0353	22.1098	Greece	leg. T. Schneider		
boltonii-group	*heros*	morphology only	38.0353	22.1098	Greece	leg. T. Schneider		
boltonii-group	*heros*	morphology only	38.0353	22.1098	Greece	leg. T. Schneider		
boltonii-group	*heros*	morphology only	38.0353	22.1098	Greece	leg. T. Schneider		
boltonii-group	*heros*	morphology only	38.0353	22.1098	Greece	leg. T. Schneider		
boltonii-group	*heros*		42.1837	19.2915	Montenegro	leg. G. De Knijf	MK779813	no data
bidentata-group	*insignis*		37.7284	26.8196	Greece	leg. T. Schneider	MW353685	MW363787
bidentata-group	*insignis*		37.5833	26.1667	Greece	leg. T. Schneider	MW353686	MW363788
bidentata-group	*insignis*		37.5833	26.1667	Greece	leg. T. Schneider	MW353687	MW363789
bidentata-group	*insignis*		39.0690	26.3468	Greece	leg. W. Lopau	MW353702	no data
bidentata-group	*insignis*		36.9287	28.7511	Turkey	Froufe et al., 2014	KF584941	KF584979
bidentata-group	*insignis*		37.0287	28.7554	Turkey	Froufe et al., 2014	KF584941	no data
bidentata-group	*insignis*		37.0287	28.7554	Turkey	Froufe et al., 2014	KF584941	no data
bidentata-group	*insignis*		37.0287	28.7554	Turkey	Froufe et al., 2014	KF584941	KF584979
bidentata-group	*insignis*		37.2196	28.3658	Turkey	Froufe et al., 2014	KF584942	no data
bidentata-group	*insignis*		37.0287	28.7554	Turkey	Froufe et al., 2014	KF584943	KF584979
bidentata-group	*insignis*		41.6771	26.5557	Turkey	leg. G.J. van Pelt	MK779846	MK861486

Table 1. *Cont.*

Group	Species Identified	Former Taxon/Notes	Latitude	Longitude	Country	Collector/Reference	Genbank COI	Genbank ITS
bidentata-group	*insignis*		39.3444	29.2579	Turkey	leg. G.J. van Pelt	MK779848	MK861487
bidentata-group	*insignis*		36.5595	30.3501	Turkey	leg. T. Schneider	MW353707	MW363797
boltonii-group	*kalkmani* sp. nov.	was *picta*, male **holotype**	40.4203	42.7446	Turkey	leg. G.J. van Pelt	MK779808	MK861450
boltonii-group	*kalkmani* sp. nov.	was *picta*, female **paratype**	39.0594	43.7540	Turkey	leg. V. Kalkman	MK779809	MK861449
bidentata-group	*mzymtae*	was sometimes treated as ssp. of *insignis*	41.6319	42.5680	Georgia	leg. T. Schneider		
bidentata-group	*mzymtae*	was sometimes treated as ssp. of *insignis*	41.6319	42.5680	Georgia	leg. T. Schneider		
bidentata-group	*mzymtae*	was sometimes treated as ssp. of *insignis*	41.6319	42.5680	Georgia	leg. T. Schneider		
bidentata-group	*mzymtae*	was sometimes treated as ssp. of *insignis*	41.6319	42.5680	Georgia	leg. T. Schneider		
bidentata-group	*mzymtae*	was sometimes treated as ssp. of *insignis*	41.6319	42.5680	Georgia	leg. T. Schneider		
bidentata-group	*mzymtae*	was sometimes treated as ssp. of *insignis*	41.6319	42.5680	Georgia	leg. T. Schneider		
bidentata-group	*mzymtae*	was sometimes treated as ssp. of *insignis*	41.6319	42.5680	Georgia	leg. T. Schneider		
bidentata-group	*mzymtae*	was sometimes treated as ssp. of *insignis*	41.6319	42.5680	Georgia	leg. T. Schneider		
bidentata-group	*mzymtae*	was sometimes treated as ssp. of *insignis*	41.6319	42.5680	Georgia	leg. T. Schneider		
bidentata-group	*mzymtae*	was sometimes treated as ssp. of *insignis*	43.4462	41.7357	Russia	leg. O. Kosterin	MN623222	no data
bidentata-group	*mzymtae*	was sometimes treated as ssp. of *insignis*	40.6797	37.9573	Turkey	leg. G.J. van Pelt	MK779841	MK861493

Table 1. *Cont.*

Group	Species Identified	Former Taxon/Notes	Latitude	Longitude	Country	Collector/Reference	Genbank COI	Genbank ITS
bidentata-group	*mzymtae*	was sometimes treated as ssp. of *insignis*	40.6797	37.9573	Turkey	leg. G.J. van Pelt	MK779842	MK861492
boltonii-group	*picta*		41.5631	24.9664	Bulgaria	leg. G.J. van Pelt	no data	no data
boltonii-group	*picta*	morphology only	42.1667	42.9833	Georgia	leg. T. Schneider		
boltonii-group	*picta*		41.3093	26.1148	Greece	Froufe et al., 2014	KF584944	KF584992
boltonii-group	*picta*		41.3093	26.1148	Greece	Froufe et al., 2014	KF584944	KF584993
boltonii-group	*picta*		40.6910	24.6209	Greece	leg. T. Schneider	MK779802	no data
boltonii-group	*picta*		37.7094	26.8178	Greece	leg. T. Schneider	MW353684	no data
boltonii-group	*picta*		44.6541	37.9347	Russia	leg. O. Kosterin	MK779799	MK861448
boltonii-group	*picta*		44.6541	37.9347	Russia	leg. O. Kosterin	MK779800	MK861447
boltonii-group	*picta*		44.6541	37.9347	Russia	leg. O. Kosterin	MK779801	no data
boltonii-group	*picta*		39.4200	29.9857	Turkey	leg. G.J. van Pelt	MK779803	no data
boltonii-group	*picta*		39.6602	35.8823	Turkey	leg. G.J. van Pelt	MK779804	MK861445
boltonii-group	*picta*		41.6281	27.5122	Turkey	leg. G.J. van Pelt	no data	MK861446
boltonii-group	*vanbrinkae*		38.6825	48.7831	Azerbaijan	leg. N. Snegovaya	MK779806	no data
boltonii-group	*vanbrinkae*	morphology only	38.6825	48.7831	Azerbaijan	leg. N. Snegovaya		
boltonii-group	*vanbrinkae*		36.0965	53.2606	Iran	leg. T. Schneider	MW353704	MW363796
boltonii-group	*vanbrinkae*		36.8741	54.8872	Iran	leg. T. Schneider	MK779805	MK861442
boltonii-group	*vanbrinkae*	larva	36.5567	51.5231	Iran	leg. M. Waldhauser	MK779807	MK861443
boltonii-group	*vanbrinkae*	morphology only	36.4700	51.5383	Iran	leg. T. Schneider		
boltonii-group	*vanbrinkae*	morphology only	36.4700	51.5383	Iran	leg. T. Schneider		
boltonii-group	*vanbrinkae*	morphology only	36.8741	54.8872	Iran	leg. T. Schneider		
boltonii-group	*vanbrinkae*	morphology only	36.8741	54.8872	Iran	leg. T. Schneider		
boltonii-group	*vanbrinkae*	morphology only	36.8741	54.8872	Iran	leg. T. Schneider		

Table 1. *Cont.*

Group	Species Identified	Former Taxon/Notes	Latitude	Longitude	Country	Collector/Reference	Genbank COI	Genbank ITS
boltonii-group	*vanbrinkae*	morphology only	36.8741	54.8872	Iran	leg. T. Schneider		
boltonii-group	*vanbrinkae*	morphology only	36.8741	54.8872	Iran	leg. T. Schneider		
boltonii-group	*vanbrinkae*	morphology only	36.8741	54.8872	Iran	leg. T. Schneider		
boltonii-group	*vanbrinkae*	morphology only	36.8741	54.8872	Iran	leg. T. Schneider		
boltonii-group	*vanbrinkae/picta*	hybrid: *vanbrinkae* × *picta*	39.3334	46.3786	Armenia	leg. V. Ananian	MK779798	MK861444
boltonii-group	*vanbrinkae/picta?*	morphology only	39.3106	46.3775	Armenia	leg. V. Ananian		
boltonii-group	*vanbrinkae/picta?*	morphology only	39.3106	46.3775	Armenia	leg. V. Ananian		

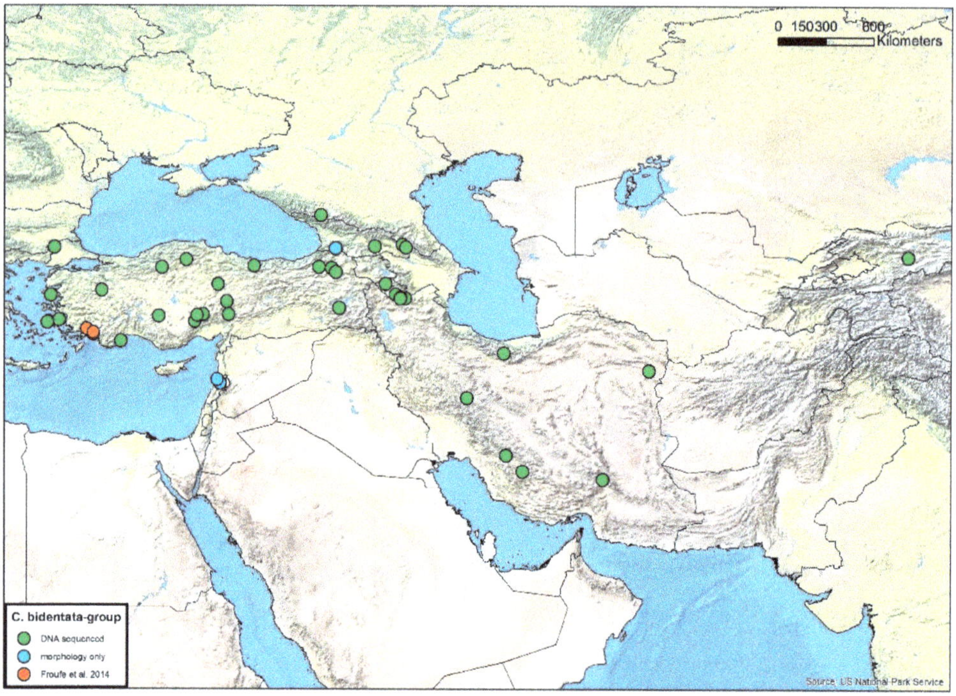

Figure 2. Map showing the origin of the investigated specimens of the *C. bidentata*-group.

At that point of the study, taxonomy of this region was no more than a guess. For example, all members of the *bidentata*-group of the region are currently treated under *C. insignis* or a subspecies of this species [11]. Furthermore, some authors treated *C. charpentieri* or *C. coronata* as separate species without specifying scientific reasons [44–47].

We used at first a molecular genetic approach analysing the COI and ITS genes for constructing bootstrap maximum likelihood trees, Bayesian inference trees and haplotype networks. In addition, evolutionary distance between a pair of sequences was measured by the *p*-distance and the Kimura 2-P distance (K2-P). The latter methods can be translated into taxonomic ranks. In zoology, about 3% difference is considered as a good limit for species difference [48]. For odonata, a difference between 2 to 3% are discussed for species discrimination and may be lower in closely related species [49,50].

A recent analysis for Western Palaearctic odonata revealed for Anisoptera, a K2-P distance above 1.96% as a good threshold [43]. This barcoding analysis on a large set of Italian odonata showed that genetic K2-*p* distance variation within morphospecies ranged from 0% to 9.17% (mean ± SD = 0.48 ± 0.62%) in Zygoptera and from 0% to 2.64% (mean ± SD = 0.33 ± 0.29%) in Anisoptera. Interspecific K2-*p* distance values ranged from 0% to 27.29% (mean ± SD = 19.41 ± 3.71%) and from 0% to 25.28% (mean ± SD = 18.25 ± 3.18%) in Zygoptera and Anisoptera, respectively [43]. Therefore, we used beside the Bayesian inference trees and haplotype networks a K2P-distance above 2% as an orientation for species allocation. We observed no relevant difference between P- and K2-P distances.

Possible hybrids were defined when specimens with their maternally inherited COI corresponding to a different species than their ITS.

2.3. Molecular Genetic Analysis

We analyzed the cytochrome c oxidase subunit I (COI) gene from the mDNA and the ITS region (ITS1, 5.8S and ITS2) from the rDNA. For details of the origin of specimens, see maps Figures 5 and 6, Figure S1 and Table 1. For the COI gene, the sequences are 568 bp long, for the ITS region the sequences are between 713 and 815 bp long, depending on the species. We used *Aeshna umbrosa* Walker, 1908 and *Anotogaster chaoi* Zhou, 1998 as outgroups.

2.4. DNA Extraction, PCR Methods, and Sequencing

Per specimen, a 1.0 mm section of a leg was transferred to a tube with 20 µL 0.05 N NaOH and 2 µL 5% Tween 20. This was heated for 15 min at 95 °C and cooled on ice. 100 µL sterile water was added to the tube and mixed. 1 to 5 µL of this solution was used in a PCR reaction. We amplified and sequenced a fragment of mDNA (the barcoding segment of the cytochrome c oxidase subunit I (COI) gene) and the entire ITS region separating the SSU and LSU region of the 18S nuclear rDNA operon comprising the ITS 1 intergenic ITS, the conserved 5.8S gene and the ITS 2 intergenic ITS, using the PCR on a 2720 Thermal Cycler of Applied Biosystems, Waltham, Massachusetts, USA. Primers used for PCR were CO1490F (5′-GGT CAA ATC ATA AAG ATA TTG G-3′) and CO2198R (5′-TAA ACT TCA GGG TGA CCA AAA AAT CA-3′) for the COI fragment [51]. Cycle conditions were 95 °C for 3 min followed by 45 cycles of 95 °C for 30 s, 48 °C for 30 s and 72 °C for 1 min. Primers used for amplifying the rDNA fragment are Vrain2F (5′-CTT TGT ACA CAC CGC CCG TCG CT-3′) and 28R1 (5′-TGA TAT GCT TAA NTT CAG CGG GT-3′) primers. Cycle conditions were 95 °C for 3 min followed by 45 cycles of 95 °C for 30 s, 54 °C for 30 s and 72 °C for 1 min. PCR products were sequenced on an ABI 3130XL automatic sequencer from Applied Biosystems with the BigDye 3.1 kit according to manufacturer instructions.

2.5. Phylogenetic Methods

MAFFT (Multiple Alignment with Fast Fourier Transform)

Multiple sequence alignments were made with the online version of MAFFT [52].

2.6. jModelTest

The model of DNA evolution that best fit the data was determined with JMODELTEST version 2.1.10 [53]. Based on the Bayesian information criteria (BIC), the best model was chosen for Bayesian inference in MRBAYES 3.2.7a [54] and maximum likelihood analysis in PAUP 4.0a168 [55].

2.7. Tree Construction with MrBayes and PAUP

The model parameters from JMODELTEST were used in MRBAYES (nst = 6 and rates = gamma for COI; nst = 1 and rates = equal for the ITS region). The settings were: 10 million generations, a sample frequency of 1000 and a burnin value of 5000 trees.

PAUP 4.0a168 was used for constructing the bootstrap maximum likelihood trees. The model parameters from JMODELTEST were used as model in PAUP (TPM2uf + G base = (0.3247 0.1447 0.1192) nst = 6 rmat = (1.8954 21.0629 1.8954 1.0000 21.0629) rates = gamma shape = 0.1580 ncat = 4 pinvar = 0 for COI, HKY + G base = (0.2011 0.2807 0.3290) nst = 2 tratio = 0.9345 rates = gamma shape = 0.3020 ncat = 4 pinvar = 0 for the ITS region). 100 bootstrap replicates (starting with a Neighbor Joining tree) were performed with a branch swapping limit set to 100,000 or 3 h (whichever comes first) per bootstrap replicate. For creating trees, *Cordulegaster* DNA-sequences from the West Palaearctic, from America and from E–Asia were used. Genbank was searched for credible sequences. Available sequences for COI and the ITS region approximately the same length as our own were added to the alignment. Their accession numbers are listed next to the names in the trees.

2.8. StarBeast

In addition, StarBeast, a multi-individual multi-locus species tree estimation program using Bayesian coalescent analysis, as implemented in the BEAST package was applied for both genes [56,57]. Xml input files were created in BEAUTI v2.6.3 [57], using the HKY + Γ + I model for both markers. The following settings were used for all analysis: base frequencies 'empirical' clock model 'Strickt clock Clock.rate = 1'; tree prior 'default values (Yule Model)'; The analyses were run in BEAST v.2.6.3 [56,57]. Analyses were run for 50 million generations, sampling every 5000th generation. Tracer v. 1.7.1 [56] was used for examining the effective sample size (ESS) for parameters and determining the burn-in Trees and posterior probabilities were summarized using TreeAnnotator v. 2.6.3 [56] and showed on the Maximum clade credibility tree, with a Posterior probability limit = 0.5 and Burnin percentage = 0. The trees were drawn in FigTree v.1.4.4 [58].

2.9. BPP

As a third program, we used BPP v. 4.3.8 for coalescent species delimitation [59]. For COI, the input species tree contained 21 species and 151 sequences. For the ITS, the input species tree contained 12 species and 53 sequences. For the combined COI and ITS, the input species tree contained 12 species and 47 sequences.

1. Running species delimitation: Settings were Speciesdelimitation = 1 0 2, Speciestree = 1, speciesmodelprior = 1, burnin = 16,000, samplefreq = 2, nsample = 500,000.
2. Running Species tree Estimation: Settings were Speciesdelimitation = 0, Speciestree = 1, speciesmodelprior = 1, burnin = 16,000, samplefreq = 2, nsample = 500,000.

2.10. Timetree

MEGA X was used to do the Timetree analysis using the RelTime method to estimate divergence times by distinct molecular dating [60]. The estimated log likelihood value is −4139.29. A discrete Gamma distribution was used to model evolutionary rate differences among sites (5 categories (+G, parameter = 0.1688)). The analysis involved 151 nucleotide sequences. Codon positions included were 1st + 2nd + 3rd + Noncoding. There were a total of 568 positions in the final dataset.

2.11. Haplotype Analysis

Haplotype networks were built in POPART with the TCS network interference method from the COI alignment [61]. A haplotype network is the evolutionary sum of mutations that defines the current haplotypes through the DNA lineages that connect the current DNA molecules to the common ancestral DNA molecule.

2.12. Evolutionary Distance Analysis

The evolutionary distance between a pair of sequences is usually measured by the number of nucleotide or amino acid substitutions between them: *p*-distance. This distance is merely the proportion (P) of nucleotide sites at which the two sequences compared are different. This is obtained by dividing the number of nucleotide differences (nd) by the total number of nucleotides compared (n). This is one of the more simple methods to evaluate differences and to calculate relationships between taxa. Another established technique for evaluating DNA sequences is the Kimura 2-parameter (K2-P) index, which is conveniently expressed in percent differences between sequences. The *p*- and K2-P distance matrix between groups was constructed in MEGA X [60]. Transitions and transversions were included, uniform rates among sites and gaps were pairwise deleted.

2.13. Morphological Analysis and Systematics

As the molecular work revealed novel taxonomic information, we had to re-describe some poorly defined and described taxa: *C. amasina* (Turkey), *C. mzymtae* (Georgia), *C. charpentieri* (East Anatolia, North-West Iran, Azerbaijan) with its variations in its large geographical range and select a neotype for *C. charpentieri* as the neotype described by

Waterston belongs to *C. picta* and should be exchanged by a new neotype [24,62]. Furthermore, we describe the new taxa *C. kalkmani* sp. nov. (East Anatolia) and *C. cilicia* sp. nov. (South–East Turkey, Lebanon). Furthermore, the phenotypic variations in characters such size and colouration of the poorly known *C. vanbrinkae* (provinces Mazandaran and Golestan in Iran and Lenkoran in Azerbaijan) were compared with the closely related species *C. picta* (Bulgaria, Greece, Turkey, Georgia). The hybrid between *C. vanbrinkae* and *C. picta* from Armenia detected by our molecular analysis is phenotypically characterized, described, depicted and compared with *C. picta* and *C. vanbrinkae*. All specimens used for molecular analysis, description, measurements and verification of characters used for the key are listed in Table 1. More detailed remarks including notes on previous taxonomy, more detailed geographical data and methods on conservation, name-bearing types, neotype, re-description, paratype, DNA isolation, access numbers for gene bank and source are provided in Figure S1. The types, and the neotype are deposited in RMNH.

Specimens were measured (total length, abdomen length, forewing length, hindwing length) with a Vernier caliper in mm (error limit: 0.03 mm). The measurements of total length, abdomen length and wing length were rounded to the nearest millimetre. For differences in the measurements a D'Agostino-Pearson Test was performed to confirm normality of the data sets followed by a two-tailed unpaired *t*-test. Values are given as individual values with means and extremes. All drawings are artwork of Ole Müller (O.M.), made on graphic tablets (Wacom Intuos, Surface Book).

For the description and for the key we used the traditional morphological features of *Cordulegaster* shown in Figure 3.

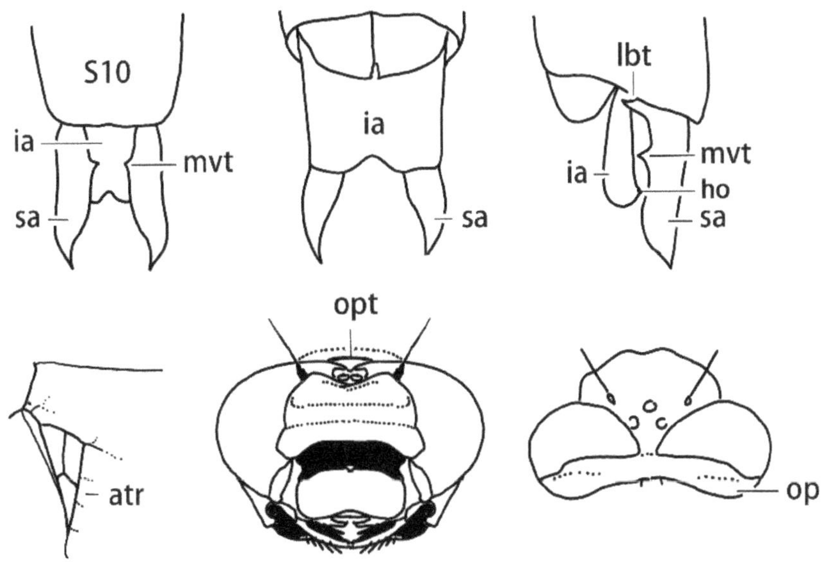

Figure 3. Main morphological characters of Codulegaster used for description and in the key. Abbreviations: ia—inferior appendage, sa—superior appendage, mvt—medioventral tooth, lbt—laterobasal tooth, ho—hook, op—occiput, opt—occipital triangle, atr—anal triangle, S10—abdominal segment 10.

2.14. Distribution Maps

The distribution maps were created as follows: Collecting localities of the specimens were first georeferenced using Google's Geocoding API (Application Programming Interface). Subsequently, the resulting coordinates were plotted in ARCGIS version 10.2.2 using the geographic coordinate system WGS 1984. The source of the background map for reference was the U.S. National Park Service, publicly available through Esri services. In

Figures 21 and 22 we have drawn the biogeographic boundary known as the Anatolian Diagonal by hand. Its position is based on literature and mountain chains visible on the background map [42].

3. Results

3.1. Molecular Genetic Analysis of the COI and ITS Genes

The COI alignment contained 151 taxa and 568 characters of which 213 were distinct data patterns. The ITS alignment contained 53 taxa and 885 characters, of which 263 were distinct data patterns. The partition alignment with two markers (COI, ITS) contained 47 taxa and 568 characters for COI, 891 for the ITS of which 150 (COI) and 296 (ITS) were distinct data patterns.

All three programs used showed for both genes in the majority-rule consensus trees four distinct major clades: *boltonii*, *bidentata*, *coronata* and a New World clade, shown are the MrBayes trees (Figures 4 and S2). Western Palaearctic *Cordulegaster* species fall in traditional *boltonii*- and *bidentata*-groups. We observed no evidence for hybridization between members of the *bidentata*- and the *boltonii*-group. *C. coronata*, until recently often treated as a subspecies of *C. insignis*, is grouped neither with *bidentata*- nor with the *boltonii*-group. All three programs with both genes and combination of both genes revealed an extra clade, which we named the *coronata*-group (Figures 4, 5 and S3–S6). For better resolution, we separated the MrBayes COI-tree in two parts and added the haplotype analysis (Figures 6 and 7). For better understanding of previously used names of the members of the genus in the region in relation to our molecular genetic results we introduced Table 2.

3.2. *Cordulegaster boltonii*-Group

In the *boltonii*-group of the Eastern region of the Western Palaearctic we recovered with molecular genetic analysis three known taxa (*C. heros*, *C. picta*, *C. vanbrinkae*) and one new taxon, which is described phenotypically below in the descriptive section (*C. kalkmani* sp. nov.). For the little known member of the *boltonii*-group *C. vanbrinkae* specimens from localities of all provinces from which this taxon has been reported so far (Armenia: Syunik; Azerbaijan: Lankaran; Iran: Mazandaran, Golestan) were included in this study (Figures 1 and S1, Table 1). In the case of specimens so far treated as *C. picta*, the whole large geographical range from East–Europe to Russian Black Sea region was covered by our molecular approach. Specimens from the East Anatolian provinces Van and Kars grouped separately from *C. picta* and *C. vanbrinkae* with both genes in MrBayes analysis (Figures 4–6). The additional two programs used for COI and a combination of COI and ITS, placed the two specimens of the new taxon apart from the other members of the *boltonii*-group (Figures S3–S6). Looking at the trees for COI, the *boltonii*-group forms two bigger clades in all three programs applied: the "western" clade with *C. boltonii*, *C. trinacriae*, and *C. princeps*; and the "eastern" clade with *C. heros*, *C. picta*, *C. vanbrinkae*, and *C. kalkmani* (Figures 4, 6, S3 and S5). The alignments suggest that *C. vanbrinkae* and *C. picta* are sister taxa (Figures 4–6 and S3–S6).

The K2-P analysis of the COI gene revealed that the proposed new taxon (*C. kalkmani*) is genetically more distant from *C. picta* than *C. vanbrinkae* is from *C. picta* (Table 3). Thus, these specimens belong, neither to *C. picta* nor to *C. vanbrinkae*. The K2-P distance between *C. picta* and *C. kalkmani* is 5.88%, and between *C. kalkmani* and *C. vanbrinkae* 4.93% (Table 3). The presence of four eastern species in the *boltonii*-group was also supported by the haplotype analysis of the COI gene (Figure 6).

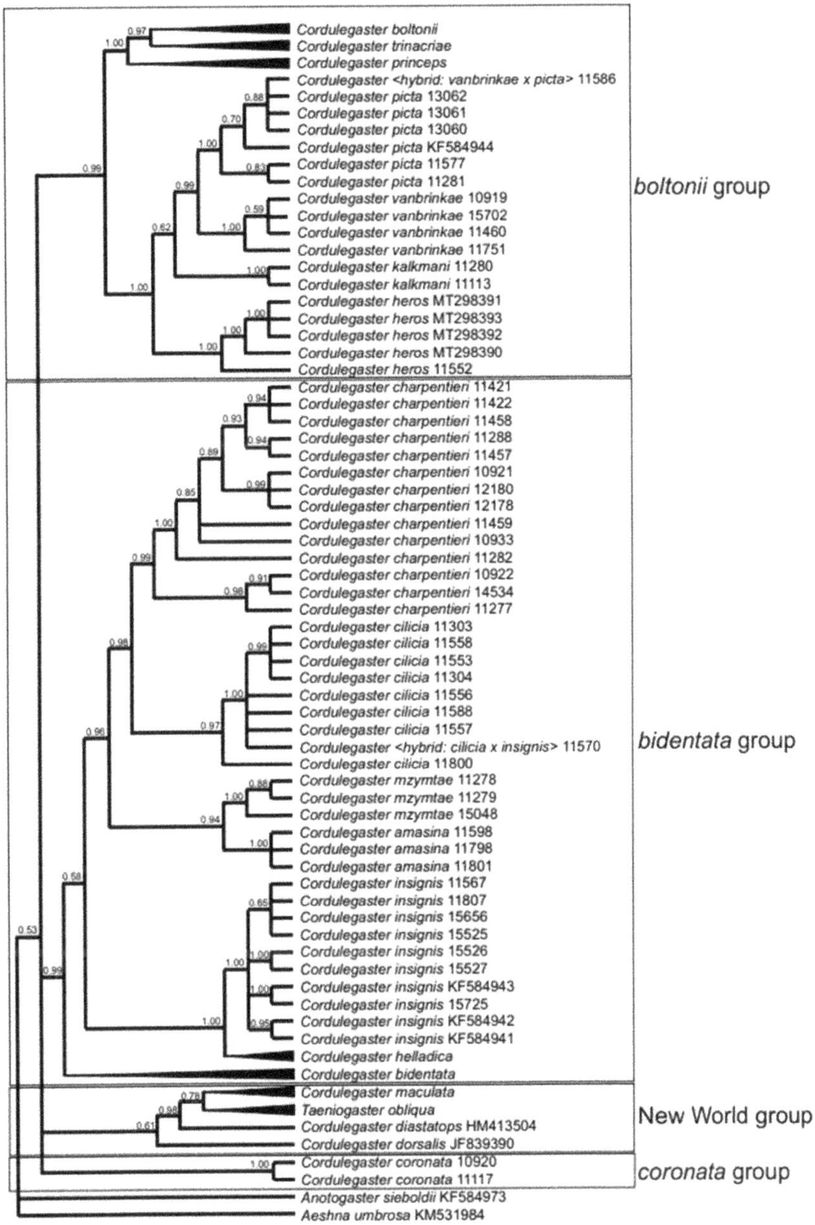

Figure 4. Overview tree for the COI gene, indicating the four major groups: *boltonii-*, *bidentata*, *coronata-*, New World-group. Maximum likelihood tree inferred with Bayesian analysis using MRBAYESs 3.2.7a using the best-fit model (GTR+I+G) identified with JMODELTEST 2.1.10. Bayesian posterior probabilities values are depicted at the nodes. Included are our own sequences (PCR number next to the name) and those retrieved from GenBank (accession numbers next to the name), if specimens identify different taxa in the COI and ITS analysis they are indicating possible hybrids).

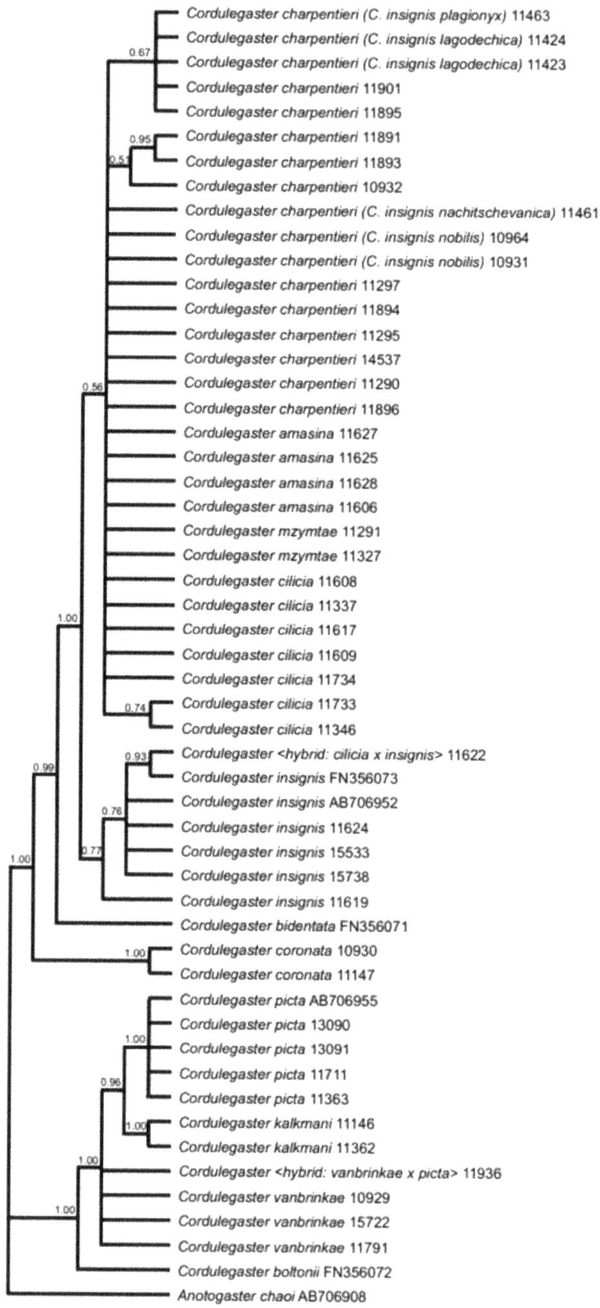

Figure 5. Overview tree from the ITS gene. Maximum likelihood tree inferred with Bayesian analysis using MRBAYES 3.2.7a using the best-fit model (HKY + G) identified with JMODELTEST 2.1.10. Bayesian posterior probabilities values are depicted at the nodes. Included are our isolated sequences (PCR number next to the name) and those retrieved from GenBank (accession numbers next to the name), if specimens identify different taxa in the COI and ITS analysis they are indicating possible hybrids.

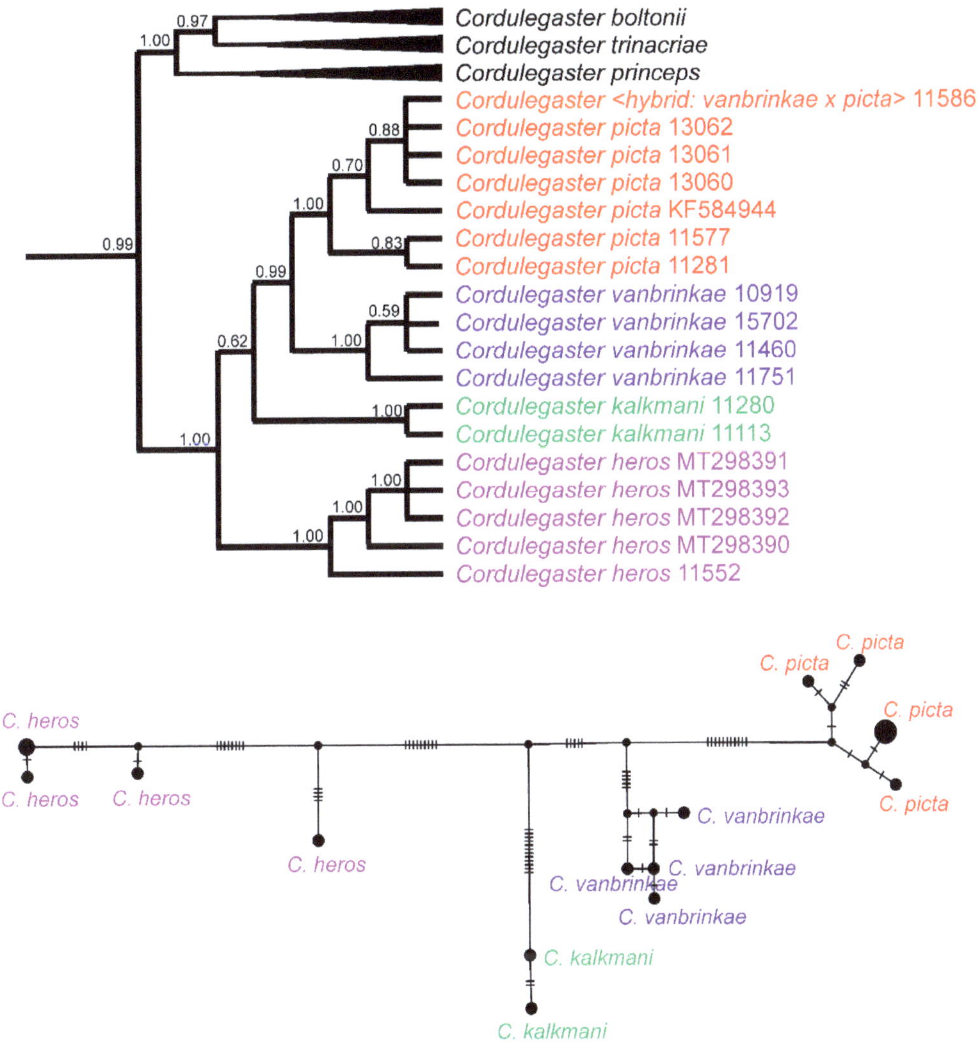

Figure 6. Tree of the COI-analysis for the *C. boltonii*-group and haplotype-analysis (TCS-network made in POPART 1.7). Bayesian posterior probability values are depicted at the nodes.

3.3. Possible Hybridisation in the Cordulegaster boltonii-Group

The two available specimens from the new taxon *C. kalkmani* showed no evidence for hybridization with other taxa. However, a specimen labeled under the name *C. vanbrinkae* from Armenia (Genbank MK779798) grouped in the COI analysis (PCR COI: 11586) with *C. picta*, while in the ITS analysis (PCR ITS: 11936) with *C. vanbrinkae* (Figures 4–6).

3.4. Cordulegaster coronata-Group

Specimens of supposed *C. coronata* from North-East Iran and Kyrgyzstan formed an extra clade in all analyses performed for both genes and the combination of the two genes (Figures 4, 5 and S3–S6). Thus, *C. coronata* does not group with members of the *bidentata*-group neither in the COI nor in the ITS analysis, which is in contrast to phenotype analysis, in which *C. coronata* is similar to *C. charpentieri* (see below in the descriptive part).

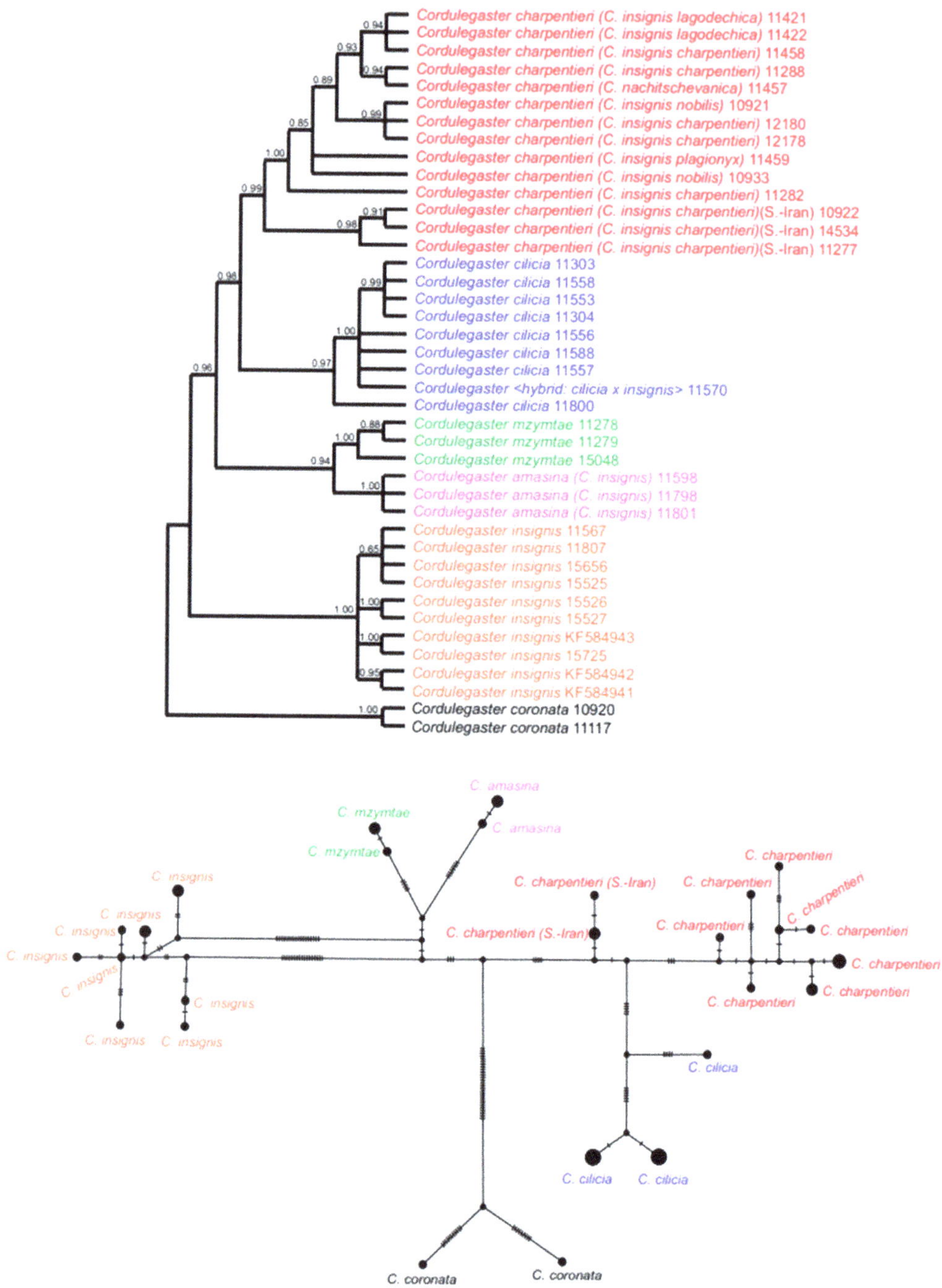

Figure 7. Tree of the COI-analysis for the *C. bidentata*-complex and *C. coronata* including haplotype-analysis (TCS-network made in POPART 1.7). Bayesian posterior probabilities are depicted at the nodes.

Table 2. (Overview Table): Actual used names and taxonomic rank as well as revised names and taxonomic rank, and the country of origin of the *Cordulegaster* specimens investigated.

Actual Used Names and Taxonomic Rank	Revised Name and Taxonomic Rank	Origin of Specimens Studied
	boltonii-group	
C. heros	*C. heros*	Bulgaria, Montenegro, Greece, Austria
C. picta	*C. picta*	Bulgaria, Turkey, Greece, Georgia, Russia
C. picta	*C. kalkmani* sp.nov.	Eastern Anatolia Turkey
C. vanbrinkae	*C. vanbrinkae*	Armenia, Azerbaijan, Iran
	bidentata-group	
C. insignis	*C. insignis*	West Turkey, East Greece
	charpentieri-complex	
C. insignis amasina	*C. amasina* stat. rev.	Turkey
C. insignis mzymtae sometimes also used on species level *C. mzymtae*	*C. mzymtae*	Turkey, Georgia, Russia
C. insignis charpentieri	*C. charpentieri* stat. rev.	Armenia, Turkey, Georgia
C. insignis lagodechica	*C. charpentieri* stat. rev. *C. insignis lagodechica* syn. nov.	Georgia
C. insignis nobilis	*C. charpentieri* stat. rev. *C. insignis nobilis* syn. nov.	Turkey, Iran
C. nachitschevanica	*C. charpentieri* stat. rev. *C. nachitschevanica* syn. nov.	Azerbaijan
C. plagionyx	*C. charpentieri* stat. rev. *C. plagionyx* syn. nov.	Azerbaijan
C. insignis	*C. cilicia* sp. nov.	Eastern Mediterranean Turkey, Lebanon
	coronata-group	
C. insignis coronata sometimes used on species level *C. coronata*	*C. coronata*	Iran, Kyrgyzstan

Table 3. Genetic distance, estimated by the Kimura 2-parameter method, between the different species in the *Codulegaster boltonii*-group and *C. coronata*.

	C. boltonii	*C. trinacriae*	*C. princeps*	*C. heros*	*C. picta*	*C. kalkmani*	*C. vanbrinkae*	Average Distance within the Group
C. boltonii								0.0077
C. trinacriae	0.0583							0.0048
C. princeps	0.0563	0.0632						0.0044
C. heros	0.0811	0.0865	0.0852					0.0151
C. picta	0.0790	0.0794	0.0865	0.0617				0.0060
C. kalkmani	0.0698	0.0884	0.0718	0.0625	0.0588			0.0036
C. vanbrinkae	0.0791	0.0838	0.0941	0.0537	0.0395	0.0493		0.0044
C. coronata	0.1108	0.1083	0.1083	0.0883	0.0950	0.1010	0.0991	0.0402

3.5. *Cordulegaster bidentata*-Group

Within the *bidentata*-group, three major clades were recognized by the majority-rule consensus trees for both genes in all three programs and with the combination analysis of COI and ITS genes with the StarBeast and BPP program (Figures 4–6 and S3–S6), which we name here *C. bidentata*, *helladica-insignis*-complex, and *charpentieri*-complex. Specimens from West Turkey grouped together with specimens from Eastern Greece islands like Samos and Ikaria. They are separated in all trees for both genes in all three programs and with the combination analysis of COI and ITS genes with the StarBeast and BPP program from all specimens collected further east, which belong to the *charpentieri*-complex (Figures 4–6 and S3–S6). Thus, our molecular analysis revealed that *C. insignis* occurs only in the Western part of Turkey further east all specimens of the *bidentata*-group belong to the *charpentieri*-complex.

3.6. *Cordulegaster charpentieri*-Complex

The COI analysis with the two programs MrBayes and StarBeast support four closely related taxa in the *charpentieri*-complex (Figures 4, 5, 7 and S3). The majority-rule consensus tree for the COI created by the BPP program revealed a reduced topology with *C. amasina* and *C. mzymtae* together, as well as *C. charpentieri* and *C. cilicia*, as well as *insignis* and *helladica*, as well as *C. picta* and *C. vanbrinkae*, as well as *C. princeps*, *C. trinacriae*, and *C. boltonii*, all in one group, respectively (Figure S5). This may indicate that the BPP program does not sufficiently distinguish more recently split taxa. Our timetree analysis supports a more recent separation of the taxa within the *charpentieri*-complex (Figure S7). For a more detailed taxonomy in this complex we followed the COI analysis suggested by the majority-rule consensus trees created by the MrBayes and StarBeast program, because these analyses also differentiated the well-established taxa *C. boltonii*, *C. trinacriae* and *C. princeps*. Finally, we suggest the differentiation of the *charpentieri*-complex in four taxa: *C. amasina* stat. rev., *C. mzymtae*, *C. cilicia* sp. nov. and *C. charpentieri* stat. rev. The phenotypical description of the suggested new taxon and the re-description of the poorly described other three taxa in this *charpentieri*-complex are presented below in the descriptive part. The trees for COI by the BPP program, and by the combination of the two genes using the StarBeast program suggest *C. amasina* and *C. mzymtae* as well as *C. charpentieri* and *C. cilicia* as sister taxa, respectively (Figures S4 and S5) supplementary y = COI-ITS combined Starbeast, z = COI BPP).

The K2-P distances in the COI analysis between the four taxa suggested here as species are 2.96–4.24% (Table 4). For Western Palaearctic Anisoptera, a recent large analysis revealed a K2-P distance above 1.96% for a good threshold [43]. The existence of four species (*C. amasina*, *C. mzymtae*, *C. cilicia* and *C. charpentieri*) in the *charpentieri*-complex is also supported by the haplotype analysis (Figure 7). To avoid further taxonomic trouble, we decided not to recognize or create subspecies within the *charpentieri*-complex.

Table 4. Genetic distance, estimated by the Kimura 2-parameter method, between the different species in the *Cordulegaster bidentata*-group and *C. coronata*.

	C. bidentata	*C. insignis*	*C. amasina*	*C. mzymtae*	*C. charpentieri*	*C. cilicia*	Average Distance within the Group
C. bidentata							0.0098
C. insignis	0.0686						0.0093
C. amasina	0.0656	0.0623					0.0012
C. mzymtae	0.0590	0.0551	0.0296				0.0012
C. charpentieri	0.0633	0.0633	0.0452	0.0424			0.0110
C. cilicia	0.0566	0.0628	0.0401	0.0421	0.0322		0.0067
C. coronata	0.1028	0.0972	0.1022	0.0966	0.0942	0.0888	0.0402

C. amasina and *C. mzymtae* occur mainly along the Black Sea. *C. amasina* occurs in the western part of this region approximately from the province Kastamonu to the province Samsun and from the Black Sea Coast to central Turkey reaching the province Ankara (Figures 2 and S1, Table 1). *C. mzymtae* the darkest member of the *bidentata*-group occurs eastwards from *C. amasina* reaching Georgia, the Russian Black Sea Coast and even until Karachay Cherkessia region in the North Caucasus (Russia) (Figures 2 and S1, Table 1).

An undescribed species, here named *C. cilicia* sp. nov., occurs in central Anatolia, and along the East Mediterranean Sea.

Surprisingly, there was evidence for a hybrid at the contact between *C. insignis and C. cilicia*. In the ITS analysis, *C. cilicia* from Turkey, Konya, Doganbey (Genbank MK779798) grouped in the ITS analysis (PCR ITS: 11622) with *C. insignis*, while in the COI analysis (PCR COI: 11570) it grouped with eight other *C. cilicia* (Figures 4, 5 and 7, Table 4).

C. charpentieri occurs in a large area including East-Central Anatolia, Georgia, Armenia, Azerbaijan and Iran, and most probably in North-East Iraq. This taxon is slightly more heterogeneous in the COI and haplotype analysis (Figures 4 and 7) compared to the other three species of the *charpentieri*-complex. However, *C. nachitschevanica* was in the COI analysis identical with Armenian *C. charpentieri*, and *C. plagionyx* was identical with North Iranian *C. charpentieri* (Figures 4 and 7). The K2-P distances are far below 2% and the haplotype analysis did not support species level (Figure 7). Therefore, we synonymized *C. nachitschevanica* syn. nov. and *C. plagionyx* syn. nov. with *C. charpentieri*. The COI and haplotype analysis of the Georgian specimens from Lagodechi (*C. i. lagodechica*) did not support a separate species or subspecies as suggested by Bartenev [30]. Therefore, we synonymized *C. insignis lagodechica* syn. nov. with *C. charpentieri*. The *C. charpentieri* specimens from South and South-East Iran (provinces Fars and Kerman) grouped extra in the COI analysis and haplotype analysis (Figures 4 and 7). However, the distance in the COI analysis between these specimens and *C. charpentieri* was less than 2%. To avoid further confusion in this group we refrain from describing further species or subspecies. For better visualization, the morphological variants were illustrated in the descriptive part below.

4. Descriptive Part

4.1. *Cordulegaster boltonii*-Group

Variation of ***Cordulegaster vanbrinkae*** and characterisation of a hybrid between *C. picta* and *C. vanbrinkae*

Material examined: *C. vanbrinkae*: 1 ♂(coll. Thomas and Elias Schneider): IRAN: Mazandaran, Veysar, 36.4700° N, 51.5383° E, 1438 m a.s.l., July 2013, leg. Thomas Schneider, July 2013; 2 ♂♂(collection coden here): IRAN: Golestan, Kaboodval waterfall/river 3 km SW Kordabad ca. 4 km SSE Aliabad, 36.8741° N, 54.8872° E, 343 m a.s.l., vii.2017, leg. Thomas Schneider.

***C. picta*:** 1 ♂(coll. Thomas and Elias Schneider): GEORGIA: Imereti, 42.0236° N, 43.4588° E, 900 m a.s.l., vii.2015, leg. Thomas Schneider. ***C.vanbrinkae* × *C. picta* (sequenced hybrid):** 1 ♂(RMNH): ARMENIA: Verin Khotanan village, 39.3334° N, 46.3786° E, 1550 m a.s.l., specimen RMNH.INS.974942, vii.2010, leg. Vasil Ananian.

C. vanbrinkae* × *C. picta (possible hybrid, no DNA could be extracted): same locality as previous specimen, RMNH.INS.975675.

Comments. In the molecular section, we detected a possible hybrid between *C. vanbrinkae* and *C. picta*. The specimen is from Armenia, geographically between known populations of *C. picta* and *C. vanbrinkae*. Therefore, we compared the morphology of this hybrid with *C. picta* and *C. vanbrinkae*.

As shown in Figure 8, by abdominal colour pattern the hybrid cannot be distinguished from *C. vanbrinkae*. *C. vanbrinkae* is usually darker than *C. picta* (Figure 8A–C,F), although more yellow specimens of *C. vanbrinkae* can be found (Figure 8C). The male appendages of the hybrid look intermediate, as the basal tooth of the superior appendages is a bit larger than in typical *C. vanbrinkae* (Figure 8J–L).

***Cordulegaster kalkmani* sp. nov.**

Type material: Holotype: 1 ♂(RMNH.INS.974903): TURKEY: Kars, 16 km NE Sarikamis, 40.4203° N, 42.7446° E, 1800 m a.s.l., 23.vii.2007, leg. Gert Jan van Pelt.

Paratype: 1 ♀(RMNH.INS.747180): TURKEY: Van, 8 km N Muradiye, near Bendimahi Selalesi 8 km, 39.0594° N, 43.7540° E, 1850 m a.s.l., 30.vi.2003, leg. Vincent Kalkman.

Etymology. The species is named after Dr. Vincent Kalkman, Naturalis Biodiversity Center, Leiden, Netherlands, who collected the female paratype.

Description of the male holotype. ***Head*** (Figure 9E,F): anteclypeus black; postclypeus yellow; labrum yellow with thicker black margin and a short black line in the middle; frons yellow, in dorsal view with black bar at the rear edge; labium yellow, mandible black; occipital triangle black with two yellow dots (Figure 9G); antenna and vertex black; eyes green (collector's observation).

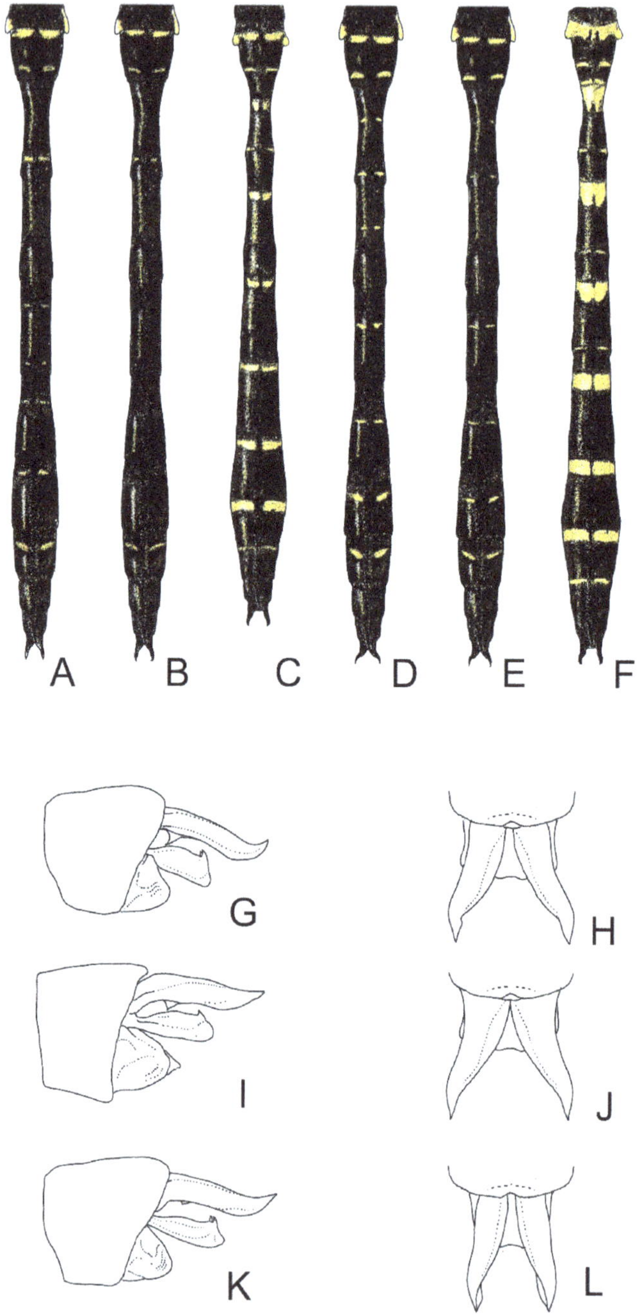

Figure 8. Abdominal variation of *C. vanbrinkae*: (**A**): Golestan, Iran. (**B**): Mazandaran, Iran. (**C**): Golestan. (**D**): Armenia, possible hybrid: *C. picta* × *C. vanbrinkae* not genetically analyzed. (**E**): Armenia, hybrid: *C. picta* × *C. vanbrinkae* genetically analyzed. (**F**): *C. picta* Georgia. (**G**,**H**): lateral and dorsal view of the male appendices from *C. picta* Georgia. (**I**,**J**): Hybrid *C. picta* × *C. vanbrinkae*, Armenia. (**K**,**L**): *C. vanbrinkae*, Mazandaran.

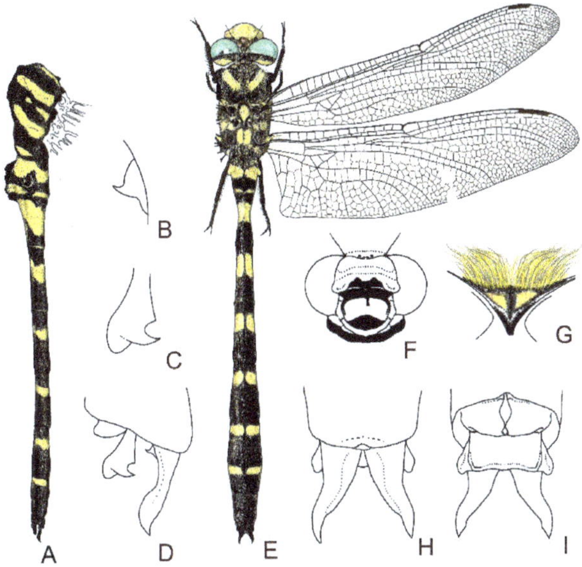

Figure 9. *Cordulegaster kalkmani* sp. nov., holotype ♂(RMNH, leg. Gert Jan van Pelt, RMNH.INS.974903, 23.07.2007, Turkey, Kars, 16 km NE Sarikamis: 40.4203° N 42.7446° E, 1800 m a.s.l.): (**A**): habitus, lateral. (**B**): medioventral tooth, lateral. (**C**): hook, lateral. (**D**): appendages, lateral. (**E**): habitus, dorsal. (**F**): head, schematically. (**G**): occipital triangle, frontal. (**H**): appendages, dorsal. (**I**): appendages, ventral.

Thorax (Figure 9A,E): anterior lobe with yellow frontal edge, median lobe of pronotum black, posterior lobe with yellow margin at the rear edge that narrows dorsolaterally; front of synthorax with broad yellow antehumeral stripes, narrowing towards pronotum leaving less than 30% black of the antehumeral region; small yellow dot on the mesanepisternum near the antealar ridge; mesepimeron and metepimeron with broad yellow bands; metepisternum black with yellow interrupted stripe-like marking reaching to one third of basal margin; metakatepisternum black; poststernum black; coxae black with yellow parts, trochanter, femora, tibiae and tarsi black.

Wing (Figure 9E hyaline; Pt black; FW 3.8 × 0.5; HW 4.5 × 0.7; anal triangle with 3 cells; all veins black, except costa.

Abdomen (Figure 9A,E): color pattern as in Figure 9A; S1 black with two yellow markings in lateral view, S1 dorsally covered with yellow hairs, yellow marking (abdominal rings) on side of S1 semilunar; S2 to S8: yellow abdominal rings of S2 and S3 dorsally connected, rings of S4 to S8 interrupted by fine black line, yellow abdominal rings leaving a bigger part black but running onto underside; S9–S10 black.

Appendages (Figure 9B–D,H,I): superior appendages in dorsal view slender and diverging, with pointed and slightly curved tips, nearly as long as S10; laterobasal tooth not visible in lateral view because it is covered by the rear edge of S10, prominent medioventral tooth, with a claw-shape tip only visible in lateral view, located at the base of the inferior near S10 (Figure 8B,D); inferior appendage in lateral view at the end with a single pointed hook-shaped tooth on each side (Figure 9C,D). Inferior appendage broad, nearly parallel-sided, the dorsally pointing lobes widen the inferior distally, hind margin not notched (Figure 9I).

Measurements (in mm): TL (inclusive appendages) 74.0, AL 58.0, HWL 46.0, FWL 46.0.

Description of female paratype. ***Head*** (Figure 10A): much as in ♂, additional small black line at crest of frons.

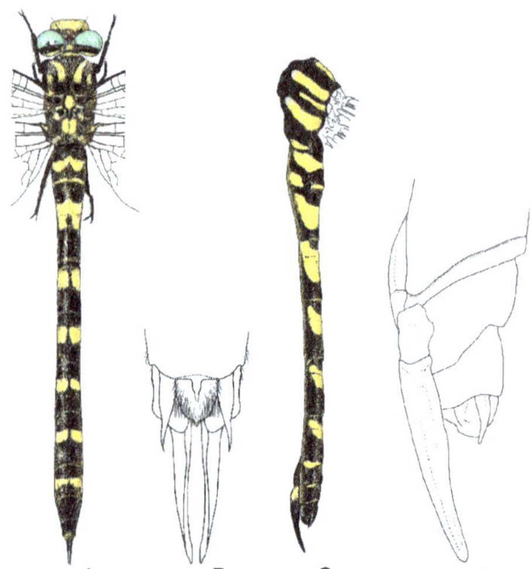

Figure 10. *Cordulegaster kalkmani* sp. nov., paratyp ♀(RMNH, leg. Vincent Kalkman, 30.06.2003, RMNH.INS. 747180, Turkey, Van, 8 km N Muradiye, near Bendimahi Selalesi, 39.0594° N 43.7540° E, 1850 m a.s.l.): (**A**): habitus, dorsal. (**B**): appendages, dorsal. (**C**): habitus, lateral. (**D**): appendages, lateral.

Thorax (Figure 10A,C): colour pattern of prothorax and synthorax similar to male; in contrast to male yellow band at metepisternum not interrupted; metakatepisternum black with a pale yellow shadow near metastigma.

Wings hyaline; pterostigmata black; costal veins with yellow leading edge.

Abdomen (Figure 10A–D): cylindrical and thicker than in male; ovipositor black, long and bent like a sword, more than three times as long as S10 as seen in lateral view.

Measurements (mm): TL (inclusive ovipositor) 82.0, AL 65.0, HWL 51.0, FWL 53.0.

Differential diagnosis

The male inferior appendage of *C. kalkmani* distally much wider than in all other members of the genus. A single hook-shaped tooth in lateral view at the apex of the inferior appendage separates it from *C. vanbrinkae* and *C. picta*. These latter two species of the Eastern *boltonii*-group have two teeth in lateral view that are often covered by bristles and are therefore difficult to see. In *C. heros* the laterobasal and medioventral teeth on superior appendages are clearly more distal than in the other three Eastern species of the *boltonii*-group. Compared to *C. vanbrinkae*, *C. kalkmani* has a stout inferior appendage in lateral view. In addition, the medioventral teeth are more clearly visible, although are closer to S10 than in *C. vanbrinkae*. In *C. picta* the medioventral teeth are more distal than in *C. kalkmani*. The medioventral teeth are more clearly hooked in *C. kalkmani* than in *C. picta*. The position of the laterobasal teeth are comparable in *C. picta* and *C. kalkmani*. In contrast to *C. vanbrinkae*, *C. kalkmani* has a yellow frons as well as *C. picta* and *C. heros*. Females within the *boltonii*-group are impossible to separate without knowing their geographical origin.

Distribution

East Anatolia and probably North–West Iran. Separated from *C. picta* through the Anatolian Diagonal, and further east by the Armenian Highland and the Caucasus Mountains. *Cordulegaster kalkmani* is separated from *C. vanbrinkae* by the Elburz Mountains, hybridization between *C. picta* and *C. vanbrinkae* occurs in South–East Armenia.

4.2. *Cordulegaster bidentata*-Group

***Cordulegaster charpentieri* (Kolenati, 1846) stat. rev.**

Cordulegaster insignis nobilis Morton, 1916, *syn. nov.*

Cordulegaster insignis lagodechica Bartenev, 1930, *syn. nov.*

Cordulegaster nachitschevanica Skvortsov and Snegovaya, 2015, *syn. nov.*

Cordulegaster plagionyx Skvortsov and Snegovaya, 2015, *syn. nov.*

Redescription. As the type is lost and the original description is poor, we selected a neotype for re-description; in addition we describe the variation of this species in its geographical range.

Material examined: ***Neotype:*** 1 ♂(RMNH.INS.974905): TURKEY: Bitlis Province, 32 km WNW Gevas, 38.3834° N, 42.7723° E, 1950 m a.s.l., 13.vii.2005, leg. Gert Jan van Pelt.

Additional material studied. 1 ♂(RMHN.INS.974907): TURKEY: Erzurum, Şenkaya, 40.5615° N, 42.3471° E, 1887 m a.s.l., 31.viii.1993, leg. Gert Jan van Pelt; 1 ♂(collection Thomas and Elias Schneider): AZERBAIJAN: Nakhichevan, Agdere, 39.11028° N, 45.9144° E, 1990 m a.s.l., 26.vii.2017, leg. Nataly Snegovaya; 1 ♂(collection Thomas and Elias Schneider): AZERBAIJAN: Balakan, 41.6761° N, 46.4931° E, 320 m a.s.l., 03.vii.2014, leg. Nataly Snegovaya; 2 ♂(collection Thomas and Elias Schneider): IRAN: Lorestan, Bagh Goije 24 km NE Aligudarz, 33.4597° N, 49.9472° E, 1817 m a.s.l., 11.vi.2015, leg. Thomas Schneider; 1 ♂(RMNH.INS.975673): IRAN: Lost Paradise, Behesht Gomshodeh, Fars, 30.3339° N, 52.1564° E, 1800 m a.s.l., 07.vi.2018, leg. Thomas Schneider; 2 ♂(collection Thomas and Elias Schneider): IRAN: Lost Paradise, Behesht Gomshodeh, Fars, 30.3339° N, 52.1564° E, 1800 m a.s.l.), 06.vi.2014, leg. Thomas Schneider; 6 ♂(collection Thomas and Elias Schneider): IRAN: Lost Paradise, Behesht Gomshodeh, Fars, 30.3339° N, 52.1564° E, 1800 m a.s.l.), 06.vi.2019, leg. Thomas Schneider; 4 ♂(collection Thomas and Elias Schneider): IRAN: Dalfard waterfalls, Kerman, 29.0022° N, 57.5869° E, 2110 m a.s.l.), 28.v.2014, leg. Thomas Schneider; 1 ♀(RMNH.INS.975674): IRAN: Lost Paradise, Behesht Gomshodeh, Fars, 30.3339° N, 52.1564° E, 1800 m a.s.l., 06.v.2017, leg. Elias and Thomas Schneider 1 ♀(collection Thomas and Elias Schneider): IRAN: Lost Paradise, Behesht Gomshodeh, Fars, 30.3339° N, 52.1564° E, 1800 m a.s.l., 06.v.2017, leg. Elias and Thomas Schneider. 2 ♀ (collection Thomas and Elias Schneider): IRAN: Lost Paradise, Behesht Gomshodeh, Fars, 30.3339° N, 52.1564° E, 1800 m a.s.l., 06.vi.2019, leg. Elias and Thomas Schneider.

Description of neotype. ***Head*** (Figure 11B,D,E): anteclypeus black; postclypeus yellow; labrum yellow with black margin; frons yellow with faint black bar; labium yellow; occiput domed and yellow, occipital triangle domed, yellow with black margins; antenna and vertex black; eyes blue (collector's observation); postocular area yellow with black upper margin.

Thorax (Figure 11A,D): anterior lobe of pronotum with yellow front edge; median lobe black with yellow dorsolateral patterns, middorsal with two small yellow dots; posterior lobe yellow with narrow middorsal line ending in front of rear edge of lobus; front of synthorax with broad big yellow antehumeral stripes, narrowing moderately towards pronotum leaving less than 20% black of antehumeral region; mesepimeron and metepimeron with broad yellow bands, metepisternum black with yellow triangle marking with base to wings and short yellow comma-shaped stripe ending near metastigma; metakatepisternum black; poststernum black; coxae black with yellow ventral parts, trochanter, femora, tibiae and tarsi black.

Wings (Figure 11D): hyaline; costal veins yellow, other veins black; pterostigma black, pterostigma FW 4.2 × 0.5, HW 4.5 × 0.7; anal triangle 3 cells; membranula narrow.

Abdomen (Figure 11A,D): colour pattern as shown in Figure 11A,B; S1 dorsal black, yellow marking on side of S2 to S9 with medium sized yellow markings leaving ca. 40% black, on dorsal view of S2 to S8 there are a pair of small yellow line-art markings on lateral distal part of segment not touching in the middle of segment, on S2 to S10 the black markings reaching lateral sides, yellow makings also present on S9 and S10.

Appendages (Figure 11C,F,G): superior appendages in dorsal view straight, slender, with pointed tips, medioventral teeth clearly visible and directed to each other, in lateral view laterobasal teeth are visible and are located near S10, in lateral view upper medioventral teeth located more ventral far away of crossing with tooth on the end of inferior appendage; inferior appendage is stout trapezoid, not notched on distal margin when viewed from above, in lateral view distally small teeth on end of the inferior appendage are visible, the three teeth form a crown-shaped complex in which the two outer teeth are fused, the inner tooth is isolated from them.

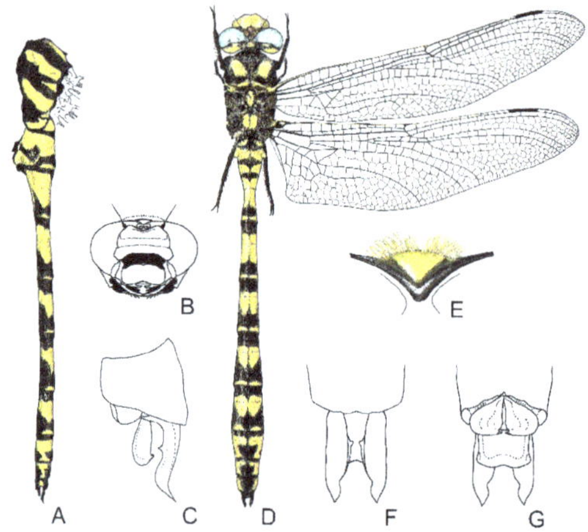

Figure 11. *Cordulegaster charpentieri* neotype ♂(RMNH, leg. Gert Jan van Pelt, RMNH.INS.974905, 13.07.2005, Turkey, Bitlis Province, 32 km WNW Gevas, 38.3834° N, 42.7723° E, 1950 m a.s.l.): (**A**): habitus, lateral. (**B**): head, schematically. (**C**): appendages, lateral. (**D**): habitus, dorsal. (**E**): occipital triangle, frontal. (**F**): appendages, dorsal. (**G**): appendages, ventral.

Measurements (mm): TL (including appendages) 65.0, AL 50.0, HWL 41.0, FWL 42.0.

Variation. *C. charpentieri* shows strong variation in colour pattern and size. We visualize this variability in figures that include yellow specimens of females (Figure 12). Individuals from South–East Iran are bigger, and those from Fars Province are more yellow (Figure 13), than typical specimens from Eastern Anatolia, the Caucasus Countries and North Iran (Figure 14). Males from Fars (n = 9): TL (including appendages) 79.0–83.0, AL: 60–63.0, HWL 47.0–50.0.

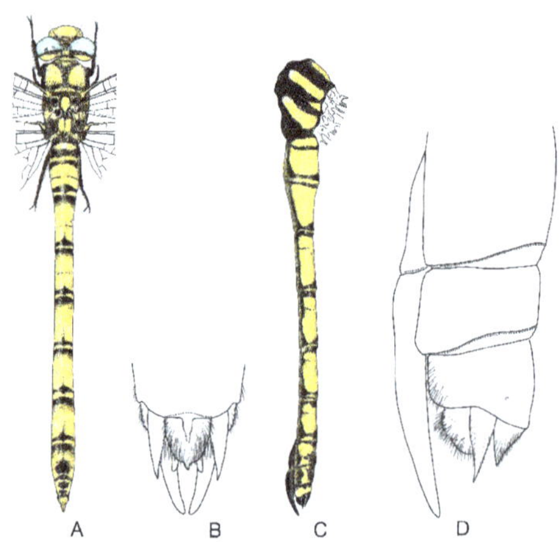

Figure 12. *Cordulegaster charpentieri*, yellow form ♀, (RMNH, leg. Elias and Thomas Schneider, RMNH.INS. 975674, 06.05.2018, Iran, Lost Paradise, Behesht Gomshodeh, Fars, 30.3339° N, 52.1564° E, 1800 m a.s.l.): (**A**): habitus, dorsal. (**B**): appendages, dorsal. (**C**): habitus, lateral. (**D**): ovipositor, lateral.

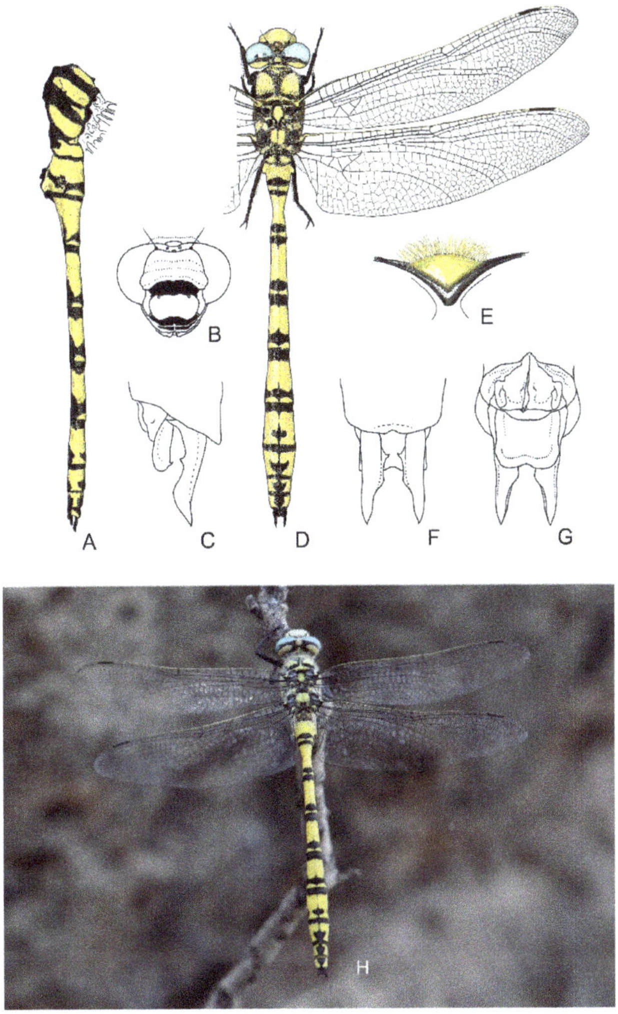

Figure 13. *Cordulegaster charpentieri* paratype ♂ (yellow form) (RMNH, leg. Thomas Schneider, RMNH.INS. 975673, 7 June 2019, Iran, Lost Paradise, Behesht Gomshodeh, Fars, 30.3339° N, 52.1563° E, 1800 m a.s.l.): (**A**): habitus, lateral. (**B**): head, schematically. (**C**): appendages, lateral. (**D**): habitus, dorsal. (**E**): occipital triangle, frontal. (**F**): appendages, dorsal. (**G**): appendages, ventral. (**H**): photo of this paratype (Dietmar Ikemeyer).

Individuals from Kerman Province are darker, more similar to typical forms, but much larger and with pterostigma extremely narrow (Figure 13). These individuals are the biggest in the *bidentata*-group, even bigger than *C. heros*, and are therefore among the largest dragonflies in the Western Palaearctic. Males from Kerman (n = 4): TL (including appendages) 81.0–85.0; AL 62.0–65.0; HWL 49.0–55.0. Individuals were found between 1700 and 2200 m a.s.l.

Females from Fars are very yellow and are impressive, heavy insects. Females from Fars (n = 4): TL (including ovipositor) 84.0–88.0; AL: 63.0–7.0; HWL 51.0–57.0.

Differential Diagnosis

The males have large abdominal markings, those on S9–10 often connect to form «7» shaped markings, apical dorsal pair of yellow spots usually present on S2–10, whereas they are less present in *C. cilicia* sp. nov. and absent in *C. mzymtae*. Individuals of *C. charpentieri* from South–East Iran are larger than all other members of the *bidentata*-group. The occiput and the postocular area are domed and yellow. The other species (*C. amasina*, *C. mzymtae* *C. cilicia*) of the *charpentieri*-complex have the occiput and postocular area more flattened and less yellow.

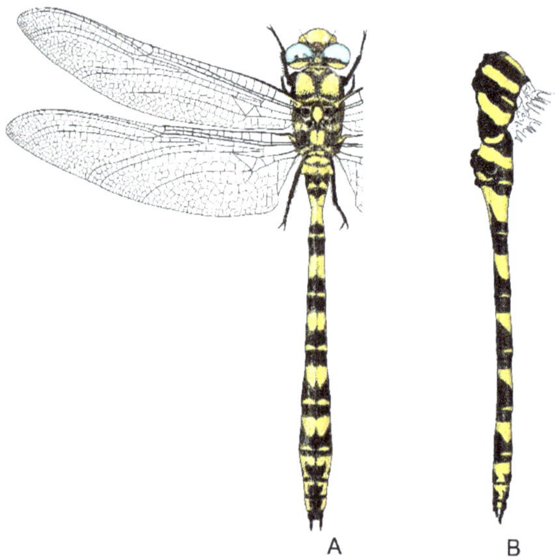

Figure 14. *Cordulegaster charpentieri*, (dark form) ♂, (in coll. Elias and Thomas Schneider, leg. Thomas Schneider, 28 May 2014, Iran Dalfard, 29.0022° N, 57.5869° E, 2110 m a.s.l.): (**A**): habitus, dorsal. (**B**): habitus, lateral.

Superior appendages in dorsal view are more straight and slender than in the other two species of the complex, inferior appendage is stout trapezoid, not rectangular as in *C. cilicia* (see below). Only the very big, yellow females of *C. charpentieri* from South–East Iran and the small and dark females of *C. mzymtae* can easily be recognized. Other females are nearly impossible to separate within the *charpentieri*-complex without knowing their geographical origin.

Distribution

Cordulegaster charpentieri has a wide distribution, from East Anatolia (Turkey), Georgia, Armenia, Azerbaijan and Iran to North–East Iraq.

***Cordulegaster cilicia* sp. nov.**

Type material. Holotype: 1 ♂(RMNH.INS.974914): TURKEY: Kahramanmaraş, Göksun, Gücüksu, Göksun River, 38.0608° N, 36.6494° E, 1350 m a.s.l., 09.vii.2008, leg. Gert Jan van Pelt.

Paratypes: 1 ♂(RMNH.INS.974912): TURKEY: Nidge, 37.9698° N, 34.6767° E, 1230 m a.s.l., 30.vi.2008, leg. van Pelt; 1 ♂(RMNH.INS.974906): TURKEY: Erzurum, Uzundere, 40.6114° N, 41.6286° E, 1010 m a.s.l., 03.viii.1996, leg. Gert Jan van Pelt; 1 ♂ (RMNH.INS.974908): TURKEY: Kayseri, Pinarbasi, 38.7205° N, 36.3950° E, 1520 m a.s.l., 19.vii.1996, leg. Gert Jan van Pelt; 1 ♂ (RMNH.INS.974909): TURKEY: Nigde, Kizilören, 38.0164° N, 36.0311° E, 1900 m a.s.l., 11.vii.1996, leg. Gert Jan van Pelt; 2 ♂(RMNH.INS.1090920 and RMNH.INS.1090922): N LEBANON: Bcharre, 34.2511° N, 36.0111° E, 1400 m a.s.l., 27.vi.1960, leg. G. A. Mavromoustakis; 1 ♀(RMNH.INS. 1090923): N LEBANON: Bcharre, 34.2511° N, 36.0111° E, 1400 m a.s.l., 27.vi.1960, leg. G. A. Mavromoustakis.

Paratype: 1 ♀(RMNH.INS.1090164): N LEBANON: Bcharre, 34.2511° N, 36.0111° E, 1400 m a.s.l., 22.vi. 1960, leg. G. A. Mavromoustakis.

Etymology. The name refers to the historical Roman province of Cilicia located on the south coast of Asia Minor. The range of this species exceeds the limits of this historical province to northeast and southeast.

Description of holotype. ***Head*** (Figure 15D–F): anteclypeus black; postclypeus yellow; labrum yellow with thicker black margin; frons yellow with faint black seam; labium yellow; occiput less domed than in *C. charpentieri*, black and yellow occipital triangle domed, yellow with pale brown margins; antenna and vertex black; eyes turquoise (collector's observation); postocular area yellow with black upper margin.

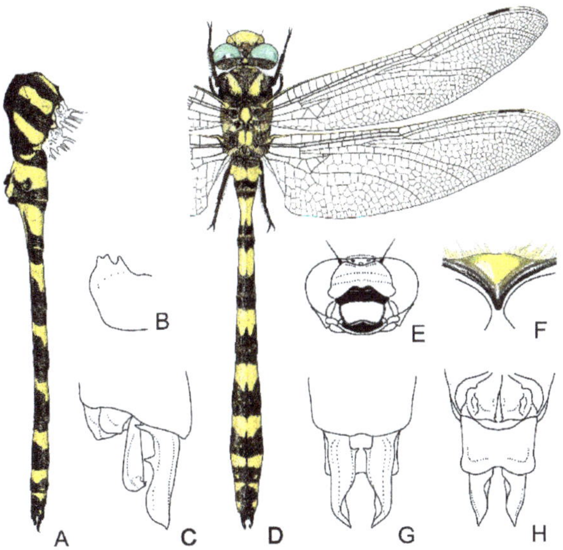

Figure 15. *Cordulegaster cilicia* sp. nov., holotype ♂, (RMNH, leg. Gert Jan van Pelt, RMNH.INS.974914, 9 July 2008, Turkey, Kahramanmaraş, Göksun, 38.0614° N, 36.4722° E, 1350 m a.s.l.): (**A**): habitus, lateral. (**B**): detailed latero-dorsal view on the hook of the inferior appendix. (**C**): appendages, lateral. (**D**): habitus, dorsal. (**E**): head, schematically. (**F**): occipital triangle, frontal. (**G**): appendages, dorsal. (**H**): appendages, ventral.

Thorax (Figure 15A,D): anterior lobe of the pronotum with yellow front edge; median lobe black with yellow dorsolateral wing-shaped patterns; posterior lobe yellow interrupted by middorsal black band; front of synthorax with broad big yellow antehumeral stripes, narrowing moderately towards pronotum leaving less than 30% black of the antehumeral region; mesepimeron and metepimeron with broad yellow bands; metepisternum black with tiny yellow markings near base of wings; metakatepisternum black; poststernum black; coxae black with yellow ventral parts, trochanter, femora, tibiae and tarsi black.

Wing (Figure 15D) hyaline; costal veins yellow, other veins black; Pt dark brown, Pt FW 4.2 × 0.5, HW 4.5 × 0.7; anal triangle 3-celled; membranula narrow.

Abdomen (Figure 15A,D): colour pattern as in Figure 15A,D; S1 dorsal black, yellow marking on side of S1 small, S2–9 with medium sized yellow markings leaving ca. 50% black, on dorsal view of S2–5 there are a pair of small, linear yellow spots present on the dorsal distal part of abdominal segments not touching in middle of the segment, on S2–10 black markings reaching lateral sides, yellow markings present on S9–10.

Appendages (Figure 15B,C,G,H): superior appendages in dorsal view broad and stout with curved and inward pointing tips, the medioventral teeth clearly visible and directed with the tips slightly in direction of S10, in the lateral view the laterobasal teeth visible and located near S10, in lateral view the medioventral teeth located more ventral and far from crossing with the tooth on end of the inferior appendage; inferior appendage in ventral view stout, nearly rectangular, slightly notched on distal margin, in lateral view distally small teeth on end of the inferior appendage are visible (Figure 15C), the three teeth form a crown-shaped complex in which the two outer teeth are fused, the inner tooth is isolated from them (Figure 15B).

Measurements (mm): TL (inclusive appendages): 71, AL 54, HWL 43, FWL 43.

Variations in males. *C. cilicia* shows some variability in size and some minor variations in colouration patterns. The eyes of some individuals are greenish, in others more blue (collector's observation). *Measurements (mm)* (paratypes, $n = 6$): TL (inclusive appendages) 70.0–73.0, AL: 53.0–55.0, HWL 40.0–46.0.

Description of female paratype. ***Head*** (Figure 16A): like in ♂.

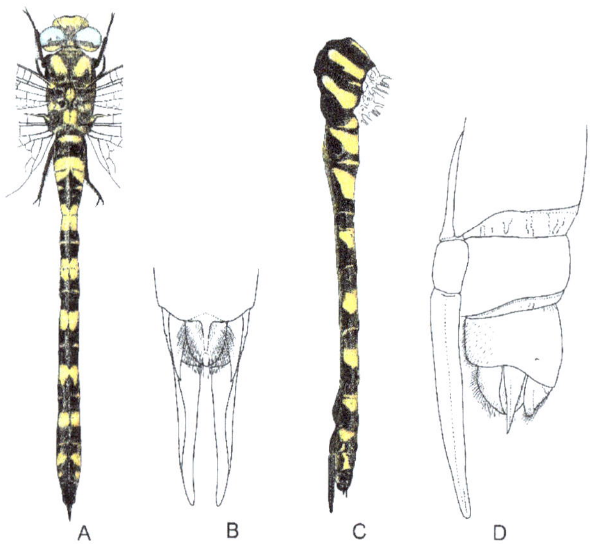

Figure 16. *Cordulegaster cilicia* sp. nov., paratype ♀, (RMNH, leg. G. A. Mavromoustakis, RMNH.INS. 1090164, 21 June 1960, North Lebanon, Bcharre, 34.2511° N, 36.0111° E, 1400 m a.s.l.): (**A**): habitus, dorsal. (**B**): appendages, dorsal. (**C**): habitus, lateral. (**D**): ovipositor, lateral.

Thorax (Figure 16A,C): much as in male, except the yellow markings of metepisternum; metepisternum black with prominent comma-shaped yellow markings.

Wings: like in male, hyaline; Pt dark brown, Pt FW 4.4 × 0.7, HW 4.8 × 0.6; membranula narrow.

Abdomen (Figure 16A,C): cylindrical and thicker than in males, ovipositor 9.0, black, only slightly curved, almost straight and long.

Measurements (mm): TL (including ovipositor) 83.0, AL 67.0, HWL 48.0, FWL 49.0.

Variations in females. Further paratype female much as the other with minor differences in measurements: TL (inclusive ovipositor) 81.0, AL 63.0, HWL 46.0, FWL 47.0.

Differential diagnosis

The males are similar in colour pattern to other species of *charpentieri*-complex, but are usually more yellow than *C. amasina* and less yellow than *C. charpentieri*, pairs of abdominal spots on apical margins of S6–10 usually absent. The occiput less domed and less yellow than in *C. charpentieri*, postocular area is less domed and mostly darker than in *C. charpentieri*; but less dark than in *C. amasina*. The superior appendages in dorsal view curved, widened at base, in contrast to *C. charpentieri*, which has nearly parallel sided superior appendages. The medioventral teeth in dorsal view located more closely to S10 than in *C. charpentieri*. The medioventral teeth in lateral view located more ventral and further from crossing with the teeth on the end of the inferior appendage, as in *C. charpentieri*. In *C. amasina* the inferior appendage slightly wider at the base, in contrast to *C. cilicia* which has a nearly rectangular shape.

Female individuals are nearly impossible to separate within the *charpentieri*-complex not knowing the geographical origin, only the dark and small *C. mzymtae*, and the large and yellow forms of *C. charpentieri* are easily identified.

Distribution

C. cilicia occurs east of Anamur along the Mediterranean coast, reaching Lebanon. It is also distributed to East-Central Turkey reaching Erzurum near Uzundere, where it meets *C. charpentieri*.

Cordulegaster amasina **Morton,1916 stat. rev.**

Material examined: 1 ♂(RMNH.INS.974910): TURKEY: Ankara, near Güvem, 40.5915° N, 32.6597° E, 1100 m a.s.l., 31.vii.2008, leg. Gert Jan van Pelt.

Additional males: 2 ♂(RMNH.INS.974929): TURKEY: Yozgat, 13 km SW of Akdagmadeni, 39.5629° N 35.7954° E, 1680 a.s.l., 27.vii.2006, leg. Gert Jan van Pelt; 1 ♂(RMNH.INS.974932): TURKEY: Kastamonu, Tosya, 20 km SE of Tosya, spring and brook, 40.9765° N, 34.1894° E 1200 m a.s.l., 05.viii.2006, leg. Gert Jan van Pelt.

Redescription of male

Head (Figure 17D,E): anteclypeus black; postclypeus yellow; labrum yellow with black margin; frons yellow and labium yellow; occipital triangle black fringed with yellowish hairs above; occiput not domed and yellow; antenna and vertex black; eyes green; postocular area yellow with black upper margin.

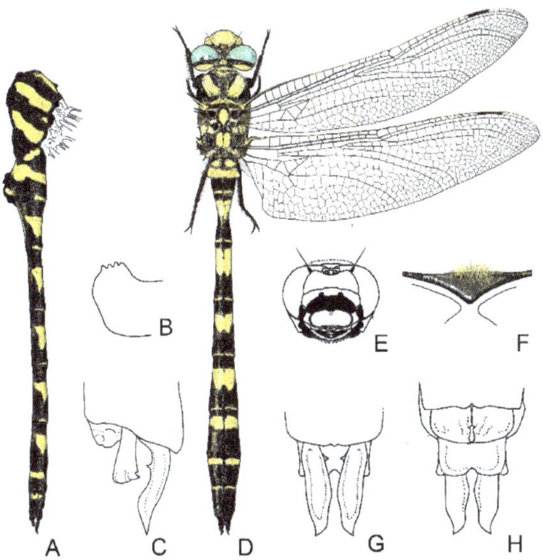

Figure 17. *Cordulegaster amasina*, re-description ♂, (RMNH, Gert Jan van Pelt, RMNH.INS.974910, 31 July 2008, Turkey, Ankara, near Güvem, 40.5915° N, 32.6596° E, 1100 m a.s.l.): (**A**): habitus, lateral. (**B**): detailed latero-dorsalview on the hook of the inferior appendix. (**C**): appendages, lateral. (**D**): habitus, dorsal. (**E**): head, schematically. (**F**): occipital triangle, frontal. (**G**): appendages, dorsal. (**H**): appendages, ventral.

Thorax (Figure 17A,D): anterior lobe of pronotum with yellow front edge; median lobe black with yellow dorsolateral ellipsoid patterns, two yellow middorsal dots; posterior lobe yellow interrupted by a middorsal black band; front of synthorax with broad big yellow antehumeral stripes, narrowing moderately towards pronotum leaving less than 20% black of antehumeral region; mesepimeron and metepimeron with broad yellow bands; metepisternum black with small yellow triangle near base of wings; metakatepisternum black; poststernum black; coxae, trochanter, femora, tibiae and tarsi black.

Wings (Figure 17D) hyaline; costal veins yellow, other veins black; Pt black, Pt FW 3.6×0.4, HW 4.0×0.5; anal triangle 3-celled; membranula narrow.

Abdomen (Figure 17A,D): colour pattern as shown in Figure 17A,D; S1 dorsally black yellow marking on side of S2–9 with medium sized yellow markings leaving ca. 40% black in dorsal view of S2–8 there are a pair of small, linear yellow spots present on the dorsal distal part of abdominal segments not touching in the middle of segment, on S2–10 black markings reaching sides, yellow markings also on S9–10.

Appendages (Figure 17B,C,G,H): superior appendages in dorsal view paddle-like with tips pointing outwards, medioventral teeth visible and directed towards each other in lateral view, laterobasal tooth visible, located near S10. In lateral view, medioventral tooth visible with a relatively broad base and a prominent hook-shaped curved tip; inferior appendage stout, inverse trapezoid, deeply notched on distal margin seen from above, in lateral view distally small teeth on end of the inferior appendage are visible (Figure 17C), the four single teeth form a small wreath that is visible from latero-dorsal view (Figure 17B), the single teeth inclined distally, in lateral view the caudal end of the inferior is slightly elongated to the rear in the ventral area (Figure 17C).

Measurements (mm): TL (inclusive appendages) 68.0, AL 53.0, FWL 39.0, HWL 39.0.

Variability: other males similar to the described one with minor variability in measurements: ***(mm)*** (males, $n = 3$): TL (including appendages) 71.0–68.0, AL 53.0–55.1, FWL 39.0–42.5, HWL 39.0–42.7.

Differential diagnosis

The males are usually smaller than in other species of the *charpentieri*-complex, *C. amasina* is only comparable in size to *C. mzymtae*. However, *C. mzymtae* is much darker than *C. amasina*. Pairs of abdominal spots on dorsal apical margin of S2–8 present in *C. amasina* are absent in *C. mzymtae*. The occipital triangle usually dark, not regular yellow as in other species of *charpentieri*-complex, in some individuals the occipital triangle of *C. amasina* and *C. mzymtae* has small yellowish markings, occiput not domed as in *C. charpentieri*, postocular area not domed and darker than in *C. charpentieri* and *C. cicilia*.

The tips of superior appendages of *C. amasina* are bent outwards (possibly an artefact of desiccation), not inwards like in *C. charpentieri*, the medioventral teeth smaller than in two other species of the group. The medioventral teeth in *C. amasina* are closer to S10 than in *C. charpentieri*. The females are smaller and darker than those of other species of *charpentieri*-complex. In contrast to the other species of the group, the small teeth at the caudal end of the inferior appendage form a regularly shaped crown consisting of four teeth. *C. amasina* shares this character with *C. mzymtae*. The size of the teeth varies among male individuals.

Distribution

C. amasina occurs approximately from the province Kastamonu to the province Samsun and from the Black Sea Coast to central Turkey reaching the province Ankara (Figure S1, Table 1).

Cordulegaster mzymtae **Bartenev, 1929**

Redescription

Material examined: 1 ♂(coll. Thomas and Elias Schneider): GEORGIA: Adjara, 5 km E Goderdzi Pass, 41.3755° N, 42.3405° E, 1800 m a.s.l., 29.vii.2015, leg. Thomas Schneider; 1 ♂ (coll. Thomas and Elias Schneider): GEORGIA: Adjara, 5 km E Goderdzi Pass, 41.6319° N, 42.5680° E, 1820 m a.s.l., 30.vii.2015, leg. Thomas Schneider; 1 ♂(coll. Thomas and Elias Schneider): GEORGIA: Adjara, 5 km E Goderdzi Pass, 41.6319° N, 42.5680° E, 1820 m a.s.l., 30.vii.2015, leg. Thomas Schneider; 1 ♂(coll. Thomas and Elias Schneider): GEORGIA: Adjara, 5 km E Goderdzi Pass, 41.6319° N, 42.5680° E, 1820 m a.s.l., 30.vii.2015, leg. Thomas Schneider; 1 ♂(RMNH.INS.974901): TURKEY: Ordu, SE of Turnasuyu, 40.6797° N, 37.9573° E, 1300 m a.s.l., 05.viii.1998, leg. Gert Jan van Pelt; 1 ♂ (RMNH.INS.974902): TURKEY: Ordu, SE of Turnasuyu, 40.6797° N, 37.9574° E, 1500 m a.s.l., 02.viii.1998, leg. Gert Jan van Pelt; 2 ♀(coll. Thomas and Elias Schneider): GEORGIA: Adjara, 5 km E Goderdzi Pass, 41.6319° N, 42.5680° E, 1820 m a.s.l., 30.vii.2015, leg. Thomas Schneider.

Description of the male. ***Head*** (Figure 18B,D): anteclypeus black; postclypeus yellow; labrum yellow with thicker black margin and a pigmented comma-shaped line in the middle; frons yellow with faint black seam; labium yellow; occiput yellow, not domed, occipital triangle black with yellowish hairs above; antenna and vertex black; eyes green (collector's observation); postocular area black.

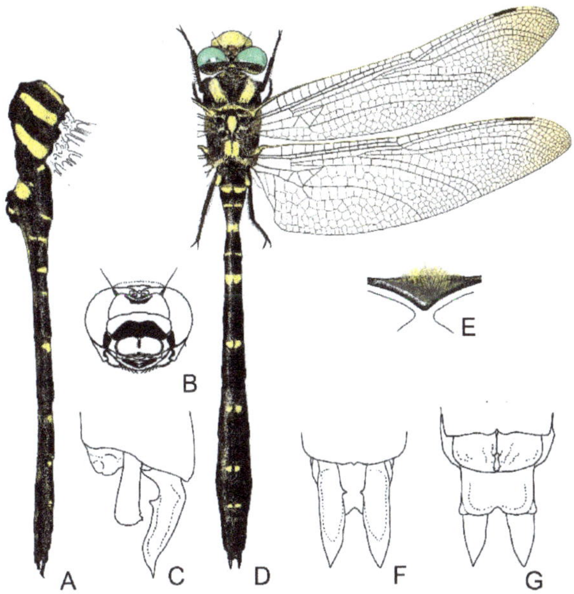

Figure 18. *Cordulegaster mzymtae*, ♂(collection Elias and Thomas Schneider, leg. Thomas Schneider, 30 July 2015, Georgia, Adjara, 5 km E Goderdzi Pass, 41.6319° N, 42.5680° E, 1820 m a.s.l.): (**A**): habitus, lateral. (**B**): head schematically. (**C**): appendages, lateral. (**D**): habitus, dorsal. (**E**): occipital triangle. (**F**): appendages, lateral. (**G**): appendages, ventral.

Thorax (Figure 18A,D): anterior lobe of pronotum black; median lobe black with yellow dorsolateral spindle-shaped patterns; posterior lobe black; front of synthorax with broad big yellow antehumeral stripes, narrowing moderately towards pronotum leaving less than 20% of antehumeral region black; mesepimeron and metepimeron with broad yellow bands; metepisternum black with small yellow dot near base of wings; metakatepisternum black; poststernum black; coxae, trochanter, femora, tibiae and tarsi black.

Wings (Figure 18D) hyaline, older individuals with semihyaline brownish wings in the distal region, costal veins yellow, other veins black; Pt black, Pt FW 3.7 × 0.4, HW 4.0 × 0.5; anal triangle 3-celled; membranula grey and narrow.

Abdomen (Figure 18A,D): dorsal yellow spots on abdomen are reduced to small pairs of semi-lunar markings leaving over 90% black; S9 and S10 completely dark in dorsal and lateral view.

Appendages (Figure 18C,F,G): superior appendages in dorsal view paddle-like, with pointed tips, medioventral teeth visible and directed towards each other, in lateral view, laterobasal teeth visible, located near S10. In lateral view, medioventral teeth visible located at the first half of the superiores, with a relatively broad base, gradually narrowing, slightly pointed; inferior appendage stout, slightly trapezoid, notched on distal margin seen from above, in lateral view distally small teeth on end of the inferior appendage are visible, the four hook-shaped single teeth form a small comb that is visible from caudal view, the single teeth inclined slightly distally, the caudal end of the inferior appendages is almost rounded in lateral view.

Measurements (mm): TL (inclusive appendages) 65.7, AL 49.6, FWL 41.4, HWL 40.8.

Variability: other males much as the described one with minor variation in measurements: (mm) (males, n = 7): TL (including appendages) 63.8–65.7, AL 49.5–49.9, FWL 41.2–42.1, HWL 40.4–42.0.

Description of female. ***Head*** (Figure 19A): much as in ♂.

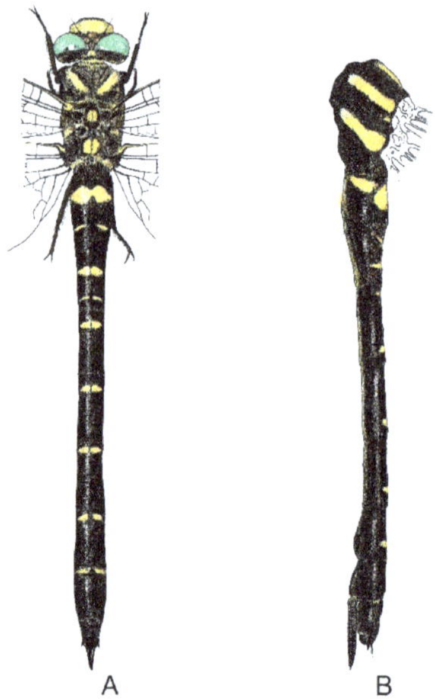

Figure 19. *Cordulegaster mzymtae*, ♀(collection Elias and Thomas Schneider, leg. Thomas Schneider, 30 July 2015, Georgia, Adjara, 5 km E Goderdzi Pass, 41.6319° N, 42.5680° E, 1820 m a.s.l.): (**A**): habitus, dorsal. (**B**): habitus, lateral.

Thorax (Figure 19A,B): much as in male, exception the yellow markings of metepisternum; metepisternum black with small comma-shaped yellow markings,

Wings: much as in male, hyaline; Pt black, Pt FW 4.0 × 0.5, HW 4.1 × 0.6.

Abdomen (Figure 19A,B): cylindrical and thicker than in males, ovipositor 5.3 mm, black, only slightly curved, almost straight and long.

Measurements (mm): TL (inclusive ovipositor) 68.5, AL 51.7, FWL 43.4, HWL 44.6.

Variability in females: second female looks much as described above, but slightly larger. ***Measurements (mm)*** (females, *n* = 2): TL (inclusive ovipositor) 68.5–68.6, AL 51.7–51.8, FWL 43.4–43.5, HWL 44.6–44.7.

Differential diagnosis

Male and female are smaller than all other species of the *C. charpentieri*-complex. They are easily recognized by their dark colour pattern. Pairs of abdominal spots on dorsal apical margin of S2–8 are absent. The occipital triangle is black, not yellow as in most other species of *charpentieri*-complex, in some individuals the occipital triangle has yellowish markings, occiput not domed as in *C. charpentieri*, postocular area not domed and darker as in all other species of this complex. Superior and inferior appendages are very similar to the appendices of *C. amasina*. As in *C. amasina*, there is also a four-tooth crown in *C. mzymtae* at the end of the inferior. In all other species of the group there are usually three teeth of different sizes visible at the end of the inferior.

Distribution

C. mzymtae and other *Cordulegaster* in it´s natural environment are shown in Figure 20.

Figure 20. Some *Cordulegaster* in their natural environment and some of their habitats: (**A**): yellow form of *C. picta* (near Köycegiz, Turkey). (**B**): dark form of *C. picta* (near Mengen, Turkey). (**C**): *C. vanbrinkae* (Hyrcanian forest near Veysar, Iran). (**D**): *C. coronata* (near Arzaneh, Iran): (**E**): *C. charpentieri* (near Dorud, Iran): (**F**): *C. mzymtae* (Georgia, Adjara, 5 km E Goderdzi Pass). (**G**): *C. mzymtae* Habitat (Georgia, Adjara, 5 km E Goderdzi Pass). (**A–E**): photos Dietmar Ikemeyer, (**F,G**): photos Elias Schneider.

C. mzymtae the darkest member of the *bidentata*-group occurs eastwards from *C. amasina* reaching Georgia, the Russian Black Sea Coast and even until Karachay Cherkessia Region in the North Caucasus (Russia) (Figure S1, Table 1).

5. Discussion

5.1. General Discussion on the Genus Cordulegaster in the Western Palaearctic

The taxonomic debate on the genus *Cordulegaster* in the Western Palaearctic has been ongoing for over one and half centuries [2,13,14,19,26,30,35,63,64]. For the Western part of the Western Palaearctic, a molecular approach exists [13]. However, for the more complex and unsettled Eastern part of the Western Palaearctic, no molecular studies have been conducted until now, and no published sequences are available so far. From the east of the Western Palaearctic, including Turkey, the Caucasus countries and Iran, several species and subspecies have been described (Table 2). All these were based on external morphology and the colour patterns of imagines, especially of the male appendices. However, colour patterns and structure of the appendices may vary and hybrids may complicate phenotypic taxonomy. *C. boltonii* and *C. trinacriae* interbreed in a broad zone in Italy [35]. In a similar way, our data support hybridization in the *boltonii*-group between *C. vanbrinkae* and *C. picta* in Armenia and in the *bidentata*-group between *C. insignis* and *C. cilicia* sp. nov. Such hybridization supports the view that pre- and post-mating barriers are inefficient, and structural differences in the genital apparatus do not prevent gene flow. It explains the limitations of morphological characters for species recognition in this genus. Only part of the variation observed is taxonomically relevant and it is sometimes impossible to identify specimens to species. As an alternative, we undertook a molecular genetic approach, which reveals the genotype, not the phenotype. It helped us to clarify the status of several doubtful subspecies and species in this region. Furthermore, an inherent limitation of the morphological method is that not all hybrids are intermediate and can only be detected by molecular genetic approach.

5.2. Cordulegaster boltonii-Group

In the *boltonii*-group of the Eastern region of the Western Palaearctic, we recovered with molecular genetic analysis three known (*C. heros*, *C. picta*, *C. vanbrinkae*) and one new species (*C. kalkmani*). These four eastern species of the *boltonii*-group (*C. heros*, *C. picta*, *C. kalkmani*, and *C vanbrinkae*) seem to be geographically separated but may meet in contact zones (Figure 21). The existence of the new taxon *C. kalkmani* was supported by the alignments of both genes separately or combined as well by the haplotype-analysis and the K2-P distances. *C. heros* is known from rather Central and Southeast Europe, reaching Ukraine in the North–East [65]. *C. picta* inhabits an area extending from North–Eastern Greece and Bulgaria, through West and North Turkey, including Samos and Lesbos in the Aegean, along the Black Sea coast as far as Russia, Georgia and Azerbaijan [16,66,67]. However, the presence of *C. picta* in Azerbaijan may be questioned, as the only voucher specimen is old and in bad condition; it is not clear whether it belongs to *C. picta*, to *C. vanbrinkae* or is a hybrid between them [16]. *Cordulegaster picta* from South-West Turkey and Samos are rather yellow, where those from the Black Sea region are much darker (Figure 20). *C. vanbrinkae* is restricted to the South Caspian Sea region [68,69]. In East Anatolia, possibly reaching Armenia and North–West Iran, a new species, *C. kalkmani* was found. The record of *C. picta* by Rastegar in North–West Iran may belong to the latter species [70]. *C. kalkmani* seems to be geographically separated from *C. picta* and *C. vanbrinkae* by the Anatolian Diagonal and further east by the Armenian Highlands and the Caucasus (Figure 21). *C. vanbrinkae* and *C. kalkmani* seem to be separated by the Armenian Highland (Figure 21). As mentioned before, we found hybridization between *C. picta* and *C. vanbrinkae* in Armenia. These hybrids may be phenotypically indistinguishable from one of their parents. This is the case of the hybrid from Armenia, which looks like *C. vanbrinkae*, causing specimens from this population to be initially described as *C. vanbrinkae* [71]. We found no evidence for hybridisation between *C. picta* and *C. kalkmani* and between *C. vanbrinkae* and *C. kalkmani*. The K2-P distances of the COI sequences revealed that *C. kalkmani* is more distant from *C. picta* (5.88%) than *C. vanbrinkae* from *C. picta* (3.95%). Correspondingly, the male appendices of *C. kalkmani* are more different from *C. picta* than those of *C. vanbrinkae* from *C. picta*.

This suggests that the latter two were separated more recently (see also timetree-analysis, Figure S7) and may be regarded as sister taxa.

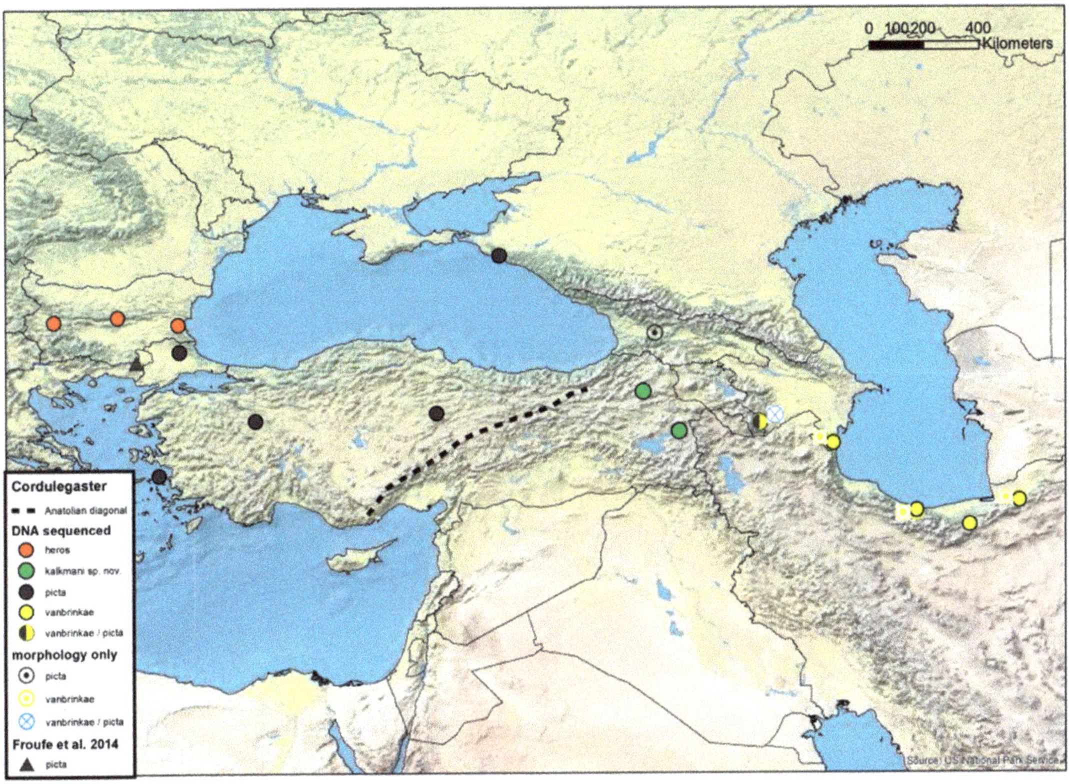

Figure 21. Map with species identified in this study for *C. boltonii*-group relative to the Anatolian Diagonal.

Hybridization may occur also between *C. boltonii* and *C. heros* as both are locally syntropic [72], and may also occur in still undocumented contact zones in Greece or Bulgaria between *C. heros* and *C. picta*, and even in Ukraine and Russia north of the Black Sea.

Thus, we highlight broad variation in colour pattern in some taxa. Male appendages, often used as diagnostic, have limits for discriminating species, and the distinction between intraspecies variation and significant differences is often hard to make. This was a source of confusion in the past. The work by Froufe showed that all European subspecies of *C. boltonii; C. b. immaculifrons* Selys, 1850; *C. b. iberica* Boudot & Jacquemin, 1995; *C. b. algirica* Morton, 1916, although statistically representative, may just be colour variations [13]. We conclude, including the data of Froufe, that the following species in the *boltonii*-group live in the West Palaearctic: *C. princeps, C. boltonii, C. trinacriae, C. heros, C. picta, C. kalkmani*, and *C. vanbrinkae*. The division of the *boltonii*-group in a "western" group, with *C. princeps, C. b. algirica* (North Africa only), *C. boltonii, C. trinacriae* and an "eastern" group with *C. heros, C. picta, C. kalkmani* and *C. vanbrinkae* is supported by our molecular data. This view was already suggested by Verschuren based on the morphology of the larvae without knowing the latter two species at that time [12].

5.3. Cordulegaster bidentata-Group

In the *bidentata*-group of the Eastern region of the Western Palaearctic, two complexes were revealed by molecular analysis: the *helladica-insignis*-complex in South-East Europe and Western Turkey, and the *charpentieri-complex* in Eastern Anatolia, Levant, Caucasus countries and Iran (Figure 22). These species are geographically separated by mountains of Anatolia and the Caucasus, although overlapping zones exist in East Anatolia (Figure 22). As widely accepted, the nominotypic *C. insignis* inhabits Western Turkey and eastern Greece islands like Samos. However, the molecular analysis of all specimens collected east of the Marmara and Aegean region grouped away from *C. insignis* using both programs for both genes separately or combining the COI and ITS genes. All the eastern specimens of the *bidentata*-group were therefore summarized here in the *charpentieri*-complex.

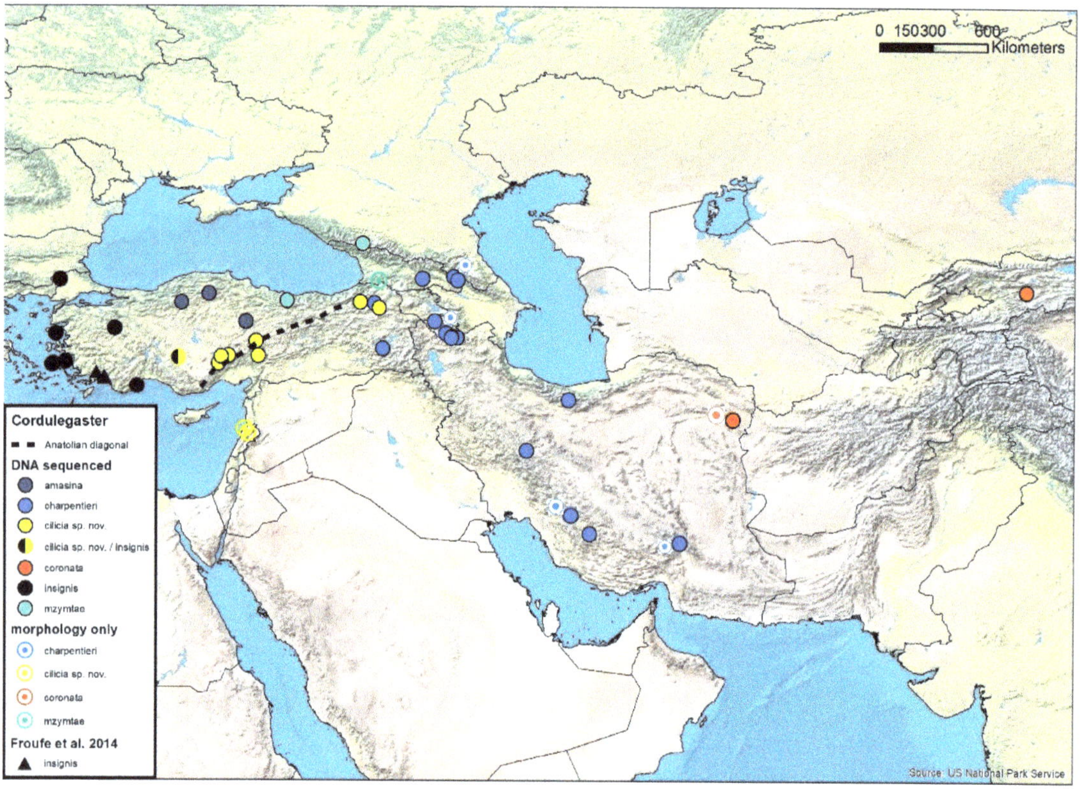

Figure 22. Map of both *C. bidentata*-complex and *C. coronata* with species identified in this study relative to the Anatolian Diagonal.

The COI analysis of the *charpentieri*-complex with two programs is in favour of four species (*C. amasina*, *C. mzymtae*, *C. cilicia*, and *C. charpentieri*). The K2-P distance between these four taxa is 2.96–4.24% (Table 4), in agreement with a full species level [43,45,46]. The ITS analysis by the MAFT program, however, put them all together indicating a more recent diversification in this taxa complex (supported by our timetree, Figure S7). This may be due to special glacial events in Anatolia as discussed below. For several aspects, including the conservation of highly endangered and unique biotopes, we would suggest to hold up these three traditional and the new taxa.

Morton introduced *C. amasina* as a race, based on five males from Amasya, Turkey [26]. As our results support the occurrence of a separate species along the western part of Black Sea Coast, we adopted the existing name for this lineage. Further east, reaching Georgia and Russia, *C. mzymtae* occurs. This taxon was sometimes treated as a subspecies of *C. insignis* [10,11]. More extensive fieldwork along the Black Sea Coast may clarify the exact distribution borders between *C. insignis*, *C. amasina*, and *C. mzymtae*. The separation between these taxa may be due to the glacial events discussed in a recent paper on banded newts in this region [73].

We did not find any published name for another clade defined by molecular data in the *charpentieri*-complex, and therefore we name it *C. cilicia*. The name refers to the historical province of Cilicia located on the south coast of Asia Minor. However, the range of this species exceeds the limits of this historical province to northeast and southeast and it inhabits East Anatolia and extends along the East–Mediterranean Sea reaching Lebanon. *C. cilicia* and *C. charpentieri* meet in Central–East Anatolia in the province of Kars (Figure 22). *C. cilicia* is well separated in the COI and haplotype analysis, and the K2-P distance to its next relative *C. charpentieri* is 3.22% and in agreement with species level for Anisoptera [43].

C. charpentieri is a widespread species occurring in Central-East Anatolia, Armenia, Georgia, and Iran. It is most likely also present in northern Iraq taking the description by Asahina into account [74]. *C. charpentieri* exhibits some variation in the colour-pattern across its range; therefore, we have depicted representative examples in this study. The size of this species is also variable, and the largest animals of this taxon exist in the South and South-Eastern Zagros Mountains. These individuals belong with *C. heros* and *Anax immaculifrons* to the largest dragonflies of the Western Palaearctic.

Anatolia emerges as a hotspot for the *charpentieri*-complex. Here, three refugia meet and interact: the Caucasus, Irano–Anatolian and the Mediterranean [40,41]. Two refugia, one in western and one in Eastern Anatolia, separated by the Anatolian diagonal, have been suggested, based on non-genetic data [75,76]. More have been found by genetic analysis of different animal groups [77,78]. The lake system was present in central Anatolia during the Pliocene and the inhabitability of the Central Anatolian Plateau in the glacial maxima of the Pleistocene probably broke up previously continuous faunal ranges leading to subspeciation [79]. Vicariant events related to the formation of the Anatolian diagonal and the orogenesis of the mountain chains in southern and Eastern Anatolia are what led to current distribution patterns of animals like *Cordulegaster*. The Anatolian diagonal was originally described for plants, but also applies to animals [41,42,77,78]. In the case of the *charpentieri*-complex it separates *C. amasina* from *C. cilicia* and the western Taurus, nominal *C. insignis* from all others. The Eastern Anatolian Mountains and the Armenian Highland separate *C. charpentieri* from *C. amasina*, and *C. cilicia* (Figure 22). However, hybridization may have occurred and may still occur in postglacial contact zones (Figure 22).

For *C. magnifica*, described by Bartenev based on a single female of unknown origin, we did not find any support in the investigated region [28]. The type is lost. The description fits to most females of 'mid-yellow' *Cordulegaster* of the *bidentata*-group in the East Mediterranean, the Middle East and the South Caucasus. In view of its unknown origin and as no other specimen or living population is known, it is not possible to ascribe it to any taxon and the name should be deleted.

Thus, in the *bidentata*-group the following species occur from West to East: *C. bidentata*, *C. helladica*, *C. insignis*, *C. amasina*, *C. mzymtae*, *C. cilicia* and *C. charpentieri*. The species *C. coronata* is genetically distant from this group, although morphologically close to the *charpentieri*-complex (Figure 23).

5.4. *Cordulegaster coronata*-Group

The *coronata*-group formed an extra clade with all three programs for both genes and also in combination of these two genes. This is the reason why this taxon was sometimes treated as a subspecies of *C. insignis* in the past [2,14]. *C. coronata* is a Middle Asian taxon, which has its Western distribution limits in North-East Iran (Razavi Khorasan) as recently documented [46,47]. In North-Khorasan, *C. coronata* and *C. charpentieri* may meet, therefore, it would be of interest to investigate specimens from this region to see if hybridisation between them is possible.

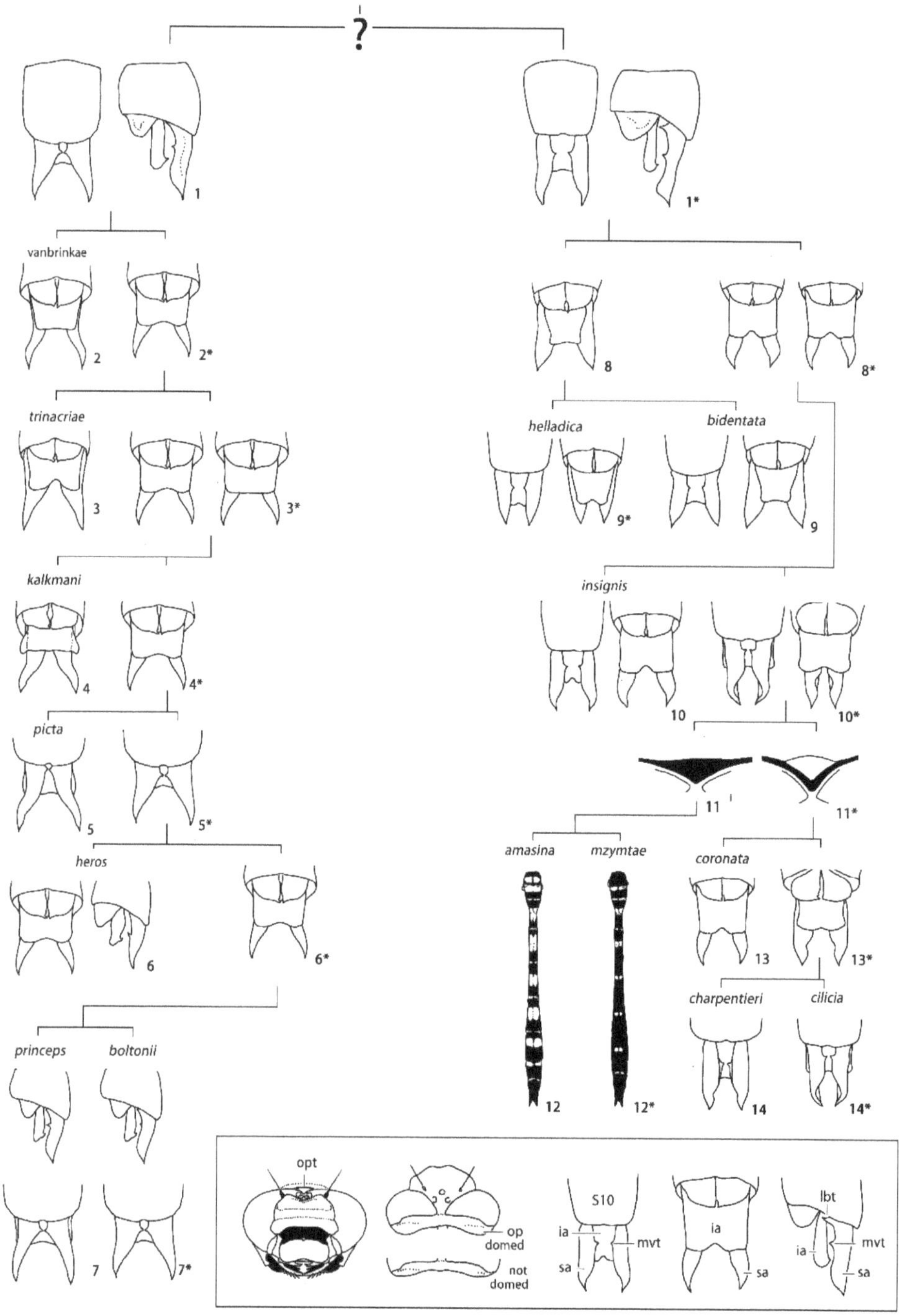

Figure 23. Dichotomous identification tree for all Western-Palaearctic *Cordulegaster* (males, graphically represented features in the text, additional features in square brackets).

5.5. Key for the Western–Palaearctic Cordulegaster

Finally, we provide a preliminary key for the Western–Palaearctic *Cordulegaster* (Figure 23). This key should be regarded as a working version for future studies. As *Cordulegaster* species seem to have separated rather recently and hybridization in contact zones is common, the phenotypical approach based upon male appendices as a discriminating character is limited. Moreover, male appendices are mobile and may become fixed at death in different positions, which may influence later interpretation. This may have led to confusion in the past. Despite all these difficulties, our key uses the known characters used for differentiation of *Cordulegaster* species (Figure 3). In contact zones this key may not work, especially where hybrids are present.

Key (only for males):

1 superior appendages in dorsal view diverging with curved outer borders and nearly close at the base, one visible tooth (medioventral tooth), laterobasal tooth not visible in lateral view ... *boltonii*-group ... 2.

1* superior appendages nearly parallel with straight outer borders and separated at the base, two teeth visible (medioventral tooth, laterobasal tooth) in lateral view ... *bidentata*-group and *coronata*-group ... 8.

2 inferior appendage trapezoid with a narrower hind margin, [frons with a thick black bar, occipital triangle black, hind margin of the inferior appendage weakly concave, narrowing distally, hooks of the inferior bifid, medioventral teeth on superior appendages very close to S10 and small] ... ***vanbrinkae***.

2* inferior appendage nearly parallel-sided ... 3.

3 hind margin of the inferior appendage not or weakly notched ... 4.

3* hind margin of the inferior appendage clearly notched, [inferior appendage clearly notched, hooks of the inferior with one single tip, 3–5 cells in anal triangle, occipital triangle mostly yellow, fons mostly unmarked] ... ***trinacriae***.

4 hind margin of the inferior appendage not notched or curved inwards, the dorsally pointing lobes of the inferior appendage widen the inferior distally, [inferior appendage at the end with a single pointed hook-shaped tooth on each side, medioventral teeth on superior appendages larger and more distal than in *C. vanbrinkae* but more proximal than in *C. picta*, 3 cells in anal triangle] ... ***kalkmani***.

4* hind margin of the inferior appendage slightly curved inwards, the dorsally pointing lobes of the inferior appendage not significantly widen the inferior distally, [4–5 cells in anal triangle] ... 5.

5 superior appendages narrow, reaching outwards, longer than S10, [hooks of the inferiorbifid, frons with a narrow black bar, occipital triangle with two yellow spots, never completely black, superior appendages long and slender, clearly divergent, longer than S10, inferior appendage slightly notched, widened distally in ventral view, 4–5 cells in anal triangle] ... ***picta***.

5* superior appendages stout, weakly pointing outwards, shorter than *S10* ... 6.

6 margins of inferior appendage not parallel-sided, widened distally, [#teeth with one single tip, 3–6 cells in anal triangle, medioventral teeth on superior appendages small, more distal than in *C. vanbrinkae*, *C. kalkmani*, and *C. picta*] ... ***heros***.

6* margins of inferior appendage nearly parallel-sided ... 7.

7 superior appendages slender,–medioventral tooth prominent, [hooks of the inferior small, dull and only with one single tip, species endemic to the High and Middle Atlas (Africa)] ... ***princeps***.

7* superior appendages stout,–medioventral tooth very small hooks of the inferiores with one single tip] ... ***boltonii***.

8 inferior appendage significantly narrower at the distal end than at the base, sometimes slightly waisted ... 9.

8* inferior appendage not significantly narrower at the distal end than at the base ... 10.

9 superior appendages reaching outwards, inferior appendage slightly waisted and with notch at the hind margin, [occipital triangle usually black although sometimes yellowish in South Italy and Sicilia] ... ***bidentata***.

9* superior appendages not reaching outwards, inferior appendage tapering towards hind margin, significantly notched at the hind margin ... ***helladica***.

10 lateral margins of the inferior appendage nearly parallel-sided over its entire length, shape rectangular, hind margin significantly notched, in dorsal view–medioventral tooth far from S10, [occiput little arched, mostly black coloured with a yellow hem, flat] ... ***insignis***.

10* lateral margins of the inferior appendage parallel-sided only at their base, inferior appendage distally getting wider ... 11.

11 occipital triangle only slightly domed, mainly black, [sometimes with yellow markings, occiput little arched, abdomen mostly black coloured] ... 12.

11* occipital triangle domed, mainly yellow ... 13.

12 abdomen [and thorax] with large extended yellow spots, [wings not coloured, hooks of the inferiores forming a 4-toothed crown] ... ***amasina***.

12* abdomen [and thorax] with only some yellow spots, hooks of the inferior appendage forming a 4-toothed crown] ... ***mzymtae***.

13 inferior appendage distally slightly smaller than at base, [occipital triangle yellow, not domed, occiput flattened, mostly black coloured] ... ***coronata***.

13* inferior appendage distally slightly wider than at base (teeth pointing outwards, this widens the inferior appendage outwards), [occipital triangle yellow and domed, occiput yellow and domed] ... 14.

14 inferior appendage longer than one half the length of superior appendages, in dorsal view medioventral teeth inserted far away from S10, [occiput mostly yellow and domed] ... ***charpentieri***.

14* inferior appendage as long as the middle of superior appendages, in dorsal view–medioventral tooth inserted nearby S10, superior appendages are shorter in relation to the inferior appendage, [short distance between laterobasal tooth and–medioventral tooth in lateral view, occiput mostly black, slightly domed] ***cilicia***.

6. Conclusions

This is the first revision of the Western Palaearctic *Cordulegaster* including the complex eastern part of the Western Palaearctic since Morton over 100 years ago [26]. We applied a two-step approach, first using molecular genetic sorting and, in a second step, morphology and description. The existence of the two traditional groups, the *boltonii*- and *bidentata*-group, is confirmed. Two new species are suggested and described, one in each group (*C. kalkmani*, *C. cilicia*). *C. coronata* Morton, 1916, however, is assigned to a different group based on its separate position in molecular trees and despite its morphological similarity to the *charpenteri-complex*. We synonymize four taxa with *C. charpentieri*: *C. insignis nobilis* Morton, 1916, *C. nachitschevanica* Skvortsov and Snegovaya, 2015, *C. plagionyx* Skvortsov and Snegovaya, 2015, and *C. insignis lagodechica* Bartenev, 1930. In contact zones of members of the *boltonii*- and the *bidentata*-group, hybridization is possible, therefore taxonomy in these areas should rule out hybrids by using not only the COI barcoding gene.

Thus, we suggest 16 taxa in *Cordulegaster* in the Western Palaearctic, eight in the *boltonii*-group (*C. princeps*, *C. b. algirica* (genetically distinct only in North Africa), *C. boltonii*, *C. trinacriae*, *C. heros*, *C. picta*, *C. kalkmani* and *C. vanbrinkae*); and seven in the *bidentata*- group (*C. bidentata*, *C. helladica*, *C. insignis*, *C. amasina*, *C. mzymtae*, *C. cilicia* and *C. charpentieri*). *C. coronata* a Middle East species with its western distribution limits in the Western Palaearctic is genetically distant from these groups ([13] and the present publication).

Supplementary Materials: The following are available online at https://www.mdpi.com/article/10.3390/d13120667/s1, Figure S1: Detailed information about the Cordulegaster specimens investigated in this study. Figure S2: CordWPalaearctis COI overview. Figure S3: COI StarBEAST. Figure S4: StarBEAST COI and SPACER. Figure S5: COI_BPP. Figure S6: COI_SPACER_BPP. Figure S7: COI timetree.

Author Contributions: T.S. designed the study and led the writing of the manuscript. A.V. conducted the molecular analysis and created the phylogenetic trees, O.M. created all drawings of the manuscript and made general contributions, G.J.v.P. and N.S. collected and analyzed data, M.C. helped finding specimens in RMNH, created the maps, H.J.D. and D.I. helped interpretation and writing the manuscript. All authors contributed critically to the drafts and gave final approval for publication. All authors have read and agreed to the published version of the manuscript.

Funding: This research received no external funding.

Informed Consent Statement: Not applicable.

Data Availability Statement: All data are provided in the paper and Supplementary Materials.

Acknowledgments: We thank Vasil Ananian, Geert De Knijf, Oleg Kosterin, Martin Lemke, and Martin Waldhauser for providing specimens. Massimo Terragni, Senckenberg Naturmuseum Frankfurt, enabled us to use legs from *Cordulegaster* specimens of Wolfgang Lopau's collection. Specimens from the collections of G. A. Mavromoustakis and Vincent Kalkman are made available by Max Caspers. Oleg Kosterin translated a paper from Bartenev. Jürgen Deckert helped us to access the Museum für Naturkunde, Berlin, Germany.

Conflicts of Interest: The authors declare no conflict of interest.

Abbreviations

Collections. RMNH—Naturalis Biodiversity Center; **Genetic; BIC**—Bayesian information criteria; **COI**—mitochondrial cytochrome coxidase subunit I; **ITS**—internal transcribed ITS; **K2-P**—Kimura 2-parameter; **PCR**—polymerase chain reaction; **Morphology. AL**—abdomen length; **FW**—forewing; **FWL**—forewing length; **HW**—hindwing; **HWL**—hindwing length; **Pt**—pterostigma; **S**—abdominal segment; **TL**—total length.

References

1. Schorr, M.; Paulson, D. World Odonata List—Revision 23. Available online: https://www2.pugetsound.edu/academics/academic-resources/slater-museum/biodiversity-resources/dragonflies/world-odonata-list2/ (accessed on 15 October 2021).
2. Fraser, F.C. A revision of the Fissilabioidea (Cordulegasteridae, Petaliidae and Petaluridae) (Order Odonata): Part 1—Cordulegasteridae. *Mem. Indian Mus.* **1929**, *9*, 69–167.
3. Zhang, H. *Dragonflies and Damselflies of China*; Chongqing University Press: Chongqing, China, 2019; Volume 1, 736p.
4. Garrison, R.W.; von Ellenrieder, N.; Louton, J.A. *Dragonfly Genera of the New World*; The Johns Hopkins University Press: Baltimore, MD, USA, 2006.
5. Boudot, J.-P.; Kalkman, V. Odonata (dragonflies and damselflies). In *The Status and Distribution of Freshwater Biodiversity in the Eastern Mediterranean*; Smith, K.G., Barrios, V., Darwall, W.R.T., Numa, C., Eds.; IUCN: Gland, Switzerland, 2014; Chapter 5; pp. 57–69.
6. Kalkman, V.J.; Boudot, J.-P.; Bernard, R.; Conze, K.-J.; De Knijf, D.; Dyatlova, E.; Ferreira, S.; Jović, M.; Ott, J.; Riservato, E.; et al. *European Red List of Dragonflies*; Publications Office of the European Union: Luxembourg, 2010.
7. Boudot, J.-P. *Cordulegaster helladica*. The IUCN Red List of Threatened Species 2020. e.T59706A140514673. Available online: https://www.iucnredlist.org/species/59706/140514673 (accessed on 15 October 2021).
8. Boudot, J.-P. *Cordulegaster helladica* ssp. *kastalia*. The IUCN Red List of Threatened Species 2020. e.T63227A188019481. Available online: https://www.iucnredlist.org/species/63227/188019481 (accessed on 15 October 2021).
9. Boudot, J.-P. *Cordulegaster vanbrinkae*. The IUCN Red List of Threatened Species 2006. E.T59709A11984778. Available online: https://www.iucnredlist.org/species/59709/11984778 (accessed on 15 October 2021).
10. Smallshire, D.; Swash, A. *Europe's Dragonflies. A Field Guide to the Damselflies and Dragonflies*; Princeton University Press & British Dragonfly Society: Woodstock, GA, USA, 2020.
11. Van Pelt, G.J. Cordulegaster. In *Field Guide to the Dragonflies of Britain and Europe*; Dijkstra, K.-D., Lewington, R., Eds.; British Wildlife Publishing: Gillingham, UK, 2020; pp. 220–231.
12. Verschuren, D. Revision of the larvae of Western–Palaearctic *Cordulegaster* Leach, 1815 (Odonata, Cordulegastridae) with a key to the considered taxa and a discussion on their affinity. *Bull. Ann. Soc. R. Belg. Entomol.* **1989**, *125*, 5–35.

13. Froufe, E.; Roufe, E.; Ferreira, S.; Boudot, J.-P.; Alves, P.C.; Harris, D.J. Molecular phylogeny of the Western Palaearctic *Cordulegaster* taxa (Odonata Anisoptera: Cordulegastridae). *Biol. J. Linn. Soc.* **2014**, *111*, 49–57. [CrossRef]
14. Boudot, J.-P. Les *Cordulegaster* du Paléarctique occidental: Identification et répartition (Odonata, Anisoptera, Cordulegastridae). *Martinia* **2001**, *17*, 3–34.
15. Corso, A. Morphological variability of *Cordulegaster trinacriae* in Italy (Odonata: Cordulegastridae). *Odonatologica* **2019**, *48*, 167–174.
16. Skvortsov, V.E.; Snegovaya, N.Y. Two new species of *Cordulegaster* Leach, 1815 from Azerbaijan (Odonata, Cordulegastridae). *J. Int. Dragonfly Fund* **2015**, *85*, 1–22.
17. Schneider, W.G. Verzeichniss der von Herrn Prof. Dr. Lowe im Sommer 1842 in der Türkei und Kleinasien gesammelten Neuroptera, nebst kurzer Beschreibung der neuen Arten. *Stettin. Entomol. Ztg.* **1845**, *6*, 110–116.
18. Schmidt, E. Auf der Spur von Kellemisch. *Entomol. Z.* **1954**, *64*, 49–62, 65–72, 74–86, 92–93.
19. Selys-Longchamps, E.; Hagen, H.A. Revue des Odonates ou Libellules d'Europe. In *Mémoires de la Société Royale des Sciences de Liège*; 1850; Volume 6, 408p, Available online: https://gallica.bnf.fr/ark:/12148/bpt6k26769q.texteImage (accessed on 15 October 2021).
20. De Selys Longchamps, E. Monographie des Gomphines. In *Mémoires de la Société Royale des Sciences de Liège*; 1858; Volume 9, 460p, plates 1–23; Available online: https://www.biodiversitylibrary.org/bibliography/39398 (accessed on 15 October 2021).
21. Kolenati, F.A. Insecta Caucasi. Meletemata Entomologica. *Imperialis Academiae Scientiarum* **1846**, *5*, 114–115.
22. De Selys-Longchamps, E. Synopsis des Gomphines. *Bull. Acad. R. Sci. Belg.* **1854**, *21*, 23–112.
23. De Selys-Longchamps, E. Odonates de l'Asie Mineure. *Ann. Soc. Entomol. Belg.* **1887**, *31*, 1–85.
24. Waterston, A.R. On the Genus *Cordulegaster* Leach, 1815 (Odonata) with special reference to the Sicilian species. *Trans. R. Soc. Edinb.* **1976**, *69*, 457–466. [CrossRef]
25. Dumont, H.J. *Aeschna charpentieri* Kolenati, 1846, a synonym of *Cordulegaster insignis* Schneider, 1845 and on the correct status of *Cordulegaster charpentieri auctorum* (Anisoptera: Cordulegastridae). *Odonatologica* **1976**, *5*, 313–321.
26. Morton, K.J. Some Palaearctic species of *Cordulegaster*. *Trans. Entomol. Soc. Lond.* **1916**, *63*, 273–290. [CrossRef]
27. Bartenev, A.N. Neue Arten und Varietäten der Odonaten des West-Kaukasus. *Zool. Anz.* **1929**, *85*, 54–68.
28. Bartenev, A.N. Die paläarktischen Arten der Untergattung *Cordulegaster* Leach. In *Travaux de la Station Biologique du Caucase du Nord*; Gorsky Institut Agronomique: Vladicaucase, Russia, 1930; Volume 3, pp. 1–32. (In Russian)
29. Theischinger, G. *Cordulegaster heros* sp. nov. und *Cordulegaster heros pelionensis* ssp. nov., zwei neue Taxa des *Cordulegaster boltonii* (Donovan)-Komplexes aus Europa (Ansioptera: Cordulegastridae). *Odonatologica* **1979**, *8*, 23–38.
30. Lohmann, H. Revision der Cordulegastridae. 2. Beschreibung neuer Arten in den Gattungen *Cordulegaster*, *Anotogaster*, *Neallogaster* und *Sonjagaster* (Anisoptera). *Odonatologica* **1993**, *22*, 273–294.
31. St. Quentin, D. Zum Vorkommen von Cordulegaster insignis Schneider in Rumänien. *Studii Communicari Sibiu* **1971**, *16*, 205–208.
32. Schmidt, E. Ergebnisse der Deutschen Afghanistan–Expedition 1956 der Landessammlungen für Naturkunde Karlsruhe sowie der Expeditionen J. Klapperich, Bonn 1952–53 und Dr. K. Lindberg, Lund (Schweden) 1957–1960. *Beiträge Nat. Forsch. SW Dtschl.* **1961**, *19*, 399–435.
33. Asahina, S. *A Revision of the Himalayan Dragonflies of the Genus Neallogaster (Odonata, Cordulegastridae)*; Bulletin of the National Science Museum: Tokyo, Japan, 1982; Volume 8, pp. 153–171.
34. Garcia-Castellanos, D.; Estrada, F.; Jiménez-Munt, I.; Gorini, C.; Fernàndez, M.; Vergés, J.; De Vicente, R. Catastrophic flood of the Mediterranean after the Messinian salinity crisis. *Nature* **2009**, *462*, 778–781. [CrossRef] [PubMed]
35. Solano, E.; Hardersen, S.; Audisio, P.; Amorosi, V.; Senczuk, G.; Antonini, G. Asymmetric hybridization in *Cordulegaster* (Odonata: Cordulegastridae): Secondary postglacial contact and the possible role of mechanical constraints. *Ecol. Evol.* **2018**, *8*, 9657–9671. [CrossRef] [PubMed]
36. Hewitt, G.M. The genetic legacy of the Quaternary ice ages. *Nature* **2000**, *405*, 907–913. [CrossRef] [PubMed]
37. Schmitt, T. Molecular biogeography of Europe: Pleistocene cycles and postglacial trends. *Front. Zool.* **2007**, *4*, 11. [CrossRef] [PubMed]
38. Hewitt, G.M. Post–glacial re–colonization of European biota. *Biol. J. Linn. Soc.* **1999**, *68*, 87–112. [CrossRef]
39. Hewitt, G.M. Speciation, hybrid zones and phylogeography—Or seeing genes in space and time. *Mol. Ecol.* **2001**, *10*, 537–549. [CrossRef] [PubMed]
40. Myers, N.; Mittermeier, R.A.; Mittermeier, C.G.; da Fonseca, G.A.; Kent, J. Biodiversity hotspots for conservation priorities. *Nature* **2000**, *403*, 853–858. [CrossRef] [PubMed]
41. Gür, H. The Anatolian diagonal revisited: Testing the ecological basis of a biogeographic boundary. *Zool. Middle East.* **2016**, *62*, 189–199. [CrossRef]
42. Davis, P.H. Distribution patterns in Anatolia with particular reference to endemism. In *Plant Life of South–West Asia*; Davis, P.H., Harper, P.C., Hedge, I.C., Eds.; Botanical Society of Edinburgh: Edinburgh, UK, 1971; pp. 15–27.
43. Galimberti, A.; Assandri, G.; Maggioni, D.; Ramazzotti, F.; Baroni, D.; Bazzi, G.; Chiandetti, I.; Corso, A.; Ferri, V.; Galuppi, M.; et al. DNA barcoding and eDNA-based biomonitoring of Italian Odonata. *Mol. Ecol. Resour.* **2020**, *21*, 183–200. [CrossRef] [PubMed]
44. Skvortsov, V.E. *The Dragonflies of Eastern Europe and Caucasus: An Illustrated Guide*; KMK Scientific Press Ltd.: Moscow, Russia, 2010.
45. Boudot, J.-P.; Borisov, S.N.; De Knijf, G.; van Grunsven, R.; Schröter, A.; Kalkman, V.J. Atlas of the dragonflies and damselflies of West and Central Asia. In *Brachytron 22*; Nederlandse Vereniging voor Libellenstudie: Den Haag, The Netherlands, 2021; pp. 4–238.

46. Schneider, T.; Ikemeyer, D.; Müller, O.; Dumont, H.J. Checklist of the dragonflies (Odonata) of Iran with new records and notes on distribution and taxonomy. *Zootaxa* **2018**, *4394*, 1–40. [CrossRef] [PubMed]
47. Schneider, T.; Ikemeyer, D. *The Damselflies and Dragonflies of Iran—Odonata Persica*; NIBUK, Natur in Buch und Kunst: Ruppichteroth, Germany, 2019; 247p.
48. Hebert, P.D.; Cywinska, A.; Ball, S.L.; de Waard, J.R. Biological identifications through DNA barcodes. *Proc. R. Soc.* **2003**, *270*, 313–321. [CrossRef] [PubMed]
49. Koroiva, R.; Pepinelli, M.; Rodrigues, M.E.; de Oliviera, R.F.; Lorenz-Lemke, A.P.; Kvist, S. DNA barcoding of odonates from the Upper Plata basin: Database creation and genetic diversity estimation. *PLoS ONE* **2017**, *12*, e0182283. [CrossRef] [PubMed]
50. Pimenta, A.L.A.; Pinto, A.P.; Takiya, D.M. Integrative taxonomy and phylogeny of the damselfly genus Forcepsioneura Lencioni, 1999 (Odonata: Coenagrionidae: Protoneurinae) with description of two new species from the Brazilian Atlantic Forest. *Arthropod Syst. Phylogenie* **2019**, *77*, 397–415.
51. Folmer, O.; Black, M.; Hoeh, W.; Lutz, R.; Vrijenhoek, R. DNA primers for amplification of mitochondrial cytochrome c oxidase subunit I from diverse metazoan invertebrates. *Mol. Mar. Biol. Biotechnol.* **1994**, 3, 294–299. [PubMed]
52. Katoh, K.; Rozewicki, J.; Yamada, K.D. MAFFT online service: Multiple sequence alignment, interactive sequence choice and visualization. *Brief. Bioform.* **2019**, *20*, 1160–1166. [CrossRef] [PubMed]
53. Darriba, D.; Taboada, G.L.; Doallo, R.; Posada, D. jModelTest 2: More models, new heuristics and parallel computing. *Nat. Methods* **2012**, *9*, 772. [CrossRef] [PubMed]
54. Ronquist, F.; Teslenko, M.; van der Mark, P.; Ayres, D.L.; Darling, A.; Höhna, S.; Larget, B.; Liu, L.; Suchard, M.A.; Huelsenbeck, J.P. MrBayes 3.2: Efficient Bayesian phylogenetic inference and model choice across a large model space. *Syst. Biol.* **2012**, *61*, 539–542. [CrossRef] [PubMed]
55. Swofford, D.L. *PAUP*. Phylogenetic Analysis Using Parsimony (*and Other Methods)*; Version 4; Sinauer Associates: Sunderland, MA, USA, 2003.
56. Drummond, A.J.; Rambaut, A. BEAST: Bayesian evolutionary analysis by sampling trees. *BMC Evol. Biol.* **2007**, *7*, 214. [CrossRef]
57. Drummond, A.J.; Suchard, M.A.; Xie, D.; Rambaut, A. Bayesische Phylogenetik mit BEAUti und dem BEAST 1.7. *Mol. Biol. Evol.* **2012**, *22*, 1185–1192. [CrossRef] [PubMed]
58. Rambout, A. GitHub. Available online: www.https://github.com/rambaut/figtree (accessed on 15 October 2021).
59. Flouri, T.; Jiao, X.; Rannala, B.; Yang, Z. Species Tree Inference with BPP using Genomic Sequences and the Multispecies Coalescent. *Mol. Biol. Evol.* **2018**, *35*, 2585–2593. [CrossRef] [PubMed]
60. Kumar, S.; Stecher, G.; Li, M.; Knyaz, C.; Tamura, K. MEGA X: Molecular Evolutionary Genetics Analysis across computing platforms. *Mol. Biol. Evol.* **2018**, *35*, 1547–1549. [CrossRef] [PubMed]
61. Leigh, J.; Bryant, D. POPART: Full-feature software for haplotype network construction. *Methods Ecol. Evol.* **2015**, *6*, 1110–1116. [CrossRef]
62. Van Pelt, G.J.; Vierstraete, A.; Dumont, H.J.; Schneider, T. *Case @@@@—Aeschna charpentieri Kolenati, 1846 (Insecta, Odonata): Proposed Replacement of the Current Neotype by a New Neotype*; Bulletin of Zoological Nomenclature; 2021; unpublished.
63. St. Quentin, D. Der Rassenkreis *Cordulegaster boltonii* (Donovan) (Odonata). *Entomol. Nachr. Österreichischer Und Schweiz. Entomol.* **1952**, *4*, 73–75.
64. Boudot, J.-P.; Jacquemin, G. Revision of *Cordulegaster boltonii* (Donovan, 1807) in southwestern Europe and northern Africa, with description of *C. b. iberica* ssp. from Spain (Anisoptera: Cordulegastridae). *Odonatologica* **1995**, *24*, 149–173.
65. Bernard, R.; Daraz, B. *Cordulegaster heros* and *Somatochlora meridionalis* in Ukraine: Solving the zoogeographical puzzle at their northern range limits (Odonata: Cordulegastridae, Corduliidae). *Odonatologica* **2015**, *44*, 255–278.
66. Kosterin, O.; Solovyev, V. Odonata found in mid-summer 2015 and 2016 at the north–westernmost Black Sea Coast of the Caucasus, with the first record of *Cordulegaster picta* Selys, 1854 in Russian Federation. *J. Int. Dragonfly Fund* **2017**, *107*, 1–43.
67. Schröter, A.; Seehausen, M.; Kunz, B.; Günther, A.; Schneider, T.; Jödicke, R. Update of the Odonata fauna of Georgia, southern Caucasus ecoregion. *Odonatologica* **2015**, *44*, 279–342.
68. Schneider, T.; Schneider, E.; Schneider, J.; Müller, O. Rediscovery of *Cordulegaster vanbrinkae* in Iran (Odonata: Cordulegastridae). *Odonatologica* **2014**, *43*, 25–34.
69. Holuša, O.; Dalecký, V.; Imanpour Namin, J. Habitat choice of *Cordulegaster vanbrinkae* in Iran (Odonata: Cordulegastridae). *Odonatologica* **2015**, *44*, 11–20.
70. Rastegar, J.; Havaskary, M.; Khodaparast, S.; Rafeii, A. A contribution to the knowledge of Odonata (Insecta) from West Azarbaijan province, northwestern Iran. *Entomofauna* **2013**, *34*, 369–375.
71. Ananian, V.; Tailly, M. *Cordulegaster vanbrinkae* Lohmann, 1993 (Odonata: Anisoptera) discovered in Armenia. *J. Int. Dragonfly Fund* **2012**, *46*, 1–11.
72. Schweighofer, W. Syntopes Vorkommen von *Cordulegaster boltonii* und *C. heros* an einem Bach im westlichen Niederösterreich (Odonata: Cordulegastridae). *Libellula* **2008**, *27*, 1–32.
73. Van Riemsdijk, I.; Arntzen, J.W.; Bogaerts, S.; Franzen, M.; Litvinchuk, S.N.; Olgun, K.; Wielstra, B. The Near East as a cradle of biodiversity: A phylogeography of banded newts (genus Ommatotriton) reveals extensive inter- and intraspecific genetic differentiation. *Mol. Phylogenet. Evol.* **2017**, *114*, 73–81. [CrossRef] [PubMed]

74. Asahina, S. The Odonata of Iraq. *Jpn. J. Zool.* **1973**, *17*, 17–36.
75. Kosswig, C. Zoogeography of the Near East. *Syst. Zool.* **1955**, *4*, 49–73. [CrossRef]
76. Çıplak, B.; Demirsoy, A.; Bozcuk, A.N. Distribution of Orthoptera in relation to the Anatolian diagonal in Turkey. *Articulata* **1993**, *8*, 1–20.
77. Bilgin, R. Back to the Suture: The distribution of intraspecific genetic diversity in and around Anatolia. *Int. J. Mol. Sci.* **2011**, *12*, 4080–4103. [CrossRef] [PubMed]
78. Kapli, P.; Botoni, D.; Ilgaz, C.; Kumlutas, Y.; Avcı, A.; Rastegar-Pouyani, N.; Fathinia, B.; Lymberakis, P.; Ahmadzadeh, F.; Poulakakis, N. Molecular phylogeny and historical biogeography of the Anatolian lizard *Apathya* (Squamata, Lacertidae). *Mol. Phylogenet. Evol.* **2013**, *66*, 992–1001. [CrossRef]
79. Erinç, S. Changes in the physical environment in Turkey since the end of the last glacial. In *The Environmental History of the Near and Middle East since the Last Ice Age*; Brice, W.C., Ed.; Academic Press: London, UK, 1978; pp. 87–110.

diversity

MDPI

Article

Molecular and Morphological Analyses Support Different Taxonomic Units for Asian and Australo-Pacific Forms of *Ischnura aurora* (Odonata, Coenagrionidae)

M. Olalla Lorenzo-Carballa *, Iago Sanmartín-Villar and Adolfo Cordero-Rivera

ECOEVO Lab, Escola de Enxeñaría Forestal, Campus A Xunqueira, Universidade de Vigo, 36005 Pontevedra, Spain; sv.iago@gmail.com (I.S.-V.); adolfo.cordero@uvigo.gal (A.C.-R.)
* Correspondence: m.o.lorenzo.carballa@gmail.com

Abstract: Despite the great technological progress that has aided taxonomical identification, taxonomical issues remain for certain species found in remote and/or understudied geographical areas. The damselfly species *Ischnura aurora* has been the subject of a long-standing taxonomical debate, focused mainly on the existence of morphological and behavioural differences between Asian and Australo-Pacific forms of this species that could justify their placement into two different species. Here, we carried out a comparative morphological analysis of specimens currently identified as *I. rubilio* from India and *I. aurora* from Asia and Oceania, combined with the analysis of mitochondrial and nuclear sequence data, both developed by us and available in public repositories. Our results split the Asian and Australo-Pacific forms of *I. aurora* into two well-differentiated taxonomic units and, hence, different (albeit closely related) species, and support the specific status of *I. rubilio*. The results of our genetic analyses suggest the existence of a third (and even fourth) taxonomic unit, stressing the need to revise all available material belonging to the different *I. aurora* subspecies that have been described. Finally, we have identified several questionable DNA sequences currently available in public repositories, upon which previous conclusions about the phylogenetic position of *I. rubilio* are based. Our study stresses the importance of being able to link available DNA sequence data with voucher specimens as well as to carry out a careful examination of DNA sequence data prior to their inclusion in taxonomical studies.

Keywords: Zygoptera; damselfly; integrative taxonomy; species delineation; barcoding; morphological analysis; DNA sequencing

Citation: Lorenzo-Carballa, M.O.; Sanmartín-Villar, I.; Cordero-Rivera, A. Molecular and Morphological Analyses Support Different Taxonomic Units for Asian and Australo-Pacific Forms of *Ischnura aurora* (Odonata, Coenagrionidae). *Diversity* **2022**, *14*, 606. https://doi.org/10.3390/d14080606

Academic Editor: Luc Legal

Received: 26 June 2022
Accepted: 26 July 2022
Published: 28 July 2022

1. Introduction

The science of taxonomy is central to many disciplines within biology, and hence being able to correctly identify species and to associate a scientific name with a particular organism is a prerequisite for ecological and conservation studies [1,2]. Until the development of alternative techniques to define species boundaries, the identification of species by means of morphological analysis was the only option available to taxonomists, even though the existence of "cryptic" species constituted a clear limitation. Taxonomy has shown great advances in recent years with the incorporation of technological advances (the most relevant being DNA sequencing) and the possibility of virtually accessing museum collections for specimen examination [3], which has led to integrative taxonomy, i.e., the study of variation in different types of datasets to delineate species boundaries more accurately [4]. Since 2003, when the DNA-based approach to taxonomy was first proposed [5–7], the so-called DNA barcode—the mitochondrial *COI* gene—has been widely used for species descriptions. However, the well-known limitations of mitochondrial DNA (e.g., incomplete lineage sorting, introgression, or the presence of nuclear copies of mitochondrial genes [8–11]) may lead to erroneous conclusions in species delimitation studies (see for example Papakostas et al. [12] or Ožana et al. [13]), and therefore the use of information from both mitochondrial and nuclear DNA markers is preferred, as it integrates information from two different genetic sources, hence allowing for a better support of taxa. Despite the acknowledged advantages of DNA sequences as a tool for species identification, species delineation is the crucial first step in taxonomy. To accurately delineate species boundaries, we need to be able to link a newly discovered species with its correct name, and describe its natural history, morphology, and behaviour [14].

Regardless of all the technological progress that has helped with taxonomic identification, the heterogeneous fieldwork effort carried out across the world may hamper the identification of specimens found in certain areas that, in some cases, host the highest diversity. Therefore, the combination of extensive fieldwork and modern technologies is necessary to clarify biodiversity, an essential matter nowadays due to the diversity loss produced because of climate change. Damselflies (Zygoptera: Odonata) constitute a good example to address taxonomic issues due to their high diversity in environments with difficult access and the high interspecific morphological similarities that exist in this insect group, sometimes only possible to unravel by genetic analysis [15]. Recent works have pointed to the genus *Ischnura* Charpentier, 1840, also known as forktails (Coenagrionidae), as a group with a non-resolved taxonomy, partly due to an incomplete knowledge of this genus in several areas of Asia [16,17]. *Ischnura* is a speciose genus with worldwide distribution, which has colonised many oceanic islands and shows great diversity in morphology, colouration, and behaviour. There are currently ca. 77 known species of *Ischnura*, which can be found in a range of diverse environments and spread across vast distribution areas [17–19]. One example of a species within this genus showing a wide geographical distribution and taxonomical issues is *Ischnura aurora* (Brauer, 1865). This species is a well-known migrant, passively dispersed throughout long distances as part of the aerial plankton [20,21] (pp. 396 and 409–411). Its ability to disperse over vast geographical areas explains its widespread distribution range that spans from the Pacific islands and Australia to the South-East Asiatic continent, India, Pakistan, and Iran [22,23] (Figure 1).

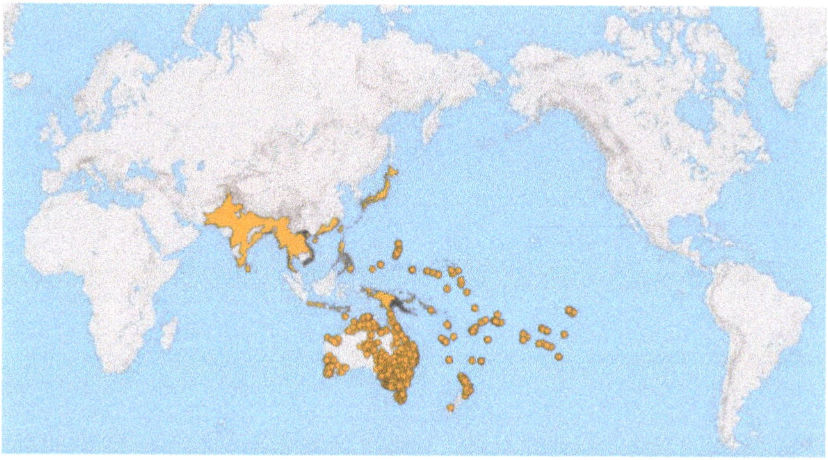

Figure 1. Map showing the geographic distribution of *Ischnura aurora* (orange dots). Modified from Dow et al. [22].

There has been a long-standing debate about which name should be used when referring to individuals of this species. In 1858, the name *Agrion delicatum* was given by Hagen to a damselfly found in Sri Lanka, Bengal (currently India and Bangladesh) and Australia [24]. In his work, Hagen included only data on body and wing sizes, but failed to provide an actual description of the species, which has remained as a *nomen nudum* since then. Later, *A. delicatum* was transferred to the genus *Ischnura* and the species *Ischnura delicata* was described using material from Asia and Australia [25]. *I. delicata* was considered a senior synonym of *Agrion (Ischnura) aurora*, a very similar damselfly earlier described by Brauer [26] using material from Tahiti. In his description of *I. delicata*, Selys mentions a male of a *"variety or juvenile"* of the species from India, which he named *Ischnura rubilio*. Selys provided information on some key morphological characteristics of *I. rubilio*: "the basal articulation of the 4–6 segments not circled in black; the 8 segment entirely blue like the 9" [25] (p. 283). Even though this taxon is still awaiting formal description, it was considered a subspecies of *I. aurora* [27] until Kalkman et al. [28] raised it to specific status. Therefore, *I. rubilio* is currently recognised as a distinct species restricted to the Indian subcontinent, whereas *I. delicata* is considered a junior synonym of *I. aurora*. At the molecular level, there has been no consensus yet about the placement of *I. rubilio* within the genus *Ischnura*. A phylogenetic analysis of the genus by Dumont [23], based on nuclear (Internal Transcribed Spacer, *ITS*) and mitochondrial (Cytochrome Oxidase I, *COI*) DNA sequence data which included specimens of *I. rubilio* collected in Bhutan and India, concluded that this species was in fact distinct from *I. aurora*, being either basal to the genus *Ischnura* or a member of the *pumilio* group. However, a more recent phylogenetic analysis by Sánchez-Guillen et al. [18], based also on nuclear (*ITS*) and mitochondrial (Cytochrome B, *CYTB* and Cytochrome Oxidase II, *COII*) sequence data, has placed *I. rubilio* as a sister species of *I. aurora*.

Beyond this debate on taxonomic hierarchy, other authors have focussed on the existence of morphological and behavioural differences between the Asian and Australo-Pacific forms of *I. aurora* that would justify their placement into two different species [27,29]. According to these authors, males of the Asian forms of the species possess large postocular spots, a completely blue eighth abdominal segment, a more pronounced dorsal tubercle in the 10th abdominal segment, acute cerci in its superior region, and narrower and pointer paraprocts, while the females mate in the adult stage. Australo-Pacific forms of *I. aurora*, on the other hand, possess small and circular postocular spots, one-third of the eighth abdominal segment with a blue colour, less pronounced dorsal tubercle in the 10th abdominal segment, rounded cerci in its superior region, and females that mate only in the teneral stage. While Papazian et al. [27] and Rowe [29] agree on the existence of morphological differences, there is no consensus on which name should be given to each of these forms. Papazian et al. [27] considered the Asian forms from the Indian subcontinent as a subspecies of *I. aurora*, named *I. aurora rubilio* (according to Selys description above), and in support of their argument, Dumont stated that "*I. aurora* is rare or totally absent West of the Wallace Line" [23] (p. 307). On the other hand, Rowe [29] suggested the existence of Asian forms of *I. aurora* in South and East Asia, as well as in Sri Lanka and India. He stated that "this 'species' should therefore be cited as *Ischnura delicata* (Hagen in Selys 1876)" [29] (p. 189), and, also, he highlighted the complex taxonomy of the Asian forms of *I. aurora*, for which three further names exist (*I. rubilio*, *I. amelia* and *I. bhimtalensis*), which have been assumed at some point to be either subspecies or junior synonyms of *I. aurora*.

Therefore, both Papazian et al. [27] and Rowe [29] coincide in restricting the "true" *Ischnura aurora* to the Australo-Pacific geographic area, but they disagree in the taxonomic classification and distribution of the Asian forms of this species. Here, we carried out a comparative morphological analysis of specimens currently identified as *I. rubilio* from India and *I. aurora* from Asia and Oceania, focusing on those characteristics traditionally used to distinguish among the Asian and Australo-Pacific forms of *I. aurora*. We combine the morphological analysis of male and female specimens with the analysis of mitochondrial and nuclear sequence data, both developed by us and available at public repositories, with the aim of clarifying the taxonomic status of these taxa.

2. Materials and Methods

2.1. Specimen Collection, DNA Extraction and Sequencing

A total of nine specimens of *I. aurora* from China (6 males, 1 female), Australia (1 male), and Fiji (1 male) belonging to ACR's collection and stored in 80% ethanol were selected for DNA extraction. Additionally, four dried-preserved specimens of *I. rubilio* (2 males, 2 females) collected in India were also used for DNA extraction (see Table 1).

Table 1. Information on *Ischnura aurora* and *I. rubilio* specimens used for DNA extraction and sequencing as described in the main text. For each specimen, we list the voucher ID, sex, and collection locality. n.a. indicates that no sequence could be obtained for a particular marker and/or specimen; Accession numbers labelled with an asterisk are those that correspond to sequences annotated as *COI*-like sequences (see Appendix A and Table 2).

Species	Specimen ID	Sex	Collection Locality	GenBank Acc. Nos	
				COI	*ITS*
Ischnura aurora	ACR819	M	Pond at Bandiana, Wodonga, Victoria, Australia.	OM964934	OM964914
I. aurora	ACR2379	M	Stream at Xi Meng, Yunnan, China.	OM964933	OM964916
I. aurora	ACR2880	M	Pond at Meng Ding, Yunnan, China.	OM964932	OM964917
I. aurora	ACR2956	M	Pond at Na Bang, Yunnan, China.	OM964931	n.a.
I. aurora	ACR3503	M	Rice fields, Huaping, Yunnan, China.	OM964930	OM964918
I. aurora	ACR3888	M	Pond in agricultural area, Mengding, Yunnan, China.	OM964929	OM964919
I. aurora	ACR3998	F	River at Meng Lun, Yunnan, China.	OM964928	OM964920
I. aurora	ACR4067	M	Stream at Meng Lun, Yunnan, China.	OM964927	OM964921
I. aurora	ACR5010	M	Somosomo Damm, Chakaudrove, Taveuni, Fiji	n.a.	OM964915
Ischnura rubilio	MB-IrbKeM	M	Trivandrum, Kerala, South India.	OM964925 *	OM964922
I. rubilio	MB-IrbKeF	F		OM964924 *	n.a.
I. rubilio	MB-IrbTam	F	Tamil Nadu, South India.	OM964926 *	n.a.
I. rubilio	MB-IrbGir	M	Unknown locality, India.	OM964923 *	n.a.

Total genomic DNA was extracted from individual legs using the GeneJet DNA extraction kit (ThermoFisher Scientific, Waltham, MA, USA), following the manufacturer's protocol. Fragments of the mitochondrial *COI* gene and the nuclear *ITS* were amplified using previously published primer pairs (COI-S0: TACCAATTATAATTGGAGGATTYGG/COI-AS0: CTTCTGGATGTCCAAARAATCA and ITS-F0: GGAAAGATGGCCAAACTTGA/ITS-28S-AS0: CCTCCGCTTATTAATATGCTTAAATTC [30]). PCR reactions were carried out at specific annealing temperatures (48 °C for *COI* and 52 °C for *ITS*) using the DreamTaq Green PCR Master Mix (ThermoFisher Scientific, Waltham, MA, USA). Before sequencing, PCR products were purified with shrimp alkaline phosphatase and exonuclease I (New England Biolabs, Ipswich, MA, USA) to remove unincorporated primers and dNTPs. Cleaned PCR products were sequenced bidirectionally using BigDye v.3.1 chemistry (Applied Biosystems, Foster City, CA, USA) at either the Macrogen Laboratories in Spain or at the CACTI genomics facility from the University of Vigo.

2.2. Genetic Analyses

DNA chromatograms were visually inspected, trimmed and automatically assembled using Geneious v. 9.1.8 (https://www.geneious.com/). Previously published *COI* and *ITS* sequences from specimens identified as *I. aurora*, *I. rubilio* or *I. delicata*; together with representative species of the Coenagrionidae genera, *Ischnura*, *Aciagrion*, *Ceriagrion*, *Coenagrion*, *Enallagma*, *Erythromma*, *Mortonagrion* and *Pseudagrion*, were downloaded from GenBank (https://www.ncbi.nlm.nih.gov/genbank/) and added to our datasets (see Appendix B Table A1). Prior to the genetic analyses, a quality control step was carried out to ensure that the sequences included in the final datasets were not derived from contaminations or, in the case of the mitochondrial DNA, also to rule out the amplification of paralogous copies of *COI* (nuclear mitochondrial DNA copies or *numts* [11,31] which were also recently described in odonates [13]). The quality control and data mining steps are described in detail in Appendix A.

After quality control, all sequences selected for inclusion in the final datasets were aligned using MAFFT [32], as implemented in Geneious v 9.1.8. Phylogenetic relationships were reconstructed using maximum likelihood (ML) and Bayesian inference (BI) approaches. ML analyses were carried out using IQTree v. 1.6.12 [33], with the best substitution model for each marker selected by ModelFinder [34]. Support of branches in the resulting tree was assessed by 10,000 ultrafast bootstrap replicates [35,36]. BI analyses were carried out using MrBayes 3.2.6 [37,38], as implemented in Geneious v 9.1.8. MCMC searches were run for 1.1 million generations, with default priors and with the GTR + I+G substitution model. Resulting phylogenetic trees were edited with FigTree v. 1.4.3. (http://tree.bio.ed.ac.uk/software/figtree/) and Inkscape v. 1.0 [39].

Genetic differentiation between *Ischnura* species (uncorrected p-distances) was estimated for each dataset in MEGA X [40] using the pairwise deletion option, which removes all ambiguous positions for each sequence pair. To further confirm the species delimitation within our datasets, we used the single-locus distance-based delimitation method implemented by the software *Assemble Species by Automatic Partitioning* (ASAP) [41]. Analyses were run at the ASAP web server (https://bioinfo.mnhn.fr/abi/public/asap/), using fasta files for each locus (*COI* and *ITS*) as the input files. The species delimitation analyses were carried out using only the sequences from *I. aurora*, *I. delicata* and *I. rubilio*, with the default options and with genetic distances computed under the Kimura (K80) model.

2.3. Morphological Analyses

For the morphological analyses, we examined a total of 31 ethanol-preserved individuals of *Ischnura aurora* (10 males and 9 females from China; 3 males and 2 females from Fiji; 3 males and 4 females from Australia; see Table A2), plus the four dried-preserved *Ischnura rubilio* specimens collected in India listed in Table 1. The aim of the morphological analysis was to determine whether the morphology of these specimens would corroborate the results of the genetic analyses and justify their placement as two different taxonomic units. Specimens were examined under an Olympus SZ60 stereoscopic microscope. Photographs of the individuals were taken using a Leica Flexacam C1 digital camera attached to the microscope at varying magnifications; and afterwards stacked and edited using the software GIMP v. 2.10 [42]. The male genital ligula of one individual of *I. rubilio* (MB-IrbkeM) and three *I. aurora* individuals (ACR2880 from China, ACR0818 from Australia and ACR5009 from Fiji) was dissected and observed under a scanning electron microscope (Philips XL30) at the CACTI microscopy service from the University of Vigo. The terminology used in the morphological descriptions follows that in Garrison et al. [43]. Abdominal segments are referred to as capital "S" plus the segment number. All measurements are given in millimetres.

3. Results

3.1. Genetic Analyses

All sequences generated in this study were deposited in GenBank with accession numbers OM964914-OM964934 (see Table 1). The *COI* sequences obtained from the *I. rubilio* samples collected in India were not included in the mtDNA dataset, as they were likely "*COI*-like" sequences or *numts*, rather than orthologous copies of the mitochondrial *COI* gene (Table 2; see also Appendix A). Regarding the sequences retrieved from GenBank, a total of thirteen sequences belonging to *I. aurora*, *I. delicata* and *I. rubilio* were also excluded from the datasets after the quality control steps carried out as described in Appendix A. Three *I. aurora COI* sequences were excluded because they corresponded to the 3′-end of the *COI* gene (Table 2 and Appendix A Figure A4). Five *COI* sequences belonging to *I. aurora* (N = 2) and *I. delicata* (N = 3) were discarded because they showed ambiguity-coded bases, which are not expected in a mitochondrial coding gene (Table 2). Three sequences belonging to *I. aurora* (N = 1) and *I. rubilio* (N = 2) were excluded from the datasets because they were likely the product of specimen misidentification or misplacement of individuals in the laboratory at the time of DNA extraction (Table 2; see also Appendix A). Finally, two *COI* sequences of *I. rubilio* were excluded from the mtDNA dataset because they showed several features consistent with these being "*COI*-like" sequences or *numts*. It is important to note that three of the *I. rubilio* sequences that were excluded from our datasets were those included in the phylogeny of the genus *Ischnura* by Dumont [23].

Table 2. Sequences of *Ischnura aurora*, *I. delicata* and *I. rubilio* that were excluded from the final genetic analyses after the quality control steps that are described in detail in the Appendix A, including the *I. rubilio* "*COI*-like" sequences generated in this study.

Taxa	Data Source	GenBank No.	Marker	Reason for Exclusion
Ischnura aurora	Nolan et al. [44]	EU219876	*COI*	Sequence corresponds to the 3′ end of the *COI* gene
I. aurora	Nolan et al [44]	EU219877	*COI*	Sequence corresponds to the 3′ end of the *COI* gene
I. aurora	Mehmood et al. (unpublished)	LC198680	*COI*	Sequence corresponds to the 3′ end of the *COI* gene
I. aurora	Ramage et al. [45]	KX053527	*COI*	Ambiguity-coded bases in sequence
I. aurora	Ramage et al. [45]	KX053531	*COI*	Ambiguity-coded bases in sequence
I. aurora	Dumont et al. [46]	FN356100	*ITS*	Specimen misidentification and/or misplacement?
Ischnura delicata	Ashfaq et al. (unpublished)	KY832433	*COI*	Ambiguity-coded bases in sequence
I. delicata	Ashfaq et al. (unpublished)	KY838304	*COI*	Ambiguity-coded bases (insertion of 3 "Ns") in sequence
I. delicata	Ashfaq et al. (unpublished)	KY844428	*COI*	Ambiguity-coded bases in sequence
Ischnura rubilio	Pavithran et al. (unpublished)	MW143324	*COI*	Both sequences are identical. No stop codons, but sequences are odd compared to references. Similar to several marine sponge genera. "*COI*-like/*COI*-numt" sequence?
I. rubilio	Dumont [23]	MH449981	*COI*	
I. rubilio	Dumont [23]	MH449992	*COI*	Specimen misidentification and/or misplacement?
I. rubilio	Dumont [23]	MH447434	*ITS*	Specimen misidentification and/or misplacement?
I. rubilio	This study	OM964923	*COI*	Ambiguity codes in sequence. Similar to *Nesobasis* spp. Annotated as "*COI*-like/*COI*-numt" sequences
I. rubilio	This study	OM964924	*COI*	
I. rubilio	This study	OM964925	*COI*	
I. rubilio	This study	OM964926	*COI*	

After the quality control steps, the final number of sequences included in each dataset was 102 for *COI* (451 bp-long alignment) and 66 for *ITS* (639 bp-long alignment). The difference in size between both datasets stems mostly from the fact that the *COI* gene (and more specifically, the Folmer region) is the marker of choice in barcoding studies, hence the higher number of sequences from this marker that are available at public repositories.

The obtained phylogenetic trees were congruent between BI and ML methods and between nuclear and mitochondrial DNA markers, with strong support for the split of *Ischnura aurora* in two clades: the Australo-Pacific and the Asian clade (see Figures 2 and 3). The Asian clade included the individuals of *I. aurora* from China that were sequenced by us, together with *I. aurora* specimens from Thailand and individuals identified as *I. delicata* and *I. rubilio* collected in Pakistan and India, respectively, whose sequences were obtained from GenBank. For the *ITS* marker, the Asian clade also included the *I. rubilio* specimen from Kerala (India) sequenced by us. The Australo-Pacific clade included all *I. aurora* individuals from Australia and Fiji sequenced by us, plus all individuals identified as *I. aurora* whose sequences were downloaded from GenBank, and which were collected in the Australo-Pacific distribution area of the species (i.e., Australia, Japan, Samoa, Tonga, French Polynesia, Fiji, Wallis and Futuna, Guam, and New Guinea; see Figures 2 and 3, Table A1). For the *COI* dataset, there were three exceptions to this pattern: the first one was an individual identified as *I. aurora* from Kerala in India (GenBank No KR149808) that falls within the Australo-Pacific clade (see Figure 2 and Table A1). The other exceptions corresponded with two other individuals identified as *I. aurora* from India (GenBank No MT511656) and New Guinea (GenBank No MH449994) which also fall outside their expected clades in the phylogenetic analyses: both individuals appear as basal to the rest of the Australo-Pacific *I. aurora* (see Figure 2 and Table A1).

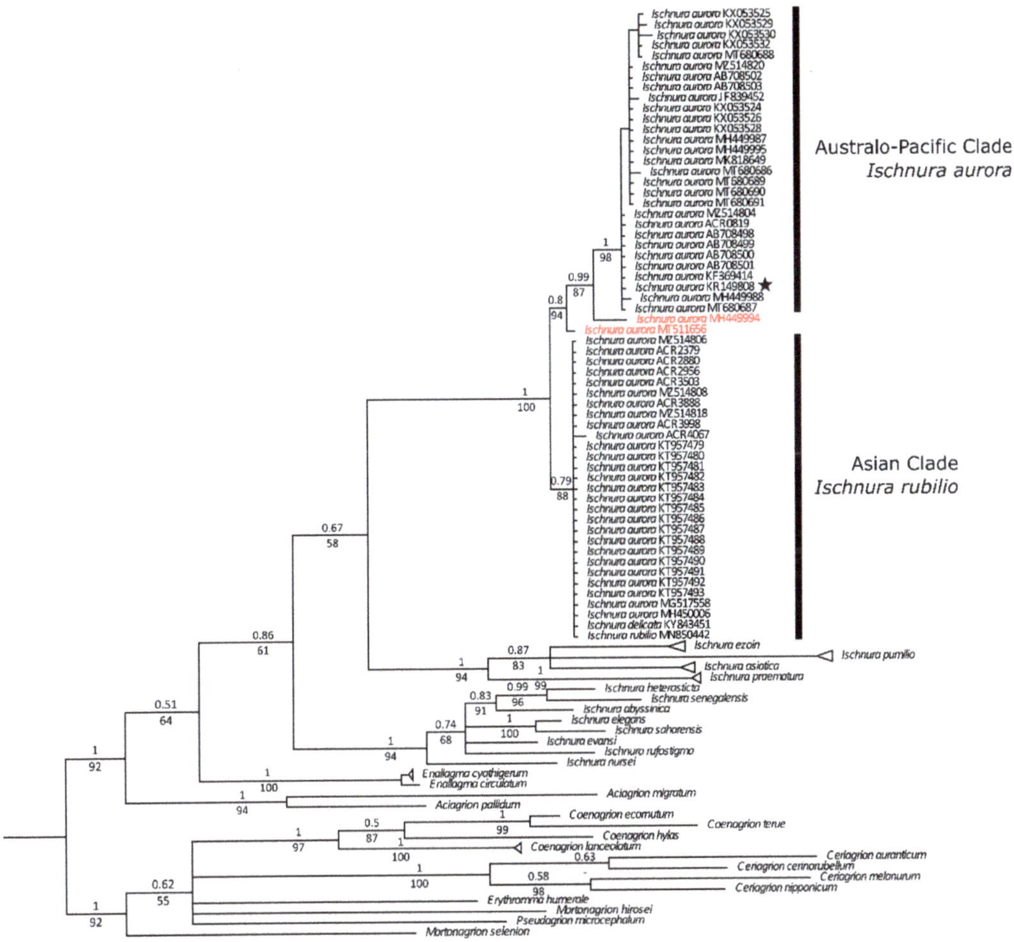

Figure 2. Tree representing the phylogenetic relationships among the *Ischnura* species analysed in this study, using mitochondrial DNA (*COI*) sequence data. Numbers above and below branches represent Bayesian posterior probability and maximum likelihood bootstrap values, respectively. Clades are labelled according to the ASAP proposed species delimitation. The star within the Australo-Pacific clade indicates the *I. aurora* individual from Kerala (India) retrieved from GenBank, while the specimens in red are the two *I. aurora* from New Guinea and India that are identified as different taxonomic units by ASAP (see also Table A1 and Figure A7a).

The ASAP species delimitation analyses were congruent with the results of the phylogenetic analyses: for both the nuclear and mitochondrial DNA datasets, partitions with the best ASAP score split the Australo-Pacific and continental Asian forms of *I. aurora* in two different taxonomic units (see Figure A7), the latter also including the specimens currently under the name *I. rubilio* and *I. delicata*. For the *COI* marker, two sequences were identified as belonging to different taxonomic units in the analysis. The first sequence was that of a specimen identified as *I. aurora* collected at Baliem River in New Guinea with accession no MH449994, which was identified as a different species in the partition with the best ASAP score. The second one was that of an *I. aurora* specimen from India with accession no MT511656, which according to the partition with the second best ASAP score would correspond to a fourth species (see Table A1 and Figure A7a).

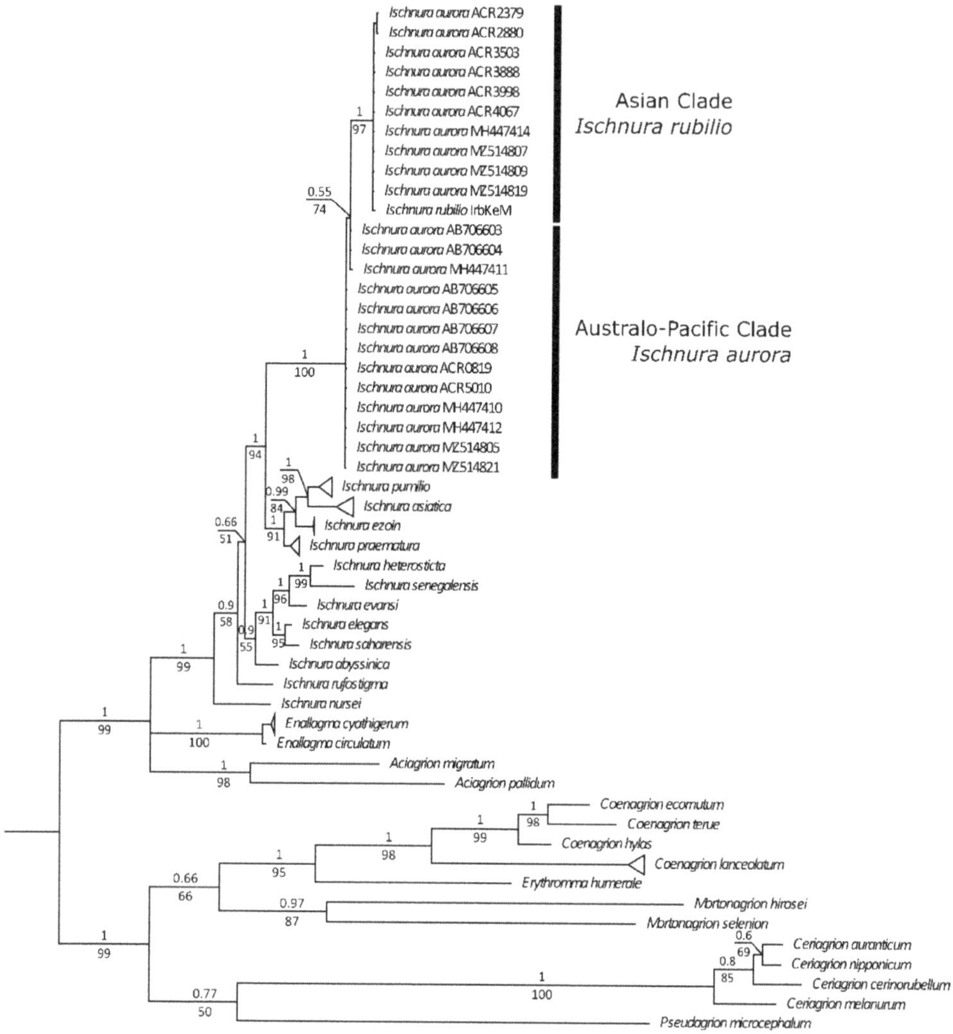

Figure 3. Tree representing the phylogenetic relationships among the *Ischnura* species analysed in this study, using nuclear DNA (*ITS*) sequence data. Numbers above and below branches represent Bayesian posterior probability and maximum likelihood bootstrap values, respectively. Clades are labelled according to the ASAP proposed species delimitation (see also Table A1 and Figure A7b).

The results presented above were further supported by the genetic distances: the average p-distance between Asian and Australo-Pacific *I. aurora* was 2% for the *ITS* marker and 3.1% for the *COI* marker. Similar values were found between the specimens labelled as *I. delicata* and *I. rubilio* and the Australo-Pacific *I. aurora*, whereas genetic differentiation between these two species and the Asian *I. aurora* was nearly zero in all cases (see Table A2). For the *COI* marker, genetic distances between the two identified clades and the *I. aurora* sequences with GenBank numbers MH449994 and MT511656 from New Guinea and India, respectively, were comparable to the distances found between the two clades (~2% in all cases; see Table A2).

3.2. Morphological Analyses

Ischnura aurora specimens from China, Australia, and Fiji all showed similar body length (males from China: 25.0 ± 0.9, N = 10; Australia: 25.3 ± 0.4, N = 2; Fiji: 24.8 ± 1.0, N = 3; females from China: 25.4 ± 1.0, N = 9; Australia: 25.7 ± 0.5, N = 5; Fiji: 23.0 ± 0.0, N = 2) and hindwing length (males from China: 11.5 ± 0.4, N = 10; Australia: 12.6 ± 0.9, N = 2; Fiji: 12.3 ± 0.4, N = 3; females from China: 13.4 ± 0.9, N = 9; Australia: 15.0 ± 0.5, N = 5; Fiji: 12.6 ± 1.3, N = 2; Figure 4). Due to their poor state of preservation, it was not possible to measure the *I. rubilio* specimens.

Figure 4. Lateral view of specimens of *Ischnura aurora* from China ((**a**) male; (**b**) female), Australia ((**c**) male, (**d**) female), and Fiji ((**e**) male, (**f**) female). Due to the poor state of preservation of the *I. rubilio* specimens, it was not possible to add an image of this species, but its morphology was similar to that observed in the *I. aurora* from China.

Head: No differences in colour patterns were observed between *I. rubilio* and the examined *I. aurora* specimens, independently of their origin. All specimens observed possessed blue small and rounded postocular spots. Some *I. aurora* females from China show a green-brownish colouration of the occipital region of the head which, in some individuals, reached the postocular spots.

Thorax: Males of both *I. rubilio* and *I. aurora* from China showed the posterior dorso-lateral part of the pronotum (greenish) to be less elevated but the dorsal part (black) to be more expanded towards the mesothorax compared to that of *I. aurora* males from Australia and Fiji (Figure 5). The posterior dorso-lateral elevations of the pronotum of females of both *I. rubilio* and *I. aurora* from China (greenish-brownish) are connected or nearly connected by an arch of the same colour (although this character shows high interindividual variability), while both elevations are restricted to the lateral border in the

I. aurora females from Australia and Fiji (Figure 6). Additionally, in females of *I. rubilio* and *I. aurora* from China, the dorsal expansion towards the mesothorax is wider in than in females from Australia and Fiji (Figure 6). Mesostigmal male protuberances and female plates show similarities between *I. rubilio* and *I. aurora* from all studied populations. Some Australian specimens show completely black mesostigmal protuberances, while tips are greenish in the other specimens. Females of *Ischnura rubilio* and *I. aurora* from China showed a narrower black humeral stripe and a less expanded black colouration in the junction between the metepisternum and metepimeron (see Figure 4d–f).

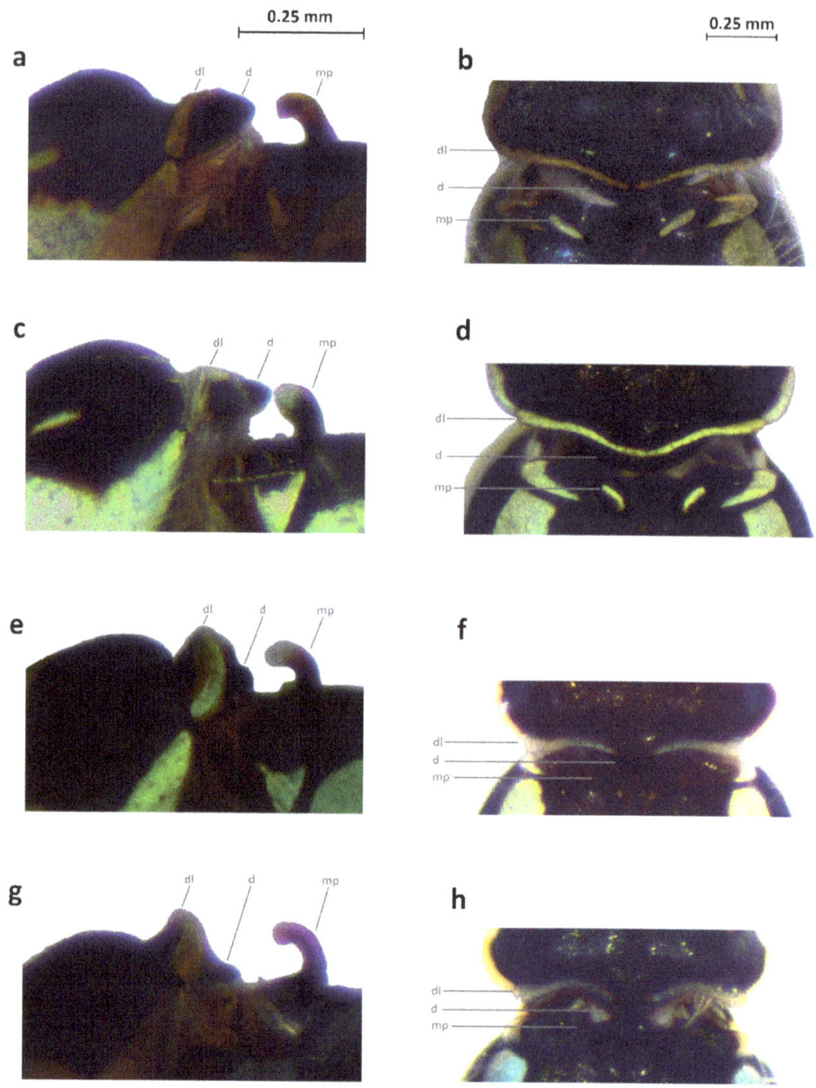

Figure 5. Posterior part of the pronotum and the anterior part of the mesothorax in males of *Ischnura rubilio* (lateral: (**a**), dorsal: (**b**)) and *Ischnura aurora* from China (lateral: (**c**), dorsal: (**d**)), Australia (lateral: (**e**), dorsal: (**f**)), and Fiji (lateral: (**g**), dorsal: (**h**)). *dl*: dorso-lateral elevation. *d*: dorsal expansion. *mp*: mesostigmal protuberance.

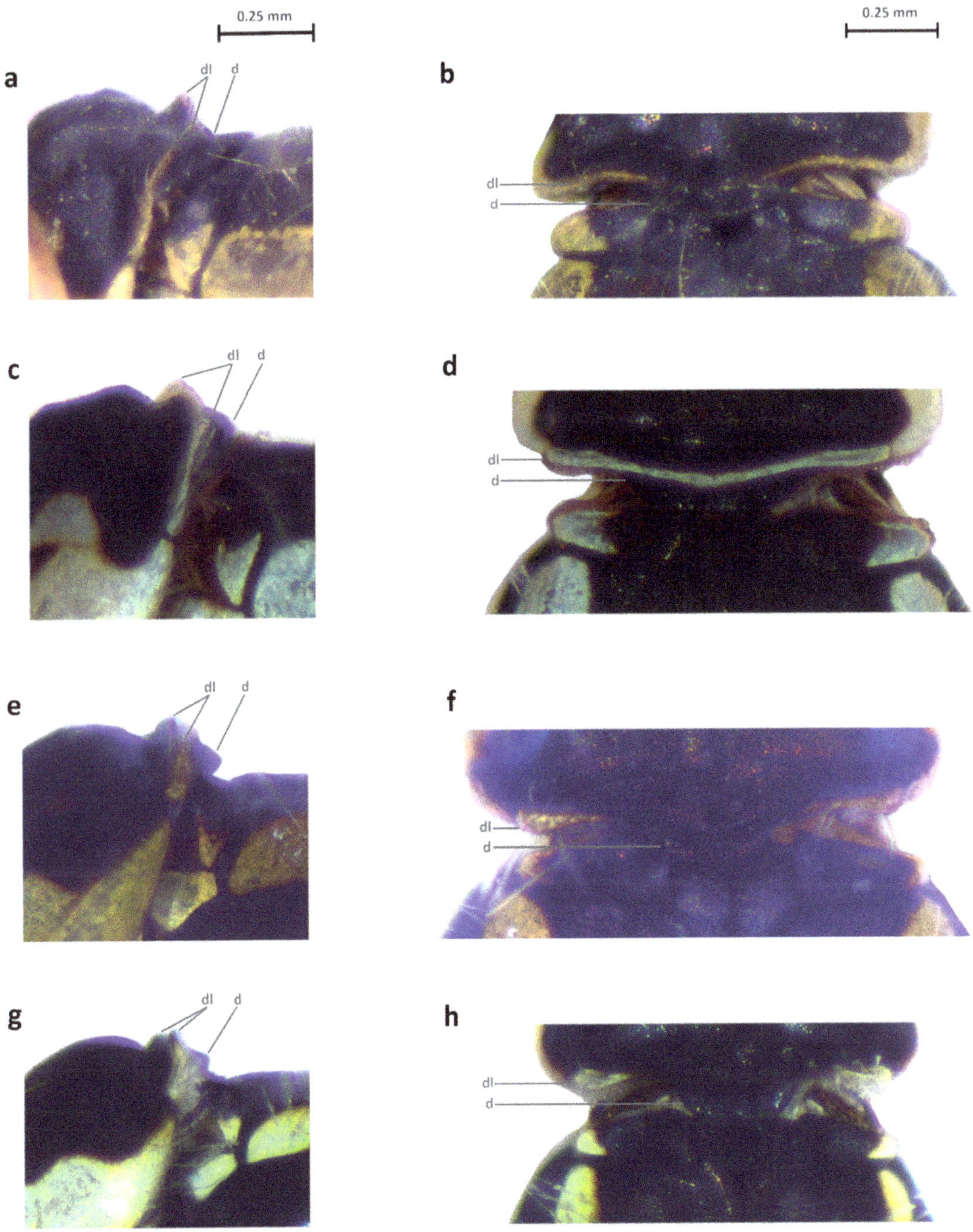

Figure 6. Posterior part of the pronotum and the anterior part of the mesothorax in females of *Ischnura rubilio* (lateral: (**a**), dorsal: (**b**)) and *Ischnura aurora* from China (lateral: (**c**), dorsal: (**d**)), Australia (lateral: (**e**), dorsal: (**f**)), and Fiji (lateral: (**g**), dorsal: (**h**)). *dl*: dorso-lateral elevation. *d*: dorsal expansion. Note that no mesostigmal protuberances exist in females.

Legs: No differences in leg colouration were found between the examined specimens.

Wings: Cell number was variable within and between all the studied populations.

Abdomen: *Ischnura rubilio* and *I. aurora* males from China show a less expanded black colouration on the dorsal part of S2 than that observed in *I. aurora* from China and Fiji. In the former, the apical stripe on S2 expands into a mid-dorsal stripe that reaches to approximately half of the segment, whereas in the *I. aurora* specimens from Australia and Fiji, this mid-dorsal stripe is much wider and may reach up to the end of S2 in some individuals (Figure 7). The basal articulation of S2 appears to be circled in black in the *I. aurora* individuals from Australia and Fiji, whereas it has a lighter colour in *I. rubilio* and *I. aurora* from China (Figure 7). S3 to S6 are citron yellow in males of *I. rubilio* and *I. aurora* collected in China and yellow-orange in males of *I. aurora* from Australia and Fiji, except for the black intersegment areas, which appear also of a lighter colour in *I. rubilio* and *I. aurora* from China (Figure 4). S6 is dorsally black in its posterior part and S7 is dorsally black in all examined individuals.

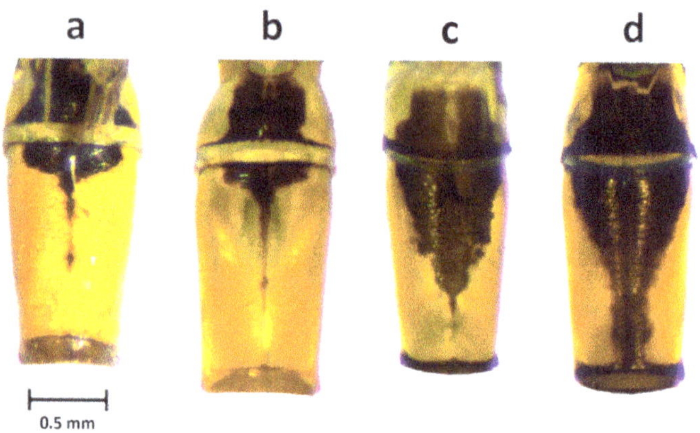

Figure 7. Dorsal view of the first and second abdominal segments of *Ischnura rubilio* (**a**) and *Ischnura aurora* from China (**b**), Australia (**c**), and Fiji (**d**).

Ischnura rubilio and *I. aurora* males from China show blue colouration in the dorsal and lateral part of S8 (except its anterior dorsal part, which shows a delta-shaped black colour) and S9, and laterally in the S10 (which vary between individuals) (Figure 7). Some individuals show blue spots in the lateral part of S7. *Ischnura aurora* males from Australia and Fiji show blue colour in the dorsal and lateral part of the S9 but only in the posterior dorso-lateral part of the S8 (1/3 or less of the segment length; see Figure 7). Individuals from Australia show certain variability in these colourations, showing black colour interrupting the blue bands of S8 and S9 but also showing blue colour in the lateral part of the S10. The dorsal tubercle of S10 is more pronounced in *I. rubilio* and *I. aurora* males from China than in *I. aurora* males from Australia and Fiji (Figure 7). All females examined showed a vulvar spine, except for one examined female from Australia. No differences were found in the length of the female genital valves.

Annal appendages: *Ischnura rubilio* and *I. aurora* males from China show more acute dorsal expression in their cerci and narrower paraprocts than males of *I. aurora* from Australia and Fiji (Figure 8). *I. aurora* from Fiji show a marked depression in the middle of the cerci under a lateral view (albeit this character shows high variability; Figure 8k), while the cerci in the rest of the specimens showed a straight border.

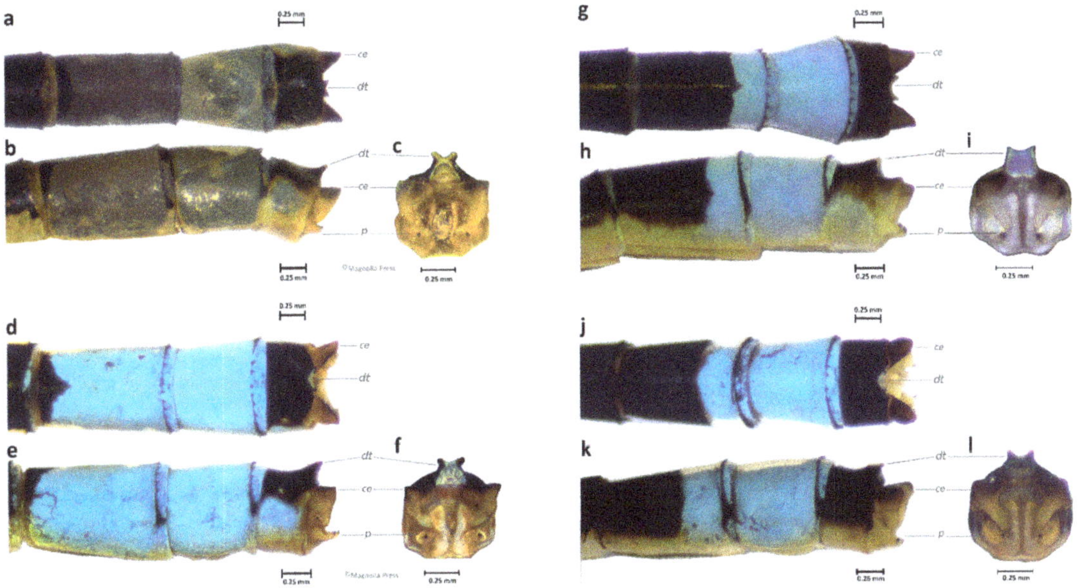

Figure 8. Dorsal, lateral, and posterior view of the abdominal appendages of *I. rubilio* (**a–c**) and *I. aurora* from China (**d–f**), Australia (**g–i**), and Fiji (**j–l**). *dt*: dorsal tubercle; *ce*: cerci; *p*: paraprocts. Images (**a–f**) are partly reproduced from Sanmartín-Villar et al. [14] with permission from the copyright holder.

Male genital ligula: No differences were observed between the male ligula of the analysed specimens (see Figure 9).

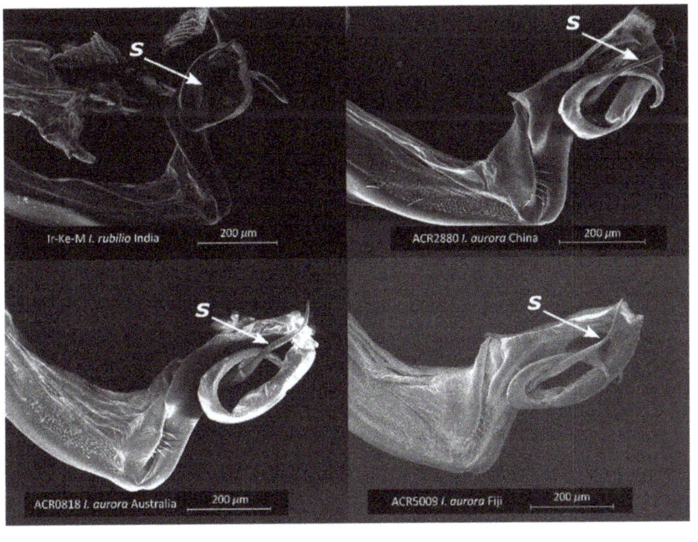

Figure 9. Scanning electron images showing the male genital ligula of *Ischnura rubilio* from India and *I. aurora* from China, Australia, and Fiji. The arrows in each image point to the spine (s) proximal to the genital ligula flexure, typical of most *Ischnura* species.

4. Discussion

The results of our genetic analyses indicate that the samples currently identified as *I. aurora* found within the Asian distribution area of this species all belong to a clade that also includes specimens currently under the names *I rubilio* and *I. delicata*. This clade appears to be closely related to that comprising the *I. aurora* individuals from the Australo-Pacific distribution area of the species. This split into two well-differentiated clades is supported by high bootstrap and posterior probability values for both nuclear and mitochondrial DNA (see Figures 2 and 3). Genetic distances between *I. rubilio, I. delicata* and *I. aurora* from China were nearly zero in all cases, whereas genetic distances between the Asian and the Australo-Pacific clade were ~3% for *COI* and ~2% for *ITS* (see Table A3). These distances found between both clades are similar to those found between other closely related species within the genus *Ischnura* (e.g., *I. elegans*/*I. saharensis*, Table A3; *I. elegans*/*I. graelsii* [47], or between closely related species in other Coenagrionidae genera (e.g., *Paracercion* [48]). The species delimitation analyses have provided further support for the placement of *I. aurora* from Asia within the same group as *I. rubilio* and *I delicata*, separated from the Australo-Pacific *I. aurora* (see Figure A7).

In agreement with the results of our molecular analyses, the morphological examination of material from Asia and Oceania points also to a closer relationship between the specimens currently under the name *I. aurora* collected in China and *I. rubilio* from India, than to *I. aurora* from Australia and Fiji. These morphological similarities also include the characteristics described above, which consistently differentiate *I. aurora* from China and *I. rubilio* from India from *I. aurora* from Australia and Fiji: in males, the less elevated but more expanded posterior part of the pronotum found in *I. rubilio* and *I. aurora* from China; and in females, the connectivity between the lateral elevations and the wider expansion, differing from the same structure in *I. aurora* females from Australia and Fiji (see Figures 5 and 6).

Beyond these morphological differences in the shape of the pronotum, all Asian and Australo-Pacific specimens examined by us differ mostly in their colouration pattern: *Ischnura rubilio* shows the same colour pattern as *I. aurora* from China, which differs from that observed in *I. aurora* from Australia and Fiji. Males of *I. rubilio* and *I. aurora* from China show a more expanded blue colouration on S8–S10. The abdominal blue colouration is highly variable in some Asian *Ischnura* [16] and similar between species (for instance, *I. praematura* Sanmartín-Villar & Zhang, 2022 shows an intermediate colouration between *I. aurora* from China and from the Australo-Pacific forms [17]), which might represent one of the main factors hindering the identification of these species. Nevertheless, we have found that all examined specimens from *I. rubilio* and *I. aurora* from China share the same colour pattern, which is different from that of *I. aurora* specimens collected in Australia and Fiji. These observations agree with those of Papazian et al. [27] and Rowe [29] regarding their examination of Asian and Australo-Pacific *I. aurora*; and with Selys in his description of *I. rubilio* [25]. Other differences that we have found between the Asian and Australo-Pacific material, and which have also been pointed out by some or all of these authors, include: (i) differences in the colour pattern of S1–S2 [27]; (ii) the paler basal articulations of the abdominal segments found in both *I. rubilio* and *I. aurora* from China when compared to Australo-Pacific *I. aurora* [25]; (iii) a more prominent dorsal tubercle in S10 in *I. rubilio* and *I. aurora* from China than in *I. aurora* from Australia and Fiji [29]; and (iv) the differences found in the anal appendages between the Asian and Australo-Pacific material that we have examined [29].

Contrary to what has been previously reported, we did not find differences in the postocular spots between *I. aurora* and *I. rubilio*. Papazian et al. [27] suggested that *I. rubilio* showed larger postocular spots and pointed out that they were "sometimes confluent across the occiput" [27] (p. 60). Even though we have observed that the occipital green-brownish colour in some females of *I. aurora* from China reaches the postocular spots, we highlight the different colouration of the postocular spots (blue) and the occipital region (green-brownish) and hence, no confluence between the postocular spots in the examined specimens. Considering the high variability that this character shows in *Ischnura* [49], including the species we examined here; and given that we found no consistent differences across the examined specimens, we conclude that postocular spots cannot be used as a trait to distinguish between *I. rubilio* and *I. aurora*.

Our morphological analysis of Asian and Australo-Pacific material allowed us to find some morphological differences in the females, which have not been addressed previously. Thus, we found differences in the thoracic colouration pattern (the black colour is less expanded in females of *I. rubilio* and *I. aurora* from China; compared to females of *I. aurora* from Australia and Fiji; Figure 4), and morphological differences in the posterior part of the pronotum that consistently separate females of *I. rubilio* and *I. aurora* from China from Australo-Pacific *I. aurora* females (see Figure 6).

Overall, the results of our analyses support the split of the Asian and Australo-Pacific forms of *I. aurora* into two well-differentiated taxonomic units and therefore different species, further supporting the specific status of *I. rubilio* [28]. The results of our genetic analyses agree with those of Sánchez-Guillén et al. [18] in placing *I. rubilio* within the *aurora* clade, but contradict those by Dumont [23], who found *I. rubilio* to be either basal to the *Ischnura* group or a member of the *pumilio* clade. The quality control of the DNA sequence data carried out by us here led to the exclusion of several questionable sequences from our datasets, among which were those belonging to *I. rubilio* from Dumont's work. These sequences are likely the product of both specimen misplacement during laboratory work and sequencing of non-orthologous copies of the mitochondrial *COI* gene (see Table 2), which led to the wrong conclusion about the phylogenetic position of this species within the *Ischnura* clade [23]. Similarly questionable, "*COI*-like" sequences were recently identified in Odonata [13], and therefore we encourage research colleagues to carefully examine the sequences they use in their studies in a similar way as we have done here or has been described elsewhere e.g., [50] to avoid the potential amplification of *numts*, which might have important consequences when addressing taxonomical questions [11,13,31].

Although wind-driven dispersal could explain the exchange of individuals between Asia and Oceania (for instance, the individual from Kerala, India with accession no KR149808 that falls within the Australo-Pacific clade; see Figure 2), long-distance dispersal may at the same time be limited by the physical barrier constituted by the sea, which will in turn restrict the contact between the identified clades and lead to the observed genetic differentiation. The observed morphological differences in the shape of the pronotum of both males and females, along with the differences in the S10 dorsal tubercle and male anal appendages found between both Asian and Australo-Pacific species might not be enough to prevent successful hybridization, which could in turn result in a morphological and/or molecular cline along the species' contact zone, with individuals showing intermediate characters between *I. rubilio* and *I. aurora*. There are several examples of hybridization between closely related *Ischnura* species [51,52]. However, as it is assumed that reproduction occurs between mature individuals in *I. rubilio* and *I. aurora* Asian forms but between mature males and teneral females in *I. aurora* from the Australo-Pacific forms [29,53], even if individuals from both species may occasionally come into contact, mating would be restricted due to these behavioural incompatibilities. Without access to the actual specimen from which the DNA sequence was obtained, we cannot establish whether the individual from Kerala that falls within the Australo-Pacific clade is a migrant individual or a hybrid.

The recent discovery of other Asian species in which teneral females mate [16] stresses the need for additional ethological studies to unravel species barriers and species population dynamics. Additionally, the presence of the elaborated mesostigmal protuberances in males but not in females suggests that these do not play a role in male–female assembly (for instance as happens in the European *Ischnura* species; see [54]) but in male–male encounters, which poses an interesting topic for future studies about the function of such structures.

Finally, the results of our genetic analyses point also to the existence of at least a third (or even a fourth) taxonomic unit identified from our *COI* dataset. These correspond to a specimen from India and another specimen from Baliem River in New Guinea, both labelled as *I. aurora*. Interestingly, Baliem River is the same locality in which the subspecies *I. aurora viduata* was first described [55]. These results stress the need to revise all available material belonging to the numerous *I. aurora* subspecies described as well as to be able to correctly link available DNA sequence data with voucher specimens.

We provide below an identification key for the separation between *I. rubilio* and *I. aurora*:

1. Pale postocular spots present, S8 partially or entirely blue dorsally, S9 blue dorsally, ... **2**

2. ♂: Posterior lobe of prothorax slightly raised laterally; expansion of the same lobe towards the mesothorax wide (approximately half of the width of the prothorax' posterior lobe), reaching the mesostigmal plates in dorsal view (Figure 5a–d); paraproct in ventral view with an accessory medial lobe in addition to the basal one; height of S10 tubercle in posterior view subequal to the width separating each tubercle (Figure 8c,f); black apical stripe in S2 expanding into a mid-dorsal stripe that reaches approximately half of the segment (Figure 7a,b); S8 entirely blue dorsally (Figure 8a,b,d,e);

♀: Posterior lobe of prothorax raised laterally; expansion of the same lobe towards the mesothorax wide (approximately half of the width of the prothorax' posterior lobe) (Figure 5a–d); black humeral stripe narrowing dorsally (Figure 4b); India [and elsewhere] .... ***rubilio***

2′. ♂: Posterior lobe of prothorax erect in its middle part; expansion of the same lobe towards the mesothorax narrow (approximately one-third of the width of the prothorax' posterior lobe), not reaching the mesostigmal plates in dorsal view (Figure 5e–h); paraproct in ventral view lacking an accessory medial lobe; distance between the bifid tubercles of S10 in posterior view greater than the height of the bifid tubercle (Figure 8i,l); black apical stripe in S2 wide, with the shape of an inverted triangle that may reach the end of S2 in some cases (Figure 7c,d); S8 with only posterior half blue dorsally (Figure 8g,h,j,k);

♀: Posterior lobe of prothorax raised laterally; expansion of the same lobe towards the mesothorax narrow (approximately one-third of the width of the prothorax' posterior lobe) (Figure 5a–d); black humeral stripe of equal width throughout (Figure 4d,f); Polynesia, Australia, New Zealand [and elsewhere] ....***aurora***

5. Conclusions

- Genetic analyses have showed that specimens currently under the names of *Ischnura rubilio* and *I. delicata* belong to a clade that also includes the *I. aurora* found within the Asian distribution area of this species.
- All the *I. aurora* found within the Australo-Pacific distribution area cluster together in a separate clade. Species delimitation analyses have identified these two clades as different taxonomic units.
- Concordant with the results of the genetic analyses, the morphology of the *I. aurora* collected in China is closer to *I. rubilio* than to *I. aurora* from Australia and Fiji.
- Given these results, we confirm the status of *I. rubilio* as a valid species and provide an identification key for its separation from *I. aurora*.
- Genetic analyses point also to the existence of at least a third taxonomic unit within the *aurora* clade, which stress the need to revise all available material belonging to the numerous subspecies of *I. aurora* that have been described.

Author Contributions: Conceptualization, M.O.L.-C., I.S.-V. and A.C.-R.; methodology, M.O.L.-C. and I.S.-V.; formal analysis, M.O.L.-C. and I.S.-V.; fieldwork and investigation, M.O.L.-C., I.S.-V. and A.C.-R.; resources, A.C.-R.; data curation, M.O.L.-C. and I.S.-V.; writing—original draft preparation, M.O.L.-C. and I.S.-V.; writing—review and editing, M.O.L.-C., I.S.-V. and A.C.-R.; supervision, M.O.L.-C. and A.C.-R.; project administration, A.C.-R.; funding acquisition, A.C.-R. All authors have read and agreed to the published version of the manuscript.

Funding: This work was funded by grants CGL2014-53140-P and PGC2018-096656-B-I00 to ACR from MCIN/AEI/10.13039/501100011033 and from "ERDF A way of making Europe", by the "European Union".

Institutional Review Board Statement: Not applicable.

Informed Consent Statement: Not applicable.

Data Availability Statement: DNA sequences generated in this study have all been submitted to GenBank (https://www.ncbi.nlm.nih.gov/genbank/). Additional data and methodology are available in Appendices A and B. All the *Ischnura* specimens examined for morphological analyses and/or used for genetics are deposited in the collections of Adolfo Cordero-Rivera and Matjas Bedjanič.

Acknowledgments: We thank Matjas Bedjanič who kindly provided us access to specimens from his personal collection. We also thank Rosser W. Garrison, who helped us with the diagnostic key included in the main text and provided valuable comments that helped us to improve our manuscript.

Conflicts of Interest: The authors declare no conflict of interest. The funders had no role in the design of the study; in the collection, analyses, or interpretation of data; in the writing of the manuscript, or in the decision to publish the results.

Appendix A. Quality Control (QC) and Curation of GenBank Sequence Data

All *COI* and *ITS* sequences deposited in GenBank that were labelled as belonging to the species *Ischnura rubilio, Ischnura aurora* and *Ischnura delicata* were downloaded from the GenBank nucleotide database (https://www.ncbi.nlm.nih.gov/genbank/), through Geneious v. 9.1.8 (https://www.geneious.com). These were added to the sequences obtained by us from samples of *I. aurora* from China, Australia and Fiji and samples of *I. rubilio* from India, as described in the Materials and Methods section of the manuscript. Additionally, sequences belonging to different species within the genus *Ischnura* and other closely related genera (*Aciagrion, Ceriagrion, Coenagrion, Enallagma, Erythromma, Mortonagrion* and *Pseudagrion*) were also downloaded from GenBank and added to each dataset. Sequences were aligned in MAFFT [32] as implemented in Geneious v. 9.1.8, with a gap open penalty of 3. Alignments were visually inspected, and tails trimmed manually before phylogenetic tree reconstruction. Phylogenetic relationships were reconstructed through maximum likelihood (ML) using FastTree 2.1.11 [56] also implemented in Geneious v. 9.1.8, with the following options: use of GTR model and optimization of the 20 Gamma Likelihood.

Appendix A.1. ITS Dataset

A total of 69 sequences were included in the preliminary *ITS* dataset. After inspection of the alignment, the sequence with accession number MZ809355, belonging to an *Ischnura rubilio* specimen from India, was excluded from further analyses because it corresponded with a partial region of the 18S ribosomal RNA gene, located upstream the region that we sequenced for our study. The tree obtained after removing this sequence and trimming the alignment tails is showed in Figure A1 below. Two of the sequences included in the dataset were placed in the tree far from their expected clades. The first one was the *Ischnura aurora* sequence with GenBank accession number FN356100, which was placed outside of the *aurora* clade. Instead, this sequence appears to be closely related to *I. nursei* within the clade that includes *I. elegans, I. senegalensis, I. heterosticta, I. evansi, I. saharensis* and *I. abyssinica* (see Figure A1). The second sequence was that of *Ischnura rubilio* with GenBank accession number MH447434, which falls outside of the *Ischnura* clade, and it is placed as a sister species to *Aciagrion migratum* (see Figure A1). Both sequences were used as queries in a BLAST search against the nr database. Searches were carried out using the MegaBLAST program implemented in Geneious v. 9.1.8 with default options.

The sequence FN356100 resulted in very similar matches with other *Ischnura* ITS sequences in the nr database, but none with *I. aurora.* The sequence MH447434 resulted in matches with *Aciagrion migratum, Proischnura subfurcata* and several *Ischnura* species, but none with *I. aurora.* The results of the phylogenetic analysis and the BLAST searches suggest that the specimens from which these sequences were obtained could have been misidentified and/or misplaced at the time of DNA extraction, and hence these sequences were removed from our datasets and excluded from the final analyses.

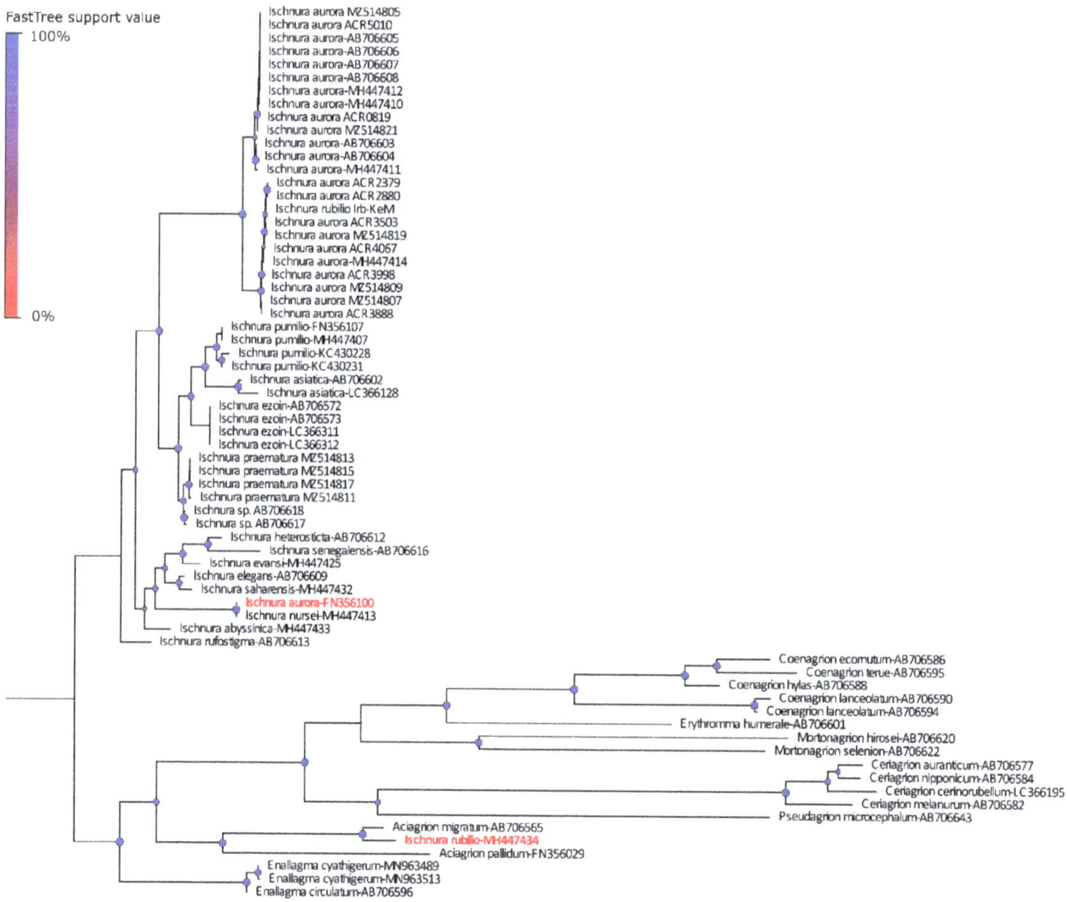

Figure A1. Maximum likelihood tree obtained with FastTree 2.1.11 for the *ITS* dataset (68 sequences; 655 bp-long). Highlighted in red are the two sequences downloaded from GenBank that were finally excluded from our dataset. The coloured dots on the nodes represent FastTree support values according to the legend on the left.

Appendix A.2. COI Dataset

The starting *COI* dataset consisted of 117 sequences and was 457 bp long. All sequences that showed ambiguities (e.g., N, M, K, Y, R . . .) were removed from the dataset: this was the case of sequences with accession numbers KY844428, KY838304, KY832433 (*Ischnura delicata*) and sequences KX053527 and KX053531 (*Ischnura aurora*). Three sequences in the alignment sequences appeared to be quite dissimilar to the rest of the sequences in the dataset, and which introduced a 6 bp-long gap in the alignment (Figure A2a). These were sequences with accession numbers EU219876, EU219877 and LC198680; all of them were labelled as *Ischnura aurora*. Translation into protein yielded nearly identical sequences without stop codons (Figure A2b).

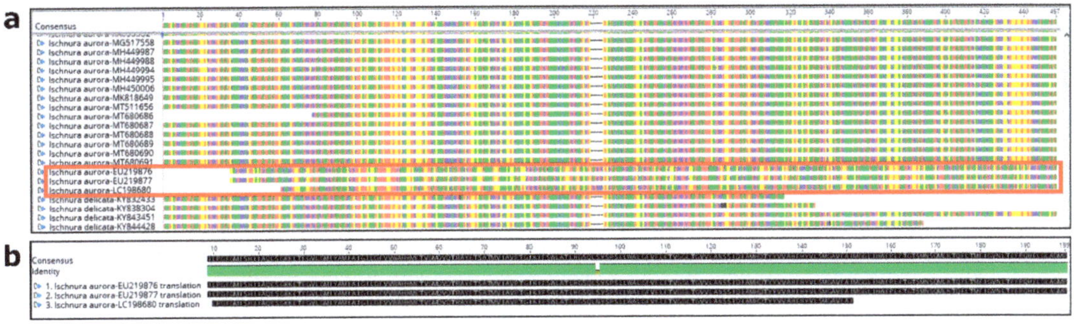

Figure A2. (**a**) Detail of the initial *COI* alignment showing the three dissimilar sequences labelled as *Ischnura aurora* (box in red). (**b**) Alignment of the amino acid sequences corresponding to these same accession number sequences.

BLAST searches were carried out also with MegaBLAST as described above using these three sequences as queries. The searches resulted in nearly the same hits for sequences EU219876 and EU219877, including several *Ceriagrion* and *Ischnura* species. The sequence LC198680 yielded matches with *Ceriagrion* and *Ischnura* species but also with Anisoptera genera such as *Anax* and *Onychogomphus*. Several of the hits obtained in these searches corresponded with sequences of *Ischnura* spp. all belonging to the same study.

The tree obtained from FastTree analysis is shown below. It can be seen how these three sequences are placed together in the same clade, at the end of a very long branch and far from the rest of the sequences included in the analysis (see Figure A3).

The alignment of these sequences with reference *COI* sequences from published complete mitochondrial genomes of *Ischnura* species (*Ischnura elegans* (MK951668), *I. senegalensis* (MT787567) and *I. pumilio* (KC878732)) revealed that they corresponded to the 3′ region of the *COI* gene, located downstream from the Folmer region that we amplified in our study (Figure A4). Therefore, these sequences were excluded from further analyses.

The *COI* sequence from *Ischnura rubilio* with accession number MH449992 falls within the *Aciagrion* clade in the FastTree ML tree (see Figure A3), similarly to the results obtained for the *ITS* sequence from this same specimen (accession number MH447434). The results of a MegaBLAST search using sequence MH449992 as a query resulted in hits with *Aciagrion*, *Ischnura pumilio* or *Ischnura elegans*, among others, which further supports our conclusion that this could be a case of misidentification and/or misplacement of individuals in the laboratory. Hence, this *COI* sequence was also removed from our dataset.

Two sequences downloaded from GenBank with accession numbers MH449981 and MW143324 labelled as *Ischnura rubilio* appear together in the ML tree, in a clade basal to *Ischnura pumilio* (see Figure A3). Both sequences can be translated into an amino acid sequence without stop codons, and both protein sequences were identical. MegaBLAST searches carried out using these two sequences as queries yielded hits with several genera of marine sponges (*Characella*, *Poecillastra* and *Theonella* among others). The sequences obtained by us from the *Ischnura rubilio* specimens from India (MB-IrbKeF, MB-IrbKeM, MB-IrbGir and MB-IrbTam; see Table 1) appear in the tree together in a clade separated from the rest of the sequences included in the analysis (see Figure A3). Translation of these sequences into the protein yielded nearly identical amino acid sequences without stop codons, and BLAST searches using these four sequences as queries returned hits against other Odonata sequences in GenBank, but the closest matches were all sequences belonging to *Nesobasis* spp., a genus in the Coenagrionidae family which is endemic to the Fiji archipelago.

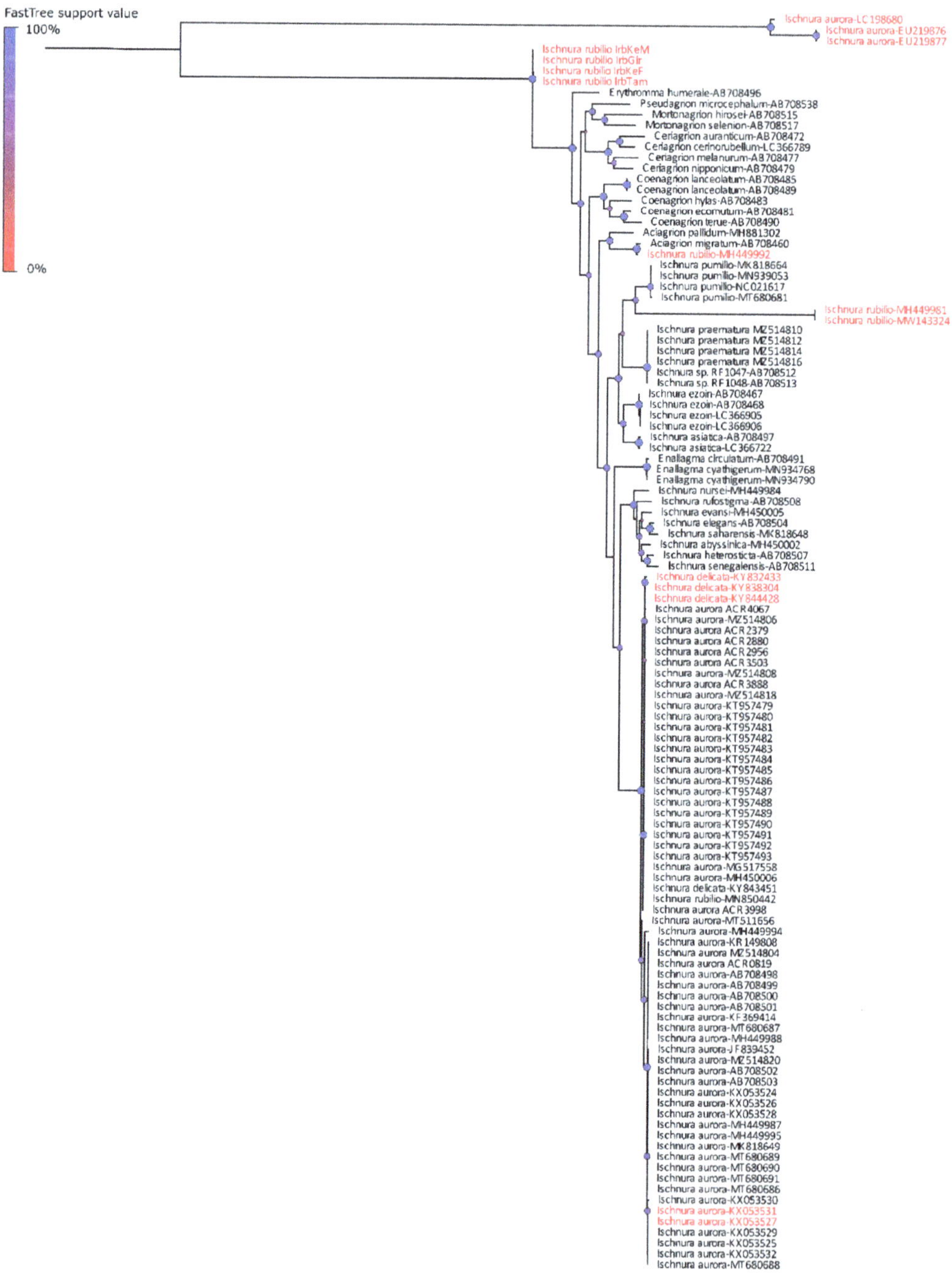

Figure A3. Maximum likelihood tree obtained with FastTree 2.1.11 for the *COI* dataset (117 sequences; 457 bp-long). The tree is rooted by the midpoint. Highlighted in red are the ten sequences downloaded from GenBank and the four sequences obtained in this study that were finally excluded from our dataset. The coloured dots on the nodes represent FastTree support values according to the legend on the left.

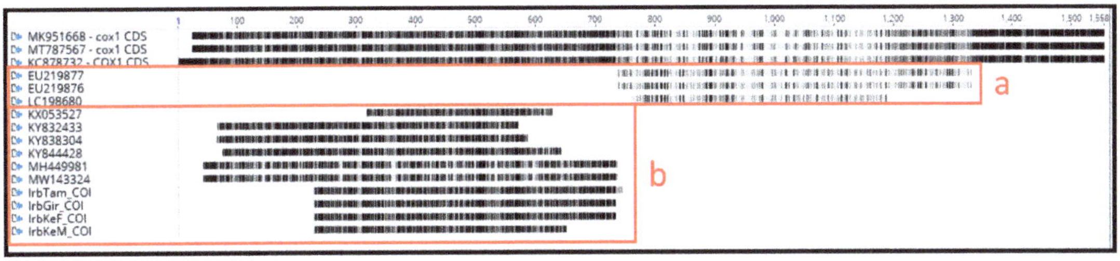

Figure A4. Alignment of sequences EU219876, EU219877 and LC198680 (box **a**) against reference *COI* sequences from *Ischnura* species (first three sequences in the alignment). Box **b** shows sequences that correspond with the Folmer region used in barcode analysis, which was the target region of the primers used in our study.

Even though the results of the BLAST searches for both sets of *I. rubilio* sequences shown mentioned above might initially lead to the conclusion that they could all be cases of sample contamination; this would in fact be quite unlikely. Regarding the *COI* sequences obtained by us from the *I. rubilio* specimens from India, none of the *Nesobasis* spp. sequences to which they are similar were obtained in our laboratory. Furthermore, the *ITS* sequence obtained from the individual MB-IrbKeM falls within the clade of *I. aurora* (see Figure A1), further supporting the fact that this is unlikely a case of contamination. In the case of sequences MH449981 and MW143324, these were produced by two independent laboratories/researchers, even at different times. Sequence MH449981 corresponds to the *I. rubilio* specimen from Kerala (India) included in the work by Dumont [23] and which was submitted to GenBank in 2018. Sequence MW143324 was submitted to GenBank in 2020 and belongs to an unpublished study on Odonata in rice ecosystems of India. Overall, the chances of getting identical (or very similar) sequences because of contamination under these circumstances are very low. Hence, we opted for a more in-depth analysis of these sequences with the aim to determine whether they should be flagged as "*COI*-like" sequences or nuclear mitochondrial copies (*numts*), rather than orthologous copies of the mitochondrial *COI* [11,31]. In fact, *numts* were recently identified in odonates of the genera *Leucorrhinia* and *Calopteryx* [13].

Examination of the chromatograms corresponding to the *I. rubilio* specimens sequenced in this study revealed the presence of some double peaks, which the program interpreted as heterozygous positions, and hence, ambiguity-coded bases in the final sequences. Additionally, the chromatograms of all these individuals showed some messy regions, which were difficult to read, and other regions that were more readable (see Figure A5), which is usually the result of the co-amplification of both *numts* and orthologous copies of the *COI* [29]. This, together with the results of the BLAST searches, suggests that these could be non-orthologous copies of the *COI* gene and, therefore, we labelled these sequences as "*COI*-like" for submission to GenBank, and excluded them from further analyses.

An examination of codon usage comparing the odd *COI* sequences of *I. rubilio* from GenBank (MH449981 and MW143324) with the apparently legitimate *COI* sequences of *I. aurora* (MT680688, MG517558 and MH449988) and the same *COI* region extracted from published complete mitochondrial genomes of other *Ischnura* species (*I. elegans* (KU958378 and MK951668), *I. pumilio* (KC878752) and *I. senegalensis* (MT787567)) is shown in Figure A6.

Amino acid composition within the selected *COI* region is identical among the reference *Ischnura* spp. included in the analysis (*I. pumilio*, *I. elegans* and *I. senegalensis*), and very similar to that of *I. aurora* (see Figure A6), whereas the two questionable sequences of *I. rubilio* show some differences in the frequency of some amino acids (e.g., F, G, I, M or S; see Figure A6). These differences in amino acid composition are not expected in a gene such as *COI*, even less in the Folmer region sequenced here, which is highly conserved across species and even across genera [11]. Even though the two *I. rubilio* sequences from GenBank showed no stop codons, the observed differences in amino acid composition between them and other *Ischnura* legitimate *COI* sequences, coupled with the fact that they appear at the end of a long branch (see Figure A3), all constitute "red flags" that make us question these as true mitochondrial *COI* sequences of *I. rubilio*, and therefore we removed them from our dataset.

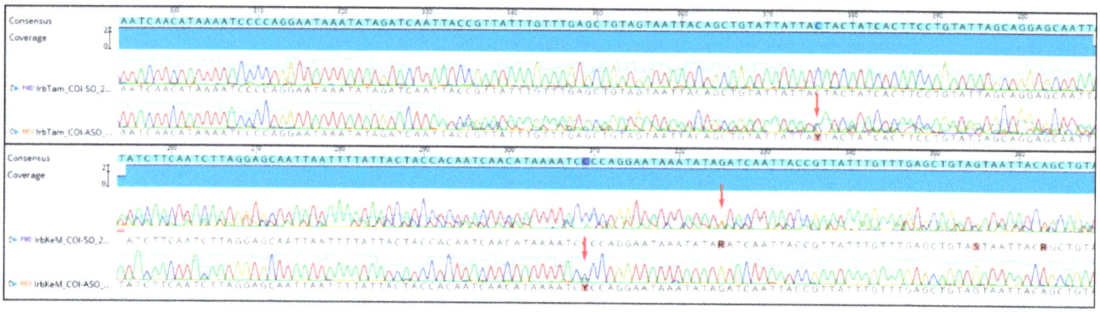

Figure A5. Screenshot of sections of two chromatograms of the *Ischnura rubilio COI* sequences obtained in this study, showing the messy but in some places readable chromatograms. The red arrows point to the double peaks which were interpreted as ambiguity-coded bases by the software.

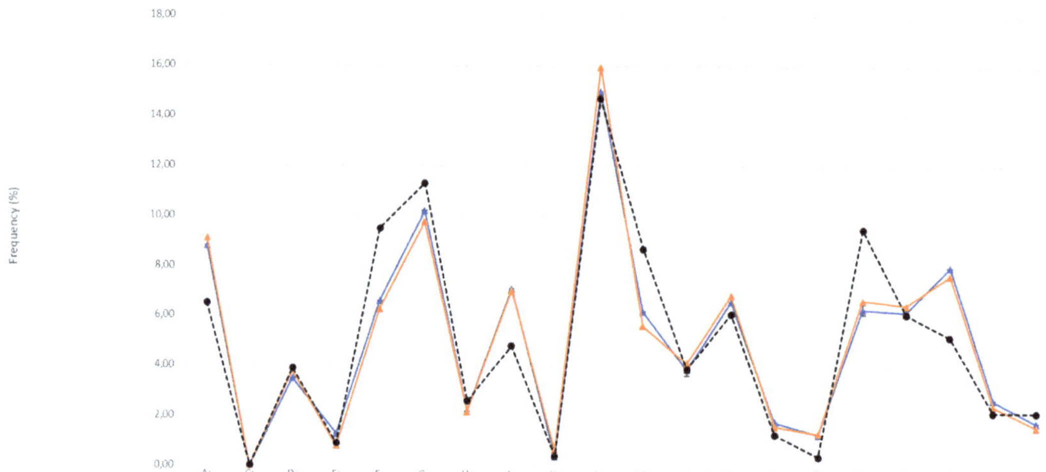

	A:	C:	D:	E:	F:	G:	H:	I:	K:	L:	M:	N:	P:	Q:	R:	S:	T:	V:	W:	Y:
Ischnura spp. mitogenomes	8,80	0,00	3,50	1,30	6,60	10,20	2,20	7,10	0,40	15,00	6,20	3,88	6,60	1,80	1,30	6,30	6,20	8,00	2,70	1,80
I. aurora	9,10	0,00	3,80	0,80	6,27	9,77	2,17	7,03	0,67	15,97	5,63	4,13	6,87	1,63	1,33	6,67	6,47	7,67	2,47	1,63
I. rubilio GenBank	6,50	0,00	3,90	0,90	9,50	11,30	2,60	4,80	0,40	14,70	8,70	3,90	6,10	1,30	0,40	9,50	6,10	5,20	2,20	2,20

Figure A6. Codon usage analysis for the questionable *COI* sequences of *Ischnura rubilio* downloaded from GenBank (MH449981 and MW143324; black dotted line with circles) and possibly true *COI* sequences of *Ischnura aurora* (orange line with triangles) and reference *COI* sequences extracted from published complete mitogenomes of several *Ischnura* species (blue line with triangles). The codon abbreviations for amino acids on the X axis are standard. The table below the X axis shows the frequency of each amino acid for each set of sequences included in the analysis.

Appendix B

Table A1. Accession numbers for the *COI* and *ITS* sequences selected for genetic analyses after quality control as described in Appendix A. For *Ischnura aurora*, *I. delicata* and *I. rubilio*, we list the information on the clade to which each individual belongs according to the genetic analyses. Highlighted in bold are the individuals that fall outside their expected clades according to the genetic and ASAP analyses (see main text and Figure A7).

Species	Clade	Collection Locality Data	GenBank Acc. Nos	
			COI	*ITS*
Aciagrion migratum	-	Yamashiro, Kyoto, Japan	AB708460	AB706565
Aciagrion pallidum	-	Thailand	MH881302	FN356029
Ceriagrion auranticum	-	Ibusuki, Kagoshima, Japan	AB708472	AB706577
Ceriagrion cerinorubellum	-	Malaysia	LC366789	LC366195
Ceriagrion melanurum	-	Kugunu, Gifu, Japan	AB708477	AB706582
Ceriagrion nipponicum	-	Suita, Osaka, Japan	AB708479	AB706584
Coenagrion ecornutum	-	Onbetsu, Hokkaido, Japan	AB708481	AB706586
Coenagrion hylas	-	Otofuke, Hokkaido, Japan	AB708483	AB706588
Coenagrion lanceolatum	-	Matsumoto, Nagano, Japan	AB708485	AB706590
C. lanceolatum	-	Abashiri, Hokkaido, Japan	AB708489	AB706594
Coenagrion terue	-	Murakami, Niigata, Japan	AB708490	AB706595
Enallagma circulatum	-	Tobetsu, Hokkaido, Japan	AB708491	AB706596
Enallagma cyathigerum	-	Bastemosen, Bornholm, Germany	MN934768	MN963489
E. cyathigerum	-	Slesvig-Holstein, Germany	MN934790	MN963513
Erythromma humerale	-	Ikeda, Hokkaido, Japan	AB708496	AB706601
Ischnura abyssinica	-	Ambo, Ethiopia	MH450002	MH447433
Ischnura asiatica	-	Fuchu, Toyama, Japan	AB708497	AB706602
I. asiatica	-	Tsukuba, Ibaraki, Japan	LC366722	LC366128
Ischnura elegans	-	Ikeda, Hokkaido, Japan	AB708504	AB706609
Ischnura evansi	-	Sarbaz Gorge, Iran	MH450005	MH447425
Ischnura ezoin	-	Anijima, Ogasawara, Tokyo, Japan	AB708467	AB706572
I. ezoin	-	Otoutojima, Ogasawara, Tokyo, Japan	AB708468	AB706573
I. ezoin	-	Mukojima, Ogasawara, Tokyo, Japan	LC366905	LC366311
I. ezoin	-	Mujkojima, Ogasawara, Tokyo, Japan	LC366906	LC366312
Ischnura heterosticta	-	Fiji	AB708507	MH447432
Ischnura nursei	-	Jaipur, India	MH449984	MH447413
Ischnura praematura	-	Yunnan, China	MZ514810	MZ514811
I. praematura	-	Yunnan, China	MZ514812	MZ514813
I. praematura	-	Yunnan, China	MZ514814	MZ514815
I. praematura	-	Yunnan, China	MZ514816	MZ514817
Ischnura pumilio	-	n.a.	MK818664	FN356107
I. pumilio	-	n.a.	MN939053	KC430228
I. pumilio	-	n.a.	MT680681	KC430231
I. pumilio	-	n.a.	NC021617	MH447407

Table A1. *Cont.*

Species	Clade	Collection Locality Data	GenBank Acc. Nos	
			COI	*ITS*
Ischnura rufostigma	-	Thailand	AB708508	AB706613
Ischnura saharensis	-	n.a.	MK818648	MH447432
Ischnura senegalensis	-	Yonaguni, Okinawa, Japan	AB708511	AB706616
Ischnura sp.	-	Yunnan, China	AB708512	AB706617
Ischnura sp.	-	Yunnan, China	AB708513	AB706618
Mortonagrion hirosei	-	Ishinomaki, Miyagi, Japan	AB708515	AB706620
M. selenion	-	Yamada, Toyama, Japan	AB708517	AB706622
Pseudagrion microcephalum	-	Yonaguni, Okinawa, Japan	AB708538	AB706643
Ischnura aurora	Australo-Pacific	New South Wales, Australia	KF369414	n.a.
Ischnura aurora	Australo-Pacific	Queensland, Australia	JF839452	n.a.
I. aurora	Australo-Pacific	Adelaide, Australia	MH449987	MH447412
I. aurora	Australo-Pacific	Perth, Australia	MH449988	MH447411
I. aurora	Australo-Pacific	Australia	MT680686	n.a.
I. aurora	Australo-Pacific	Japan	MT680687	n.a.
I. aurora	Australo-Pacific	American Samoa	MK818649	n.a.
I. aurora	Australo-Pacific	American Samoa	MT680690	n.a.
I. aurora	Australo-Pacific	Tonga	MT680689	n.a.
I. aurora	Australo-Pacific	French Polynesia	MT680688	n.a.
I. aurora	Australo-Pacific	French Polynesia	MT680691	n.a.
I. aurora	Australo-Pacific	Fiji	AB708502	AB706607
I. aurora	Australo-Pacific	Fiji	AB708503	AB706608
I. aurora	Australo-Pacific	Maroe Bay, Huahine island, French Polynesia	KX053530	n.a.
I. aurora	Australo-Pacific	Afareaitu, Moorea island, French Polynesia	KX053529	n.a.
I. aurora	Australo-Pacific	Paopao river, Moorea island, French Polynesia	KX053532	n.a.
I. aurora	Australo-Pacific	Paopao river, Moorea island, French Polynesia	KX053524	n.a.
I. aurora	Australo-Pacific	Pihaena, Moorea island, French Polynesia	KX053528	n.a.
I. aurora	Australo-Pacific	Mount Mauru, Tahiti island, French Polynesia	KX053526	n.a.
I. aurora	Australo-Pacific	Mount Mauru, Tahiti island, French Polynesia	KX053525	n.a.
I. aurora	Australo-Pacific	Wallis and Futuna	n.a.	MH447410
I. aurora	Australo-Pacific	Guam	AB708500	AB706605
I. aurora	Australo-Pacific	Guam	AB708501	AB706606
I. aurora	Australo-Pacific	Iojima, Ogasawara, Tokyo, Japan	AB708498	AB706603
I. aurora	Australo-Pacific	Iojima, Ogasawara, Tokyo, Japan	AB708499	AB706604
I. aurora	Australo-Pacific	Baliem valley New Guinea	MH449995	n.a.
I. aurora	**Australo-Pacific**	**Malappuram, Kerala, India**	**KR149808**	**n.a.**
I. aurora	Asian	Ugani Sahib, Rajpura, Patial (Punjab), India	MG517558	n.a.
I. aurora	Asian	Nakhon Sawan, Thailand	MH450006	MH447414

Table A1. *Cont.*

Species	Clade	Collection Locality Data	GenBank Acc. Nos	
			COI	*ITS*
I. aurora	Asian	Thailand	KT957479	n.a.
I. aurora	Asian	Thailand	KT957480	n.a.
I. aurora	Asian	Thailand	KT957481	n.a.
I. aurora	Asian	Thailand	KT957482	n.a.
I. aurora	Asian	Thailand	KT957483	n.a.
I. aurora	Asian	Thailand	KT957484	n.a.
I. aurora	Asian	Thailand	KT957485	n.a.
I. aurora	Asian	Thailand	KT957486	n.a.
I. aurora	Asian	Thailand	KT957487	n.a.
I. aurora	Asian	Thailand	KT957488	n.a.
I. aurora	Asian	Thailand	KT957489	n.a.
I. aurora	Asian	Thailand	KT957490	n.a.
I. aurora	Asian	Thailand	KT957491	n.a.
I. aurora	Asian	Thailand	KT957492	n.a.
I. aurora	Asian	Thailand	KT957493	n.a.
Ischnura delicata	Asian	Islamabad, Pakistan	KY843451	n.a.
Ischnura rubilio	Asian	India	MN850442	n.a.
I. aurora	**3rd taxonomic unit in ASAP**	**Baliem Valley New Guinea**	**MH449994**	**n.a.**
I. aurora	**Asian? 4th taxonomic unit in ASAP**	**India**	**MT511656**	**n.a.**

Table A2. List of *Ischnura aurora* material belonging to Adolfo Cordero-Rivera's personal collection used for morphological examination (see main text). Listed are the voucher ID, sex and collection details for each specimen. Collector's names are as follows: Adolfo Cordero-Rivera (ACR), Iago Sanmartín-Villar (ISV) and Haomiao Zhang (HZ).

Voucher ID	Sex	Collection Date	Collection Locality
ACR-00738	male	01/12/2013	Long Swamp; Nelson, Victoria, Australia. ACR leg and det.
ACR-00776	female	02/12/2013	Ming Ming Swamp, Grampians National Park, Victoria, Australia. ACR leg and det.
ACR-00793	female	04/12/2013	Ming Ming Swamp, Grampians National Park, Victoria, Australia. ACR leg and det.
ACR-00794	female	04/12/2013	Ming Ming Swamp, Grampians National Park, Victoria, Australia. ACR leg and det.
ACR-00795	female	04/12/2013	Ming Ming Swamp, Grampians National Park, Victoria, Australia. ACR leg and det.
ACR-00818	male	09/12/2013	pond at Bandiana, Wodonga, Victoria, Australia. ACR leg and det.

Table A2. *Cont.*

Voucher ID	Sex	Collection Date	Collection Locality
ACR-00819	male	09/12/2013	pond at Bandiana, Wodonga, Victoria, Australia. ACR leg and det.
ACR-02338	male	08/05/2015	River at Si Fang Jing, Yunnan, China. ISV leg and det.
ACR-02379	male	13/05/2015	River at Xi Meng, Yunnan, China. ISV leg and det.
ACR-02880	male	19/06/2015	Meng Ding, Yunnan China. ISV leg and det.
ACR-02956	male	26/06/2015	Na Bang, Yunnan, China. ISV leg and det.
ACR-02957	female	26/06/2015	Na Bang, Yunnan, China. ISV leg and det.
ACR-03503	male	02/07/2015	Rice field at Huaping. Yunnan, China. ISV leg and det.
ACR-03504	male	02/07/2015	Rice field at Huaping. Yunnan, China. ISV leg and det.
ACR-03505	male	02/07/2015	Rice field at Huaping. Yunnan, China. ISV leg and det.
ACR-03506	male	02/07/2015	Rice field at Huaping. Yunnan, China. ISV leg and det.
ACR-03507	female	02/07/2015	Rice field at Huaping. Yunnan, China. ISV leg and det.
ACR-03508	female	02/07/2015	Rice field at Huaping. Yunnan, China. ISV leg and det.
ACR-03509	female	02/07/2015	Rice field at Huaping. Yunnan, China. ISV leg and det.
ACR-03510	female	02/07/2015	Rice field at Huaping. Yunnan, China. ISV leg and det.
ACR-03511	female	02/07/2015	Rice field at Huaping. Yunnan, China. ISV leg and det.
ACR-03513	female	02/07/2015	Rice field at Huaping. Yunnan, China. ISV leg and det.
ACR-03888	male	10/06/2016	Pond in agricultural area. Mengding, Yunnan, China. ACR leg and det.
ACR-03917	female	11/06/2016	Pond in agricultural area. Mengding, Yunnan, China. HZ leg and det.
ACR-03998	female	19/06/2016	River at Meng Lun, Yunnan, China. ACR leg and det.
ACR-04067	male	24/06/2016	Stream at Meng Lung, Yunnan, China. ACR leg and det.
ACR-05007	male	06/06/2018	Somosomo damm, Chakaudrove, Taveuni, Fiji.
ACR-05008	female	06/06/2018	Somosomo damm, Chakaudrove, Taveuni, Fiji.
ACR-05009	male	06/06/2018	Somosomo damm, Chakaudrove, Taveuni, Fiji.
ACR-05010	male	06/06/2018	Somosomo damm, Chakaudrove, Taveuni, Fiji.
ACR-05091	female	11/06/2018	Korovuli, Seqaqa, Labasa, Vanua Levu, Fiji.

Table A3. Evolutionary divergence (uncorrected p-distances) estimated with MEGA X from nuclear (*ITS*, above) and mitochondrial (*COI*, below) DNA sequence data.

	Australo-Pacific *aurora*	**Asian *aurora***	***I. delicata***	***I. rubilio***	***I. aurora* MH449994**	***I. aurora* MT511656**	***I. abyssinica***	***I. asiatica***	***I. elegans***	***I. evansi***	***I. ezoin***	***I. heterosticta***	***I. nursei***	***I. praematura***	***I. pumilio***	***I. rufostigma***	***I. saharensis***	***I. senegalensis***
Australo-Pacific *aurora*		0.020	n.a.	0.017	n.a.	n.a.	0.081	0.094	0.088	0.101	0.082	0.095	0.116	0.069	0.083	0.087	0.091	0.109
Asian *aurora*	0.031		n.a.	0.001	n.a.	n.a.	0.084	0.094	0.090	0.104	0.079	0.094	0.110	0.070	0.081	0.081	0.097	0.106
I. delicata	0.031	0.000			n.a.	n.a.	n.a.	n.a.	n.a.	n.a.	n.a.	n.a.	n.a.	n.a.	n.a.	n.a.	n.a.	n.a.
I. rubilio	0.031	0.000	0.000		n.a.	n.a.	0.041	0.052	0.047	0.052	0.047	0.052	0.064	0.050	0.052	0.041	0.047	0.058
I. aurora MH449994	0.024	0.027	0.027	0.027		n.a.	n.a.	n.a.	n.a.	n.a.	n.a.	n.a.	n.a.	n.a.	n.a.	n.a.	n.a.	n.a.
I. aurora MT511656	0.022	0.013	0.013	0.013	0.022		n.a.	n.a.	n.a.	n.a.	n.a.	n.a.	n.a.	n.a.	n.a.	n.a.	n.a.	n.a.
I. abyssinica	0.135	0.126	0.126	0.126	0.131	0.131		0.079	0.036	0.045	0.062	0.057	0.074	0.054	0.069	0.042	0.042	0.066
I. asiatica	0.132	0.128	0.127	0.127	0.132	0.123	0.150		0.079	0.094	0.044	0.084	0.103	0.042	0.040	0.075	0.083	0.101
I. elegans	0.138	0.129	0.129	0.129	0.133	0.124	0.064	0.145		0.036	0.066	0.043	0.068	0.058	0.065	0.053	0.014	0.055
I. evansi	0.131	0.126	0.126	0.126	0.131	0.124	0.058	0.141	0.058		0.079	0.034	0.073	0.065	0.080	0.051	0.038	0.057
I. ezoin	0.130	0.125	0.125	0.125	0.136	0.116	0.149	0.089	0.147	0.145		0.077	0.086	0.026	0.033	0.064	0.076	0.090
I. heterosticta	0.129	0.120	0.120	0.120	0.122	0.120	0.058	0.145	0.069	0.064	0.132		0.074	0.068	0.079	0.060	0.045	0.040
I. nursei	0.134	0.130	0.130	0.130	0.132	0.128	0.087	0.158	0.078	0.070	0.159	0.083		0.080	0.092	0.076	0.075	0.083
I. praematura	0.148	0.144	0.144	0.144	0.144	0.144	0.140	0.112	0.156	0.142	0.118	0.144	0.153		0.030	0.055	0.064	0.083
I. pumilio	0.150	0.154	0.154	0.154	0.161	0.154	0.160	0.125	0.147	0.148	0.115	0.154	0.148	0.129		0.070	0.075	0.092
I. rufostigma	0.122	0.113	0.113	0.113	0.118	0.109	0.078	0.127	0.069	0.071	0.136	0.078	0.090	0.138	0.154		0.056	0.073
I. saharensis	0.134	0.137	0.137	0.137	0.137	0.137	0.069	0.154	0.035	0.062	0.155	0.080	0.094	0.149	0.147	0.075		0.061
I. senegalensis	0.131	0.133	0.133	0.133	0.129	0.133	0.069	0.145	0.084	0.078	0.154	0.051	0.096	0.136	0.174	0.082	0.086	

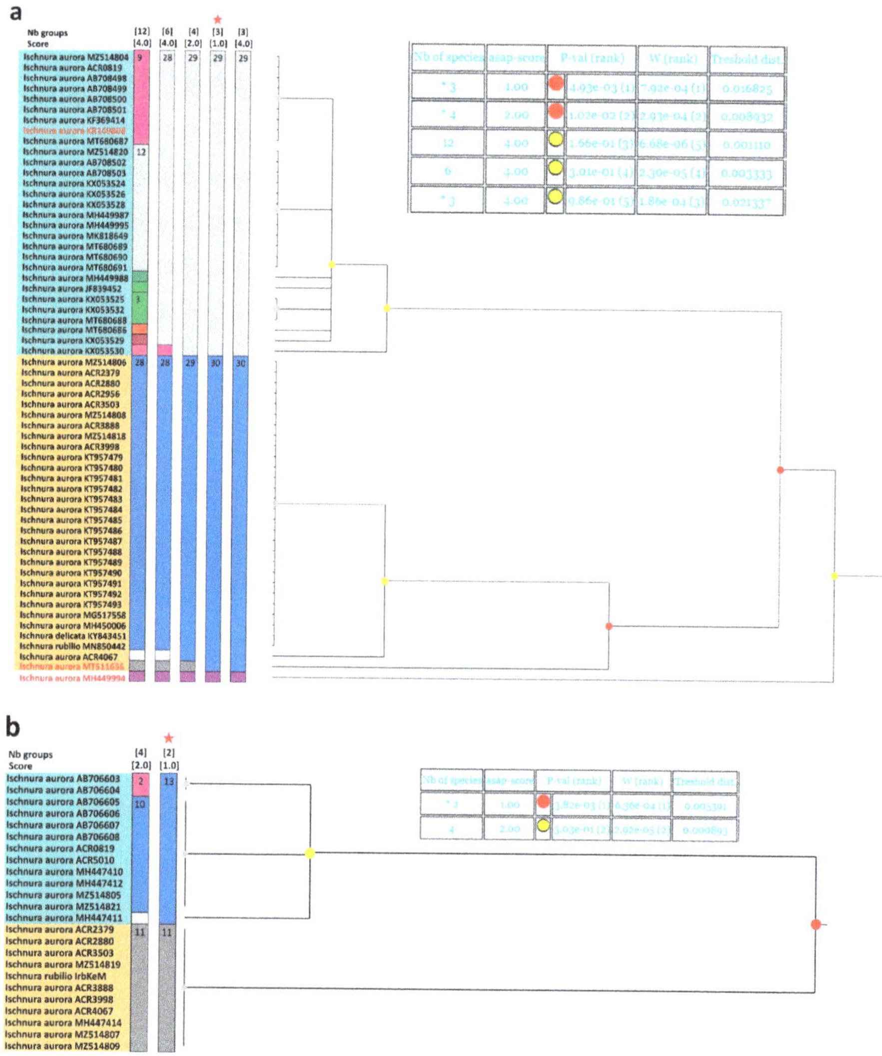

Figure A7. Graphical output of ASAP, showing the results of the species delimitation analysis using the *Ischnura aurora*; *I. delicata* and *I. rubilio* COI sequences and (**b**) the *I. aurora* and *I. rubilio* ITS sequences. Each node of the dendrogram is coloured depending on its probability of being a panmictic species: darker colours for nodes that may be split into smaller groups. The tables show the best partitions found by ASAP in each case, scored and sorted by their *p*-value (the smallest p-value has rank 1) and their rank of relative barcode gap width (the largest gap has rank 1). The asap-score is the average of both ranks: the smaller the asap-score, the better. The predicted number of groups in each partition, together with their asap-score, are also indicated in the dendrograms, with the red star indicating the best partition according to the analysis in each case. Australo-Pacific and Asian clades are highlighted with the same colours used in Figure 2 in the main text. In (**a**), the individual highlighted in red inside the Australo-Pacific clade corresponds to a specimen of *I. aurora* from India. The individuals highlighted in red at the bottom of the dendrogram correspond to the specimens of *I. aurora* from India and New Guinea with COI sequences that are ~2% divergent from the rest of the sequences included in the analyses and identified as belonging to a different species in the ASAP analysis.

References

1. Wheeler, Q. Taxonomic triage and the poverty of phylogeny. *Philos. Trans. R. Soc. Lond. B* **2004**, *359*, 571–583. [CrossRef] [PubMed]
2. Wilson, E.O. Taxonomy as a fundamental discipline. *Philos. Trans. R. Soc. Lond. B* **2004**, *359*, 739. [CrossRef] [PubMed]
3. Padial, J.M.; Miralles, A.; De la Riva, I.; Vences, M. The integrative future of taxonomy. *Front. Zool.* **2010**, *7*, 16. [CrossRef] [PubMed]
4. Dayrat, B. Towards integrative taxonomy. *Biol. J. Linn. Soc.* **2005**, *85*, 407–415. [CrossRef]
5. Hebert, P.D.; Cywinska, A.; Ball, S.L.; Dewaard, J. Biological identification through DNA barcodes. *Proc. R. Soc. B Biol. Sci.* **2003**, *270*, 313–321. [CrossRef]
6. Tautz, D.; Arctander, P.; Minelli, A.; Thomas, R.H.; Vogler, A.P. A plea for DNA taxonomy. *Trends in Ecology and Evolution* **2003**, *18*, 70–74. [CrossRef]
7. Baker, S.; Dalebout, M.L.; Lavery, S.; Ross, H.A. www.DNA-surveillance: Applied molecular taxonomy for species conservation and discovery. *Trends Ecol. Evol.* **2003**, *18*, 271–272. [CrossRef]
8. Moritz, C.; Cicero, C. DNA barcoding: Promise and pitfalls. *PLoS Biol.* **2004**, *2*, e354. [CrossRef]
9. Bachtrog, D.; Andolfatto, P. Selection, recombination and demographic history in *Drosophila miranda*. *Genetics* **2006**, *174*, 2045–2059. [CrossRef]
10. Rubinoff, D. Utility of mitochondrial DNA barcodes in species conservation. *Conserv. Biol.* **2006**, *20*, 1026–1033. [CrossRef]
11. Song, H.; Buhay, J.E.; Whiting, M.F.; Crandall, K.A. Many species in one: DNA barcoding overestimates the number of species when nuclear mitochondrial pseudogenes are coamplified. *Proc. Natl. Acad. Sci. USA* **2008**, *105*, 13486–13491. [CrossRef] [PubMed]
12. Papakostas, S.; Michaloudi, E.; Proios, K.; Brehm, M.; Verhage, L.; Rota, J.; Peña, C.; Stamou, G.; Pritchard, V.L.; Fontaneto, D.; et al. Integrative taxonomy recognizes evolutionary units despite widespread mitonuclear discordance: Evidence from a rotifer cryptic species complex. *Syst. Biol.* **2016**, *65*, 508–524. [CrossRef] [PubMed]
13. Ožana, S.; Dolný, A.; Pánek, T. Nuclear copies of mitochondrial DNA as a potential problem for phylogenetic and population genetic studies of Odonata. *Syst. Entomol.* **2022**, 1–12. [CrossRef]
14. Goulding, T.C.; Dayrat, B. Integrative taxonomy: Ten years of practice and looking into the future. *Arch. Zool. Mus. Lomonosov Mosc. State Univ.* **2016**, *54*, 116–133.
15. Carvalho, F.G.; Duarte, L.S.; Seger, G.D.S.; Nakamura, G.; Guillermo-Ferreira, R.; Cordero-Rivera, A.; Juen, L. Detecting Darwinian shortfalls in the Amazonian Odonata. *Neotrop. Entomol.* 2022, *in press*. [CrossRef]
16. Sanmartín-Villar, I.; Zhang, H.; Cordero-Rivera, A. Colour polymorphism and ontogenetic colour changes in *Ischnura rufostigma* (Odonata: Coenagrionidae). *Odonatologica* **2016**, *45*, 77–86. [CrossRef]
17. Sanmartín-Villar, I.; Lorenzo-Carballa, M.O.; Zhang, H.; Cordero-Rivera, A. *Ischnura praematura* sp. nov. (Odonata: Zygoptera: Coenagrionidae): A species from Yunnan (China) whose females mate in the teneral state. *Zootaxa* **2022**, *5087*, 59–74. [CrossRef]
18. Sánchez-Guillén, R.; Ceccarelli, S.F.; Villalobos, F.; Neupane, S.; Rivas-Torres, A.; Sanmartín-Villar, I.; Wellenreuther, M.; Bybee, S.M.; Velásquez-Vélez, M.I.; Realpe, E.; et al. The evolutionary history of colour polymorphism in *Ischnura* damselflies (Odonata: Coenagrionidae). *Odonatologica* **2020**, *49*, 333–370. [CrossRef]
19. Blow, R.; Willink, B.; Svensson, E.I. A molecular phylogeny of forktail damselflies (genus Ischnura) reveals a dynamic macroevolutionary history of female colour polymorphisms. *Mol. Phylogenetics Evol.* **2021**, *160*, 107–134. [CrossRef]
20. O'Farrell, A.F. Odonata. In *The Insects of Australia*; Mackerras, I.M., Ed.; Melbourne University Press: Melbourne, Australia, 1973; pp. 241–261.
21. Corbet, P.S. *Dragonflies: Behaviour and Ecology of Odonata*; Harley Books: Colchester, UK, 1999; p. 882. ISBN 0-946-58964-X.
22. Dow, R.A.; Rowe, R.; Marinov, M. *Ischnura aurora*. The IUCN Red List of Threatened Species 2020: E.T167375A83371053. 2020. Available online: https://www.iucnredlist.org/species/167375/83371053 (accessed on 2 November 2021).
23. Dumont, H.J. Phylogeny of the genus *Ischnura*, with emphasis on the old world taxa (Zygoptera: Coenagrionidae). *Odonatologica* **2013**, *42*, 301–308.
24. Hagen, H.A. Synopsis der Neuroptera Ceylons [Pars I]. *Verhandlungen der Kaiserlich-Königlichen Zoologisch-Botanischen Gesellschaft in Wien* **1858**, *8*, 471–488.
25. Sélys-Longchamps, M.E. Synopsis des Agrionines, 5me légion: *Agrion* (suite). *Bulletins de l'Académie Royale des Sciences, des Lettres et des Beaux-Arts de Belgique* **1876**, *41*, 247–322.
26. Brauer, F. Dritter Bericht über die auf der Weltfahrt der kais. Fregatte Novara gesammelten Libellulinen. *Verhandlungen der K. K. Zoologisch-Botanischen Gesellschaft* **1865**, *15*, 501–512.
27. Papazian, M.; Dumont, H.J.; Mary-Sasal, N.J. The Odonata of the Pacific Ocean Islands of Wallis and Futuna, with special reference to speciation in *Ischnura aurora* (Brauer). *Odonatologica* **2007**, *36*, 53–62.
28. Kalkman, V.J.; Babu, R.; Bedjanič, M.; Connif, K.; Gyeltshen, T.; Khan, M.K.; Subramanian, K.A.; Zia, A.; Orr, A.G. Checklist of the dragonflies and damselflies (Insecta: Odonata) of Bangladesh, Bhutan, India, Nepal, Pakistan and Sri Lanka. *Zootaxa* **2020**, *4849*, 1–84. [CrossRef]
29. Rowe, R.J. *Ischnura aurora* (Brauer 1865) (Zygoptera: Coenagrionidae), an Australo-Pacific species. *N. Z. J. Zool.* **2010**, *37*, 189–192. [CrossRef]
30. Futahashi, R.; Okude, G.; Sugimura, M.; Ugai, S. Interspecific hybrids in Japanese Odonata. *Tombo* **2018**, *60*, 1–49.

31. Buhay, J.E. "COI-like" sequences are becoming problematic in molecular systematic and DNA barcoding studies. *J. Crustacean Biol.* **2009**, *29*, 96–110. [CrossRef]
32. Katoh, K.; Standley, D.M. MAFFT multiple sequence alignment software version 7: Improvements in performance and usability *Mol. Biol. Evol.* **2013**, *30*, 772–780. [CrossRef]
33. Nguyen, L.T.; Schmidt, H.A.; von Haeseler, A.; Minh, B.Q. IQ-TREE: A fast and effective stochastic algorithm for estimating maximum likelihood phylogenies. *Mol. Biol. Evol.* **2015**, *32*, 268–274. [CrossRef]
34. Kalyaanamoorthy, S.; Minh, B.Q.; Wong, T.K.F.; von Haeseler, A.; Jermiin, L.S. ModelFinder: Fast model selection for accurate phylogenetic estimates. *Nat. Methods* **2017**, *14*, 587–589. [CrossRef] [PubMed]
35. Minh, B.Q.; Nguyen, M.A.T.; von Haeseler, A. Ultrafast approximation for phylogenetic bootstrap. *Mol. Biol. Evol.* **2013**, *30*, 1188–1195. [CrossRef] [PubMed]
36. Hoang, D.T.; Chernomor, O.; von Haeseler, A.; Minh, B.Q.; Vinh, L.S. UFBoot2: Improving the ultrafast bootstrap approximation *Mol. Biol. Evol.* **2018**, *35*, 518–522. [CrossRef] [PubMed]
37. Huelsenbeck, J.P.; Ronquist, F. MRBAYES: Bayesian inference of phylogeny. *Bioinformatics* **2001**, *17*, 754–755. [CrossRef] [PubMed]
38. Ronquist, F.; Huelsenbeck, J.P. MRBAYES 3: Bayesian phylogenetic inference under mixed models. *Bioinformatics* **2003**, *19*, 1572–1574. [CrossRef] [PubMed]
39. Inkscape. Inkscape Project. 2020. Available online: https://inkscape.org (accessed on 25 June 2022).
40. Kumar, S.; Stecher, G.; Li, M.; Knyaz, C.; Tamura, K. MEGA X: Molecular Evolutionary Genetics Analysis across computing platforms. *Mol. Biol. Evol.* **2018**, *35*, 1547–1549. [CrossRef] [PubMed]
41. Puillandre, N.; Brouillet, S.; Achaz, G. ASAP: Assemble species by automatic partitioning. *Mol. Ecol. Resour.* **2021**, *21*, 609–620. [CrossRef]
42. The GIMP Development Team. GIMP. 2019. Available online: https://www.gimp.org (accessed on 25 June 2022).
43. Garrison, R.W.; von Ellenrieder, N.; Louton, J.A. *Damselfly Genera of the New World*; Johns Hopkins University Press: Baltimore, MD, USA, 2010; p. 490.
44. Nolan, L.; Hogg, I.D.; Sutherland, D.L.; Stevens, M.I.; Schnabel, K.E. Allozyme and mitochondrial DNA variability within the New Zealand damselfly genera *Xanthocnemis*, *Austrolestes*, and *Ischnura* (Odonata). *N. Z. J. Zool.* **2007**, *34*, 371–380. [CrossRef]
45. Ramage, T.; Martins-Simoes, P.; Mialdea, G.; Allemand, R.; Duplouy, A.; Rousse, P.; Davies, N.; Roderick, G.K.; Charlat, S. A DNA barcode-based survey of terrestrial arthropods in the Society Islands of French Polynesia: Host diversity within the SymbioCode Project. *Eur. J. Taxon.* **2017**, *272*, 1–13. [CrossRef]
46. Dumont, H.J.; Vierstraete, A.; Vanfleteren, J.R. A molecular phylogeny of the Odonata (Insecta). *Syst. Entomol.* **2010**, *35*, 6–18. [CrossRef]
47. De Knijf, G.; Sparrow, D.; Dimitriou, A.; Kent, R.; Kent, H.; Siedle, K.; Lewis, J.; Crossley, L. Distribution, ecology and status of a threatened species *Ischnura intermedia* (Insecta: Odonata), new for Europe. *Int. J. Odonatol.* **2016**, *19*, 257–274. [CrossRef]
48. Zhang, H.; Ning, X.; Yu, X.; Bu, W.-J. Integrative species delimitation based on *COI*, *ITS*, and morphological evidence illustrates a unique evolutionary history of the genus *Paracercion* (Odonata: Coenagrionidae). *PeerJ* **2021**, *9*, e11459. [CrossRef] [PubMed]
49. Cordero-Rivera, A. Ciclomorfosis y fenología en *Ischnura graellsii* Rambur, 1842 (Odonata: Coenagrionidae). *Actas II Congr. Ibérico Entomol.* 1998; 419–430.
50. Ribeiro Leite, L.A. Mitochondrial pseudogenes in insect DNA barcoding: Differing points of view on the same issue. *Biota Neotrop.* **2012**, *12*, 301–308. [CrossRef]
51. Galindo-Ruiz, N.; Velasquez-Velez, M.I.; Cano-Cobos, Y.; Sánchez-Guillén, S.A.; Realpe, E. Description of a putative hybrid between *Ischnura cyane* and *I. capreolus* from Colombia (Odonata: Coenagrionidae). *Not. Odonatol.* **2019**, *9*, 144–151. [CrossRef]
52. Monetti, L.; Sánchez Guillén, R.A.; Cordero-Rivera, A. Hybridization between *Ischnura graellsii* (Vander Linder) and *I. elegans* (Rambur) (Odonata: Coenagrionidae): Are they different species? *Biol. J. Linn. Soc.* **2002**, *76*, 225–235. [CrossRef]
53. Rowe, R.J. *Ischnura aurora* (Brauer), a dragonfly with unusual mating behaviour (Zygoptera: Coenagrionidae). *Odonatologica* **1978**, *7*, 375–383.
54. Sánchez-Guillén, R.A.; Wellenreuther, M.; Cordero-Rivera, A. Strong asymmetry in the relative strengths of prezygotic and postzygotic barriers between two damselfly sister species. *Evolution* **2012**, *66*, 690–707. [CrossRef]
55. Lieftinck, M.A. The dragonflies (Odonata) of New Guinea and neighbouring islands. Part VII. Results of the Third Archbold expedition 1938-1939 and of the Le Roux Expedition 1939 to Netherlands New Guinea (II. Zygoptera). *Nova Guin.* **1949**, *5*, 1–271.
56. Price, M.N.; Dehal, P.S.; Arkin, A.P. FastTree 2-Approximately Maximum-Likelihood Trees for Large Alignments. *PLoS ONE* **2010**, *5*, e9490. [CrossRef]

diversity

MDPI

Article

DNA Barcoding and New Records of Odonates (Insecta: Odonata) from Paraíba State, Brazil

Ricardo Koroiva [1,2,*], Vanessa Gabrielle Nóbrega Gomes [2] and Diogo Silva Vilela [3]

[1] Departamento de Sistemática e Ecologia, Universidade Federal da Paraíba, João Pessoa 58051-900, Paraíba, Brazil
[2] Núcleo de Biodiversidade, Instituto Nacional do Semiárido, Campina Grande 58434-700, Paraíba, Brazil; vnobrega.gomes@gmail.com
[3] Departamento de Ciências Biológicas, Faculdade de Ciências e Letras de Assis, Universidade Estadual Paulista Júlio de Mesquita Filho, São Paulo 19806-900, Brazil; deeogoo@gmail.com
* Correspondence: ricardo.koroiva@insa.gov.br

Abstract: Odonates (Insecta: Odonata) are important insects in the food chains of freshwater environments around the world, being used as a model species for areas of behavior and analysis of environmental quality. In Brazil, especially in the Northeastern region, both knowledge about the distribution and molecular information of odonate species found in the two main biomes of the region is still limited. Aiming to improve these issues, here, we carried out an Odonata survey in two locations and built a DNA barcode database for species from the state of Paraíba. In total, 15 first records were reported for this Brazilian state and 142 specimens from 27 genera and 45 species had their 'Folmer' cytochrome c oxidase subunit I (COI) fragment evaluated. The database we generated includes data for 70% of the Odonata species found in Paraíba state. For 16 species, this is the first DNA barcode available in public sequence repositories. Our results demonstrate that using the COI in the regional scale can help identify and delimit those evaluated. Eight species (17%) showed a low percentage of differentiation (<2%) compared to other species currently deposited in the GenBank or BOLD System; nevertheless, we present morphological traits that reaffirm our identifications. Barcode data provide new insights into Neotropical diversity and deliver basic information for taxonomic analyses.

Keywords: dragonfly; damselflies; DNA barcode; Brazilian northeast

Citation: Koroiva, R.; Gomes, V.G.N.; Vilela, D.S. DNA Barcoding and New Records of Odonates (Insecta: Odonata) from Paraíba State, Brazil. *Diversity* **2022**, *14*, 203. https://doi.org/10.3390/d14030203

Academic Editor: Luc Legal

Received: 21 February 2022
Accepted: 8 March 2022
Published: 10 March 2022

1. Introduction

Odonates (Insecta: Odonata) are a group of fascinating insects that can be found all over the world (except for Antarctica) [1]. They have been used as model animals in several areas, such as behavioral studies and the analysis of environmental quality (e.g., [2–4]). This group plays an important role in the trophic network of freshwater environments, both as an efficient predator of invertebrates and as a prey for several vertebrates [5]. Over the last decade, scientists have increasingly discovered new species, which has improved the taxonomic knowledge of odonates in the Neotropical realm; however, information on their taxonomy ('Linnean' shortfall), distribution ('Wallacean' shortfall) and genetic diversity remains scarce (see [6,7]).

The traditional identification of odonates based on their phenotypic traits is often assumed to be difficult, considering that the older depictions are based on brief descriptions and rare illustrations (see examples in [8]). Even the most recent definitions present limited information to facilitate the identification of both larvae and adults. In general, in adults, anal appendages, wing venation and genitalia are often used to identify and classify odonates [9,10]. Despite the possibility of identifying some species at the larval level, the number of this type of descriptions is still scarce, especially in the Neotropical region [9].

Since Hebert and colleagues [11,12] suggested using cytochrome oxidase subunit I (COI) sequences as a global bioidentification system for animals in 2003, there has been great progress in using it for species identification and species discovery of odonates. This system, which is called DNA barcoding, has proven to be highly effective for delimiting and identifying different groups (e.g., [13,14]). As a result of an initiative by researchers at the University of Guelph (Ontario, Canada), a worldwide database has been maintained for the deposit and public identification of sequences; this database is called the BOLD system (the DNA Barcode of Life Data System) [15]. Among the advantages of this system compared with other public sequence repositories (e.g., GenBank) is that it is easy to access and download specimen data and sequences. For Neotropical odonates, the use of DNA barcoding is still limited, and there has been some discussion regarding its effectiveness by Koroiva et al. [16]. Moreover, Vilela et al. [17,18] have substantiated its direct application through integrative taxonomy, i.e., the framework to delimit and describe taxa by integrating information from multiple and complementary perspectives [19], using molecular and morphological data.

Currently, there are about 872 species of odonates in Brazil [20]. Several biomes can be found in this country, including the Atlantic Forest and 'Caatinga'. The former is considered a biodiversity hotspot, and it is one of the most threatened tropical forests in the world [21]. Meanwhile, the latter is the only biome exclusive to Brazil. It is characterized by a semi-arid climate and has been recognized as an affected biome as a result of anthropic actions, such as free-living livestock and fuelwood extraction [22,23]. These two biomes are present in the state of Paraíba, a small territory state localized in Northeastern Brazil. The region's list of Odonata species was recently published by Koroiva et al. [8]. In this work, the authors identified 49 species living in the region; this provides a good sampling of the diversity of the state, despite the few collections present in the Atlantic Forest. Nonetheless, it is essential to elaborate upon a broad and diverse species collection in order to realize representative genetic databases. In turn, using them to construct a reference database is the next step to enable the use of new molecular tools (e.g., DNA metabarcoding) in this region.

Keeping in mind the difficulties in the morphological identification of dragonflies, molecular tools present a promising way to solve this impediment in Neotropical species [16]. In this study, we present a DNA barcode library for the Odonata identification in the Paraíba state. This work aims (i) to improve the information on the species present in the Atlantic Forest of Paraíba; (ii) to establish DNA barcode libraries for the odonatofauna of Paraíba based on the COI gene and (iii) to evaluate the accuracy of the DNA barcodes in defining species in relation to both the regional and global databases.

2. Materials and Methods

2.1. Ethics Statement

This study was conducted with the appropriate permission (SISBIO license number 74324-5 and JBBM license number 002/2021/JBBM/SUDEMA).

2.2. Data Collection

Between May 2020 and November 2021, 740 specimens were collected from 10 municipalities in the state of Paraíba. Information about climate classification, precipitation, vegetation types and the geology of the sampling area are available in the work of Koroiva et al. [8]. In addition, two sites not considered by Koroiva et al. [8] were sampled during the 10 sampling campaigns between October 2020 and November 2021: João Pessoa Botanical Garden (JBBM), which is located in the municipality of João Pessoa (−7.135867, −34.860025; datum WGS84), and the "Banho do Jair" stream, found in the municipality of Santa Rita (−7.000965, −34.98836; datum WGS84). Notably, both areas are located in the Atlantic Forest fragments. The morphological identification of all specimens was done with the help of experts (see Acknowledgements) in Odonata taxonomy and by using the taxonomic keys of Lencioni [9,24,25] and Garrison et al. [10,26]. Our collection of specimens followed the methodology presented in Vilela et al. [27]. In terms of the classification, we followed Paulson et al. [28]. Voucher specimens were deposited into the Entomological Collection of the Department of Systematics and Ecology at the Federal University of Paraíba (DSEC/UFPB).

2.3. Extracting, Amplifying and Sequencing

All the DNA from the samples was extracted using the Blood & Tissue DNA Mini Kit (Ludwig Biotec, Alvorada, Brazil) from one leg, and it was preserved in ethanol. We amplified the genetic material of 44 specimens of 21 species collected in the two sampled sites mentioned above. We also used another 70 specimens of the 27 species previously collected by Koroiva et al. [8] and deposited in DSEC/UFPB (see Figure 1). For most species, we used specimens from different municipalities in an effort to analyze different populations (see BOLD dataset on http://dx.doi.org/10.5883/DS-ODOPB, accessed on 11 February 2022). In total, 658 bp were amplified from the 5′ region of the Cox1 gene using the M13-tailed primers OdoF1_t1 (5′–TGTAAAACGACGGCCAGTATTCAACHAATCATAARGATATTG G-3′) and OdoR1_t1 (5′–CAGGAAACAGCTATGACTAAACTTCTGGATGYCCRAARA AYCA-3′) (Semotok, unpublished, BOLD Systems http://www.boldsystems.org/index.php/Public_Primer_PrimerSearch, accessed on 27 January 2022). When it was not possible to amplify them, an approximately 421 bp long fragment at the 3′ end of the barcoding region was amplified by using the forward primer BF2 (5′-GCHCCHGAYATRGCHTTYCC-3′) and the reverse primer BR2 (5′-TCDGGRTGNCCRAARAAYCA-3′) described by Elbrecht and Leese [29]. This procedure was performed with consideration that the primers commonly used in COI amplification (OdoF1_t1-OdoR1_t1, HCO2198-LCO1490 [30], HCO2198_t1-LCO1490_t1 [31], LepF1-LepR1 [32] and LepF1_t1-LepR1_t1 [33]) were not successful in the amplification for many species, especially in Zygoptera, with exceptions of *Telebasis corallina* (Selys, 1876) and *Hetaerina rosea* Selys, 1853.

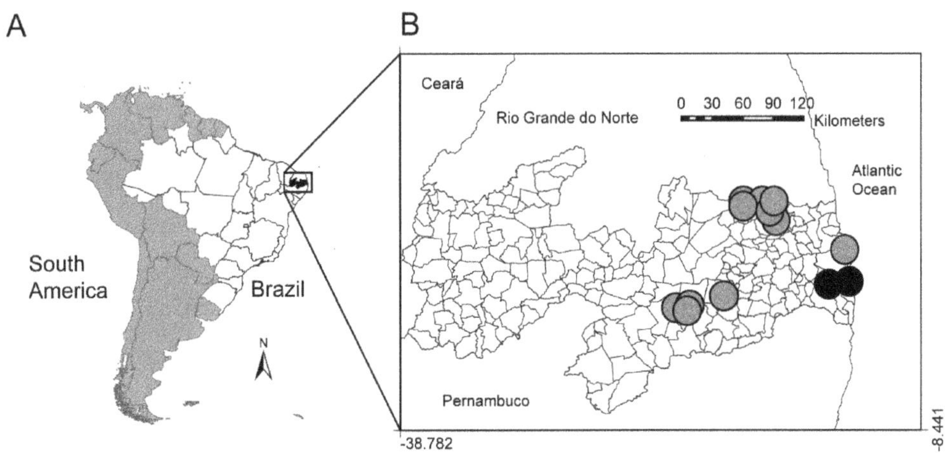

Figure 1. Geographical location of sampled specimens of odonates. (**A**) Map of South America (dark grey) highlighting the geopolitical division of Brazil (white); (**B**) municipality division of Paraíba with the locality of our two sampling sites (black dots) and the other localities for specimens obtained in DSEC/UFPB (grey dots).

The PCR conditions for amplification consisted of 1 × buffer (Colorless GoTaq® Flexi Buffer; Promega Corp., Madison, WI, USA), 0.2 mM dNTP mix, 0.2 μM of each primer, 2 mM $MgCl_2$, 1U Taq polymerase (GoTaq® G2 hot start polymerase, Promega Corp., Madison, WI, USA) and 2 μL of template DNA; these materials were placed in a total reaction volume of 25 μL. The PCR cycling program to OdoF1_t1 and OdoR1_t1 followed Vilela et al. [18]. For the BF2-BR2 primers, the PCR cycling program was run as follows: initial denaturation step with 3 min at 95 °C, 35 cycles of denaturation for 30 s at 95 °C, annealing for

45 s at 50 °C and extension for 1 min at 72 °C, and final extension for 5 min at 72 °C. The PCR products were purified with ethanol/sodium acetate and sequenced in an ABI 3130 Genetic Analyzer (Applied Biosystems, Waltham, MA, USA). The OdoF1_t1 and OdoR1_t1 sequencing were performed using M13 Forward (5′-TGTAAAACGACGGCCAGT-3′) and M13 Reverse primers (5′-CAGGAAACAGCTATGAC-3′), respectively. The sequence data were uploaded to GenBank (accession numbers OL806732 to OL806735 and OL806621 to OL806730) and were made available on BOLD as a dataset (http://dx.doi.org/10.5883/DS-ODOPB, accessed on 11 February 2022).

2.4. Data Analysis

To check the sequence quality of both strands and to assemble and edit them if necessary, we used GENEIOUS v 9.0.5 [34]. Furthermore, we aligned the sequences for each gene loci using Muscle v3.8.425 [35] (module implemented in GENEIOUS v 9.0.5) at the default setting. Five species that had less than three individuals were found to have sequenced in our database (i.e., singletons and doubletons; see on http://dx.doi.org/10.5883/DS-ODOPB, accessed on 11 February 2022). Using the sequences from the Brazilian specimens deposited in the BOLD system (accessed on 27 January 2022), we determined that our database had the incorporation of all sequences of these species (28 specimens); in turn, this allowed the analysis of intraspecific variation, totaling 142 specimens. The genetic distances between and within species were estimated using the Kimura's two-parameter substitution model (K2P) (but see Srivathsan and Meier [36]); these were calculated using the MEGA X software [37]. To increase the robustness of the homology statement and to elevate the matrix occupancy, long sequences were truncated to cover only the 'Folmer' region of the COI gene. This is the most commonly used region for DNA barcoding, as it covers 658 nt of the 5′-end of the gene. For insects, the region can be amplified using the 'Folmer' primer pair (HCO2198 and LCO1490; [30]); subsequently, truncation was carried out following the positioning of these primers.

Next, we calculated the mean and maximum genetic divergence values and the lowest genetic distance on our (regional) database to the nearest neighbor in MEGA X [37]. We then plotted the empirical K2P values associated with intra- and interspecific comparisons against each other following the methods detailed by Koroiva and Kvist [38], in order to highlight and visualize any potential "barcode gap" (but see discussion on Wiemers and Fiedler [39]). To evaluate them in the global databases, we used the default settings of Web BLAST (Basic Local Alignment Search Tool; https://blast.ncbi.nlm.nih.gov/Blast.cgi, accessed on 11 February 2022) on GenBank [40] in order to identify the nearest matching sequences and the Ident and E(xpect) values. In addition, we used the Species Level Barcode Records option on BOLD (http://www.boldsystems.org/index.php/IDS_OpenIdEngine; accessed on 11 February 2022, [41]) to obtain a list of the sequences with the highest similarity. With consideration of previous works (e.g., [16]), we examined the species closest to those that showed less than 2% genetic divergence.

2.5. Species Delimitation

We used two methods to delimit the species: distance-based and tree-based methods. For a distance-based method, we performed ABGD (Automatic Barcode Gap Discovery analysis) online (http://wwwabi.snv.jussieu.fr/public/abgd/, accessed on 11 February 2022) [42] because the program is optimized for the COI gene; we used the values of Pmin = 0.001 and Pmax = 0.10, steps = 20, relative gap width (X) = 0.75 and Nb bins = 20 and K2P. The ABGD resulted in a stable genetic group count with a range of prior intraspecific values (P = 0.0113–0.0483) and the results of these grouping are presented. ASAP (Assemble Species by Automatic Partitioning) [43] analyses were performed on the website (https://bioinfo.mnhn.fr/abi/public/asap/, accessed on 11 February 2022) using K2P; we considered only the partition showing the lowest ASAP score.

For tree-based methods, we used the Poisson tree process (PTP) as implemented in the PTP (http://species.h-its.org/, accessed on 11 February 2022) using the default settings [44,45] and the generalized mixed Yule coalescent (GMYC) method. To run the PTP analysis, we first built a tree with RAxML (v 8.2.12) [46] using a GTR GAMMA I model and 1000 bootstrap replicates. The resulting tree was used as the input tree to run on the web server.

A single-threshold GMYC analysis was conducted in the species delimitation service (https://species.h-its.org/gmyc/, accessed on 11 February 2022) [47]. An ultrametric single-locus gene tree was obtained using BEAST v.2.6.6 [48] with 1.5×10^8 MCMC generations under relaxed lognormal clock, the Yule process tree model, and a burn-in of the first 10% generations of the final consensus tree. The posterior distributions (ESS > 200) were examined in Tracer v. 1.6 [49]. The best fitting available model was identified through the jModelTest v. 2.1.7 (AIC & BIC, GTR + I + G) [50,51].

Finally, we created a neighbor-joining (NJ) dendrogram to provide a graphic representation of the divergence pattern between species. An NJ tree was inferred using MEGA X software [37]. In all the tree-based analysis, three species were used as the out group (HQ941355 *Baetis adonis*, HQ943539 *Baetis phoebus* and HQ987969 *Ephemerella mucronata*). It should be noted that the tree presented here is only intended to represent the distance matrix and it should not be interpreted as a phylogenetic hypothesis.

3. Results

3.1. Sampling

Our samples added 15 new species records to the odonatofauna of Paraíba: *Dasythemis venosa* (Burmeister, 1839), *Dythemis nigra* Martin, 1897, *Epipleoneura metallica* Rácenis, 1955, *Erythrodiplax* cf. *fervida* (Erichson in Schomburgk, 1848), *Idioneura ancilla* Selys, 1860, *Macrothemis imitans* Karsch, 1890, *Metaleptobasis bicornis* (Selys, 1877), *Micrathyria didyma* (Selys in Sagra, 1857), *Micrathyria mengeri* Ris, 1919, *Nephepeltia berlai* Santos, 1950, *Orthemis flavopicta* Kirby, 1889, *Perithemis lais* (Perty, 1834), *Tauriphila australis* (Hagen, 1867), *Telebasis griffinii* (Martin, 1896) and *Triacanthagyna septima* (Selys in Sagra, 1857).

After performing the morphological analysis, all specimens (including those deposited in the DSEC/UFPB) that had previously been identified as *Anatya guttata* (Erichson in Schomburgk, 1848) were now identified as *Anatya januaria* Ris, 1911.

3.2. Genetic Variation

A total of 142 mitochondrial COI barcode sequences were obtained from five families, 27 genera and 45 species (mean by species 3.15; max = 15, min = 1, Table 1, Figure 2). All of the analyzed sequences were larger than 353 bp. The average base pairs of the sequences were 524 bp (SD = 116.54) and the median was 592 bp. Thirteen singletons were registered in the database (Table 1). In the regional database (Table 1), the greatest intraspecific variation was the Erythrodiplax fusca (Rambur, 1842) with 1.85% and the smallest inter-specific variation occurred between the Erythrodiplax leticia Machado, 1996 and the Erythrodiplax cf. fervida, with 4.87%.

Table 1. Similarity match approach of the generated sequences of collected specimens with the regional database sequences and global database sequences of GenBank and BOLD (Species Level Barcode Records) for species identification. N, number of barcode sequences; Mean(%) = average of intraspecific genetic distance value (expressed as percent); Max(%) = maximum intraspecific genetic distance value (expressed as percent); Nearest neighbor = most closely related species retrieved; DNN = lowest genetic distance to the nearest neighbor (expressed as percent).

Species	N	Mean (%)	Max (%)	Similarity Match Regional Database Nearest Neighbor	DNN (%)	GenBank id. (%)	GenBank E-Value	Similarity Match Global Database GenBank Species	Bold Similarity	Bold Species
Acanthagrion gracile	1	-	-	*Lestes forficula*	19.01	92.17	1.00×10^{-149}	*Acanthagrion cuyabae*	no match	no match
Anatya januaria	3	0.36	0.57	*Erythrodiplax umbrata*	15.44	99.73	0	*Anatya* sp.	99.74	*Anatya guttata*
Argia reclusa	3	0.16	0.25	*Orthemis discolor*	18.88	98.74	0	*Argia reclusa*	98.73	*Argia reclusa*
Brachymesia furcata	2	0.85	0.85	*Brachymesia herbida*	10.55	99.67	0	*Brachymesia furcata*	99.67	*Brachymesia furcata*
Brachymesia herbida	1	-	-	*Brachymesia furcata*	10.55	90.05	0	*Brachymesia furcata*	99.33	*Brachymesia herbida*
Diastatops obscura	3	0.86	1.3	*Orthemis aequilibris*	12.2	99.83	0	*Diastatops obscura*	99.83	*Diastatops obscura*
Dythemis nigra	3	0.16	0.25	*Orthemis discolor*	13.87	100	0	*Dythemis multipunctata*	100	*Dythemis nigra*
Epipleoneura metallica	1	-	-	*Dythemis nigra*	20.59	85.4	0	*Argia fumipennis*	no match	no match
Erythemis plebeja	6	0	0	*Erythemis carmelita*	10.64	90.95	0	*Erythemis peruviana*	100	*Erythemis plebeja*
Erythemis vesiculosa	5	0.32	0.50	*Erythrodiplax umbrata*	14.62	87.50	0	*Acisoma attenboroughi*	99.85	*Erythemis vesiculosa*
Erythemis peruviana	2	0.33	0.33	*Erythemis plebeja*	13.02	88.39	0	*Agrionoptera insignis*	100	*Erythemis peruviana*
Erythemis carmelita	3	0.10	0.16	*Erythemis plebeja*	10.64	97.90	0	*Erythemis* sp.	98.67	*Erythemis mithroides*
Erythrodiplax avittata	2	0	0	*Erythrodiplax basalis*	11.04	86.93	0	*Erythrodiplax paraguayensis*	no match	no match
Erythrodiplax umbrata	3	1.14	1.37	*Orthemis discolor*	13.74	99.84	0	*Erythrodiplax umbrata*	99.83	*Erythrodiplax umbrata*
Erythrodiplax leticia	2	0.16	0.16	*Erythrodiplax cf fervida*	4.87	92.53	0	*Erythemis* sp.	no match	no match
Erythrodiplax fusca	15	0.89	1.85	*Erythrodiplax avittata*	15.75	94.53	0	*Erythrodiplax connata*	99.85	*Erythrodiplax fusca*
Erythrodiplax cf. *fervida*	2	0	0	*Erythrodiplax leticia*	4.87	91.71	0	*Erythemis* sp.	no match	no match
Erythrodiplax basalis	6	0.11	0.75	*Erythrodiplax avittata*	11.04	87.24	6.00×10^{-118}	*Erythrodiplax paraguayensis*	99.74	*Erythrodiplax paraguayensis*
Erythrodiplax cf. *unimaculata*	1	-	-	*Erythrodiplax leticia*	8.58	95.86	6.00×10^{-163}	*Libellulidae* sp.	97.11	*Erythrodiplax kimminsi*
Hetaerina rosea	3	0.52	0.78	*Ischnura capreolus*	19.6	98.96	0	*Hetaerina sanguinea*	98.95	*Hetaerina sanguinea*
Ischnura fluviatilis	4	0.12	0.26	*Telebasis filiola*	15.96	99.22	0	*Ischnura fluviatilis*	98.68	*Ischnura fluviatilis*
Ischnura capreolus	4	0.25	0.49	*Lestes forficula*	13.2	100	0	*Ischnura capreolus*	100	*Ischnura capreolus*
Lestes forficula	3	0.69	1.03	*Ischnura capreolus*	13.2	100	0	*Lestes forficula*	98.78	*Lestes forficula*
Macrothemis griseofrons	6	0.1	0.17	*Erythrodiplax umbrata*	15.39	88.07	0	*Agrionoptera insignis*	no match	no match
Mecistogaster kesselringi	1	-	-	*Lestes forficula*	14.76	95.92	6.00×10^{-178}	*Mecistogaster amalia*	no match	no match
Metaleptobasis bicornis	4	0.12	0.25	*Telebasis corallina*	13.75	93.42	6.00×10^{-163}	*Metaleptobasis selysi*	no match	no match
Miathyria marcella	6	0.53	0.92	*Perithemis lais*	13.19	87.84	0	*Agrionoptera insignis*	100	*Miathyria marcella*
Micrathyria hesperis	4	0.38	0.51	*Uracis imbuta*	14.37	87.98	4.00×10^{-125}	*Micrathyria stawiarskii*	no match	no match
Micrathyria ocellata	3	0.11	0.16	*Nephepeltia belai*	13.76	100	0	*Micrathyria ocellata*	100	*Micrathyria ocellata*
Nephepeltia berlai	1	-	-	*Micrathyria ocellata*	13.76	98.11	0	*Nephepeltia aequisetis*	98.1	*Nephepeltia aequisetis*
Orthemis schmidt	4	0.62	0.78	*Orthemis discolor*	5.00	99.48	0	*Orthemis schmidti*	99.48	*Orthemis schmidti*
Orthemis aequilibris	1	-	-	*Orthemis discolor*	4.92	95.25	0	*Orthemis discolor*	99.83	*Orthemis aequilibris*
Orthemis discolor	7	0.51	0.91	*Orthemis aequilibris*	4.92	99.03	0	*Orthemis* sp.	100	*Orthemis discolor*
Orthemis flavopicta	1	-	-	*Orthemis discolor*	9.69	93.79	4.00×10^{-176}	*Orthemis cultriformis*	no match	no match
Pantala flavescens	3	0.65	0.65	*Tramea cophysa*	14.9	99.84	0	*Pantala flavescens*	99.83	*Pantala flavescens*
Perithemis tenera	3	0.87	1.27	*Brachymesia furcata*	13.28	91.1	0	*Libellulidae* sp.	99.83	*Perithemis icteroptera*
Perithemis lais	1	-	-	*Brachymesia furcata*	13.14	88.21	0	*Libellulinae* sp.	99.25	*Perithemis lais*
Progomphus dorsopallidus	1	-	-	*Perithemis tenera*	18.93	86.22	4.00×10^{-171}	*Tanypteryx hageni*	no match	no match
Tauriphila australis	1	-	-	*Orthemis flavopicta*	14.64	100	0	*Tauriphila australis*	100	*Tauriphila australis*
Telebasis corallina	4	0.47	0.8	*Orthemis schmidt*	13.08	88.28	0	*Telebasis willinki*	97.9	*Telebasis corallina*
Telebasis filiola	4	0.26	0.51	*Telebasis corallina*	13.26	98.72	0	*Telebasis willinki*	98.96	*Telebasis willinki*
Telebasis griffinii	3	0.52	0.78	*Uracis imbuta*	17.12	88.89	8.00×10^{-127}	*Telebasis digiticollis*	no match	no match
Tramea cophysa	4	0.83	1.02	*Brachymesia furcata*	12.88	99.11	0	*Tramea cophysa*	98.34	*Tramea binotata*
Uracis imbuta	1	-	-	*Nephepeltia berlai*	14.35	94.08	1.00×10^{-179}	*Uracis imbuta*	99.04	*Uracis imbuta*
Zenithoptera lanei	1	-	-	*Telebasis corallina*	19.33	98.24	8.00×10^{-166}	*Zenithoptera* sp.	no match	no match

Figure 2. Examples of odonates (Insecta: Odonata) collected and sequenced from Paraíba State, Brazil. (**A**) *Anatya januaria*, (**B**) *Perithemis tenera*, (**C**) *Tramea cophysa*; (**D**) *Hetaerina rosea*, (**E**) *Telebasis filiola*, (**F**) *Lestes forficula*; (**G**) *Micrathyria hesperis*, (**H**) *Erythrodiplax basalis* and (**I**) *Erythemis plebeja*.

Figure 3 shows the clear separation of intraspecific and interspecific distances and the so-called "barcoding gap" on the regional DNA barcode library. Considering the global databases, eight species showed close proximity (<2%) to records of other deposited species (Table 1): *A. januaria*, *Erythemis carmelita* Williamson, 1923, *Erythrodiplax basalis* (Kirby, 1897), *H. rosea*, *N. berlai*, *Perithemis tenera* (Say, 1840), *Telebasis filiola* (Perty, 1834) and *Tramea cophysa* Hagen, 1867.

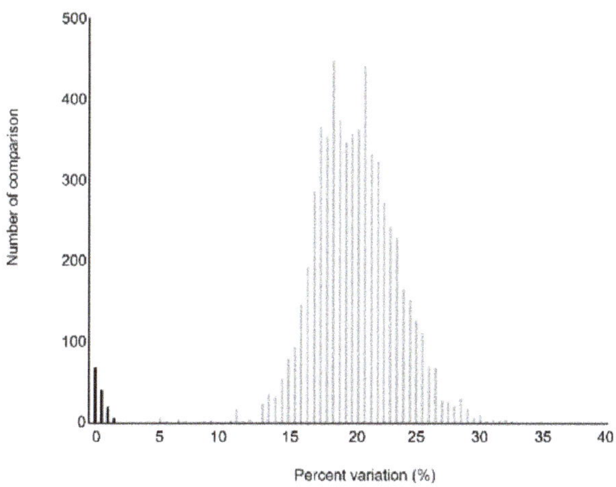

Figure 3. Frequency distribution of intraspecific (black) and interspecific (grey) genetic divergence in the sampled odonates (regional database).

3.3. Species Delimitation

Species delimitation analyses provided 45 species defined by morphological delimitation. Using molecular methods to delimit species, the results were 45 for both the ABGD-initial partition and ABGD-recursive partition, 39 for ASAP, 46 for PTP and GMYC (Figure 4).

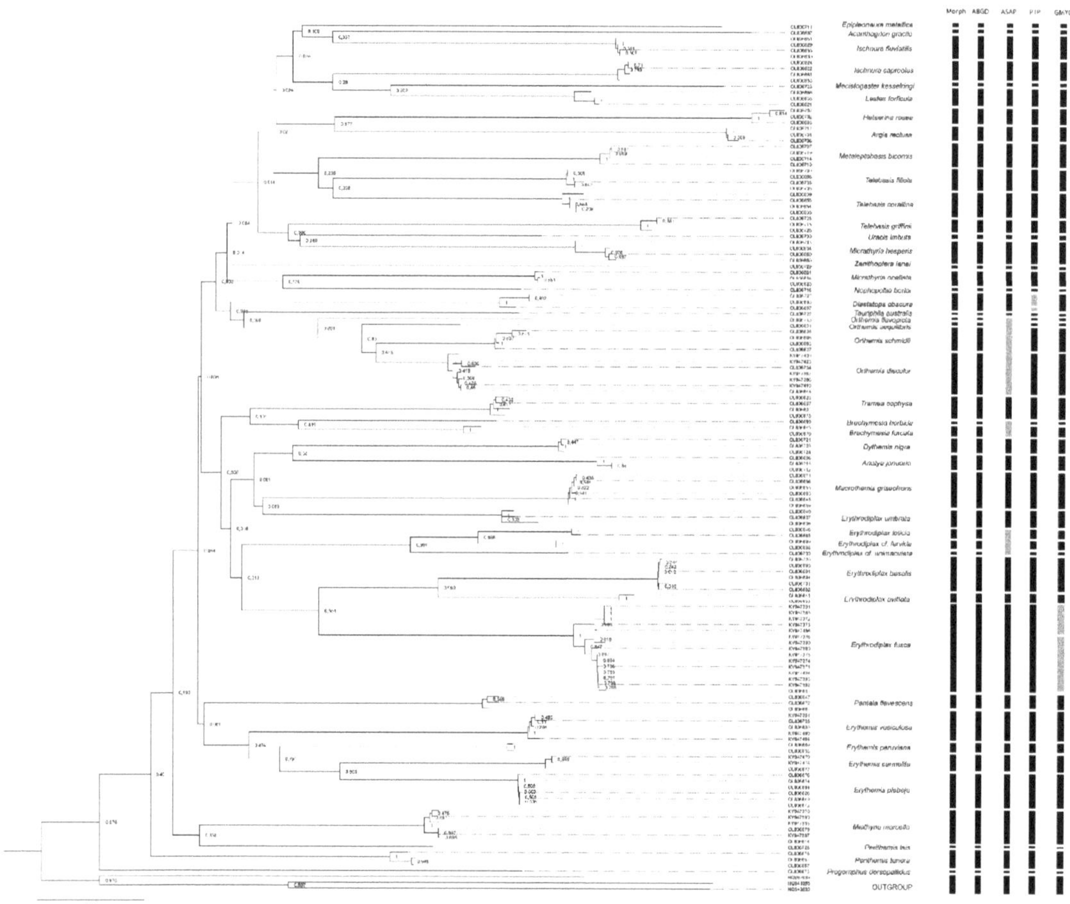

Figure 4. Molecular species delimitation of odonates from the Paraíba state, Brazil, based on DNA barcodes. Black bars indicate congruent results between the molecular and morphological identifications, and grey bars indicate divergence results from morphological identification. It should be noted that the neighbor-joining tree presented here is only intended to represent the distance matrix, and it should not be interpreted as a phylogenetic hypothesis.

4. Discussion

This study adds 15 Odonata species to the state of Paraíba. As a result, this state now has 64 species recorded; thus, it is third in number of species in the Northeast region of Brazil, behind only Bahia and Ceará (174 species [52] and 73 species [53], respectively). This library of DNA sequences is the first publication of DNA barcoding for the 16 Odonata species in public sequence repositories: *Acanthagrion gracile* (Rambur, 1842), *A. januaria*, *Ep. metallica*, *E. avittata* Borror, 1942, *E. leticia*, *E.* cf. *fervida*, *E. basalis*, *E.* cf. *unimaculata* (De Geer, 1773), *M. griseofrons* Calvert, 1909, *Mecistogaster kesselringi* Soldati and Machado, 2019, *Metaleptobasis bicornis* (Selys, 1877), *N. berlai*, *O. flavopicta*, *Progomphus dorsopallidus*

Byers, 1934, *T. filiola* and *Zenithoptera lanei* Santos, 1941. The database generated for the present study provides data for 70% (45 species) of the odonate species found in Paraíba state. Despite all these numbers, we recognize that this work is a starting point for the study of odonates in Paraíba state, considering that the likely diversity of the region must be far greater than our estimates.

It is important to highlight the methodological limitations that we faced for the amplification of Neotropical odonates. Although many studies indicate the effectiveness of the tested primers on several continents (e.g., [54,55]), they did not have the amplification capacity for all the species amplified in this study, most notably in Zygoptera. Problems with amplifying Neotropical species using primers commonly used in the world is not new, and it is not exclusive to odonates (see [13] for frogs and [14] for fishes). Jennings and collaborators [14] highlight the importance of this type of information considering that the trial and error nature for the choice of primers wastes labor and reagents.

In addition to demonstrating that the two pairs of primers used in this study are capable of amplifying the COI fragment in the two suborders present in the Neotropical region, this study shows that it is also important to consider the usefulness of this marker to discriminate between different species in terms of their DNA barcode and metabarcoding studies. The regional database does not present the overlapping of inter- and intraspecific genetic variation; however, in the global database analysis (using Genbank and the BOLD System), eight species showed close genetic proximity to the species we examined (<2%). Below, we discuss the most likely hypothesis for each taxon as well as the disagreements we noticed in the global dataset evaluation.

Our results showed that the analyzed specimens of *A. januaria* from Paraíba state are 99.74% similar to the available sequences for *A. guttata* (Erichson, 1848). Problems with determining the species of *Anatya* are largely recognized, mainly due to some specific variable characteristics being wrongly interpreted (see [10], p. 224). Our specimens had subtle differences among them, including the size of cerci and body length; however, based on the comparison of these structures drawn by Ris ([56], p. 424) and Garrison et al. [10], the shape of the posterior hamule left almost no doubt that our specimens belong to *A. januaria* (Figure 5A).

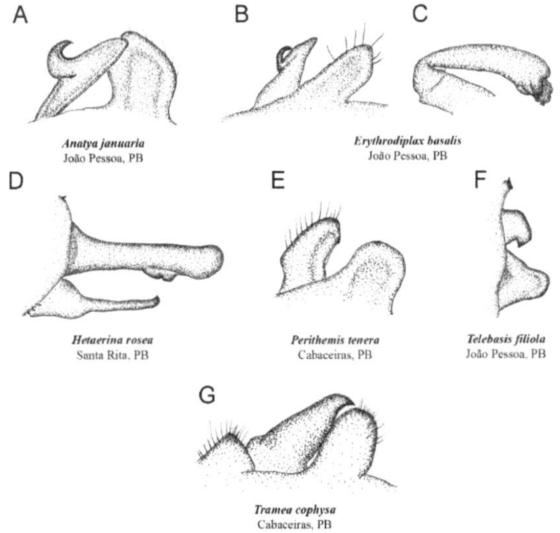

Figure 5. Morphology of male Odonata adults collected in Paraíba state, Brazil (lateral view of): (**A**) secondary genitalia of *Anatya januaria*; (**B**) secondary genitalia and (**C**) vesica spermalis of *Erythrodiplax basalis*; (**D**) caudal appendages of *Hetaerina rosea*; (**E**) secondary genitalia of *Perithemis tenera*; (**F**) caudal appendages of *Telebasis filiola*; (**G**) secondary genitalia of *Tramea cophysa*.

In turn, the *E. carmelita* we analyzed was 98.67% similar to a sequence identified as the species *E. mithroides*. We are confident with regard to our morphological identification because, despite their color resemblance, these two species are very easily distinguishable because of their abdominal characteristics. *Erythemis carmelita* belongs to the group of *Erythemis* species that possesses greatly swollen basal abdominal segments (which are slightly swollen in *E. mithroides*) and narrow remainder segments (which are broad in *E. mithroides*) ([10], p. 240–241).

In addition, we found a high similarity between the sequences of our *E. basalis* and the deposited sequences of *E. paraguayensis* (Förster, 1905). Borror [57] states that specimens of *E. paraguayensis* are "apt to be confused with small individuals of *basalis*" (p. 153); however, morphologically, these species are difficult to be confused with one another because of multiple characteristics: *E. basalis* belongs to Borror's basalis group, which consists of species that have hamules that are usually slender and have an outer branch a little longer than their inner branch (evident in lateral view, Figure 5B), a terminal segment of a slender penis, small and rounded lateral lobes (Figure 5C)—characteristics that we can easily observe in *E. basalis* specimens [57].

In contrast, *E. paraguayensis* belongs to the connata group, in which males present hamules that are moderately robust, their outer branch is almost equal in size to inner branch, and the median process of the penis is very prominent and slender, similar to what we can observe in the males of *E. paraguayensis*. In addition, *E. paraguayensis* is one of the smallest species of the genus and the extent of the small spots of the hind wings nearly trespasses upon the cubital space. In contrast, *E. basalis* is a much larger species, its hind wing spots reaching (or in some cases, trespassing upon) the first antenodal vein. Based on these characters, we argue that our morphological identification is correct.

Although our sequences of *H. rosea* are 98.95% similar to the ones assigned to *H. sanguinea* Selys, 1853, those two species are highly unlikely to be confused with one another. This is mainly because of the paraprocts (Figure 5D) are long and well developed in *H. rosea*, while they are vestigial in *H. sanguinea* (see [58] for a revision). Another character that can easily separate the two species is the median lobe of cercus. It is bilobed in *H. rosea* and entirely so in *H. sanguinea* (see [24], pp. 66–67).

Nephepeltia berlai shares several morphological characters with *N. aequisetis* Calvert, 1909, with the most prominent among them being the tubercle on the venter of thorax, the length and placement of the hind tibiae spurs, and also the vesica spermalis (see [59] for a revision). However, there are some characters on the vesica spermalis (medio-ectal distal process) and the cercus (level of the distal end of ventral toothed carina at about distal fourth of cercus length) that allow for a separation of the two species. Based on those characters, we believe that our specimens are *N. berlai*.

We based our identifications of the genus *Perithemis* largely on the study of von Ellenrieder and Muzón [60], which was the first study to separate the species of this genus using characteristics other than coloration and wing venation. The males we identified as *P. tenera* (Say, 1840) (regarded as *P. mooma* Kirby, 1889 in [60]) present wings that are uniformly colored (as in *P. icteroptera* Selys in Sagra, 1857), have a tip of hamuli (Figure 5E) at least 0.40 of the ventral margin (at the level of ventral margin in *P. icteroptera*), and a penis with a first segment trapezoidal (rounded in *P. icteroptera*). Our sequences were 99.83% similar to the sequences assigned to *P. icteroptera*; however, the characteristics described above led us to identify our specimens as *P. tenera*.

The sequences of our specimens identified as *T. filiola* were 98.96% identical to the specimens deposited assigned as *T. willinki* Fraser, 1948. These species are very similar morphologically, as it is stated in the revision of the genus [61]. However, despite the great resemblance of these taxa, Garrison [61] properly diagnosed the two species, showing that the cercus of *T. filiola* (Figure 5F) is distinctly shorter than the paraproct (it is subequal to the paraproct in *T. willinki*). In turn, we followed the same diagnosis to assign our specimens as *T. filiola*.

Lastly, we identified some males of *T. cophysa* showing agreement with the diagnosis presented in DeMarmels and Rácenis [62]. However, our sequences were 98.34% similar to the sequences assigned to *T. binotata* (Rambur, 1842) in the BOLD System. These two taxa, although they belong to the same genus, are quite different. Due to their color and morphological differences, they were placed in different groups within *Tramea*. The cophysa group, to which *T. cophysa* belongs, is composed of four species that share "only one constant characteristic common to all four species and, peculiar to the "cophysa-group", are two oblique pale lateral bands on the synthorax" [62]. Such a characteristic is present in our specimens while it is absent in *T. binotata*, as this is a species with an overall blue-grey coloration, contrasting with the reddish coloration of the species of the cophysa group. Additionally, as a result of our comparison of morphological features, such as the shape of the posterior hamule and the length and shape of the cercus (Figure 5G), we were able to make a safe distinction between our specimens to other taxa within *Tramea*.

The molecular species delimitation results were identical to the morphological results for most of the species we examined. The presence of divergent results is commonly used to indicate possible cryptic species (e.g., [63]); in turn our identical results in terms of morphological, ABGD, and the occasional divergences suggest that COI has the ability to delimit in the evaluated species.

In summary, our results demonstrate that DNA barcoding can be used to delimit and differentiate odonates on a regional scale. Of the 45 species evaluated in this study, only eight species (17%) showed any disagreement with the global databases and all species (100%) could be identified when we consider only our regional DNA barcode database. Considering the issues with the global databases, our results for the number of species that can be readily identified using their DNA barcoding (83%) are close to those found in other regions of the world. In a genetic database for the Central and North European odonates, the effectiveness was 88% in a set of 103 species [64]. Values between 79% and 89% were also found in datasets in countries such as the Philippines (in a set of 38 species; [54]), Italy (in a set of 88 species; [65]) and Malta (in a set of 10 species, [55]). In Brazil, the only evaluation (in a set of 38 species) performed indicated a success rate between 79% and 94% depending on the analysis criteria [16].

Subsequently, all these results indicate the importance of performing careful morphological analysis (see this kind of problem in [66]). Moreover, as indicated by Koroiva and Kvist [38], it reinforces that the use of the global database does not allow the correct establishment of all molecular identifications to be correct. Many reasons justify the discrepancy between the morphological and molecular results. Among in odonates, three main causes are the presence of cryptic species, rapid and/or recent radiation events and errors in identifying the deposited specimens [38]. Regarding this last issue, the description and presentation of the key structures for identification presented above aimed to facilitate future comparisons in this sense.

5. Conclusions

The establishment of an Odonata DNA barcode library for the Paraíba state is a milestone that will improve the taxonomy and biodiversity conservation for Neotropical species. Despite the difficulties of using traditional primers for amplifying the Neotropical species, our results demonstrate that using the COI in the regional scale can help identify and delimit those evaluated. Our results for the number of species that can be readily identified using their DNA barcoding (83%) are close to the results found in other regions of the world. Keeping in mind the problems of using public genetic databases for identification, here, we present morphological evidence for our identifications in the cases of disagreement. In turn, this facilitates comparisons and allows for new questions to arise about the genetic diversity of tropical species.

Author Contributions: Conceptualization, R.K.; methodology, R.K., V.G.N.G. and D.S.V.; formal analysis, R.K., V.G.N.G. and D.S.V.; data curation, R.K. and D.S.V.; writing—original draft preparation, R.K., V.G.N.G. and D.S.V.; writing—review and editing, R.K., V.G.N.G. and D.S.V. All authors have read and agreed to the published version of the manuscript.

Funding: This research was partially funded by PROPESQ/PRPG/UFPB, grant number PVA13280-2020 and Coordenação de Aperfeiçoamento de Pessoal de Nível Superior—Brasil (CAPES), finance Code 001. DSV thanks Fundação de Amparo à Pesquisa do Estado de São Paulo (FAPESP) for a postdoctoral fellowship grant (Proc. 2019/26438-9).

Institutional Review Board Statement: Not applicable.

Data Availability Statement: Data are contained within the article; specimens analyzed in this study are deposited in Entomological Collection of the Department of Systematics and Ecology of the Federal University of Paraíba (DSEC/UFPB) and are available on request to the collection managers. The sequences are available at GenBank (accession numbers OL806732 to OL806735 and OL806621 to OL806730) and BOLD system (http://dx.doi.org/10.5883/DS-ODOPB, accessed on 11 February 2022).

Acknowledgments: We are thankful to Jardim Botânico de João Pessoa (JBBM/SUDEMA) for the assistance in sampling. We are grateful to Frederico Augusto de Atayde Lencioni and Rosser W. Garrison for confirmation of species identifications. We are also grateful to the academic editor and the anonymous reviewers for their valuable comments.

Conflicts of Interest: The authors declare no conflict of interest.

References

1. Trueman, J.W.H. A brief history of the classification and nomenclature of Odonata. *Zootaxa* **2007**, *1668*, 381–394. [CrossRef]
2. Cordoba-Aguilar, A. *Dragonflies and Damselflies: Model Organisms for Ecological and Evolutionary Research*, 1st ed.; Oxford University Press: Oxford, UK, 2009.
3. Miguel, T.B.; Oliveira-Junior, J.M.B.; Ligeiro, R.; Juen, L. Odonata (Insecta) as a tool for the biomonitoring of environmental quality. *Ecol. Indic.* **2017**, *81*, 555–566. [CrossRef]
4. Veras, D.S.; Lustosa, G.S.; Moura, L.P.; Ferreira, M.F.R.; Juen, L. Differences in land use modify Odonata assemblages in the Cerrado-Caatinga ecotone. *Acta Limnol. Bras.* **2020**, *32*, e15. [CrossRef]
5. May, M. Odonata: Who They Are and What They Have Done for Us Lately: Classification and Ecosystem Services of Dragonflies. *Insects* **2019**, *10*, 62. [CrossRef] [PubMed]
6. Bried, J.; Ries, L.; Smith, B.; Patten, M.; Abbott, J.; Ball-Damerow, J.; Cannings, R.; Cordero-Rivera, A.; Córdoba-Aguilar, A.; De Marco, P.; et al. Towards Global Volunteer Monitoring of Odonate Abundance. *Bioscience* **2020**, *70*, 914–923. [CrossRef]
7. Gómez-Tolosa, M.; Rivera-Velázquez, G.; Rioja-Paradela, T.M.; Mendoza-Cuenca, L.F.; Tejeda-Cruz, C.; López, S. The use of Odonata species for environmental assessment: A meta-analysis for the Neotropical region. *Environ. Sci. Pollut. Res.* **2021**, *28*, 1381–1396. [CrossRef] [PubMed]
8. Koroiva, R.; Pereira-Colavite, A.; Batista, F.R.D.C.; Vilela, D.S. Checklist and contribution to the knowledge of the odonatofauna of Paraíba state, Brazil. *Biota Neotrop.* **2021**, *21*, e20211196. [CrossRef]
9. Lencioni, F.A.A. *Damselflies of Brazil—An Illustrated Identification Guide—Southeast Region*; F.A.A. Lencioni: Jacareí, Brazil, 2017.
10. Garrison, R.W.; von Ellenrieder, N.; Louton, J.A. Dragonfly Genera of the New Word: An Illustrated and Annotated Key to the Anisoptera. Available online: http://entomologia.net/L_Odonata/Odonata_Dragonfly_Genera_of_the_New_World__An_Illustrated_and_Annotated_Key_to_the_Anisoptera.pdf (accessed on 5 March 2016).
11. Hebert, P.D.N.; Cywinska, A.; Ball, S.L.; de Waard, J.R. Biological identifications through DNA barcodes. *Proc. Biol. Sci.* **2003**, *270*, 313–321. [CrossRef]
12. Hebert, P.D.N.; Ratnasingham, S.; de Waard, J.R. Barcoding animal life: Cytochrome c oxidase subunit 1 divergences among closely related species. *Proc. R. Soc. B Biol. Sci.* **2003**, *270*, S96–S99. [CrossRef]
13. Lyra, M.L.; Haddad, C.F.B.; de Azeredo-Espin, A.M.L. Meeting the challenge of DNA barcoding Neotropical amphibians: Polymerase chain reaction optimization and new COI primers. *Mol. Ecol. Resour.* **2017**, *17*, 966–980. [CrossRef]
14. Jennings, W.B.; Ruschi, P.A.; Ferraro, G.; Quijada, C.C.; Silva-Malanski, A.C.G.; Prosdocimi, F.; Buckup, P.A. Barcoding the Neotropical freshwater fish fauna using a new pair of universal COI primers with a discussion of primer dimers and M13 primer tails. *Genome* **2019**, *62*, 77–83. [CrossRef] [PubMed]
15. Ratnasingham, S.; Hebert, P.D.N. BOLD: The Barcode of Life Data System (www.barcodinglife.org). *Mol. Ecol. Notes* **2007**, *7*, 355–364. [CrossRef] [PubMed]
16. Koroiva, R.; Pepinelli, M.; Rodrigues, M.E.; Roque, F.O.; Lorenz-Lemke, A.P.; Kvist, S. DNA barcoding of odonates from the Upper Plata basin: Database creation and genetic diversity estimation. *PLoS ONE* **2017**, *8*, e0182283. [CrossRef] [PubMed]

17. Vilela, D.S.; Koroiva, R.; Cordero-Rivera, A.; Guillermo-Ferreira, R. A further study on *Franciscobasis* Machado & Bedê, 2016 (Odonata: Coenagrionidae), a newly described genus from Minas Gerais, Brazil. *PLoS ONE* **2019**, *14*, e0223241. [CrossRef]
18. Vilela, D.S.; Anjos-Santos, D.; Koroiva, R.; Cordero-Rivera, A.; Guillermo-Ferreira, R. Revision of the genus Minagrion Santos, 1965 (Odonata: Coenagrionidae). *Zootaxa* **2020**, *4786*, 176–198. [CrossRef] [PubMed]
19. Pante, E.; Schoelinck, C.; Puillandre, N. From integrative taxonomy to species description: One step beyond. *Syst. Biol.* **2015**, *64*, 152–160. [CrossRef] [PubMed]
20. IUCN. The IUCN Red List of Threatened Species: Version 2021-3. Available online: https://www.iucnredlist.org/search?permalink=7d7c85cb-009c-4c4d-83ae-25e4675dcd32 (accessed on 18 February 2022).
21. Laurance, W.F. Conserving the hottest of the hotspots. *Biol. Conserv.* **2009**, *142*, 1137. [CrossRef]
22. Marinho, F.P.; Mazzochini, G.G.; Manhães, A.P.; Weisser, W.W.; Ganade, G. Effects of past and present land use on vegetation cover and regeneration in a tropical dryland forest. *J. Arid Environ.* **2016**, *132*, 26–33. [CrossRef]
23. da Silva, J.M.C.; Barbosa, L.C.F. Impact of Human Activities on the Caatinga. In *Caatinga*; Silva, J.M.C., da Tabarelli, M., Leal, I.R., Eds.; Springer International Publishing: Cham, Switzerland, 2017; pp. 359–368.
24. Lencioni, F.A.A. *Damselflies of Brazil: An Illustrated Identification Guide, Volume 1: Non-Coenagrionidae Families*; All Print Editora: São Paulo, Brazil, 2005; ISBN 8598310565.
25. Lencioni, F.A.A. *Damselflies of Brazil: An Illustrated Identification Guide, Volume 2: Coenagrionidae*; All Print Editora: São Paulo, Brazil, 2006; ISBN 8577180344.
26. Garrison, R.W.; von Ellenrieder, N.; Louton, J.A. *Damselfly Genera of the New World: An Illustrated and Annotated Key to the Zygoptera*; Johns Hopkins University Press: Baltimore, MD, USA, 2010; ISBN 0801896703.
27. Vilela, D.S.; Koroiva, R.; Tosta, T.H.A.; Novaes, M.C.; Guillermo-Ferreira, R. Dragonflies and damselflies from the West of Minas Gerais, Brazil: Checklist and new records. *Biota Neotrop.* **2020**, *20*, e20190851. [CrossRef]
28. Paulson, D.; Schorr, M.; Deliry, C. World Odonata List. Available online: https://www2.pugetsound.edu/academics/academic-resources/slater-museum/biodiversity-resources/dragonflies/world-odonata-list2/ (accessed on 11 February 2021).
29. Elbrecht, V.; Leese, F. Validation and Development of COI Metabarcoding Primers for Freshwater Macroinvertebrate Bioassessment. *Front. Environ. Sci.* **2017**, *5*, 11. [CrossRef]
30. Folmer, O.; Black, M.; Hoeh, W.; Lutz, R.; Vrijenhoek, R. DNA primers for amplification of mitochondrial cytochrome c oxidase subunit I from diverse metazoan invertebrates. *Mol. Mar. Biol. Biotechnol.* **1994**, *3*, 294–299. [PubMed]
31. Foottit, R.G.; Maw, H.E.L.; Havill, N.P.; Ahern, R.G.; Montgomery, M.E. DNA barcodes to identify species and explore diversity in the Adelgidae (Insecta: Hemiptera: Aphidoidea). *Mol. Ecol. Resour.* **2009**, *9*, 188–195. [CrossRef] [PubMed]
32. Hebert, P.D.N.; Penton, E.H.; Burns, J.M.; Janzen, D.H.; Hallwachs, W. Ten species in one: DNA barcoding reveals cryptic species in the neotropical skipper butterfly Astraptes fulgerator. *Proc. Natl. Acad. Sci. USA* **2004**, *101*, 14812–14817. [CrossRef] [PubMed]
33. Zhang, J.; Hanner, R. Molecular Approach to the Identification of Fish in the South China Sea. *PLoS ONE* **2012**, *7*, e30621. [CrossRef] [PubMed]
34. Kearse, M.; Moir, R.; Wilson, A.; Stones-Havas, S.; Cheung, M.; Sturrock, S.; Buxton, S.; Cooper, A.; Markowitz, S.; Duran, C.; et al. Geneious Basic: An integrated and extendable desktop software platform for the organization and analysis of sequence data. *Bioinformatics* **2012**, *28*, 1647–1649. [CrossRef] [PubMed]
35. Edgar, R.C. MUSCLE: Multiple sequence alignment with high accuracy and high throughput. *Nucleic Acids Res.* **2004**, *32*, 1792–1797. [CrossRef]
36. Srivathsan, A.; Meier, R. On the inappropriate use of Kimura-2-parameter (K2P) divergences in the DNA-barcoding literature. *Cladistics* **2012**, *28*, 190–194. [CrossRef] [PubMed]
37. Kumar, S.; Stecher, G.; Li, M.; Knyaz, C.; Tamura, K. MEGA X: Molecular Evolutionary Genetics Analysis across Computing Platforms. *Mol. Biol. Evol.* **2018**, *35*, 1547–1549. [CrossRef]
38. Koroiva, R.; Kvist, S. Estimating the barcoding gap in a global dataset of cox1 sequences for Odonata: Close, but no cigar. *Mitochondrial DNA Part A* **2018**, *29*, 765–771. [CrossRef]
39. Wiemers, M.; Fiedler, K. Does the DNA barcoding gap exist?—A case study in blue butterflies (Lepidoptera: Lycaenidae). *Front. Zool.* **2007**, *4*, 8. [CrossRef]
40. Benson, D.A.; Cavanaugh, M.; Clark, K.; Karsch-Mizrachi, I.; Lipman, D.J.; Ostell, J.; Sayers, E.W. GenBank. *Nucleic Acids Res.* **2013**, *41*, D36–D42. [CrossRef]
41. Ratnasingham, S.; Hebert, P.D.N. A DNA-Based Registry for All Animal Species: The Barcode Index Number (BIN) System. *PLoS ONE* **2013**, *8*, e66213. [CrossRef]
42. Puillandre, N.; Lambert, A.; Brouillet, S.; Achaz, G. ABGD, Automatic Barcode Gap Discovery for primary species delimitation. *Mol. Ecol.* **2012**, *21*, 1864–1877. [CrossRef]
43. Puillandre, N.; Brouillet, S.; Achaz, G. ASAP: Assemble species by automatic partitioning. *Mol. Ecol. Resour.* **2021**, *21*, 609–620. [CrossRef]
44. Zhang, J.; Kapli, P.; Pavlidis, P.; Stamatakis, A. A general species delimitation method with applications to phylogenetic placements. *Bioinformatics* **2013**, *29*, 2869–2876. [CrossRef]
45. Kapli, P.; Lutteropp, S.; Zhang, J.; Kobert, K.; Pavlidis, P.; Stamatakis, A.; Flouri, T. Multi-rate Poisson Tree Processes for single-locus species delimitation under Maximum Likelihood and Markov Chain Monte Carlo. *Bioinformatics* **2017**, *33*, 1630–1638. [CrossRef]

46. Stamatakis, A. RAxML version 8: A tool for phylogenetic analysis and post-analysis of large phylogenies. *Bioinformatics* **2014**, *30*, 1312–1313. [CrossRef]
47. Fujisawa, T.; Barraclough, T.G. Delimiting species using single-locus data and the generalized mixed yule coalescent approach: A revised method and evaluation on simulated data sets. *Syst. Biol.* **2013**, *62*, 707–724. [CrossRef]
48. Bouckaert, R.; Vaughan, T.G.; Barido-Sottani, J.; Duchêne, S.; Fourment, M.; Gavryushkina, A.; Heled, J.; Jones, G.; Kühnert, D.; De Maio, N.; et al. BEAST 2.5: An advanced software platform for Bayesian evolutionary analysis. *PLoS Comput. Biol.* **2019**, *15*, e1006650. [CrossRef]
49. Rambaut, A.; Suchard, M.A.; Xie, D.; Drummond, A.J. Tracer, v.1.6. Available online: http://beast.bio.ed.ac.uk/Tracer (accessed on 6 May 2016).
50. Darriba, D.; Taboada, G.L.; Doallo, R.; Posada, D. jModelTest 2: More models, new heuristics and parallel computing. *Nat. Methods* **2012**, *9*, 772. [CrossRef]
51. Guindon, S.; Gascuel, O. A Simple, Fast, and Accurate Algorithm to Estimate Large Phylogenies by Maximum Likelihood. *Syst. Biol.* **2003**, *52*, 696–704. [CrossRef] [PubMed]
52. Ribeiro, C.; Firme, B.; Araujo, S.A.; de Sá, A.; Zander, F.; Teixeira, K.; Santos, L.R.; Rodrigues, M.E. Check-list of Odonata from the state of Bahia, Brazil: Ecological information, distribution, and new state records. *Odonatologica* **2021**, *50*, 161–186. [CrossRef]
53. Takiya, D.; Santos, A.P.; Pinto, Â.; Henriques-Oliveira, A.L.; Carvalho, A.; Sampaio, B.; Clarkson, B.; Moreira, F.; Avelino-Capistrano, F.; Gonçalves, I.; et al. Aquatic Insects from the Caatinga: Checklists and diversity assessments of Ubajara (Ceará State) and Sete Cidades (Piauí State) National Parks, Northeastern Brazil. *Biodivers. Data J.* **2016**, *4*, e8354. [CrossRef] [PubMed]
54. Casas, P.A.S.; Sing, K.-W.; Lee, P.-S.; Nuñeza, O.M.; Villanueva, R.J.T.; Wilson, J.-J. DNA barcodes for dragonflies and damselflies (Odonata) of Mindanao, Philippines. *Mitochondrial DNA Part A* **2018**, *29*, 206–211. [CrossRef] [PubMed]
55. Rewicz, T.; Móra, A.; Tończyk, G.; Szymczak, A.; Grabowski, M.; Calleja, E.J.; Pernecker, B.; Csabai, Z. First records raise questions: DNA barcoding of Odonata in the middle of the Mediterranean. *Genome* **2021**, *64*, 196–206. [CrossRef]
56. Ris, F. Libellulinen. In *Collections Zoologiques du Baron Edm. De Selys Longchamps—Catalogue Systématique et Descriptif*; Hayez: Bruxelles, Belgium, 1911; pp. 385–528.
57. Borror, D.J. *A revision of the Libelluline genus Erythrodiplax (Odonata)*; The Ohio State University: Columbus, OH, USA, 1942.
58. Garrison, R.W. A synopsis of the genus Hetaerina with descriptions of four new species (Odonata: Calopterygidae). *Trans. Am. Entomol. Soc.* **1990**, *116*, 175–259.
59. von Ellenrieder, N. A synopsis of the Neotropical genus Nephepeltia (Odonata: Libellulidae), including description of a new species, synonymies, and a key to males. *Zootaxa* **2014**, *3796*, 121–146. [CrossRef]
60. von Ellenrieder, N.; Muzón, J. The Argentinean species of the genus Perithemis Hagen (Anisoptera: Libellulidae). *Odonatologica* **1999**, *28*, 1–14.
61. Garrison, R.W. A synopsis of the genus Telebasis (Odonata: Coenagrionidae). *Int. J. Odonatol.* **2009**, *12*, 1–121. [CrossRef]
62. De Marmels, J.; Rácenis, J. An analysis of the cophysa-group of Tramea Hagen, with descriptions of two new species (Anisoptera: Libellulidae). *Odonatologica* **1982**, *11*, 109–128.
63. Koroiva, R.; Rodrigues, L.R.R.; Santana, D.J. DNA barcoding for identification of anuran species in the central region of South America. *PeerJ* **2020**, *8*. [CrossRef] [PubMed]
64. Geiger, M.; Koblmüller, S.; Assandri, G.; Chovanec, A.; Ekrem, T.; Fischer, I.; Galimberti, A.; Grabowski, M.; Haring, E.; Hausmann, A.; et al. Coverage and quality of DNA barcode references for Central and Northern European Odonata. *PeerJ* **2021**, *9*, e11192. [CrossRef] [PubMed]
65. Galimberti, A.; Assandri, G.; Maggioni, D.; Ramazzotti, F.; Baroni, D.; Bazzi, G.; Chiandetti, I.; Corso, A.; Ferri, V.; Galuppi, M.; et al. Italian odonates in the Pandora's box: A comprehensive DNA barcoding inventory shows taxonomic warnings at the Holarctic scale. *Mol. Ecol. Resour.* **2021**, *21*, 183–200. [CrossRef] [PubMed]
66. Vega-Sánchez, Y.M.; Lorenzo-Carballa, M.O.; Vilela, D.S.; Guillermo-Ferreira, R.; Koroiva, R. Comment on "Molecular identification of seven new Zygopteran genera from South China through partial cytochrome oxidase subunit I (COI) gene". *Meta Gene* **2020**, *25*, 100759. [CrossRef]

diversity

MDPI

Article

Climate Change Is Driving Shifts in Dragonfly Species Richness across Europe via Differential Dynamics of Taxonomic and Biogeographic Groups

Kent Olsen [1,*], Jens-Christian Svenning [2,3,4] and Henrik Balslev [2]

1 Research and Collections, Natural History Museum Aarhus, Wilhelm Meyers Allé 10, DK-8000 Aarhus, Denmark
2 Section for Ecoinformatics and Biodiversity, Department of Biology, Aarhus University, Ny Munkegade 116, DK-8000 Aarhus, Denmark
3 Center for Biodiversity Dynamics in a Changing World, Department of Biology, Aarhus University, Ny Munkegade 116, DK-8000 Aarhus, Denmark
4 Center for Sustainable Landscapes under Global Change, Department of Biology, Aarhus University, Nordre Ringgade 1, DK-8000 Aarhus, Denmark
* Correspondence: kent@nathist.dk; Tel.: +45-40272030

Abstract: Understanding how changes in species richness pattern correlate with range changes in different taxonomic and biogeographic groups is important for conservation because it allows for generalizations about which species are at greatest risk. Here, we assessed whether changes in species richness patterns result from generalized range shifts across taxonomic and biogeographic groups or from changes in specific subsets of species. Using data from 1988 and from 2010, we studied changes in distributional range of European dragonfly species, using outline distribution maps for all dragonflies combined and separately for taxonomic suborders (Zygoptera and Anisoptera) and biogeographic groups (Boreo-alpine, Eurasian, Mediterranean, and Tropical). The results demonstrated differing range dynamics for Zygoptera and Anisoptera, with Anisoptera driving local turnover in species richness to a greater extent than Zygoptera. The distributional range of Tropical and Mediterranean species had expanded to a much greater extent than that of Eurasian and Boreo-alpine species. Large-scale changes in species richness arose from several divergent, group-specific processes. Overall, local diversity especially declined in parts of southern and south-eastern Europe, reflecting local losses in multiple species rather than major range contractions among Mediterranean or Eurasian species. In fact, among the biogeographic groups, overall range declines were most prominent among Boreo-alpine species, highlighting the particular threat from climate change to this group.

Keywords: biodiversity; geographic range expansion; Odonata; range dynamics; range size; species distribution; species richness; zoogeography

Citation: Olsen, K.; Svenning, J.-C.; Balslev, H. Climate Change Is Driving Shifts in Dragonfly Species Richness across Europe via Differential Dynamics of Taxonomic and Biogeographic Groups. *Diversity* **2022**, *14*, 1066. https://doi.org/10.3390/d14121066

Academic Editors: M. Olalla Lorenzo-Carballa and Ricardo Koroiva

Received: 19 September 2022
Accepted: 2 December 2022
Published: 3 December 2022

Publisher's Note: MDPI stays neutral with regard to jurisdictional claims in published maps and institutional affiliations.

1. Introduction

The latitudinal gradient of species richness is well documented for most higher taxa in both terrestrial and aquatic environments [1]. Current environmental changes are causing shifts in geographical distributions of species, leading to new patterns of species richness and assemblages at regional and local scales. Human-driven climate change is already profoundly affecting species distributions, causing substantial range shifts and expansions in species that can keep pace with changes in climate and resources [2] or that can adapt to new resource conditions and exploit formerly unsuitable habitats [3,4]. Species that cannot do either will experience range contraction or local extinction [5]. Habitat loss and degradation may cause range shifts to lag behind changes in climate and resources if unsuitable habitat restricts or blocks emigration by spatially isolated and small populations [6]. In such cases, small local populations can become more susceptible to extinction [7] because of stochastic events such as extreme weather or climatic variation at range boundaries where individuals live at the limits of their physiological tolerances [8]. The overall consequences of these range shift dynamics are changes in local species richness and assemblages, which eventually may lead to shifts in the latitudinal richness gradient [9]. Climate change is expected to reduce the number of species globally [10], but species richness at regional and local scales could increase or decrease.

Different species from a variety of ecological systems are showing poleward range expansion in the Northern Hemisphere that is consistent with climate change [2], and this has had important effects on distributions and regional species richness [11]. Commonly used scenarios for future changes predict an increase in global temperature of 1.8–6.4 °C in this century [12]. Phenotypic plasticity may play a key role in surviving a changing climate [3,4], but such climatic changes will probably stress insects, which likely have insufficient adaptive potential to keep pace with the rate of change [13]. In principle, evolutionary adaptation could be a response [14,15], but niche conservatism in some insects including dragonflies suggests a limited scope for this strategy [16,17]. Most species therefore are expected to show altered distribution rather than adaptation to warmer temperatures in situ [18].

The ranges of several European dragonfly species (Odonata) have expanded or moved northward (e.g., [19–22]. In this taxonomic group, temperature is a major determinant of species distribution [21], life cycle regulation and larvae growth responses [23,24], shifts in voltinism and seasonal regulation [25,26], and phenology [25,27–29], as well as immune function capacity [30,31] and pigment production for thermoregulation [32]. These different effects and responses originate in the facts that (1) dragonflies are flying insects that lay eggs in aquatic habitats, with larvae strictly tied to water for months or even years prior to emergence; (2) climate changes influence distribution in space and time of habitats and food resources; and (3) their metabolic and physiological processes are temperature dependent [33].

Knowing the range dynamics of species affected by climate warming is imperative for understanding which species are most likely to experience expansion or contraction of their range in response to global climate change. In conservation, information about trends in species range shifts is needed for setting priorities and assigning threat status. Change in occurrence have been used by International Union for Conservation of Nature (IUCN) to determine species status in the European Red List, which shows that 15% of European dragonfly species are threatened, with 2% being critically endangered, 4% endangered, and 9% classified as vulnerable. A further 11% are considered to be near threatened within Europe [34]. Additionally, a representative global assessment of conservation status has been completed and analyzed for dragonflies, currently the only insect group for which that has been conducted [35]. Only a few studies have concentrated on more integrated measures of change, such as species richness and local species assemblage [36]. However, although species are expected to respond individually, the overall consequences of environmental changes will likely be shifts in local species richness and assemblage composition.

The combined latitudinal and altitudinal species richness gradient for European dragonflies, ranges from many species in warmer southern regions to fewer species in colder northern regions, and Europe overall is species-poor compared with the tropics [37]. The higher diversity of dragonflies in the mountains is influenced not only by temperature and rainfall but also by the greater diversity of habitats in these areas [38]. Nevertheless, the legacies of past climate may be important for understanding current species distributions. Traditionally, it has been thought that three major Pleistocene refugia on the Iberian, Italian, and Balkan peninsulas were the source of recolonization of most of the temperate part of Europe after the last ice age [39,40]. Recent studies based on plants, terrestrial vertebrates, and butterflies have revealed a much more complicated situation, however, and postglacial recolonization may have been sourced from the east and from small ice age refugia in Europe north of the Alps (e.g., Simonsen and Huemer [41], Ursenbacher et al. [42], Brochmann et al. [43], and Schmitt et al. [44]). For highly mobile species, present-day ranges are believed to be primarily governed by current environmental conditions rather than changes in environmental conditions over time. In contrast, current ranges of less mobile species may represent only partially incomplete post-glacial recolonization [45]. Dragonflies are believed to have high dispersal capacities in general [33], so that present day climate warming and resources should mainly drive changes from past distribution and geographical richness patterns; restrictions by physical barriers such as high-altitude ranges are expected to have a lesser influence by having prevented postglacial recolonization in certain regions.

The order Odonata consists of the two suborders: true dragonflies (Anisoptera) and damselflies (Zygoptera). True dragonflies are capable of using thoracic muscle vibration to heat their body and to a certain degree regulate their hemolymph circulation, adaptations which damselflies lack and as a consequence they are considered thermoconform [33]. Hence, true dragonflies and damselflies are affected differently by temperature [46] where true dragonflies in general are more tolerant to high and low temperatures than damselflies, despite living in the same latitudes or altitudes [17,33]. Furthermore, true dragonflies are generally large, robust, and physically strong, and their hind wings have a broad base and are larger than the front pair. Damselflies are typically smaller and therefore do not fly as fast as true dragonflies in active flight, and their front and hind wings are similar in shape. Most European dragonfly species are strong fliers that are able to move between suitable habitats. Commuting between roosting, foraging, and reproductive sites up to several kilometers apart does not lead to relocation of next generations into a different habitat. Dispersal, in contrast, is unidirectional and may be a response to unfavorable habitat conditions, mass emergence after unusual weather, or population increases following favorable weather conditions [47]. True dragonflies and damselflies may occupy suitable habitats for several generations and then move to other suitable regions when the original habitat deteriorates [33]. This mobility helps species to maintain continuity of reproduction in the face of discontinuous habitat suitability, and as a result, their distributional range and biogeographic species assemblage becomes dynamic.

Environmental change can have strong effects on the population dynamics, distribution, and diversity of dragonflies [22,23,37,48], making them well suited for evaluating the mechanisms of these changes. Their sensitivity to habitat quality, amphibious life cycle, and ease of identification combined with the substantial knowledge about their distribution and ecological requirements uniquely suit them for studies of the effects of environmental changes in the short term (water pollution, structural changes in running and standing water) and the long term (species conservation and biogeography).

For our work here, we used distribution maps of European dragonflies from 1988 [49] and 2010 [34] to identify how their species ranges have changed during those 22 years. Our scope was to track temporal changes in distribution and to highlight patterns in the latitudinal and longitudinal movements at the margins of the dragonfly ranges. For this purpose, we analyzed geographic patterns of change in species richness with the aim of identifying species groups sharing functional and biogeographic traits that primarily drive local turnover in species assemblages and cause geographical shifts in species richness patterns. The five specific questions we addressed are as follows: (1) Are increases in ranges of true dragonflies greater than those of damselflies? (2) Have the ranges of southern species increased more than the ranges of continental and northern species? (3) Will northern species and high-elevation species experience reduced overall ranges as their realized climatic envelopes shrink because of global warming? (4) Are northwards range shifts greater than movements at other range margins reflecting a directional poleward shift rather than a non-directional range expansion? (5) Are geographical shifts in species richness patterns driven mainly by southern species rather than by species from a more continental and northern origin? We argue that these measures would be particularly useful for detecting effects of environmental changes and highlighting the importance of using insects—and especially dragonflies—as first-level indicators of environmental health.

We expect the current findings to support conservation efforts by providing additional means of determining species and species groups most at risk.

2. Materials and Methods

2.1. Study Area

Our study area covered 6,331,488 km^2 of the westernmost peninsula of Eurasia (=Europe) limited by the Arctic Ocean to the north, by the Atlantic Ocean to the west, and by the Mediterranean Sea to the south (see Figure 2 in Olsen et al. [22]). The eastern border of the study area followed a combination of the 35° E longitude and the eastern margin used in outline range maps in Askew [49]. All larger European islands in the Mediterranean Sea were included as in Olsen et al. [22].

2.2. Data

2.2.1. Species Distribution Data

Distributional ranges of dragonflies in Europe were obtained from two points taken 22 years apart, based on outline maps in Askew [49] and Kalkman et al. [34]. The maps do not always present the full distributional range and sometimes represent only the part that falls within the westernmost Eurasian peninsula. We excluded data from east of 35° E and south of the Mediterranean Sea because dragonfly occurrences in these regions are not well documented (e.g., Dijkstra and Lewington [50]).

Of the 130 species of dragonflies known to occur within the study area, we constructed outline range maps for 123, after excluding vagrant species, species new to science since 1988, and species without a range map in Askew [49] or Kalkman et al. [34] (see Table S1 for a list of excluded species, species that in 2010 were included as new in Europe, taxonomic and nomenclatural changes, and modifications to species ranges). Of the 123 included species, 4 colonized Europe during the 22-year period, whereas 119 species occurred in both data sets (see Table A1 and Table S1 for a list of the 4 and 119 species).

Maps from Askew [49] were georeferenced in ArcGIS 10.2 [51] based on scanned TIFF images, whereas maps from [34] were provided as shape files from the Freshwater Biodiversity Unit under the IUCN Global Species Program. All species ranges categorized as extant in Kalkman et al. [34] were included, whereas all ranges with a signature of extinction were omitted. The distribution maps in 1988 and 2010 were cut with the same European coastline layer in ArcGIS 10.2 [51] to ensure that species ranges followed the same extent of land cover and to facilitate direct comparison.

2.2.2. Species Classification

As functional traits, we used the morphological characteristics that distinguish the taxonomic suborders of European dragonflies—damselfly (Zygoptera) (n = 41) and true dragonfly (Anisoptera) (n = 82) species (see Table A1 for a list of species in each suborder). For biogeographic traits, we used four groups—Tropical (n = 14), Mediterranean (n = 56), Eurasian (n = 36), and Boreo-alpine (n = 17) species (see Table A1 for a list of species in each group). These subdivisions by functional and biogeographic traits based on Dijkstra and Lewington [50], Sternberg [52], and Beschovski et al. [53] allowed us to distinguish species responses and differential changes in species richness patterns arising from southern Mediterranean fauna elements (Tropical and Mediterranean groups) from the species responses with a more continental distribution in central and northern Europe (Eurasian and Boreo-alpine groups). It also allowed us to capture effects caused by species with an Afrotropical and Oriental origin (Tropical group) from the more extreme habitat specialists, such as the Boreo-alpine species.

2.3. Data Analysis

2.3.1. Range Shifts

In ArcGIS 10.2 [51], we transformed the outline distributions into gridded maps with 880 cells of 100 × 100 km to estimate distributional range as an occupancy of grid cells. The large grid resolution allowed us to minimize artefacts from outline range maps, including false absences or more commonly false presences, and thus avoid overestimating the extent of occurrence of species [22]. To address questions 1–3 of our study, we calculated differences in total species range (ΔR) between 1988 and 2010 for stable (ΔR = 0), contracting (negative ΔR), and expanding (positive ΔR) species as relative change, giving the percentage change in the number of occupied cells.

To address question 4 of our study, we measured change in distributional range as shifts in northern, southern, eastern, and western boundaries, calculated by subtracting minimum and maximum latitude and longitude for each species in 1988 from the values in 2010. All range shift distances were standardized so that expansions and contractions were expressed with positive and negative values, respectively.

To further address question 4, we determined directionality in range shifts by calculating direction (0–360°) and compass distance of range centroid shifts. All distances were calculated using an equidistant projection in ArcGIS 10.2 [51].

2.3.2. Species Richness

To address question 5 of our study, we calculated latitudinal species richness (number of species in 100-km latitudinal intervals) to evaluate changes in species richness in each biogeographic group between 1988 and 2010. We subdivided our study area into 10 × 10 km grid cells (total 67.374) and calculated local species richness (number of species in each grid cell) by overlaying the grid onto the outline distribution maps. We then subtracted the number of species in each grid cell in 1988 from the number in 2010 to evaluate geographic patterns in diversity changes over the 22-year period. When plotting geographical patterns of species richness, we chose to reduce the grid cell size to 10 × 10 km because higher resolution allowed us visually to detect patterns at a more local scale than if we used the 100 × 100 km grid as applied in the statistical analysis.

2.4. Statistical Analysis

The various measures of range shifts (overall range shift; range shift at the four range margins: north, south, east, and west; and shift in range centroid) in European damselfly (Zygoptera) and true dragonfly (Anisoptera) species were analyzed with the taxonomic suborders—Zygoptera and Anisoptera—used as unmatched test groups in a Mann–Whitney–Wilcoxon test.

The various measures of range shifts (overall range shift; range shift at the four range margins: north, south, east, and west; and shift in range centroid) in biogeographical groups of European dragonfly species were analyzed with the biogeographic groups—Tropical, Mediterranean, Eurasian, and Boreo-alpine—used as unmatched test groups in a Kruskal–Wallis test.

Range shifts at range margins for all species, for damselfly species and true dragonfly species, and for species in the four biogeographical groups—Tropical, Mediterranean, Eurasian, and Boreo-alpine—were analyzed with the northern margin and the southern, eastern, and western margins combined (=other margins) used as unmatched test groups in a Mann–Whitney–Wilcoxon test.

All statistical tests were performed using R [54], and polar plots were made with the plotrix package [55].

3. Results

3.1. Range Shift Pattern

On average, range sizes increased between 1988 and 2010. Median change in overall range size was 254,965 km^2, median percentage change in range was 18%, and median change in number of 100 × 100 km grid cells was 35. Of the 123 species, 106 had expanding ranges (including 4 that colonized Europe between 1988 and 2010), 3 species (all damselflies) had stable range sizes, and 14 species experienced range contractions (7 damselflies and 7 true dragonflies) (Tables A1 and S2).

Percentage change in range was significantly larger in true dragonflies than in damselflies (Figure 1A, Table S3). The magnitude of the range changes differed significantly among biogeographic groups, with the largest increase in Tropical and Mediterranean species compared with Eurasian and Boreo-alpine taxa. With separate analyses for the two suborders, however, the changes were significant only for true dragonflies (Figure 1A, Table S4). Within the Boreo-alpine group, 13 species had expanding ranges, 1 species had a stable range size, and 3 species showed range contractions (Tables A1 and S2).

Latitudinal shifts are much larger than longitudinal shifts. The latitudinal shift at the northern range margin was significantly larger than shifts at the other three margins combined, and when accounting for suborder, the differences were significant for both damselflies and true dragonflies (Table S5). When accounting for biogeographic group, shifts in the northern range margin were significantly larger than shifts in the other three directions for Tropical, Mediterranean, and Eurasian species, but not for Boreo-alpine species (Table S5). When shifts at the four range margins were analyzed separately rather than together, we found a significant difference between damselfly and true dragonfly species only at the western border (Figure 1C–F, Table S3). In addition, we found a significant difference among the four biogeographic groups at the eastern border, but when suborder was considered, the differences were significant only for true dragonflies (Figure 1C–F, Table S4).

The centroid of dragonfly ranges shifted by 176 km on average (median 138 km), with a significant difference between damselfly and true dragonfly species (Figure 1B, Table S3) and among the four biogeographic groups (Figure 1B, Table S4). However, when accounting for suborder, the differences between biogeographical groups were not significant for either damselflies or true dragonflies (Table S4). Centroids of dragonfly ranges moved in all directions, with a mean shift towards south–southwest (202°) (Figure 2). Damselfly and true dragonfly species did not differ significantly in the direction of the range shifts (Figure 2, Table S3), but the biogeographic groups did show differences (Figure 2, Table S4).

3.2. Species Richness Pattern

On average, the latitudinal species richness (number of species in 100 km latitudinal intervals) increased from 1988 to 2010 by 5.4 species (median 5.0 species), representing 1.3 damselfly and 4.1 true dragonfly species (Figure 3). The largest increase was between the Mediterranean Sea and 46° N, with an average of 7.1 species (1.3 damselfly and 6.3 true dragonfly), and between 52° N and 63° N, with an average of 7.4 species (2.8 damselfly and 4.6 true dragonfly).

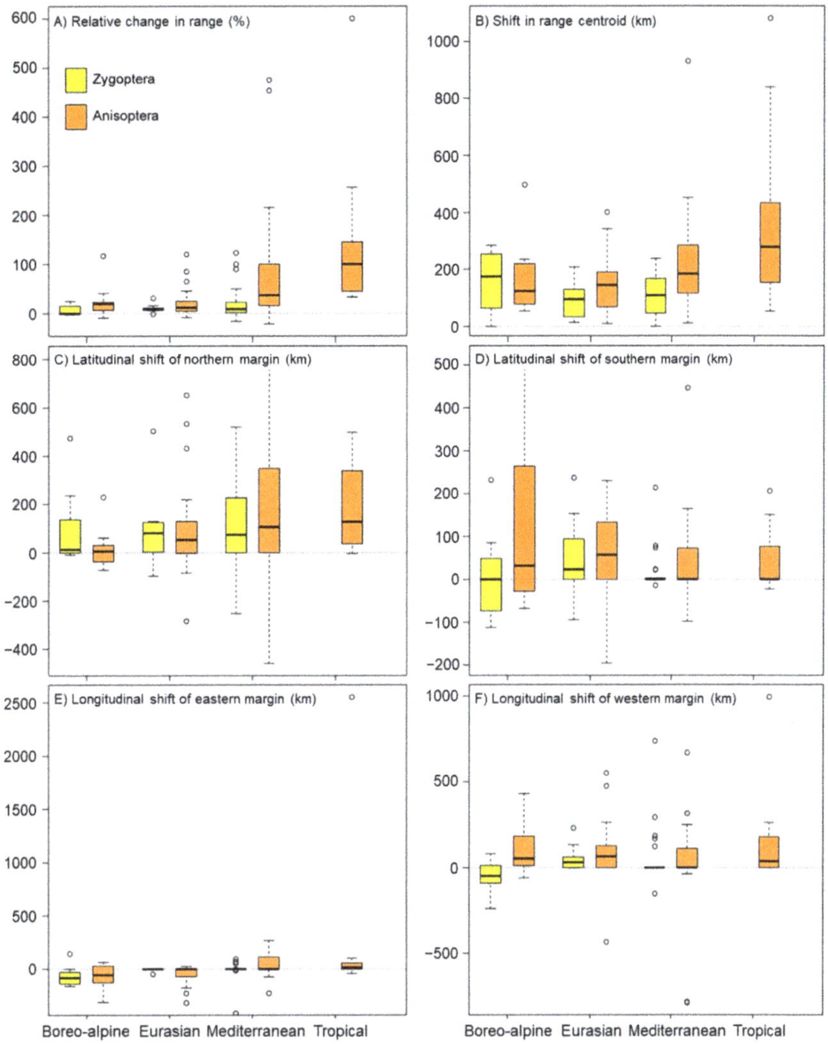

Figure 1. Range shifts in European dragonfly (Odonata) species from 1988 to 2010. (**A**) Relative change in range (percent change in number of occupied 100 × 100 km grid cells). (**B**) Distance shift of range centroid. (**C**) Latitudinal shift of northern range margin. (**D**) Latitudinal shift of southern range margin. (**E**) Longitudinal shift of eastern range margin. (**F**) Longitudinal shift of western range margin. Data separated by biogeographical groups (Tropical, Mediterranean, Eurasian, and Boreo-alpine) for damselflies (Zygoptera, yellow) and true dragonflies (Anisoptera, orange). The box-and-whisker plots illustrate the spread and skewness of the data through their quartiles and the median (thick black middle line). The whiskers extending from the box show data variability outside the upper and lower quartiles. Outlier points that differed significantly from the rest of the dataset are plotted as individual points (empty circles) beyond the whiskers.

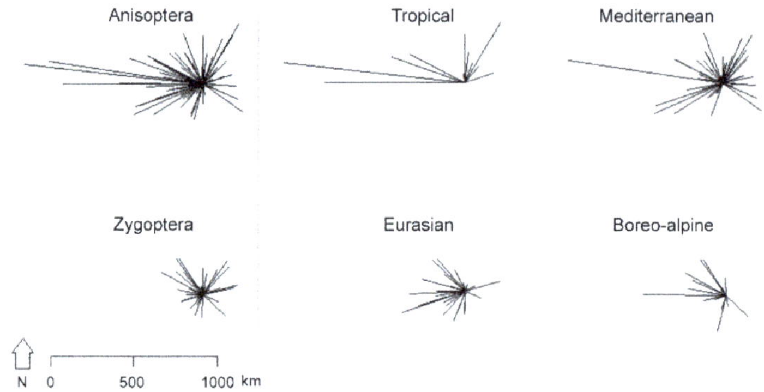

Figure 2. Plots showing shifts (1988–2010) in the range centroids of European (Odonata) dragonflies. Distance (km) and compass direction (°) of range shifts in range centroids in species from 1988 to 2010. Data separated by taxonomic suborder (damselflies [Zygoptera] and true dragonflies [Anisoptera]) and biogeographical groups (Tropical, Mediterranean, Eurasian, and Boreo-alpine).

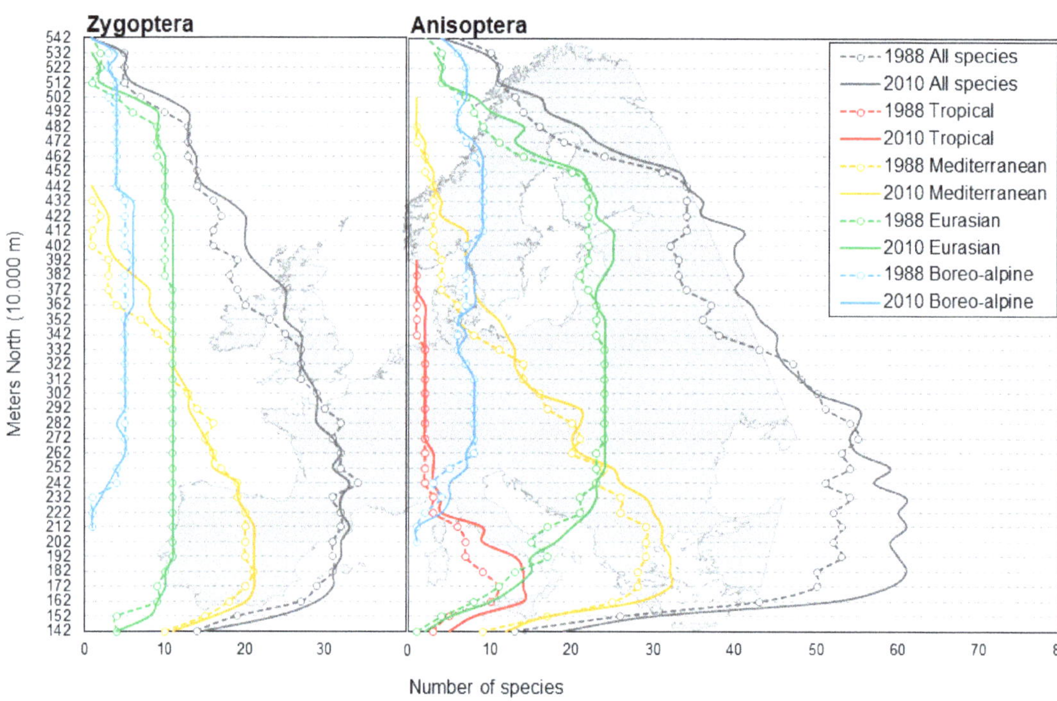

Figure 3. Shift in latitudinal species richness of European dragonfly (Odonata) species, plotted against latitude. Species richness presented as number of species in 100-km latitudinal intervals in species from 1988 to 2010. Data separated by biogeographical groups (Tropical—*red*, Mediterranean—*yellow*, Eurasian—*green*, and Boreo-alpine—*blue*) for damselflies (Zygoptera, *left*) and true dragonflies (Anisoptera, *right*).

The biogeographic group of damselflies with the largest increase in latitudinal species richness was the Mediterranean, where richness increased most at Scandinavian latitudes, so that the expansion of their northern range included a shift from Central Europe into the Scandinavian zone (Figure 3). Additionally, the true dragonfly fauna that accounted for the largest increase were the Tropical and Mediterranean in Southern Europe, and Mediterranean and Eurasian in Central and Northern Europe (Figure 3).

Average local species richness (number of species in 10 km $\times$ 10 km grid cells) increased by 7.3 species (median 7.0 species), with 2.0 damselfly and 5.3 true dragonfly species. The highest values for local species richness were observed in eastern and central Europe, with a maximum (>64) in the lowlands north of the Alps and Carpathian Mountains, and in the region west of the western Alps (Figure 4). Local species richness of the two suborders followed a similar geographic pattern, although a hotspot west of the Alps was more pronounced in damselflies compared with true dragonflies, and true dragonflies were more species rich than damselflies in lowlands north of the Alps and Carpathian Mountains (Figure 4). The diversity center for damselflies was located around the "Massif Central" in France (28 species), whereas the diversity center for true dragonflies (46 species) was located west of the northwestern pre-Alps and areas in northern Slovakia along the Carpathian Mountains (Figure 4).

The geographic pattern of changes in local species richness differed among biogeographic groups (Figure S1), which followed variation in the percentages of 10 km $\times$ 10 km grid cells in which the number of Tropical, Mediterranean, Eurasian, or Boreo-alpine species increased, decreased, or remained unchanged between 1988 and 2010. The Boreo-alpine group showed the largest decline in the percentage of 10 $\times$ 10 km grid cells where species from that group occurred in 1988 compared with the similar decline in species from Tropical, Mediterranean, and Eurasian groups. In contrast, the Mediterranean and Eurasian species showed the largest percentage of cells with increasing diversity (Figure 5).

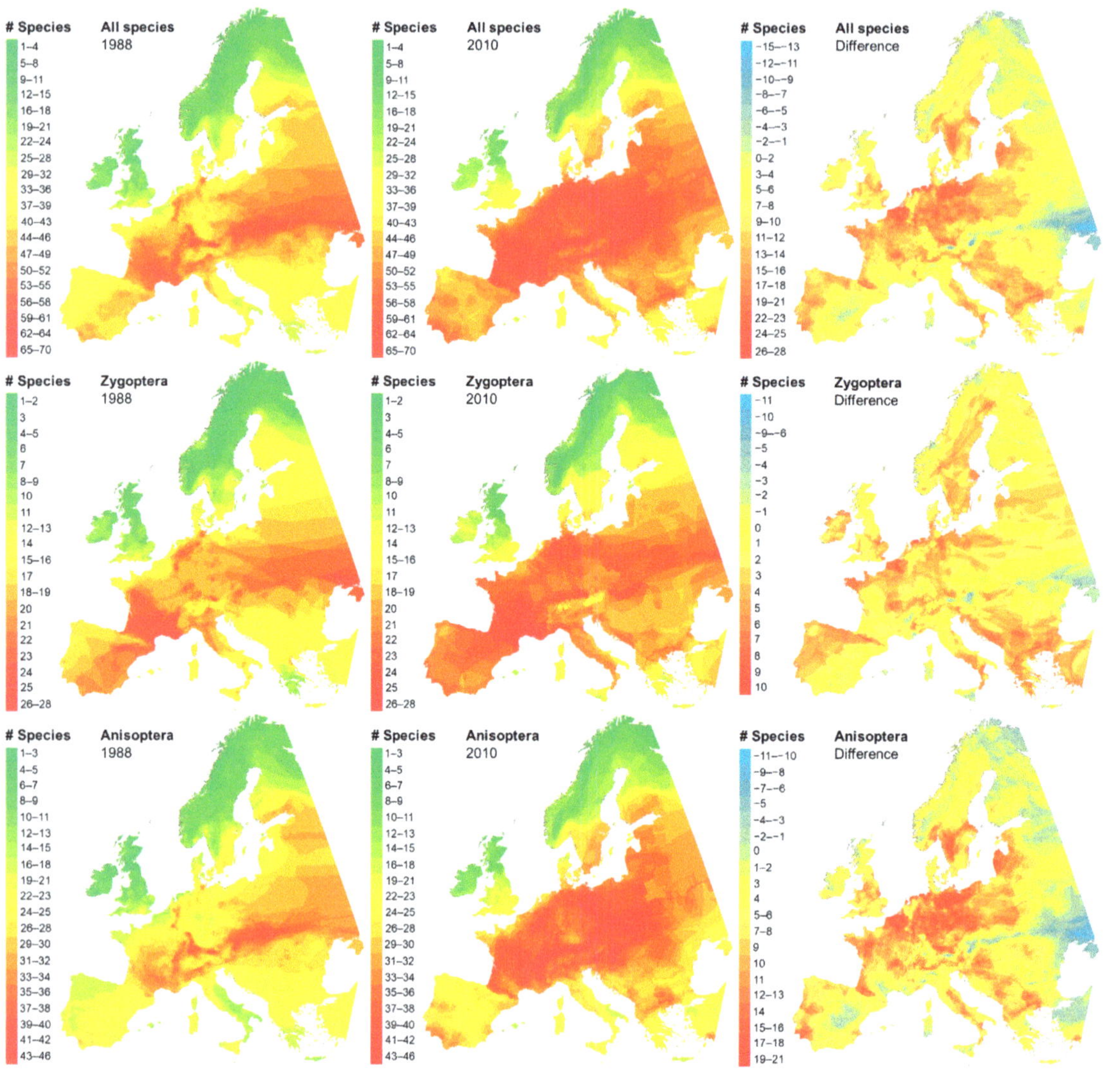

Figure 4. Geographical patterns of European dragonfly (Odonata) species richness and changes by taxonomic suborder. Observed species richness in 1988 according to Askew [49] (left column), observed species richness in 2010 according to Kalkman et al. [34] (middle column), and difference in observed species richness between 1988 and 2010 (negative value = decrease, positive value = increase) (right column). European data presented for all species (upper row), damselfly (Zygoptera) species (middle row), and true dragonfly (Anisoptera) species (lower row) at 10 km × 10 km grid resolution.

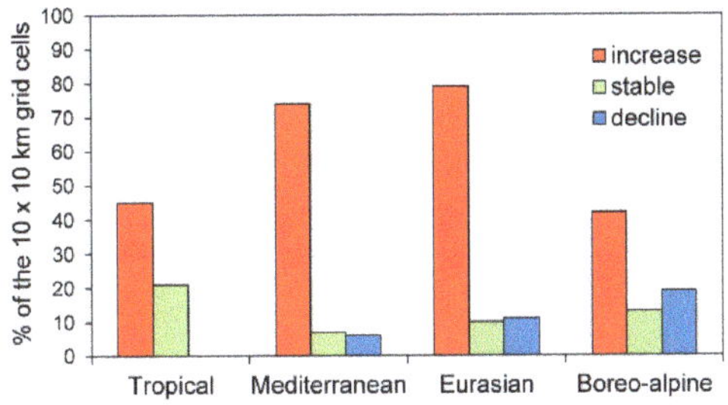

Figure 5. Percentages of 10 × 10 km grid cells (total 67.374 grid cells) where the number of European dragonfly (Odonata) species increased, decreased, or remained unchanged from 1988 to 2010, separated into biogeographic groups (Tropical, Mediterranean, Eurasian, and Boreo-alpine).

4. Discussion

4.1. Data Quality

We used expert-drawn outline range maps to address how species ranges have changed on a continental scale. Even though they are expert-drawn maps, one drawback of outline range maps is that species do not occur uniformly within their range [56], so that these maps can include false absences or presences [57]. Consequently, because of ignorance about the internal range structure [58], such maps may overestimate species occurrence [59], as has been addressed in the macroecological literature (e.g., Graham and Hijmans [36] and Hurlbert and White [60]). Moreover, a common critique is that outline maps represent only knowledge about the distribution that the respective authors have, rather than giving the true species distribution. If so, any analysis based on these maps could reflect changes in what the authors know rather than the true range patterns. Nevertheless, they represent the best currently available data on European dragonflies for addressing macroecological questions such as ours and multiple comparisons of outline distributions to find differences in species ranges or species richness have been published on various taxonomic groups within plants, vertebrates, and invertebrates (e.g., Hawkins et al. [61] and references therein), including macroecological studies on European dragonflies similar to ours (e.g., Olsen et al. [22], Grewe et al. [62], Hof et al. [63], Hof et al. [64], and Kalkman et al. [65]. Furthermore, of the 85 reviewed analyses of species richness in Hawkins et al. [61], 69% were based on range maps. We acknowledge that our maps represent rough approximations of the distribution of European dragonflies, but as Hurlbert and Jetz [66] demonstrated, using a sufficiently large grid resolution can surmount these problems.

Even though dragonflies are among the taxa with the best data record in space and time across Europe [5,65,67], an important concern about studies focusing on range shift is that expansions could simply be the outcome of a higher number of records. Although there will be some sampling heterogeneity on continental scale [65], the most significant northern range border shifts we found were for species that colonized Central Europe from the Mediterranean or extended their previous northern range border in Central Europe northwards into areas know to be well studied historically [34,67]. We found no general indication that the ranges shifts have been caused by false expansions due to an increase of knowledge. Hence, we assume that a lower sampling intensity in parts of Europe did not affect the observed range shifts on a 100 × 100 km grid level. Moreover, a bias in the distribution estimates should matter only if there were strong differences in mapping accuracy between damselflies and true dragonflies or among the four biogeographic groups. Finally, studies relying on true observations of range shifts rather than outcomes based on range maps yield results that support our findings on range shift in European dragonfly species, both on a more local scale (e.g., Hickling et al. [19], Knijf and Anselin [20], Ott [21], Suhling et al. [37], Hassall and Thompson [48], Riservato et al. [68], and Termaat et al. [69]), and continental scale (e.g., Kalkman et al. [65] and Boudot and Kalkman [67].

Because our geographic scope is Europe, the eastern border does not follow natural boundaries as the other range margins do. We will therefore have underestimated any shift eastwards for species distributed along the eastern border of the study area. As long as the bias causes underestimation rather than overestimation, however, we argue that we still obtained biologically meaningful and valuable information on range shift directionality by including the eastern margin.

4.2. Are Increases in Ranges of True Dragonflies Greater Than Those of Damselflies?

Compared with damselflies, the true dragonflies were more prone to overall range increases independent of biogeographic origin. The geographic differences between poorly dispersing damselflies and easily dispersing true dragonflies revealed that distribution pattern and ranges seemed to be regulated differently between the two suborders. Moreover, range expansion and successful establishment are subject to physical constraints. Despite the relatively weak dispersal ability of damselflies, their passive flight across land areas should still be sufficient to confer on them sufficient geographic plasticity to keep pace with shifts in climatic envelope and resources, but they have less ability to cross wide physical barriers such as the Mediterranean Sea or the North Sea. For dispersing species, distance between suitable habitats is important with regard to their chances of tracking climate and environmental change. If suitable dispersal corridors are absent, species responses to climate change may not be realized [6,22,70]. For tropical species, which currently are represented only by true dragonflies, it is reasonable to believe that the Sahara and the Mediterranean Sea together constitute a barrier preventing Afro-Tropical damselfly species from colonizing southern Europe or at least causing them to fall behind true dragonflies. Of the four species identified as new for Europe since 1988, the three tropical species (*Orthetrum sabina*, *Orthetrum taeniolatum* and *Trithemis kirbyi*) are true dragonflies, whereas only the Mediterranean species, *Ischnura fountaineae*, which arrived from North Africa to the Italian island Panteleria southwest of Sicily, is a damselfly [50]. With regard to damselfly species, which may not be able to keep up with the dispersal capacity of tropical true dragonflies, the latitudinal range centers of damselfly and true dragonfly species within the Mediterranean group did not differ significantly in 1988 ($p > 0.05$). In contrast, in the Eurasian and Boreo-alpine groups, true dragonflies on average showed a more northerly located range center. We cannot rule out that the difference in some species could have been caused by true dragonflies being able re-colonize Europe faster after the last glaciations than some damselflies, and we have not been able to find any studies supporting directly that damselflies exhibit postglacial dispersal limitation to the present day. However, because body size and the ability to vibrate thoracic muscles to heat their body matters for survival in a colder environment, we suggest that this difference in range center could also have resulted from a synergistic effect of the larger size, thermoregulation abilities and better flight capacity of true dragonflies being generally more robust and physically strong than the thermoconform damselflies.

4.3. Have the Ranges of Southern Species Increased More Than the Ranges of Continental and Northern Species?

Distributions of southern species (Tropical and Mediterranean groups) expanded to a larger extent than those of northern species (Eurasian and Boreo-alpine groups), which is consistent with other studies, at least in the temperate part of the world (e.g., Hickling et al. [19], Knijf and Anselin [20], and Ott [21]. This finding supports the expectation that it is especially in the species adapted to a warmer climate that we see the greatest range expansions.

The northward range expansions and the Afro-Tropical species entering southern Europe indicate that the range expansion of southern dragonfly species in Europe in particular is ongoing. However, the Sahara and the Mediterranean Sea together seem to constitute a barrier that cause the initial colonization to occur at a relatively low rate and range shifts are geographically skewed, with most range expansions occurring in central and northern Europe, whereas changes in the south are fewer and smaller. Here, the only species with a relatively large change in range were the Afro-Tropical species entering Europe. For most lowland species in southern Europe, temperature is not as much the constraining factor for their range as is the occurrence of freshwater and suitable habitat [71].

4.4. Will Northern Species and High-Elevation Species Experience Reduced Overall Ranges as Their Realized Climatic Envelopes Shrink Because of Global Warming?

Most Boreo-alpine species had expanded their distributional range, and we find no indications that these are false expansions due to increase of knowledge. Hence, we argue that any shifts in species range are not simply constrained by the availability of suitable habitat but may to some degree also be explained by constraints in the realized thermal niche and distributions at the range margins. The change in range centroids in Boreo-alpine species showed a mean longitudinal shift towards the west mainly into Fennoscandia and Central Europe, highlighting that turnover in species richness may be driven not only by northbound range expansions, but also by westward colonization. This pattern of longitudinal shift towards the west is counterintuitive to what we would expect for climate-driven range shift [2]. However, as the realized niche is not necessarily the same as the fundamental niche of the species, the thermal niche of some of the Boreo-alpine species could have been wider than the temperature, which used to be available in the area they are occupying, so when climate warms, they are still inside their fundamental niche and can perform better, causing the range expansion with the largest towards the west. Furthermore, it has been suggested that more dispersal (gene flow) may occur from central to peripheral populations (asymmetric migration) than the reverse [19]. If gene flow is stronger than selection along the range margins, core populations represent sources and the peripheral populations are sinks where genetic variation is continuously replenished [19]. In this way, local evolutionary adaptation at range margins may be prevented even though climate change triggers a different selective pressure than in the core range. When global warming eventually reaches the area with the genetically more diverse central populations, it may cause a shift in the species' realized niche. In response, rather than persisting only with a narrow range of habitat characteristics, the species may gradually adapt in a way that allows for the exploitation of formerly unsuitable habitats [3,4]. Following this, we suggest that some Boreo-alpine species with westward range expansion may not yet have been able to colonize all available areas and persist within their climatic envelope or may show some degree of thermal release during a climate-driven range expansion. The result could be a shift in their thermal niche, making them able to adapt to new resource conditions and exploit formerly unsuitable habitats. This pattern may explain why it was mostly Boreo-alpine species that expanded westward, especially so during the last decades, when global warming has been affecting boreal forest and taiga at increasing rates [12].

Evidence of a strong negative impact of climate change on dragonflies is lacking, although local examples of desiccation of bog habitats have been described [72]. We found three Boreo-alpine species that showed range contraction, namely the relatively widespread *Coenagrion lunulatum* and *Leucorrhinia albifrons* and the much rarer *Nehalennia speciosa*. We suggest that these trends could be consequences of global warming, and if so, they provide a negative signal for selection of oligotrophic freshwater species. Range contraction may not necessarily be driven by a decrease in their realized climatic envelopes as much as by habitat loss and degradation. For some dragonfly species, available habitat continues to decline because of global warming and drainage of wetland areas, but pollution and overgrowth of habitats may also threaten them [34]. Loss and degradation of habitat will cause local populations to go extinct and simultaneously escalate the degree of fragmentation. This pattern is believed to explain why *Nehalennia speciosa* is declining and has already become regionally extinct in many areas across its European range [73].

4.5. Are Northwards Range Shifts Greater Than Movements at Other Range Margins Reflecting a Directional Poleward Shift Rather Than Non-Directional Range Expansion?

Even though latitudinal shift at the northern range margin was larger on average than shifts at any other margins, we expected that range shift resulting from climate warming would occur not only at the northern margins, but also as poleward shifts of southern range margins [74]. We did not find significant contraction of the southern range boundaries, which could be explained by the fact that southern boundaries of distributional ranges for most species lie outside Europe, but also to some degree because the southern limit has been historically poorly defined. Nevertheless, especially in Mediterranean species, local diversity had declined in regions on the Iberian Peninsula and along the Mediterranean Sea and the Black Sea [68,71]. This is likely due to habitat destruction and degradation, pollution, mismanagement of water bodies due to increased water demand and a lower level of precipitation due to climate change [68]. These dynamics jointly illustrate that species occurrences may decline or that species may go locally extinct without causing a current contraction in overall range [2].

Tropical species highlight the colonization corridors for Afro-Tropical species entry into southern Europe, by crossing the narrowest straits of the Mediterranean Sea through the Iberian or Italian peninsulas, by using Mediterranean islands as stepping stones, or through the Near East. Although the initial colonization of Europe occurred as northbound movement from Africa [75], it is important to note that the colonization did not immediately follow a direct northerly route. When species have crossed the Mediterranean Sea, they may occur in fragmented ranges with populations scattered along various southern latitudes. Increase in range and subsequent colonization by these subpopulations means that the direction in their range shift appears stochastic rather than following a recognizable pattern. This appearance is illustrated by the tropical species *Paragomphus genei*, which had the largest change in range centroid in a 1081-km shift westward, resulting from recent range expansion across the Iberian Peninsula, where the species has begun to reproduce in watering pools constructed for sheep [76]. This example offers a possible explanation for why, in contrast to expectations, shift patterns in range centroid and overall range were mixed. Expanding species had an overall northbound directionality in their range expansion, and we expected this trend to manifest not only at the northern range boundary but also in the direction of range centroid shifts. However, as the plots illustrate (Figure 2), there was no common directionality because of differences in habitat requirements, habitat dispersal abilities, knowledge in distribution and other factors [22]

4.6. Are Geographical Shifts in Species Richness Patterns Driven Mainly by Southern Species Rather Than by Species from a More Continental and Northern Origin?

Since 1988, average range size for species in all biogeographic groups has increased, with southern species especially having expanded their ranges to the north, some increasing by hundreds of kilometers. This expansion, in turn, is driving increased overall species richness to a higher extent than expansions in continental and northern species, especially so across all of Central Europe. For tropical species, the most species-rich regions in Europe are currently situated in the southern part of the Iberian Peninsula and in southern Turkey. In contrast, local species richness of Mediterranean species has declined in southern Europe, but increased in northwestern, central, and eastern Europe, which is a general trend observed across animals and plants [7]. This overall pattern of increase in local species richness in dragonflies is mirrored almost completely by the Eurasian species, except for an additional increase in central Scandinavia and on the Balkan and Italian peninsulas. There is high diversity of Eurasian species throughout Central Europe, but with a clear decline along the northern Pyrenees, indicating that for the continental species associated with lower elevation land areas, the mountain areas may constitute physical barriers that limit range expansion from the northeast into the Iberian Peninsula. The diversity in Boreo-alpine species has increased in most of Scandinavia and in various isolated, rather fragmented parts of Central Europe, in contrast to declines in species richness north of the Black Sea.

Recent data on continental scale support that the expansion of ranges to the north, which in general was observed across all biogeographic groups between 1988 and 2010, seem to continue [67], which, in turn, drives shifts in dragonfly species richness across Europe even further [65]. We would argue that the shift in European species richness seen in our study and supported by more recent data could be accelerated further by very warm summers such as observed in the last few years.

The distribution of dragonfly species in Europe exemplifies a distinct biogeographical pattern. The lowest diversity occurs in the northern parts of mainland Europe and on islands such as Great Britain and Ireland, whereas the highest is found in Central Europe, where tropical and Mediterranean species co-occur with species from more temperate and boreal climates. As Central Europe is an area of confluence of multiple expansion routes, a higher overall diversity is to be expected. Confirmation of this region as a hotspot for dragonfly diversity underlines the high ecological importance of the Central European wetlands for water-linked species. In contrast, the Mediterranean glacial refuge areas are not as species rich as the central part of Europe. Recent studies based on plants, terrestrial vertebrates, and butterflies have demonstrated that postglacial recolonization of most of temperate Europe has come not only from the three major Pleistocene refugia on the Iberian, Italian, and Balkan peninsulas, but also from the east and from small ice age refugia in Europe north of the Alps (e.g., Simonsen and Huemer [41], Ursenbacher et al. [42], Brochmann et al. [43], Schmitt et al. [44]). Dragonflies are believed to have high dispersal capacities [33], so that changes in distribution and geographical richness pattern should mainly be driven by present-day climate warming and availability of suitable habitat [16] and only to a lesser extent by historical legacies of past climate; however, physical barriers such as high-altitude mountain ranges may prevent postglacial recolonization in certain regions. Such barriers could explain why the Mediterranean region is not as rich in species as the central part of Europe. Moreover, especially for the less mobile dragonfly species, it could indicate that assemblages of species in the westernmost peninsula of Eurasia and other geographically isolated regions such as Fennoscandia could reflect stepwise reductions in species immigration from glacial refuge regions during the postglacial recolonization process due to geographic distance [77] and physical barriers [45].

5. Conclusions

The results of this study demonstrate that large-scale changes in patterns of dragonfly species richness are the result of several divergent dynamics that differ for the taxonomic suborders and biogeographic groups of dragonflies. In addition to showing an overall increase in species range for the whole order, true dragonflies were more prone than damselflies to exhibit overall range expansions, independent of biogeographic origin. Consequently, true dragonflies had more local turnover than did damselflies. Even though damselflies are strong enough to keep pace with shifts in the climatic envelope by passive flight across land, they are usually not strong enough to cross wide physical barriers such as the Sahara, Mediterranean Sea, or the North Sea. Tropical and Mediterranean species had much more expanded ranges than did Eurasian and Boreo-alpine species. The greatest range expansions were found in warm-adapted species, with the most prominent diversity changes shaped by southern species. However, several Boreo-alpine species also expanded their ranges, especially westward. This pattern suggests that thermal release during climate-mediated range expansion may shift local species richness across Europe. The Central European hotspot for dragonfly diversity documents the high conservation value of the Central European wetlands for water-linked species. Local species richness declined in the Iberian Peninsula, the Mediterranean Sea, and the Black Sea areas, suggesting a negative impact of climate change on dragonfly species in warm regions. Range contractions of three Boreo-alpine habitat specialists in oligotrophic freshwater should be given special conservation attention to avoid regional extinction. This is especially true for *Nehalennia speciosa*, which face severe challenges from climate change declines.

We have provided an assessment on a European scale of how dragonfly species richness patterns have changed over a 22-year period. Our focus was on range shifts between groups of species with shared traits that we consider to be important drivers of changes in species richness. In this way, we document that understanding range changes and tracking changes in diversity patterns are important tools for conservation of dragonflies, and at the same time, dragonflies emerge as important first-level indicators of environmental health and conservation needs.

Supplementary Materials: The following supporting information can be downloaded at: https://www.mdpi.com/article/10.3390/d14121066/s1. Figure S1: Geographical patterns of European dragonfly (Odonata) species richness and changes by biogeographical groups; Table S1: List of European dragonfly (Odonata) species excluded from the analysis, where modifications to species range have been applied, that colonized Europe after 1988, and with taxonomic and nomenclatural changes between 1988 and 2010; Table S2: Data for the analysis of range shift in European dragonfly (Odonata) species. The list includes taxonomic suborder and scientific names of all species included in the analysis of range shift and changes in species richness 1988–2010. Also presented are the per species biogeographic group, range shift trend, relative change in range (percent change in number of occupied 100 × 100 km grid cells), distance of shift in northern, southern, eastern, and western range margins, and direction of shift in range centroid and distance of shift in range centroid; Table S3: Summary statistics of various measures of range shifts between 1988 and 2010 in European damselfly (Zygoptera) and true dragonfly (Anisoptera) species; Table S4: Summary statistics of various measures of range shifts between 1988 and 2010 in all European dragonfly (Odonata), damselfly (Zygoptera), and true dragonfly (Anisoptera) species, and species in the four biogeographical groups—Tropical, Mediterranean, Eurasian, and Boreo-alpine; and Table S5: Summary statistics of range margin shifts between 1988 and 2010 at the northern margin and the southern, eastern, and western margins combined in all European dragonfly (Odonata), damselfly (Zygoptera), and true dragonfly (Anisoptera) species, and species in the four biogeographical groups—Tropical, Mediterranean, Eurasian, and Boreo-alpine [78,79].

Author Contributions: Conceptualization, K.O., J.-C.S. and H.B.; methodology, K.O.; formal analysis, K.O.; investigation and data curation, K.O.; writing—original draft preparation, K.O.; writing—review and editing, K.O., J.-C.S. and H.B. All authors have read and agreed to the published version of the manuscript.

Funding: This study was supported by the Danish Ministry of Culture (TAKT2010-035), the Graduate School of Science and Technology of Aarhus University, and the Natural History Museum, Aarhus. This study is a contribution to JCS's VILLUM Investigator project "Biodiversity Dynamics in a Changing World" funded by VILLUM FONDEN (grant 16549) and the Center for Informatics Research on Complexity in Ecology (CIRCE), funded by Aarhus University and the Aarhus University Research Foundation under the AU IDEAS program.

Institutional Review Board Statement: Not applicable.

Informed Consent Statement: Not applicable.

Data Availability Statement: Data are provided in the paper, Supplementary Figure S1, and Supplementary Tables S1–S5.

Acknowledgments: We thank Brody Steven Sandel for help with R scripts and Peder Klith Bøcher and Lonnie Mikkelsen for ArcGIS assistance. We are particularly grateful to Melanie Bilz and Jemma Able from the Freshwater Biodiversity Unit under the IUCN Global Species Program for providing range maps from 2010 as ArcGIS shapefiles.

Conflicts of Interest: The authors declare no conflict of interest. The funders had no role in the design of the study; in the collection, analyses, or interpretation of data; in the writing of the manuscript; or in the decision to publish the results.

Appendix A

List of European dragonfly (Odonata) species included in the analysis of range shift and changes in species richness 1988–2010 with information on taxonomic suborder, scientific name including author and year of species description, biogeographic group, and range shift trend. * Indicate the four species that in 2010 were included as new in Europe.

Table A1. List of European dragonfly (Odonata) species (n = 123) included in the analysis of range shift and changes in species richness 1988–2010 with information on taxonomic suborder, scientific name including author and year of species description, biogeographic group, and range shift trend. * Indicate the four species that in 2010 were included as new in Europe. Nomenclature and taxonomy follow IUCN [80].

Suborder	Species	Biogeographic Group	Range Shift Trend
Zygoptera	*Calopteryx haemorrhoidalis* Vander Linden, 1825	Mediterranean	Expansion
Zygoptera	*Calopteryx splendens* Harris, 1780	Eurasian	Expansion
Zygoptera	*Calopteryx virgo* Linnaeus, 1758	Eurasian	Expansion
Zygoptera	*Calopteryx xanthostoma* Charpentier, 1825	Mediterranean	Contraction
Zygoptera	*Ceriagrion georgifreyi* Schmidt, 1953	Mediterranean	Expansion
Zygoptera	*Ceriagrion tenellum* De Villers, 1789	Mediterranean	Expansion
Zygoptera	*Chalcolestes viridis* Vander Linden, 1825	Mediterranean	Expansion
Zygoptera	*Coenagrion armatum* Charpentier, 1840	Boreo-alpine	Expansion
Zygoptera	*Coenagrion caerulescens* Fonscolombe, 1838	Mediterranean	Expansion
Zygoptera	*Coenagrion hastulatum* Charpentier, 1825	Boreo-alpine	Expansion
Zygoptera	*Coenagrion hylas* Trybom, 1889	Boreo-alpine	Stable
Zygoptera	*Coenagrion intermedium* Lohmann, 1990	Mediterranean	Stable
Zygoptera	*Coenagrion johanssoni* Wallengren, 1894	Boreo-alpine	Expansion
Zygoptera	*Coenagrion lunulatum* Charpentier, 1840	Boreo-alpine	Contraction
Zygoptera	*Coenagrion mercuriale* Charpentier, 1840	Mediterranean	Expansion
Zygoptera	*Coenagrion ornatum* Selys, 1850	Mediterranean	Expansion
Zygoptera	*Coenagrion puella* Linnaeus, 1758	Mediterranean	Expansion
Zygoptera	*Coenagrion pulchellum* Vander Linden, 1825	Eurasian	Expansion
Zygoptera	*Coenagrion scitulum* Rambur, 1842	Mediterranean	Expansion
Zygoptera	*Enallagma cyathigerum* Charpentier, 1840	Eurasian	Expansion
Zygoptera	*Epallage fatime* Charpentier, 1840	Mediterranean	Expansion
Zygoptera	*Erythromma lindenii* Selys, 1840	Mediterranean	Expansion
Zygoptera	*Erythromma najas* Hansemann, 1823	Eurasian	Expansion
Zygoptera	*Erythromma viridulum* Charpentier, 1840	Mediterranean	Expansion
Zygoptera	*Ischnura elegans* Vander Linden, 1820	Eurasian	Expansion
Zygoptera	*Ischnura fountaineae* * Morton, 1905	Mediterranean	Expansion
Zygoptera	*Ischnura genei* Rambur, 1842	Mediterranean	Stable
Zygoptera	*Ischnura graellsii* Rambur, 1842	Mediterranean	Expansion
Zygoptera	*Ischnura pumilio* Charpentier, 1825	Mediterranean	Expansion
Zygoptera	*Lestes barbarus* Fabricius, 1798	Mediterranean	Expansion
Zygoptera	*Lestes dryas* Kirby, 1890	Eurasian	Expansion
Zygoptera	*Lestes macrostigma* Eversmann, 1836	Mediterranean	Contraction
Zygoptera	*Lestes sponsa* Hansemann, 1823	Eurasian	Expansion
Zygoptera	*Lestes virens* Charpentier, 1825	Eurasian	Expansion
Zygoptera	*Nehalennia speciosa* Charpentier, 1840	Boreo-alpine	Contraction
Zygoptera	*Platycnemis acutipennis* Selys, 1841	Mediterranean	Expansion
Zygoptera	*Platycnemis latipes* Rambur, 1842	Mediterranean	Contraction
Zygoptera	*Platycnemis pennipes* Pallas, 1771	Eurasian	Contraction
Zygoptera	*Pyrrhosoma nymphula* Sulzer, 1776	Eurasian	Expansion
Zygoptera	*Sympecma fusca* Vander Linden, 1820	Mediterranean	Contraction
Zygoptera	*Sympecma paedisca* Brauer, 1877	Boreo-alpine	Expansion
Anisoptera	*Aeshna affinis* Vander Linden, 1820	Mediterranean	Expansion

Table A1. *Cont.*

Suborder	Species	Biogeographic Group	Range Shift Trend
Anisoptera	*Aeshna caerulea* Ström, 1783	Boreo-alpine	Expansion
Anisoptera	*Aeshna crenata* Hagen, 1856	Boreo-alpine	Expansion
Anisoptera	*Aeshna cyanea* Müller, 1764	Eurasian	Contraction
Anisoptera	*Aeshna grandis* Linnaeus, 1758	Eurasian	Expansion
Anisoptera	*Aeshna isoceles* (Müller, 1767)	Mediterranean	Expansion
Anisoptera	*Aeshna juncea* Linnaeus, 1758	Boreo-alpine	Expansion
Anisoptera	*Aeshna mixta* Latreille, 1805	Mediterranean	Expansion
Anisoptera	*Aeshna serrata* Hagen, 1856	Eurasian	Expansion
Anisoptera	*Aeshna subarctica* Walker, 1908	Boreo-alpine	Expansion
Anisoptera	*Aeshna viridis* Eversmann, 1836	Eurasian	Expansion
Anisoptera	*Anax imperator* Leach, 1815	Mediterranean	Expansion
Anisoptera	*Anax parthenope* Selys, 1839	Mediterranean	Expansion
Anisoptera	*Boyeria cretensis* Peters, 1991	Mediterranean	Expansion
Anisoptera	*Boyeria irene* Fonscolombe, 1838	Mediterranean	Expansion
Anisoptera	*Brachythemis impartita* Karsch, 1890	Tropical	Expansion
Anisoptera	*Brachytron pratense* Müller, 1764	Eurasian	Expansion
Anisoptera	*Caliaeschna microstigma* Schneider, 1845	Mediterranean	Expansion
Anisoptera	*Cordulegaster bidentata* Selys, 1843	Mediterranean	Expansion
Anisoptera	*Cordulegaster boltonii* Donovan, 1807	Eurasian	Contraction
Anisoptera	*Cordulegaster helladica* Lohmann, 1993	Mediterranean	Expansion
Anisoptera	*Cordulegaster heros* Theischinger, 1979	Mediterranean	Expansion
Anisoptera	*Cordulegaster insignis* Schneider, 1845	Mediterranean	Contraction
Anisoptera	*Cordulegaster picta* Selys, 1854	Mediterranean	Expansion
Anisoptera	*Cordulegaster trinacriae* Waterston, 1976	Mediterranean	Expansion
Anisoptera	*Cordulia aenea* Linnaeus, 1758	Eurasian	Expansion
Anisoptera	*Crocothemis erythraea* Brullé, 1832	Tropical	Expansion
Anisoptera	*Diplacodes lefebvrii* Rambur, 1842	Tropical	Expansion
Anisoptera	*Epitheca bimaculata* Charpentier, 1825	Eurasian	Expansion
Anisoptera	*Gomphus flavipes* Selys, 1837	Eurasian	Expansion
Anisoptera	*Gomphus graslinii* Rambur, 1842	Mediterranean	Expansion
Anisoptera	*Gomphus pulchellus* Selys, 1840	Mediterranean	Expansion
Anisoptera	*Gomphus schneiderii* Selys, 1850	Mediterranean	Expansion
Anisoptera	*Gomphus simillimus* Selys, 1840	Mediterranean	Expansion
Anisoptera	*Gomphus vulgatissimus* Linnaeus, 1758	Eurasian	Expansion
Anisoptera	*Leucorrhinia albifrons* Burmeister, 1839	Boreo-alpine	Contraction
Anisoptera	*Leucorrhinia caudalis* Charpentier, 1840	Boreo-alpine	Expansion
Anisoptera	*Leucorrhinia dubia* Vander Linden, 1825	Eurasian	Expansion
Anisoptera	*Leucorrhinia pectoralis* Charpentier, 1825	Eurasian	Expansion
Anisoptera	*Leucorrhinia rubicunda* Linnaeus, 1758	Boreo-alpine	Expansion
Anisoptera	*Libellula depressa* Linnaeus, 1758	Eurasian	Expansion
Anisoptera	*Libellula fulva* Müller, 1764	Eurasian	Expansion
Anisoptera	*Libellula quadrimaculata* Linnaeus, 1758	Eurasian	Expansion
Anisoptera	*Lindenia tetraphylla* Vander Linden, 1825	Mediterranean	Contraction
Anisoptera	*Macromia splendens* Pictet, 1843	Mediterranean	Expansion
Anisoptera	*Onychogomphus costae* Selys, 1885	Mediterranean	Expansion
Anisoptera	*Onychogomphus forcipatus* Linnaeus, 1758	Mediterranean	Expansion
Anisoptera	*Onychogomphus uncatus* Charpentier, 1840	Mediterranean	Expansion
Anisoptera	*Ophiogomphus cecilia* Fourcroy, 1785	Eurasian	Expansion
Anisoptera	*Orthetrum albistylum* Selys, 1848	Mediterranean	Expansion
Anisoptera	*Orthetrum brunneum* Fonscolombe, 1837	Mediterranean	Expansion
Anisoptera	*Orthetrum cancellatum* Linnaeus, 1758	Mediterranean	Expansion
Anisoptera	*Orthetrum chrysostigma* Burmeister, 1839	Tropical	Expansion
Anisoptera	*Orthetrum coerulescens* Fabricius, 1798	Mediterranean	Expansion
Anisoptera	*Orthetrum nitidinerve* Selys, 1841	Mediterranean	Expansion

Table A1. *Cont.*

Suborder	Species	Biogeographic Group	Range Shift Trend
Anisoptera	*Orthetrum sabina* * Drury, 1773	Tropical	Expansion
Anisoptera	*Orthetrum taeniolatum* * Schneider, 1845	Tropical	Expansion
Anisoptera	*Orthetrum trinacria* Selys, 1841	Tropical	Expansion
Anisoptera	*Oxygastra curtisii* Dale, 1834	Mediterranean	Contraction
Anisoptera	*Pantala flavescens* Fabricius, 1798	Tropical	Expansion
Anisoptera	*Paragomphus genei* Selys, 1841	Tropical	Expansion
Anisoptera	*Selysiothemis nigra* Vander Linden, 1825	Mediterranean	Expansion
Anisoptera	*Somatochlora alpestris* Selys, 1840	Boreo-alpine	Expansion
Anisoptera	*Somatochlora arctica* Zetterstedt, 1840	Boreo-alpine	Expansion
Anisoptera	*Somatochlora flavomaculata* Vander Linden, 1825	Eurasian	Expansion
Anisoptera	*Somatochlora meridionalis* Nielsen, 1935	Mediterranean	Expansion
Anisoptera	*Somatochlora metallica* Vander Linden, 1825	Eurasian	Contraction
Anisoptera	*Somatochlora sahlbergi* Trybom, 1889	Boreo-alpine	Expansion
Anisoptera	*Sympetrum danae* Sulzer, 1776	Eurasian	Expansion
Anisoptera	*Sympetrum depressiusculum* Selys, 1841	Eurasian	Expansion
Anisoptera	*Sympetrum flaveolum* Linnaeus, 1758	Eurasian	Expansion
Anisoptera	*Sympetrum fonscolombii* Selys, 1840	Tropical	Expansion
Anisoptera	*Sympetrum meridionale* Selys, 1841	Mediterranean	Expansion
Anisoptera	*Sympetrum pedemontanum* O.F.Müller, 1766	Eurasian	Expansion
Anisoptera	*Sympetrum sanguineum* Müller, 1764	Eurasian	Expansion
Anisoptera	*Sympetrum sinaiticum* Dumont, 1977	Mediterranean	Expansion
Anisoptera	*Sympetrum striolatum* Charpentier, 1840	Eurasian	Expansion
Anisoptera	*Sympetrum vulgatum* Linnaeus, 1758	Eurasian	Expansion
Anisoptera	*Trithemis annulate* Palisot de Beauvois, 1807	Tropical	Expansion
Anisoptera	*Trithemis festiva* Rambur, 1842	Tropical	Expansion
Anisoptera	*Trithemis kirbyi* * Selys, 1891	Tropical	Expansion
Anisoptera	*Zygonyx torridus* Kirby, 1889	Tropical	Expansion

References

1. Brown, J.; Lomolino, M. *Biogeography*; Sinauer Associates: Sunderland, MA, USA, 1998.
2. Lenoir, J.; Svenning, J.C. Climate-Related Range Shifts—A Global Multidimensional Synthesis and New Research Directions. *Ecography* **2015**, *38*, 15–28. [CrossRef]
3. Thomas, C.D.; Bodsworth, E.J.; Wilson, R.J.; Simmons, A.D.; Davies, Z.G.; Musche, M.; Conradt, L. Ecological and Evolutionary Processes at Expanding Range Margins. *Nature* **2001**, *411*, 577–581. [CrossRef] [PubMed]
4. Lancaster, L.T.; Dudaniec, R.Y.; Hansson, B.; Svensson, E.I. Latitudinal Shift in Thermal Niche Breadth Results from Thermal Release during a Climate-Mediated Range Expansion. *J. Biogeogr.* **2015**, *42*, 1953–1963. [CrossRef]
5. Channell, R.; Lomollno, M.V. Dynamic Biogeography and Conservation of Endangered Species. *Nature* **2000**, *403*, 84–86. [CrossRef] [PubMed]
6. Fahrig, L. Effects of Habitat Fragmentation on Biodiversity. *Annu. Rev. Ecol. Evol. Syst.* **2003**, *34*, 59–71. [CrossRef]
7. Wiens, J.J. Climate-Related Local Extinctions Are Already Widespread among Plant and Animal Species. *PLoS Biol.* **2016**, *14*, e2001104. [CrossRef]
8. Opdam, P.; Wascher, D. Climate Change Meets Habitat Fragmentation: Linking Landscape and Biogeographical Scale Levels in Research and Conservation. *Biol. Conserv.* **2004**, *117*, 285–297. [CrossRef]
9. Chaudhary, C.; Richardson, A.J.; Schoeman, D.S.; Costello, M.J. Global Warming Is Causing a More Pronounced Dip in Marine Species Richness around the Equator. *Proc. Natl. Acad. Sci. USA* **2021**, *118*, e2015094118. [CrossRef]
10. Thomas, C.D.; Cameron, A.; Green, R.E.; Bakkenes, M.; Beaumont, L.J.; Collingham, Y.C.; Erasmus, B.F.N.; de Siqueira, M.F.; Grainger, A.; Hannah, L.; et al. Extinction Risk from Climate Change. *Nature* **2004**, *427*, 145–148. [CrossRef]
11. Flenner, I.; Sahlén, G. Dragonfly Community Re-Organisation in Boreal Forest Lakes: Rapid Species Turnover Driven by Climate Change? *Insect Conserv. Divers.* **2008**, *1*, 169–179. [CrossRef]
12. IPCC. *Climate Change 2007: The Physical Science Basis: Contribution of Working Group I to the Fourth Assessment*; Report of the Intergovernmental Panel on Climate Change; Cambridge University Press: Cambridge, UK; New York, NY, USA, 2007.
13. Gienapp, P.; Teplitsky, C.; Alho, J.S.; Mills, J.A.; Merilä, J. Climate Change and Evolution: Disentangling Environmental and Genetic Responses. *Mol. Ecol.* **2008**, *17*, 167–178. [CrossRef]
14. Davis, M.B.; Shaw, R.G. Range Shifts and Adaptive Responses to Quaternary Climate Change. *Science* **2001**, *292*, 673–679. [CrossRef] [PubMed]

15. Johansson, H.; Stoks, R.; Nilsson-Örtman, V.; Ingvarsson, P.K.; Johansson, F. Large-Scale Patterns in Genetic Variation, Gene Flow and Differentiation in Five Species of European Coenagrionid Damselfly Provide Mixed Support for the Central-Marginal Hypothesis. *Ecography* **2013**, *36*, 744–755. [CrossRef]
16. Kellermann, V.; Loeschcke, V.; Hoffmann, A.A.; Kristensen, T.N.; Fløjgaard, C.; David, J.R.; Svenning, J.C.; Overgaard, J. Phylogenetic Constraints in Key Functional Traits Behind Species' Climate Niches: Patterns of Desiccation and Cold Resistance across 95 Drosophila Species. *Evolution* **2012**, *66*, 3377–3389. [CrossRef] [PubMed]
17. Bota-Sierra, C.A.; García-Robledo, C.; Escobar, F.; Novelo-Gutiérrez, R.; Londoño, G.A. Environment, Taxonomy and Morphology Constrain Insect Thermal Physiology along Tropical Mountains. *Funct. Ecol.* **2022**, *36*, 1924–1935. [CrossRef]
18. Hickling, R.; Roy, D.B.; Hill, J.K.; Fox, R.; Thomas, C.D. The Distributions of a Wide Range of Taxonomic Groups Are Expanding Polewards. *Glob. Chang. Biol.* **2006**, *12*, 450–455. [CrossRef]
19. Hickling, R.; Roy, D.B.; Hill, J.K.; Thomas, C.D. A Northward Shift of Range Margins in British Odonata. *Glob. Chang. Biol.* **2005**, *11*, 502–506. [CrossRef]
20. de Knijf, G.; Anselin, A. When South Goes North: Mediterranean Dragonflies (Odonata) Conquer Flanders (North-Belgium). *BioRisk* **2010**, *5*, 141–153. [CrossRef]
21. Ott, J. Dragonflies and Climatic Change—Recent Trends in Germany and Europe. *BioRisk* **2010**, *5*, 253–286. [CrossRef]
22. Olsen, K.; Svenning, J.-C.; Balslev, H. Niche Breadth Predicts Geographical Range Size and Northern Range Shift in European Dragonfly Species (Odonata). *Diversity* **2022**, *14*, 719. [CrossRef]
23. Flenner, I.; Richter, O.; Suhling, F. Rising Temperature and Development in Dragonfly Populations at Different Latitudes. *Freshw. Biol.* **2010**, *55*, 397–410. [CrossRef]
24. Johansson, F. Latitudinal Shifts in Body Size of Enallagma Cyathigerum (Odonata). *J. Biogeogr.* **2003**, *30*, 29–34. [CrossRef]
25. Dingemanse, N.J.; Kalkman, V.J. Changing Temperature Regimes Have Advanced the Phenology of Odonata in the Netherlands. *Ecol. Entomol.* **2008**, *33*, 394–402. [CrossRef]
26. Braune, E.; Richter, O.; Söndgerath, D.; Suhling, F. Voltinism Flexibility of a Riverine Dragonfly along Thermal Gradients. *Glob. Chang. Biol.* **2008**, *14*, 470–482. [CrossRef]
27. Hassall, C.; Thompson, D.J.; French, G.C.; Harvey, I.F. Historical Changes in the Phenology of British Odonata Are Related to Climate. *Glob. Chang. Biol.* **2007**, *13*, 933–941. [CrossRef]
28. Richter, O.; Suhling, F.; Müller, O.; Kern, D. A Model for Predicting the Emergence of Dragonflies in a Changing Climate. *Freshw. Biol.* **2008**, *53*, 1868–1880. [CrossRef]
29. Johansson, F.; Śniegula, S.; Brodin, T. Emergence Patterns and Latitudinal Adaptations in Development Time of Odonata in North Sweden and Poland. *Odonatologica* **2010**, *39*, 97–106.
30. Stoks, R.; de Block, M.; McPeek, M.A. Physiological Costs of Compensatory Growth in a Damselfly. *Ecology* **2006**, *87*, 1566–1574. [CrossRef]
31. de Block, M.; Stoks, R. Compensatory Growth and Oxidative Stress in a Damselfly. *Proc. R. Soc. B Biol. Sci.* **2008**, *275*, 781–785. [CrossRef]
32. Sternberg, K. Colours, Colour Change, Colour Patterns and 'Cuticular Windows' as Light Traps. Their Thermoregulatoric and Ecological Significance in Some Aeshna Species (Odonata: Aeshnidae). *Zool. Anz.* **1996**, *235*, 77–88.
33. Corbet, P.S. *Dragonflies: Behavior and Ecology of Odonata*; Harley Books: Colchester, UK, 2004.
34. Kalkman, V.J.; Boudot, J.-P.; Bernard, R.; Conze, K.-J.; de Knijf, G.; Dyatlova, E.; Ferreira, S.; Jović, M.; Ott, J.; Riservato, E.; et al. *European Red List of Dragonflies*; Publications Office of the European Union: Luxembourg, 2010.
35. Clausnitzer, V.; Kalkman, V.J.; Ram, M.; Collen, B.; Baillie, J.E.M.; Bedjanič, M.; Darwall, W.R.T.; Dijkstra, K.D.B.; Dow, R.; Hawking, J.; et al. Odonata Enter the Biodiversity Crisis Debate: The First Global Assessment of an Insect Group. *Biol. Conserv.* **2009**, *142*, 1864–1869. [CrossRef]
36. Graham, C.H.; Hijmans, R.J. A Comparison of Methods for Mapping Species Ranges and Species Richness. *Glob. Ecol. Biogeogr.* **2006**, *15*, 578–587. [CrossRef]
37. Suhling, F.; Sahlén, G.; Gorb, S.; Kalkman, V.; Dijkstra, K.-D.; van Tol, J. Order Odonata. In *Ecology and General Biology: Thorp and Covich's Freshwater Invertebrates*; Thorp, J., Rogers, D., Eds.; Academic Press: San Diego, CA, USA, 2015; pp. 893–932.
38. Kalkman, V.J.; Clausnitzer, V.; Dijkstra, K.-D.B.; Orr, A.G.; Paulson, D.R.; van Tol, J. Global Diversity of Dragonflies (Odonata) in Freshwater. In *Freshwater Animal Diversity Assessment*; Balian, E., Lévêque, C., Segers, H., Martens, K., Eds.; Springer: Dordrecht, The Netherlands, 2007; pp. 351–363.
39. de Lattin, G. *Grundriss Der Zoogeographie*; Fischer: Stuttgart, Germany, 1967.
40. Husemann, M.; Schmitt, T.; Zachos, F.E.; Ulrich, W.; Habel, J.C. Palaearctic Biogeography Revisited: Evidence for the Existence of a North African Refugium for Western Palaearctic Biota. *J. Biogeogr.* **2014**, *41*, 81–94. [CrossRef]
41. Simonsen, T.J.; Huemer, P. Phylogeography of *Hepialus Humuli* (L.) (Lepidoptera: Hepialidae) in Europe: Short Distance vs. Large Scale Postglacial Expansions from Multiple Alpine Refugia and Taxonomic Implications. *Insect Syst. Evol.* **2014**, *45*, 209–250. [CrossRef]
42. Ursenbacher, S.; Carlsson, M.; Helfer, V.; Tegelström, H.; Fumagalli, L. Phylogeography and Pleistocene Refugia of the Adder (Vipera Berus) as Inferred from Mitochondrial DNA Sequence Data. *Mol. Ecol.* **2006**, *15*, 3425–3437. [CrossRef]
43. Brochmann, C.; Gabrielsen, T.M.; Nordal, I.; Landvik, J.Y.; Elven, R. Glacial Survival or Tabula Rasa? The History of North Atlantic Biota Revisited. *Taxon* **2003**, *52*, 417–450. [CrossRef]

44. Schmitt, T.; Rákosy, L.; Abadjiev, S.; Müller, P. Multiple Differentiation Centres of a Non-Mediterranean Butterfly Species in South-Eastern Europe. *J. Biogeogr.* **2007**, *34*, 939–950. [CrossRef]
45. Baselga, A.; Lobo, J.M.; Svenning, J.C.; Aragón, P.; Araújo, M.B. Dispersal Ability Modulates the Strength of the Latitudinal Richness Gradient in European Beetles. *Glob. Ecol. Biogeogr.* **2012**, *21*, 1106–1113. [CrossRef]
46. Castillo-Pérez, E.U.; Suárez-Tovar, C.M.; González-Tokman, D.; Schondube, J.E.; Córdoba-Aguilar, A. Insect Thermal Limits in Warm and Perturbed Habitats: Dragonflies and Damselflies as Study Cases. *J. Therm. Biol.* **2022**, *103*, 103164. [CrossRef]
47. Dumont, H.J.; Desmet, K. Trans-Sahara and Trans-Mediterranean Migratory Activity of Hemianax Ephippiger (Burmeister) in 1988 and 1989 (Anisoptera: Aeshnidae). *Odonatologica* **1990**, *19*, 181–185.
48. Hassall, C.; Thompson, D.J. The Effects of Environmental Warming on Odonata: A Review. *Int. J. Odonatol.* **2008**, *11*, 131–153. [CrossRef]
49. Askew, R.R. *The Dragonflies of Europe*; Harley Books (BH & A Harley Ltd.): Colchester, UK, 1988.
50. Dijkstra, K.-D.; Lewington, R. *Field Guide to the Dragonflies of Britain and Europe*; British Wildlife Publishing: Dorset, UK, 2006.
51. ESRI. *ArcGIS Desktop [GIS Software]*, version 10.2; Environmental Systems Research Institute: Redlands, CA, USA, 2010.
52. Sternberg, K. Die Postglaziale Besiedlung Mitteleuropas Durch Libellen, Mit Besonderer Berucksichtigung Sudwestdeutschlands (Insecta, Odonata). The Postglacial Colonization of Central Europe by Dragonflies, with Special Reference to Southwestern Germany (Insecta, Odonata). *J. Biogeogr.* **1998**, *25*, 319–337. [CrossRef]
53. Beschovski, V.; Marinov, M. Fauna, Ecology, and Zoogeography of Dragonflies (Insecta: Odonata) of Bulgaria. In *Biogeography and Ecology of Bulgaria*; Fet, V., Popov, A., Eds.; Springer: Dordrecht, The Netherlands, 2007; pp. 199–231.
54. R Core Team. *R: A Language and Environment for Statistical Computing*; R Core Team: Vienna, Austria, 2022.
55. Lemon, J. Plotrix: A Package in the Red Light District of R. *R-News* **2006**, *6*, 8–12.
56. Pulliam, H.R. On the Relationship between Niche and Distribution. *Ecol. Lett.* **2000**, *3*, 349–361. [CrossRef]
57. Rondinini, C.; Wilson, K.A.; Boitani, L.; Grantham, H.; Possingham, H.P. Tradeoffs of Different Types of Species Occurrence Data for Use in Systematic Conservation Planning. *Ecol. Lett.* **2006**, *9*, 1136–1145. [CrossRef]
58. Gaston, K.J. *The Structure and Dynamics of Geographic Ranges*; Oxford University Press: Oxford, UK, 2003.
59. Jetz, W.; Sekercioglu, C.H.; Watson, J.E.M. Ecological Correlates and Conservation Implications of Overestimating Species Geographic Ranges. *Conserv. Biol.* **2008**, *22*, 110–119. [CrossRef]
60. Hurlbert, A.H.; White, E.P. Disparity between Range Map- and Survey-Based Analyses of Species Richness: Patterns, Processes and Implications. *Ecol. Lett.* **2005**, *8*, 319–327. [CrossRef]
61. Hawkins, B.A.; Field, R.; Cornell, H.V.; Currie, D.J.; Guégan, J.F.; Kaufman, D.M.; Kerr, J.T.; Mittelbach, G.G.; Oberdorff, T.; O'Brien, E.M.; et al. Energy, Water, and Broad-Scale Geographic Patterns of Species Richness. *Ecology* **2003**, *84*, 3105–3117. [CrossRef]
62. Grewe, Y.; Hof, C.; Dehling, D.M.; Brandl, R.; Brändle, M. Recent Range Shifts of European Dragonflies Provide Support for an Inverse Relationship between Habitat Predictability and Dispersal. *Glob. Ecol. Biogeogr.* **2013**, *22*, 403–409. [CrossRef]
63. Hof, C.; Brändle, M.; Dehling, D.M.; Munguía, M.; Brandl, R.; Araújo, M.B.; Rahbek, C. Habitat Stability Affects Dispersal and the Ability to Track Climate Change. *Biol. Lett.* **2012**, *8*, 639–643. [CrossRef]
64. Hof, C.; Brändle, M.; Brandl, R. Lentic Odonates Have Larger and More Northern Ranges than Lotic Species. *J. Biogeogr.* **2006**, *33*, 63–77. [CrossRef]
65. Kalkman, V.J.; Boudot, J.-P.; Futahashi, R.; Abbott, J.C.; Bota-Sierra, C.A.; Guralnick, R.; Bybee, S.M.; Ware, J.; Belitz, M.W. Diversity of Palaearctic Dragonflies and Damselflies (Odonata). *Diversity* **2022**, *14*, 966. [CrossRef]
66. Hurlbert, A.H.; Jetz, W. Species Richness, Hotspots, and the Scale Dependence of Range Maps in Ecology and Conservation. *Proc. Natl. Acad. Sci. USA* **2007**, *104*, 13384–13389. [CrossRef] [PubMed]
67. Boudot, J.-P.; Kalkman, V.J. *Atlas of the European Dragonflies and Damselflies*; KNNV-Publishing: Utrecht, The Netherlands, 2015.
68. Riservato, E.; Boudot, J.; Ferreira, S.; Jović, M.; Kalkman, V.; Schneider, W.; Samraoui, B.; Cuttelod, A. *The Status and Distribution of Dragonflies of the Mediterranean Basin*; IUCN: Gland, Switzerland; Malaga, Spain, 2009.
69. Termaat, T.; Kalkman, V.; Bouwman, J. Changes in the Range of Dragonflies in the Netherlands and the Possible Role of Temperature Change. *BioRisk* **2010**, *5*, 155–173. [CrossRef]
70. Hill, J.K.; Thomas, C.D.; Huntley, B. Climate and Habitat Availability Determine 20th Century Changes in a Butterfly's Range Margin. *Proc. R. Soc. B Biol. Sci.* **1999**, *266*, 1197–1206. [CrossRef]
71. Boudot, J.-P.; Kalkman, V.J.; Azpilicueta Amorin, M.; Bogdanović, T.; Cordero Rivera, A.; Degabriele, G.; Dommanget, J.-L.; Ferreira, S.; Garrigos, B.; Jović, M.; et al. Atlas of the Odonata of the Mediterranean and North Africa. *Libellula Suppl.* **2009**, *9*, 1–256.
72. de Knijf, G.; Flenker, U.; Vanappelghem, C.; Manci, C.O.; Kalkman, V.J.; Demolder, H. The Status of Two Boreo-Alpine Species, Somatochlora Alpestris and S. Arctica, in Romania and Their Vulnerability to the Impact of Climate Change (Odonata: Corduliidae). *Int. J. Odonatol.* **2011**, *14*, 111–126. [CrossRef]
73. Bernard, R.; Wildermuth, H. Nehalennia Speciosa (Charpentier, 1840) in Europe: A Case of a Vanishing Relict (Zygoptera: Coenagrionidae). *Odonatologica* **2005**, *34*, 335–378.
74. Hampe, A.; Petit, R.J. Conserving Biodiversity under Climate Change: The Rear Edge Matters. *Ecol. Lett.* **2005**, *8*, 461–467. [CrossRef]

75. Simonsen, T.J.; Olsen, K.; Djernæs, M. The African-Iberian Connection in Odonata: MtDNA and NcDNA Based Phylogeography of Aeshna Cyanea (MüLler, 1764) (Odonata: Aeshnidae) in Western Palaearctic. *Arthropod Syst. Phylogeny* **2020**, *78*, 309–320. [CrossRef]
76. Boudot, J.-P.; Clausnitzer, V.; Samraoui, B.; Suhling, F.; Dijkstra, K.-D.B.; Schneider, W. *Paragomphus genei*. The IUCN Red List of Threatened Species 2016: E.T59976A83861834. Available online: https://www.iucnredlist.org/species/59976/83861834 (accessed on 26 August 2022).
77. Jiménez-Alfaro, B.; Girardello, M.; Chytrý, M.; Svenning, J.-C.; Willner, W.; Gégout, J.-C.; Agrillo, E.; Campos, J.A.; Jandt, U.; Kącki, Z.; et al. History and Environment Shape Species Pools and Community Diversity in European Beech Forests. *Nat. Ecol. Evol.* **2018**, *2*, 483–490. [CrossRef] [PubMed]
78. Aguesse, P. Les Odonates de l'Europe Occidentale, Du Nord de l'Afrique et Des Îles Atlantiques. In *Faune de l'Europe et du Bassin Méditerranéen*; Masson et Cie: Paris, France, 1968; Volume 4, pp. 1–258.
79. Pilgrim, E.M.; von Dohlen, C.D. Molecular and Morphological Study of Species-Level Questions within the Dragonfly Genus Sympetrum (Odonata: Libellulidae). *Ann. Entomol. Soc. Am.* **2007**, *100*, 688–702. [CrossRef]
80. IUCN The IUCN Red List of Threatened Species. Version 2015-3. Available online: www.iucnredlist.org (accessed on 22 August 2016).

diversity

MDPI

Article

Niche Breadth Predicts Geographical Range Size and Northern Range Shift in European Dragonfly Species (Odonata)

Kent Olsen [1,*], Jens-Christian Svenning [2,3,4] and Henrik Balslev [2]

1 Research and Collections, Natural History Museum Aarhus, Wilhelm Meyers Allé 10, DK-8000 Aarhus, Denmark
2 Section for Ecoinformatics and Biodiversity, Department of Biology, Aarhus University, Ny Munkegade 116, DK-8000 Aarhus, Denmark
3 Center for Biodiversity Dynamics in a Changing World, Department of Biology, Aarhus University, Ny Munkegade 116, DK-8000 Aarhus, Denmark
4 Center for Sustainable Landscapes under Global Change, Department of Biology, Aarhus University, Nordre Ringgade 1, DK-8000 Aarhus, Denmark
* Correspondence: kent@nathist.dk; Tel.: +45-40272030

Abstract: We studied how range sizes and shifts in species ranges depend on niche breadth in European dragonflies. We measured range sizes and shifts over a 22-year period (1988–2010) and grouped species into those reproducing in permanent running (perennial lotic) water, permanent standing (perennial lentic) water, and temporary (running or standing) water. Running water species are more specialized and have narrower niches with a more fixed niche position than standing water species. Temporary water species are more generalist and have broader niches without a fixed niche position as clear as permanent water species because they may utilize both temporary and permanent habitats. Running water species have smaller ranges, and some of them have contracted their ranges more than species reproducing in standing or temporary waters; that is, they are especially at risk of habitat loss and climate change because of the joint effects of their narrow niches and small range sizes. Temporary water species track climate changes better than permanent water species. This suggests that ecological specialization may cause contemporary range shifts to lag behind changes in climate and resources. Furthermore, it indicates that recent changes in climate and human land use cause biotic homogenization, where specialists are outperformed and replaced by generalists.

Keywords: aquatic invertebrates; biodiversity; freshwater ecology; geographic range; habitat preference; Odonata; odonatology; range dynamics; range size; species distribution

Citation: Olsen, K.; Svenning, J.-C.; Balslev, H. Niche Breadth Predicts Geographical Range Size and Northern Range Shift in European Dragonfly Species (Odonata). *Diversity* **2022**, *14*, 719. https://doi.org/10.3390/d14090719

Academic Editors: M. Olalla Lorenzo-Carballa and Ricardo Koroiva

Received: 20 July 2022
Accepted: 25 August 2022
Published: 30 August 2022

1. Introduction

Climate change strongly impacts the distributions of dragonflies, and it seems that, at least at temperate latitudes, it has become one of the most important driving forces behind distributional changes in this insect order [1–16]. Climate change causes dramatic shifts in geographical distributions of species. Species that are limited by their lower thermal tolerance threshold and are capable of at least partially tracking changes in climate experience significant range expansions [17]. Species from a variety of ecological systems [17], across [18] and within broad taxonomic groups [10], exhibit northward range expansions, consistent with a warming climate. Apart from their climatic envelopes, the occurrence of species is also determined by the availability of suitable habitats [19,20]. Species that cannot keep pace with climate change [17] or adapt to new environmental conditions [21,22] by utilizing formerly unavailable habitats or new habitats will lag behind changes in climate and resources and eventually experience range contraction or local to regional extinction [23,24].

The ongoing climate changes cause habitat alterations and threaten certain aquatic habitats [25]. A 20% decrease in summer precipitation has been projected for all areas around the Mediterranean Basin, and increased evapotranspiration due to rising air and water temperatures will also reduce water availability [25]. Higher proportions of winter precipitation will be received as rain, and snow will accumulate in smaller quantities and melt earlier in the season, leading to reduced summer flows [25]. Such shifts in the form and timing of precipitation and runoff may disrupt the life cycle of species adapted to permanent running water by causing unstable flow regimes and summertime desiccation of streams and rivers [12]. Drying has already been observed in southern Europe, where increasing numbers of standing water bodies dry out or become temporarily dry [2]. Human population growth and industrial and agricultural development further intensify the pressures on aquatic habitats and increase the levels of water scarcity [12]. In this context, change in climate and human land use are expected to have a negative impact on the distribution of species dependent on aquatic habitats, especially species adapted to permanent running water habitats.

Dragonflies reproduce in most running and standing water habitats, and their larvae cannot freely migrate in search of suitable conditions in the same way as adults can [26]. Many dragonfly species inhabit water bodies that dry out periodically, and dragonflies' adaptations to that condition include prolonged egg stage [27], drought-resistant eggs [28,29], or larvae that can survive drought periods for as long as 8 months in moist cracks in the sediment [30]. Some species have a rapid larval development and a shortened larval stage [27], and some are able to accelerate larval development in response to rapid water loss [31]. However, not all species have these adaptations, and the development of most dragonfly larvae lasts from 1 to 3 years, but depending on the latitude/altitude and habitat suitability, it may last up to 6 years [15,26]. One of the effects of climate change is therefore caused by the selective pressure that shift in water permanence constitutes on species assemblages.

A major challenge in conservation ecology is to understand how different species respond to climate and environmental changes; that is, what enables some species to persist while others decline? Several studies have linked responses to climate change to species traits, such as niche breadth [8,11,32–38]. Local species diversity often increases when the temperature rises mainly due to the arrival of ecological generalists that can respond quickly because they can utilize a large variety of habitats [34,36,39]. This is, however, accompanied by loss of habitat specialists that only have the ability to utilize a narrow range of habitats [19,36]. Consequently, ecological specialists with narrow niches are replaced by broad niche generalists, and this in turn causes biotic homogenization, where species from a variety of taxonomic groups and geographic regions are lost due to recent climate change [19,40–42]. Hence, the dynamics of range shifts are constrained not only by climatic boundaries but also by the niche breadth of species. Functional species traits are therefore important when determining and understanding how species respond to climate change [8,11,33,36–38]. This has led to the hypothesis that key traits, such as niche breadth, can be used to predict temporal changes in species range.

We study how niche breadth in a larval habitat influences the response of European dragonfly species to climate change. The larvae of species adapted to permanent running (perennial lotic) waters have a narrower niche breadth with a more fixed niche position than permanent standing (perennial lentic) water and temporary (running or standing) water species. Species adapted to life in temporary waters have the broadest niche breadth without a fixed niche position as they may be found not only in temporary water but also in various permanent habitats, sometimes including both running and standing waters (Figure 1). According to the niche breadth of a larvae habitat, we divided species into three groups: species reproducing in (1) permanent running water, (2) permanent standing water, and (3) temporary water. Instead of a single focus on the running- to standing-water gradient [8,11,33], we chose to include the complete range of habitat resources by also implementing the temporary water species as an independent functional group. First, we chose this because the larvae of many dragonfly species may utilize different aquatic habitats along both a running- to standing-water gradient and a permanent- to temporary-water gradient [26]. As the transitions from permanent to temporary water bodies affects different dragonfly species in different ways [9], it specifically allows us to test how species with different niche breadths within the complete range of habitat resources react to climatic change. Second, climate change is known to cause biotic homogenization in terrestrial organisms [19,40–42], and as the three ecological habitat categories also represent a gradient from habitat specialist to habitat generalist, it allows us to evaluate whether this also applies to organisms whose larvae are strictly aquatic. The larvae of species adapted to permanent running waters are more specialized and habitat constrained than species that can also live in standing waters. Species adapted to life in temporary waters can be considered as more generalist than permanent water species as they may reproduce not only in temporary water but also in various permanent habitats, sometimes including both running and standing waters (Figure 1). Third, by including temporary habitats, it allows us to evaluate the relative importance of temporary ponds that are expected to have an increasing functional significance as an aquatic resource. This is important because among the gradients of water bodies, temporary habitats constitute an essential freshwater ecosystem as underlined by the inclusion of the Mediterranean temporary ponds as a priority habitat for conservation (code 3170) in the European Union (EU) Habitats Directive (92/43/EEC) [43]. Due to their heterogeneity, temporary waters support high diversity and act as stepping stones for the dispersal of species [44].

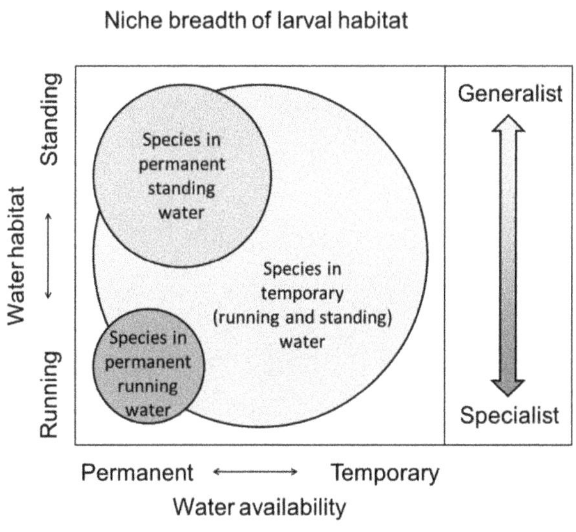

Figure 1. Size of the ecological niche space of the aquatic larvae of the European dragonfly (Odonata) species separated in three habitat categories: permanent running (perennial lotic) water, permanent standing (perennial lentic) water, and temporary (running or standing) water. Color scale to the right represents a gradient going from habitat specialist with a narrow niche breadth to habitat generalist with a broad niche breadth as also illustrated with the size of the ecological niche space of three habitat categories.

We use distribution maps for European dragonflies from 1988 [45] and 2010 [12] to determine differences in range sizes and follow how both latitudinal and northern range borders of the species have changed over a 22-year period. As documented above, there is a gap in our knowledge of how a key species trait, such as niche breadth, of the larval habitat of European dragonfly species influences their response to climate change. Especially, studies that evaluate the specialist–generalist gradient and test for climate-driven homogenization in aquatic communities and organisms that represent both aquatic and terrestrial ecosystems are important. In this context, we specifically ask: (1) Do species reproducing in permanent standing water and temporary water have larger ranges than species reproducing in permanent running water? (2) Do species reproducing in temporary habitats track changes better than species in permanent habitats? (3) Do species reproducing in permanent running water contract more or expand less than species in permanent standing water and temporary water? (4) Do generalist species outperform specialists with a narrow niche breadth? Answering these questions will help explain how environmental changes affect species differently according to their ability to utilize various types of aquatic habitats.

2. Materials and Methods

2.1. Study Area

Our study area included Europe, which, as the westernmost peninsula of Eurasia, is limited to the north by the Arctic Ocean, to the west by the Atlantic Ocean, and to the south by the Mediterranean Sea. The eastern border of the study area follows a combination of the 35° E longitude and the eastern margin used in outline range maps in Askew [45]. All larger European islands in the Mediterranean were included (Figure 2).

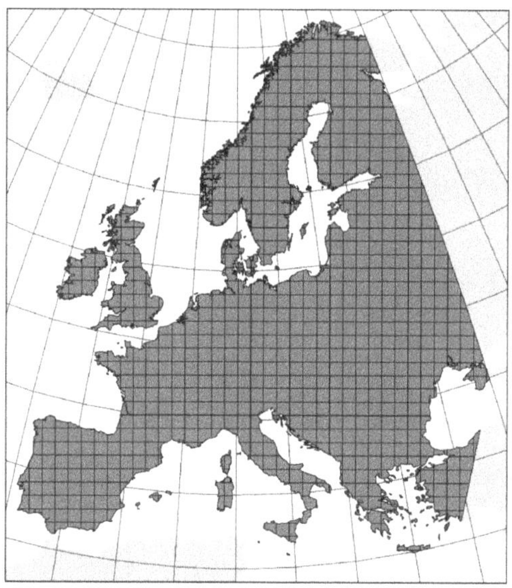

Figure 2. Study area covers the westernmost peninsula of Eurasia. The gridded map with 880 cells of 100 × 100 km highlighted in grey.

2.2. Data

Distribution ranges of dragonflies in Europe were taken from outline maps in Askew [45] and in the online species summary in Kalkman et al. [12]. Maps from Askew [45] were georeferenced in ArcGIS 10.2 [46] based on scanned TIFF images, whereas maps from Kalkman et al. [12] were provided as shape files from the Freshwater Biodiversity Unit under the IUCN Global Species Program. The maps do not always represent the full distribution of a species, but only the part that falls within the Eurasian peninsula, including western Russia and Africa north of the Sahara. We excluded data from east of 35° E and south of the Mediterranean Sea because dragonfly occurrences in these regions are not well documented (e.g., Dijkstra and Lewington) [47]. Each distribution map from the 1988 and 2010 data was cut with the same European coastline layer in ArcGIS 10.2 [46] to ensure that species ranges followed the same extent of land cover to make them directly comparable.

Of the 130 species of dragonflies known to occur within the study area, we constructed outline range maps for 123 species after excluding vagrant species and species without range maps in one or both data sets (see Table S1 for a list of excluded species, taxonomic and nomenclatural notes, and modifications to species ranges). Of the 123 species included,

4 colonized Europe between 1988 and 2010, whereas 119 species occurred in both data sets (see Table S1 for a list of the 4 and 119 species, respectively). All parts of species ranges categorized as extant were included, whereas range parts where the species was scored as extinct were omitted.

Based on a combination of scientific and popular species names and the wordings—*ephemeral water, ephemeral pond, temporary water, temporary pond, seasonal water*, and *seasonal pond*—we searched in the key book references on Eurasian dragonfly species [45,47–51] and peer-reviewed literature found in, for example, Web of Science [27,52–55] (per March 2016), for which species can occur in temporary water bodies in Europe, regions of origin for species colonizing Europe, or similar latitudes east of Europe and adjacent southern regions. The larval habitat of species not reported to sometimes occur in temporary water bodies was subsequently reviewed to determine whether they had a preference for standing over running water. By this, we assigned a niche breadth of the larval habitat for each species with a primary emphasis on whether they are able to reproduce successfully in temporary (running and standing) water bodies and a secondary emphasis on the remaining species' ability to utilize permanent standing waters over permanent running ones. The 123 species were divided into three ecological habitat categories—permanent running waters (n = 37), permanent standing waters (n = 49), and temporary (running or standing) waters (n = 37)—based on references in Table S2.

2.3. Data Analysis

We transformed the outline distributions into gridded maps with 880 cells of 100 × 100 km in ArcGIS 10.2 [46] to estimate distributional range as an occupancy of grid cells. The large grid resolution allowed us to minimize artefacts from outline range maps, including false absences or more commonly false presences, thus overestimating the extent of occurrence of species. For each species, we calculated both overall range size within the study area expressed as the number of occupied grid cells and latitudinal range expressed as the latitudinal extent of distributional range between northern and southern range borders in 1988 and 2010, respectively.

Change in distributional range was measured for each species as shifts in both latitudinal range and northern range border. The former was calculated by subtracting the latitudinal extent of distributional range in 1988 from the range in 2010 and the latter by subtracting the maximum latitude in 1988 from the values in 2010. Range expansions and contractions were expressed with positive and negative values, respectively. Species experiencing contraction in latitudinal range or northern range border were divided into two subsets to test for correlations between the extent of their range contraction and their ability to utilize habitats.

2.4. Statistical Analysis

Differences within a year in overall range size and latitudinal range were analyzed in two-way combinations of the three unmatched test groups—permanent running water, permanent standing water, and temporary (running or standing) water—with Mann–Whitney–Wilcoxon tests.

Shift in latitudinal range and northern range border was analyzed in two-way combinations of the three unmatched test groups—permanent running water, permanent standing water, and temporary (running or standing) water—with Mann–Whitney–Wilcoxon tests.

All statistical tests were performed using R [56].

3. Results

3.1. Range Size

Species reproducing in permanent standing water and temporary water had, on average, larger ranges in 1988 and 2010, than permanent running water species (Figure 3, Table S3). Differences between species in permanent running water and permanent standing water were highly significant for both overall range and latitudinal range in 1988 and 2010, respectively (Figure 3, Table S3). In contrast, differences between species in permanent running water and temporary water were only significant for overall range in 2010 and not latitudinal range (Figure 3, Table S3).

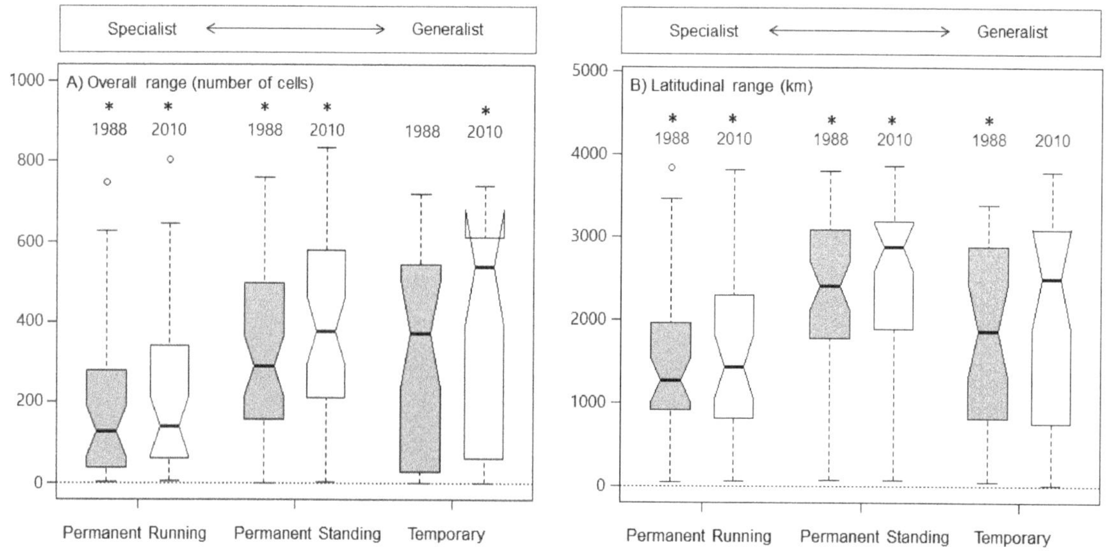

Figure 3. Range sizes in European dragonfly (Odonata) species adapted to permanent running (perennial lotic) water, permanent standing (perennial lentic) water, and temporary (running and standing) water habitats in 1988 (grey) and 2010 (white), respectively, with overall range size expressed as number of occupied 100 × 100 km grid cells, and latitudinal range expressed as the latitudinal extent of distributional range between northern and southern range borders separated by habitat utilization. The box-and-whisker plots illustrate the spread and skewness of the data through their quartiles. The whiskers extending from the box show data variability outside the upper and lower quartiles. Outlier points that differ significantly from the rest of the dataset are plotted as individual points (empty circles) beyond the whiskers. The box notches narrows around the median (thick black middle line) and offers a visual guide on the significance of the difference of medians; if the notches of two boxes do not overlap, this will provide evidence of a statistically significant difference between the medians. Significant variables at $\alpha = 0.05$ in the within-year analysis of differences in overall range size and latitudinal range of the three unmatched test groups—permanent running water, permanent standing water, and temporary (running or standing) water—are indicated with *. Top scale represents a gradient going from habitat specialist with a narrow niche breadth to habitat generalist with broad niche breadth.

Differences between species reproducing in permanent standing water and temporary water were only significant for latitudinal range in 1988, despite the latter having a larger median overall range and a smaller median latitudinal range in 1988 and 2010 compared with permanent standing water species (Figure 3, Table S3).

3.2. Range Shifts

On average, the overall range and latitudinal range expanded, and the northern range border shifted farther north from 1988 to 2010 in all the three ecological habitat categories (Table S4). The median change in the overall range in the number of 100 × 100 km grid cells was 35, the median change in the latitudinal range was 130 km, and the median shift in the northern range margin was 61 km for all species.

Species reproducing in permanent standing water and temporary water had, on average, expanded their latitudinal range and northern range border between 1988 and

2010 to a larger extent than species in permanent running water (Figure 4, Table S4). Differences between species in permanent running water and permanent standing water were, however, not significant for a shift in either latitudinal range or northern range border (Figure 4, Table S4). In contrast, differences between species in permanent running water and temporary water and between permanent standing water and temporary water were highly significant for both range shift measures; that is, temporary water species had expanded both their latitudinal range and northern range border to a much larger extent than species in permanent (both running and standing) waters (Figure 4, Table S4).

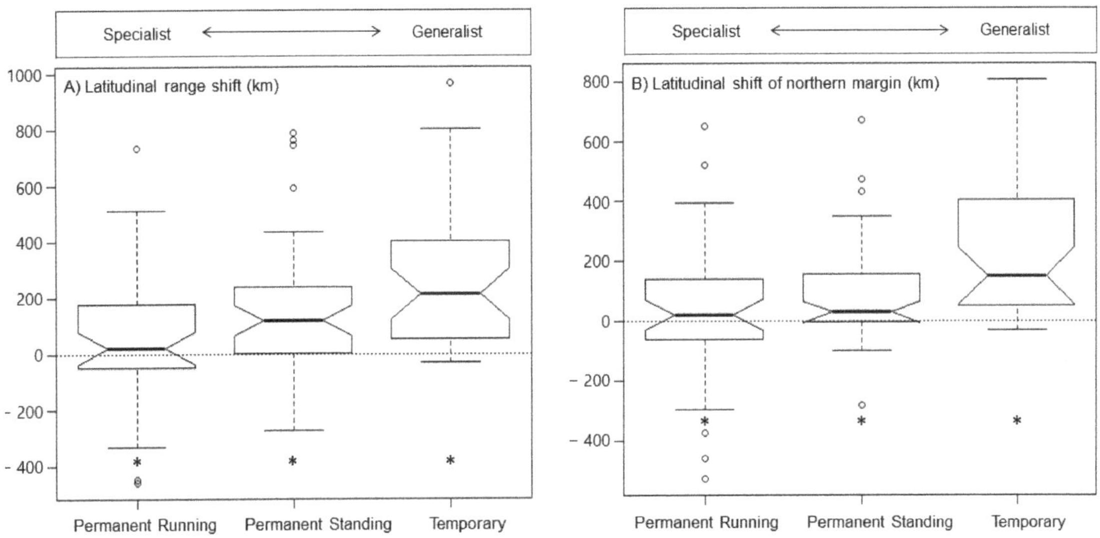

Figure 4. Range shift in European dragonfly (Odonata) species adapted to permanent running (perennial lotic) water, permanent standing (perennial lentic) water, or temporary (running and standing) water habitats between 1988 and 2010 with a shift in latitudinal range expressed as the change in latitudinal extent of distributional range and a shift in northern range border expressed as the change in the maximum of distributional range latitude separated by habitat utilization. Range expansion and contraction is expressed with positive and negative values, respectively. The box-and-whisker plots illustrate the spread and skewness of the data through their quartiles. The whiskers extending from the box show data variability outside the upper and lower quartiles. Outlier points that differ significantly from the rest of the dataset are plotted as individual points (empty circles) beyond the whiskers. The box notches narrow around the median (thick black middle line) and offer a visual guide of the significance of the difference of medians; if the notches of two boxes do not overlap, this will provide evidence of a statistically significant difference between the medians. Significant variables at $\alpha = 0.05$ in the analysis of a shift in latitudinal range and northern range border of the three unmatched test groups —permanent running water, permanent standing water, and temporary (running or standing) water—are indicated with *. Top scale represents a gradient going from habitat specialist with a narrow niche breadth to habitat generalist with a broad niche breadth.

Of the species experiencing contraction in either latitudinal range or northern range border between 1988 and 2010, those in permanent running water had, on average, contracted latitudinal range and northern range border to a much larger extent than species in permanent standing water and temporary water (Table S4). Differences in range contraction between the three habitat categories were, however, not significant for either of the range shift measures (Table S4). Nevertheless, the percentages of species in each category that had experienced range contraction were different, with relatively many more range contracting species in the permanent running water group than in the other two groups (Figure 5).

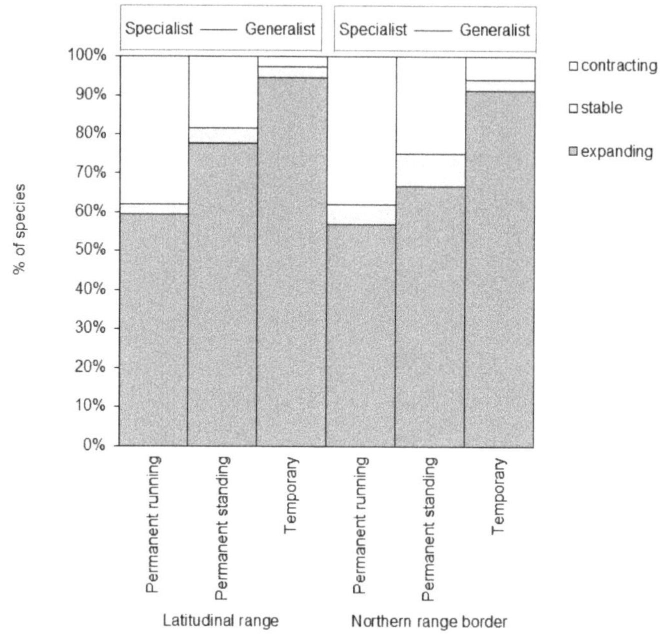

Figure 5. Percentage of European dragonfly (Odonata) species in three ecological habitat categories—permanent running (perennial lotic) water, permanent standing (perennial lentic) water, and temporary (running and standing) water habitats—that have experienced expanding, stable, or contracting latitudinal range or northern range border between 1988 and 2010. Top scale represents a gradient going from habitat specialist with a narrow niche breadth to habitat generalist with a broad niche breadth.

4. Discussion

4.1. Data Quality

Dragonflies are among the taxa with the best data record in space and time across Europe [47], and range maps that are expert-drawn outline maps are the best currently available data for addressing macroecological questions. One drawback of range maps is that such maps may include false absences or more commonly false presences [57], thus overestimating the extent of occurrence of species [58]. Moreover, a common critique is that outline maps only represent the knowledge on distribution of the respective authors rather than the true species distribution. Some critics suggest that analysis will not show true patterns, but only changes in authors' knowledge. Nevertheless, multiple studies on various taxonomic groups within plants, vertebrates, and invertebrates have used outline distribution to find differences in species ranges or species richness (e.g., Hawkins et al. [59]), including macroecological studies on European dragonflies similar to ours (e.g., Grewe et al. [8], Hof et al. [11], and Hof et al. [33]). Furthermore, of the 85 reviewed analyses of species richness in Hawkins et al. [59], 69% were based on range maps. We acknowledge that our maps represent rough approximations of the distribution of European dragonflies, but as demonstrated in Hurlbert et al. [60], this problem can be counteracted by using a sufficiently large grid resolution. Other range map skeptics suggest that range expansions are simply the outcome of a higher number of records. Although we cannot rule out sampling heterogeneity on continental scales, the most significant northern range border shifts were found for species that invaded Central Europe from the Mediterranean or extended their previous northern range border in Central Europe northwards. As Central Europe is one of the best sampled regions [47], we assume that a lower sampling intensity in other parts of Europe should not affect these observed range shifts. Moreover, a bias in the distribution estimates should only matter if there were strong differences in mapping accuracy between species in the three ecological habitat categories. Finally, more local studies that have reported smaller-scale range shifts in European dragonfly species based on point data of range shifts support our findings (e.g., Riservato et al. [2], Knijf et al. [3], Hassall et al. [9], Hickling et al. [10], Ott [13], Suhling et al. [15], and Termaat et al. [16]).

4.2. Do Species Reproducing in Permanent Standing Water and Temporary Water Have Larger Ranges Than Species Reproducing in Permanent Running Water?

Dragonflies in general have high dispersal ability [26], and as a result, their distribution should mainly be driven by changes in current climate and resources because their flying capabilities allow them to track changing climatic and environmental conditions. However, species distributions are not maintained in a quasi-equilibrium state with climate, and we found that species that have adapted to permanent standing water and temporary water, and that are thus more generalists in their habitat utilization than species in permanent running water, have much larger ranges than species in permanent running water. That species with broad niches have larger ranges than species with narrow niches clearly demonstrates that range dynamics in European dragonflies are heavily influenced by habitat specialization or linked dispersal limitation. The fact that permanent running water species have relatively small ranges compared with generalists could arise from their extreme habitat specialization rather than from their reduced dispersal ability. As proposed by Baselga et al. [61], habitat specialization and dispersal ability may, however, be like 'two sides of the same coin', where dispersal ability and habitat specialization may represent a relationship filtering species response to changes in climate and resources [62–64].

Compared with generalist species that reproduce in standing waters, larval adaptations to running water, which occasionally can be fast-flowing and scouring during the wet season, may include a modified body shape and a modified morphology of body parts [26]. This includes a flattened body and an enlarged contact surface with the substrate, narrow and long caudal appendages, reduced or lost swimming ability, reduced number of antenna segments, and leg spurs that allow the larvae to burrow into sandy-stone substrates [15,26]. Due to these adaptations, some running water species are strictly linked to certain microhabitats within different types of permanent running water [26]. This produces more separate and less mobile populations, thus reducing dispersal success and range expansion and thereby causing range filling in these specialist species to lag behind changes in climate and resources [8,11]. In contrast, the landscape has been, and still is, more permeable to generalist species that are adapted to either permanent standing water or temporary water habitats because migration is not restricted or blocked by unsuitable habitats as in specialists [23]. Generalists have a larger proportion of suitable habitats available and, therefore, have larger ranges that are more interconnected, which should further increase both dispersal success and range expansion. Further, as standing waters and especially temporary waters are less persistent in time compared with running waters, these habitats may also have selected for a higher dispersal ability in dragonflies adapted to standing water [11]. Our results clearly support this as generalist species with the broadest niche breadth tend to have larger ranges than permanent standing water species at the same time as they exhibit the most climate-driven response by being most prone to increase their ranges. Similarly, Bota-Sierra et al. [65] found that generalist species in the western Andes in Colombia colonized ecosystems recently created by human activities, while specialist species remain in forest ecosystems.

Due to the European topography, postglacial invasions of dragonflies may have been confined to only a few routes where especially mountain chains acted as barriers, while geotectonic depressions and wide river plains likely played important roles as corridors [66]. Current ranges of dispersal limited organisms often represent partially incomplete postglacial colonization from southern glacial refugia [61] because habitat specialization has prevented colonization in species with narrow niches. Here, the historical landscape may have filtered distributions differentially through the continuous glacial races [66–68]. However, besides a stepwise loss of species with increasing latitude, we find little evidence of postglacial colonization lag of European dragonflies on a continental scale [68,69]. The proportion of standing water species increases with latitude, and in addition to their larger latitudinal ranges, they have more northern distribution centers and range boundaries than running water species [33]. The differences in range size between generalist and specialist species may therefore have ecological explanations, namely, dispersal limitations connected to habitat specialization and the availability of a suitable habitat. This is especially relevant for those dragonflies that have adapted to life in permanent running water and that rely on rivers and streams for successful dispersal among water as their adaptations prevent them from utilizing other habitat types, even only as a stepping stone during colonization.

Some species that we categorized as being able to utilize temporary habitats have relatively small ranges compared with permanent standing water species and are even similar in size to some permanent running water species. This is due to the fact that temporary water species may include not only species adapted to standing water, but also species with a preference for running water that occasionally are also found in seasonal habitats. Furthermore, especially temporary water species include several Afrotropical species, which have only recently established populations north of the Mediterranean Sea and, therefore, only have a restricted current distribution from where they are colonizing southern Europe. At the moment, they are strongly dispersal limited as an effect of their small range size and not so much because they experience habitat loss and are vulnerable to climate change.

4.3. Do Species Reproducing in Temporary Habitats Track Changes Better Than Species in Permanent Habitats?

Besides an overall increase in ranges, species adapted to temporary habitats expanded their latitudinal ranges and northern range borders the most. Consequently, temporary water species migrate north much more than species bound to permanent (both running and standing) waters. This is illustrated by some of the most significant shifts in northern range borders seen in European dragonflies, such as in *Anax imperator* Leach, 1815; *Anax parthenope* (Selys, 1839); *Aeshna affinis* Vander Linden, 1820; *Sympetrum fonscolombii* (Selys, 1840); and *Crocothemis erythraea* (Brullé, 1832) which are all generalist species with the ability to utilize temporary water and which are known to have either invaded Central Europe from the Mediterranean or extended their previous northern range boundaries in Central Europe northwards [13,14]. Common to them, they utilize not only temporary waters, but also various permanent habitats, sometimes including both running and standing waters. Further, they can respond quickly because they can utilize the broadest range of niches, which allows them to persist during times of climate change and habitat loss through the efficient relocation to suitable habitats. The ephemeral nature of temporary habitats selects for high dispersal ability, where their shortened larval stage and rapid larval development [27] clearly provide an advantage in a rapidly changing environment compared with larval stages lasting from 1 to 6 years as in permanent water species [15,26].

4.4. Do Species Reproducing in Permanent Running Water Contract More or Expand Less Than Species in Permanent Standing Water and Temporary Water?

Of those species that experience range contraction, we found an overweight of running water species, which, on average, also showed larger contractions than species that inhabit permanent standing water or temporary water. We suggest that these larger contractions are not driven by a decrease in their realized climatic envelopes as much as by habitat loss and degradation; that is, the declines in habitat specialists are not believed to be linked to climate as much as to differential land use or eutrophication effects. Life forms are even and ever more directly affect by a continuous decline in the availability of a suitable habitat [12]. Changes (of which climate effects are just a part) due to human land use, drainage of wetland areas, pollution, eutrophication, and overgrowth of habitats may cause synergistic treats, where local populations are driven to extinction and habitat fragmentation is prompted to escalate [12]. This is believed to be the reason why *Nehalennia speciosa* is declining and has already gone regionally extinct in many areas across its European range [70]. Contemporary land use is believed to produce more separate and less mobile populations. Especially, distance between suitable habitats in dragonfly specialists is important for their chances of at least partially tracking changes in habitats and resources. However, migratory routes already seem to be partly interrupted over long distances due to loss of habitats, and the connections to distribution centers are also broken in some species [66]. Without suitable dispersal corridors, species responses to changes in climate and resources may not be realized [20,23]. Combined with the dramatic changes to running water environments that prompt them to dry up, especially during the dry season [12], our results suggest that permanent running water species, in particular, could face severe challenges in response to such changes. However, this effect may apply not only to species with a small range size and a narrow niche breadth, but also to less specialized species that cannot persist in temporary dry habitats.

4.5. Do Generalist Species Outperform Specialists with a Narrow Niche Breadth?

We found that specialist species in permanent running waters could be most vulnerable to habitat loss and climate change, likely due to interacting effects of a narrow niche and a small range size. Furthermore, the fact that generalist species have the largest ranges and probably track climate changes the best, and that relatively more specialist than generalist species contract ranges, could indicate that generalists are outperforming specialists. This could provide an early warning of biotic homogenization in European dragonflies similar to what has been found in several studies in which climate change response has been related to niche breadth. It is seen in plants [35], butterflies [34,37,38], frogs [32], reptiles [40], birds [19], and mammals [42], but also in other studies on dragonflies, as in North America [71] and the British Isles [36].

5. Conclusions

Habitat-defined groups of dragonfly species differ in their response to climate and environmental changes, and we suggest that species with narrow habitat requirements will be most affected by ongoing climate change and differential land use due to synergistic effects of a narrow niche and small range size. We find that species that are adapted to habitats that become temporarily dry move north to a much larger extent than species in permanent water habitats, including both running and standing waters. By this, we suggest that temporary waters support dragonfly diversity and act as stepping stones for the dispersal of generalist species. Species adapted to permanent standing water or temporary water habitats, which are less persistent in time and space than running water, disperse better than species adapted to permanent running water habitats. Species in permanent standing water and temporary water have larger ranges than specialist species in permanent running water. Additionally, relatively more species with a small range size and a narrow niche breadth living in permanent running water contract their ranges than less specialized species. This suggests that ecological specialization or dispersal limitations connected to ecological specialization may have prevented postglacial colonization in certain regions. It also suggests that specializations may cause contemporary range shifts to lag behind changes in climate and resources because the landscape may be less permeable to specialized species. Furthermore, this could provide an early indication that recent changes in climate and human land use cause biotic homogenization in the European dragonflies, where ecological specialists are outperformed and replaced by generalists.

Supplementary Materials: The following supporting information can be downloaded at: https://www.mdpi.com/article/10.3390/d14090719/s1; Table S1: List of species excluded from the analysis, where modifications to species range have been applied, which have colonized Europe after 1988, and with taxonomic and nomenclatural changes between 1988 and 2010 [71–74]; Table S2: List of species included in the analysis with individual measures of range size and range shift and ecological habitat category with references. See methods for detailed descriptions of the entries on species range in this table; Table S3: Different measures of range size and test results by ecological habitat category; Table S4: Different measures of range shift and test results for all species and range contracting species by ecological habitat category.

Author Contributions: Conceptualization, K.O., J.-C.S. and H.B.; methodology, K.O.; formal analysis, K.O.; investigation and data curation, K.O.; writing—original draft preparation, K.O.; writing—review and editing, K.O., J.-C.S. and H.B. All authors have read and agreed to the published version of the manuscript.

Funding: This study was supported by the Danish Ministry of Culture (TAKT2010-035), the Graduate School of Science and Technology of Aarhus University, and the Natural History Museum, Aarhus. This study is a contribution to JCS's VILLUM Investigator project "Biodiversity Dynamics in a Changing World" funded by VILLUM FONDEN (grant 16549) and the Center for Informatics Research on Complexity in Ecology (CIRCE), funded by Aarhus University and the Aarhus University Research Foundation under the AU IDEAS program.

Institutional Review Board Statement: Not applicable.

Data Availability Statement: Data are provided in the paper and Supplementary Tables S1–S4.

Acknowledgments: We thank Brody Steven Sandel for the help with R scripts and Peder Klith Bøcher and Lonnie Mikkelsen for the ArcGIS assistance. We are particularly grateful to Melanie Bilz and Jemma Able from the Freshwater Biodiversity Unit under the IUCN Global Species Program for providing range maps from 2010 as ArcGIS shapefiles.

Conflicts of Interest: The authors declare no conflict of interest. The funders had no role in the design of the study; in the collection, analyses, or interpretation of data; in the writing of the manuscript; or in the decision to publish the results.

References

1. Bell, N.; Dylmer, E.; Olsen, K. Status of Aeshna Serrata in Denmark. *Brachytron* **2014**, *16*, 38–42.
2. Riservato, E.; Boudot, J.; Ferreira, S.; Jović, M.; Kalkman, V.; Schneider, W.; Samraoui, B.; Cuttelod, A. *The Status and Distribution of Dragonflies of the Mediterranean Basin*; IUCN: Gland, Switzerland; Malaga, Spain, 2009.
3. de Knijf, G.; Anselin, A. When South Goes North: Mediterranean Dragonflies (Odonata) Conquer Flanders (North-Belgium). *BioRisk* **2010**, *5*, 141–153. [CrossRef]
4. de Kniff, G.; Anselin, A.; Goffart, P. Changes in Ranges: Invertebrates on the Move. In Proceedings of the 13th International Colloquium of the European Invertebrate Survey, Leiden, Germany, 2–5 September 2001; Reemer, M., van Helsdingen, P., Kleukers, R., Eds.; European Invertebrate Survey: Leiden, Germany, 2003; pp. 33–38.
5. de Knijf, G.; Flenker, U.; Vanappelghem, C.; Manci, C.O.; Kalkman, V.J.; Demolder, H. The Status of Two Boreo-Alpine Species, Somatochlora Alpestris and S. Arctica, in Romania and Their Vulnerability to the Impact of Climate Change (Odonata: *Corduliidae*). *Int. J. Odonatol.* **2011**, *14*, 111–126. [CrossRef]
6. Dingemanse, N.J.; Kalkman, V.J. Changing Temperature Regimes Have Advanced the Phenology of Odonata in the Netherlands. *Ecol. Entomol.* **2008**, *33*, 394–402. [CrossRef]
7. Gonseth, Y.; Monnerat, C. Recent Changes in Distribution of Dragonflies in Switzerland (Odonata). In Proceedings of the 13th International Colloquium of the European Invertebrate Survey, Leiden, Germany, 2–5 September 2001; Reemer, M., van Helsdingen, P., Kleukers, R., Eds.; European Invertebrate Survey: Leiden, Germany, 2003; pp. 23–31.
8. Grewe, Y.; Hof, C.; Dehling, D.M.; Brandl, R.; Brändle, M. Recent Range Shifts of European Dragonflies Provide Support for an Inverse Relationship between Habitat Predictability and Dispersal. *Glob. Ecol. Biogeogr.* **2013**, *22*, 403–409. [CrossRef]
9. Hassall, C.; Thompson, D.J. The Effects of Environmental Warming on Odonata: A Review. *Int. J. Odonatol.* **2008**, *11*, 131–153. [CrossRef]
10. Hickling, R.; Roy, D.B.; Hill, J.K.; Thomas, C.D. A Northward Shift of Range Margins in British Odonata. *Glob. Change Biol.* **2005**, *11*, 502–506. [CrossRef]
11. Hof, C.; Brändle, M.; Dehling, D.M.; Munguía, M.; Brandl, R.; Araújo, M.B.; Rahbek, C. Habitat Stability Affects Dispersal and the Ability to Track Climate Change. *Biol. Lett.* **2012**, *8*, 639–643. [CrossRef]
12. Kalkman, V.J.; Boudot, J.-P.; Bernard, R.; Conze, K.-J.; de Knijf, G.; Dyatlova, E.; Ferreira, S.; Jović, M.; Ott, J.; Riservato, E.; et al. *European Red List of Dragonflies*; Publications Office of the European Union: Luxembourg, 2010.
13. Ott, J. Dragonflies and Climatic Change—Recent Trends in Germany and Europe. *BioRisk* **2010**, *5*, 253–286. [CrossRef]
14. Ott, J. Expansion of Mediterranean Odonata in Germany and Europe—Consequences of Climatic Changes. In *Fingerprints of Climate Change—Adapted Behaviour and Shifting Species Ranges*; Walther, G., Burga, C., Edwards, P., Eds.; Kluwer Academic; Plenum Publication: New York, NY, USA, 2001; pp. 89–111.

15. Suhling, F.; Sahlén, G.; Gorb, S.; Kalkman, V.; Dijkstra, K.-D.; van Tol, J. Order Odonata. In *Ecology and General Biology: Thorp and Covich's Freshwater Invertebrates*; Thorp, J., Rogers, D., Eds.; Academic Press: San Diego, CA, USA, 2015; pp. 893–932.
16. Termaat, T.; Kalkman, V.; Bouwman, J. Changes in the Range of Dragonflies in the Netherlands and the Possible Role of Temperature Change. *BioRisk* **2010**, *5*, 155–173. [CrossRef]
17. Lenoir, J.; Svenning, J.C. Climate-Related Range Shifts—A Global Multidimensional Synthesis and New Research Directions. *Ecography* **2015**, *38*, 15–28. [CrossRef]
18. Hickling, R.; Roy, D.B.; Hill, J.K.; Fox, R.; Thomas, C.D. The Distributions of a Wide Range of Taxonomic Groups Are Expanding Polewards. *Glob. Change Biol.* **2006**, *12*, 450–455. [CrossRef]
19. Davey, C.M.; Chamberlain, D.E.; Newson, S.E.; Noble, D.G.; Johnston, A. Rise of the Generalists: Evidence for Climate Driven Homogenization in Avian Communities. *Glob. Ecol. Biogeogr.* **2012**, *21*, 568–578. [CrossRef]
20. Hill, J.K.; Thomas, C.D.; Huntley, B. Climate and Habitat Availability Determine 20th Century Changes in a Butterfly's Range Margin. *Proc. R. Soc. B Biol. Sci.* **1999**, *266*, 1197–1206. [CrossRef]
21. Lancaster, L.T.; Dudaniec, R.Y.; Hansson, B.; Svensson, E.I. Latitudinal Shift in Thermal Niche Breadth Results from Thermal Release during a Climate-Mediated Range Expansion. *J. Biogeogr.* **2015**, *42*, 1953–1963. [CrossRef]
22. Thomas, C.D.; Bodsworth, E.J.; Wilson, R.J.; Simmons, A.D.; Davies, Z.G.; Musche, M.; Conradt, L. Ecological and Evolutionary Processes at Expanding Range Margins. *Nature* **2001**, *411*, 577–581. [CrossRef]
23. Fahrig, L. Effects of Habitat Fragmentation on Biodiversity. *Annu. Rev. Ecol. Evol. Syst.* **2003**, *34*, 59–71. [CrossRef]
24. Channell, R.; Lomollno, M.v. Dynamic Biogeography and Conservation of Endangered Species. *Nature* **2000**, *403*, 84–86. [CrossRef]
25. *IPCC Climate Change 2021: The Physical Science Basis: Contribution of Working Group I to the Sixth Assessment*; Report of the Intergovernmental Panel on Climate Change; Cambridge University Press: Campridge, UK; New York, NY, USA, 2021.
26. Corbet, P.S. *Dragonflies: Behavior and Ecology of Odonata*; Harley Books: Colchester, UK, 2004.
27. Corbet, P.S.; Suhling, F.; Soendgerath, D. Voltinism of Odonata: A Review. *Int. J. Odonatol.* **2006**, *9*, 1–44. [CrossRef]
28. Cham, S.A. Ovipositing Behaviour and Observations on Eggs and Prolarva of Ischnura Pumilio. *J. Br. Dragonfly Soc.* **1992**, *8*, 6–10.
29. de Block, M.; McPeek, M.A.; Stoks, R. Life-History Evolution When Lestes Damselflies Invaded Vernal Ponds. *Evol. Ecol.* **2008**, *62*, 485–493. [CrossRef] [PubMed]
30. Wiggins, G.; Mackay, R.; Smith, I. Evolutionary and Ecological Strategies of Animals in Annual Temporary Pools. *Arch. Für Hydrobiol. Suppl.* **1980**, *58*, 97–206.
31. Fischer, Z. Food Composition and Food Preference in Larvae of *Lestes sponsa* (L.) in Astatic Water Environment. *Pol. Arch. Hydrobiol.* **1967**, *14*, 59–71.
32. Botts, E.A.; Erasmus, B.F.N.; Alexander, G.J. Small Range Size and Narrow Niche Breadth Predict Range Contractions in South African Frogs. *Glob. Ecol. Biogeogr.* **2013**, *22*, 567–576. [CrossRef]
33. Hof, C.; Brändle, M.; Brandl, R. Lentic Odonates Have Larger and More Northern Ranges than Lotic Species. *J. Biogeogr.* **2006**, *33*, 63–77. [CrossRef]
34. Menéndez, R.; Megías, A.G.; Hill, J.K.; Braschler, B.; Willis, S.G.; Collingham, Y.; Fox, R.; Roy, D.B.; Thomas, C.D. Species Richness Changes Lag behind Climate Change. *Proc. R. Soc. B Biol. Sci.* **2006**, *273*, 1465–1470. [CrossRef]
35. Ozinga, W.A.; Colles, A.; Bartish, I.V.; Hennion, F.; Hennekens, S.M.; Pavoine, S.; Poschlod, P.; Hermant, M.; Schaminée, J.H.J.; Prinzing, A. Specialists Leave Fewer Descendants within a Region than Generalists. *Glob. Ecol. Biogeogr.* **2013**, *22*, 213–222. [CrossRef]
36. Powney, G.D.; Cham, S.S.A.; Smallshire, D.; Isaac, N.J.B. Trait Correlates of Distribution Trends in the Odonata of Britain and Ireland. *PeerJ* **2015**, *3*, e1410. [CrossRef]
37. Pöyry, J.; Luoto, M.; Heikkinen, R.K.; Kuussaari, M.; Saarinen, K. Species Traits Explain Recent Range Shifts of Finnish Butterflies. *Glob. Chang Biol.* **2009**, *15*, 732–743. [CrossRef]
38. Warren, M.S.; Hill, J.K.; Thomas, J.A.; Asher, J.; Fox, R.; Huntley, B.; Roy, D.B.; Telfer, M.G.; Jeffcoate, S.; Harding, P.; et al. Rapid Responses of British Butterflies to Opposing Forces of Climate and Habitat Change. *Nature* **2001**, *414*, 65–69. [CrossRef]
39. Angert, A.L.; Crozier, L.G.; Rissler, L.J.; Gilman, S.E.; Tewksbury, J.J.; Chunco, A.J. Do Species' Traits Predict Recent Shifts at Expanding Range Edges? *Ecol. Lett.* **2011**, *14*, 677–689. [CrossRef] [PubMed]
40. Foufopoulos, J.; Kilpatrick, A.M.; Ives, A.R. Climate Change and Elevated Extinction Rates of Reptiles from Mediterranean Islands. *Am. Nat.* **2011**, *177*, 119–129. [CrossRef] [PubMed]
41. Hill, J.K.; Thomas, C.D.; Fox, R.; Telfer, M.G.; Willis, S.G.; Asher, J.; Huntley, B. Responses of Butterflies to Twentieth Century Climate Warming: Implications for Future Ranges. *Proc. R. Soc. B Biol. Sci.* **2002**, *269*, 2163–2171. [CrossRef] [PubMed]
42. Myers, P.; Lundrigan, B.L.; Hoffman, S.M.G.; Haraminac, A.P.; Seto, S.H. Climate-Induced Changes in the Small Mammal Communities of the Northern Great Lakes Region. *Glob. Change Biol.* **2009**, *15*, 1434–1454. [CrossRef]
43. Ruiz, E. *Management of Natura 2000 Habitats. 3170 *Mediterranean Temporary Ponds. Directive 92/43/EEC on the Conservation of Natural Habitats and of Wild Fauna and Flora*; European Commission: Luxembourg, 2008.
44. Pérez-Bilbao, A.; Benetti, C.; Garrido, J. Biodiversity and Conservation of Temporary Ponds—Assessment of the Conservation Status of "Veiga de Ponteliñares", NW Spain (Natura 2000 Network), Using Freshwater Invertebrates. In *Biodiversity in Ecosystems—Linking Structure and Function*; Blanco, J., Lo, Y.-H., Eds.; InTech: London, UK, 2015; pp. 241–269.
45. Askew, R.R. *The Dragonflies of Europe*; Harley Books (BH & A Harley Ltd.): Colchester, UK, 1988.

46. *ESRI ArcGIS Desktop v. 10.2 [GIS Software]*; Environmental Systems Research Institute: Redlands, CA, USA, 2010.
47. Dijkstra, K.-D.; Lewington, R. *Field Guide to the Dragonflies of Britain and Europe*; British Wildlife Publishing: Totnes, UK, 2006.
48. IUCN. The IUCN Red List of Threatened Species. Version 2015-3. Available online: www.iucnredlist.org (accessed on 17 May 2016).
49. Sternberg, K.; Buchwald, R. *Die Libellen Baden-Württembergs: Grosslibellen (Anisoptera)*; Eugen Ulmer: Stuttgart, Germany, 2000; Volume 2.
50. Sternberg, K.; Buchwald, R. *Die Libellen Baden-Württembergs: Allgemeiner Teil, Kleinlibellen (Zygoptera)*; Eugen Ulmer: Stuttgart, Germany, 1999; Volume 1.
51. Beschovski, V.; Marinov, M. Fauna, Ecology, and Zoogeography of Dragonflies (Insecta: *Odonata*) of Bulgaria. In *Biogeography and Ecology of Bulgaria*; Fet, V., Popov, A., Eds.; Springer: Dordrecht, Germany, 2007; pp. 199–231.
52. Carchini, G.; della Bella, V.; Solimini, A.G.; Bazzanti, M. Relationships between the Presence of Odonate Species and Environmental Characteristics in Lowland Ponds of Central Italy. *Ann. Limnol. Int. J. Limnol.* **2007**, *43*, 81–87. [CrossRef]
53. Koli, V.K.; Bhatnagar, C.; Shekhawat, D.S. Diversity and Species Composition of Odonates in Southern Rajasthan, India. *Proc. Zool. Soc.* **2015**, *68*, 202–211. [CrossRef]
54. Suhling, F.; Jödicke, R.; Schneider, W. Odonata of African Arid Regions—Are There Desert Species? *Cimbebasia* **2003**, *18*, 207–224.
55. Florencio, M.; Díaz-Paniagua, C. Presencia de Lestes Macrostigma (Eversmann, 1836) (Odonata: *Lestidae*) En Las Lagunas Temporales Del Parque Nacional de Doñana (Sudoeste de España). *Boletín Soc. Entomológica Aragonesa* **2012**, *50*, 579–581.
56. R Core TeAm. *R: A Language and Environment for Statistical Computing*; R Core TeAm: Vienna, Austria, 2022.
57. Rondinini, C.; Wilson, K.A.; Boitani, L.; Grantham, H.; Possingham, H.P. Tradeoffs of Different Types of Species Occurrence Data for Use in Systematic Conservation Planning. *Ecol. Lett.* **2006**, *9*, 1136–1145. [CrossRef]
58. Jetz, W.; Sekercioglu, C.H.; Watson, J.E.M. Ecological Correlates and Conservation Implications of Overestimating Species Geographic Ranges. *Conserv. Biol.* **2008**, *22*, 110–119. [CrossRef]
59. Hawkins, B.A.; Field, R.; Cornell, H.V.; Currie, D.J.; Guégan, J.F.; Kaufman, D.M.; Kerr, J.T.; Mittelbach, G.G.; Oberdorff, T.; O'Brien, E.M.; et al. Energy, Water, and Broad-Scale Geographic Patterns of Species Richness. *Ecology* **2003**, *84*, 3105–3117. [CrossRef]
60. Hurlbert, A.H.; Jetz, W. Species Richness, Hotspots, and the Scale Dependence of Range Maps in Ecology and Conservation. *Proc. Natl. Acad. Sci. USA* **2007**, *104*, 13384–13389. [CrossRef] [PubMed]
61. Baselga, A.; Lobo, J.M.; Svenning, J.C.; Aragón, P.; Araújo, M.B. Dispersal Ability Modulates the Strength of the Latitudinal Richness Gradient in European Beetles. *Glob. Ecol. Biogeogr.* **2012**, *21*, 1106–1113. [CrossRef]
62. Jocque, M.; Field, R.; Brendonck, L.; de Meester, L. Climatic Control of Dispersal-Ecological Specialization Trade-Offs: A Metacommunity Process at the Heart of the Latitudinal Diversity Gradient? *Glob. Ecol. Biogeogr.* **2010**, *19*, 244–252. [CrossRef]
63. Abbott, J.C.; Bota-Sierra, C.A.; Guralnick, R.; Kalkman, V.; González-Soriano, E.; Novelo-Gutiérrez, R.; Bybee, S.; Ware, J.; Belitz, M.W. Diversity of Nearctic Dragonflies and Damselflies (Odonata). *Diversity* **2022**, *14*, 575. [CrossRef]
64. Bota-Sierra, C.A.; García-Robledo, C.; Escobar, F.; Novelo-Gutiérrez, R.; Londoño, G.A. Environment, Taxonomy and Morphology Constrain Insect. Thermal Physiology along Tropical Mountains. *Funct. Ecol.* **2022**, *36*, 1924–1935. [CrossRef]
65. Bota-Sierra, C.A.; Flórez-V, C.; Escobar, F.; Sandoval-H, J.; Novelo-Gutiérrez, R.; Londoño, G.A.; Cordero-Rivera, A. The Importance of Tropical Mountain Forests for the Conservation of Dragonfly Biodiversity: A Case from the Colombian Western Andes. *Int. J. Odonatol.* **2021**, *24*, 233–247. [CrossRef]
66. Sternberg, K. Die Postglaziale Besiedlung Mitteleuropas Durch Libellen, Mit Besonderer Berucksichtigung Sudwestdeutschlands (Insecta, Odonata). The Postglacial Colonization of Central Europe by Dragonflies, with Special Reference to Southwestern Germany (Insecta, Odonata). *J. Biogeogr.* **1998**, *25*, 319–337. [CrossRef]
67. Corser, J.D.; White, E.L.; Schlesinger, M.D. Odonata Origins, Biogeography, and Diversification in an Eastern North American Hotspot: Multiple Pathways to High Temperate Forest Insect. Diversity. *Insect. Conserv. Divers.* **2014**, *7*, 393–404. [CrossRef]
68. Heiser, M.; Schmitt, T. Do Different Dispersal Capacities Influence the Biogeography of the Western Palearctic Dragonflies (Odonata)? *Biol. J. Linn. Soc.* **2010**, *99*, 177–195. [CrossRef]
69. Keil, P.; Simova, I.; Hawkins, B.A. Water-Energy and the Geographical Species Richness Pattern of European and North African Dragonflies (*Odonata*). *Insect. Conserv. Divers.* **2008**, *1*, 142–150. [CrossRef]
70. Bernard, R.; Wildermuth, H. Nehalennia Speciosa (Charpentier, 1840) in Europe: A Case of a Vanishing Relict (Zygoptera: *Coenagrionidae*). *Odonatologica* **2005**, *34*, 335–378.
71. Ball-Damerow, J.E.; M'Gonigle, L.K.; Resh, V.H. Changes in Occurrence, Richness, and Biological Traits of Dragonflies and Damselflies (Odonata) in California and Nevada over the Past Century. *BioDivers. Conserv.* **2014**, *23*, 2107–2126. [CrossRef]
72. Boudot, J.-P.; Kalkman, V.J.; Azpilicueta Amorin, M.; Bogdanović, T.; Cordero Rivera, A.; Degabriele, G.; Dommanget, J.-L.; Ferreira, S.; Garrigos, B.; Jović, M.; et al. Atlas of the Odonata of the Mediterranean and North Africa. *Libellula Suppl.* **2009**, *9*, 1–256.
73. Aguesse, P. Les Odonates de l'Europe Occidentale, Du Nord de l'Afrique et Des Îles Atlantiques. In *Faune de l'Europe et du Bassin Méditerranéen*; Masson et Cie: Paris, France, 1968; Volume 4, pp. 1–258.
74. Pilgrim, E.M.; von Dohlen, C.D. Molecular and Morphological Study of Species-Level Questions within the Dragonfly Genus Sympetrum (Odonata: *Libellulidae*). *Ann. Entomol. Soc. Am.* **2007**, *100*, 688–702. [CrossRef]

diversity

MDPI

Article

Population Density and Abundance of the Northernmost Population of *Cordulegaster heros* (Anisoptera: Cordulegastridae) in Europe (Czech Republic) with Notes on Its Biogeographical Range

Otakar Holuša [1,*] and Kateřina Holušová [2]

[1] Department of Environmental Science and Natural Resources, Faculty of Regional Development and International Studies, Mendel University in Brno, Tř. Gen Píky 7, CZ-613 00 Brno, Czech Republic
[2] Department of Forest and Wood Product Economics and Policy, Faculty of Forestry and Wood Technology, Mendel University in Brno, Zemědělská 3, CZ-613 00 Brno, Czech Republic
* Correspondence: holusao@email.cz

Abstract: *Cordulegaster heros* is a Balkan species with a disjunctive area extending into Central Europe. The population in the Chřiby Mts. in the southeastern Czech Republic is the northernmost population, and this population was intensively studied from 2010 to 2021 to establish basic data on its abundance. In the territory, the geomorphological characteristics of streams, characteristics of sediment in streams, habitat, emergence time, and period of flight were recorded, and population viability was evaluated. Larvae were recorded in 10 small forest streams (altitude of 235–426 m a.s.l.), with an average minimum width of 51.9 cm, an average maximum width of 177.7 cm, an average minimum depth of 6.5 cm, an average maximum depth (in pools) of 21 cm, and an average stream gradient of 1.9 grades. The sediments in each stream exhibited a grain size distribution with an average fraction less than 0.05 mm represented by 6.3%, a fraction of 0.05–0.1 mm represented by 21.1%, a fraction of 0.1–2 mm represented by 52.1%, a fraction of 2–5 mm represented by 12.1%, a fraction of 5–20 mm represented by 8%, and a fraction of 20+ mm represented by 0.3%. The larval abundance was 0.1–6.7 larvae per 1 m^2 of suitable sediment. The emergence period was recorded from 28 May to 1 July. The emergence site was categorized as larvae-dominated plant leave (57% of cases), plant stalks (21%), and tree trunks (17%). Exuviae occurred at an average of 154 cm at horizontal distance from the shore and an average vertical height of 77 cm above the ground. The average total distance of larval movement was 205 cm. The flight period in 2021 was recorded from 15 June to 11 August with peak flight activity noted in the third week of June. The northernmost population of *C. heros* was evaluated as viable and stable.

Keywords: *Cordulegaster heros*; Cordulegastridae; Odonata; northern border distribution; population abundance; biogeography; Chřiby Hills; Czech Republic

Citation: Holuša, O.; Holušová, K. Population Density and Abundance of the Northernmost Population of *Cordulegaster heros* (Anisoptera: Cordulegastridae) in Europe (Czech Republic) with Notes on Its Biogeographical Range. *Diversity* **2022**, *14*, 854. https://doi.org/10.3390/d14100854

Academic Editors: M. Olalla Lorenzo-Carballa and Ricardo Koroiva

Received: 17 September 2022
Accepted: 6 October 2022
Published: 9 October 2022

1. Introduction

Cordulegaster heros was described by Theischinger [1] with type locality in Austria, Lower Austria-Sankt Andrä vor dem Hagnethale in central Europe. The species is found on the Balkan Peninsula (including Greece, North Macedonia, Montenegro, Albania, Croatia, Slovenia, Serbia, Bosnia and Herzegovina, Bulgaria, northeastern Italy, and Romania) and central Europe, including the foothills of the Alps in eastern Austria [2–5], the foothills of the Carpathians Mountains in Slovakia [6,7], the Czech Republic [8], and outside of the Carpathians Mountains in Ukraine [9]. Separate areas of growth have been reported from the hills in the middle of the Pannonian lowland in Hungary [10,11].

Since the species was described approximately 45 years ago, its range has seen great progress in the last few years in terms of knowledge and boundary refinement. In 1988, practically nothing was known about its distribution, as the species occurrence was known from a few localities in Austria, western Hungary, the western part of the former Yugoslavia [12], and Greece [1]. Later, Boudot [2] reported its distribution over a large area of the Balkans Peninsula, but the range often follows national boundaries and includes "geometric" connections between the northern and southern parts of the range. Later, van Pelt [3] removed some inconsistencies and reported a more or less disjunctive range throughout the Balkans with the northernmost extension in southern Slovakia. The most up-to-date range has been recently reported [13], where the areal distribution throughout the western Balkans is already shown, although with "gaps" of knowledge in Albania and Bosnia and Herzegovina. Unfortunately, there are still insufficient data on the northern and northeastern borders, as evidenced by the findings for the territory of Ukraine [9]. The center of the species range is located in Slovenia, eastern Austria, and northwestern Croatia, where the species is common, with large populations reported ([14,15], Holuša unpubl.). Until the description of *C. heros* in 1979, its occurrence was confused with that of the congeneric species *C. boltonii* and *C. bidentata*. A more precise eastern limit of the distribution of *C. boltonii* in Central Europe was reported by Holuša [16]. Although *C. boltonii* is reported in Romania [17] and Slovakia [18–20], this was possibly due to confusion with *C. heros* and *C. bidentata* [16].

The possible occurrence of *C. heros* in the territory of the Czech Republic was first reported by Bernard (pers. Comm.). Later, Holuša [21] introduced the idea that the species could occur in the southeastern part of the territory of the Czech Republic on the southern slopes of the Carpathian Mountains on the border with the Pannonian lowland. However, extensive surveys of this area since 1998 confirmed the occurrence as late as 2009 (one dead female) [8]. This record has been the absolute northernmost occurrence reported to date. In contrast, Staufer and Holuša [8] "classified" it as an occasional "visit"; nevertheless, they conceded the potential occurrence of a small permanent population. The occurrence of larvae was established in 2011 from one specimen at the locality of the Kudlovický potok stream in the Kudlovická valley near the village of Kudlovice and at the Jankovický potok stream in the village of Jankovice [22].

The aim of this paper is to present a complete overview of all findings of *C. heros* in the Czech Republic during intensive research carried out in 2010–2021 in order to provide a first evaluation of population abundance and to evaluate its stability. Knowledge of the occurrence and ecology of this species is also very important, as this species is listed as requiring protection (Annex II and IV of Council Directive 92/43/EHS).

2. Materials and Methods

A detailed survey was performed in the area of the Chřiby Hills and Ždánický les Hills in the southeastern part of the Czech Republic. In both areas, all watercourses were repeatedly surveyed (grid squares of the Central European Mapping Network 6867, 6868, 6967, and 6968 for Ždánický les Hills [23] and 6769, 6869, 6870, 6968, 6969, 6970, and 6770 for Chřiby Hills). After finding larvae in 2011 in the northern area of the Chřiby Hills, some streams were selected for intensive surveys. Several sites were then selected on each stream to survey larvae in detail with individual sites located approximately 500–700 m from each other, depending on the nature of the stream. The following number of sites was then selected on each stream (Table 1). Only sites where *C. heros* was detected are listed in the site descriptions.

Table 1. Description of watersheds with localities with *Cordulegaster heros* records in the Czech Republic (only localities with detected occurrence are listed, localities are sorted upstream).

Number of Watershed	Locality Name (Cadastral Territory)	Code of Locality *	Altitude (m a.s.l.)	Geographical Coordinates (N/E)		Grid Mapping Square	Larvae Density (per 1 m^2 of Sediment)
I.	Cvrčovický potok stream (Cvrčovice u Zdounek)	Ia.	235	49°13′24.71″	17°20′28.89″	6770	0.7
		Ib.	248	49°13′17.51″	17°20′51.26″	6770	0.4
		Ic.	265	49°12′59.78″	17°21′08.39″	6770	0.8
		Id.	270	49°12′45.78″	17°21′15.83″	6770	0.3
		Ie.	294	49°12′19.27″	17°21′35.88″	6770	0.7
		If.	307	49°12′06.05″	17°21′23.53″	6770	0.1
II.	Divocký potok stream (Divoky)	IIa.	285	49°12′11.69″	17°19′42.62″	6769	0.4
		IIb.	298	49°11′54.36″	17°19′50.56″	6869	0.2
		IIc.	312	49°11′53.77″	17°19′51.13″	6869	4.0
III.	Roštínský potok stream (Roštín)	IIIa.	371	49°10′46.72″	17°18′44.81″	6869	0.1
IV.	Vašákův potok stream (Roštín)	IVa.	340	49°10′34.67″	17°17′23.07″	6869	0.2
V.	stream of Litava (Zástřizly)	Va.	361	49°08′52.72″	17°14′54.34″	6869	0.3
		Vb.	426	49°09′05.37″	17°16′06.35″	6869	0,1
VI.	Kudlovický potok stream (Košíky, Lubná u Kroměříže, Kostelany)	VIa.	255	49°10′11.52″	17°25′51.31″	6870	0.7
		VIb.	295	49°11′01.59″	17°24′41.22″	6870	4.5
		VIc.	320	49°11′39.34″	17°23′54.94″	6870	0.2
		VId.	328	49°11′41.59″	17°23′26.90″	6870	0.1
		VIe.	320	49°11′18.41″	17°23′59.90″	6870	0.3
	Habešský potok stream–right-side tributary of Kudlovický potok stream	VIf.	346	49°11′07.23″	17°23′40.57″	6870	0.3
		VIg.	356	49°10′59.78″	17°23′20.28″	6870	6.7
VII.	Jankovický potok stream (Jankovice u Uherského Hradiště)	VIIa.	309	49°09′44.99″	17°22′36.64″	6870	0.5
		VIIb.	316	49°09′49.38″	17°22′32.72″	6870	6.3
		VIIc.	334	49°09′59.90″	17°22′28.30″	6870	1.3
		VIId.	352	49°10′07.97″	17°22′19.96″	6870	0.4
		VIIe.	355	49°10′16.87″	17°22′15.19″	6870	0.1
		VIIf.	362	49°10′26.88″	17°22′04.71″	6870	0.2
		VIIg.	383	49°10′35.61″	17°21′56.09″	6870	1.3
	Upper left-side tributary of Jankovický potok stream	VIIh.	354	49°10′21.73″	17°22′14.47″	6870	0.3
	Upper right-side tributary of Jankovický potok stream	VIIi.	371	49°10′21.12″	17°21′59.64″	6870	0.2
VIII.	Bunčovský potok stream (Velehrad)	VIIIa.	352	49°09′53.34″	17°20′53.31″	6870	0.3
		VIIIb.	355	49°10′07.99″	17°20′42.39″	6870	0.1
		VIIIc.	368	49°10′05.02″	17°20′49.50″	6870	0.2
		VIIId.	403	49°10′24.00″	17°20′55.09″	6870	0.1
		VIIIe.	286	49°09′08.41″	17°21′46.88″	6870	4.0
		VIIIf.	389	49°10′15.71″	17°20′58.41″	6870	0.1
IX.	stream of Salaška (Salaš u Velehradu)	IXa.	298	49°08′42.51″	17°20′15.25″	6870	0.8
		IXb.	319	49°09′00.89″	17°19′49,33″	6869	0.1
		IXc.	323	49°09′06.93″	17°19′09.09″	6869	0.5
		IXd.	342	49°09′05.95″	17°19′19.44″	6869	1.5
	Upper left-side tributary of stream of Salaška	IXe.	332	49°09′07.25″	17°19′53.29″	6869	5.0
		IXf.	344	49°09′20.63″	17°18′51.26″	6869	0.1
X.	Buchlovický potok stream (Buchlovice)	Xa.	310	49°05′48.06″	17°19′01.18″	6969	4.0

* roman numeral - designation of the watershed of watercourse, alphabetical letter - location on the watercourse

For the watersheds, individual sites (locations) were selected on the main stream and tributaries, where larvae were surveyed over a 50 m transect. In this reach, all suitable sediment deposits where larvae were suspected to occur were examined in detail. Where larvae were found, basic stream characteristics (stream width—minimum/maximum), stream depth (minimum/maximum in pools), and stream gradient were recorded. Phytocenological notes were made on vegetation and tree cover composition, and other dragonfly species were monitored. After the total larvae were found and recorded, they were again released into the stream. At the site where larvae were detected, a sediment sample weighing approximately 700–1000 g was collected, to determine the sediment grain fraction representation. Grain size was determined on the dried sample, which was divided into fines and skeleton fractions. The skeleton was separated into grain size fractions (2–5 mm, 5–20 mm, 20 mm or greater) using sieves and the fine-earth into grain size fractions (less than 0.05 mm, 0.05–0.1 mm, and 0.1–2 mm) using the floating method [24]. For the 50 m transect, the suitable sediment area was estimated using the average length and average width.

Emergence monitoring was conducted on all streams from 2013–2014, and then 2020–2021. A 100 m transect was selected where larvae were detected. Exuviae were collected from the beginning of May to the beginning of July in the riparian parts of the sites studied. The search for larvae occurred approximately 10 m from the shoreline and up to 5 m on tree trunks. The distance from the shoreline (vertical projection of the emergence place) and the height above the ground (vertical distance from the place where the shoreline ends) were measured. The total distance to the emergence place is the sum of the height and the distance from the shore, as travelled by the larvae from the shoreline to the place where the adult emerged. The position of its thorax was used to identify the position of the exuviae, and distances were measured from this point. The site of emergence was evaluated according to where the exuviae was attached, i.e., tree roots, tree trunks, tree sticks, tree leaves, plant stalks, and plant leaves.

In 2021, the flight period of the adults at the Habešský potok stream (locality VIf) stream site was studied in detail. The flight activity of adults was monitored from 2 June to 18 August 2021. The site was monitored in suitable weather (partly cloudy to clear sky, midday temperature, approximately 20 °C or greater) from 8:30 a.m. to 7 p.m. CET from 2 to 15 June, from 15 July to 18 August 2021 (sunrise on these days at this latitude is at 4:55 and 5:20 a.m. EST, respectively), and from 5:30 a.m. to 9:15 p.m. CET from 19 June to 11 July 2021 (sunrise on these days at this latitude ranges from 4:50 to 5:05 a.m. CET). The passing adult was caught on the first pass and marked with a number on the wings. On the next overflight, only the number was detected, and the overflight was noted. The total number of adults within a day and the passage of non-marked adults were recorded. The number of adults observed each day was determined based on the number of marked individuals and the number of passes of unmarked individuals.

The distribution of the species in Europe was processed in a grid map of 12 × 11 km squares to correspond to the KFME (Kartierung der Flora Mitteleuropas) squares for each country. Previously published data [7,25,26] and our own data [Holuša unpubl.] for Slovakia, Hungary [27], and Austria [14] were used to show the occurrence in individual countries. The Pannonian biogeographical boundaries are modified for Slovenia and Austria based on state boundaries but are based on the works of [28,29]. The maps were processed in ESRI 2020 ArcGIS ArcMap 10.8 software.

3. Results

3.1. *Locations of Codulegaster Heros Populations in the Czech Republic*

Cordulegaster heros was found at 42 localities (Table 1) in 10 forest stream catchments in the northern Chřiby region of southern Moravia in the southeastern part of the Czech Republic (Figure 1) in an area approximately 100 km^2 in size. These localities are located in five faunistic squares, 6769, 6770, 6869, 6870, and 6969, at altitudes ranging from 235 to 426 m a.s.l. with an average of 328 m a.s.l. (Figure 2). The species was not detected in the adjacent area, i.e., Ždánický les Hills.

The area of occurrence of *C. heros* in Chřiby Hills in the territory of the Czech Republic constitutes a separate area within its disjunctive area. The nearest known occurrence in Austria is 135 km to the southwest ([30], Holuša unpbl.), and the range in the Záhorie lowland and in the Little Carpathians (Malé Karpaty Mts.) in Slovakia [7] (Figure 1) is 50 km to the south. The area in Chřiby Hills is the northernmost occurrence of the species, and the locality Cvrčovický potok stream (Cvrčovice village cadastral territory), with coordinates of 49°13′24.71″ N, 17°20′28.89″ E, is the absolutely northernmost point of occurrence in Europe and the world.

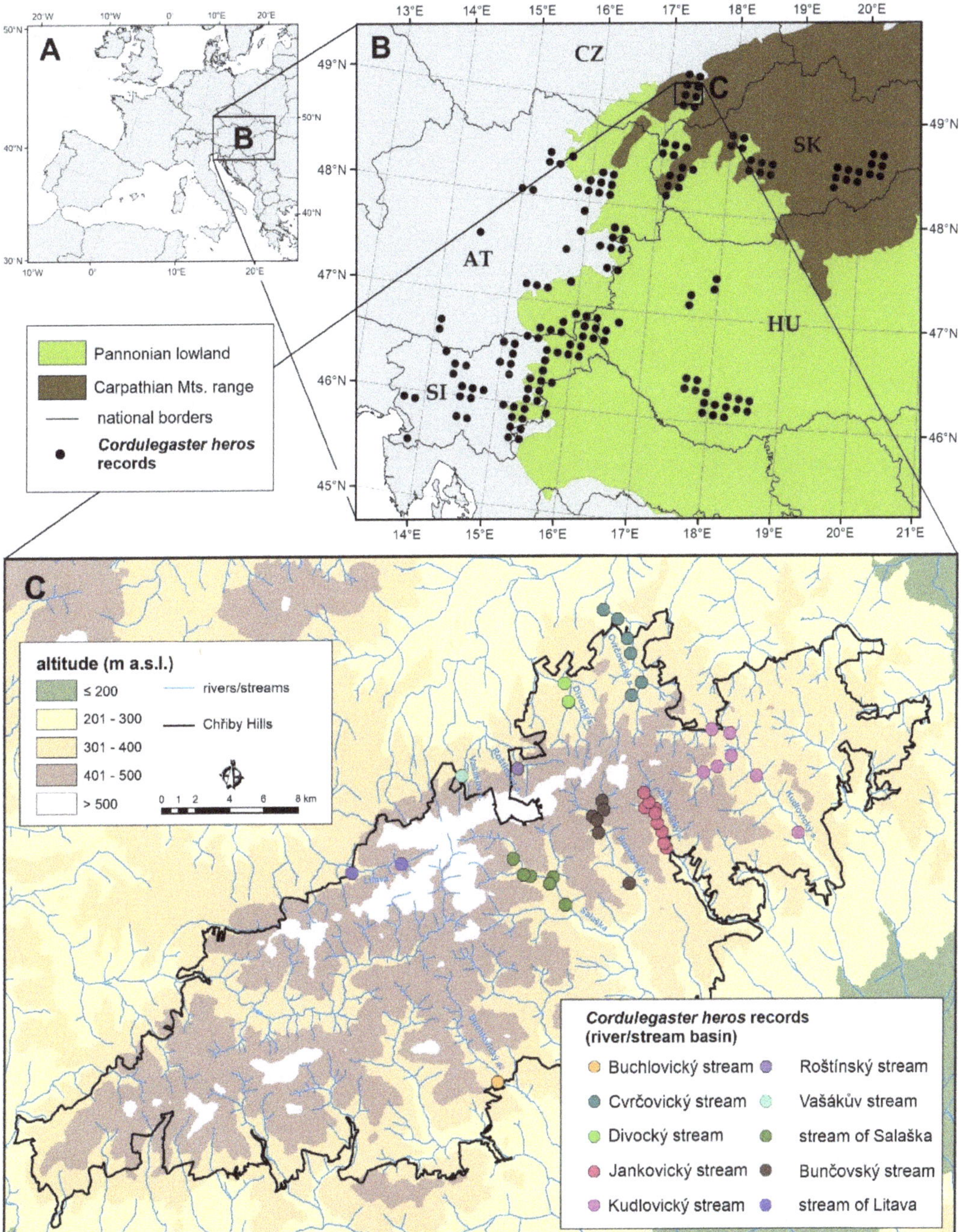

Figure 1. Known records of *Cordulegaster heros* in the Czech Republic (state as of 30 June 2022). (**A**) Map of western, south and central Europe (**B**) Map of central Europe with records of *C. heros* (AT—Austria, CZ—Czech Republic, HU—Hungary, SI—Slovenia, SK—Slovakia) (**C**) Region of the Chřiby Hills in the Czech Republic with *C. heros* records in individual watershed.

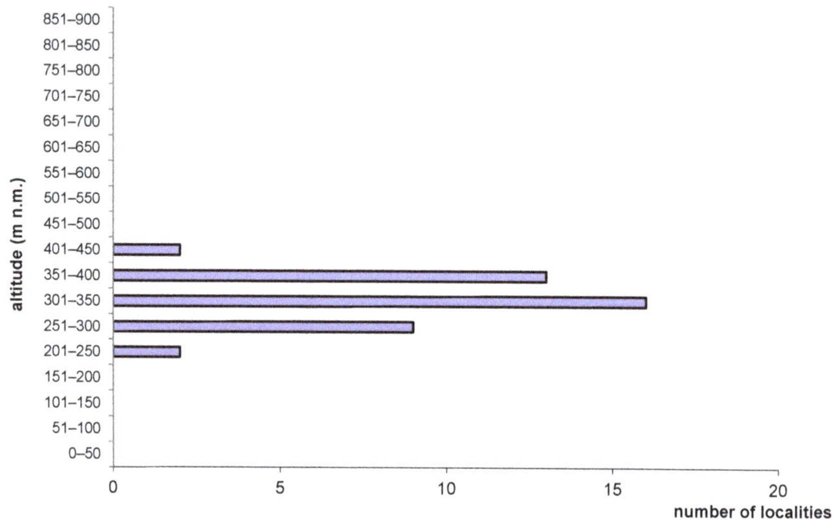

Figure 2. Number of localities with *Cordulegaster heros* populations in the Czech Republic based on altitude (all authors (n = 42)).

3.2. Habitat Characteristics

The species habitat is characterized by forest streams, with an average minimum width of 51.9 cm, an average maximum width of 177.7 cm, an average minimum depth of 6.5 cm, an average maximum depth (in pools) of 21 cm (Figure 3a), and an average stream gradient of 1.9 grades (range 1–3.5 grades). All streams are natural or seminatural, and more or less meandering. In some locations, the stream courses have been straightened due to the construction of forest roads, and the adjacent bank is strengthened in places by stone flatwork. The banks are gradual but also straight where the stream bed meets the adjacent slope. Most of the stream alluvium is covered by a variable vegetation cover (average 46% with values ranging from 5 to 100% based on tree cover).

The vegetation is most frequently composed of the following species listed in order of frequency of occurrence: *Aegopodium podagraria*, *Athyrium filix-femina*, *Carex remota*, *C. sylvatica*, *Dryopteris filix-mas*, *Geranium robertianum*, *Glechoma hederacea*, *Impatiens noli tangere*, *Lamium maculatum*, *Urtica dioica*, *Stachys sylvatica*, *Petasites albus*, *Pulmonaria officinalis*, *Rubus fruticosus*, *R. hirtus*, *Carex brizoides*, *Brachypodium sylvaticum*, *Carex pendula*, *Lycopus europaeus*, *Lysimachia numularia*, *Mercurialis perennis*, *Oxalis acetosella*, *Impatiens parviflora*, and others. All localities are located in the forest complex of Chřiby Hills (Figure 4a,b). The alluvia are covered by forest stands characterized by the following two types of composition: 1. *Alnus glutinosa* and *Fraxinus excelsior* with an admixture of *Fagus sylvatica* and *Acer pseudoplatanus*, and 2. dominated by *Fagus sylvatica* and *Acer pseudoplatanus* with an admixture of *Carpinus betulus* and *Alnus glutinosa*. Individually, *Tilia cordata*, *Quercus petraea*, *Sambucus nigra*, nonnative *Picea excelsa*, and *Larix decidua* may also be represented.

The presence of suitable sediment is necessary for larval survival. Based on grain size characteristics (Figure 3b), 6.3% of samples exhibit an average fraction of less than 0.05 mm, 21.1% with a fraction of 0.05–0.1 mm, 52.1% with a fraction of 0.1–2 mm, 12.1% with a fraction of 2–5 mm, 8% with a fraction of 5–20 mm, and 0.3% with a fraction of 20+ mm. Thus, the average ratio of fine (fraction less than 2 mm grain size) to skeleton (fraction greater than 2 mm) sediment was 79.6:20.4%. This finding shows that the sediment in these habitats is sandy and dominated by medium sand with an admixture of fine sand and a mixture of gravel. Stones (from 20 mm) and clay are very poorly represented.

In habitats with *C. heros*, seven additional dragonfly species were found. The most frequent three species included *Calopteryx virgo*, *Cordulegaster bidentata* (larvae of both species were found), and *Aeshna cyanea*. Only single adult specimens of other species, including *Ophigomphus cecilia*, *Somatochlora metallica*, *Platycnemis pennipes*, and *Lestes viridis*, were recorded. The larvae of *C. bidentata* were recorded syntopically at several sites at the small side tributaries of the main streams, where *C. bidentata* occurs. The larvae enter the main stream by flushing at higher water levels in the watercourse.

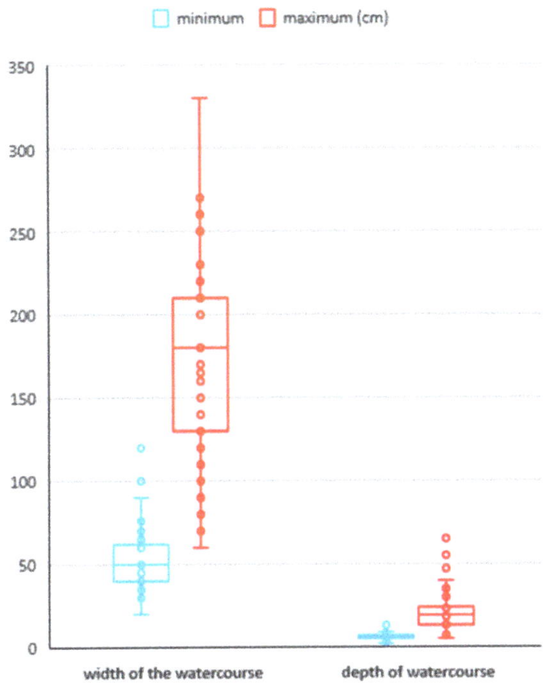

(a)

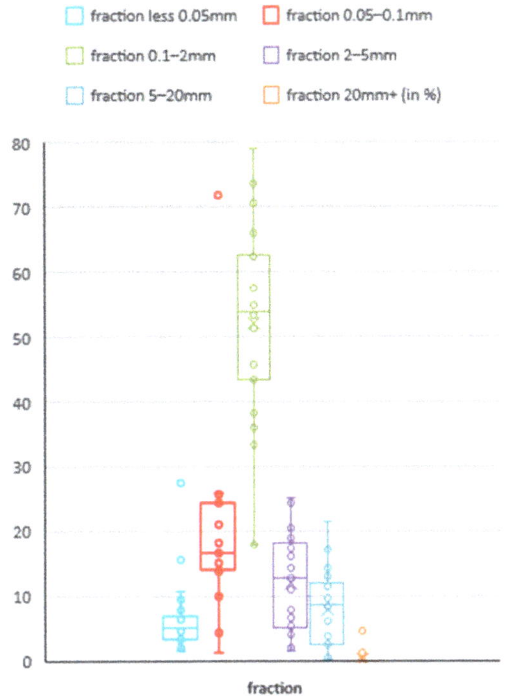

(b)

Figure 3. Characteristics of habitat at localities with *Cordulegaster heros* larvae (n = 42): (**a**) width and depth of watercourses, (**b**) grain composition of the sediment.

3.3. Population Abundance

The larvae of *C. heros* occurred in areas with suitable sediment, which varied from single point locations (e.g., locality Xa) to beds that were approximately 100% composed of alluvial sediment (e.g., locality VIIIb). The average larval abundance (all instars combined) was 2.8 larvae/50 m section. With respect to sediment area, the density reached 0.1–6.7 larvae per 1 m^2 of sediment (Table 1). The maximum number of larvae found per 50 m section of watercourse was 10 with a maximum of 4 larvae per site (15 × 15 cm). It is clear from the distribution of sites that most larvae were found in the upper parts of the streams (e.g., Kudlovický potok stream); thus, larvae found in the lower parts of the streams are due to larval flushing during high flow conditions. In these lower parts, the occurrence of adults (both males and females) was recorded, but no ovipositions were observed. Females most often prefer streams with specific characteristics (Figure 3a,b).

(a) (b)

Figure 4. Habitat of *Cordulegaster heros* in the Czech Republic: (**a**) locality of Habešský potok stream (Kostelany cadastral territory, southern Czech Republic) (locality VIe) (18 July 2013, photo Otakar Holuša), (**b**) locality of Cvrčovický potok stream (Cvrčovice cadastral territory, southern Czech Republic) (locality Id) (4 June 2016, photo Otakar Holuša).

3.4. Time and Place of Emergence

The emergence period occurred from 28 May (the earliest date of detection of exuviae in 2014) to 1 July (the latest date of detection of exuviae in 2013). A total of 186 exuviae were found (several exuviae were in varying degrees of damage, such as the abdomen or head were missing). The site of emergence was evaluated with respect to location, i.e., the place where the adult emerged: (a) tree roots (exposed roots of living trees in the vertical banks of the streams), (b) tree trunks, (c) tree sticks, (d) tree leaves, (e) plant stalks, and (f) plant leaves (Figure 5). The presence of these emergence places is based on the characteristics of the habitat that the species inhabits, namely, watercourses shaded with rich vegetation and tree layers. Plant leaves were the dominant site of emergence (57%, Figure 5) followed by plant stalks (21%) and tree trunks (17%), tree leaves (15%), tree sticks (5%), and tree roots (4%) (Figure 6a,b). Even the roots of old trees, which were uncovered and formed "overhangs" from the soil due to water erosion, were selected by larvae for emergence. It is obvious that the larva at the end of its "journey" climbed as if on the ceiling and continued horizontally on the overhanging root.

Emergence sites were located at different distances from the shoreline and at different heights. Specifically, exuviae were noted from 0 to 790 cm ($\ddot{x}$ = 154 cm) at a horizontal distance from the shore and from 10 to 310 cm ($\ddot{x}$ = 77 cm) at a vertical height above the ground. The total distance of larval movement ranged from 10 to 1080 cm ($\ddot{x}$ = 205 cm) (Figures 7 and 8).

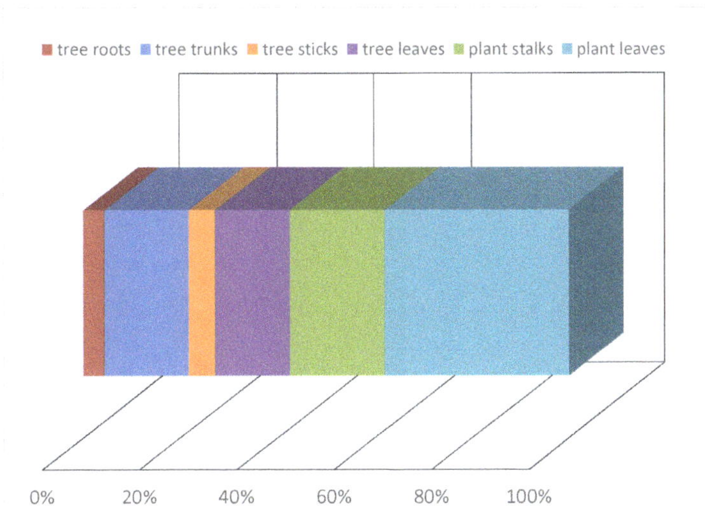

Figure 5. Proportion of different types of emergence sites of *Cordulegaster heros*.

(**a**)

(**b**)

Figure 6. Position of exuviae of *Cordulegaster heros*: (**a**) on tree roots at steep bank, locality Cvrčovický potok stream (Cvrčovice cadastral territory, southern Czech Republic) (locality Ie), (4 June 2016, photo by Otakar Holuša); (**b**) on trunk of *Alnus glutinosa*, locality Cvrčovický potok stream (Cvrčovice cadastral territory, southern Czech Republic) (locality Ie), (8 July 2016, photo by Otakar Holuša).

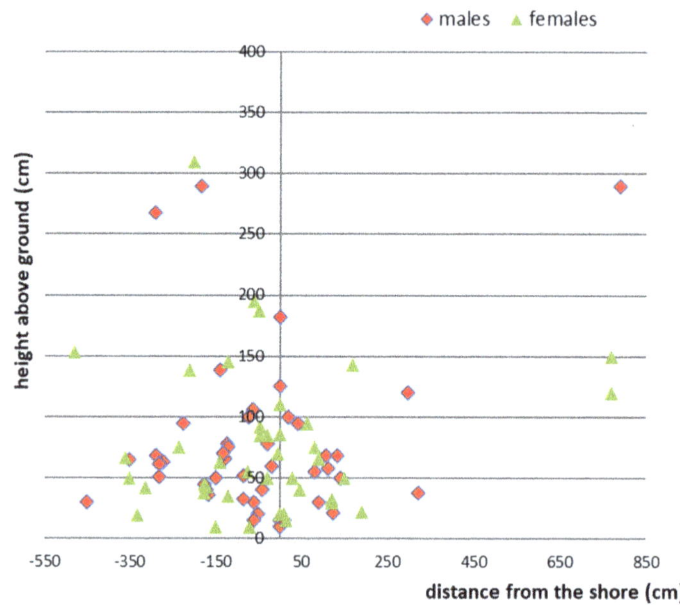

Figure 7. Position of exuviae of *Cordulegaster heros* in riparian parts of the habitat (x-axes: (−) left bank, (+) right bank).

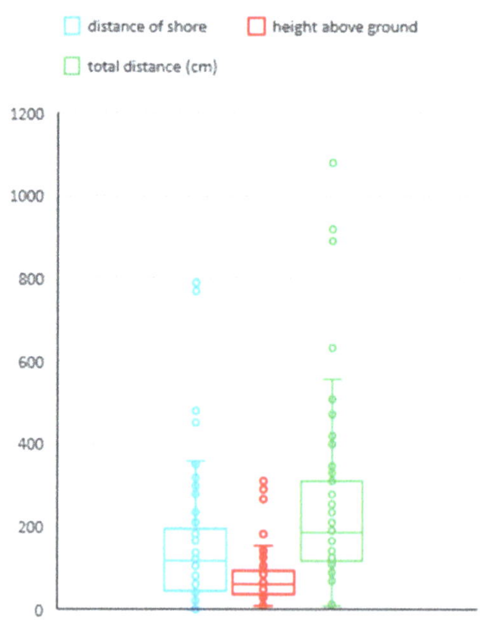

Figure 8. Variability of position of exuviae of *Cordulegaster heros* (distance from shore, height above ground, and total distance; all reported in cm).

3.5. Flight Time of Adults

The flight period of *C. heros* was identified from 15 June 2021 (the earliest date of detection of adults outside the year of detailed flight period monitoring) to 11 August 2021 (the latest date of detection of flying adults was 13 August 2013; one male observed)

(Figure 9). Flying activity increased sharply in the second half of June, with a peak in the second half of June. Specifically, on 23 June 2021, 103 individuals were recorded during the day plus another 150 flights unmarked adults. Thus, the potential number of adults in one day at one site was approximately 200. In the second half of June, flight activity began to decline gradually, with ca. 40 individuals (marked) and an additional 30 overflights observed in early July. In the second half of July, a further decline was noted with only 3 to 10 individuals (marked) recorded per day with 10 additional overflights. In the first ten days of August, only a few individuals were observed (four to six individuals in total). The flight period was completed at the end of the first third of August. No additional flying individuals were recorded after 13 August. Based on the comparison of the emergence and flight periods, it is obvious that adults spend approximately 2 weeks out of the habitat after emergence before returning and starting their activity by flying in their "home" habitat.

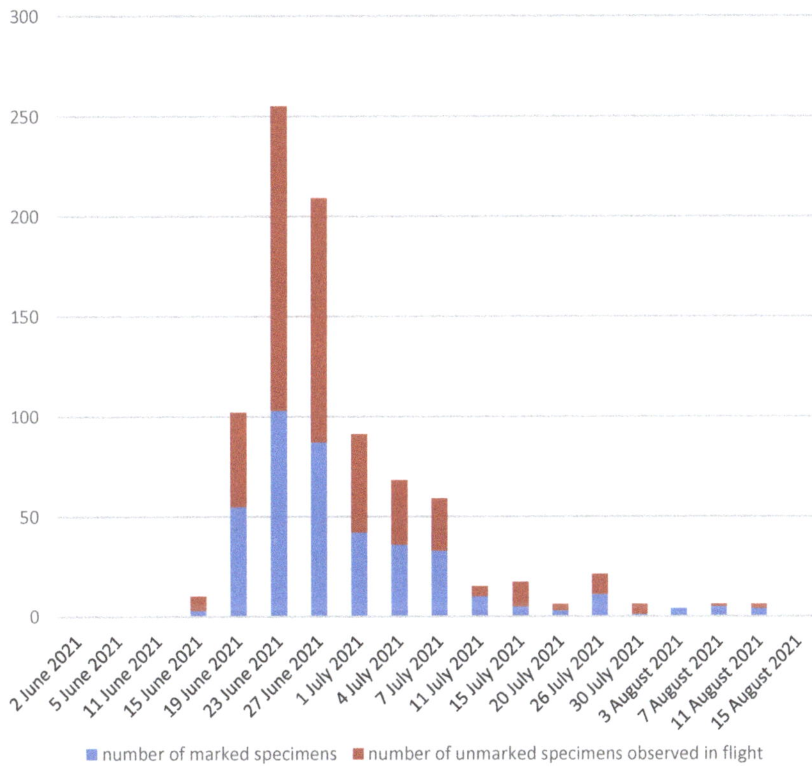

Figure 9. Flight time of *Cordulegaster heros* in the Habešský potok stream (Kostelany cadastral territory, southern Czech Republic) (locality VIe) in 2021.

4. Discussion

Cordulegaster heros is a European endemic species with a habitat that extends throughout the Balkan Peninsula. In the literature, the species is placed in the "east Mediterranean species" group [7], which is more associated with the zoogeographical division of the *boltonii* group of *Cordulegaster* species. As noted in [9], the designation "east Mediterranean" is rather broad given that the eastern part of the Mediterranean is occupied by *Cordulegaster* species related to *C. heros*, i.e., *C. picta*, and further east by *C. vanbrinkae* [31] and the currently least known *C. kalkmani* [32].

The species' area is unlikely to experience the same "evolutionary" changes that have occurred over the past 45 years, i.e., the period from the discovery of *C. heros* to the present day. If we compare the state of knowledge in 1988 [12] and today [13], we note great progress in the knowledge of its distribution. The species' area, thus, mostly includes the countries of the Balkan Peninsula with its boundaries lying in the northeastern area of Italy, eastern Austria, southeastern Czech Republic, southern Slovakia, western Ukraine, and eastern Romania. On this boundary, the occurrence is not continuous, comprising several separate areas. The northernmost area is the Chřiby Hills area in the Czech Republic (Figure 1), with the closest population occurring in Austria at a distance of 160 km and the easternmost population occurring in Slovakia at a distance of 70 km. Thus, the absolute northernmost point of occurrence is the locality of the Cvrčovický potok stream in Cvrčovice village at GPS 49°13′24.71″ N and 17°20′28.89″ E. The species is not thought to live outside of the surveyed area in the Chřiby Hills, as the landscape further afield is an agricultural landscape without forest belts along the streams.

Taking into account the distribution of the species and its center of the area, which is due to its numerous occurrences in Slovenia ([15], Bedjanič pers. comm., Holuša unpubl.), northern Croatia (Holuša unpubl.), and the Illyrian area, *C. heros* can be included in the Illyrian faunal element. Bernard and Daraż [9] refer to this species as part of the Ponto-Mediterranean faunal element, which does not fully correspond to its distribution or its postglacial distribution. From the mentioned Illyrian refugium, it is assumed that the main species of central European forests, such as *Quercus petraea*, *Carpinus betulus*, *Fagus sylvatica*, and *Abies alba*, especially the spreading of vegetation belts sensu Schmid [33,34], used this refugium to survive the ice age. As described by Bernard and Daraż [9], *C. heros* used three migratory directions to inhabit the central European area in the postglacial period: western (going north along the Austrian–Hungarian border from the refugium), central (going northeast along the eastern foothills of the Carpathian Mountains in western Romania), and eastern (going north along the eastern foothills of the Carpathian Mountains in eastern Romania towards the Ukraine) directions. The territory of the Czech Republic was, therefore, colonized by the western migration direction, which went from the refugium directly northwards along the eastern foothills of the Alps and the western edge of the Pannonian lowland. This migration stream continued to southern Slovakia, although the *C. heros* area in the space, including eastern Austria, the Czech Republic, and western Slovakia, is disjunctive.

The occurrence of *C. heros* in the Czech Republic has long been presumed by Holuša [16,21], and it has been expected in regions with suitable habitats, i.e., in the belt of regions on the forested southern slopes of the Carpathians Mountains, especially regions without interruption of large lowlands. Although research has been carried out since 1998, the first occurrence of the species in this country was recorded as late in 2009. This first record of *C. heros* adult in 2009 was considered to be a vagrant specimen [8], but the authors did not exclude the occurrence of a permanent population in the Czech Republic. The fact that the population of *C. heros* escaped attention until the beginning of the 21st century can be attributed to two main facts: first, very limited information about the occurrence of central European species of the genus *Cordulegaster* was available, i.e., *C. boltonii* and *C. bidentata*, in the Czech Republic at the end of the century and these species were, therefore, evaluated as "rare" [35–38]. Forest complexes, especially forest springs, were not the focus of attention for entomologists and odonatological research. It was only discovered in the last 20 years that the species are common in the Czech Republic and occupy large areas there (Holuša unpubl.). In addition, there is confusion regarding species determination. The second is the so-called the "paradox of Chřiby Hills". Specifically, the area of the Ždánický Forest Hills and Chřiby Hills has not been and is not currently the focus of attention of entomologists. This area includes small hills with complexes of "monotonous" forests, and the most interesting species are mostly found at in the southern foothills of this area, where sub-Mediterranean species occur, e.g., *Tibicina haematodes* near Archlebov [39]. Moreover, the highest positions of this area do not reach higher altitudes to host montane Carpathian species, so the interest of entomologists has been concentrated in the neighboring areas. It was not until the end of the 20th century that the first odonatological surveys were performed here [40]; however, *C. heros* was not found here, or, it is possible, it was confused with *C. bidentata*.

The area of Central Europe and the territory of the Czech Republic represents the locations where three species of *Cordulegaster*—*C. boltonii* and *C. heros* of the *boltonii* group and *C. bidentata* as a representative of the *bidentata* group—meet. The occurrence of *C. boltonii* extends its eastern limit of distribution into the Waldviertel in Lower Austria in Austria [14], and localities with syntopic occurrences of *C. boltonii* and *C. heros* are known in this area around Melk [30]. The eastern boundary of the range of *C. boltonii* runs northeast into the Czech Republic in the area of the Českomoravská vrchovina Hills, and then continues along the edge of the Nízký Jeseník Hills (Holuša unpubl.) and continues into Poland, where it follows the foothills of the Carpathian Mountains and continues eastwards towards Ukraine [9]. All the revised data on *C. boltonii* obtained from the territory of Slovakia [20] were assessed as erroneous, as these samples were actually *C. bidentata* [16]. Similarly, *C. boltonii* was reported in the territory of Hungary [41] and Romania [17,42]. In Hungary, it was confused with *C. heros* [43], which is similar to that noted in Romania. In these territories, there is no overlap of the ranges of *C. boltonii* and *C. heros*, and their area boundaries are at least 70 km apart.

The locations of *C. heros* in the Czech Republic have similar characteristics to those in neighboring countries. The altitude fully corresponds to the data from Slovakia, where the species occurs between 160 and 516 m a.s.l., with its main occurrence noted in the range of 201–300 m a.s.l. (43% of all localities) [7]. In Austria, *C. heros* inhabits localities from 180 m a.s.l. to 720 m a.s.l. [14]. In Slovenia, *C. heros* inhabits localities from 0 to 800 m a.s.l. with a range of 200–300 m a.s.l. [15].

The geomorphological characteristics of the streams are comparable to those reported in other areas. In Slovakia, biotopes are reported [7,44,45] as shaded streams and streamlets in forest areas in hills with clean water, with widths from 20 to 420 cm, water depths from 2 to 18 cm, and sandy-gravel sediments. Similarly, Bernard and Daraż [9] from Ukraine describe them as streams with widths of 0.8–1.2 m, sandy-gravel sediment (sand 79.6%, gravel 15.6%), a small admixture of silt (4.8%), and a dominant soil fraction of very fine sand (11.8%) and fine sand (25.6%). From sites in Austria [46], the most favorable microbiotope is characterized as streams with fine, medium, and coarse sand substrate, whereas the less favorable microbiotope is characterized as streams with fine, medium, and coarse gravel substrate. Detailed characteristics of the habitat from the southern part of the species area are not currently available.

With regard to larval density, data are available from Lang [46] and Boda et al. [47]; however, these data are reported in larval density per 10 m of stream length with Hungarian data, indicating larval density per m^2. It is not clear what area of suitable sediment was available in these habitats, so comparisons are not possible. The highest local density of 25 larvae per 0.25 m^2 of sediment (majority of last larval instar-14) when a group of larvae was collected in a suitable pool was reported from the Little Carpathians (Malé Karpaty Mts.) by Holuša and Kúdela [7].

The observed emergence period is similar to data obtained from more southern areas (Austria) with emergence occurring from the beginning of June to the end of June [14] and southern Hungary from the beginning of the last half of May until 31 July [48]. At sites in Hungary, the locations where exuviae were found are clearly dominated by trees followed by shrubs and plants [46], which is due to the nature of the vegetation of the streambeds. Completely identical results are available for the location of exuviae in the vertical and horizontal directions from the shoreline, including the total distance reported by Boda et al. [48], even with a maximum total distance of 1030 cm, which is the distance the larva had to travel from the shoreline to the emergence site.

The flight activity of *C. heros* is significantly shorter than that of other central European *Cordulegaster* species. However, the data available from Austria [30] and Slovakia [49] are similar to those reported for the northernmost population; thus, the first adults appear in mid-June and fly until the end of July. However, in our case, the adults did not appear until mid-August. The same peak of flight activity was found by Balász [49] in southern Slovakia. Specifically, in the third week of June, up to 100 individuals per day appeared at the locality.

5. Conclusions

Although the northernmost distribution of *C. heros* has an area of approximately 100 km^2, it includes 10 individual forest streams and brook watersheds. The populations in the individual watersheds are numerous and do not represent individual records for larvae or adults. However, their numbers are similar to those of populations found further south in Austria and Hungary. In most streams, there is a sufficient abundance of suitable sediment for larval development. The sites are located in forest complexes with a predominance of *Fagus sylvatica*, *Carpinus betulus*, *Quercus petraea*, *Acer pseudoplatanus*, and *Alnus glutinosa*, and thus are forests with a high degree of naturalness. Thus, minimal potential threats are encountered. Existing forests are managed in a manner that does not affect the population status; therefore, a minimal threat is observed. With forest management, only minimal impacts can be expected from timber harvesting and transport across the stream bed. The potential for the construction of water works (small dams or ponds) or the construction or repair of forest roads along streambeds represent significant impacts. The habitat condition is, therefore, excellent, providing suitable conditions for the future occurrence of the population. However, it is necessary to monitor the response of the population to any major interventions in the catchment areas of individual streams and, where appropriate, to monitor the trends in population abundance.

Author Contributions: Conceptualization, O.H. and K.H.; methodology, O.H. and K.H.; software, O.H.; validation, O.H.; formal analysis, O.H. and K.H.; investigation, O.H. and K.H.; resources, O.H. and K.H.; data curation, O.H. and K.H.; writing—original draft preparation, O.H. and K.H.; writing—review and editing, O.H. and K.H.; visualization, O.H. and K.H.; supervision, O.H.; project administration, O.H. and K.H.; funding acquisition, O.H. and K.H. All authors have read and agreed to the published version of the manuscript.

Funding: This research was funded by Agency for Nature and Landscape Conservation of the Czech Republic of in frame of monitoring NATURA 2000 species.

Institutional Review Board Statement: Not applicable.

Data Availability Statement: Not applicable.

Acknowledgments: We thank to Jiří Trombík (Czech Agricultural University in Prague) for helping with creating map images.

Conflicts of Interest: The authors declare no conflict of interest. The funders had no role in the design of the study; in the collection, analyses, or interpretation of data; in the writing of the manuscript; or in the decision to publish the results.

References

1. Theischinger, G. *Cordulegaster heros* sp. nov. und *Cordulegaster heros pelionensis* ssp. nov., zwei neue taxa des *Cordulegaster boltoni* (Donovan)-komplexes aus Europa (Anisoptera: Cordulegsteridae). *Odonatologica* **1979**, *8*, 23–28.
2. Boudot, J.P. Les *Cordulegaster* du Paléarctique occidental: Identification et répartition (Odonata, Anisoptera, Cordulegastridae). *Martinia* **2001**, *17*, 1–34.
3. Van Pelt, G.J. Cordulegaster Leach, 1815—Goldenrings. In *Field Guide to the Dragonflies of Britain and Europe Including Western Turkey and North-Western Africa*; Dijkstra, K.-D.B., Lewington, R., Eds.; British Wildlife Publishing: Gillingham, UK, 2006; pp. 210–221.
4. Boudot, J.P.; Kalkman, V.; Azpilicueta Amorín, M.; Bogdanović, T.; Rivera, A.C.; Degabriele, G.; Dommanget, J.L.; Ferreira, S.; Garrigós, B.; Jović, M.; et al. *Atlas of the Odonata of the Mediterranean and North Africa*; Libellula; Gesellschaft deutschsprachiger Odonatologen: Börnsen, Germany, 2009; pp. 1–256.
5. Bedjanič, M.; Šalamun, A. Large golden-ringed dragonfly *Cordulegaster heros* Theischinger 1979, new for the fauna of Italy (Odonata: Cordulegastridae). *Nat. Slov.* **2003**, *5*, 19–29.
6. Janský, V.; David, S. Výskyt vážky *Cordulegaster heros* ssp. *heros (Odonata: Cordulegastridae) na Slovensku. Acta Rerum Nat. Musei Natl. Slov.* **2008**, *54*, 61–68.

7. Holuša, O.; Kúdela, M. New records of occurrence of *Cordulegaster heros* Theischinger, 1979 in Slovakia. *Acta Mus. Beskid.* **2010**, *2*, 75–87.
8. Staufer, M.; Holuša, O. First record of *Cordulegaster heros* in the Czech Republic, with notes on *Cordulegaster* spp. in southern Moravia (Odonata: Cordulegastridae). *Libellula* **2010**, *29*, 197–204.
9. Bernard, R.; Daraż, B. *Cordulegaster heros* and *Somatochlora meridionalis* in Ukraine: Solving the zoogeographical puzzle at their northern range limits (Odonata: Cordulegastridae, Corduliidae). *Odonatologica* **2015**, *44*, 255–278.
10. Tóth, S. A ritka hegyiszitakötő (*Cordulegaster heros* Theischinger, 1979) előfordulása a Zselicben. *Nat. Som.* **2006**, *9*, 141–144. [CrossRef]
11. Csordás, L.; Ferincz, A.; Lőkkös, A.; Rozner, G. New data on the distribution of Large Golden Ringed Dragonfly (*Cordulegaster heros* Theischinger, 1979) (Odonata) in Zselic hills. *Nat. Som.* **2009**, *15*, 53–56.
12. Askew, R.R. *The Dragonflies of Europe*; Harley Books: Colchester, UK, 1988; pp. 140–141.
13. Boudot, J.-P.; Holuša, O. Cordulegaster heros Theischinger, 1979. In *Atlas of the European Dragonflies and Damselflies*; Boudot, J.-P., Kalkman, V.J., Eds.; KNNV Publishing: Zeist, The Netherlands, 2015; pp. 216–218.
14. Raab, R.; Chovanec, A.; Pennerstorfer, J. *Libellen Österreichs*; Springer: Wien, Austria, 2006; pp. 180–181.
15. Kotarac, M. *Atlas Kačjiv Pastirjev (Odonata) Slovenije z Rdečim Seznamom: Project Slovenskega Odonatološkega Društva*; Atlas Faune et Florae Sloveniae 1. Center za Kartogarfio Favne in Flore: Miklavž na Dravskem polju, Slovenia, 1997; pp. 94–95.
16. Holuša, O. Notes on the distribution of *Cordulegaster* spp. in Central Europe. In Proceedings of the Abstracts 5th WDA International Symposium of Odonatology, Swakopmund, Namibia, 16–20 April 2007.
17. Cîrdei, F.; Bulimar, F. *Odonata. Fauna Republici Populare Române, Insecta Vol. 7., Fasc. 5*; Academia Republici Populare Române: Bucureşti, Romania, 1965; pp. 1–277.
18. Balthasar, V. Limnologické výskumy v slovenských vodách. *Učená Spoloč. Šafárikova Bratisl.* **1936**, *34*, 50–51.
19. Buláňková, E. Výskyt niektorých ohrozených druhov vážok na Slovensku. *Spr. Slov. Ent. Spoloč.* **1993**, *5*, 23–26.
20. Straka, V. Vážky (Odonata) Slovenska. *Zbor. Slov. Nár. Múz. Prír. Vedy* **1990**, *36*, 121–147.
21. Holuša, O. The occurrence of *Cordulegaster* sp. in Czech Republic—Result of influence of habitat ecological factors in different biogeographical regions? In Proceedings of the Program and Abstracts, 4th WDA International Symposium of Odonatology, Pontevedra, Spain, 26–30 July 2005.
22. Holuša, O.; Dalecký, V.; Holušová, K. First record of larvae of *Cordulegaster heros* (Theischinger, 1979) (Odonata: Cordulegastridae) in the Czech Republic. *Acta Mus. Beskid.* **2011**, *3*, 65–69.
23. Pruner, L.; Míka, P. Seznam obcí a jejich částí v České republice s čísly mapových polí pro síťové mapování fauny. *Klapalekiana* **1996**, *32*, 1–175.
24. Rejšek, K. *Lesnická Pedologie: Cvičení*; Mendelova zemědělská a lesnická Univerzita v Brně: Brno, Czech Republic, 1999; pp. 1–152.
25. David, S.; Šácha, D. Komentovaný seznam vážek (Odonata) Slovenké republiky. *Ochr. Prír.* **2019**, *3*, 49–78.
26. Buczyński, P.; Michoński, G.; Zawal, A. New site of *Cordulegaster heros* Theischinger, 1979 in western Slovakia—The confirmation of its northern border of range of occurrence in Europe (Odonata: Cordulegastridae). *Ecol. Montenegrina* **2020**, *28*, 20–22. [CrossRef]
27. Ambrus, A.; Danyik, T.; Kovács, T.; Olajos, P. *Magyarország Szitakötőinek Kézikönyve*; Magyar Természettudományi Múzeum, Herman Ottó Intézet Nonprofit Kft.: Budapet, Hungary, 2018; pp. 1–290.
28. Kilian, W.; Müller, F.; Starlinger, F. Die forstlichen Wuchsgebiete Österreichs. Eine Naturraumgliederung nach waldökologischen Gesichtspunkten. The forestation areas in Austria. A natural space breakdown by forest ecological aspects. *FBVA/BFW Ber.* **1994**, *82*, 1–60.
29. Ciglič, R.; Perko, D. Slovenia in geographical typifications and regionalizations of Europe. *Geogr. Vestn.* **2012**, *84*, 23–37.
30. Schweigofer, W. Syntopes Vorkommen von *Cordulegaster boltonii* und *C. heros* an einem Bach im westlichen Niederösterreich (Odonata: Cordulegastridae). *Libellula* **2008**, *27*, 1–32.
31. Holuša, O. Description of the last instar larva of *Cordulegaster vanbrinkae* and emergence place from northern Iran (Odonata: Cordulegastridae). *Int. J. Odonatol.* **2022**, *25*, 72–79. [CrossRef]
32. Schneider, T.; Vierstraete, A.; Müller, O.; van Pelt, G.J.; Caspers, M.; Ikemeyer, D.; Snegovaya, N.; Dumont, H.J. Taxonomic Revision of Eastern Part of Western Palaearctic *Cordulegaster* Using Molecular Phylogeny and Morphology, with the Description of Two New Species (Odonata: Anisoptera: Cordulegastridae). *Diversity* **2021**, *13*, 667. [CrossRef]
33. Schmid, E. Die Vegetationskartierung der Schweiz. *Ber. Geobot. Forsch. Rübel Zürich* **1939**, *41*, 76–85.
34. Schmid, E. Vegetationsgürtel und Biocoenose. *Ber. Schweiz. Bot. Geselschaft* **1949**, *51*, 461–474.
35. Teyrovský, V. Řád Vážky—Odonata. In *Klíč Zvířeny ČSR III*; ČSAV, ČSSR: Praha, Czech Republic, 1959; pp. 183–193.
36. Zelený, J. Vážky—Odonata. In *Klíč k Určování Vodních Larev Hmyzu*; Academia, ČSSR: Praha, Czech Republic, 1980; pp. 68–85.
37. Holuša, O. Cordulegaster boltonii. In *Vážky České Republiky. Ekologie, Ochrana a Rozšíření*; Český Svaz Ochránců Přírody: Vlašim, Czech Republic, 2007; pp. 462–465.
38. Holuša, O. Cordulegaster bidentata. In *Vážky České Republiky. Ekologie, Ochrana a Rozšíření*; Český Svaz Ochránců Přírody: Vlašim, Czech Republic, 2007; pp. 458–461.
39. Rozšíření a biologie cikády viničné (Tibicina haematodes) v České republice. Available online: http://botzool.sci.muni.cz/cikada-vinicna (accessed on 5 May 2021).

40. Bezděčka, P.; Vážky, C. Dragonflies (Odonata) of the Chřiby Highlands (Moravia, Czech Republic). In Proceedings of the Vážky 2000. Sborník Referátů III. Celostátního Semináře Odonatologů, Chráněné krajinné oblasti Třeboňsko, Czech Republic, 15–18 June 2000.
41. Dévai, G.; Bodnár, G.; Benedek, P. A szitakötők (Odonata) magyarországi előfordulási adatainak elemzése. *Acta Biol. Debrecina* **1976**, *13* (Suppl. 1), 9–92.
42. Lehrer, A.Z.; Bulimar, F. Sinteze cartografice ale patrimoniului natural al României II. Ordinul Odonata Fabricius, 1792. *Nymphaea: Folia Nat. Bihariae* **1979**, *7*, 343–393.
43. Ambrus, A.; Bánkuti, K.; Kovács, T. The Odonata fauna of Kisalföld and the West-Hungarian marginal zone. *Győr-Moson-Sopron M. Múz. Kiadv. Tanulm.* **1992**, *2*, 1–81.
44. Holuša, O. The analysis of the habitat of *Cordulegaster heros* on its northern area border in Slovakia. In Proceedings of the Abstracts, 6th WDA International Congress of Odonatology, Xalapa, Mexico, 7–12 June 2009.
45. Holuša, O.; Holušová, K. Kolik toho víme o rozšíření největší evropské vážky páskovce velkého *Cordulegaster heros* (Odonata: Cordulegastridae) ve střední Evropě? In Proceedings of the Zoologické dny 2012. Sborník Abstraktů z Conference, Olomouc, Czech Republic, 9–10 February 2007.
46. Lang, C.; Müller, H.; Waringer, J.A. Larval habitats and longitudinal distribution patterns of *Cordulegaster heros* Theischinger and *C. bidentata* Selys in an Austrian forest stream (Anisoptera: Cordulegastridae). *Odonatologica* **2001**, *30*, 395–409.
47. Boda, R.; Bereczki, C.; Pernecker, B.; Mauchart, P.; Csabai, Z. Life history and multiscale habitat preferences of the redlisted Balkan Goldenring, *Cordulegaster heros* Theischinger, 1979 (Insecta, Odonata), in South-Hungarian headwaters: Does the species have mesohabitat-mediated microdistribution? *Hydrobiol.* **2015**, *760*, 121–132. [CrossRef]
48. Boda, R.; Bereczki, C.; Ortmann-Ajkai, A.; Mauchart, P.; Pernecker, B.; Csabai, Z. Emergence behaviour of the red listed Balkan Goldenring (*Cordulegaster heros* Theischinger, 1979) in Hungarian upstreams: Vegetation structure affects the last steps of the larvae. *J. Ins. Conserv.* **2015**, *19*, 547–557. [CrossRef]
49. Balász, A.; Fric, Z.F.; Holuša, O. Flying activity and population dynamics of *Cordulegaster heros* Theischinger, 1979 (Insecta: Odonata: Cordulegastridae) in Slovakia. *Int. J. Odonatol.* **2020**, *23*, 1–9.

diversity

MDPI

Article

Detection and Monitoring of Riverine Dragonfly of Community Interest (Insecta: Odonata): Proposal for a Standardised Protocol Based on Exuviae Collection

Loan Arguel [1], Alice S. Denis [1,2], Samuel Danflous [2], Nicolas Gouix [2], Frédéric Santoul [1], Laëtitia Buisson [1] and Laurent Pelozuelo [1,*]

1 Laboratoire d'Ecologie Fonctionnelle et Environnement, CNRS—INPT-UPS, Université Paul Sabatier, Bâtiment 4R1, 118 Route de Narbonne, CEDEX 09, 31062 Toulouse, France

2 Conservatoire d'Espaces Naturels d'Occitanie, BP 57611, 75 Voie du TOEC, CEDEX 3, 31076 Toulouse, France

* Correspondence: laurent.pelozuelo@univ-tlse3.fr

Abstract: Collecting quantitative data on insect species occurrence and abundance is a major concern to document population trends. This is especially the case to assess the conservation status of species listed in the European Habitats Directive and to assess the efficiency of mitigation measures with a view to achieve the "no net loss of biodiversity" goal for protected species. However, at present, populations of riverine dragonflies listed in the Habitats Directive and protected under French national law are poorly quantified and monitored. Exuviae collection could be used for such monitoring but a standardised protocol is lacking. We here proposed and tested such a protocol to monitor riverine dragonfly populations through exhaustive exuviae collection along river bank transects. To define the optimal transect size and number of visits, ninety-eight 100 m-long transects divided into 10 m-long plots were monitored on three rivers in southern France. Each transect was visited three times over the emergence period. In the course of each visit, all the exuviae along transects were collected and identified. From our results, we recommend collecting exuviae along 100 m of river bank in the course of two visits in order to both maximise the species detection and minimise the monitoring cost.

Keywords: riverine community; survey method; sampling issues; field protocol; conservation; splendid cruiser; pronged clubtail; orange-spotted emerald

Citation: Arguel, L.; Denis, A.S.; Danflous, S.; Gouix, N.; Santoul, F.; Buisson, L.; Pelozuelo, L. Detection and Monitoring of Riverine Dragonfly of Community Interest (Insecta: Odonata): Proposal for a Standardised Protocol Based on Exuviae Collection. *Diversity* **2022**, *14*, 728. https://doi.org/10.3390/d14090728

Academic Editors: M. Olalla Lorenzo-Carballa and Ricardo Koroiva

Received: 23 July 2022
Accepted: 30 August 2022
Published: 2 September 2022

1. Introduction

Among the ongoing challenges in conservation entomology stands the necessity to acquire quantitative data to document the trends of insect populations [1,2]. Based on long-term quantitative studies, alerts on massive decline of insect populations have multiplied in recent years [3–7]. This situation clearly urges the need for quantitative datasets allowing (i) to detect and monitor local populations, particularly those of threatened or protected species and (ii) to monitor national or supra-national population trends. However, for several dragonfly species, protocols suitable for the production of such quantitative data are simply lacking. This is the case in south-western Europe of the riverine community composed of the Splendid Cruiser *Macromia splendens* (Pictet, 1843), the Pronged Clubtail *Gomphus graslinii* (Rambur, 1842) and the Orange-spotted Emerald *Oxygastra curtisii* (Dale, 1834) [8]. These three species have an unfavourable conservation status at the European community level (Table 1), they are listed in annexes II and IV of the Habitat Directive of the European community, and are strictly protected in Spain, Portugal and France. Moreover, they are southwestern Europe endemic species, their area of distribution being almost limited to France and the Iberian Peninsula [9–12]. This imposes to assess the conservation status of populations at local scale within the Natura 2000 sites and at the national scale [13] (art. 17). It also imposes to offset negative impacts in the case of habitat alteration or destruction. However, in the context of high anthropogenic pressure on rivers (i.e., increasing demand for infrastructure allowing flood-control, providing irrigation water or producing hydro-electricity), quantitative assessment of both national population trends and local population response to habitat alteration (or restoration) is not possible due to a lack of standardised data. Those species are thus often neglected in the Natura 2000 sites where they are present [14] and the French action plans in favour of dragonflies call for the development of effective standardised protocols [12,15]. To date, national and local environmental authorities, biodiversity consultants in charge of impact assessment studies and nature conservation organisations mostly rely on presence/absence datasets obtained through the compilation of opportunistic sightings. Here, we aimed at providing a standardised protocol to detect those protected species and provide an indicator of their population size.

Table 1. The Splendid Cruiser *Macromia splendens*, Pronged Clubtail *Gomphus graslinii*, and Orange-spotted Emerald *Oxygastra curtisii* conservation status and legal status.

Species	Conservation Status		French Region [3]	Legal Status	
	Europe [1]	France [2]		Europe [4]	France [5]
M. splendens	VU	VU	VU (Occitanie) EN (Aquitaine) VU (Rhônes-Alpes)	Annexes II et IV	Art. 2
G. graslinii	NT	LC	NT (Occitanie) LC (Aquitaine) VU (Rhônes-Alpes)	Annexes II et IV	Art. 2
O. curtisii	NT	LC	LC (Occitanie) LC (Aquitaine) LC (Rhônes-Alpes)	Annexes II et IV	Art. 2

[1] Kalkman et al., 2010 [16]; [2] UICN France et al., 2016 [17]; [3] Charlot et al., 2018; Barneix et al., 2016; Deliry et al., 2014 [18–20]; [4] European Union Council Directive 92/43/EEC 1992 [13]; [5] Arrêté du 23 April2007 fixant les Listes des Insectes Protégés sur l'ensemble du Territoire et les Modalités de leur Protection; 2007 [21].

Population size can hardly be assessed through adult observation because these dragonflies mature away from emergence sites, males exhibit exacerbated territoriality (especially *O. curtisii*), and imagoes are highly dependent on meteorological conditions for their flying activity. Exuviae collection was investigated as a mean to monitor species of conservation interest, specific richness and odonata community composition since the 2000's. Foster and Soluk [22] first proved the usefulness of exuviae collection to monitor the population of the endangered Hine's emerald dragonfly, *Somatochlora hineana* Williamson 1931, and Oertli [23] recommended to "prioritize exuviae collection, then larvae and only lastly the adults" to sample Odonata. Hardersen and collaborators in particular compared exuviae collection to larvae collection and adult survey in lotic and lentic habitats [24–26]. Raebel et al. [27] and da Silva-Méndez et al. [28], they showed the different sampling methods are not interchangeable: each has its own advantages and drawbacks. Contrary to adult survey, exuviae provide an unequivocal proof of autochthony and habitat suitability, their collection is poorly or not invasive, and they can easily be used in standardised methods to produce quantitative indicators. Whether or not exuviae collection reduce statistical bias is still controversial [27–29] but it is gaining popularity. There are first attempts to monitor populations of *M. splendens* in Catalunya [11,30,31] through exuviae collection and recently, da Silva-Méndez et al. [28] investigated exuviae persistence time for several riverine species, including *M. splendens*, *G. graslinii* and *O. curtisii*. They confirmed including exuvia collection is essential to assess riverine communities in north-western Iberian peninsula. We thus took advantage of the increased knowledge on exuviae identification—user-friendly identification keys now available, such as Doucet [32] and Boudot & Grand [33]—and increased interest for exuviae collection as a detection and monitoring tool [22,27,34–40] to propose a monitoring protocol based on exuviae collection.

We tested this protocol on three rivers in south-western France (Tarn, Lot and Dourdou de Camarès), during which the sampling distance and the number of visits per season were calibrated. We also provide information about the cost of such a protocol.

2. Materials and Methods

2.1. Monitoring Sites and Transects Location

Forty-nines sites were sampled on three rivers in south-western France, in the Occitanie region: twenty-one on the river Lot, twenty-six on the river Tarn, and one on the river Dourdou de Camarès (Figures 1 and 2, Table 1 and Table S1). Presence of reproductive populations of *M. splendens*, *G. graslinii*, and *O. curtisii* in at least some stretches of those three rivers was known prior to sampling [9,40,41].

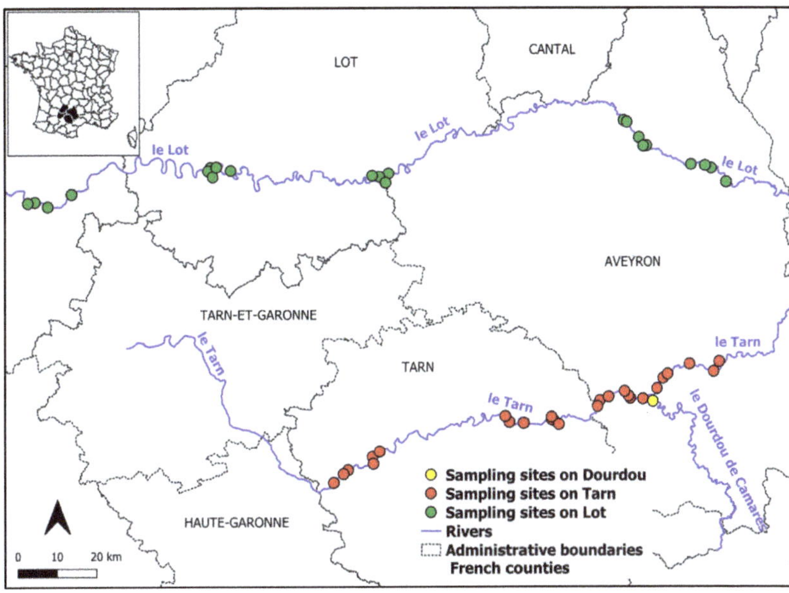

Figure 1. Location of study sites in south-western France. Blue lines and blue names indicate the rivers Tarn, Lot and Dourdou de Camarès. Red, green and yellow circles show sampling sites on Tarn, Lot, and Dourdou de Camarès, respectively. The names in capital letters correspond to the French county and their limits are shown in black dotted lines.

Monitoring sites were chosen for their easy access to water. Then, on each monitoring site, two 100 m long river bank stretches (there after named transects) were positioned, one on the left river bank, one on the right river bank. Transects were chosen in a way to fit, as far as possible, the description of the favourable habitats for the targeted species: deep waters, low stream velocity, presence of a dense riparian vegetation with shaded places, or rocky river bank [41–43]. Additionally, we avoided obstacles preventing access to the river bank, such as fallen trees, mud, and sand-banks. A global positioning system handheld (Garmin GPSMAP 65s model) was used to geolocalise the position of each transect and a 20 m long rope with marks every 10 m was used to measure and divide each of them into ten 10 m long plots. For the time of the study, pink warning tape was used to show the beginning and the end of each 100 m transect and each 10 m plot.

Figure 2. Overview of the three rivers sampled: (**a**) Dourdou de Camarès, (**b**) Lot, and (**c**) Tarn.

2.2. Exuviae Collection

Along each transect, all Anisoptera exuviae within reach were collected from each 10 m long plot and stored for later identification. Exuviae were collected from a short and easily maneuverable kayak (Mojito model by Rotomod, 250 cm long, 76 cm width and 16 kg weight). The operator visually inspected the riverbank's substrate, herbaceous vegetation and trees roots, trunks, and branches. As emerging larvae can walk high to complete emergence (Arguel, Pelozuelo and Denis, personal observation), trunks, branches and mineral cliffs and rocks were scouted from their base to approximately 3 m high. Exuviae found at such height were collected using the 2.2 m long kayak paddle. On each transect, the total sampling time and at each 10 m portion was recorded.

Each site was sampled three times from June to August 2015 to cover the entire expected emergence period of the riverine community in south-western France (Table S1). The first visit to each transect took place between 9 June and 14 July, the second visit between 9 July and 7 August, and the third one between 27 July and 26 August (Table S1). The date for the first visit were chosen to be around the peak of emergence, based on information's available at that time on Vère and Aveyron rivers (Denis and Pelozuelo, personal observation) and later confirmed [44]. However, it is noteworthy that local phenology is not described on those rivers, and emergence can even occur later due to deep and thus cold waters of the Lot and Tarn rivers, as observed on Tarn river in 2020 and 2021 [45]. The delay between each visit on the same site was on average, 25 days between the first and the second visit (Min = 15, Max = 42, Median = 23) and 18 days between the second and the third visits (Min = 11, Max = 23, Median = 20). We could not have shorter delays between the visits due to rainy weeks.

Given that *M. splendens*, *G. graslinii*, and *O. curtisii* are protected species in France, the exuviae collection was authorised by prefectural decrees n°81-2014-05 and n°82-2014-05. Collected exuviae were identified in the laboratory, using a binocular microscope (Leica Zoom 220 model) and the identification key of dragonfly exuviae of France [32]. Numbers of exuviae per species, per 10 m plot of each transect were then obtained.

2.3. Practical Considerations: Time and Cost Required for Such a Protocol

The time required to sample each 100 m long transect was measured, from the beginning to the end of each transect. The cost of such a study has also been estimated based on the prices of the different material items required, as given on several websites. The costs of the equipment specifically required to this study (navigation equipment, roof bars, etc.) were estimated separately from the cost of the basic equipment generally present in an ecology laboratory (binocular microscope, car, identification key, etc.).

2.4. Data Analysis

For each 10 m plot, 100 m transect, or site (left plus right river bank transects), species richness and abundance were calculated. We then calibrated the sampling effort (i.e., number of visits and length of transects) that would be required to maximise abundance, species richness, and detection of the target species (i.e., *M. splendens*, *O. curtisii*, and *G. graslinii*). We calculated the number of new species detected and the abundance of exuviae collected at each visit to assess the efficient number of visits. We compared them using a Friedman rank sum test, followed by a non-parametric Wilcoxon signed-rank test with Bonferroni adjustment to find post-hoc statistical differences. We also calculated the cumulated richness detected by plots. Cumulated species richness was plotted to calibrate the effective length of transects. We finally tested the correlation between exuviae abundance by transect and sampling duration using the Spearman correlation test. All analyses were performed with R 4.0.3 (R Core Team, 2020) and all maps were made with QGIS Desktop 3.12.3.

Species that were poorly present in our samples of exuviae (i.e., less than 10 exuviae across all exuviae collected over the 98 transects) and exuviae that we could not identify were not included in the analysis. As *M. splendens* was one of the species of community interest we focused on, its exuviae number are shown even if they were below the fixed threshold. For each species, densities of exuviae per 100 m for first plus second visit were calculated.

3. Results

In total, 5831 exuviae from 11 Anisoptera species (with more than 10 exuviae each) were collected. Six species are typically riverine according to the description of their habitats [10,33], two are occasionally riverine, and three are rather associated with standing waters. Besides the species targeted by this study (i.e., *M. splendens*, *G. graslinii* and *O. curtisii*), the community found in this region included *Onychogomphus forcipatus* (Linnaeus, 1758), *Boyeria irene* (Boyer de Fonscolombe, 1838), *Gomphus vulgatissimus* (Linnaeus, 1758) as riverine species, plus *G. pulchellus* (Selys, 1840), and *Somatochlora metallica* (Vander Linden, 1825) as occasionally riverine and *Orthetrum cancellatum* (Linnaeus, 1758), *Anax imperator* (Leach, 1815), and *A. parthenope* (Selys, 1839) as rather associated with standing waters. *Aeshna mixta* (Latreille, 1805) (6 exuviae), *Cordulegaster boltonii* (Donovan, 1807) (3 exuviae), *Gomphus similimus* (Selys, 1840) (1 exuvia), *Libellula fulva* (O.F Müller, 1764) (1 exuvia), *Orthetrum albistylum* (Selys, 1848) (1 exuvia), *Orthetrum brunneum* (Boyer de Fonscolombe, 1837) (1 exuvia), *Sympetrum sanguineum* (OF Müller, 1764) (1 exuvia), *Sympetrum striolatum/meridionalis* (Charpentier, 1840 and Selys, 1841) (10 exuviae), and *Trithemis annulata* (Palisot de Beauvois, 1807) (4 exuviae) were also detected but given the low number of exuviae collected from each of these species, they were discarded from further analysis. The rivers sampled are relatively wide and deep (around 90 m for the Lot, 100 m for the Tarn and 30 m for the Dourdou-de-Camarès) and thus the scarcity of *C. boltonii* is not surprising as it prefers smaller and shallower tributaries.

Furthermore, there were some exuviae that could not be identified to the species level and thus were not considered in the analyses: *Anax* sp. (28 exuviae), *Aeshna* sp. (1 exuvia), *Gomphus* sp. (10 exuviae), and *Sympetrum* sp. (3 exuviae).

The presence of the three species of community interest in different places of the studied rivers (Tarn, Lot and Dourdou de Camarès) was confirmed. *O. curtisii* and *G. graslinii* were found on the Tarn and the Lot, with relatively greater numbers of the downstream part of these two rivers (Tables 2 and 3 and Figure 3). *M. splendens* was detected on two sites on the Tarn and one on the Lot. Density of *M. splendens* was much lower compared to *O. curtisii* and *G. graslinii*. On the Dourdou de Camarès river banks, only *O. curtisii* was detected.

Table 2. List of dragonflies species with more than 10 exuviae detected on each river (Lot, Tarn, and Dourdou de Camarès) with the total number of transects where the species was detected at least once and the associated percentage of positive transects (between parentheses). Riverine species according to Dijkstra [10] and Boudot and Grand [33] are indicated by asterisks and this study's target species (i.e., protected ones) are indicated in bold.

Odonata Species	Rivers		
	Lot (n = 44)	Tarn (n = 52)	Dourdou de Camarès (n = 2)
O. curtisii	22 (50)	16 (30.8)	1 (50)
M. splendens	1 (2.3)	2 (3.8)	-
G. graslinii	27 (61.4)	24 (46.2)	-
**G. vulgatissimus*	23 (52.3)	31 (59.6)	2 (100)
**O. forcipatus*	34 (72.3)	45 (86.5)	2 (100)
**B. irene*	26 (59.1)	27 (51.9)	1 (50)
S. metallica	12 (27.3)	6 (11.5)	-
O. cancellatum	16 (36.4)	1 (1.9)	-
G. pulchellus	5 (11.4)	6 (11.5)	-
A. imperator	4 (9.1)	2 (3.8)	-
A. parthenope	4 (9.1)	2 (3.8)	-

Table 3. Average exuviae density per 100 m long transects obtained during the first plus second visits for each species detected on each river (Lot, Tarn, and Dourdou de Camarès). The corresponding standard deviations are shown between parentheses. Riverine species according to Dijkstra [10] and Boudot and Grand [33] are indicated by asterisks and this study's target species (i.e., protected ones) are indicated in bold.

Odonata Species	Rivers		
	Lot (n = 44)	Tarn (n = 52)	Dourdou de Camarès (n = 2)
O. curtisii	29.4 (91.5)	11.7 (26.7)	3 (4.2)
M. splendens	0.02 (0.2)	0.1 (0.3)	-
G. graslinii	8.8 (13.8)	10.7 (24.5)	-
**G. vulgatissimus*	6.3 (18.1)	5.9 (11)	8.5 (6.4)
**O. forcipatus*	10.6 (14.9)	17.6 (17.2)	94.5 (36.1)
**B. irene*	5.2 (13.3)	2.2 (3.2)	2.5 (0.7)
S. metallica	0.7 (1.4)	0.2 (0.6)	-
O. cancellatum	1.8 (5.4)	0.02 (0.1)	-
G. pulchellus	0.3 (1.1)	0.2 (0.6)	-
A. imperator	0.05 (0.2)	0.04 (0.2)	-
A. parthenope	0.02 (0.2)	0.1 (0.3)	-

3.1. Number of Visits

On the three visits carried out, it appears very clearly that the first visit was the most informative in terms of both specie richness and abundance (Figures 4 and 5. Indeed, the number of new species detected on a site was significantly much lower during the second (Friedman test, $p < 0.001$; Wilcoxon signed-rank test, P1–2 < 0.001) and the third visits (Friedman test, $p < 0.001$; Wilcoxon signed-rank test, P1–3 < 0.001, P2–3 = 0.01; Figure 4).

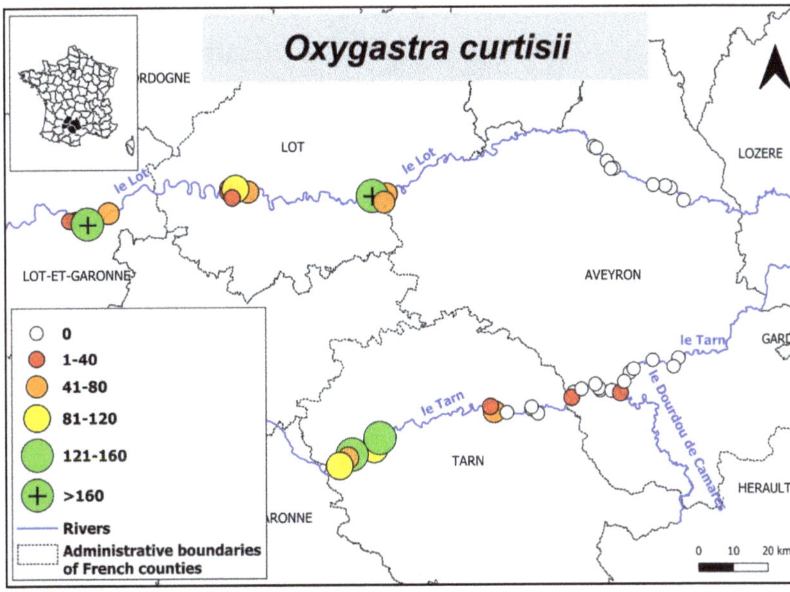

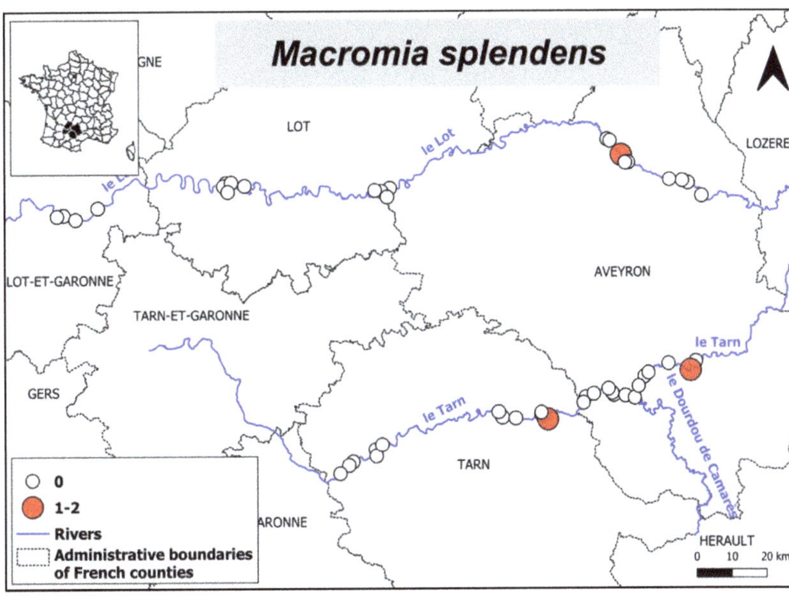

Figure 3. *Cont.*

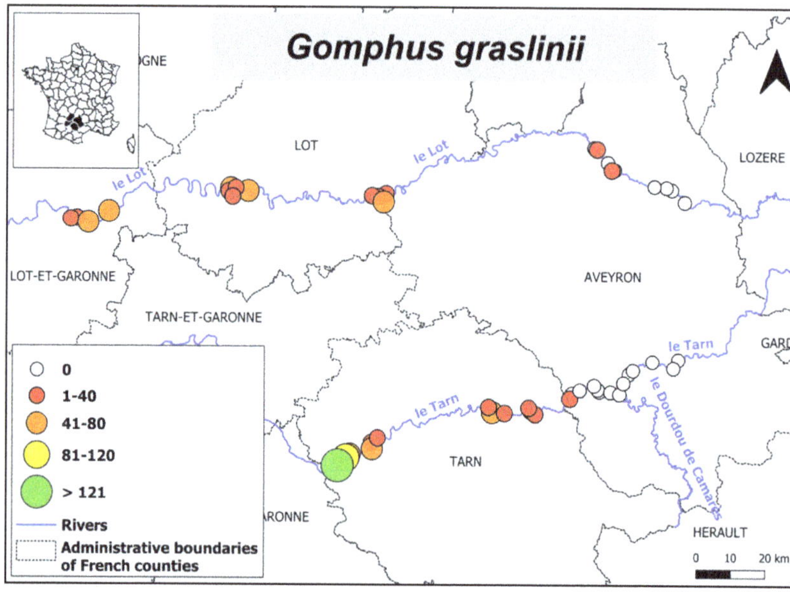

Figure 3. Distribution maps of exuviae of the three species of community interest *Oxygastra curtisii, Macromia splendens* and *Gomphus graslinii*. Bold lines and bold names indicate the rivers Tarn, Lot, and Dourdou de Camarès. The gradient of colour and size of the circles highlights the differences in numbers of exuviae.

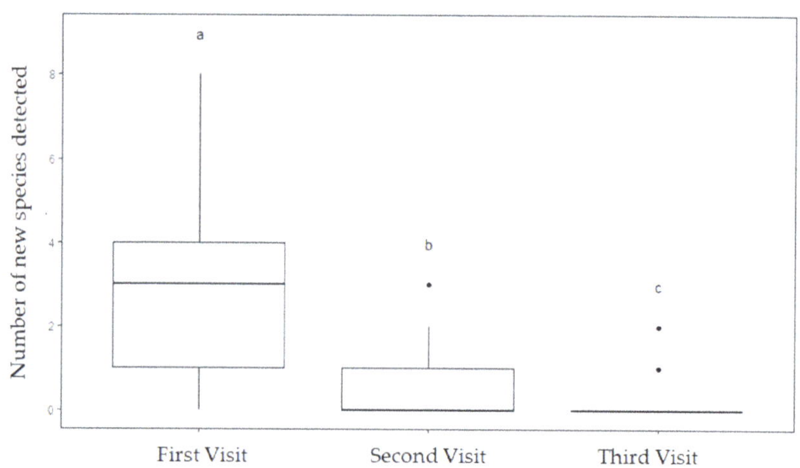

Figure 4. Number of new species detected calculated by transect ($n = 98$) depending on the number of visits. Box plots indicate median, range, and first and third quartiles. Points indicate outliers. Significant differences are indicated by different letters.

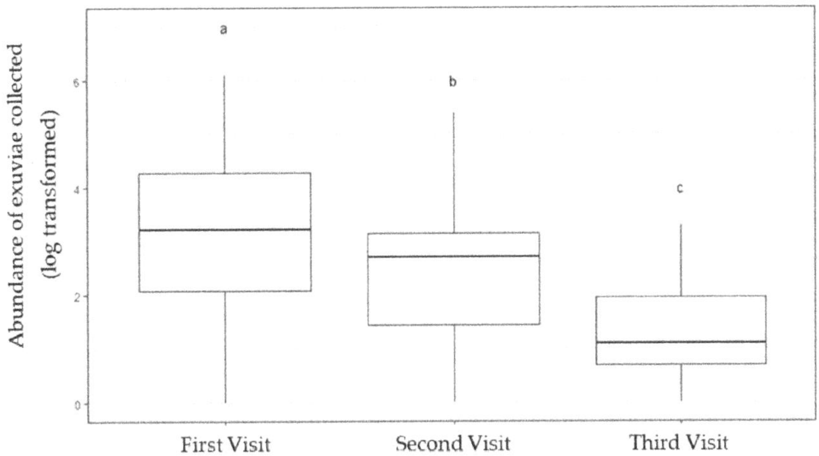

Figure 5. Abundance of exuviae collected (all species pooled) per 100 m long transects (n = 98 transects) depending on the visit rank. Abundances are log-transformed. Box plots indicate median, range, and first and third quartiles. Significant differences are indicated by different letters.

The species richness detected during the first visit represented on average 82.8% of the total detected richness on a site. This percentage fell respectively to 12.8% and 4.4% during the second and the third visits. Moreover, the total abundance of exuviae was significantly higher during the first visit than during the other two (Friedman test, $p < 0.001$; Wilcoxon signed-rank test, P1–2 < 0.001, P1–3 < 0.001, P2–3 < 0.001; Figure 5).

The three species of community interest were mostly detected during the first and second visits. For *O. curtisii*, detection occurred during the first visit on 94.9% of the transects where the species was detected at least once during the entire sampling period (n = 39), for *G. graslinii* on 86.3% of transects (n = 51) and for *M. splendens* on 66.7% of transects (n = 3) (Figure 6). The second visit allowed detecting *G. graslinii* for the first time in 11.8% of cases, *M. splendens* in 33.3% of cases, and *O. curtisii* in 5.1% of cases. In the third visit, only *G. graslinii* was detected on a single transect (on the Lot river) among the 51 sampled (Figure 6).

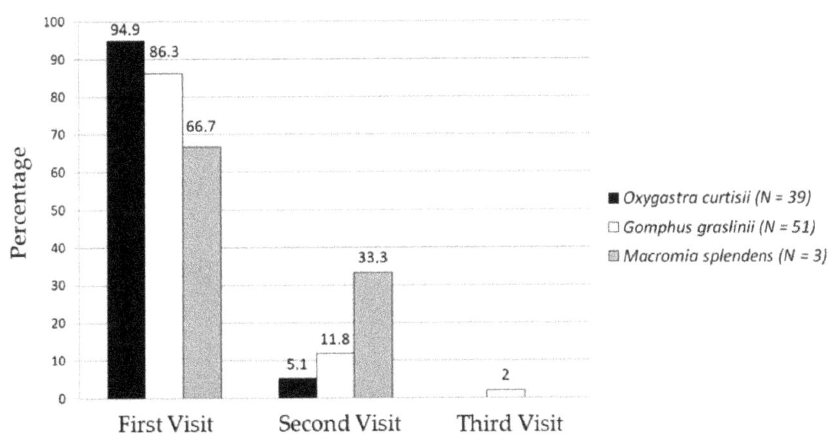

Figure 6. Distribution of the first detection for the three species of community interest, *Gomphus graslinii*, *Macromia splendens*, and *Oxygastra curtisii*, across the first, second, and third visits.

The first visit enabled us to collect 69.4% of the total exuviae collected, while the second and the third ones represented respectively 25.5% and 5.1% of the total. With the two first visits, we thus detected about 95.6% of the total species richness and collected 94.9% of the total abundance.

Regarding the three species of community interest, the abundance of exuviae was significantly higher during the first passage and it significantly decreased at the second and third visit (for *G. graslinii*: Friedman test, $p < 0.001$; Wilcoxon signed-rank test, P1–2 < 0.01, P1–3 < 0.001, P2–3 < 0.001; for *O. curtisii*: Friedman test, $p < 0.001$; Wilcoxon signed-rank test, P1–2 < 0.001, P1–3 < 0.001, P2–3 < 0.001) (Figure 7). The number of *M. splendens* exuviae was too small (four exuviae) to carry any analysis.

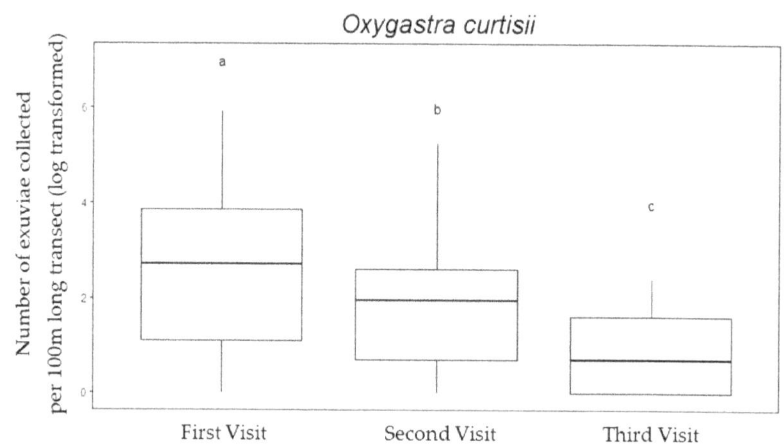

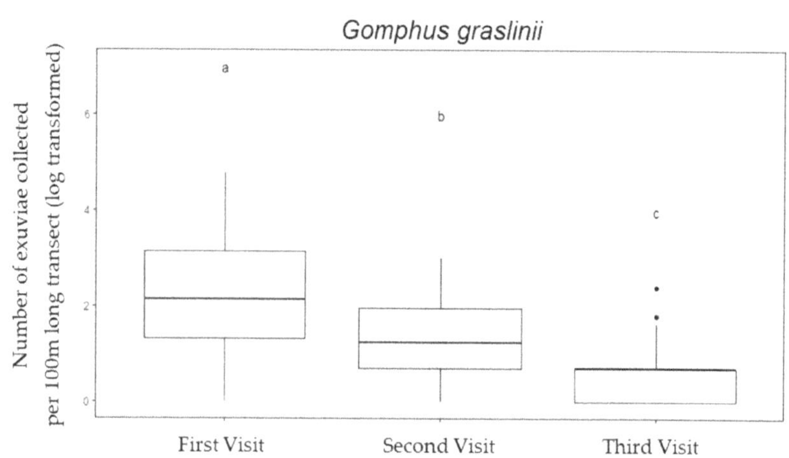

Figure 7. Abundance of exuviae of *Gomphus graslinii* and *Oxygastra curtisii* collected along 100 m-long transects (n = 98 transects) during the first, second, and third visits. Abundances are log-transformed. Box plots indicate median, range, and first and third quartiles. Points indicate outliers. Significant differences are indicated by different letters.

3.2. Transect Length

Along the 100 m transects, the number of new species detected mostly increased within the first 70 m then tended to stabilise. The first four plots allowed, on average, detection of over 70% of the species. The arbitrary threshold of 90% of detected species richness was reached at, on average, between 60 and 70 m (Figure 8).

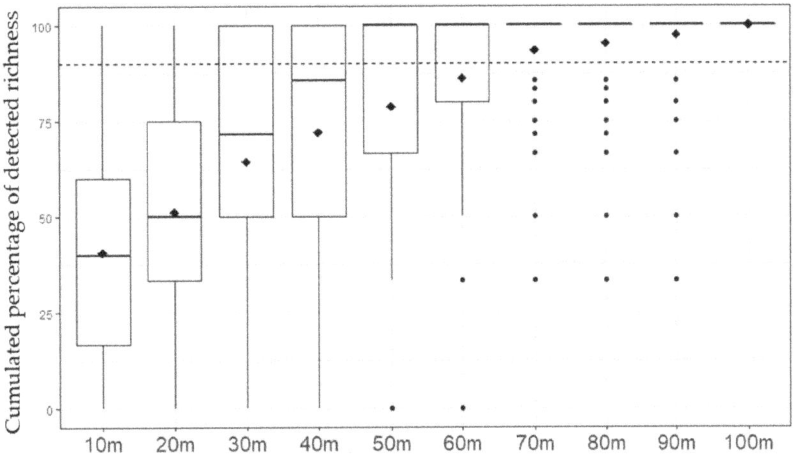

Figure 8. Cumulated percentage of detected richness depending on number of 10 m plots scouted for exuviae along each transect (n = 98 transects). Box plots indicate median, range, and first and third quartiles. Points indicate outliers. Black diamonds indicate the mean of each section. Dashed line indicates the 90% threshold.

Regarding the three species of community interest, detection generally took place within the first 10 m plots. *G. graslinii* and *O. curtisii*, are detected in the first 10 m plot in 47.1% and 59% of positive transects respectively (Figure 9). Beyond 60 m, the first detection of *G. graslinii* and *O. curtisii* was very low (<8%). The number of *M. splendens* was too low to make any conclusion.

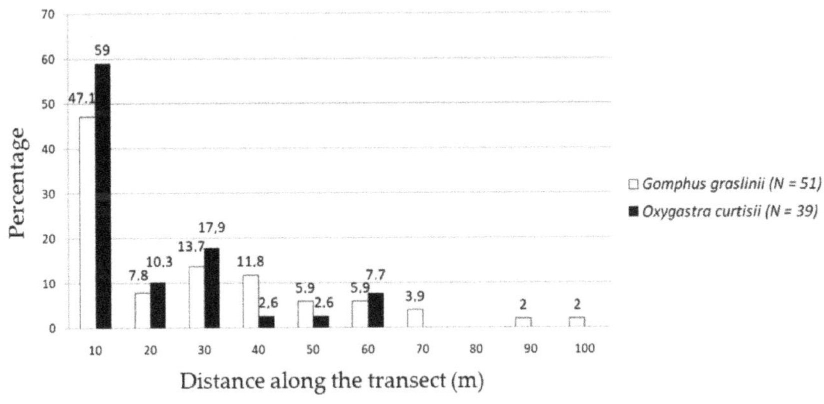

Figure 9. Distribution of the first detection of *Gomphus graslinii* and *Oxygastra curtisii* along the 100 m long transects (n = number of positive transects for each species).

3.3. Practical Considerations: How Long Did It Take and How Much Did It Cost?

On average (n = 98), exuviae collection along one 100 m long transect required 1 h 14 min (Max = 2 h 11 min and Min = 29 min) during the first visit, and, respectively, 52 min (Max = 1 h 37 min and Min = 28 min) and 39 min (Max = 1 h 7 min and Min = 14 min) during the second and the third visits (Table S2). The duration of exuviae collection is significantly and positively correlated with exuviae abundance (Spearman's rank correlation rho = 0.78, $p < 0.001$). This is the raw time necessary to collect exuviae but the time to handle, load, and unload the kayak on the car roof must be added (approximately 20 min) as well as the time necessary to drive from the lab to the different sites. Within normal conditions, two persons on two separate kayaks can sample three sites per day (i.e., six transects per day).

Equipment cost for exuviae collection (one kayak and its accessories such as a paddle, a lifejacket, a 5 L waterproof container, containers for exuviae collection, one handheld GPS, and a 20 m-long rope) is around €1000 per person and two persons are required for security reasons. The material cost for exuviae identification (binocular microscope, identification key, etc.) is around €850.

4. Discussion

This study aimed to propose a standardised protocol for monitoring riverine dragonfly communities. We investigated the optimal number of visits to pay to each transect and the optimal length of transect to accurately describe the species composition and abundance of the local Anisoptera community, with a special focus on species of community interest, *O. curtisii*, *M. splendens*, and *G. graslinii*. Regarding the number of visits per transect, our results showed that two visits are required to detect the majority of species and specifically our three species of community interest. Even if most species were detected during the first visit, a second visit brought valuable additional information as *O. curtisii* and *G. graslinii* would have been not detected in 5.1% and 11.8% of our positive transects without a second visit. On the contrary, there is no need for a third visit as the number of new species detected during this visit was nearly zero and, if not, new species detected were species with no conservation issue, e.g., *A. parthenope*, *T. annulata*, or *L. fulva*. Our conclusion is similar for abundance: two visits allowed us to collect around 94.9% of the total number of exuviae. On the contrary, the proportion of exuviae collected during the third visit was low (5.1%). This pattern might be the result of our visit schedule: the first visit probably occurred around the emergence peak, the second visit at the end of the emergence period—allowing collection of the exuviae of the individuals emerging lately—and the third visit may have occurred when the emergence period was already finished, and thus only exuviae unseen during the first and second visit were remaining.

Therefore, we conclude that it is not useful to carry out a third visit, especially since it would occur lately in the season and would probably not allow the detection of species of community interest which are all early species [41,46]. As sampling effort is always a trade-off between the search for exhaustiveness and time and money allocated, we recommend two visits rather than three.

We did not investigate the delay between each visit. Even if we recognise that this would deserve more attention, the 18- up to 25-day-long interval between first and second visit seems appropriate. Such a delay allows to stop sampling and to wait for new emergence in case of unpredictable meteorological event such as a storm or a flood washing away the exuviae. During this study, we had to face periods of heavy rains which obliged us to stop exuviae sampling and increased the delay between the first and second visit. This probably had a negative impact on the exuviae densities [27]; however, this was not quantified in our case and, hence, we cannot provide a solution to manage the impact of such events.

We also investigated the effect of transect length on our ability to detect species. Our results showed that a 70 m-long transect would be enough to detect, on average, 90% of the species locally present on each transect. This particularly applies for *O. curtisii* and *G. graslinii* which were detected within the first 10 m plots in a vast majority of positive transects, 87% and 69% respectively within the first 30 m. Concerning *M. splendens*, we cannot draw any conclusion because this species was too scarce on our transects (only three positive transects, with a total of four exuviae), even though the species had been previously observed in various location of the Tarn and Lot rivers [41,47–49]. We would recommend to sample 100 m-long transects rather than 70 m-long ones. First, sampling 100 m-long transects would only last 15 min more on average. Second, a 100 m-long distance is easy to remember and handle in our decimal metric system. Third, we can expect a 100 m-long transect to increase our chances to detect *M. splendens*.

Compared with the results of da Silva-Méndez in the NW Iberian Peninsula [28] and those of samplings we have carried out in other French rivers (Aveyron, Vère, Viaur, unpublished data), exuviae densities were surprisingly low on some sites, particularly for *M. splendens*, whose exuviae were occasionally collected by hundreds on Tarn river in the 1980's [41]. Such a situation has poor chance to be due to late collection date as our first visit took place around the date 50% of emergence occurred on close by rivers [44]. Furthermore, in a recent survey [45], around fifty *M. splendens* exuviae could be found on Tarn river between mid-July and early August. As adults damselflies were also few (Denis, Pelozuelo and Danflous, personal observation), we rather think it is due to a low level of Odonata populations at that time. Dam emptying operation at the Pinet dam in 2003 and 2009, with the water level moving down by 9 m and 15 m respectively [50], could have heavily impacted Odonata populations on the upper Tarn river for several years. Heavy rains that occurred at the end of June could also washed away an important part of the exuviae. Anyway, locally low densities of exuviae do not undermine our results and both the transect length and the two visit we recommend would be enough for an accurate description of dragonflies community in rivers with higher densities.

Differences in detection between adults and exuviae of Odonata have been shown in recent studies and the exuviae collection is today one of the most reliable methods for monitoring species [25,29]. Moreover, exuviae detection offers numerous benefits: on one hand, exuviae are "the most important indicators of resident populations" [51], i.e., the best cue of on-site reproduction and development [27,41,52] and on the other hand, the number of exuviae provides the most reliable estimate of population density [23]. In addition, exuviae collection is a non-invasive method, an essential quality when dealing with protected or red-listed species [22]. Indeed, exuviae collection has become in recent years a popular sampling method to inventory Odonata in lentic [22,26] and lotic habitats [26,40,53,54]. However, few standardised protocols based on the exuviae collection have been investigated. There are examples of exhaustive exuviae collection during a defined duration [40], of exhaustive exuviae collection over transects randomly drawn each year [53,55,56] or over chosen "sentinel" sites to be monitored every year ([8], this study). Number of visits and transect length differ from one protocol to another; however, such differences are adaptations to local conditions. In our case, riverbanks are not drastically modified from one year to another in the rivers monitored and our protocol can easily be applied. With a good knowledge of the target species' emergence periods, two visits on identical 100 m-long transects every year are enough to monitor riverine dragonfly communities in a way that allows to obtain trends after several years of monitoring. Thanks to the use of "sentinel" sites, sampled each year, spatial and temporal variations in exuviae densities can be highlighted. As important interannual variation might be expected, several years of monitoring would be recommended to establish a local reference.

However, for this type of protocol, additional recommendations can be made. First, as a feedback from our own field-work experience with kayak for exuviae collection, the use of short individual kayaks would be recommended, even on small rivers, since it allows good access to the river banks regardless of water levels, without trampling the aquatic habitat and larvae. It should however be noted that in some rare cases kayaks may become impractical in small Mediterranean rivers, restricted to pools during part of the summer. Additionally, cleaning the kayak and accessories between sites/sampling dates may be required to prevent the spread of invasive aquatic species if such a risk is identified.

For safety purposes, surveys should always be carried out in pairs. Kayaks also enable sampling to be carried out in a comfortable position (sampling sessions can sometimes last all day) and to access areas that are difficult to access on foot (e.g., steep banks, sections of isolated rivers). Then, it is important to plan sampling according to meteorological and hydrological conditions. Exuviae can be washed by storms (rainfall and gusts of wind) or when water levels rise (because of dam water release or precipitations) [57]. Thus, we recommend leaving a few days after those occasional disturbances, to allow larvae to emerge.

Standardised protocols based on exuviae collection are currently the most reliable and relevant methods for detecting and monitoring riverine dragonflies. They should be implemented to improve knowledge of targeted species population trends and ensure their conservation, as recommended by the latest French National Action Plan for dragonflies [12]. In the future, they should be used to test and develop other methods such as quantitative environmental DNA approaches. Efforts to produce genetic barcodes for species identification [58,59] and to understand the persistence and accumulation of dragonflies DNA in their ecosystems [60] would probably soon make quantitative environmental DNA approaches available. Using standardised exuviae collection with such genetic methods will thus be useful to calibrate and validate the use of genetic methods to monitor odonate populations in nature.

Finally, we should emphasise that this protocol has been used every year since 2018. Except for the impossibility to quantify how much storms may impact exuviae densities, no particular drawback has been identified since, and the dataset obtained is being analysed in order to describe temporal and spatial variability for these three targeted species and identify rivers and rivers portion of highest conservation value.

5. Conclusions

The extensive sampling effort set up during this study has allowed us to propose and calibrate a relevant protocol for surveying riverine dragonfly communities. Even though sample sites were only located on three rivers in south-western France, this method may be suited to abroad range of temperate rivers. Thus, we propose riverine dragonfly surveys to be conducted by two observers in kayaks (i.e., one along each bank) in the course of two visits during the emergence period to collect exuviae of all species along a 100 m transect of river bank. According to our results, this method maximises detected richness while minimising the duration of sampling. These results are a major issue since exuviae collection is rarely undertaken when surveying riverine dragonflies, especially as no standardised monitoring program currently exists for the three protected species in France (i.e., *O. curtisii*, *M. splendens*, and *G. graslinii*) and only recently effort were also dedicated to develop a protocol for the monitoring of *Ophiogomphus cecilia* (Geoffroy in Fourcroy, 1785) and *Stylurus flavipes* (Charpentier, 1825) [12,53]. This is also particularly true regarding environmental impact assessments, which aim to avoid, mitigate, and offset adverse impacts on biodiversity, and particularly on protected species. Methods for detecting riverine dragonflies and quantifying their populations should be relevant and robust to ensure that decision-makers' judgments are well-founded. Thus, we expect our protocol proposal to raise awareness among the experts involved in impact assessment and the administration to review their studies.

Supplementary Materials: The following supporting information can be downloaded at: https://www.mdpi.com/article/10.3390/d14090728/s1, Table S1: Location and sampling dates of the ninety-eight transects. GPS points are provided in WGS84 geographic format; Table S2: Details of survey durations on the different rivers (average, standard deviation, maximum duration and minimum duration).

Author Contributions: Conceptualization, A.S.D., S.D. and L.P.; methodology (field data collection), A.S.D. and L.P.; methodology (analysis data collection), S.D.; formal analysis, A.S.D. and L.B.; figures preparation, L.A.; writing—original draft preparation, A.S.D. and L.P.; editing, L.A.; final draft writing, L.A. and L.P.; supervision, N.G., F.S. and L.P. All authors have read and agreed to the published version of the manuscript.

Funding: This research was funded by Agence Nationale de la Recherche et de la Technologie (Cifre PhD grant n° 2015/0051), Electricité de France (EDF), European Regional Development Fund (FEDER), Agence de l'Eau Adour-Garonne and a donation by Mrs. MANN. The APC was funded by Laboratoire d'Ecologie Fonctionnelle et Environnement.

Institutional Review Board Statement: Not applicable.

Data Availability Statement: Not applicable.

Acknowledgments: We thank the following people for their respective contributions to this work: Baptiste Charlot and Laetitia Pêcheur for their great help to identify the exuviae and also Marine Valet and Maxime Sacré for their help in the data collection.

Conflicts of Interest: The authors declare no conflict of interest. The funders had no role in the design of the study; in the collection, analyses, or interpretation of data; in the writing of the manuscript, or in the decision to publish the results.

References

1. Didham, R.K.; Barbero, F.; Collins, C.M.; Forister, M.L.; Hassall, C.; Leather, S.R.; Packer, L.; Saunders, M.E.; Stewart, A.J.A. Spotlight on Insects: Trends, Threats and Conservation Challenges. *Insect Conserv. Divers.* **2020**, *13*, 99–102. [CrossRef]
2. Didham, R.K.; Basset, Y.; Collins, C.M.; Leather, S.R.; Littlewood, N.A.; Menz, M.H.M.; Müller, J.; Packer, L.; Saunders, M.E.; Schönrogge, K.; et al. Interpreting Insect Declines: Seven Challenges and a Way Forward. *Insect Conserv. Divers.* **2020**, *13*, 103–114. [CrossRef]
3. Hallmann, C.A.; Sorg, M.; Jongejans, E.; Siepel, H.; Hofland, N.; Schwan, H.; Stenmans, W.; Müller, A.; Sumser, H.; Hörren, T.; et al. More than 75 Percent Decline over 27 Years in Total Flying Insect Biomass in Protected Areas. *PLoS ONE* **2017**, *12*, e0185809. [CrossRef] [PubMed]
4. Vogel, G. Where Have All the Insects Gone? *Science* **2017**, *356*, 576–579. [CrossRef] [PubMed]
5. Sánchez-Bayo, F.; Wyckhuys, K.A.G. Worldwide Decline of the Entomofauna: A Review of Its Drivers. *Biol. Conserv.* **2019**, *232*, 8–27. [CrossRef]
6. Van Klink, R.; Bowler, D.E.; Gongalsky, K.B.; Swengel, A.B.; Gentile, A.; Chase, J.M. Meta-Analysis Reveals Declines in Terrestrial but Increases in Freshwater Insect Abundances. *Science* **2020**, *368*, 417–420. [CrossRef]
7. Wagner, D.L. Insect Declines in the Anthropocene. *Annu. Rev. Entomol.* **2020**, *65*, 457–480. [CrossRef]
8. Denis, A.S. Impacts de l'anthropisation sur la Diversité Odonatologique au Sein des cours d'eau: Vers une Meilleure Prise en Compte des Espèces de la Directive Habitats Faune Flore. Ph.D. Thesis, Université de Toulouse, Toulouse, France, 2018.
9. Dommanget, J.-L. *Etude Faunistique et Bibliographique des Odonates de France*; INRA-Muséum National d'Histoire Naturelle: Paris, France, 1987; p. 277.
10. Dijkstra, K.-D.B. *Guide Des Libellules de France et d'Europe*; Guide Delachaux; Delachaux et Niestlé: Paris, France, 2015; ISBN 978-2-603-02153-8.
11. de Las Heras, M.; Cordero-Rivera, A.; Cabana, M.; Romeo, A.; Rey-Muñiz, X.L.; Mezquita, I.; Gainzarain, J.A.; Salvador Vilariño, V.; Evangelio-Pinach, J.M.; Díaz, C.; et al. Distribución Ibérica de Gomphus Graslinii, Oxygastra Curtisii y Macromia Splendens (Insecta: Odonata), Especies Protegidas Por La Directiva Hábitats. *Bol. Rola* **2017**, *9*, 15–54.
12. Houard, X. *Plan National d'actions En Faveur Des Libellules-Agir Pour La Préservation Des Odonates Menacés et de Leurs Habitats 2020–2030*; Office pour les Insectes et leur Environnement-DREAL Hauts-de-France-Ministère de la transition écologique: Toulouse, France, 2020; p. 66.
13. European Union. Council Directive 92/43/EEC 1992 on the Conservation of Natural Habitats and of Wild Flora and Fauna. *Off. J. Eur. Communities* **1992**, *206*, 7–49.
14. Costes, A.; Danflous, S.; Laurent, B.; Delpon, G. Premier retour d'expérience sur la prise en compte des odonates dans les DOCOB des sites Natura 2000 de Midi-Pyrénées. In *Les Invertébrés dans la Conservation et la Gestion des Espaces Naturels*; Muséum national d'Histoire naturelle: Toulouse, France, 2015; pp. 22–25.
15. Dupont, P. *Plan national d'actions en faveur des Odonates-Libellules et Demoiselles menacées 2011–2015*; Office pour les Insectes et leur Environnement/Société Française d'Odonatologie-Ministère de l'Ecologie, de l'Energie et du Développement durable et de la Mer: Toulouse, France, 2010; p. 170.
16. Kalkman, V.; Boudot, J.-P.; Bernard, R.; Conze, K.-J.; De Knijf, G.; Dyatlova, E.; Ferreira, S.; Jović, M.; Ott, J.; Riservato, E.; et al. *European Red List of Dragonflies*; Publications Office: Luxembourg, 2010; p. 28.
17. UICN France, MNHN, OPIE & SFO. *La Liste Rouge Des Espèces Menacées En France*; Chapitre Libellules de France Métropolitaine; UICN France, MNHN, OPIE & SFO: Paris, France, 2016; p. 5.
18. Charlot, B.; Danflous, S.; Louboutin, B.; Jaulin, S. *Liste Rouge Des Odonates d'Occitanie-Rapport d'évaluation*; CEN Midi-Pyrénées & OPIE: Toulouse, France, 2018; p. 102.
19. Barneix, M.; Bailleux, G.; Soulet, D. *Liste Rouge régionale des Odonates d'Aquitaine*; Observatoire Aquitain de la Faune Sauvage (Coordination): Pessac, France, 2016; p. 40.
20. Deliry, C. Le Groupe Sympetrum Liste Rouge des Odonates de la région Rhône-Alpes. In *Histoires Naturelles*; Flammarion: Paris, France, 2014; p. 35.
21. Ministre de l'Ecologie et du Développement Durable Arrêté Du 23/04/07 Fixant Les Listes Des Insectes Protégés Sur l'ensemble Du Territoire et Les Modalités de Leur Protection. 2007. Available online: https://www.legifrance.gouv.fr/loda/id/JORFTEXT000000465500/ (accessed on 25 June 2022).
22. Foster, S.E.; Soluk, D.A. Evaluating Exuvia Collection as a Management Tool for the Federally Endangered Hine's Emerald Dragonfly, Somatochlora Hineana Williamson (Odonata: Cordulidae). *Biol. Conserv.* **2004**, *118*, 15–20. [CrossRef]

23. Oertli, B. The Use of Dragonflies in the Assessment and Monitoring of Aquatic Habitats. In *Dragonflies and Damselflies: Model Organisms for Ecological and Evolutionary Research*; Oxford Biology: New York, NY, USA, 2008; pp. 79–95. ISBN 978-0-19-923069-3.
24. Hardersen, S. Dragonfly (Odonata) Communities at Three Lotic Sites with Different Hydrological Characteristics. *Ital. J. Zool.* **2008**, *75*, 271–283. [CrossRef]
25. Giugliano, L.; Hardersen, S.; Santini, G. Odonata Communities in Retrodunal Ponds: A Comparison of Sampling Methods. *Int. J. Odonatol.* **2012**, *15*, 13–23. [CrossRef]
26. Hardersen, S.; Corezzola, S.; Gheza, G.; Dell'Otto, A.; La Porta, G. Sampling and Comparing Odonate Assemblages by Means of Exuviae: Statistical and Methodological Aspects. *J. Insect Conserv.* **2017**, *21*, 207–218. [CrossRef]
27. Raebel, E.M.; Merckx, T.; Riordan, P.; Macdonald, D.W.; Thompson, D.J. The Dragonfly Delusion: Why It Is Essential to Sample Exuviae to Avoid Biased Surveys. *J. Insect Conserv.* **2010**, *14*, 523–533. [CrossRef]
28. Da Silva-Méndez, G.; Riso, S.; Lorenzo-Carballa, M.O.; Cordero-Rivera, A. Sampling Larvae, Exuviae or Adults of Odonata for Ecological Studies: A Test of Methods in Permanent Rivers in the Iberian Peninsula. *Odonatologica* **2022**, *51*, 63–81.
29. Bried, J.T.; D'Amico, F.; Samways, M.J. A Critique of the Dragonfly Delusion Hypothesis: Why Sampling Exuviae Does Not Avoid Bias: Critiquing the Dragonfly Delusion. *Insect Conserv. Divers.* **2012**, *5*, 398–402. [CrossRef]
30. Luque, P. Macromia Splendens i Gomphus Graslinii, Dues Noves Espècies d'odonats per a Catalunya. *Butlletí Inst. Catalana Història Nat.* **2006**, *74*, 113–116.
31. Martínez-Martínez, D.; Olmo-Vidal, J.M.; Luque Pino, P.; Querol, J.M.; del Mai, M.J.M.; Casanova, C.B.; Pujol, A.G. Seguimiento de La Población de Macromia Splendens Mediante El Muestreo de Exuvias En La Terra Alta, Cataluña. In Proceedings of the Simposio Ibérico de Odonatología SIO-2015, Córdoba, Spain, 1–3 May 2015; Asociación de Educación Ambiental El Bosque Animado/Red de Observadores de Libélulas en Andalucía: Málaga, Spain, 2015.
32. Doucet, G. *Clé de Détermination Des Exuvies Des Odonates de France*, 3rd ed.; Société Française d'Odonatologie; Fondation Nature & Découvertes: Bois d'Arcy, France, 2016; ISBN 979-10-94955-00-0.
33. Boudot, J.-P.; Grand, D.; Wildermuth, H.; Monnerat, C. *Les Libellules de France, Belgique et Luxembourg*, 2nd ed.; Parthenope; Biotope: Mèze, France, 2017; ISBN 978-2-36662-176-1.
34. Ferreras-Romero, M.; Corbet, P.S. Seasonal Patterns of Emergence in Odonata of a Permanent Stream in Southwestern Europe. *Aquat. Insects* **1995**, *17*, 123–127. [CrossRef]
35. Suhling, F. Temporal Patterns of Emergence of the Riverine Dragonfly Onychogomphus Uncatus (Odonata: Gomphidae). *Hydrobiologia* **1995**, *302*, 113–118. [CrossRef]
36. Farkas, A.; Jakab, T.; Tóth, A.; Kalmár, A.F.; Dévai, G. Emergence Patterns of Riverine Dragonflies (Odonata: Gomphidae) in Hungary: Variations between Habitats and Years. *Aquat. Insects* **2012**, *34*, 77–89. [CrossRef]
37. Zebsa, R.; Khelifa, R.; Kahalerras, A. Emergence Pattern, Microhabitat Choice, and Population Structure of the Maghribian Endemic Gomphus Lucasii Selys, 1849 (Odonata: Gomphidae) in Northeastern Algeria. *Aquat. Insects* **2014**, *36*, 245–255. [CrossRef]
38. Maingeot, M.; Motte, G.; Goffart, P. Première Étude de l'émergence de La Cordulie à Corps Fin (Oxygastra Curtisii) Le Long de l'Ourthe. *Nat. Belg.* **2015**, *96*, 57–83.
39. Golfieri, B.; Hardersen, S.; Maiolini, B.; Surian, N. Odonates as Indicators of the Ecological Integrity of the River Corridor: Development and Application of the Odonate River Index (ORI) in Northern Italy. *Ecol. Indic.* **2016**, *61*, 234–247. [CrossRef]
40. Hardersen, S.; Toni, I. Proposal for a Time-Based Standard Sampling Method for the Monitoring of Gomphus Flavipes (Charpentier, 1825) and Ophiogomphus Cecilia (Fourcroy, 1785) (Odonata: Gomphidae). *Fragm. Entomol.* **2019**, *51*, 55–62. [CrossRef]
41. Dommanget, J.-L. *Etude de Macromia splendens (Pictet, 1843) dans la vallée du Tarn (Tarn, Aveyron) et statut national de l'espèce (Odonata, Anisoptera, Macromiidae)*; Société Française d'Odonatologie-Ministère de l'Aménagement du Territoire et de l'Environnement Direction de la Nature et des Paysages: Paris, France, 2001; p. 134.
42. Cordero-Rivera, A. Distribution, Habitat Requirements and Conservation of Macromia Splendens Pictet (Odonata: Corduliidae) in Galicia (Nw Spain). *Int. J. Odonatol.* **2000**, *3*, 73–83. [CrossRef]
43. Lohr, M. *Sur l'habitat et la répartition de Macromia splendens (Pictet, 1843) et Gomphus graslinii (Dale, 1834) dans la rivière de l'Hérault (département de l'Hérault)*; Société Française d'Odonatologie: Vallet, Loire-Atlantique, France, June 2005; pp. 115–124.
44. Denis, A.S.; Payet, O.; Danflous, S.; Gouix, N.; Santoul, F.; Buisson, L.; Pelozuelo, L. Intraspecific Variability of the Phenology and Morphology of Three Protected Dragonflies between Natural and Artificial Habitats. *J. Insect Conserv.* **2018**, *22*, 419–431. [CrossRef]
45. Danflous, S.; Norel, H.; Charlot, B.; Lim, M. *Inventaire & Suivi Des Odonates d'intérêt Communautaire Sur Le Barrage Du Pinet–Résultats 2021*; CEN Occitanie: Montpellier, France, 2021; p. 57.
46. Cordero-Rivera, A.; Utzeri, C.; Santolamazza Carbone, S. Emergence and Adult Behaviour of Macromia Splendens (Pictet) in Galicia, Northwestern Spain (Anisoptera: Cordullidae). *Odonatologica* **1999**, *28*, 333–342.
47. Lieftinck, M.A. Macromia Splendens (Pictet, 1843) in Europe with Notes on Its Habits, Larva, and Distibution (Odonata). *Tijdschr. Voor Entomol.* **1965**, *108*, 41–59.
48. Bilek, A. Ergänzende Beobachtungen Zur Lebensweise von Macromia Splendens (Pictet 1843) Und Einingen Anderen in Der Guyenne Vorkommenden Odonata-Arten. *Entomol. Z.* **1969**, *79*, 117–124.
49. Pelozuelo, L.; Costes, A.; Delpon, G.; Calvignac, R.; Alquier, D.; Haber, E.; Polisset, P. *Macromia Splendens en Midi Pyrénées: Enfin des Nouvelles en 2012!* OPIE Midi-Pyrénées & LPO Tarn: Labruguière, France, 2012; pp. 1–15.

50. Danflous, S.; Norel, H.; Charlot, B.; Lim, M. *Inventaire & Suivi Des Odonates d'intérêt Communautaire Sur le Barrage Du Pinet*; CEN Occitanie: Montpellier, France, 2020; p. 53.
51. Chovanec, A.; Waringer, J. Ecological Integrity of River-Floodplain Systems-Assessment by Dragonfly Surveys (Insecta: Odonata). *Regul. Rivers Res. Manag.* **2001**, *17*, 493–507. [CrossRef]
52. Corbet, P.S. Are Odonata Useful as Bioindicators? *Libellula* **1993**, *12*, 91–102.
53. Baeta, R.; Bard, D.; Chantereau, M.; Fritsch, B.; Herbrecht, F.; Hudin, S.; Itrac-Bruneau, R.; Multeau, D.; Paillat, R.; Rambourdin, M.; et al. Protocole de Suivi Diachronique Des Populations Ligériennes de Gomphus Flavipes et d'Ophiogomphus Cecilia. 2015. 6p + annexes. Available online: http://odonates.pnaopie.fr/wp-content/uploads/2015/02/GomphesdeLoire_Protocole_avril2015.pdf (accessed on 25 June 2022).
54. DuBois, R.B. Detection Probabilities and Sampling Rates for Anisoptera Exuviae along River Banks: Influences of Bank Vegetation Type, Prior Precipitation, and Exuviae Size. *Int. J. Odonatol.* **2015**, *18*, 205–215. [CrossRef]
55. Baeta, R. Suivi Diachronique Des Populations Ligériennes de Gomphus Flavipes et d'Ophiogomphus Cecilia En Région Centre Val-de-Loire (Saison 2016-Deuxième Année de Suivi à l'échelle Régionale). 2017, p. 13. Available online: http://www.anepe-caudalis.fr/wa_files/GomphesLoire_IndreetLoire_Saison2016_VF.pdf (accessed on 25 June 2022).
56. Baeta, R.; Herbrecht, F.; Fierimonte, B. Stylurus flavipes et Ophiogomphus cecilia deux mystérieuses libellules. *Loire Terroirs* **2020**, *105*, 10–17.
57. Aliberti Lubertazzi, M.A.; Ginsberg, H.S. Persistence of Dragonfly Exuviae on Vegetation and Rock Substrates. *Northeast. Nat.* **2009**, *16*, 141–147. [CrossRef]
58. Rach, J.; DeSalle, R.; Sarkar, I.N.; Schierwater, B.; Hadrys, H. Character-Based DNA Barcoding Allows Discrimination of Genera, Species and Populations in Odonata. *Proc. R. Soc. B Biol. Sci.* **2008**, *275*, 237–247. [CrossRef]
59. Damm, S.; Schierwater, B.; Hadrys, H. An Integrative Approach to Species Discovery in Odonates: From Character-Based DNA Barcoding to Ecology. *Mol. Ecol.* **2010**, *19*, 3881–3893. [CrossRef]
60. Schmidt, K.J.; Soluk, D.A.; Maestas, S.E.M.; Britten, H.B. Persistence and Accumulation of Environmental DNA from an Endangered Dragonfly. *Sci. Rep.* **2021**, *11*, 18987. [CrossRef]

diversity

MDPI

Article

Similar Response of a Range Expanding Dragonfly to Low- and High-Elevation Predators

Rabah Zebsa [1,2], Hayat Mahdjoub [2] and Rassim Khelifa [2,3,4,*]

1 Laboratoire Biologie, Eau et Environment (LBEE), Université 8 Mai 1945, Guelma 24000, Algeria; zebsarabah@gmail.com
2 Department of Evolutionary Biology and Environmental Studies, University of Zurich, Winterthurerstrasse 190, CH-8057 Zurich, Switzerland; hayatmahdjoub@gmail.com
3 Institute for Resources, Environment and Sustainability, University of British Columbia, Vancouver, 429-2202 Main Mall, Vancouver, BC V6T 1Z4, Canada
4 Department of Biological Sciences, Simon Fraser University, Burnaby, BC V5A 1S6, Canada
* Correspondence: rassimkhelifa@gmail.com

Abstract: Recent range expansion of many species northward and upward in elevation suggests that the expanding species are able to cope with new biotic interactions in the leading edge. To test this hypothesis, we used a common garden experiment expanding the elevation range of an obligatorily univoltine dragonfly (*Sympetrum striolatum*) to investigate whether the growth, behavioral (food intake), and morphological (8th and 9th abdominal lateral spine) responses differed when confronted with dragonfly predators that dominate low-elevation (*Aeshna cyanea*) and high-elevation (*A. juncea*) lentic freshwater systems under two temperature treatments (20 °C and 24 °C). Growth rate and growth efficiency increased at higher temperature. Overall, low- and high-elevation predators induced a similar increase in growth rate and growth efficiency but a decrease in food intake at 24 °C. Lateral abdominal spines were longer only in low-elevation dragonflies at 18 °C. Our study suggests that range-expanding species may have been successful in colonizing new areas at higher elevations because they respond to dominant high-elevation predators in a similar way to the more familiar low-elevation predators.

Keywords: biotic responses; freshwater; Odonata; life history; chemical cues

Citation: Zebsa, R.; Mahdjoub, H.; Khelifa, R. Similar Response of a Range Expanding Dragonfly to Low- and High-Elevation Predators. *Diversity* **2022**, *14*, 302. https://doi.org/10.3390/d14040302

Academic Editors: M. Olalla Lorenzo-Carballa and Ricardo Koroiva

Received: 16 February 2022
Accepted: 12 April 2022
Published: 15 April 2022

1. Introduction

To comprehend and predict future climate impacts on freshwater communities, there is growing interest in the biotic response of species to new colonizer species and its consequences on life history traits [1–5]. The ability to recognize and respond to different types of predators is a key trait that contributes to the successful establishment of new populations [6,7]. The response of growth rate to predatory stress has been widely investigated because of its intimate relationship with fitness [8]. Because the magnitude of biotic interactions (e.g., predation) can change with temperature, recent studies have focused on the thermal dynamics of predator-prey interaction to understand the outcome of range shift events that occur across geographic gradients such as latitude and elevation [9–11]. Both consumptive (visual cues) and non-consumptive (chemical cues) effects of predation can affect growth rate and induce changes in fitness-related traits [3,12,13]; however, studies on non-consumptive effects of cold-adapted (unfamiliar) predators on warm-adapted prey that gradually expand its geographic range to cooler areas are still lacking [4].

There are different physiological and behavioral mechanisms that determine growth rate, and understanding these mechanisms unravels some underlying plastic responses [5,14]. Usually, growth rate is a function of the amount of food ingested (food intake) and the amount of energy allocated to growth (growth efficiency) [15,16]. Thus, although a constant growth rate across an environmental gradient may be interpreted as an absence of the adaptive response, it may hide dynamic plasticity in the behavioral and physiological components that contribute to the growth rate [17,18]. On one hand, an increased food intake likely increases individual activity and foraging behavior for prey acquisition, which often results in a higher mortality rate due to increased detectability by predators [7]. On the other hand, an accelerated growth efficiency leads to faster energy allocation and results in lower energy reserves and immune function [5,19]. Thus, it is important to understand whether behavioral and physiological components of the growth rate change in response to different types of predators and different temperatures to reveal the selective forces acting against the evolution of growth rate and predict the population dynamics of plasticity in new habitats after range expansion.

Apart from physiology and behavior, predators may induce plastic morphological responses such as defensive weaponry that reduces successful predation attempts [20,21]. As with growth efficiency, investment in defensive weaponry has survival costs and leads to a lower allocation of energy to other functional traits [22]. It has been shown that there is a negative relationship between behavioral and morphological defenses, so-called trait compensation [21,23]. Spending energy on defensive weaponry results in reduced vigilance while foraging. It is still not clear how species with time constraints respond to various types of predators, and which morphological, physiological, and/or behavioral mechanisms are prioritized.

While recent experimental studies have investigated the latitudinal patterns in predator–prey interactions [4,5], similar studies on the elevational gradient are still lacking. In the current study, we investigate the elevational range shift in Switzerland of *Sympetrum striolatum*, a widespread obligatorily univoltine species in North Africa, Europe, and Asia. To understand whether the species is able to cope with new biotic interactions at higher elevations, we assessed in a common garden experiment the response of growth rate, growth efficiency, food intake, and morphology (larval lateral spines) to temperature and different predation treatments. We exposed the species to two predators, a familiar (*Aeshna cyanea*: mainly at low elevations) dragonfly predator and an unfamiliar (*A. juncea*: mainly at high elevations) dragonfly predator. The low-high elevation dragonfly comparison reveals how the species would respond in a scenario of range shift to new habitats with a new predator, a phenomenon that has occurred repeatedly during the last decades [24,25]. Documenting the mechanisms controlling variation in growth rate across different treatments of predation and temperature is essential to understand the evolution of plasticity and predict future population dynamics in freshwater ecosystems [17].

2. Materials and Methods

2.1. Study Species and Distribution Data

Sympetrum striolatum is an obligatorily univoltine dragonfly that inhabits various types of wetlands, mostly stagnant freshwater, where it lays eggs from mid-summer to autumn and occupies an intermediate trophic level (as both predator and prey) in the freshwater food web. The species has shown a northward range expansion in the UK in recent decades [25], which makes it appropriate for the research question. In Switzerland, the species is historically considered a low-elevation species [26]. Long-term observations of adult *S. striolatum* carried out in Switzerland between 1990 and 2013 were obtained from CSCF (Centre Suisse de Cartographie de la Faune. The data included not only adult observations but also sex and reproductive states. To assess range shift across elevation, we included only records that showed a reproductive state such as copulation, oviposition, and emergence. The 95th percentile of the elevational distribution (leading edge) of *S. striolatum* and *A. cyanea*, and the 5th percentile (trailing edge) of the elevational distribution of *A. juncea* were calculated for each year to assess their historical changes during 1990–2013.

2.2. Study Site and Treatments

To investigate the effect of predation risk on the behaviour and physiology of dragonflies as well as the underlying proximate mechanisms, the response of larvae of *Sympetrum striolatum* to two different predators was assessed in a common garden experiment in the laboratory at the University of Zurich, Switzerland. Eggs of *S. striolatum* were collected from two sites in Zurich, Switzerland (Irchel pond: 47.3977° N, 8.5448° E, 480 m elevation; Chatzensee pond: 47.4264° N, 8.4878°, 440 m elevation). Both sites occurred within or near urban areas. We obtained the eggs from copulating females that were captured with a hand net in the middle of the day in early October 2015. The female abdomen was immersed in a vial with water, which led to the oviposition (about a few hundred eggs per female) [27]. Five different females from each site were sampled. Eggs were brought to the laboratory within two hours and put at room temperature (21 °C) and natural light conditions. Eggs of *S. striolatum* had a direct development and hatched within 2–3 weeks. After hatching, 20 larvae were placed in each of the six treatments (2 temperatures [18 °C, 24 °C] × 3 predation treatments [control, low-elevation, and high-elevation]) and raised individually in 180-mL plastic cups filled with dechlorinated tap water and floating on water tubs. Using polystyrene foam, the cups were able to remain floating on the surface of the water.

A randomized full factorial design using eggs from each of the two populations which were exposed to the combination of two predation treatments (low-elevation [*A. cyanea*] and high-elevation [*A. juncea*]) and two constant temperatures (20 and 24 °C). Temperatures were maintained by setting heaters at the bottom of water tubs. The low-elevation dragonfly was *Aeshna cyanea* and the high-elevation dragonfly was *A. juncea*, whose larvae were collected from Irchel forest (47.3893° N, 8.5611° E, 660 m) and Flumserberg (47.0661° N, 9.2546° E, 1904 m), respectively. Although the two species' elevational distribution overlap, the largest populations of each do not coexist at a similar elevation [26]. The two temperatures used in the experiment are far below the thermal critical maximum of *S. striolatum* [28].

Larvae under predation treatments received only chemical (no visual) cues from predators. The predator medium containing chemical cues was acquired by individually setting the predators (four large dragonfly larvae [size > 4 cm] in containers filled with dechlorinated tap water [12 × 12 cm filled to the height of 4 cm]). All predators were fed every day with a single *S. striolatum* larva (body length: 0.5–1 cm). We mixed together the chemical cues of four individuals of the same predator species, then 1 mL of the solution was provided daily to the prey. All *S. striolatum* larvae were fed once a day with *Artemia* sp. nauplii (~80–110 individuals in 5 mL).

2.3. Growth Efficiency and Food Intake

During 60 days after hatching, growth rates were quantified by measuring the head width of larvae every four days, taking pictures with a stereoscope, and estimating the distance between the edges of the eyes to the nearest 0.01 mm using ImageJ v1.52. To estimate the growth efficiency, a 4-day experiment as described by Stoks, Swillen and De Block [5] was used on 60-day-old larvae. This experiment consists of weighing larvae from different treatments to the nearest 0.01 mg using an electronic balance at the start and the end of the 4-day period. Then growth rate was measured as [log(final mass)–log(initial mass)]/4 days, which accounts for differences in mass at the start of the experiment [18]. Given the widespread distribution of *Daphnia* in freshwater ecosystems worldwide, and its importance as prey to dragonflies [29], we provided *S. striolatum* larvae with 20 individual *Daphnia* of similar size each day. The number of *Daphnia* eaten per day was counted and replaced. The number of eaten daphnia was converted to total dry mass ingested using an equation [dry mass = a wet mass + b] established through weighing (to the nearest µg using an electronic balance) *Daphnia* before and after drought-treatment in the oven at 60 °C during 48 h. Dry mass of larvae was also estimated in the same way to obtain a conversion factor (dry mass = 0.114 × wet mass + 0.478).

2.4. Morphological Defense

Predation avoidance in freshwater prey might involve morphological modifications that reduce the probability of being eaten [1,20,21]. To test whether there is variability in the morphological response to different predator types, the 8th and 9th lateral abdominal spines of larvae were measured at the end of the experiments, 65 days after hatching. The size of lateral abdominal spines shows plasticity when exposed to predators [21]. The length of the spines was corrected for the larval size by dividing it by the head width of larvae; a widely used measure of body size which describes the developmental stage of larvae [30].

2.5. Statistical Analyses

All statistical analyses were carried out using R 3.5.1 [31]. We calculated the leading and trailing edge of the distribution of all three species using 95th and 5th percentile of the elevation of the occurrence data, respectively. Then, elevational range shift was assessed using simple linear regressions including our estimate of the elevational distribution edge (leading or trailing) as a response variable and years as an explanatory variable. To determine differences in growth rate among treatments, linear mixed-effects (LME) model including head width as a response variable and time (days), predation and temperature as explanatory variables, and individual identifier as a random effect, was carried out using *lme4* package [32]. We used the lstrends function from lsmeans package [33] on our LME model to calculate for each treatment average growth rate with 95% confidence intervals. Since we had a single measure per individual (no repeated measures), we used a linear model to determine the difference in growth efficiency, food intake and lateral spine length between temperature and predation treatments.

3. Results

3.1. Elevational Range Shift

The 95th margin of the elevational distribution (leading edge) shifted by 4.6 m/year (Figure 1; linear model: slope = 4.66 m/yr, R^2 = 0.21, p = 0.02). The trailing edge (5th percentile) of the high-elevation predator *A. juncea* did not show a significant shift (slope = −2.63 m/yr, R^2 = 0.03, p = 0.40), nor the leading edge of the low-elevation predator *A. cyanea* (slope = −12.2 m/yr, R^2 = 0.14, p = 0.06) during the same period.

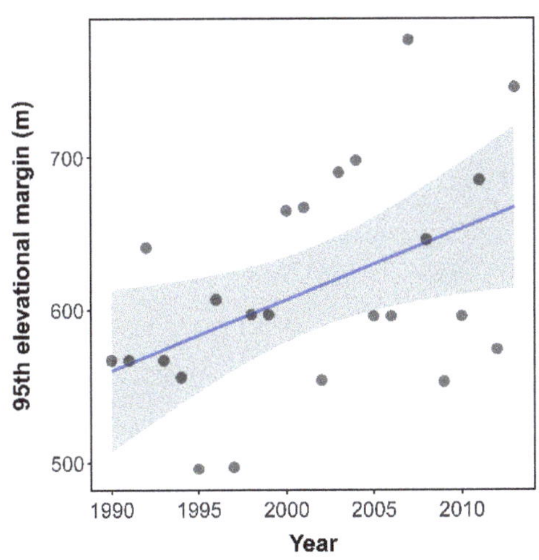

Figure 1. Historical changes of the leading elevational edge (95th percentile) of *Sympetrum striolatum* in Switzerland between 1990 and 2013. The linear regression was positively significant (R^2 = 0.21, p = 0.02), suggesting a historical shift towards higher elevations.

3.2. Larval Growth

Larvae grew faster in water at 24 °C than at 18 °C (LME: $p = 0.0001$; Table S1). Growth rate was significantly different among predation treatments (LME: $p = 0.002$; Table S1), revealing a faster growth in predation treatment compared to the control treatment (Figure 2). The significant interaction between time and temperature and the non-significant three-way interaction between time, temperature, and predation indicate that larvae responded similarly to the different predators at both temperatures ($p > 0.05$; Table S1).

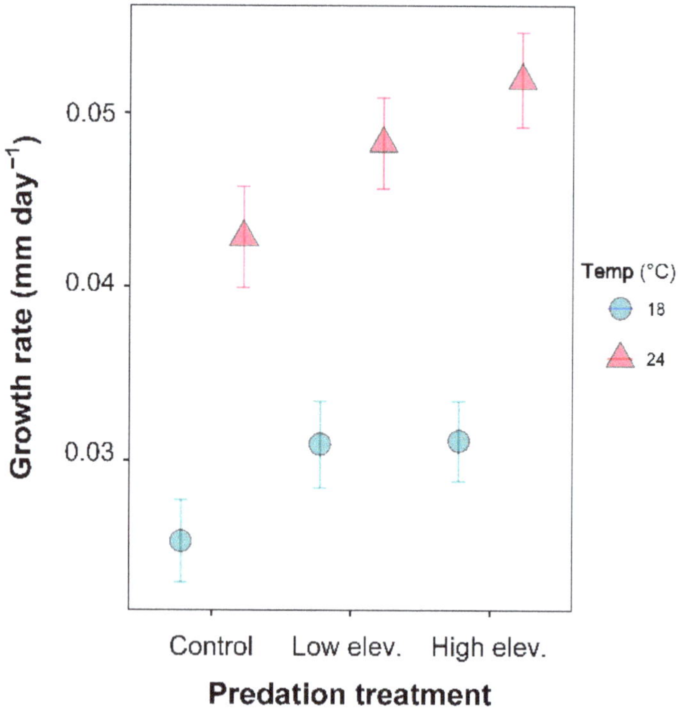

Figure 2. Growth rate of *Sympetrum striolatum* larvae in the combination of two temperature and three predator treatments. Error bars are 95% confidence intervals.

Growth efficiency was significantly different between temperature (LME: $p = 0.0002$) and predator treatments (LME: $p < 0.002$) (Figure 3). Growth efficiency was 81.6% greater at higher temperature than at lower temperature. Larvae in predation treatments showed a higher growth efficiency than in the control treatment, but there was no significant difference between low and high-elevation predator (Figure 3). There was a small significant interaction between temperature and predation treatment (Table S2).

3.3. Behavioral Response

There was a significant interaction between temperature and predation (Figure 4, Table S3). At 18 °C, larvae showed a lower food intake in high-elevation predator treatment ($p = 0.002$). At 24 °C, larval food intake was lower in both low- and high-elevation predator treatment than in the control, although marginal for the high-elevation predator (Table S3, $p = 0.01$; $p = 0.07$, respectively).

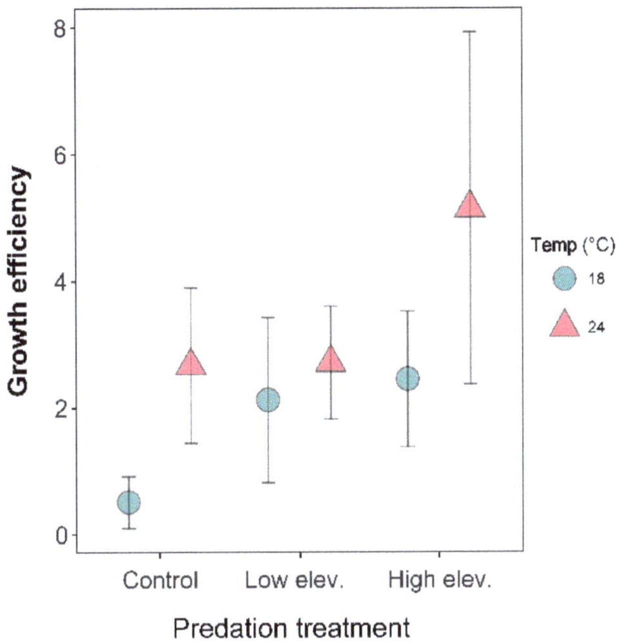

Figure 3. Growth efficiency of *Sympetrum striolatum* larvae in the combination of two temperature and three predator treatments. Error bars are 95% confidence intervals.

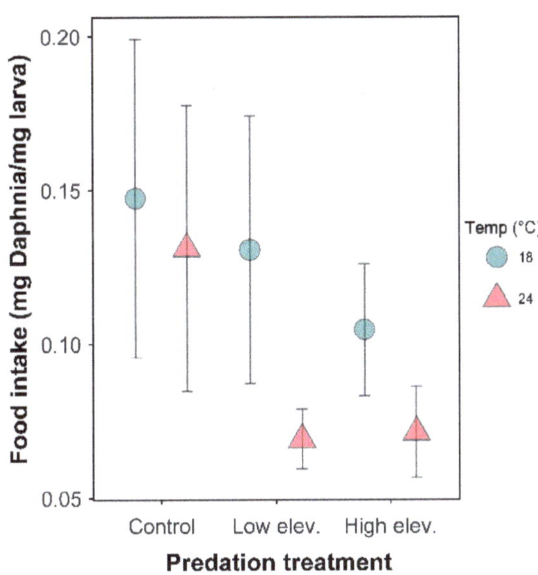

Figure 4. Food intake (corrected for body mass) of *Sympetrum striolatum* larvae in the combination of two temperature and three predator treatments. Error bars are 95% confidence intervals.

3.4. Morphological Response

The length of the 8th lateral spine was longer in the low-elevation dragonfly treatment ($p = 0.01$; Table S4a, Figure 5a), but neither temperature nor the interaction of temperature with predation showed a significant effect (Table S4a). The length of the 8th lateral spine was 32% longer in the low-elevation predator treatment than in the control (Figure 5a, Table S4a). Similar to the 8th lateral spine, there was no significant effect of temperature on the length of the 9th lateral spine (Table S4b, $p = 0.64$). However, larvae in the low-elevation dragonfly treatment under 18 °C had significantly longer 9th spines than the other treatments ($p = 0.001$; Table S4b, Figure 5b). The average length of the 9th spine in the low-elevation predator treatment was 33% longer than that of the control.

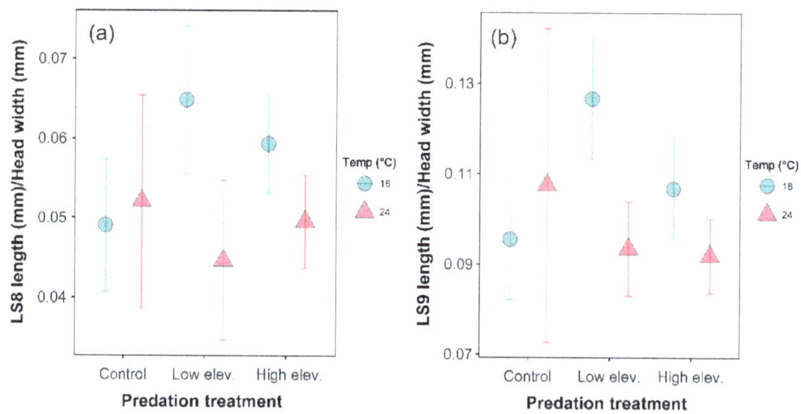

Figure 5. Length of the 8th (**a**) and 9th (**b**) lateral spine of *Sympetrum striolatum* larvae in the combination of two temperature and three predator treatments. Error bars are 95% confidence intervals.

4. Discussion

In this study, we showed that the low-elevation obligatorily univoltine *S. striolatum* expanded its range by 4.6 m/yr across elevation during 1990–2013, which suggests that the species was able to adapt to the new biotic environment. The common garden experiments assessing the non-consumptive response of growth (rate and efficiency), behavior (food intake), and morphology (abdominal spine) of *S. striolatum* to different predators including low- (familiar) and high-elevation (non-familiar) dragonflies showed that overall dragonfly larvae responded similarly in growth and food intake to both predators, but they responded distinctly in morphology to the familiar dragonfly predator.

4.1. Range Shift

Sympetrum striolatum has shown an elevational range shift of 4.6 m/year between 1990 and 2013. The same species has been reported to expand its northern range limit in the UK by 346 km between 1960 and 1995, which is equivalent to a rate of 8.6 km/year [25]. Given that this species is still flourishing at low elevations and in the southern range limit, it can be inferred that the range has expanded both in latitude and elevations. Discrepancies between latitudinal and elevational range shift are probably due to the steeper cline in environmental conditions across elevation, which restricts the speed at which new populations are established at higher elevations. However, the successful establishment of higher elevation populations is probably due to recent climate warming in the Swiss Alps, but it also suggests that the species was able to cope with new biotic conditions. Furthermore, the absence of elevational shift of the trailing distribution edge of *A. juncea* suggests that the encounter probability between *S. striolatum* and *A. juncea* has increased. We suggest that the overlap will increase even further in the following years.

4.2. Temperature Effects

Temperature had various effects on growth, food intake, and length of lateral spines of larvae. Higher temperatures led to an increased growth rate and growth efficiency. In many taxa, growth rate is expected to increase with temperature when temperatures are below the thermal optimum [34], which is typically higher than 24 °C in temperate Libellulidae [35].

However, unlike studies on odonates [5,36] and butterflies [37] which found that the increase in growth rate was associated with an increase in food intake, our study showed that it was associated with an increase in growth efficiency [16]. Thus, it is likely that this is an adaptive mechanism that allows obligatorily univoltine species such as *S. striolatum* to respond to an increase in temperature by increasing growth efficiency despite lowering food intake. Such a strong physiological response probably has considerable fitness trade-offs for adults [38]. In a climate warming scenario, temperature-induced changes in physiology and behavior might have a major impact on predator-prey interaction and community structure [4,39,40]. For example, an increase of 4 °C (difference in temperature treatments in this study) causes an increase in growth rate and a decrease in food intake which may lead to an earlier emergence (shorter exposure of prey to predators) and a decline in predation rates (less prey eaten during the short exposure time). These effects may have direct [41] and indirect consequences [42] on the freshwater food web. Interestingly, a positive response in anti-predatory morphological defenses was observed only at low temperatures where food intake was higher, suggesting that at higher temperature larvae invested in growth rather than anti-predatory weaponry.

4.3. Predation Effects

The analysis of growth rate showed a significant difference among predator treatments. Unlike the commonly reported pattern of growth reduction under predation treatments [17,43], we found a predatory-induced acceleration of growth which was also reported in odonates [5,44–46] and other vertebrates [47,48]. One explanation of the increased growth rate under predation is the 'escape theory' which is predicted by some optimal models when non-predatory costs of growth exist [49,50]. This theory is further supported by the fact that many *Sympetrum* species are obligatorily univoltine [51], and thus environmental limitations such as predation or time constraints result in a fastening of growth rates [5,52]. This finding highlights the fact that the recent range shift of many large species of odonates, invertebrates, and vertebrates might alter the life history of many prey taxa of freshwater ecosystems, which might result in major evolutionary and ecological changes.

There was a lower food intake under dragonfly predation (both low- and high-elevation) than in the control, which suggests that larvae detected the predator through chemical cues and subsequently lowered their food intake probably to avoid being detected and eaten [2]. A similar decline in food intake under dragonfly predation was detected in damselflies [5,17] and dragonflies [53]. Furthermore, there was a higher growth efficiency under dragonfly treatment than in the control treatment, which means that larvae grew fast with respect to the amount of food consumed. A similar increase in growth efficiency has been observed in damselflies in the presence of predation [5]. The lower food intake and higher growth efficiency for a species adapted to time constraints could be theoretically explained by optimality models [50]. Food intake involves at least the movement of mouthpieces (mentum), which makes prey larvae detectable by large dragonfly predators. Thus, decreasing food intake is a common anti-predatory behavioral mechanism that individuals adopt to increase their survival probability [54]. High growth efficiency is a physiological anti-predatory response that allows larvae to minimize their exposure to predation and reach the adult stage as early as possible [55]. This hypothesis is supported by previous studies on many taxa carried out in freshwater ecosystems on odonates [45,46], frogs [48], and fish [3,47]. The costs of growth acceleration have been documented and could involve a weakening of the immune system and reduction in energy reserves [15], and could even carry over until the adult stage where individuals encounter shorter longevity and reproductive success [56].

Morphological responses to predators have been well documented in odonates, but mostly against fish [20,21,57]. Interestingly, our study showed a morphological response of the lateral abdominal spines when exposed to chemical cues of dragonfly predators. Studies have shown that fish reject larval dragonflies which have long abdominal spines, thus allowing dragonflies to increase their survival probabilities [20]. It is unclear how longer spines benefit larvae from dragonfly predation. One likely hypothesis is that longer spines might be a by-product of physiological responses rather than an adaptive response to dragonfly predation [47]. In fact, higher growth efficiency might result in morphological changes that might include variation in the length or morphology of abdominal structures such as lateral spines. However, this hypothesis does not explain the observed longer lateral spines in low-elevation dragonfly treatment but not in high-elevation dragonfly treatment. This intriguing response to the more familiar predator warrants further investigations of the role of abdominal lateral spines against dragonfly predation.

While larvae used to estimate physiological and behavioral parameters had similar ages, there might be a difference in larval instars among individuals that might influence the response of larvae. Nevertheless, the observed physiological and behavioral responses were similar to other studies [5,45,48]. Since antipredator response can be costly in invertebrates in general [58], one would expect that selection should favor accuracy in detecting the predators that induce mortality or damage [59]. The fact that low- and high-elevation predators (*A. cyanea* and *A. juncea*) are closely related species with potentially similar diet could explain the similar behavioral and physiological response of *S. striolatum* larvae. Studies have linked the composition of chemical signals (kairomones) and predator diet [60–62]. The ability to detect a non-familiar predator is a dispersal asset for species, allowing larvae to develop, persist, and successfully establish a viable population in new habitats. For instance, many libellulids such as *S. striolatum* are invaders of ponds and lakes of northern and high-elevation areas [63], causing changes in species composition and probably competing with rare specialist species [35,64]. It could be that their success in invading new territories is due to their ability to respond similarly to new predators as their familiar predator. However, whether populations of *S. striolatum* at low elevations living in different types of landscapes (e.g., natural, suburban, or urban) respond similarly to different predators remains to be investigated. Our results suggest that species that successfully expanded their range to higher elevation habitats might be preadapted to the new biotic interactions. Future studies should investigate the implication of odonate range expansion on the dynamics of freshwater and terrestrial foodweb [65].

Supplementary Materials: The following supporting information can be downloaded at: https://www.mdpi.com/article/10.3390/d14040302/s1, Table S1: Summary results of the linear mixed-effects model regressing head width of *Sympetrum striolatum* against time, temperature and predation treatments. Table S2: Summary results of the linear mixed-effects model regressing growth efficiency of *Sympetrum striolatum* against temperature and predation treatments. Table S3: Summary results of the linear mixed-effects model regressing food intake of *Sympetrum striolatum* against temperature and predation treatments. Table S4: Summary results of the linear mixed-effects model regressing the length of the 8th (a) and 9th (b) lateral spine corrected for body size of *Sympetrum striolatum* against temperature and predation treatments.

Author Contributions: Conceptualization, R.K.; methodology, R.K., R.Z. and H.M.; software, R.K., R.Z. and H.M.; validation, R.K.; formal analysis, R.Z.; investigation, R.K., R.Z. and H.M.; data curation, H.M.; writing—original draft preparation, R.K. and R.Z.; writing—review and editing, R.K., R.Z. and H.M.; visualization, H.M.; supervision, R.K.; funding acquisition, R.K. All authors have read and agreed to the published version of the manuscript.

Funding: This research was funded by the Swiss National Science Foundation, grant number P2ZHP2_175028. The APC was funded by Swiss National Science Foundation.

Institutional Review Board Statement: Not applicable.

Informed Consent Statement: Not applicable.

Data Availability Statement: The data presented in this study are available on request from the corresponding author.

Acknowledgments: We would like to thank the Centre suisse de cartographie de la faune and Christian Monnerat for providing the Swiss atlas data of odonates.

Conflicts of Interest: The authors declare no conflict of interest.

References

1. Sih, A. Predators and prey life style. In *Predation: Direct and Indirect Impacts on Aquatic Communities*; Kerfoot, W.C., Sih, A., Eds.; University Press of New England: Hanover, UK, 1987.
2. Werner, E.E.; Anholt, B.R. Ecological consequences of the trade-off between growth and mortality rates mediated by foraging activity. *Am. Nat.* **1993**, *142*, 242–272. [CrossRef] [PubMed]
3. Bell, A.; Dingemanse, N.; Hankison, S.; Langenhof, M.; Rollins, K. Early exposure to nonlethal predation risk by size-selective predators increases somatic growth and decreases size at adulthood in threespined sticklebacks. *J. Evol. Biol.* **2011**, *24*, 943–953. [CrossRef] [PubMed]
4. De Block, M.; Pauwels, K.; Van Den Broeck, M.; De Meester, L.; Stoks, R. Local genetic adaptation generates latitude-specific effects of warming on predator–prey interactions. *Glob. Chang. Biol.* **2013**, *19*, 689–696. [CrossRef] [PubMed]
5. Stoks, R.; Swillen, I.; De Block, M. Behaviour and physiology shape the growth accelerations associated with predation risk, high temperatures and southern latitudes in Ischnura damselfly larvae. *J. Anim. Ecol.* **2012**, *81*, 1034–1040. [CrossRef]
6. Richardson, J.M. A comparative study of phenotypic traits related to resource utilization in anuran communities. *Evol. Ecol.* **2002**, *16*, 101–122. [CrossRef]
7. Stoks, R.; McPeek, M.; Mitchell, J. Evolution of prey behavior in response to changes in predation regime: Damselflies in fish and dragonfly lakes. *Evolution* **2003**, *57*, 574–585. [CrossRef]
8. Conover, D.O.; Schultz, E.T. Phenotypic similarity and the evolutionary significance of countergradient variation. *Trends Ecol. Evol.* **1995**, *10*, 248–252. [CrossRef]
9. Wellborn, G.A.; Skelly, D.K.; Werner, E.E. Mechanisms creating community structure across a freshwater habitat gradient. *Annu. Rev. Ecol. Evol. Syst.* **1996**, *27*, 337–363. [CrossRef]
10. Parmesan, C.; Ryrholm, N.; Stefanescu, C.; Hill, J.K.; Thomas, C.D.; Descimon, H.; Huntley, B.; Kaila, L.; Kullberg, J.; Tammaru, T. Poleward shifts in geographical ranges of butterfly species associated with regional warming. *Nature* **1999**, *399*, 579. [CrossRef]
11. Hassall, C.; Thompson, D.J. Accounting for recorder effort in the detection of range shifts from historical data. *Methods Ecol. Evol.* **2010**, *1*, 343–350. [CrossRef]
12. Lima, S.L.; Dill, L.M. Behavioral decisions made under the risk of predation: A review and prospectus. *Can. J. Zool.* **1990**, *68*, 619–640. [CrossRef]
13. Antoł, A.; Sniegula, S. Damselfly eggs alter their development rate in the presence of an invasive alien cue but not a native predator cue. *Ecol. Evol.* **2021**, *11*, 9361–9369. [CrossRef] [PubMed]
14. Blanckenhorn, W.; Demont, M. Bergmann and converse Bergmann latitudinal clines in arthropods: Two ends of a continuum? *Integr. Comp. Biol.* **2004**, *44*, 413–424. [CrossRef]
15. Stoks, R.; Block, M.D.; McPeek, M.A. Physiological costs of compensatory growth in a damselfly. *Ecology* **2006**, *87*, 1566–1574. [CrossRef]
16. Stoks, R.; Block, M.D.; Van De Meutter, F.; Johansson, F. Predation cost of rapid growth: Behavioural coupling and physiological decoupling. *J. Anim. Ecol.* **2005**, *74*, 708–715. [CrossRef]
17. McPeek, M.A. The growth/predation risk trade-off: So what is the mechanism? *Am. Nat.* **2004**, *163*, E88–E111. [CrossRef]
18. McPeek, M.A.; Grace, M.; Richardson, J.M. Physiological and behavioral responses to predators shape the growth/predation risk trade-off in damselflies. *Ecology* **2001**, *82*, 1535–1545. [CrossRef]
19. Thaler, J.S.; McArt, S.H.; Kaplan, I. Compensatory mechanisms for ameliorating the fundamental trade-off between predator avoidance and foraging. *Proc. Natl. Acad. Sci. USA* **2012**, *109*, 12075–12080. [CrossRef]
20. Mikolajewski, D.J.; Rolff, J. Benefits of morphological defence demonstrated by direct manipulation in larval dragonflies. *Evol. Ecol. Res.* **2004**, *6*, 619–626.
21. Mikolajewski, D.J.; Johansson, F. Morphological and behavioral defenses in dragonfly larvae: Trait compensation and cospecialization. *Behav. Ecol.* **2004**, *15*, 614–620. [CrossRef]
22. Kishida, O.; Trussell, G.C.; Mougi, A.; Nishimura, K. Evolutionary ecology of inducible morphological plasticity in predator–prey interaction: Toward the practical links with population ecology. *Popul. Ecol.* **2010**, *52*, 37–46. [CrossRef]
23. Dmitriew, C.; Rowe, L. Resource limitation, predation risk and compensatory growth in a damselfly. *Oecologia* **2005**, *142*, 150–154. [CrossRef] [PubMed]
24. Angert, A.L.; Crozier, L.G.; Rissler, L.J.; Gilman, S.E.; Tewksbury, J.J.; Chunco, A.J. Do species' traits predict recent shifts at expanding range edges? *Ecol. Lett.* **2011**, *14*, 677–689. [CrossRef] [PubMed]
25. Hickling, R.; Roy, D.B.; Hill, J.K.; Thomas, C.D. A northward shift of range margins in British Odonata. *Glob. Chang. Biol.* **2005**, *11*, 502–506. [CrossRef]
26. Wildermuth, H.; Gonseth, Y.; Maibach, A. Die Libellen der Schweiz. *Fauna Helv.* **2005**, *12*, 398.
27. Khelifa, R.; Zebsa, R.; Kahalerras, A.; Mahdjoub, H. Clutch size and egg production in *Orthetrum nitidinerve* Selys, 1841 (Anisoptera: Libellulidae): Effect of body size and age. *Int. J. Odonatol.* **2012**, *15*, 51–58. [CrossRef]

28. Khelifa, R. Spatiotemporal Pattern of Phenology across Geographic Gradients in Insects. Ph.D. Thesis, University of Zurich, Zurich, Switzerland, 2017.
29. Thompson, D.J. Towards a realistic predator-prey model: The effect of temperature on the functional response and life history of larvae of the damselfly, Ischnura elegans. *J. Anim. Ecol.* **1978**, *47*, 757–767. [CrossRef]
30. Corbet, P.S. *Dragonflies: Behaviour and Ecology of Odonata*; Harley Books: Colchester, UK, 1999.
31. R Development Core Team. *R: A Language and Environment for Statistical Computing*; R Foundation for Statistical Computing: Vienna, Austria, 2019.
32. Bates, D.; Mächler, M.; Bolker, B.; Walker, S. Fitting Linear Mixed-Effects Models Using lme4. *J Stat. Softw.* **2015**, *67*, 1–48. [CrossRef]
33. Lenth, R. Least-Squares Means: The R package (lsmeans). *J Stat. Softw.* **2016**, *69*, 1–33. [CrossRef]
34. Angilletta, M.J. *Thermal Adaptation: A Theoretical and Empirical Synthesis*; Oxford University Press: Oxford, UK, 2009.
35. Suhling, F.; Suhling, I.; Richter, O. Temperature response of growth of larval dragonflies—An overview. *Int. J. Odonatol.* **2015**, *18*, 15–30. [CrossRef]
36. Van Doorslaer, W.; Stoks, R. Growth rate plasticity to temperature in two damselfly species differing in latitude: Contributions of behaviour and physiology. *Oikos* **2005**, *111*, 599–605. [CrossRef]
37. Karl, I.; Fischer, K. Why get big in the cold? Towards a solution to a life-history puzzle. *Oecologia* **2008**, *155*, 215–225. [CrossRef]
38. De Block, M.; Stoks, R. Fitness effects from egg to reproduction: Bridging the life history transition. *Ecology* **2005**, *86*, 185–197. [CrossRef]
39. Hunt, R.J.; Swift, M. Predation by larval damselflies on cladocerans. *J. Freshw. Ecol.* **2010**, *25*, 345–351. [CrossRef]
40. Raczyński, M.; Stoks, R.; Johansson, F.; Sniegula, S. Size-mediated priority effects are trait-dependent and consistent across latitudes in a damselfly. *Oikos* **2021**, *130*, 1535–1547. [CrossRef]
41. Traill, L.W.; Lim, M.L.; Sodhi, N.S.; Bradshaw, C.J. Mechanisms driving change: Altered species interactions and ecosystem function through global warming. *J. Anim. Ecol.* **2010**, *79*, 937–947. [CrossRef]
42. Yang, L.H.; Rudolf, V. Phenology, ontogeny and the effects of climate change on the timing of species interactions. *Ecol. Lett.* **2010**, *13*, 1–10. [CrossRef]
43. Altwegg, R. Predator-induced life-history plasticity under time constraints in pool frogs. *Ecology* **2002**, *83*, 2542–2551. [CrossRef]
44. Slos, S.; Stoks, R. Behavioural correlations may cause partial support for the risk allocation hypothesis in damselfly larvae. *Ethology* **2006**, *112*, 143–151. [CrossRef]
45. Wohlfahrt, B.; Mikolajewski, D.J.; Joop, G.; Vamosi, S.M. Ontogenetic changes in the association between antipredator responses and growth variables. *Ecol. Entomol.* **2007**, *32*, 567–574. [CrossRef]
46. Mikolajewski, D.J.; Brodin, T.; Johansson, F.; Joop, G. Phenotypic plasticity in gender specific life-history: Effects of food availability and predation. *Oikos* **2005**, *110*, 91–100. [CrossRef]
47. Johansson, F.; Andersson, J. Scared fish get lazy, and lazy fish get fat. *J. Anim. Ecol.* **2009**, *78*, 772–777. [CrossRef] [PubMed]
48. Benard, M.F. Predator-induced phenotypic plasticity in organisms with complex life histories. *Annu. Rev. Ecol. Evol. Syst.* **2004**, *35*, 651–673. [CrossRef]
49. Abrams, P.A.; Rowe, L. The effects of predation on the age and size of maturity of prey. *Evolution* **1996**, *50*, 1052–1061. [CrossRef] [PubMed]
50. Higginson, A.; Ruxton, G. Adaptive changes in size and age at metamorphosis can qualitatively vary with predator type and available defenses. *Ecology* **2010**, *91*, 2756–2768. [CrossRef] [PubMed]
51. Corbet, P.S.; Suhling, F.; Soendgerath, D. Voltinism of Odonata: A review. *Int. J. Odonatol.* **2006**, *9*, 1–44. [CrossRef]
52. Johansson, F.; Stoks, R.; Rowe, L.; De Block, M. Life history plasticity in a damselfly: Effects of combined time and biotic constraints. *Ecology* **2001**, *82*, 1857–1869. [CrossRef]
53. Pierce, C. Predator avoidance, microhabitat shift, and risk-sensitive foraging in larval dragonflies. *Oecologia* **1988**, *77*, 81–90. [CrossRef]
54. Sih, A.; Bell, A.; Johnson, J.C. Behavioral syndromes: An ecological and evolutionary overview. *Trends Ecol. Evol.* **2004**, *19*, 372–378. [CrossRef]
55. Dmitriew, C.M. The evolution of growth trajectories: What limits growth rate? *Biol. Rev.* **2011**, *86*, 97–116. [CrossRef]
56. Metcalfe, N.B.; Monaghan, P. Growth versus lifespan: Perspectives from evolutionary ecology. *Exp. Gerontol.* **2003**, *38*, 935–940. [CrossRef]
57. Johansson, F. Reaction norms and production costs of predator-induced morphological defences in a larval dragonfly (*Leucorrhinia dubia*: Odonata). *Can. J. Zool.* **2002**, *80*, 944–950. [CrossRef]
58. Van Buskirk, J. Phenotypic lability and the evolution of predator-induced plasticity in tadpoles. *Evolution* **2002**, *56*, 361–370. [CrossRef]
59. Belden, L.; Wildy, E.; Hatch, A.; Blaustein, A. Juvenile western toads, Bufo boreas, avoid chemical cues of snakes fed juvenile, but not larval, conspecifics. *Anim. Behav.* **2000**, *59*, 871–875. [CrossRef]
60. Kiesecker, J.M.; Chivers, D.P.; Anderson, M.; Blaustein, A.R. Effect of predator diet on life history shifts of red-legged frogs, Rana aurora. *J. Chem. Ecol.* **2002**, *28*, 1007–1015. [CrossRef]
61. Li, D.; Jackson, R.R. Influence of diet-related chemical cues from predators on the hatching of egg-carrying spiders. *J. Chem. Ecol.* **2005**, *31*, 333–342. [CrossRef]

62. Ferland-Raymond, B.; Murray, D.L. Predator diet and prey adaptive responses: Can tadpoles distinguish between predators feeding on congeneric vs. conspecific prey? *Can. J. Zool.* **2008**, *86*, 1329–1336. [CrossRef]
63. Ott, J. The expansion of *Crocothemis erythraea* (Brullé, 1832) in Germany-an indicator of climatic changes. In *Odonata: Biology of Dragonflies*; Scientific Publishers: Jodhpur, India, 2007; pp. 201–222.
64. Korkeamäki, E.; Suhonen, J. Distribution and habitat specialization of species affect local extinction in dragonfly Odonata populations. *Ecography* **2002**, *25*, 459–465. [CrossRef]
65. Trussell, G.C.; Ewanchuk, P.J.; Matassa, C.M. The fear of being eaten reduces energy transfer in a simple food chain. *Ecology* **2006**, *87*, 2979–2984. [CrossRef]

diversity

MDPI

Article

First Record of Microsporidia Infection in the Damselfly *Ischnura elegans* Larvae: Temperature and Predator Cue Effects on the Host's Life History

Andrzej Antoł [1], Anna Maria Labecka [2], J. I. Ronny Larsson [3] and Szymon Sniegula [1,*]

1 Institute of Nature Conservation, Polish Academy of Sciences, al. Adama Mickiewicza 33, 31-120 Kraków, Poland; andrzejantol@gmail.com
2 Institute of Environmental Sciences, Jagiellonian University, ul. Gronostajowa 7, 30-387 Kraków, Poland; anna.labecka@uj.edu.pl
3 Department of Biology, University of Lund, Sölvegatan 35, SE-223 62 Lund, Sweden; ronny.larsson@biol.lu.se
* Correspondence: szymon.sniegula@gmail.com or sniegula@iop.krakow.pl

Abstract: Here, we report, for the first time, a microsporidian infection in laboratory-reared larvae of the damselfly *Ischnura elegans*. Infected larvae originated from field-collected adult females, which were caught in southern Poland in August 2020 (the second half of the flight season). Higher rearing temperatures and the presence of predator cues from the invasive alien signal crayfish (*Pacifastacus leniusculus*) increased the number of infected larvae. Infected larvae had distorted wing development, and all individuals died before emergence. Hence, microsporidian infection in *I. elegans* larvae impacted damselfly morphology and life history. We propose that warming temperature and stress caused by non-consumptive effects triggered by invasive alien predators are possible factors that produce negative fitness consequences following microsporidian infection in a key amphibious ectotherm.

Keywords: microsporidia; parasitism; predator–prey interaction; invasive alien species; temperature; *Ischnura elegans*; *Pacifastacus leniusculus*

Citation: Antoł, A.; Labecka, A.M.; Larsson, J.I.R.; Sniegula, S. First Record of Microsporidia Infection in the Damselfly *Ischnura elegans* Larvae: Temperature and Predator Cue Effects on the Host's Life History. *Diversity* **2022**, *14*, 428. https://doi.org/10.3390/d14060428

Academic Editors: M. Olalla Lorenzo-Carballa and Ricardo Koroiva

Received: 27 March 2022
Accepted: 24 May 2022
Published: 27 May 2022

1. Introduction

Parasitism is one of the most common and important interspecific interactions. It occurs in the kingdoms of all living organisms and takes different forms, from occasional ectoparasitism to close obligatory endoparasitism. Parasitic infections have various outcomes for the hosts, ranging from temporary mild sickness to death [1]. Parasites can affect host physiology, immunology (e.g., by causing tissue damage or increasing metabolism) [1–3], behaviour (e.g., changes in attraction to light) [2,4] and life history (e.g., decreases in fecundity) [5]. A high prevalence of parasitic infection may change host population density or even lead to population extinction [6], especially if host individuals are weakly resistant to a particular parasite or if a parasite is highly transmissible [2,6].

Researchers have focused on parasite–host interactions and have claimed that insects—taxonomically the most numerous and diverse group of animals on Earth [7]—are probably infected by equal numbers of parasites and parasitoids [8–10]. An important group among these invertebrates are dragonflies (Odonata), which are amphibious and hemimetabolic insects that have aquatic larvae and terrestrial imago. Both stages of odonates can be infected by parasites belonging to several systematic groups, including ectoparasitic water mites (arachnids) and endoparasitic gregarines (protists) [11,12]. Water mites decrease the survival of infected adults [13]. Gregarines invade both larval and adult odonate stages [14]. Gregarine infection leads to a shorter imago lifespan [15] or lower adult fat content, which negatively affects host reproduction [16]. Nematodes (Nematoda) and plathelminths (Digenea and Cestoda) are other endoparasites that infect odonates [11].

Odonates have also been recorded as hosting endoparasitic microsporidia fungal-related protists [17–19]. In such infections, there is a scarcity of information regarding parasitic effects on odonate fitness-related traits [17,18,20].

Microsporidia, which constitute approximately 1500 named species, are unicellular parasites that reproduce through spores. As parasites, they are limited to animal hosts, including insects, crustaceans, fishes and humans [21]. Microsporidia can be host- and tissue-specific. These species are transferred vertically and horizontally and often have multiple spore types and sometimes intermediate hosts. Alternatively, microsporidia can be opportunistic (towards hosts and tissues) in situations with horizontal transmission and one host [22]. In extreme cases, parasites can take over host cells and change host metabolism and reproduction [23,24]. In odonates, infection with microsporidia occurs in the fat body, where the parasite can be found at various developmental stages. The infected odonate larvae are often whiter or paler than noninfected ones [18,19]. However, there is no information on whether and to what extent microsporidia affect odonate life history. Data from other groups of insects indicate that infections caused by microsporidia might negatively affect host fitness-related traits, e.g., fecundity [5,8,25].

Here, we present the first record of microsporidian infection in laboratory-reared damselfly *Ischnura elegans* (Odonata: Zygoptera) larvae, a model species for eco-evolutionary studies [26–29]. This endoparasite affected larval survival, larval size and emergence success in the host.

2. Materials and Methods

Some of the data from our long-term experiment on damselflies are presented, in which we reared individuals from the egg stage through the larval stage until emergence or pre-emergence death.

Adult female *I. elegans* in copula were collected from two ponds in the city of Kraków, Poland, at two time points during summer: 5 July 2020 (Staw Płaszowski, 50.042908, 19.967240, and Staw Dąbski, 50.064735, 19.987438) and 8 August 2020 (Staw Płaszowski). After collection, females were placed in plastic containers equipped with wet filter paper for egg laying. The plastic containers were placed in a Styrofoam box with ice packs to keep the temperature low. Under such conditions, females were transported by car to the Institute of Nature Conservation at the Polish Academy of Sciences (PAS), where the laboratory experiment was run. Females were kept in containers in a room with a natural photoperiod and temperature until they laid eggs.

The temperatures at which eggs and larvae were reared were then changed once a week to follow seasonal changes in the mean weekly temperatures in shallow-water habitats (optimal for damselfly larvae [11]). As in previous laboratory experiments of damselflies' life histories [30,31], the temperature was derived from lake model FLake [32]. The experiment consisted of two temperature treatments: a reference temperature that mimicked the actual temperature at the collection site and an elevated temperature treatment, where the temperature was elevated by 4 °C to mimic the predicted temperature change by the end of the 21st century [33]. Except for the overwintering conditions (see below), we followed weekly seasonal changes in daylight (photoperiod) according to civil twilight length in Kraków. In the wild, *I. elegans* overwinter in the larval stage, so that individuals experienced winter diapause. During the experiment, we simulated winter conditions. We programmed a constant temperature of 6 °C in the reference temperature treatment and 10 °C in the elevated-temperature treatment. No light was provided during the overwintering conditions. For experimental temperature and photoperiod distributions and ranges, see the Supporting Information (Supplementary Material Table S1).

Every egg clutch laid by individual females was divided into six and placed in separate plastic containers (15 cm × 11 cm × 7.5 cm) filled with 600 mL of water. The water consisted of ¾ dechlorinated tap water and ¼ dechlorinated tap water with or without predator cues originating from European perch (*Perca fluviatilis*) and signal crayfish (*P. leniusculus*). Eggs from every female were present in every treatment: the three predation cues (none, perch, or crayfish) and the two temperatures (reference or elevated). Perch and crayfish cues were not mixed; hence, damselflies from different predator treatment groups experienced these predator cues independently. At hatching, every predator treatment group during the egg stage was further divided into two new groups: a group that experienced predator cues during the egg stage and the larval stage and a control group that did not experience predator cues during the larval stage (only the predator effects from the egg stage) (Figure 1). The water was refilled every other day to keep the predator cue approximately constant, considering the length of cue biodegradation [34]. Additionally, previous experiments have demonstrated that predator cue refill every other day affects damselfly life history traits [28,29]. Hatching occurred two to three weeks after the eggs had been laid. After hatching, the larvae were transferred to other containers (19 cm × 12 cm × 9 cm) filled with 1 l of water and kept in groups of 15–20 individuals for the first 14 days. Keeping individuals in groups at this stage increases larval survival. Fourteen days after hatching, every larva was individually placed in a plastic cup (height = 9 cm, diameter = 4 cm, volume = 200 mL) filled with 100 mL of water. The water refill in every container that held 15–20 larvae and the water refill in cups holding individual larvae were analogous (¾ dechlorinated tap water and ¼ of dechlorinated tap water with or without predator cue, ¼ changed every second day) to that previously described. The larvae were fed twice a day (morning and afternoon) during weekdays and once a day during weekends with 1 mL of *Artemia* sp. Nauplii solution (mean = 198.5, SD = 92.4 nauplii/1 mL, N = 38). During the first 14 days, larvae kept in groups received 10 mL of solution. During the winter, the larvae were fed once a day. When the larvae reached the prefinal instar before emergence (F-1), they were additionally provided with one live *Chironomidae* larvae three times a week (Monday, Wednesday and Friday).

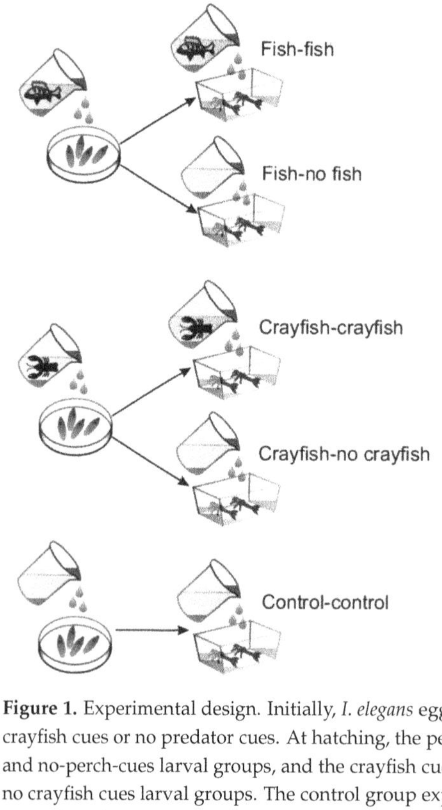

Figure 1. Experimental design. Initially, *I. elegans* eggs were assigned to groups receiving perch cues, crayfish cues or no predator cues. At hatching, the perch cues egg group was divided into perch-cues and no-perch-cues larval groups, and the crayfish cues egg group was divided into crayfish cues and no crayfish cues larval groups. The control group experienced no perch or crayfish cues in either the egg or larval stage.

When the larvae reached the prefinal and final instars (F-1 and F-0), we noticed that some individuals had white structures inside of their body. For these individuals, we recorded the length of larval development (in days) between hatching and emergence or between hatching and death. We photographed the larvae to measure larval head width and wing pad length. In odonates, head width reflects overall body size and is used together with wing pad length to identify the larval instar. Live larvae were photographed with a binocular microscope (SMZ 745T; Nikon, Tokyo, Japan) with a camera (DFK 23UP031; Nikon). Larvae were placed in a drop of water on a Petri dish and the photographer took a picture. We photographed larvae on 1 April 2021, when the vast majority of noninfected larvae had already emerged, and on 22 April 2021, to estimate their growth. The head widths and wing pads lengths were measured using ImageJ software (NIH, Bethesda, MD, USA).

2.1. Histology

Three randomly selected larvae were chosen for microscopic analyses. Damselflies were cut into two halves, and both were used to prepare fresh smears and fixed in Bouin's solution histological slides. Following fixation, the samples were dehydrated in a graded ethyl alcohol series (Linegal Chemicals, Warszawa, Poland), cleared with isopropyl alcohol (Leica, Wetzlar, Germany) and Clearene (Leica), and embedded in Paraplast Plus (Leica). Serial cross-sections (4–6 μm thick) were cut with a rotary microtome (Hyrax M55, Zeiss, Oberkochen, Germany). Histological slides were deparaffinized in Clearene (Leica), rehydrated in a graded ethyl alcohol series (Linegal Chemicals), and stained with Ehrlich haematoxylin (Carl Roth, Karlsruhe, Germany) and with a 1% ethanol solution of eosin Y (Analab, Warszawa, Poland). Then, the slides were dehydrated in 96% ethanol (Linegal Chemicals) and isopropyl alcohol (Leica), cleared in Clearene (Leica), and embedded in CV Ultra (Leica). Dried and nonstained fresh smears as well as fixed and stained tissue cross-sections were analysed under a light microscope (Eclipse 80i, Nikon) in a bright field and photographed using a digital camera (Axio Cam 305 colour, Zeiss) and image acquisition software (ZEN 3.3. blue edition, Zeiss).

2.2. Statistical Analysis

We found infections only in larvae originating from one population (Staw Płaszowski, later collection date) and only in individuals who were reared in the elevated temperature treatment. Therefore, we limited our analyses to this group.

As we noticed the infected larvae in the middle of the experiment, it was possible that some of them would die before being classified as infected. Thus, comparisons of the larval development duration of infected and noninfected individuals would be biased. Therefore, we reported only raw values for larval development duration.

All the analyses were performed in R software [35] with the following packages: emmeans [36], lme4 [37] and lmerTest [38]; ggplot2 [39] was used for graph preparation.

We analysed the data on the presence of infection among predator cue treatments using a generalized linear model with a binomial distribution and a logit link function. We plotted the distribution of larval head widths and wing pad lengths, and for comparison, we added head width and wing pad length ranges for F-1 instars based on observation of the larvae originating from females collected at the same site [40]. In independent tests, we compared the head widths and wing pad lengths between two measurement dates via a general linear mixed model. Head widths or wing pad lengths were response variables, measurement date was a fixed predictor, and individual ID was a random factor. We used Tukey HSD tests for pairwise level comparisons.

3. Results

3.1. General View

Infected larvae had white structures along the long axis of their bodies (thorax and abdomen) that were visible through the body shell. In the analysed subset, where infections were present (Staw Płaszowski, later collection date), there were 182 larvae in total, of which 33 were classified as infected, 87 as not infected that emerged with success and 62 as not infected that died before emergence. Larval infection was significantly affected by predator cue treatment ($\chi^2 = 15.12$, $p = 0.001$). Pairwise comparisons revealed significant differences between the perch (egg and larva) vs. signal crayfish (egg and larva) groups ($p = 0.04$) and the control vs. signal crayfish (egg and larva) groups ($p = 0.05$). In both cases, infected larvae were more abundant in the signal crayfish group. A higher number of infected larvae was also found in the signal crayfish (egg)–control (larva) group, but the difference was not significant (the lowest $p = 0.2$ from the pairwise comparison with the perch (egg and larva) group) (Figure 2).

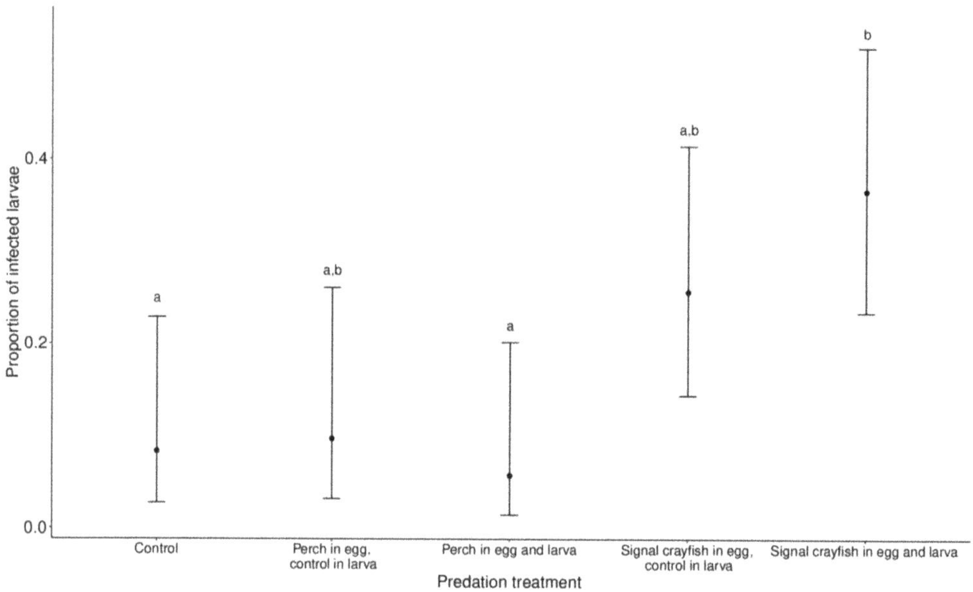

Figure 2. The proportion of microsporidian infections in *I. elegans* larvae in the different predator treatment groups. *Ischnura elegans* eggs and larvae were exposed to perch or signal crayfish cues either during egg or egg and larval stages. The control group was not exposed to predator cues during either stage. Data are displayed as the marginal means ±95% CI. Different letters at the top of the error bars indicate significant differences between experimental groups.

3.2. *Histological Description*

Histological analysis showed that larvae were infected with microsporidia (Figure 3A). These parasites filled the thorax and abdomen (Figure 3B). On the cross-sections of the body, hypertrophied cells of the fat body lobes were filled in by the microsporidia and surrounding organs (Figure 3B–F). Germ gonads with active gametogenesis were visible (Figure 3F).

3.3. *Life History*

Infected larvae did not metamorphose, and so did not emerge. The longest development duration between hatching and emergence of noninfected larvae was 231 days. The longest development duration of noninfected larvae that died before emergence was 251 days. Six infected larvae lived longer than 251 days; of these, the longest larval lifespan was 261 days.

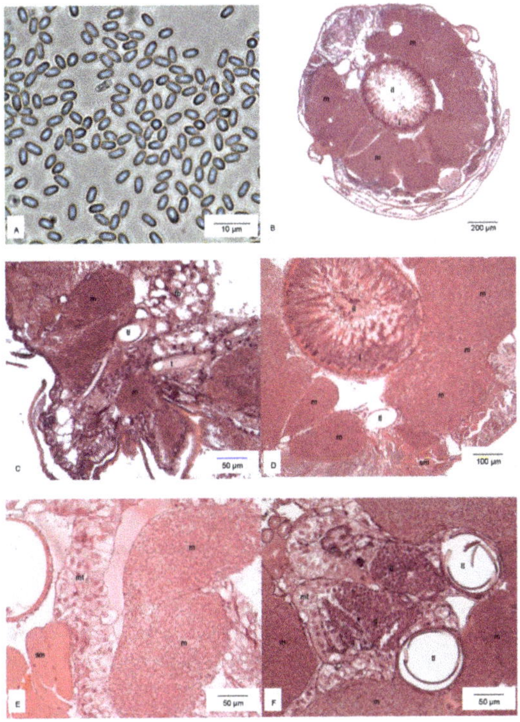

Figure 3. Microsporidia (**A**). Cross-section through the thorax (**B**). Hypertrophied cells of the fat body filled in by microsporidia and the surrounding noninfected fat body (**C**), intestine (**D**), Malpighian tubules (**E**), and germ gonads (**F**). Abbreviations: fb, noninfected fat body; g, germ gonads; i, intestine; il, intestinal lumen; m, hypertrophied cells of the fat body with microsporidia; mt, Malpighian tubules; sm, striated muscles; t, tracheal tube; tl, tracheal lumen.

In infected larvae, the range of head widths overlapped with the head widths of reference larvae, for both F-1 (first measurement date) and F-0 (second measurement date). The head widths increased significantly between the two measurement dates ($t = 4.46$, $p = 0.003$), as did the wing pad lengths ($t = 2.26$, $p = 0.04$). However, after removing two outliers from the data from the second measurement date, the wing pad length did not differ between measurement dates ($t = 1.04$, $p = 0.32$).

4. Discussion

We report, for the first time, the negative effects of endoparasitic microsporidian infection on life history in a damselfly model for eco-evolutionary research, *I. elegans*. Other insects [8,22,25,41], including odonates (suborder Anisoptera: *Aeschna viridis* [17], *Orthetrum albistylum speciosum* [20], *Aeschna grandis* and *Libellula quadrimaculata* [18,42], *Tholymis tillarga* [43], *Tramea limbata* [44]; suborder Zygoptera: *Calopteryx virgo* [45], *Coenagrion pulchellum* [46] *C. hastulatum* [19]) were reported to carry this pathogen; however, these results were based mainly on observational, not experimental, studies. Hence, previous results could not directly answer the question of whether parasites or other factors caused reductions in host fitness. The results of our experimental approach provide insights into pathogenic effects on fitness-related traits in the studied organism and how these pathogenic effects might modify damselfly population dynamics.

We could not identify the microsporidia to the species level, which would otherwise extend the list of parasites in odonates (and *I. elegans*, in particular). However, our identification at the phylum level indicates that microsporidia are likely as common in odonates as in other groups of insects [25].

None of the infected *I. elegans* larvae emerged, indicating that microsporidian infection has a lethal effect on the premature damselfly. Similar negative effects on life history traits have been reported in other insects. For example, in the mosquito *Aedes aegypti*, microsporidian infection led to a reduction in the number of eggs laid by adult females and lower offspring hatching success [5]. A similar infection effect on the host was found in *Muscidifurax raptor*, a parasitoid wasp of the house fly (*Musca domestica*) [8]. In *Drosophila melanogaster*, artificial microsporidian infection of adults led to death caused by the parasites' use of the fat storage [41]. If the pathogen is highly transmissible, it might negatively affect the host population. However, we think it is possible that damselfly larvae carrying fewer endoparasites, which could not be identified as infected 'by eye', successfully emerged. In such situations, it would be interesting to study delayed or carry-over pathogenic effects from the larval to imago stage.

The considerable increase in head width in larvae carrying microsporidia could suggest that these individuals had reached the minimal size for emergence; however, minimal wet mass for emergence is also important [40]. During the experiment, we did not weigh the larvae. Nonetheless, the largest infected larvae, estimated 'by eye', reached a body size similar to that of the noninfected individuals that eventually successfully emerged. Distorted wing development during last instars would likely prevent the damselflies from unfolding their wings and flying if they emerged. Notably, in mosquitoes, microsporidia did not affect wing development [8]. Individuals with head widths that indicated that they were F-1 or F-0 instars tended to have wing lengths that would classify them as previous instars. Moreover, hardly any wing development occurred between the two measurement dates. Over the same period, head width significantly increased. We suggest that delayed wing development led to prolonged larval development and finally larval death. Nonetheless, measuring a larger number of morphological traits would probably show other aspects of abnormal development in the infected larvae.

We found infections only in animals originating from one population and at one time, sampled later in the season. The larvae from the first collection date, which included both collection sites, were held in the same climatic chamber. This suggests that the parasites did not transfer between the cups, or that earlier-collected larvae were more resistant to infection and did not show symptoms. The infection of individuals from just one population and one sampling date indicates that, in our laboratory, microsporidia were transferred horizontally. Some microsporidia genera can also be transferred vertically [47]. The higher experimental temperatures in the elevated temperature condition probably increased pathogen development and, thus, damselfly sickness because only larvae grown at higher temperatures showed signs of infection. This might also explain the presence of parasites only in larvae whose mothers were field-sampled later in the summer, when the ambient temperature was higher than during the earlier sampling point.

According to pathogen transfer, microsporidian spores could be disseminated among damselflies through water exchange between the aquaria that housed predators and cups holding damselfly eggs and larvae. However, this explanation is less likely since infected damselfly larvae were reported, although with lower frequency, in all experimental groups.

Infections were generally more frequent in the signal crayfish cue group than in the perch and control groups. This suggests that chemicals released by the crayfish promoted the infection. Animals experiencing predator cues are more stressed [48,49]. Stress can cause changes in a number of traits, including physiological traits such as metabolic rate [48] and immune function, in potential prey [50–52]. Reduced immune function can, in turn, increase the exposure of host animals to pathogens. The results from previous studies on *I. elegans* indicated that exposing damselfly egg and larval stages to predator cues delayed egg and larval development [28,29,40], thus indicating that this damselfly is sensitive to predator-risk cues. Hence, a possible explanation for the highest microsporidian infection experienced by the crayfish cue group could be stress reactions in the host. Interestingly, larval infection was only elevated in the signal crayfish cue group, indicating that a stronger stress is imposed by alien invasive allopatric predators (the signal crayfish cue group) than native sympatric predators (the perch cue group). This discrepancy is in accordance with our previous results, where we showed prolonged egg development in groups treated with signal crayfish cues, but not perch cues, when compared to the control group [29].

5. Conclusions

Our results broaden our knowledge of the effects of microsporidian infection on life history traits in a key amphibious organism. We underline the importance of possible microsporidian infection in damselflies that are used as model organisms in eco-evolutionary studies. Overlooking this factor may impact the experimental results. Future studies should focus on the mechanism of infection and the impact of microsporidia on different host traits and conditions that increase the risk of parasitism. Such studies would shed light on the association between microsporidian infection and hosts exposed to ecologically challenging conditions.

Supplementary Materials: The following supporting information can be downloaded at: https://www.mdpi.com/article/10.3390/d14060428/s1; Table S1. Experimental temperature and photoperiod distributions and ranges.

Author Contributions: Conceptualization, A.A. and S.S.; methodology, A.A., A.M.L. and S.S.; software, A.A., A.M.L. and S.S.; validation, A.A., A.M.L., J.I.R.L. and S.S.; formal analysis, A.A. and S.S.; investigation, A.A., A.M.L. and S.S.; resources, S.S. and A.M.L.; data curation, A.A.; writing—A.A., A.M.L., J.I.R.L. and S.S.; writing—original draft preparation, A.A., A.M.L., J.I.R.L. and S.S.; writing—review and editing, A.A., A.M.L., J.I.R.L. and S.S.; visualization, A.A., A.M.L. and S.S.; supervision, A.A. and S.S.; project administration, S.S.; funding acquisition, S.S. All authors have read and agreed to the published version of the manuscript.

Funding: The research leading to these results was funded by the Norwegian Financial Mechanism 2014–2021, project no. 2019/34/H/NZ8/00683 (ECOPOND). A.A. and S.S. were supported by the National Science Centre, Poland (grant 2019/33/B/NZ8/00521) and the Institute of Nature Conservation. A.M.L. was supported by the Institute of Environmental Sciences, Jagiellonian University (N18/DBS/000003).

Institutional Review Board Statement: The animal study protocol was approved by the Local Ethical Committee (ref. 394/2020), Regional Directorate for Environmental Protection in Białystok (ref. WPN.6205.21.2020.ML) and Regional Directorate for Environmental Protection in Kraków (ref. OP-158 I.672.8.2020.MK1).

Informed Consent Statement: Not applicable.

Data Availability Statement: The data presented in this study are available in open repository https://zenodo.org (accessed on 1 May 2021).

Acknowledgments: We would like to thank E. Sieńko, M. Nowicki, and P. Laskowski for catching the crayfish and U. Norling for help in identifying the parasites we found in our culture.

Conflicts of Interest: The authors declare no conflict of interest.

References

1. Gunn, A.; Pitt, S.J. *Parasitology: An Integrated Approach*; John Willey & Sons: Hoboken, NJ, USA, 2012; ISBN 9780470684245.
2. Goater, T.M.; Goater, C.P.; Esch, G.W. *Parasitism, the Diveristy and Ecology of the Animal Parasites*; Cambridge University Press: Cambridge, UK, 2014; ISBN 9788578110796.
3. Sadd, B.M.; Siva-Jothy, M.T. Self-harm caused by an insect's innate immunity. *Proc. R. Soc. B Biol. Sci.* **2006**, *273*, 2571–2574. [CrossRef] [PubMed]
4. Bethel, W.M.; Holmes, J.C. Altered evasive behavior and responses to light in amphipods harboring acanthocephalan cystacanths. *J. Parasitol.* **1973**, *59*, 945–956. [CrossRef]
5. Becnel, J.J.; Garcia, J.J.; Johnson, M.A. Edhazardia aedis (Microspora: Culicosporidae) effects on the reproductive capacity of Aedes aegypti (Diptera: Culicidae). *J. Med. Entomol.* **1995**, *32*, 549–553. [CrossRef] [PubMed]

6. Ebert, D.; Lipsitch, M.; Mangin, K.L. The Effect of Parasites on Host Population Density and Extinction: Experimental Epidemiology with Daphnia and Six Microparasites. *Am. Nat.* **2000**, *156*, 459. [CrossRef]
7. Grimaldi, D.; Engel, M.S. *Evolution of the Insects*; Cambridge University Press: Cambridge, UK, 2005; ISBN 9780521821490.
8. Geden, C.J.; Long, S.J.; Rutz, D.A.; Becnel, J.J. Nosema disease of the parasitoid muscidifurax raptor (Hymenoptera: Pteromalidae): Prevalence, patterns of transmission, management, and impact. *Biol. Control* **1995**, *5*, 607–614. [CrossRef]
9. Fredensborg, B.L.; Fossdal Í Kálvalíð, I.; Johannesen, T.B.; Stensvold, C.R.; Nielsen, H.V.; Kapel, C.M.O. Parasites modulate the gut-microbiome in insects: A proof-of-concept study. *PLoS ONE* **2020**, *15*, e0227561. [CrossRef]
10. Humber, R.A. Fungal Pathogens and Parasites of Insects. In *Applied Microbial Systematics*; Priest, F.G., Goodfellow, M., Eds.; Springer Science+Business Media: Dordrecht, The Netherlands, 2000; pp. 203–231, ISBN 1599051494.
11. Corbet, P.S. *Dragonflies—Behaviour and Ecology of Odonata*; Harley Books: Colchester, UK, 1999.
12. Forbes, M.R.; Robb, T. Testing hypotheses about parasite-mediated selection using odonate hosts. In *Dragonflies and Damselflies: Model Organisms for Ecological and Evolutionary Research*; Córdoba-Aguilar, A., Ed.; Oxford University Press: Oxford, UK, 2008; pp. 175–188.
13. Braune, P.; Rolff, J. Parasitism and survival in a damselfly: Does host sex matter? *Proc. R. Soc. B Biol. Sci.* **2001**, *268*, 1133–1137. [CrossRef]
14. Hecker, K.R.; Forbes, M.R.; Léonard, N.J. Parasitism of damselflies (*Enallagma boreale*) by gregarines: Sex biases and relations to adult survivorship. *Can. J. Zool.* **2002**, *80*, 162–168. [CrossRef]
15. Åbro, A. Gregarines: Their Effects on Damselflies (Odonata: Zygoptera). *Entomol. Scandanavi* **1971**, *2*, 294–300. [CrossRef]
16. Siva-Jothy, M.T.; Plaistow, S.J. A fitness cost of eugregarine parasitism in a damselfly. *Ecol. Entomol.* **1999**, *24*, 465–470. [CrossRef]
17. Sokolova, Y.Y.; Kryukova, N.A.; Glupov, V.V.; Fuxa, J.R. Systenostrema alba Larsson 1988 (Microsporidia, Thelohaniidae) in the dragonfly Aeshna viridis (Odonata, Aeshnidae) from South Siberia: Morphology and molecular characterization. *J. Eukaryot. Microbiol.* **2006**, *53*, 49–57. [CrossRef] [PubMed]
18. Larsson, J.I.R. On the taxonomy of the genus Systenostrema Hazard & Oldacre, 1975 (Microspora, Thelohaniidae), with description of two new species. *Syst. Parasitol.* **1988**, *11*, 3–17. [CrossRef]
19. Larsson, J.I.R. On the Cytology and Taxonomic Position of Nudispora biformis N. G., N. Sp. (Microspora, Thelohaniidae), a Microsporidian Parasite of the Dragon Fly Coenagrion hastulatum in Sweden. *J. Protozool.* **1990**, *37*, 310–318. [CrossRef]
20. Nakamura, H.; Kurimoto, N.; Imura, Y.; Hatakeyama, Y. The First Isolation of Microsporidia from Dragonflies in Japan. *Jpn. J. Appl. Entomol. Zool.* **2021**, *65*, 29–34. [CrossRef]
21. Weiss, L.M.; Becnel, J.J. *Microsporidia Pathogens of Opportunity*; John Willey & Sons: Hoboken, NJ, USA, 2014.
22. Becnel, J.J.; White, S.E.; Shapiro, A.M. Review of microsporidia-mosquito relationships: From the simple to the complex. *Folia Parasitol.* **2005**, *52*, 41–50. [CrossRef]
23. Senderskiy, I.V.; Timofeev, S.A.; Seliverstova, E.V.; Pavlova, O.A.; Dolgikh, V.V. Secretion of Antonospora (Paranosema) locustae proteins into infected cells suggests an active role of microsporidia in the control of host programs and metabolic processes. *PLoS ONE* **2014**, *9*, 93585. [CrossRef]
24. Tamim El Jarkass, H.; Reinke, A.W. The ins and outs of host-microsporidia interactions during invasion, proliferation and exit. *Cell. Microbiol.* **2020**, *22*, e13247. [CrossRef]
25. Becnel, J.J.; Andreadis, T.G. Microsporidia in Insects. In *Microsporidia: Pathogens of Opportunity*; Weiss, L.M., Becnel, J.J., Eds.; John Willey & Sons: Hoboken, NJ, USA, 2014; pp. 521–571, ISBN 9781118395264.
26. Córdoba-Aguilar, A. *Dragonflies and Damselflies: Model Organisms for Ecological and Evolutionary Research*; Córdoba-Aguilar, A., Ed.; Oxford University Press: Oxford, UK; New York, NY, USA, 2008; ISBN 9789896540821.
27. Stoks, R.; Córdoba-Aguilar, A. Evolution ary Ecology of Odonata: A Complex Life Cycle Perspective. *Annu. Rev. Entomol.* **2012**, *57*, 249–265. [CrossRef]
28. Sniegula, S.; d'Amour Nsanzimana, J.; Johansson, F. Predation risk affects egg mortality and carry over effects in the larval stages in damselflies. *Freshw. Biol.* **2019**, *64*, 778–786. [CrossRef]
29. Antoł, A.; Sniegula, S. Damselfly eggs alter their development rate in the presence of an invasive alien cue but not a native predator cue. *Ecol. Evol.* **2021**, *11*, 9361–9369. [CrossRef]
30. Sniegula, S.; Golab, M.J.; Johansson, F. Cannibalism and activity rate in larval damselflies increase along a latitudinal gradient as a consequence of time constraints. *BMC Evol. Biol.* **2017**, *17*, 167. [CrossRef] [PubMed]
31. Johansson, F.; Watts, P.C.; Sniegula, S.; Berger, D. Natural selection mediated by seasonal time constraints increases the alignment between evolvability and developmental plasticity. *Evolution* **2021**, *75*, 464–475. [CrossRef] [PubMed]
32. Lake Model Flake FLake. Available online: http://www.flake.igb-berlin.de/ (accessed on 26 March 2022).
33. Parry, M.L.; Canziani, O.F.; Palutikof, J.P.; van der Linden, P.J.; Hanson, C.E. (Eds). *IPCC Climate Change Impacts, Adaptation and Vulnerability. Contribution of Working Group II to the Fourth Assessment Report of the Intergovernmental Panel on Climate Change*; Cambridge University Press: Cambridge, UK, 2007.
34. Van Buskirk, J.; Krügel, A.; Kunz, J.; Miss, F.; Stamm, A. The Rate of Degradation of Chemical Cues Indicating Predation Risk: An Experiment and Review. *Ethology* **2014**, *120*, 942–949. [CrossRef]
35. R Core Team. *R: A Language and Environment for Statistical Computing*; R Foundation for Statistical Computing: Vienna, Austria, 2021.
36. Lenth, R.V. Emmeans: Estimated Marginal Means, aka Least-Squares Means. *R Package Version* **2021**, *163*, 1–91.

37. Bates, D.; Mächler, M.; Bolker, B.M.; Walker, S.C. Fitting Linear Mixed-Effects Models Using lme4. *J. Stat. Softw.* **2015**, *67*, 1–48. [CrossRef]
38. Kuznetsova, A.; Brockhoff, P.B.; Christensen, R.H.B. General rights lmerTest Package: Tests in Linear Mixed Effects Models. *J. Stat. Softw.* **2017**, *82*, 1–26. [CrossRef]
39. Wickham, H. *Ggplot2: Elegant Graphics for Data Analysis*, 2nd ed.; Springer: New York, NY, USA, 2016.
40. Sniegula, S.; Raczyński, M.; Golab, M.J.; Johansson, F. Effects of predator cues carry over from egg and larval stage to adult life-history traits in a damselfly. *Freshw. Sci.* **2020**, *39*, 804–811. [CrossRef]
41. Franchet, A.; Niehus, S.; Caravello, G.; Ferrandon, D. Phosphatidic acid as a limiting host metabolite for the proliferation of the microsporidium Tubulinosema ratisbonensis in Drosophila flies. *Nat. Microbiol.* **2019**, *4*, 645–655. [CrossRef]
42. Larsson, J. Ultrastructural Investigation of 2 Microsporidia with Rod-Shaped Spores, with Descriptions of Cylindrospora-Fasciculata Sp-Nov and Resiomeria-Odonatae Gen Et Sp-Nov (microspora, Thelohaniidae). *Protistologica* **1986**, *22*, 379–398.
43. Kalavati, C.; Narasimhamurti, C.C. New microsporidian parasite, *Toxoglugea tillargi* Sp-N from an odonate, *Tholymis tillarga*. *Acta Protozool.* **1978**, *17*, 279–283.
44. Narasimhamurti, C.C.; Ahamed, S.N.; Kalavati, C. Two new species of microsporidia from the larvae ofTramea limbata (Odonata: Insecta). *Proc. Anim. Sci.* **1980**, *89*, 531–535. [CrossRef]
45. Sprague, V. Classification and Phylogeny of the Microsporidia. In *Comparative Pathobiology: Volume 2 Systematics of the Microsporidia*; Bulla, L.A., Cheng, T.C., Eds.; Springer US: Boston, MA, USA, 1977; pp. 1–30, ISBN 978-1-4613-4205-2.
46. Larsson, J. A revisionary study of the taxon Tuzetia Maurand, Fize, Fenqick and Michel, 1971, and related forms (Microspora, Tuzetiidae). *Protistologica* **1983**, *19*, 323–355.
47. Cali, A.; Takvorian, P.M. Developmental Morphology and Life Cycles of the Microsporidia. In *Microsporidia: Pathogens of Opportunity*; Weiss, L.M., Becnel, J.J., Eds.; John Willey & Sons: Hoboken, NJ, USA, 2014; pp. 71–135.
48. Antoł, A.; Kierat, J.; Czarnoleski, M. Sedentary prey facing an acute predation risk: Testing the hypothesis of inducible metabolite emission suppression in zebra mussels, Dreissena polymorpha. *Hydrobiologia* **2018**, *810*, 109–117. [CrossRef]
49. Jermacz, Ł.; Dzierżyńska-Białończyk, A.; Kobak, J. Predator diet, origin or both? Factors determining responses of omnivorous amphipods to predation cues. *Hydrobiologia* **2017**, *785*, 173–184. [CrossRef]
50. Mikolajewski, D.J.; Stoks, R.; Rolff, J.; Joop, G. Predators and cannibals modulate sex-specific plasticity in life-history and immune traits. *Funct. Ecol.* **2008**, *22*, 114–120. [CrossRef]
51. Adamo, S.A.; Easy, R.H.; Kovalko, I.; MacDonald, J.; McKeen, A.; Swanburg, T.; Turnbull, K.F.; Reeve, C. Predator exposure-induced immunosuppression: Trade-off, immune redistribution or immune reconfiguration? *J. Exp. Biol.* **2017**, *220*, 868–875. [CrossRef] [PubMed]
52. Raczyński, M.; Stoks, R.; Johansson, F.; Sniegula, S. Size-mediated priority effects are trait-dependent and consistent across latitudes in a damselfly. *Oikos* **2021**, *130*, 1535–1547. [CrossRef]

diversity

MDPI

Article

Evaluating Potential Distribution and Niche Divergence among Populations of the World's Largest Living Damselfly, *Megaloprepus caerulatus* (Drury, 1782)

Alondra Encarnación-Luévano, Jaime Antonio Escoto-Moreno and Giovanna Villalobos-Jiménez *

Colección Zoológica, Departamento de Biología, Centro de Ciencias Básicas, Universidad Autónoma de Aguascalientes, Av. Universidad #940, Ciudad Universitaria, Aguascalientes C.P. 20131, Ags., Mexico; alondra.encarnacion@gmail.com (A.E.-L.); jerjaem2002@yahoo.es (J.A.E.-M.)

* Correspondence: villalobos.giovanna@gmail.com

Abstract: *Megaloprepus caerulatus* is a Neotropical species with a highly specialised niche, found from Mexico to Bolivia, primarily in mature tropical forests lower than 1500 masl. It is also the damselfly with the largest wingspan in the world. Recent studies found strong genetic isolation among populations of *M. caerulatus*. Further studies found genetic and morphological divergence, but ecological divergence was not tested. Here, we test for ecological divergence by evaluating niche differences among populations of *M. caerulatus* in Los Tuxtlas (Mexico), Corcovado (Costa Rica), Barro Colorado (Panama), and La Selva (Costa Rica). We used Ecological Niche Modelling (ENM) to compare potential distribution ranges, and we estimated the breadth and overlap of the ecological niche using equivalence and similarity tests. The potential distributions estimated with ENM were heavily fragmented and we found no geographic overlap of potential distributions among populations. However, we found geographic correspondence between populations with a close phylogenetic relationship. Even though all similarity tests were non-significant, the results of the equivalence tests suggest niche divergence between Corcovado and the other three populations, but also between Barro Colorado (Panama) and La Selva. These results show evidence of strong ecological divergence in Corcovado and Barro Colorado populations.

Keywords: divergence; ecological niche modelling; niche equivalence; niche similarity; Odonata; helicopter damselfly

Citation: Encarnación-Luévano, A.; Escoto-Moreno, J.A.; Villalobos-Jiménez, G. Evaluating Potential Distribution and Niche Divergence among Populations of the World's Largest Living Damselfly, *Megaloprepus caerulatus* (Drury, 1782). *Diversity* **2022**, *14*, 84. https://doi.org/10.3390/d14020084

Academic Editors: M. Olalla Lorenzo-Carballa and Ricardo Koroiva

Received: 31 December 2021
Accepted: 23 January 2022
Published: 26 January 2022

1. Introduction

The niche is integral in ecology for the study of species and as tools for conservation. Many contributions have been made on niche conceptualization, all of which relate to the basic taxonomic unit, the species. Since Grinnell [1], niche theory has been constantly developed and multiple studies at fine (i.e., physiological) and coarse (geographical) scales have been explored. Although direct evaluation of physiological constraints would be ideal to niche reconstruction, specific data are unavailable for many species in all taxonomic groups. Thus, Ecological Niche Modelling (ENM) can be used as a proxy. Tolerance limits reflect the fundamental ecological niche of a species and can be studied via their coarse-resolution associations with environments manifested across their geographic distributions [1–4]. Under certain assumptions and limitations, the correlational ENM approaches can estimate limits that approach the fundamental niche [5].

Within the fundamental niche, the existing niche is the one that corresponds to environments represented within the species' distributional area [2,6]. This existing niche can be estimated using correlative modelling approaches [5]. The ENM has been a widely used approach for estimating niches by searching for associations between species localities and their corresponding environmental conditions [5,7–14]. Estimating abiotically suitable, potential, and occupied distribution areas has been the oldest and most widely used application for the ENM. Moreover, when expressing the existing fundamental niche geographically [1], the constancy or change of the space in which the niche is manifested through time becomes apparent [6,15].

Environmental changes in the area that has been historically accessible to a species can take place over time spans of a few thousand years or even much shorter, over centuries or even decades [16]. Therefore, to avoid extinction, they must either geographically track the existing fundamental niche, or be able to change it via evolutionary responses in physiological or behavioural traits [17]. Currently, several studies have matched evolutionary information with the adaptation to climate and other aspects of environment to improve the understanding of species' niche evolution e.g., [18–24].

One way to search for niche evolution is testing whether niche characteristics are constant or divergent among genetic relatedness throughout the historically accessible area. Exploring the degree of overlap between them, environmental niche similarity, and its geographic correspondence as a function of environmental suitability has been the focus of recent studies [25]. Currently, there is an increasing interest in addressing this issue below the species level [10,12,23,25–29]. The existence of subspecies, ecotypes, and locally adapted populations suggests there is genetically-based geographic variation in physiological traits, which conveys adaptation to climate and other environmental aspects. In species with a highly specialised niche, geographic events that disrupt large primary habitats lead to isolated populations which can either become extinct or divergent [30]. Hence, populations or any significant evolutionary unit could be modelled to identify their environmental affinities [22,25].

The helicopter damselfly *Megaloprepus caerulatus* (Odonata: Coenagrionidae), the world's largest living zygopteran, shows a highly specialised ecological niche throughout its geographic range [31]. It is a Neotropical forest specialist distributed from Mexico to Bolivia inhabiting only mature, moist forests with a closed canopy up to 1500 m elevation [32–35]. Moreover, this species represents one of the few odonates that reproduce in water-filled tree holes [36–38]. This is an important limiting resource for *Megaloprepus*, since only a few tree species are able to form the water-filled holes needed for reproduction [39]. Compared to other helicopter damselflies, *M. caerulatus* is more common in primary forests than secondary forests, hence it is extremely susceptible to forest conversion [34]. One of the factors that affect the behaviour of *M. caerulatus* is the morphotype. This species can have a sexually monomorphic or dimorphic wing band [31]. In monomorphic populations, both males and females present an iridescent dark blue band, followed by a milky white spot near the wing tip, although females show a much more conspicuous white tip than males [31]. Dimorphic males have a distinct white UV reflective band proximal to the blue band [40,41]. The blue band shared by males and females facilitates conspecific detection under direct sunlight of treefall gaps, the white tip allows males to recognise females and initiates a sexual response, and the white band in dimorphic males identifies them as rivals, which elicits an aggressive response from other males [40]. Stronger UV reflectance from the white band in dimorphic males increases the likelihood of winning a territorial contest [41].

De Selys Longchamps [42] described three subspecies based on wing characteristics: (1) *M. c. caerulatus*, distributed from Central America to Colombia, Guyana, Ecuador, and Bolivia; (2) *M. c. brevistigma*, found in Colombia east of the Andes, Venezuela, and Peru, and (3) *M. c. latipennis*, from Mexico and Guatemala. *M. c. caerulatus* shows a sexually dimorphic wing pattern, whereas the others are monomorphic. However, recent research on the population structure and genetic diversity of *M. caerulatus* has shown a limited gene flow between morphotypes of the species [31,35]. Feindt et al. [35] suggested the existence of three distinct genetic clusters considering four study populations: Barro Colorado Island (BCI) from Panamá, La Selva Biological Research Station (SELVA) from Costa Rica (both *M. c. caerulatus*, dimorphic), Corcovado National Park (CNP-SIR) from Costa Rica (*M. c.* subsp. nov., monomorphic), and Los Tuxtlas Biosphere Reserve (TUX) from Mexico (*M. c. latipennis*, monomorphic). These clusters were genetically as different from each other as other odonate species, suggesting these subspecies in fact are more likely different species. Subsequently, Fincke et al. [31] identified genetic and morphological clades among the aforementioned study populations, in addition to the following populations: La Bartola Reserve (BART) from Nicaragua, Canandé Reserve (CAN) in Ecuador (both dimorphic), El Jaguar (EJ) in Nicaragua, and Cusuco National Park (HON) in Honduras (both monomorphic). Monomorphic populations showed lower adult density, lower resource defence, fewer male-male interactions and, consequently, lower sexual selection on males, suggesting sexual selection is a diverging mechanism between monomorphic and dimorphic populations [31].

However, ecological divergence among populations or subspecies of *M. caerulatus* has not been tested. In this study we evaluate whether niche characteristics of *Megaloprepus* remain constant or instead are divergent. We analyse how intraspecific genetic-level ENMs of *Megaloprepus* might differ in terms of (1) potential distribution ranges, and (2) the breadth and overlap of the ecological niche between populations or subspecies.

2. Materials and Methods

2.1. Biological and Environmental Data

In order to have a complete database of the historical records of *M. caerulatus*, an exhaustive search was carried out from literature [31–33,35,43–49] and from the GBIF database [50], which is an online portal providing access to primary biodiversity data. Even though only "research grade" records were selected, all GBIF records were verified whenever possible through photographs of the specimens to make sure they were correctly identified to species and morphotype level. Records lacking GPS coordinates but with detailed locations were georeferenced with Google Earth Pro. The resulting database had 139 records (80 from GBIF, 59 from taxonomic records) ranging from Mexico to Ecuador (Figure 1a).

Our study groups were chosen from the eight genetically and morphologically distinct populations from the subspecies *M. c. caerulatus*, *M. c. latipennis*, and *M. c.* subsp. nov. suggested by Feindt et al. [35] and Fincke et al. [31]. For this purpose, we analysed the 139 records in our database to find geographic (based on terrestrial ecoregions; [51]) and morphological (i.e., dimorphic and monomorphic) correspondence with the populations of the former studies (Figure 1a). We prioritised those populations with a strong reference of genetic and morphological difference and with an adequate number of records to complete the ENM. Therefore, our study groups are four, the northernmost in Mexico and the southernmost in Panama (Figure 1b). There are two monomorphic groups, Los Tuxtlas (16 data points; Figure 1c) and Corcovado (15 data points; Figure 1d), as well as two dimorphic groups, La Selva (44 data points; Figure 1d), and Barro Colorado (9 data points; Figure 1d). For simplicity, we refer to our study groups as "populations", although the monomorphic groups Los Tuxtlas and Corcovado are different subspecies (*M. c. latipennis* and *M. c.* subsp. nov., respectively), while dimorphic groups La Selva and Barro Colorado are populations of the same subspecies, *M. c. caerulatus*.

We used spatial environmental data relevant to the biology of the species to produce the dataset that would represent the ecological niches [52–55]. Climatic data were obtained from WorldClim Version 1.4 [56] (http://www.worldclim.org/ (accessed on 19 May 2018)). Land use was obtained from EarthEnv [57] (http://www.earthenv.org/ (accessed on 3 October 2019)) and soil evapotranspiration from the Global High-Resolution Soil-Water Balance dataset [58]. Topographic data were obtained from GMTED2010 [59] (https://on.doi.gov/30VfqfR (accessed on 10 October 2019)) and HydroSHEDS [60] (https://hydrosheds.org/ (accessed on 24 November 2019)). All spatial data were downloaded at a spatial resolution of 30 s (~1 km^2). To avoid duplication of environmental information, a correlation analysis was performed. The environmental values associated with the occurrence data (one per pixel) were used to perform the Spearman non-parametric correlation test with the "Hmisc" package [61] and a principal component analysis in R [62]. The variables that were correlated with a greater number of other variables were eliminated.

We also conducted an exploratory Jackknife test with the niche modelling algorithm of Maximum Entropy (Maxent, v3.3.3.e; [63]) in order to identify the variables that contribute the least to the construction of the model, and thus eliminate them prior to modelling the definitive ecological niche model and the similarity analyses. This resulted in the selection of 15 environmental variables: mean diurnal range, isothermality, temperature seasonality, temperature annual range, annual precipitation, precipitation of wettest month, precipitation of driest month, precipitation seasonality, precipitation of wettest quarter, precipitation of driest quarter, precipitation of warmest quarter, evergreen broadleaf tree cover, mean annual evapotranspiration, drainage direction, and flow accumulation. In fact, many of these variables have also been of importance in other odonate ENMs [11,14].

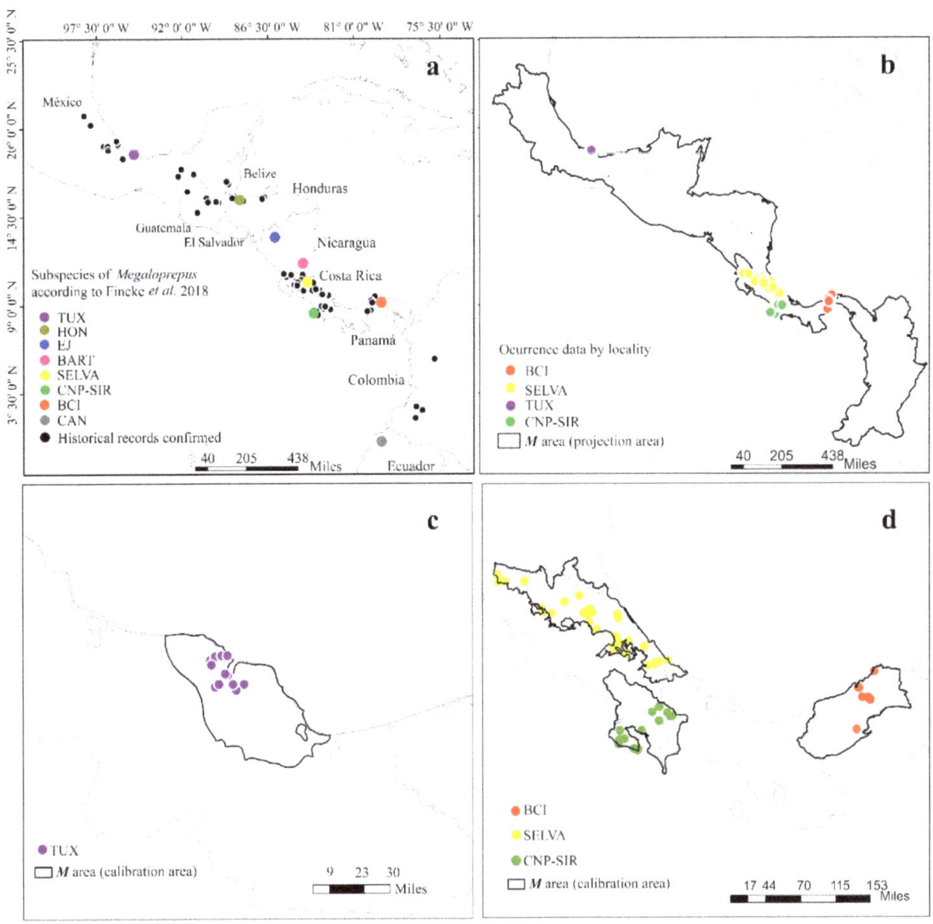

Figure 1. Geographic location of *Megaloprepus* records. (**a**) Populations that were studied in Feindt et al. [35] and Fincke et al. [31] as well as all historical records for the species (black circles). This study focused on Los Tuxtlas (TUX), La Selva (SELVA), Corcovado (CNP-SIR), and Barro Colorado (BCI); (**b**) Occurrence data used for niche modelling and similarity tests for TUX (purple), SELVA (yellow), CNP-SIR (green), and BCI (orange); their ecological niches were projected to a greater geographic extension (black solid line polygon) which encompasses all known records for the species; (**c**,**d**) Occurrence data and models for each population, which were calibrated in M regions delimited by biological and geographic conditions, also shown as black solid line polygons.

2.2. Ecological Niche Modelling

In ENM, three major factors are considered to explain distributions of species: biotic (*B*), abiotic (*A*), and mobility (*M*) constraints (***BAM***; [2]). Biotic factors are denoted by ***B***; however, at coarse resolutions, the biotic component is frequently diffuse and non-limiting, in contrast to how it is manifested at finer spatial resolutions [64]. On the other hand, the remaining two components have broad-scale effects. The abiotic factors, called ***A***, represent the geographic region presenting favourable conditions, and, finally, ***M*** is the area that has historically been accessible to the species via dispersal over relevant periods [5]. Although this approach is simplified based on static approximations to the three classes of factors [65], it has proven to be a useful heuristic test. Since our hypothesis was tested on geographic extents, a coarse resolution, the component ***B*** was not considered. Evaluations of niche similarity were made in terms of whether two niches are more similar than expected given the set of environments accessible to each population across their ***M*** [66,67]. We tested the niche similarity under environmental space [68] to compare the niches of *Megaloprepus* populations and test for divergence.

For niche model calibration, we applied the Maximum Entropy algorithm implemented in Maxent [63], which fits a distribution of probabilities across the study area subject to the constraints of the environmental characteristics of known occurrences. Evaluation data were separated a priori, therefore no data were assigned for evaluation within Maxent. Other settings included: regularization value = 1, maximum number of points for the background = 10,000, maximum iterations = 500 and convergence threshold = 0.00001. We turned off the clamping and extrapolation options, following Owens et al. [69] to avoid artificial extrapolations of extreme values of environmental variables. To convert the output into a binary map, we used the maximum training sensitivity plus specificity threshold, because it has been shown to be an optimal method for presence-only data [70].

Only in the case of the potential distribution of SELVA, occurrences were divided into random subsets: 80% for model calibration and 20% for model evaluation. Conversely, for the rest of the *Megaloprepus* populations, fewer than 25 occurrence points were available, therefore models were calibrated with all data. Models were calibrated across regions posited as historically accessible to each population (Figure 1b). ***M*** regions were delimited considering aspects of the distribution and life history of the population [65], using the limits of the surrounding ecoregions [51] and watershed boundaries and sub-basin delineation [71] as a guide (Figure 1c,d). Once the models were calibrated, they were transferred to a broader region, including the union of the ***M*** hypotheses across all of the *Megaloprepus* study populations (Figure 1b).

The model evaluation was performed differently according to the sample sizes of populations. The potential distribution of SELVA had a large sample size ($N = 44$), therefore the model was evaluated using a modification of the area under the curve of the receiver operating characteristic (partial ROC AUC ratios; [72]) using the graphical interphase Niche Toolbox [73]. This test only evaluates over the spectrum of the prediction and allows for differential weighting of the two error components (omission and commission; [72]). Thus, AUCs were limited to the proportional areas over which the models truly made predictions, and we only considered models that presented omission errors of less than 5% [72]. In the case of BCI, CNP-SIR and TUX, evaluation was accomplished via the Jackknife strategy developed for small sample-sizes by Pearson et al. [74]: significance was evaluated over n models, each excluding one locality from among the n available and evaluating the success of the model in terms of anticipating the excluded locality. The probability of these observed levels of success and failure was calculated using scripts provided by Pearson et al. [74]. This test was applied to binary models created by applying Minimum Training Presence (MTP) approaches [74].

2.3. Niche Similarity Tests

The niche similarity tests were performed in an environmental space as proposed by Broennimann et al. [68]. This method uses Schoener's D metric [75] as a measure of environmental niche overlap and includes a statistical framework to test for niche similarity as proposed by Warren et al. [66]. The value of D ranges between 0, when two populations have no overlap in the environmental space, and 1 when two populations share the same environmental space. With this method, a multivariate environment grid is created using the first two axes of a PCA that summarise all the environmental variables previously selected (PCA-env). A Gaussian kernel density is applied to estimate the occupancy of each cell (z_{ij}), and the D metric is calculated based on the different z_{ij} values obtained [68]. We tested for niche equivalence to assess whether the ecological niches of a pair of *Megaloprepus* populations were significantly different from each other and if the two niche spaces were interchangeable. This test only assesses if the two populations are identical in their niche space by using the environmental data of their exact locations and does not consider the surrounding ***M*** space, unlike the similarity test [76]. Equivalency test compares the niche overlap values (D) of a pair of populations to a null distribution of 100 overlap values. On the other hand, niche similarity test assesses if the ecological niches of any pair of populations are less similar than expected by chance, accounting for the differences in the surrounding environmental conditions in the ***M*** regions, which are the geographic areas where the populations are distributed [66]. We determined that ecological niches in comparison were less equivalent and/or less similar if the niche overlap value of the populations being compared was significantly lower than the overlap values form the null distribution ($p \leq 0.05$). This analysis was performed using the Ecospat package [77] in R. Since our goal was to test for niche divergence, we selected alternative = "lower". Thus, we present six cross-comparisons between *Megaloprepus* populations.

3. Results

The niche model performance of all populations was high and better than expected by chance. In the case of SELVA, the Partial ROC value was 1.606 ($p = 0$) and in the case of BCI, CNP-SIR and TUX the jackknife test (used for populations with fewer occurrence data) showed high success rates (0.77, 0.87, 0.94, respectively), as well as statistical significance ($p = 0$).

The environmental variables used contributed differently to the ENM of each population. For example, in the case of CNP-SIR, the variables that contributed the most were precipitation of the wettest month (45%) and mean diurnal range (10.3%), whereas precipitation seasonality contributed with 8.6%, mean annual evapotranspiration with 8.2%, isothermality with 6.3% and temperature annual range with 5.7%. For BCI, mean diurnal range also contributed significantly to the model with 57%, as well as evergreen broadleaf tree cover with 26.5%, precipitation of the driest quarter with 6.3% and flow accumulation with 5.1%. For SELVA, the two major contributors to the model were mean annual evapotranspiration (30.6%) and precipitation of the driest quarter (20.6%), although other variables that contributed significantly were mean diurnal range (15.5%), temperature annual range (12.2%) and flow accumulation (9.2). For TUX, mean diurnal range was the variable with the most significant contribution of 58.1%, others were mean annual evapotranspiration (11.7%) and precipitation seasonality (9.6%). Other variables had a contribution of <5% in each model.

The niche projection resulted in the potential distribution of the four populations studied. We found that the potential distributions of the four *Megaloprepus* populations do not overlap. Additionally, all potential distributions are fragmented and, in most cases, discontinuous (Figure 2). The potential distribution of TUX is restricted to Sierra Los Tuxtlas in Veracruz at the coast of the Gulf of Mexico (Figure 2a). Interestingly, the model does not predict suitable conditions towards other regions to the north or south where nearby historical records for *Megaloprepus* are located (see biological and environmental data in methods; Figure 1a). The potential distribution of SELVA is mainly located in Eastern Costa Rica and North Panama (next to the Atlantic Coast), but partly extends to Colombia, relatively close to the Canandé population of Fincke et al. [31], where other historical records in Colombia are found (Figure 2b). The potential distribution of CNP-SIR is predominantly found in Western Costa Rica, except for a few small, patchy areas towards the Pacific coast of Guatemala and the Southern Mexican border (Figure 2c). The potential distribution of BCI is predominantly found in Panama but extends discontinuously towards the east of Nicaragua along the Atlantic Coast (Figure 2d). This potential distribution corresponds with La Bartola population studied by Fincke et al. [31], but not with La Selva, which is located between the two (Figure 2d).

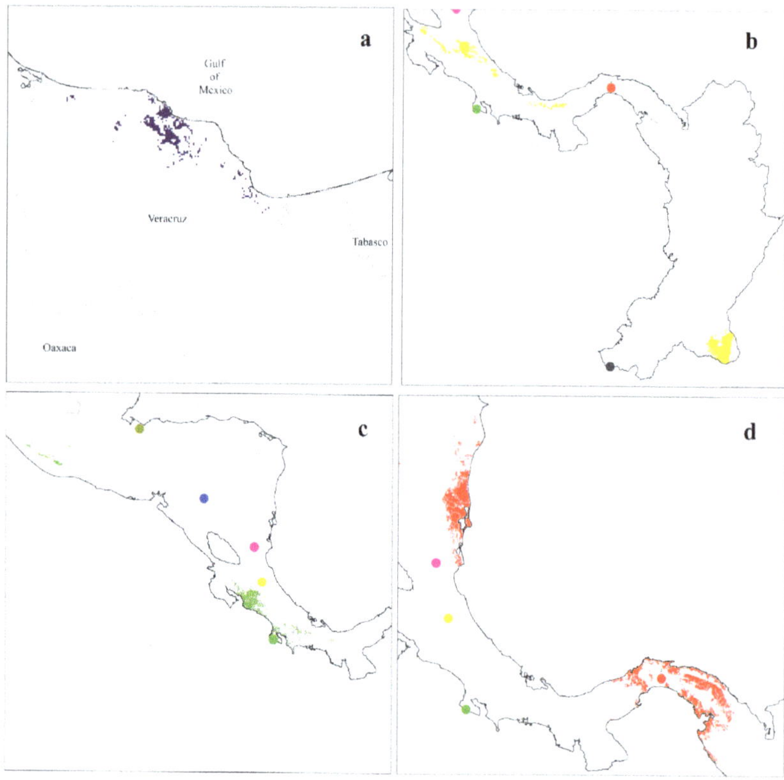

Figure 2. Potential distributions of *Megaloprepus* populations studied. Populations studied by Fincke et al. [31] are shown in the same symbology as Figure 1a for reference. (**a**) TUX potential distribution (purple); (**b**) SELVA (yellow); (**c**) CNP-SIR (green); (**d**) BCI (orange).

In order to compare the niches at the environmental space, we performed pairwise tests for equivalency and similarity, therefore there are six comparisons for the four study populations. What we found can be summarised in three points. First, in all cases the PCA-env plots show low overlap between ***M*** environments ($D = 0$, except in Figure 3c) (Figure 3, left graphs). Second, the equivalence tests show evidence of niche divergence in four of the comparisons: TUX and CNP-SIR, SELVA and CNP-SIR, SELVA and BCI and between CNP-SIR and BCI ($p \leq 0.05$; Figure 3a,d–f). It is important to keep in mind that this evidence of divergence occurs when comparing the niche with environmental data of exact locations and without considering the surrounding space, contrary to the analysis of similarity. Third, no evidence of niche divergence was found in the similarity tests, since all pairwise comparisons were non-significant ($p \geq 0.05$) (Figure 3a–f).

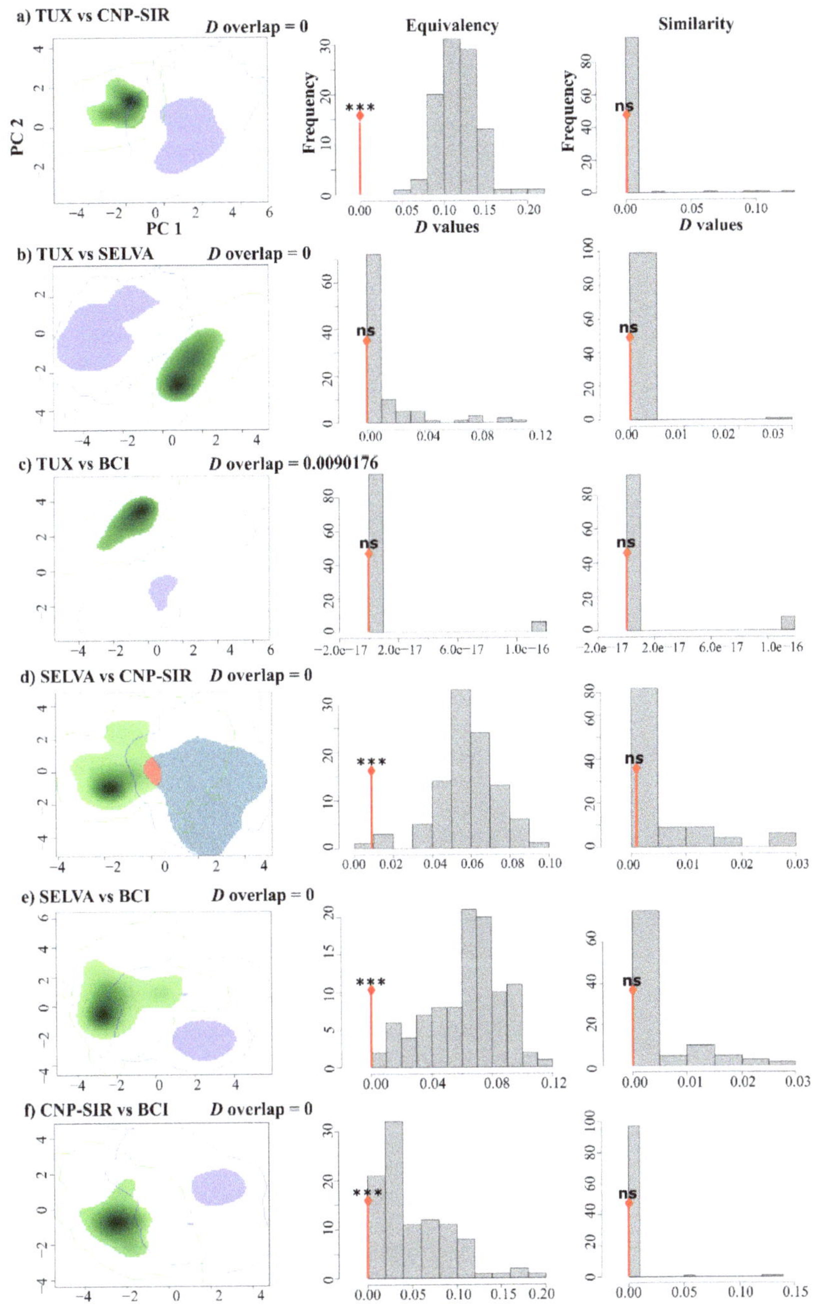

Figure 3. Similarity test analysis for *Megaloprepus* populations. We present the six pairwise comparisons between populations studied, (**a**) TUX vs. CNP-SIR; (**b**) TUX vs. SELVA; (**c**) TUX vs. BCI; (**d**) SELVA vs. CNP-SIR; (**e**) SELVA vs. BCI; (**f**) CNP-SIR vs. BCI. The PCA-env plots (left) are shown according to the similarity test of Broennimann et al. [68]. Shaded areas in each plot show the density of the occurrences of the populations by cell. The solid contour lines illustrate 100% of the available

(background) environment. The green colour represents the niche of the first population and the blue colour the niche of the second population. The red shaded areas are the niche intersections among kernel densities of occurrences. The histograms correspond to the results of equivalency (centre) and similarity tests (right) to test for niche divergence. Histograms show the observed niche overlap D between the two ranges (red lines with a diamond) and simulated niche overlaps (grey bars) on which tests of niche divergence are calculated from 100 iterations. Test significance is shown (ns, non-significant; *** $p < 0.001$).

4. Discussion

This study shows that the potential distributions of the *Megaloprepus* populations studied in this work are formed by highly fragmented areas of suitable habitat ranging from South Mexico to South Colombia. According to Fincke et al. [31] no *Megaloprepus* subspecies are known to occur sympatrically. Our results support this fact, since the potential distributions do not present any geographic overlap (Figure 2a–d). However, we found geographic correspondence between populations with a close phylogenetic relationship [31,35] despite being rather separated populations (e.g., La Selva and Canandé; Barro Colorado and La Bartola).

Our results suggest that the potential distribution of Los Tuxtlas (*M. c. latipennis*), which is monomorphic, is heavily isolated. There is no potential distribution nearby areas where historical records have been found, such as localities in the state of Chiapas, Mexico [48] (Figures 1a and 2a). This may be explained by the biogeographic history of the region to which the subspecies is restricted, the Sierra Los Tuxtlas. This is an isolated mountainous area on the Gulf of Mexico that was originated due to intense volcanic activity in the Miocene [78]. It is also the northernmost relict of moist tropical forest in the continent [79], in which vertical colonization [78], divergent evolutionary processes [80], and high rates of endemism have been documented [45].

The CNP-SIR potential distribution from Costa Rica (*M. c.* subsp. nov.), also monomorphic, is predominantly found on the Pacific coast of Costa Rica, but despite a large distance, it also reaches small areas in the north of Guatemala and the state of Chiapas (Mexico) next to the Pacific (Figure 2c). This northern range of suitable conditions for Corcovado (Figure 2c) almost coincides with several records of *Megaloprepus* found in Guatemala (Figure 1a). This can be partially explained because the Sierra Madre of Chiapas and the Central Chiapas Massif constitute the northern projection of the Central American mountain system [79]. These elevations yield a specific floristic composition ranging from Tropical-Subtropical moist forest to Tropical-Subtropical coniferous forest [81].

We also found interesting results for dimorphic populations (*M. c. caerulatus*). Despite the geographic proximity between the populations of La Selva and Barro Colorado, their potential distributions do not overlap (Figure 2b,d). All the records used in the niche modelling for both populations are located in the East Central America biogeographic province sensu [82]. The disruption of the potential distribution of La Selva may have originated from biogeographic barriers formed during the Pleistocene, which is a long period wherein landscapes had a dynamic history of dramatic changes, due to geological and climatic processes that impacted a wide variety of taxa through fragmentation and displacement of populations [83]. The potential distribution of the La Selva population shows the Napo province [84] in South America as an ideal habitat if individuals could have access (Figure 2b), which corresponds with the Napo Pleistocene Refuge proposed by Haffer [85]. However, it is important to mention that there is a debate regarding the validity of the Refugial Hypothesis in South America [86]. Therefore, further research is needed to unveil the underlying mechanisms driving the potential distribution of La Selva.

Additionally, we found significant ecological divergence between La Selva and Barro Colorado, despite being very similar populations, behaviourally and morphologically [31]. Perhaps the main difference between La Selva and Barro Colorado is their environmental conditions, particularly rainfall and seasonality. La Selva is a wet tropical forest where the larval habitat of the species seems not to dry up, whereas Barro Colorado is a tropical moist forest where larval habitats dry up for 2–3 months each year [31]. Adults from Barro Colorado population could become inactive during the dry season, which would imply an ecological adaptation via seasonal changes in behaviour. Behavioural traits have been shown to affect niche similarity or divergence in other species at inter and intraspecific level [24,29]. Even though many of the environmental variables that contributed significantly to the niche models are related to rainfall and seasonality, we cannot conclude from our data that these factors play a role in the ecological divergence found, because the PCA-env used in similarity and equivalence tests summarises all environmental variables in a multivariate environment grid using two PCs. This makes it difficult to ascertain which factors are driving ecological divergence. Alternatively, the ecological divergence between La Selva and Barro Colorado could be due to possible physiological differences in the larvae or adults, which would need to be verified in future studies.

We must mention that this study largely focuses on the abiotic factors driving niche divergence. However, *Megaloprepus* depends entirely on water-filled tree holes for reproduction. These holes are formed from indentations in the tree bole or buttress—particularly in tree species that are susceptible to burl formation—or in fallen trees in which flutings are filled with water [38,39]. Trees with smooth boles, such as Bursera simarouba, are unlikely to form tree holes [39]. *Megaloprepus* reproduces predominantly in tree holes found in gaps where other trees or large branches have fallen [87]. Moreover, large holes (>1 L) are relatively rare but support a greater number of emerging adults per season and provide more resources for the larvae, producing larger adults than small holes (<1 L) which usually only support one emerging adult [38,87]. In fact, this is the reason why territorial males primarily defend large tree holes as breeding sites [87]. Only a subset of available tree species can provide water-filled holes (especially >1 L), such as *Ceiba pentandra, Dipteryx panamensis, Platypodium elegans, Ficus trigonata,* and *Alseis blackiana,* among others [39]. Hence, tree holes are a limiting population resource for *Megaloprepus* [38]. Even though we included the general habitat type in the "Evergreen Broadleaf Tree Cover" variable, we did not consider the distribution of the tree species that yield the water-filled holes needed for *Megaloprepus* to thrive. This is a crucial biotic factor that could play a key role in their distribution and divergence, and should be accounted for in future studies.

One of the most important findings in this study is the niche differences between *Megaloprepus* populations in the geographic and environmental space in the equivalence test, particularly in Corcovado (*M. c.* subsp. nov.) and Barro Colorado populations (*M. c. caerulatus*) (Figure 3f). However, we must emphasize that ecological divergence is rare or very slow-occurring in species with highly specialised niches [30]. Numerous studies have found that niches generally remain constant throughout phylogeny, at least in the short-to-medium term [18,24,66,88,89]. Therefore, ecological divergence could be a result of an ancient and complex geographic event. Toussaint et al. [90] tested three different clock partitioning schemes and two different tree models, which suggest an ancient origin for the diversification of Neotropical giant damselflies in the Paleogene-Eocene (50–40 Ma). Feindt et al. [35] assume that the historical distribution of helicopter damselflies might have been in the northern portion of South America. According to morphological and phylogenetic studies, *Megaloprepus* is closely related to *Anomisma* [91] and *Microstigma* [90]. These taxa present an exclusive current distribution in South America with a northern limit in Colombia, without having any presence in Central America [92]. It has been suggested that the closing of the land bridge connecting south Nicaragua and Colombia was during the late Pliocene 3 Ma [93,94], although new palaeontological studies suggest this occurred 13–15 Ma [95,96]. Until the closing of this land bridge, migration of montane entomofauna between the Central American Nucleus and South America represented a very difficult undertaking and was only possible through the mountains of Talamanca [97]. To this day, fauna from Costa Rica and Panama show a higher affinity towards South American biodiversity than fauna from Mexico and other countries in Central America [97]. The mountains of Talamanca are over 1500 masl and, considering *M. caerulatus* is only found in areas below 1500 masl, these mountains may have constituted an important barrier for dispersal of *Megaloprepus* between East Central America and the Pacific region. Geographic barriers limit gene diffusion and populations become geographically isolated [30]. This could be driving divergence in the Corcovado population, located west to the mountains of Talamanca.

On the other hand, Fincke et al. [31] found the Tuxtlas population as the most ancestral, suggesting the species dispersed in a N-S direction. This would change the perception of the dispersion of *Megaloprepus* described thus far, which highlights the need to investigate the areas of endemism, along with the primary biogeographic homology and if possible, the relationships between the areas of secondary biogeographic homology sensu [98]. Due to the complex history of Central America, another scenario is also possible where dispersion was first S-N, then extinction occurred in intermediate fragments, and when the temperature and other environmental conditions changed in southern Mexico [97], the populations dispersed from N-S along the continuous fragments remaining. If our results of geographic correspondence between populations with a close phylogenetic relationship [31,35] are evidence of an ancient, continuous distribution, then they should be subject to discussion and testing. This is particularly necessary if the distribution eventually became fragmented as a result of natural events that occurred long ago, perhaps during the Pleistocene, which is likely given the correspondence of La Selva potential distribution with a Pleistocene refuge in South America [85,99].

In order to decipher the biogeographic history of *M. caerulatus*, it is paramount to identify more populations representative of South America, particularly monomorphic populations found in the northern portion of South America (*M. caerulatus brevistigma*; [42]), since very few records were found in this area (Figure 1a). It is also important to characterise populations found north of Los Tuxtlas [43,47], Chiapas (Mexico) and Guatemala to verify if their phylogenetic relationship is closer to Los Tuxtlas (*M. c. latipennis*) or Corcovado (*M. c.* subsp. nov.), as suggested by the potential distributions found in this study. This would shed light on the divergence of Corcovado (*M. c.* subsp. nov.) and provide more information to characterise monomorphic populations from Los Tuxtlas (*M. c. latipennis*) or from South America (*M. c. brevistigma*) as the most basal node.

Conversely, no difference was found in niche similarity among the populations studied. This approach not only considers the record's environmental data, but also the surrounding area, which can be heterogeneous in fragmented landscapes [100]. *Megaloprepus caerulatus* is a species with a highly specialised niche [35] which shows limited dispersal in open areas [39] and is susceptible to forest conversion [34]. Landscape fragmentation is a well-known issue especially in Los Tuxtlas [39], although fragmentation rates and connectivity to other forested areas vary considerably among the populations studied [35]. Therefore, environmental heterogeneity may constitute an excluding factor, which might explain the non-significant niche similarity and significant niche equivalence found in this study. Nevertheless, the relevance of surrounding environments (i.e., ***M*** environments) in highly specialised species must be tested in order to relate niche theory to niche conservatism or divergence.

So far, we have speculated about the possible drivers of the ecological divergence found in this study, which is likely a result of complex processes that occurred long ago. However, *Megaloprepus* was found to be fairly sensitive to recent land use change [13], which could be due to various factors. Firstly, odonates—particularly zygopterans—with a large body size, such as *Megaloprepus*, are more prone to extinction [101,102]. This could be due to the long development period required to achieve such large size, which makes them more vulnerable to predators [101]. On the other hand, it may also be related to their thermal tolerance. Larger water-breathing ectotherms may be more susceptible to impaired heat tolerance by oxygen limitation [103], which could restrict the capacity of large damselflies to obtain oxygen from the environment [102].

Additionally, their highly specialised niche can make them particularly vulnerable to current anthropogenic land use change. As previously mentioned, *Megaloprepus* is highly dependent on mature, moist Neotropical forests and it is most common in primary forests [34]. Although they can disperse considerable distances in forest understorey, they show relatively low flight endurance in open areas (almost 1 km in open water) and their dispersal largely depends on tree-hole species [39]. Therefore, they are extremely susceptible to forest conversion [34,39]. Land use change is perhaps the most important factor constricting and/or fragmenting their distribution range, as has been shown in other odonate species [13,54,104]. In addition, climate change may be a more unpredictable threat and could exacerbate the effects of forest fragmentation [39]. This is important because we are barely beginning to understand the patterns of divergence in *Megaloprepus*, and while we do not know for certain how these populations will respond to land use change, these divergent populations might differ in their response, or in other words, some populations could be more prone to extinction than others. Further research is essential to identify the impact of human activities on these populations and increase conservation efforts.

5. Conclusions

In conclusion, we found that *Megaloprepus* populations show some potential ecological divergence, which is in line with previous studies that found genetic and morphological divergence in these populations. This prepares the basis for examining the specific biotic and abiotic factors limiting the distribution of these populations and driving niche divergence. Further studies are also encouraged to (1) disentangle the biogeographic history of *Megaloprepus*, and (2) evaluate the impact of current anthropogenic land use changes and climate change in the distribution and conservation of *Megaloprepus*. This will allow a better understanding of the divergence in these populations and predict their future under anthropogenic disturbance.

Author Contributions: Conceptualization, J.A.E.-M. and A.E.-L.; methodology, A.E.-L.; software, A.E.-L.; validation, A.E.-L.; formal analysis, A.E.-L.; investigation, A.E.-L., J.A.E.-M. and G.V.-J.; resources, A.E.-L., J.A.E.-M. and G.V.-J.; data curation, G.V.-J.; writing—original draft preparation, A.E.-L., J.A.E.-M. and G.V.-J.; writing—review and editing, A.E.-L., J.A.E.-M. and G.V.-J.; visualization, A.E.-L.; supervision, J.A.E.-M.; project administration, J.A.E.-M. and G.V.-J.; funding acquisition, J.A.E.-M. All authors have read and agreed to the published version of the manuscript.

Funding: This research was funded by the Universidad Autónoma de Aguascalientes, grant number PIB19-1 and PIB20-4N. The APC was funded by the Universidad Autónoma de Aguascalientes, grant number PIB22-2.

Institutional Review Board Statement: Not applicable.

Informed Consent Statement: Not applicable.

Data Availability Statement: Publicly available datasets from the Global Biodiversity Information Facility (GBIF) were analysed in this study. This data can be found at https://doi.org/10.15468/dl.dqiufw (accessed on 30 December 2021).

Acknowledgments: The authors would like to thank Christoffer Fägerström from the Lund Museum of Zoology (MZLU), Max Caspers from the Naturalis Biodiversity Center, James E. Berrian from the San Diego Natural History Museum (SDNHM), Bill Mauffray from the Florida State Collection of Arthropods (FSCA), and Dennis Paulson for providing photographs of the *Megaloprepus* specimens, as well as Loren Bowman, Christopher Hoffman, and the anonymous reviewers for their valuable feedback on the early versions of this manuscript.

Conflicts of Interest: The authors declare no conflict of interest. The funders had no role in the design of the study; in the collection, analyses, or interpretation of data; in the writing of the manuscript, or in the decision to publish the results.

References

1. Grinnell, J. The niche-relationships of the California Thrasher. *Auk* **1917**, *34*, 427–433. [CrossRef]
2. Soberón, J.; Peterson, A.T. Interpretation of models of fundamental ecological niches and species' distributional areas. *Biodivers. Inform.* **2005**, *2*, 1–10. [CrossRef]
3. Soberón, J. Grinnellian and Eltonian niches and geographic distributions of species. *Ecol. Lett.* **2007**, *10*, 1115–1123. [CrossRef]

4. Barve, N.; Martin, C.; Brunsell, N.A.; Peterson, A.T. The role of physiological optima in shaping the geographic distribution of Spanish moss. *Glob. Ecol. Biogeogr.* **2014**, *23*, 633–645. [CrossRef]
5. Peterson, A.T.; Soberón, J.; Pearson, R.G.; Anderson, R.P.; Martínez-Meyer, E.; Nakamura, M.; Araújo, M.B. *Ecological Niches and Geographic Distributions*; Princeton University Press: Princeton, NJ, USA, 2011; p. 314.
6. Jackson, S.T.; Overpeck, J.T. Responses of plant populations and communities to environmental changes of the late Quaternary. *Paleobiology* **2000**, *26*, 194–220. [CrossRef]
7. Peterson, A.T. Predicting species' geographic distributions based on ecological niche modeling. *Condor* **2001**, *103*, 599–605. [CrossRef]
8. Ortega-Huerta, M.A.; Peterson, A.T. Modeling ecological niches and predicting geographic distributions: A test of six presence-only methods. *Rev. Mex. Biodivers.* **2008**, *79*, 205–216.
9. Elith, J.; Leathwick, J.R. Species distribution models: Ecological explanation and prediction across space and time. *Annu. Rev. Ecol. Evol. Syst.* **2009**, *40*, 677–697. [CrossRef]
10. Smith, A.B.; Godsoe, W.; Rodríguez-Sánchez, F.; Wang, H.H.; Warren, D. Niche estimation above and below the species level. *Trends Ecol. Evol.* **2019**, *34*, 260–273. [CrossRef]
11. Collins, S.D.; McIntyre, N.E. Modeling the distribution of odonates: A review. *Freshw. Sci.* **2015**, *34*, 1144–1158. [CrossRef]
12. Costa Bastos, R.; Schlemmer Brasil, L.; Oliveira-Junior, J.M.B.; Geraldo Carvalho, F.; Lennox, G.D.; Barlow, J.; Juen, L. Morphological and phylogenetic factors structure the distribution of damselfly and dragonfly species (Odonata) along an environmental gradient in Amazonian streams. *Ecol. Indic.* **2021**, *122*, 107257. [CrossRef]
13. Rodríguez-Tapia, G.; Rocha-Ortega, M.; Córdoba-Aguilar, A. An index to estimate the vulnerability of damselflies and dragonflies (Insecta: Odonata) to land use changes using niche modeling. *Aquat. Insects* **2020**, *41*, 254–272. [CrossRef]
14. Boys, W.A.; Siepielski, A.M.; Smith, B.D.; Patten, M.A.; Bried, J.T. Predicting the distributions of regional endemic dragonflies using a combined model approach. *Insect Conserv. Divers.* **2021**, *14*, 52–66. [CrossRef]
15. Ackerly, D.D. Community assembly, niche conservatism, and adaptive evolution in changing environments. *Int. J. Plant Sci.* **2003**, *164*, S165–S184. [CrossRef]
16. Balanyá, J.; Oller, J.M.; Huey, R.B.; Gilchrist, G.W.; Serra, L. Global genetic change tracks global climate warming in *Drosophila subobscura*. *Science* **2006**, *313*, 1773–1775. [CrossRef] [PubMed]
17. Holt, R.D. The microevolutionary consequences of climate change. *Trends Ecol. Evol.* **1990**, *5*, 311–315. [CrossRef]
18. Peterson, A.T.; Soberón, J.; Sánchez-Cordero, V. Conservatism of ecological niches in evolutionary time. *Science* **1999**, *285*, 1265–1267. [CrossRef]
19. Martínez-Meyer, E.; Peterson, A.T.; Hargrove, W.W. Ecological niches as stable distributional constraints on mammal species, with implications for Pleistocene extinctions and climate change projections for biodiversity. *Glob. Ecol. Biogeogr.* **2004**, *13*, 305–314. [CrossRef]
20. Jakob, S.S.; Ihlow, A.; Blattner, F.R. Combined ecological niche modelling and molecular phylogeography revealed the evolutionary history of *Hordeum marinum* (Poaceae)—Niche differentiation, loss of genetic diversity, and speciation in Mediterranean Quaternary refugia. *Mol. Ecol.* **2007**, *16*, 1713–1727. [CrossRef]
21. Pearman, P.B.; Guisan, A.; Broennimann, O.; Randin, C.F. Niche dynamics in space and time. *Trends Ecol. Evol.* **2008**, *23*, 149–158. [CrossRef]
22. Pearman, P.B.; D'Amen, M.; Graham, C.H.; Thuiller, W.; Zimmermann, N.E. Within-taxon niche structure: Niche conservatism, divergence and predicted effects of climate change. *Ecography* **2010**, *33*, 990–1003. [CrossRef]
23. Brown, J.L.; Carnaval, A.C. A tale of two niches: Methods, concepts, and evolution. *Front. Biogeogr.* **2019**, *11*, e44158. [CrossRef]
24. Encarnación-Luévano, A.; Peterson, A.T.; Rojas-Soto, O.R. Burrowing habit in *Smilisca* frogs as an adaptive response to ecological niche constraints in seasonally dry environments. *Front. Biogeogr.* **2021**, *13*, e50517. [CrossRef]
25. Serra-Varela, M.J.; Grivet, D.; Vincenot, L.; Broennimann, O.; Gonzalo-Jiménez, J.; Zimmermann, N.E. Does phylogeographical structure relate to climatic niche divergence? A test using maritime pine (*Pinus pinaster* Ait.). *Glob. Ecol. Biogeogr.* **2015**, *24*, 1302–1313. [CrossRef]
26. Banta, J.A.; Ehrenreich, I.M.; Gerard, S.; Chou, L.; Wilczek, A.; Schmitt, J.; Kover, P.X.; Purugganan, M.D. Climate envelope modelling reveals intraspecific relationships among flowering phenology, niche breadth and potential range size in *Arabidopsis thaliana*. *Ecol. Lett.* **2012**, *15*, 769–777. [CrossRef] [PubMed]
27. Marcer, A.; Méndez-Vigo, B.; Alonso-Blanco, C.; Picó, F.X. Tackling intraspecific genetic structure in distribution models better reflects species geographical range. *Ecol. Evol.* **2016**, *6*, 2084–2097. [CrossRef]
28. Mota-Vargas, C.; Rojas-Soto, O.R. Taxonomy and ecological niche modeling: Implications for the conservation of wood partridges (genus *Dendrortyx*). *J. Nat. Conserv.* **2016**, *29*, 1–13. [CrossRef]
29. Bried, J.T.; Siepielski, A.M. Predator driven niches vary spatially among co-occurring damselfly species. *Evol. Ecol.* **2019**, *33*, 243–256. [CrossRef]
30. Wiens, J.J.; Ackerly, D.D.; Allen, A.P.; Anacker, B.L.; Buckley, L.B.; Cornell, H.V.; Damschen, E.I.; Jonathan Davies, T.; Grytnes, J.A.; Harrison, S.P.; et al. Niche conservatism as an emerging principle in ecology and conservation biology. *Ecol. Lett.* **2010**, *13*, 1310–1324. [CrossRef]
31. Fincke, O.M.; Xu, M.; Khazan, E.S.; Wilson, M.; Ware, J.L. Tests of hypotheses for morphological and genetic divergence in *Megaloprepus* damselflies across Neotropical forests. *Biol. J. Linn. Soc.* **2018**, *125*, 844–861. [CrossRef]

32. Eaton, A.E.; Calvert, P.P. *Biologia Centrali-Americana: Insecta, Neuroptera, Ephemeridæ & Odonata*; Taylor & Francis: London, UK, 1892–1908; p. 420.
33. Hedström, I.; Sahlén, G. A key to the adult Costa Rican "helicopter" damselflies (Odonata: Pseudostigmatidae) with notes on their phenology and life zone preferences. *Rev. Biol. Trop.* **2001**, *48*, 1037–1056.
34. Fincke, O.M.; Hedström, I. Differences in forest use and colonization by Neotropical tree-hole damselflies (Odonata: Pseudostigmatidae): Implications for forest conversion. *Stud. Neotrop. Fauna Environ.* **2008**, *43*, 35–45. [CrossRef]
35. Feindt, W.; Fincke, O.; Hadrys, H. Still a one species genus? Strong genetic diversification in the world's largest living odonate, the Neotropical damselfly *Megaloprepus caerulatus*. *Conserv. Genet.* **2014**, *15*, 469–481. [CrossRef]
36. Young, A.M. Feeding and oviposition in the Giant Tropical Damselfly *Megaloprepus coerulatus* (Drury) in Costa Rica. *Biotropica* **1980**, *12*, 237. [CrossRef]
37. Fincke, O.M. Giant damselflies in a tropical forest: Reproductive biology of *Megaloprepus caerulatus* with notes on *Mecistogaster* (Zygoptera: Pseudostigmatidae). *Adv. Odonatol.* **1984**, *2*, 13–27.
38. Fincke, O.M. Interspecific competition for tree holes: Consequences for mating systems and coexistence in Neotropical damselflies. *Am. Nat.* **1992**, *139*, 80–101. [CrossRef]
39. Fincke, O.M. Use of forest and tree species, and dispersal by giant damselflies (Pseudostigmatidae): Their prospects in fragmented forests. In *Forests and Dragonflies, Proceedings of the 4th WDA International Symposium of Odonatology, Pontevedra, Spain, 26–30 July 2005*; Cordero-Rivera, A., Ed.; Pensoft Publishers: Sofia, Bulgaria, 2006; pp. 103–125.
40. Schultz, T.D.; Fincke, O.M. Structural colours create a flashing cue for sexual recognition and male quality in a Neotropical giant damselfly. *Funct. Ecol.* **2009**, *23*, 724–732. [CrossRef]
41. Xu, M.; Fincke, O.M. Ultraviolet wing signal affects territorial contest outcome in a sexually dimorphic damselfly. *Anim. Behav.* **2015**, *101*, 67–74. [CrossRef]
42. de Selys Longchamps, M. Révision du synopsis des Agrionines. Première partie comprenant les légions Pseudostigma—Podagrion—Platycnemis et Protoneura. *Mem. Couronnés Académie R. Belg.* **1886**, *38*, 1–233.
43. Escoto-Moreno, J.A.; Hernández-Hernández, A.; Hernández-Hernández, J.A.; Márquez, J.; Silva-Briano, M.; Novelo-Gutiérrez, R. El registro más septentrional de la libélula gigante neotropical *Megaloprepus caerulatus* (Drury, 1782) (Odonata: Coenagrionidae) en el continente Americano. *Gayana* **2018**, *82*, 90–93. [CrossRef]
44. Measey, G.J. Some Odonata from Belize, Central America. *Not. Odonatol.* **1994**, *4*, 40–46.
45. González-Soriano, E.; Dirzo, R.; Vogt, R.C. *Historia Natural de los Tuxtlas*; Instituto de Biología, UNAM: Mexico City, Mexico, 1997; p. 647.
46. Machado, J. Inventario y Estudio Comparativo de la Fauna de Odonata en tres Áreas de Honduras. Bachelor Thesis, Universidad Zamorano, San Antonio de Oriente, Honduras, November 2001.
47. Cuevas-Yáñez, K. Los odonatos (Insecta: Odonata) de la Hidroeléctrica de Patla (El Pozo) y del Río Tecpatlán, Zihuateutla, Puebla, México. *Dugesiana* **2007**, *14*, 83–91.
48. González-Soriano, E.; Paulson, D.R. Los odonatos de Chiapas. In *Chiapas: Estudios Sobre su Diversidad Biológica*; Álvarez-Noguera, F., Ed.; Instituto de Biología, UNAM: Mexico City, Mexico, 2011; pp. 299–314.
49. Esquivel, C. Las libélulas de la zona de El Rodeo, cantón de Mora, San José, Costa Rica. *Brenesia* **2012**, *77*, 329–342.
50. GBIF. GBIF Occurrence Download. Available online: https://doi.org/10.15468/dl.dqiufw (accessed on 19 August 2019).
51. Olson, D.M.; Dinerstein, E.; Wikramanayake, E.D.; Burgess, N.D.; Powell, G.V.N.; Underwood, E.C.; D'amico, J.A.; Itoua, I.; Strand, H.E.; Morrison, J.C.; et al. Terrestrial ecoregions of the World: A new Map of Life on Earth: A new global map of terrestrial ecoregions provides an innovative tool for conserving biodiversity. *Bioscience* **2001**, *51*, 933–938. [CrossRef]
52. Corbet, P.S. *Dragonflies: Behaviour and Ecology of Odonata (Revised Edition)*, 2nd ed.; Comstock Publishing Associates: New York, NY, USA, 2004; p. 829.
53. Corbet, P.S.; Brooks, S.J. *Dragonflies*; Collins: London, UK, 2008; p. 454.
54. Cuevas-Yáñez, K.; Rivas, M.; Muñoz, J.; Córdoba-Aguilar, A. Conservation status assessment of *Paraphlebia* damselflies in Mexico. *Insect Conserv. Divers.* **2015**, *8*, 517–524. [CrossRef]
55. Rangel-Sánchez, L.; Nava-Bolaños, A.; Palacino-Rodríguez, F.; Córdoba-Aguilar, A. Estimating distribution area in six *Argia* damselflies (Insecta: Odonata: Coenagrionidae) including *A. garrisoni*, a threatened species. *Rev. Mex. Biodivers.* **2018**, *89*, 921–926. [CrossRef]
56. Hijmans, R.J.; Cameron, S.E.; Parra, J.L.; Jones, P.G.; Jarvis, A. Very high resolution interpolated climate surfaces for global land areas. *Int. J. Climatol.* **2005**, *25*, 1965–1978. [CrossRef]
57. Tuanmu, M.N.; Jetz, W. A global 1-km consensus land-cover product for biodiversity and ecosystem modelling. *Glob. Ecol. Biogeogr.* **2014**, *23*, 1031–1045. [CrossRef]
58. Trabucco, A.; Zomer, R.J. Global High-Resolution Soil-Water Balance. Figshare Fileset. Available online: https://figshare.com/articles/Global_High-Resolution_Soil-Water_Balance/7707605/3 (accessed on 18 November 2019).
59. Danielson, J.J.; Gesch, D.B. *Global Multi-Resolution Terrain Elevation Data 2010 (GMTED2010): U.S. Geological Survey Open-File Report 2011–1073*; U.S. Geological Survey: Reston, VA, USA, 2011; p. 26.
60. Lehner, B.; Verdin, K.; Jarvis, A. New global hydrography derived from spaceborne elevation data. *Eos* **2008**, *89*, 93–94. [CrossRef]
61. Harrell, F.E. Hmisc: Harrell Miscellaneous. R Package Version 4.3-0. Available online: https://CRAN.R-project.org/package=Hmisc (accessed on 13 September 2019).

62. R Core Team. R: A Language and Environment for Statistical Computing. R Foundation for Statistical Computing, Vienna, Austria. Available online: http://www.R-project.org/ (accessed on 30 May 2020).
63. Phillips, S.J.; Anderson, R.P.; Schapire, R.E. Maximum entropy modeling of species geographic distributions. *Ecol. Modell.* **2006**, *190*, 231–259. [CrossRef]
64. Peterson, A.T.; Soberón, J. Species distribution modeling and ecological niche modeling: Getting the Concepts Right. *Nat. Conserv.* **2012**, *10*, 102–107. [CrossRef]
65. Barve, N.; Barve, V.; Jiménez-Valverde, A.; Lira-Noriega, A.; Maher, S.P.; Peterson, A.T.; Soberón, J.; Villalobos, F. The crucial role of the accessible area in ecological niche modeling and species distribution modeling. *Ecol. Modell.* **2011**, *222*, 1810–1819. [CrossRef]
66. Warren, D.L.; Glor, R.E.; Turelli, M. Environmental niche equivalency versus conservatism: Quantitative approaches to niche evolution. *Evolution* **2008**, *62*, 2868–2883. [CrossRef] [PubMed]
67. Peterson, A.T. Ecological niche conservatism: A time-structured review of evidence. *J. Biogeogr.* **2011**, *38*, 817–827. [CrossRef]
68. Broennimann, O.; Fitzpatrick, M.C.; Pearman, P.B.; Petitpierre, B.; Pellissier, L.; Yoccoz, N.G.; Thuiller, W.; Fortin, M.J.; Randin, C.; Zimmermann, N.E.; et al. Measuring ecological niche overlap from occurrence and spatial environmental data. *Glob. Ecol. Biogeogr.* **2012**, *21*, 481–497. [CrossRef]
69. Owens, H.L.; Campbell, L.P.; Dornak, L.L.; Saupe, E.E.; Barve, N.; Soberón, J.; Ingenloff, K.; Lira-Noriega, A.; Hensz, C.M.; Myers, C.E.; et al. Constraints on interpretation of ecological niche models by limited environmental ranges on calibration areas. *Ecol. Modell.* **2013**, *263*, 10–18. [CrossRef]
70. Liu, C.; White, M.; Newell, G. Selecting thresholds for the prediction of species occurrence with presence-only data. *J. Biogeogr.* **2013**, *40*, 778–789. [CrossRef]
71. Lehner, B.; Grill, G. Global river hydrography and network routing: Baseline data and new approaches to study the world's large river systems. *Hydrol. Process.* **2013**, *27*, 2171–2186. [CrossRef]
72. Peterson, A.T.; Papeş, M.; Soberón, J. Rethinking receiver operating characteristic analysis applications in ecological niche modeling. *Ecol. Modell.* **2008**, *213*, 63–72. [CrossRef]
73. Osorio-Olvera, L.; Lira-Noriega, A.; Soberón, J.; Peterson, A.T.; Falconi, M.; Contreras-Díaz, R.G.; Martínez-Meyer, E.; Barve, V.; Barve, N. ntbox: An r package with graphical user interface for modelling and evaluating multidimensional ecological niches. *Methods Ecol. Evol.* **2020**, *11*, 1199–1206. [CrossRef]
74. Pearson, R.G.; Raxworthy, C.J.; Nakamura, M.; Townsend Peterson, A. Predicting species distributions from small numbers of occurrence records: A test case using cryptic geckos in Madagascar. *J. Biogeogr.* **2007**, *34*, 102–117. [CrossRef]
75. Schoener, T.W. Nonsynchronous spatial overlap of lizards in patchy habitats. *Ecology* **1970**, *51*, 408–418. [CrossRef]
76. Aguirre-Gutiérrez, J.; Serna-Chavez, H.M.; Villalobos-Arambula, A.R.; Pérez de la Rosa, J.A.; Raes, N. Similar but not equivalent: Ecological niche comparison across closely-related Mexican white pines. *Divers. Distrib.* **2015**, *21*, 245–257. [CrossRef]
77. Di Cola, V.; Broennimann, O.; Petitpierre, B.; Breiner, F.T.; D'Amen, M.; Randin, C.; Engler, R.; Pottier, J.; Pio, D.; Dubuis, A.; et al. ecospat: An R package to support spatial analyses and modeling of species niches and distributions. *Ecography* **2017**, *40*, 774–787. [CrossRef]
78. Halffter, G.; Morrone, J.J. An analytical review of Halffter's Mexican transition zone, and its relevance for evolutionary biogeography, ecology and biogeographical regionalization. *Zootaxa* **2017**, *4226*, 1–46. [CrossRef]
79. Morrone, J.J. Biogeographic regionalization and biotic evolution of Mexico: Biodiversity's crossroads of the New World. *Rev. Mex. Biodivers.* **2019**, *90*, 1–68. [CrossRef]
80. Rzedowski, J. El endemismo en la flora fanerogámica mexicana: Una apreciación analítica preliminar. *Acta Bot. Mex.* **1991**, 47–64. [CrossRef]
81. Dinerstein, E.; Olson, D.; Joshi, A.; Vynne, C.; Burgess, N.D.; Wikramanayake, E.; Hahn, N.; Palminteri, S.; Hedao, P.; Noss, R.; et al. An ecoregion-based approach to protecting half the terrestrial realm. *Bioscience* **2017**, *67*, 534–545. [CrossRef]
82. Morrone, J.J. *Biogeografía de América Latina y el Caribe, M&T–Manuales & Tesis SEA, Vol. 3*; CYTED, UNESCO-ORCYT & SEA: Zaragoza, Spain, 2001; 148p.
83. Avise, J.C.; Walker, D. Pleistocene phylogeographic effects on avian populations and the speciation process. *Proc. R. Soc. B Biol. Sci.* **1998**, *265*, 457–463. [CrossRef]
84. Morrone, J.J. Biogeographical regionalisation of the neotropical region. *Zootaxa* **2014**, *3782*, 1–110. [CrossRef]
85. Haffer, J. Speciation in amazonian forest birds. *Science* **1969**, *165*, 131–137. [CrossRef] [PubMed]
86. Bush, M.B.; de Oliveira, P.E. The rise and fall of the Refugial Hypothesis of Amazonian speciation: A paleoecological perspective. *Biota Neotrop.* **2006**, *6*, bn00106012006. [CrossRef]
87. Fincke, O.M. Consequences of larval ecology for territoriality and reproductive success of a neotropical damselfly. *Ecology* **1992**, *73*, 449–462. [CrossRef]
88. Eaton, M.D.; Soberón, J.; Peterson, A.T. Phylogenetic perspective on ecological niche evolution in american blackbirds (Family Icteridae). *Biol. J. Linn. Soc.* **2008**, *94*, 869–878. [CrossRef]
89. Petitpierre, B.; Kueffer, C.; Broennimann, O.; Randin, C.; Daehler, C.; Guisan, A. Climatic niche shifts are rare among terrestrial plant invaders. *Science* **2012**, *335*, 1344–1348. [CrossRef] [PubMed]
90. Toussaint, E.F.A.; Bybee, S.M.; Erickson, R.J.; Condamine, F.L. Forest giants on different evolutionary branches: Ecomorphological convergence in helicopter damselflies. *Evolution* **2019**, *73*, 1045–1054. [CrossRef]

91. Ingley, S.J.; Bybee, S.M.; Tennessen, K.J.; Whiting, M.F.; Branham, M.A. Life on the fly: Phylogenetics and evolution of the helicopter damselflies (Odonata, Pseudostigmatidae). *Zool. Scr.* **2012**, *41*, 637–650. [CrossRef]
92. Paulson, D.R. Middle American Odonata By Country. Slater Museum of Natural History, University of Puget Sound. Available online: https://www.pugetsound.edu/academics/academic-resources/slater-museum/biodiversity-resources/dragonflies/middle-american-odonata/ (accessed on 3 September 2020).
93. Marshall, L.G.; Webb, S.D.; Sepkoski, J.J.; Raup, D.M. Mammalian evolution and the great American interchange. *Science* **1982**, *215*, 1351–1357. [CrossRef]
94. Rich, P.V.; Rich, T.H. The Central American dispersal route: Biotic history and palaeogeography. In *Costa Rican Natural History*; Janzen, D.H., Ed.; The University of Chicago Press: Chicago, IL, USA, 1983; pp. 12–34.
95. Montes, C.; Cardona, A.; Jaramillo, C.; Pardo, A.; Silva, J.C.; Valencia, V.; Ayala, C.; Pérez-Angel, L.C.; Rodriguez-Parra, L.A.; Ramirez, V.; et al. Middle Miocene closure of the Central American Seaway. *Science* **2015**, *348*, 226–229. [CrossRef]
96. Hoorn, C.; Flantua, S. An early start for the Panama land bridge. *Science* **2015**, *348*, 186–187. [CrossRef]
97. Halffter, G. Biogeography of the Montane Entomofauna of Mexico and Central America. *Annu. Rev. Entomol.* **1987**, *32*, 95–114. [CrossRef]
98. Morrone, J.J. *Homología Biogeográfica: Las Coordenadas Espaciales de la Vida. Cuadernos del Instituto de Biología 37*; Instituto de Biología, UNAM: Mexico City, Mexico, 2004; p. 199.
99. Brown, K.S. Areas where humid tropical forest probably persisted. In *Biogeography and Quaternary History in Tropical America*; Whitmore, T.C., Prance, G.T., Eds.; Clarendon Press: Oxford, UK, 1987; pp. 44–45.
100. Hiebeler, D. Populations on fragmented landscapes with spatially structured heterogeneities: Landscape generation and local dispersal. *Ecology* **2000**, *81*, 1629. [CrossRef]
101. Suárez-Tovar, C.M.; Rocha-Ortega, M.; González-Voyer, A.; González-Tokman, D.; Córdoba-Aguilar, A. The larger the damselfly, the more likely to be threatened: A sexual selection approach. *J. Insect Conserv.* **2019**, *23*, 535–545. [CrossRef]
102. Rocha-Ortega, M.; Rodríguez, P.; Bried, J.; Abbott, J.; Córdoba-Aguilar, A. Why do bugs perish? Range size and local vulnerability traits as surrogates of Odonata extinction risk. *Proc. R. Soc. B Biol. Sci.* **2020**, *287*, 20192645. [CrossRef] [PubMed]
103. Leiva, F.P.; Calosi, P.; Verberk, W.C.E.P. Scaling of thermal tolerance with body mass and genome size in ectotherms: A comparison between water- and air-breathers. *Philos. Trans. R. Soc. B Biol. Sci.* **2019**, *374*, 20190035. [CrossRef]
104. Rocha-Ortega, M.; Rodríguez, P.; Córdoba-Aguilar, A. Can dragonfly and damselfly communities be used as bioindicators of land use intensification? *Ecol. Indic.* **2019**, *107*, 105553. [CrossRef]

diversity

MDPI

Article

Land Uses for Pasture and Cacao Cultivation Modify the Odonata Assemblages in Atlantic Forest Areas

Laís R. Santos [1,2] and Marciel E. Rodrigues [1,2,*]

[1] Laboratory of Aquatic Organisms ("LOA"), Department of Biological Sciences, Santa Cruz State University (UESC), Ilheus 45662-900, Bahia, Brazil
[2] Graduate Program in Tropical Aquatic Systems ("PPGSAT"), Department of Biological Sciences, Santa Cruz State University (UESC), Ilheus 45662-900, Bahia, Brazil
* Correspondence: merodrigues@uesc.br

Abstract: Tropical forests such as the Atlantic Forest are under constant threats from the impact of human activities, mostly being caused by the loss of native forest areas for other land uses. This study aimed to evaluate the effect of changes in land use for pasture and cacao cultivation on the richness and composition of Odonata assemblages in comparison to native forest areas. We also evaluated the species as possible indicators of these different land uses. In total, 64 streams were sampled in southern Bahia, Brazil. A total of 84 species were recorded. The results indicated that changes in land use modify the richness and composition of Odonata assemblages. Regarding composition, our results indicated a difference among the assemblages in the three land use areas and that the native areas maintain more stable assemblages. According to the indicator species analysis, 13 species were recorded as possible bioindicators for different land uses. Changes in aquatic ecosystems and their surroundings caused by different land uses a select group of different species groups, modifying Odonata diversity among these areas. Notably, land uses that maintain a certain integrity of the environment, as in the case of cacao cultivation, are the best alternatives for conserving Odonata biodiversity in comparison with pasture.

Keywords: bioindicators; dragonflies; cacao cultivation; pasture; aquatic ecosystems

Citation: Santos, L.R.; Rodrigues, M.E. Land Uses for Pasture and Cacao Cultivation Modify the Odonata Assemblages in Atlantic Forest Areas. *Diversity* **2022**, *14*, 672. https://doi.org/10.3390/d14080672

Academic Editors: M. Olalla Lorenzo-Carballa, Ricardo Koroiva and Michael Wink

Received: 23 July 2022
Accepted: 11 August 2022
Published: 18 August 2022

1. Introduction

Land-use changes for agriculture and livestock grazing along with urbanization have intensified over the years, modifying the dynamics of ecosystems and causing the loss of biodiversity [1–4]. In Brazil, the Atlantic Forest, considered a biodiversity hotspot, has been the most widely affected by changes in land uses, including extensive and disorderly timber extraction [1,4–6]. Therefore, it is critical to understand how the type of land use affects ecosystems and their biodiversity [4,7], especially in aquatic ecosystems [8–10].

Among the different types of land uses in Atlantic Forest areas, farming and livestock grazing stand out the most due to the amount (number and size) of modified areas [4]. However, in some regions, other types of land uses are also important. This is the case of areas used for cacao cultivation in the southern region of Bahia, in which cacao is cultivated in the understory of the forest and part of the native vegetation is maintained. This cultivation system is regionally called *Cabruca* [11–13] and is viewed as a sustainable model of production within the remnant Atlantic Forest areas. However, all of these different land uses in Atlantic Forest fragments make them priority areas for conservation actions due to the high loss and fragility of their remaining ecosystems and local biodiversity [2,14], mainly within aquatic ecosystems [10,15,16].

The changes caused by different land uses in the aquatic ecosystems are numerous and affect the chemical conditions as well as the physical and biological structure. These degrade the water quality of water bodies, favor siltation, alter the hydrological regime, and increase the incidence of light [8,17,18]. These factors lead to alterations in the environmental conditions and reduce the variability of habitats by homogenization, which reduces the availability of food resources and oviposition sites and increases intra-and interspecific competition [18–21]. Moreover, these effects compromise the ability of these environments to maintain their natural communities, thus causing the local extinction of more sensitive species [7,10,16,22–27]

Since species of the Odonata order depend on aquatic ecosystems, they are widely used in studies evaluating the effects of land-use changes on aquatic ecosystems and their surroundings. The species are also used as surrogates of various groups of aquatic insects [28], especially when assessing the impacts caused by different land uses (agriculture, pasture, and urban development) [21,28–33]. However, there are still few studies that have evaluated these effects within the Atlantic Forest and with different types of land use changes (especially in cacao cultivation areas).

Among dragonflies, both the larvae and adults exhibit morphological, ecophysiological, and behavioral characteristics that are closely related to habitat, such as diet and reproduction, oviposition, flight behavior, dispersal ability, and thermoregulation capacity [34–36]. These characteristics divide species into groups that can reflect the quality and integrity of the ecosystems in which they are found [21,29,35], and also help classify species as forest specialists, open area specialists or habitat generalists [33]. For this reason, they can be used as bioindicators of changes in land use [21,28].

The Odonata species considered forest specialists are extremely dependent on the integrity of aquatic ecosystems and surrounding areas. Therefore, they are sensitive to environmental changes and highly susceptible to local extinction when habitats are modified. Species considered open area specialists belong to a group that is adapted to non-forested environments with high levels of solar incidence. They are generally found in natural open areas and in aquatic environments with intermediate levels of alterations of the surroundings. Lastly, species considered habitat generalists are more tolerant of modifications in natural environments and different levels of human impact. For this reason, they are found in areas with various levels of anthropization, including areas where other species cannot develop [21,33].

This study was designed to evaluate the effects of different land uses (pasture, cacao cultivation, and native forest) on the richness and composition of Odonata species within Atlantic Forest areas. In addition, we aimed to identify the possible existence of species that can be regarded as bioindicators of these land uses. We predicted that the pasture areas would have greater species richness than the *cabruca* and native vegetation areas. Modifications in natural ecosystems allowed habitat generalists and open area specialists to colonize these areas and increase local richness [21,33,37–39]. Regarding composition, the prediction was that different land uses would select different species groups. The composition of native forest and *cabruca* areas is similar since it consists of species classified as forest specialists, while pastures tend to be inhabited by species considered to be habitat generalists or open area specialists [33,38]. Considering bioindicator species, we expected the different land uses to be associated with selected group of species that might be considered bioindicators of the native, cacao cultivation, and pasture areas [39].

2. Material and Methods

2.1. Study Area

This study was conducted in the southern region of the state of Bahia, in the municipalities of Ilhéus, Una, Uruçuca, Itacaré, Buerarema, São José da Vitória, Porto Seguro, and Santa Cruz Cabrália (Figure 1), located in the Atlantic Forest domain. Regional climate according to Köppen-Geiger is classified as Tropical Forest Climate Af (tropical super humid) with evenly distributed rainfall throughout the year. A total of 64 streams were sampled twice of which 24 were in native forest areas, 17 were in in pasture areas, and 23 were in areas of cacao cultivation. The streams are considered low-order (first to third), with an average width of 2 m.

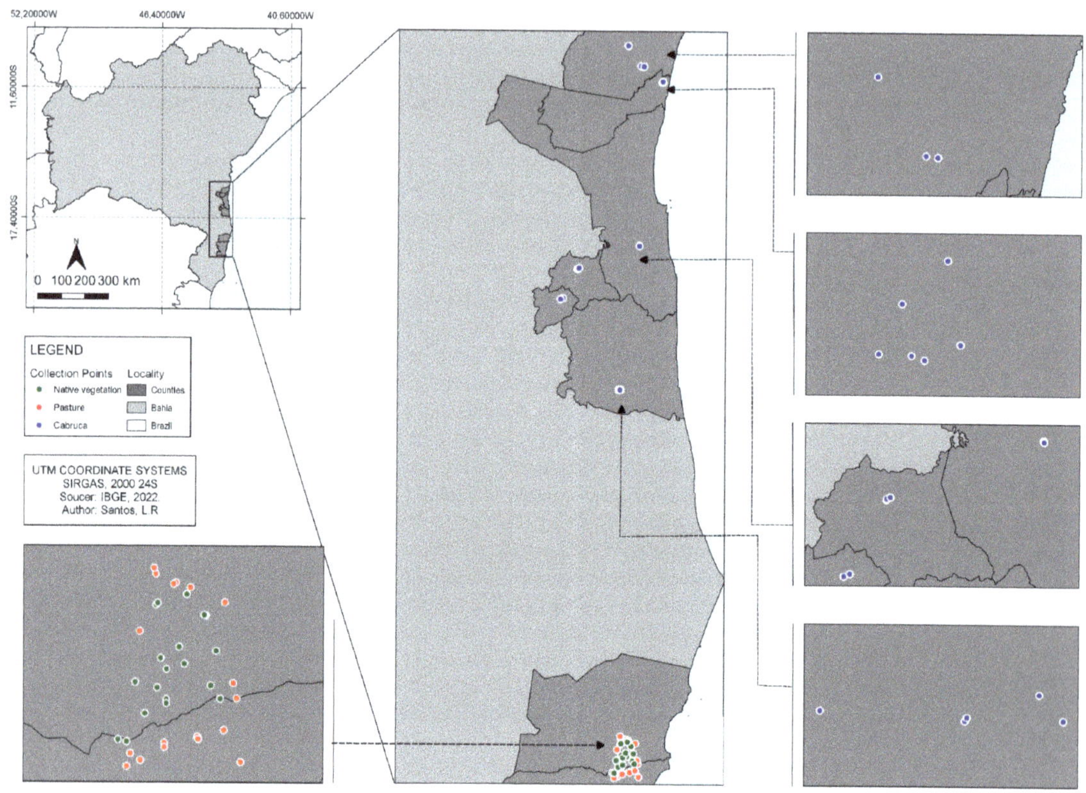

Figure 1. Map of the municipalities with the sampling points in areas of native vegetation (green), *cabruca* (blue), and pasture (red).

Collections in native areas were carried out in the municipalities of Porto Seguro and Santa Cruz Cabrália, specifically inside the Estação Veracel Private Natural Heritage Reserve (RPPN), between September and October 2018 and between February and March 2019. In the municipalities of Una and Uruçuca, collections were carried out in permanent preservation areas ("APPs") on private properties between October and November 2019 and July and August 2020. In general, the streams within native vegetation areas have margins with riparian forest and canopy cover, without evidence of physical pollution (disposal of waste discharge of effluents) and with greater physical integrity of the channels (stable margins, little or no evident silting). The streams had widths varying between 78 cm the 501 cm (mean 276 cm and sd 135) and depths varying between 8 cm and 35 cm (mean 28 cm and sd 12).

In pasture areas, collections were carried out in the municipalities of Porto Seguro and Santa Cruz, on private properties and settlements in the region surrounding the Veracel RPPN, during September and October 2018 and in February and March 2019. In general, the streams had little or no riparian vegetation, although some had channel dams to form ponds for drinking water for animals and other purposes on the property. At some points, the margins were unstable and the channel bed was silted. A small amount of household waste was also frequently observed, usually plastic containers. The streams had the width varying between 74 cm and 540 cm (mean 265 cm and sd 128) and depth varying between 12 cm and 63 cm (mean 19 cm and sd 15).

In cabruca areas, collections were carried out on properties of organic cacao producers belonging to the Cabruca Cooperative in the municipalities of Ilhéus, Itacaré, Uruçuca, Buerarema, São José da Vitória and Una, between the months of September and November

2019 and July and August 2020. These sampling sites generally had stable margins with little or no evidence of silting. The riparian vegetation had a slightly closed canopy and many of the sampled areas are used for cacao cultivation until the margins of the streams, forming a canopy over the aquatic ecosystems. The streams had width varying between 31 cm and 292 cm (mean 124 cm and sd 0.70) and depth varying between 5 cm and 96 cm (mean 24 cm and sd 24).

2.2. Sampling Method

Adult specimens were sampled within a 100-m section on both margins of the streams. They were collected using an entomological net, with a total sampling effort of 1.30 h for each site. The samples were collected from 9:00 a.m. to 3 p.m., always on sunny days [40]. The collected individuals were sent to the Laboratory of Aquatic Organisms of Santa Cruz State University for identification, which was carried out with the aid of keys [41–45] and other more specific taxonomic keys, along with expert help (see acknowledgments). The collected material is deposited in the collection of aquatic insects of the Santa Cruz State University—UESC.

2.3. Data Analysis

To evaluate the response of Odonata richness in the areas with different land uses, the data were tested with a generalized linear method (GLM) using a log-linear [46]. For this analysis, richness was used as a response variable and land uses as predictor variables (native, cabruca, and pasture). To evaluate the relationship of assemblages with the different land uses, principal coordinate analysis (PcoA) and PERMANOVA [47] were performed. For the PcoA, a distance matrix was generated to determine the difference between the assemblages in each land use area and the generated cluster was tested by means of PERMANOVA, from 999 repetitions.

The indicator species analysis (IndVal) for each land use and among the land uses was calculated according to De Cáceres [48]. With this analysis, it is possible to create combinations and evaluate species associated with each type of land use, resulting in association values for each type of use. IndVal also calculates specificity and fidelity. Specificity identifies the probability that a species belongs to a given land use while fidelity indicates how many times the species was recorded for the total sampled points in a particular land use area [48]. All analyses were conducted with the R software and the packages "vegan" [49], "RT4Bio", and "Indicspecies" [50].

3. Results

In total, 1558 individuals belonging to 84 species were collected, of which 1217 specimens belong to 34 species were of the suborder Zygoptera and 341 specimens to 50 species were of the suborder Anisoptera. In native forest areas, 514 specimens and 38 species were collected; in the pasture areas, 332 specimens and 43 species were collected; and in the cabruca areas, 712 specimens and 54 species were collected.

Of the 84 species collected, 13 were common to the three types of land uses. A total of 8 species were exclusive in native forest areas, 26 species only occurred in cabruca areas, and 12 were only found in pasture areas. In total, ten species were recorded and found both in the pasture and native forest areas, seven species were found in the native forest and cabruca areas, and eight species were found in both the cabruca and pasture areas (Appendix A).

Species richness among the areas with the different land uses, according to the GLM test, showed a significant difference (AIC = 301.84, df = 61, $p = 0.0003$), with the model explaining 24% of the data variation, resulting in mean richness of 8.13 for cabruca areas, 4.835 for native areas, and 7.05 for pasture areas (Figure 2)

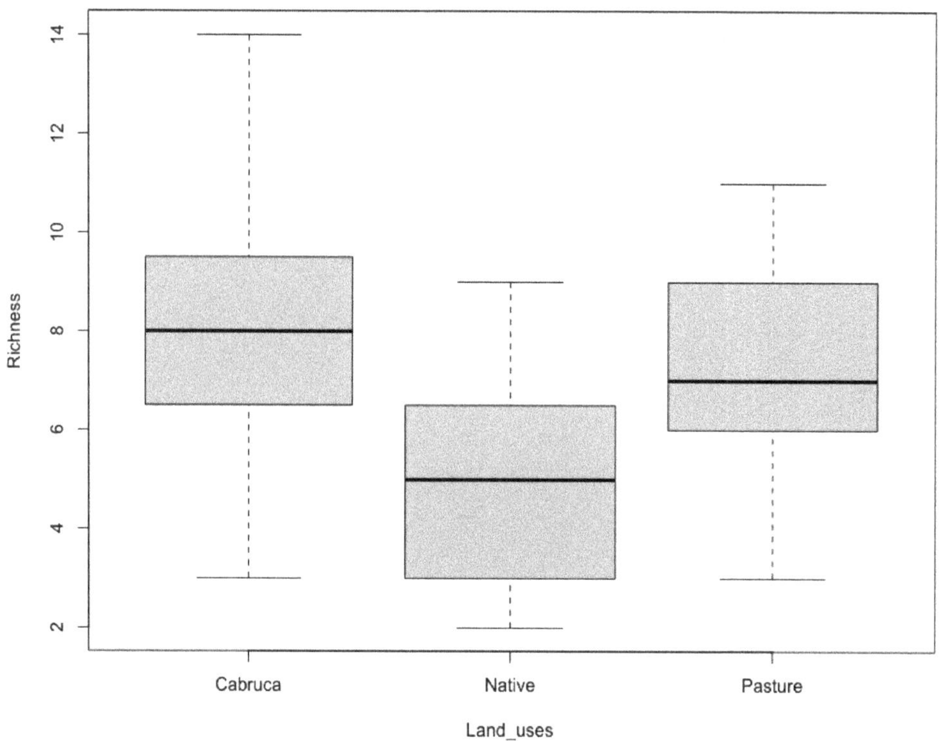

Figure 2. Richness of the order Odonata among the different land use areas: cabruca, native, and pasture. The bold line indicates the mean values (8.13 cabruca, 4.835 native, and 7.05 pasture).

Regarding composition, the ordination analysis showed that the assemblages found in streams with different land uses were different from each other (PERMANOVA $p = 0.001$, R = 0.5092). In summary, the species composition differed for each land use (i.e., the assemblages found in the collection sites in each of the land uses were more similar to each other when compared to the different land uses). The assemblages found in native forest areas were most similar to each other, showing a lower variation between the species collected in areas with native vegetation. In the cabruca and pasture areas, the assemblages had the greatest variation in species composition among the sampled sites (Figure 3).

According to the IndVal analysis, some species were identified as possible bioindicators for each of the land uses (Table 1). A total of 20 species were selected. In the cabruca areas, the species were *Acanthagrion aepiolum* Tennessen, 2004, *Argia chapadae* Calvert, 1909, *Aceratobasis nathaliae* Lencioni, 2004, *Epipleoneura metallica* Rácenis, 1955, *Heteragrion consors* Hagen in Selys, 1862, *Erythrodiplax castanea* Burmeister, 1839, and *Perithemis thais* Kirby, 1889. For the native forest areas, the only species was *Heliocharis amazona* Selys, 1853. In the pasture areas, the species were *Ischnura capreolus* Hagen, 1861, *Telebasis corallina* Selys, 1876, *Erythrodiplax paraguayensis* Förster, 1905, *Erythrodiplax leticia* Machado, 1996, *Planiplax phoenicura* Ris, 1912, *Acanthagrion gracile* Rambur, 1842, *Perithemis lais* Perty, 1834 and *Erythemis credula* Hagen, 1861. Moreover, the following indicator species were obtained for area pairs: cabruca-native forest: *Heteragrion aurantiacum* Selys, 1862; cabruca-pasture: *Erythrodiplax fusca* Rambur, 1842; and native forest-pasture: *Argia hasemani* Calvert, 1909, and *Epipleoneura machadoi* Rácenis, 1960. High specificity values, greater than 0.8, were obtained for all species except *Perithemis lais*, with 0.63. These species always occurred in a single type of land use. In contrast, mostly intermediate or low fidelity values were obtained (almost all of them being less than 0.6). These species had an intermediate or low frequency in the total of each land use.

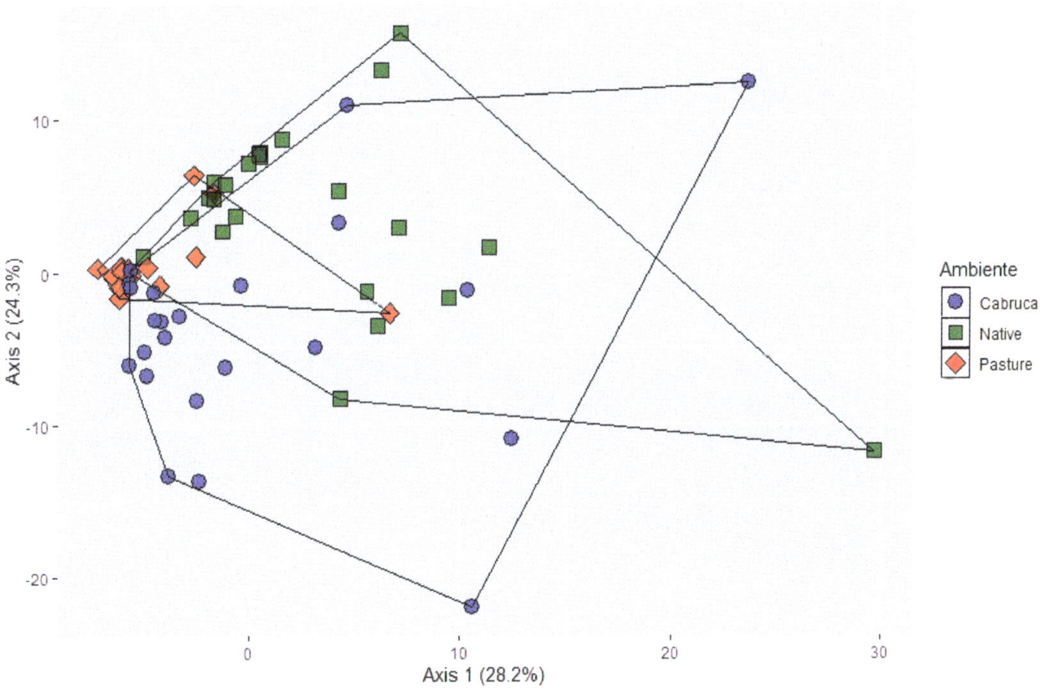

Figure 3. PcoA chart demonstrating the similarity of the composition of Odonata assemblages among the land uses: native forest, cabruca, and pasture areas. The dots in green or with the letter N are associated with areas of native vegetation, dots in blue or with the letter C are associated with areas of cacao cultivation, and dots in red or with the letter P are associated with pasture areas.

Table 1. Indicator species for the different land uses (cabruca, native forest and pasture) based on the IndVal test.

Species	Cabruca	Native Forest	Pasture	Index Value	*p*-Value	Specificity (A)	Fidelity (B)
Acanthagrion aepiolum	x			0.718	0.001	0.9889	0.5217
Acanthagrion gracile			x	0.453	0.024	0.8712	0.2353
Aceratobasis nathaliae	x			0.417	0.017	1.000	0.1739
Argia chapadae	x			0.830	0.001	0.8339	0.8261
Argia hasemani		x	x	0.733	0.001	1.000	0.5366
Epipleoneura machadoi		x	x	0.494	0.027	1.000	0.2439
Epipleoneura metallica	x			0.417	0.011	1.000	0.1739
Erythrodiplax castanea	x			0.417	0.024	1.000	0.1739
Erythemis credula			x	0.420	0.024	1.000	0.1765
Erythrodiplax fusca	x		x	0.712	0.003	0.9224	0.5500
Erythrodiplax leticia			x	0.485	0.004	1.000	0.2353

Table 1. *Cont.*

Species	Cabruca	Native Forest	Pasture	Index Value	*p*-Value	Specificity (A)	Fidelity (B)
Erythrodiplax paraguayensis			x	0.531	0.002	0.9600	0.2941
Heliocharis amazona		x		0.456	0.025	1.000	0.2083
Heteragrion aurantiacum	x	x		0.824	0.001	0.9123	0.7447
Heteragrion consors	x			0.659	0.001	1.000	0.4348
Ischnura capreolus			x	0.737	0.001	0.9231	0.5882
Perithemis lais			x	0.432	0.0216	0.6358	0.2941
Perithemis thais	x			0.674	0.001	0.9495	0.4783
Planiplax phoenicura			x	0.485	0.005	1.000	0.2353
Telebasis corallina			x	0.554	0.003	0.8698	0.3529

4. Discussion

Our results revealed that changes in natural landscapes for other land uses modify the richness and composition of the Odonata assemblages. The cabruca and pasture areas had a greater number of species than the native forest areas. Moreover, the composition differed among the assemblages in the three land uses, which partly corroborates our predictions. Studies assessing the effect of different land uses on Odonata richness have revealed an increase in richness in altered environments when compared to native forest areas [21,33,37]. Modifications in native forest areas cause disturbances of different magnitudes and favor the entry and colonization of Odonata species in these ecosystems. The transformations alter the physical environmental characteristics of the surroundings and aquatic ecosystems and facilitate the colonization of species which are considered open area specialists and habitat generalists [33]. Thus, the different land uses evaluated here (pasture and cabruca) may be maintaining a high richness of species that tolerate slight disturbances when compared to the native forest areas. Of the three evaluated land uses, the cabruca areas exhibited the greatest richness. The cabruca cultivation system caused less severe changes to the ecosystems than the other land uses (namely pasture, agriculture, and urban development). The cabruca areas maintain some of the characteristics found in preserved environments, such as a greater presence of trees, leading to increased canopy cover and, consequently, to greater physical integrity of the channels (stable margins and little or no silting). These characteristics protect aquatic ecosystems from more extensive alterations, while maintaining the physical integrity and quality of water bodies and their surroundings. Thus, the cabruca areas maintain a part of the forest specialist species such as *Forcepcioneura serrabonita, Heteragrion consors, and Perilestes fragilis* [51–53] Moreover, some open area specialist species benefit from several of these changes, such as partial canopy opening, especially *Perithemis thais, Erythrodiplax paraguayensis* and *Orthemis discolor* [33,54], which increases the richness in these areas compared to native areas.

Although the native areas generally exhibit lower richness than other land use areas, they maintain species that are more sensitive to environmental changes, such as forest specialist species [33]. In our study, the species *Heteragrion aurantiacum, Heliocharis amazona, Leptagrion macrurum, Kiautagrion acutum, Gomphidae* sp1, and *Aceratobasis cornicauda* were only found in the native areas (or exhibited greater abundance in these areas). This finding stresses the importance of native areas to preserving and maintaining the diversity of species that are more sensitive to anthropogenic changes [21,33]. In particular, these environments can maintain highly specific habitats such as phytotelmata, and the loss of these habitats can lead to the local extinction of associated species, such as *K. acutum* and *L. macrurum* recorded in this study. These species are endemic to the Atlantic Forest with few occurrence records [55,56]. Furthermore, *K. acutum* is included in the Red List of Threatened Species as being critically endangered [57].

In terms of composition, more drastic changes were observed in the pasture areas than in the cabruca and native areas. Loss of integrity of the aquatic ecosystem, mainly caused by the removal of vegetation and use of these areas by animals, more abruptly alters the physical structure and quality of these ecosystems and their surroundings and homogenizes the habitats for both the larvae and adults of Odonata [21,33,38]. In this regard, these areas can benefit some species and impair others. Open area specialist species and habitat generalist species that benefit from pastures have been observed in these areas, as in the case of *Ischnura capreolus, Erythrodiplax paraguayensis, Perithemis lais,* and *Erythrodiplax fusca* [32,33,52].

Our results indicated a difference between the assemblages in the three land use areas. The native forest areas showed less variation between the assemblages, while the cabruca and pasture areas differed more from each other. This result reveals that native forest maintains more stable assemblages than other land uses. As they are subject to different levels of anthropic modifications, they also exhibit less similar assemblages. These results have been reported in other studies comparing the composition of Odonata assemblages in native areas with palm trees, pastures, and urban areas [21,33,38].

Composition is a good measure to assess the effect of changes in natural environments on Odonata assemblages. Moreover, it has proved effective in studies such as that of Carvalho [33] for evaluating the effects of extensive palm tree cultivation areas in relation to native and pasture areas in streams of the Cerrado biome [38] and in studies on the impacts of vegetation removal on Odonata assemblages [21,32,37,58]. The different land uses modify the composition of the Odonata assemblages due to changes in the environmental variables of the aquatic ecosystems and their surroundings, which allow species with different ecological and behavioral characteristics to remain and colonize these areas [21,33,35,59].

Previous studies in the Amazon and Atlantic Forest have used a "zygopteran/anisopteran" ratio as an index of anthropogenic effect. Native forests are usually dominated by specialist zygopterans, whereas altered environments with more light contain more anisopterans which lower this ratio [60,61]. Among the species considered as bioindicators, our results revealed the existence of indicator species for each of the three land uses. Among the selected species, almost all had high specificity values (A), thus revealing that species had a high correlation with their respective land uses. However, in relation to fidelity (B), the values were relatively low. The species had low representativeness among the total number of sampled points for each of the land uses. Among the recorded species, *Heliocharis amazona* was classified as an indicator of forest areas. This species is always associated with more pristine environments [37,42] and it is considered a forest specialist [33].

In the cabruca areas, the species *Acanthagrion aepiolum, Aceratobasis nathaliae, Argia chapadae, Epipleoneura metallica, Erythrodiplax castanea, Erythrodiplax fusca, Heteragrion consors,* and *Perithemis thais* were considered possible bioindicators. Notably, some of these species are commonly recorded in more forested areas, as in the case of *Heteragrion aurantiacum, Heteragrion consors,* while others are common in more open or anthropogenic areas, as in the case of *Erythrodiplax fusca* and *Perithemis thais* [21,29,32,33,38]. Furthermore, the cabruca areas help protect more sensitive species, such as those classified as forest specialists, and favors some species considered open area specialists or generalists, thus increasing total richness in these areas.

In the pasture areas, the species *Acanthagrion gracile, Erythemis credula, Erythrodiplax leticia, Erythrodiplax paraguayensis, Ischnura capreolus, Perithemis lais, Planiplax phoenicura,* and *Telebasis corallina* were identified as indicators. In general, these species are found in environments with low canopy cover over the channel, consequently resulting in higher solar incidence, as well as in lentic environments or those with slow flow. Therefore, they are considered open area specialists. Moreover, they are commonly recorded in other studies in natural open areas or areas transformed into pastures [21,29,32,33].

Among the three evaluated land uses, cabruca farming stands out as a sustainable production model within the Atlantic Forest domain in southern Bahia. According to Cassano [62], cabruca agroforestry systems effectively contribute to the conservation of fauna and flora. Moreover, this contribution is directly related to factors of composition, structure, and management of cacao plantations. Studies carried out with terrestrial invertebrates, birds, and mammals [12,26,63] have shown that cacao-cabruca areas are important for feeding and reproduction and serve as corridors between forest remnants for these species.

In this regard, it is critical to understand how changes in land use affect ecosystems and their biodiversity. Today, changes in land use are highlighted as one of the main anthropogenic problems worldwide, especially in Brazil [4]. Therefore, further knowledge on these effects can support decision-making and proposals for management methods and sustainable use practices to help protect aquatic ecosystems and associated biodiversity [16]. Moreover, ecosystem balance can be maintained without causing interference in local and regional diversity. Thus, production systems that minimize human impacts, as in the case of cabruca systems, are gaining increasing attention. The importance of compliance with current legislation such as the Forest Code [27] should also be stressed. As shown in the present study, riparian vegetation must be maintained in areas of agriculture and pasturing to protect the physical integrity of aquatic ecosystems, their surroundings, and the biodiversity associated with these environments.

Author Contributions: Conceptualization, M.E.R. and L.R.S.; methodology, M.E.R. and L.R.S.; formal analysis, M.E.R. and L.R.S.; investigation, M.E.R. and L.R.S.; resources, M.E.R.; data curation, M.E.R. and L.R.S.; writing—original draft preparation, M.E.R. and L.R.S.; writing—review and editing, M.E.R. and L.R.S.; supervision, M.E.R.; project administration, M.E.R.; funding acquisition, M.E.R. All authors have read and agreed to the published version of the manuscript.

Funding: This research was funded by Fundação de Amparo à Pesquisa do Estado da Bahia (FAPESB) for fellowships for the first author LRS (process number 073.6787.2020.0007415-43). The Santa Cruz State University—UESC and to the National Council for Scientific and Technological Development—CNPQ, for funding the research project (registered number UESC/PROPP 0220.1100.1693; registered number CNPQ 423737/2018-0).

Institutional Review Board Statement: Not applicable.

Data Availability Statement: Not applicable.

Acknowledgments: We are grateful to the Bahia State Research Foundation (FAPESB) for fellowships for the first author; Santa Cruz State University (UESC) and National Council for Scientific and Technological Development (CNPq); the Graduate Program in Tropical Aquatic Systems (PPGSAT/UESC) and Department of Research and Graduate Studies (PROPP/UESC) for payment of the translation of the manuscript; Estação Veracel RPPN and Cabruca Cooperative for allowing research and for support in field activities and accommodation; and Francisco Valente Neto and Fernando Carvalho for the suggestions and criticisms about this manuscript. To the friends of the Odonatólogos da Bahia group for their support during collections, sorting, material identification and friendship. To the post-graduate program in Tropical Aquatic Systems—PPGSAT/UESC. The Dean of Research and Graduate Studies (PROPP/UESC), for payment of the translation of the manuscript.

Conflicts of Interest: The authors declare no conflict of interest. They also state that the funders had no role in the study design; in the collection, analysis or interpretation of data; in writing the manuscript; or in the decision to publish the results.

Appendix A. Species Recorded for Different Land Uses in Cabruca, Native Forest, and Pasture Areas in the Sampled Streams of an Atlantic Forest Region in Southern Bahia, Brazil

SUBORDEM	Family/Species	Abundance			
ZYGOPTERA	**CALOPTERYGIDAE**	**Cabruca**	**Native Forest**	**Pasture**	**Total**
	Hetaerina longipes Hagen in Selys, 1853	23	25	17	65
	Hetaerina rosea Selys, 1853	87	113	18	218
	COENAGRIONIDAE				
	Acanthagrion aepiolum Tennessen, 2004	85	1	0	86
	Acanthagrion cuyabae Calvert, 1909	0	0	2	2
	Acanthagrion gracile (Rambur, 1842)	1	0	5	6
	Aceratobasis cornicauda (Calvert, 1909)	0	1	0	1
	Aceratobasis macilenta (Rambur, 1842)	1	0	0	1
	Aceratobasis nathaliae (Lencioni, 2004)	5	0	0	5
	Argia chapadae Calvert, 1909	154	32	0	186
	Argia hasemani Calvert, 1909	0	42	24	66
	Epipleoneura machadoi Rácenis, 1960	0	10	14	24
	Epipleoneura metallica Rácenis, 1955	7	0	0	7
	Forcepsioneura sancta (Hagen in Selys, 1860)	1	4	3	8
	Forcepsioneura serrabonita Pinto & Kompier, 2018	12	1	0	13
	Idioneura ancilla Selys, 1860	6	1	4	11
	Ischnura capreolus (Hagen, 1861)	4	1	44	49
	Kiautagrion acutum Santos, 1961	0	3	0	3
	Leptagrion macrurum (Burmeister, 1839)	0	10	0	10
	Metaleptobasis selysi Santos, 1956	4	0	0	4
	Neoneura ethela Williamson, 1917	4	2	0	6
	Neoneura sylvatica Hagen in Selys, 1886	0	0	5	5
	Nehalennia minuta (Selys in Sagra, 1857)	0	0	4	4
	Telagrion longum Selys, 1876	3	1	1	5
	Telebasis corollina (Selys, 1876)	2	7	43	52
	Telebasis willinki Fraser, 1948	1	0	0	1
	DICTERIADIDAE				
	Heliocharis amazona Selys, 1853	0	8	0	8
	LESTIDAE				
	Archilestes exoletus (Hagen in Selys, 1862)	4	0	0	4
	Lestes forficula Rambur, 1842	0	0	8	8
	Lestes tricolor Erichson in Schomburgk, 1848	0	0	1	1
	HETERAGRIONIDAE				
	Heteragrion aurantiacum Selys, 1862	87	203	20	310
	Heteragrion consors Hagens in Selys, 1862	34	0	0	34
	Heteragrion gracile Machado, 2006	0	2	0	2
	PERILESTIDAE				
	Perilestes fragilis Hagen in Selys, 1862	6	4	2	12
ANISOPTERA	**GOMPHIDAE**				
	Gomphoides praevia St. Quentin, 1967	1	0	0	1
	Gomphoides sp1	0	1	0	1
	Progomphus sp	0	1	1	2
	Progomphus montanus Belle, 1973	0	0	2	2
	Phyllogomphoides sp	0	1	1	2
	Zonophora calippus Selys, 1869	0	1	2	3
	LIBELULIDAE				
	Anatya guttata (Erichson in Schomburgk, 1848)	5	0	2	7

Anatya januaria Ris, 1911	2	0	0	2
Dasythemis essequiba Ris, 1919	1	0	0	1
Dasythemis venosa (Burmeister, 1839)	1	0	0	1
Diastatops obscura (Fabricius, 1775)	3	0	8	11
Diastatops nigra Montgomery, 1940	9	3	0	12
Elasmothemis alcebiadesi (Santos, 1945)	6	0	0	6
Elasmothemis cannacrioides (Calvert, 1906)	0	7	3	10
Elga leptostyla Ris, 1909	1	0	0	1
Erythemis carmelita Williamson, 1923	1	0	0	1
Erythemis credula (Hagen, 1861)	0	0	4	4
Erythemis vesiculosa (Fabricius, 1775)	0	1	0	1
Erythrodiplax avittata Borror, 1942	0	1	3	4
Erythrodiplax castanea (Burmeister, 1839)	14	0	0	14
Erythrodiplax famula (Erichson in Schomburgk, 1848)	1	0	0	1
Erythrodiplax funerea (Hagen, 1861)	0	3	0	3
Erythrodiplax fusca (Rambur, 1842)	62	9	30	101
Erythrodiplax latimaculata Ris, 1911	1	0	0	1
Erythrodiplax leticia Machado, 1996	0	0	6	6
Erythrodiplax lygaea Ris, 1911	1	0	1	2
Erythrodiplax maculosa (Hagen, 1861)	3	0	0	3
Erythrodiplax media Borror, 1942	4	0	0	4
Erythrodiplax paraguayensis (Förster, 1905)	0	1	17	18
Erythrodiplax umbrata (Linnaeus, 1758)	5	2	3	10
Erythrodiplax sp1	1	0	0	1
Erythrodiplax sp2	2	0	0	2
Erythrodiplax sp3	1	0	0	1
Macrothemis tenuis Hagen, 1868	4	0	0	4
Micrathyria atra (Martin, 1897)	0	1	1	2
Micrathyria artemis Ris, 1911	8	0	2	10
Micrathyria catenata Calvert, 1909	1	0	2	3
Micrathyria mengeri Ris, 1919	0	0	1	1
Micrathyria ungulata Förster, 1907	12	0	2	14
Nephepeltia phryne (Perty, 1833)	1	0	0	1
Oligoclada abbreviata (Rambur, 1842)	1	0	0	1
Oligoclada umbricola Borror, 1931	1	0	2	3
Orthemis attenuata (Erichson in Schomburgk, 1848)	3	4	2	9
Orthemis discolor (Burmeister, 1839)	4	1	0	5
Perithemis lais (Perty, 1833)	1	3	5	9
Perithemis thais Kirby, 1889	18	1	0	19
Planiplax phoenicura Ris, 1912	0	0	9	9
Tauriphila argo (Hagen, 1869)	0	0	1	1
Uracis infumata (Rambur, 1842)	2	0	0	2
Zenithoptera viola Ris, 1910	0	0	5	5
Total Abundance	**712**	**514**	**332**	**1558**
Zygoptera Abundance	**531**	**471**	**215**	**1217**
Anisoptera Abundance	**181**	**43**	**117**	**341**

References

1. Mittermeier, R.A.; Turner, W.R.; Larsen, F.W.; Brooks, T.M.; Gascon, C. Global Biodiversity Conservation: The Critical Role of Hotspots. In *Biodiversity Hotspots*; Zachos, F., Habel, J., Eds.; Springer: Berlin/Heidelberg, Germany, 2011; pp. 3–22. [CrossRef]
2. Barlow, J.; França, F.; Gardner, T.A.; Hicks, C.; Lennox, G.D.; Berenguer, E.; Castello, L.; Economo, E.P.; Ferreira, J.; Guénard, B.; et al. The future of hyperdiverse tropical ecosystems. *Nature* **2018**, *559*, 517–526. [CrossRef] [PubMed]
3. Pereira, P.; Bogunovic, I.; Muñoz-Rojas, M.; Brevik, E.C. Soil ecosystem services, sustainability, valuation and management. *Curr. Opin. Environ. Sci. Health* **2018**, *5*, 7–13. [CrossRef]
4. de Mello, K.; Taniwaki, R.H.; de Paula, F.R.; Valente, R.A.; Randhir, T.O.; Macedo, D.R.; Leal, C.G.; Rodrigues, C.B.; Hughes, R.M. Multiscale land use impacts on water quality: Assessment, planning, and future perspectives in Brazil. *J. Environ. Manag.* **2020**, *270*, 110879. [CrossRef] [PubMed]

5. Almeida, D.S. *Recuperação Ambiental da Mata Atlêntica*, 3rd ed.; Editus: Luxembourg, 2016.
6. SOS Mata Atlântica, 2018. Relatório Anual 2018. São Paulo—SP. Available online: https://www.sosma.org.br/wpcontent/uploads/2019/07/RA_SOSMA_2018DIGITAL.pdf (accessed on 12 August 2019).
7. Albert, J.S.; Destouni, G.; Duke-Sylvester, S.M.; Magurran, A.E.; Oberdorff, T.; Reis, R.E.; Winemiller, K.O.; Ripple, W.J. Scientists' warning to humanity on the freshwater biodiversity crisis. *Ambio* **2020**, *50*, 85–94. [CrossRef]
8. Allan, J.D. Landscapes and Riverscapes: The Influence of Land Use on Stream Ecosystems. *Annu. Rev. Ecol. Evol. Syst.* **2004**, *35*, 257–284. [CrossRef]
9. Yu, S.; Xu, Z.; Wu, W.; Zuo, D. Effect of land use types on stream water quality under seasonal variation and topographic characteristics in the Wei River basin, China. *Ecol. Indic.* **2016**, *60*, 202–212. [CrossRef]
10. Dala-Corte, R.B.; Melo, A.S.; Siqueira, T.; Bini, L.M.; Martins, R.T.; Cunico, A.M.; Pes, A.M.; Magalhães, A.L.B.; Godoy, B.S.; Leal, C.G.; et al. Thresholds of freshwater biodiversity in response to riparian vegetation loss in the Neotropical region. *J. Appl. Ecol.* **2020**, *57*, 1391–1402. [CrossRef]
11. Lobão, D.E.; Setenta, W.C.; Valle, R.R. Sistema agrossilvicultural cacaueiro-modelo de agricultura sustentável. *Agrossilvicultura* **2004**, *1*, 163–173.
12. Cassano, C.R.; Schroth, G.P.; De Faria, D.E.; Delabie, J.H.C.; Bede, L. Landscape and farm scale management to enhance biodiversity conservation in the cocoa producing region of southern Bahia, Brazil. *Biodivers. Conserv.* **2009**, *18*, 577–603. [CrossRef]
13. Kersul, M.G.; Costa, N.A.; Boullosa, R.G.; Silva, A.A.; Rios, O.; Munhoz, A.D.; Andrade-Silva, B.E.; Maldonado, A.; Gentile, R.; Alvarez, M.R. Helminth communities of sigmodontine rodents in cocoa agroforestry systems in Brazil. *Int. J. Parasitol. Parasites Wildl.* **2020**, *11*, 62–71. [CrossRef]
14. Vanwalleghem, T.; Gómez, J.; Amate, J.I.; de Molina, M.G.; Vanderlinden, K.; Guzmán, G.; Laguna, A.; Giráldez, J. Impact of historical land use and soil management change on soil erosion and agricultural sustainability during the Anthropocene. *Anthropocene* **2017**, *17*, 13–29. [CrossRef]
15. Leal, C.G.; Lennox, G.D.; Ferraz, S.F.B.; Ferreira, J.; Gardner, T.A.; Thomson, J.R.; Berenguer, E.; Lees, A.C.; Hughes, R.M.; Mac Nally, R.; et al. Integrated terrestrial-freshwater planning doubles conservation of tropical aquatic species. *Science* **2020**, *370*, 117–121. [CrossRef] [PubMed]
16. Tonello, G.; Decian, V.S.; Restello, R.M.; Hepp, L.U. The conversion of natural riparian forests into agricultural land affects ecological processes in Atlantic forest streams. *Limnologica* **2021**, *91*, 125927. [CrossRef]
17. Teresa, F.B.; Casatti, L. Importância da vegetação ripária degradada em região intensamente desmatada no sudeste do Brasil: Um estudo com peixes de riacho. *Panamjas* **2010**, *in press*.
18. Lange, M.; Weisser, W.W.; Gossner, M.M.; Kowalski, E.; Türke, M.; Joner, F.; Fonseca, C.R. The impact of forest management on litter-dwelling invertebrates: A subtropical–temperate contrast. *Biodivers. Conserv.* **2011**, *20*, 2133–2147. [CrossRef]
19. Ometo, J.P.H.B.; Martinelli, L.A.; Ballester, M.V.; Gessner, A.; Krusche, A.V.; Victoria, R.L.; Williams, M. Effects of land use and water chemistry and macroinvertebrates in two streams of the Piracicaba river basin, south-east Brazil. *Freshw. Biol.* **2000**, *44*, 327–337. [CrossRef]
20. Benstead, J.P.; Douglas, M.M.; Pringle, C.M. Relationships of Stream Invertebrate Communities to Deforestation in Eastern Madagascar. *Ecol. Appl.* **2003**, *13*, 1473–1490. [CrossRef]
21. Rodrigues, M.E.; Roque, F.D.O.; Guillermo-Ferreira, R.; Saito, V.S.; Samways, M.J. Egg-laying traits reflect shifts in dragonfly assemblages in response to different amount of tropical forest cover. *Insect Conserv. Divers.* **2018**, *12*, 231–240. [CrossRef]
22. Sala, O.E.; Chapin, F.S.; Armesto, J.J.; Berlow, E.; Bloomfield, J.; Dirzo, R.; Huber-Sanwald, E.; Huenneke, L.F.; Jackson, R.B.; Kinzig, A.; et al. Global Biodiversity Scenarios for the Year 2100. *Science* **2000**, *287*, 1770–1774. [CrossRef]
23. Ferreira, W.; Hepp, L.; Ligeiro, R.; Macedo, D.; Hughes, R.; Kaufmann, P.; Callisto, M. Partitioning taxonomic diversity of aquatic insect assemblages and functional feeding groups in neotropical savanna headwater streams. *Ecol. Indic.* **2017**, *72*, 365–373. [CrossRef]
24. Casatti, L. Alterações no Código Florestal Brasileiro: Impactos potenciais sobre uma ictiofauna. *Biota Neotropica* **2010**, *10*, 31–34. [CrossRef]
25. Reid, A.J.; Carlson, A.K.; Creed, I.F.; Eliason, E.J.; Gell, P.A.; Johnson, P.T.J.; Kidd, K.A.; MacCormack, T.J.; Olden, J.D.; Ormerod, S.J.; et al. Emerging threats and persistent conservation challenges for freshwater biodiversity. *Biol. Rev.* **2019**, *94*, 849–873. [CrossRef] [PubMed]
26. Cabral, J.P.; Faria, D.; Morante-Filho, J.C. Landscape composition is more important than local vegetation structure for understory birds in cocoa agroforestry systems. *For. Ecol. Manag.* **2021**, *481*, 118704. [CrossRef]
27. Brazil. Lei n° 12.651—Institui o novo Código Florestal Brasileiro. 2012. Available online: http://www.planalto.gov.br/ccivil_03/_ato2011-2014/2012/lei/l12651.htm (accessed on 27 January 2022).
28. Gómez-Tolosa, M.; Rivera-Velázquez, G.; Rioja-Paradela, T.M.; Mendoza-Cuenca, L.F.; Tejeda-Cruz, C.; López, S. The use of Odonata species for environmental assessment: A meta-analysis for the Neotropical region. *Environ. Sci. Pollut. Res.* **2021**, *28*, 1381–1396. [CrossRef] [PubMed]
29. De Oliveira, J.M.B., Jr.; Shimano, Y.; Gardner, T.A.; Hughes, R.M.; De Marco, P.D., Jr.; Juen, L. Neotropical dragonflies (Insecta: Odonata) as indicators of ecological condition of small streams in the eastern Amazon. *Austral Ecol.* **2015**, *40*, 733–744. [CrossRef]
30. Rodrigues, M.; Roque, F.D.O.; Quintero, J.M.O.; Pena, J.C.C.; de Sousa, D.C.; De Marco, P., Jr. Nonlinear responses in damselfly community along a gradient of habitat loss in a savanna landscape. *Biol. Conserv.* **2016**, *194*, 113–120. [CrossRef]

31. De Oliveira, J.M.B., Jr.; De Marco, P., Jr.; Dias-Silva, K.; Leitão, R.P.; Leal, C.; Pompeu, P.S.; Gardner, T.A.; Hughes, R.M.; Juen, L. Effects of human disturbance and riparian conditions on Odonata (Insecta) assemblages in eastern Amazon basin streams. *Limnogical* **2017**, *66*, 31–39. [CrossRef]
32. Miguel, T.B.; Oliveira-Junior, J.M.B.; Ligeiro, R.; Juen, L. Odonata (Insecta) as a tool for the biomonitoring of environmental quality. *Ecol. Indic.* **2017**, *81*, 555–566. [CrossRef]
33. Carvalho, F.G.; Roque, F.D.O.; Barbosa, L.; Montag, L.F.D.A.; Juen, L. Oil palm plantation is not a suitable environment for most forest specialist species of Odonata in Amazonia. *Anim. Conserv.* **2018**, *21*, 526–533. [CrossRef]
34. Corbet, P.S.; May, M.L. Fliers and perchers among Odonata: Dichotomy or multidimensional continuum? A provisional reappraisal the flier/percher template. *Int. J. Odonatol.* **2008**, *11*, 155–171. [CrossRef]
35. De Marco, P., Jr.; Batista, J.D.; Cabette, H.S.R. Community Assembly of Adult Odonates in Tropical Streams: An Ecophysiological Hypothesis. *PLoS ONE* **2015**, *10*, e0123023. [CrossRef] [PubMed]
36. Valente-Neto, F.; Roque, F.D.O.; Rodrigues, M.; Juen, L.; Swan, C.M. Toward a practical use of Neotropical odonates as bioindicators: Testing congruence across taxonomic resolution and life stages. *Ecol. Indic.* **2016**, *61*, 952–959. [CrossRef]
37. Júnior, C.D.S.M.; Juen, L.; Hamada, N. Analysis of urban impacts on aquatic habitats in the central Amazon basin: Adult odonates as bioindicators of environmental quality. *Ecol. Indic.* **2015**, *48*, 303–311. [CrossRef]
38. Calvão, L.B.; Juen, L.; Junior, J.M.B.D.O.; Batista, J.D.; Júnior, P.D.M. Land use modifies Odonata diversity in streams of the Brazilian Cerrado. *J. Insect Conserv.* **2018**, *22*, 675–685. [CrossRef]
39. Rodrigues, M.E.; Moura, E.B.; Roque, F.D.O. Dragonflies as indicators of the environmental conditions of veredas in a region of central-western Brazil. *Oecol. Aust.* **2019**, *23*, 969–978. [CrossRef]
40. Batista, J.D.; Ferreira, V.R.S.; Cabette, H.S.R.; de Castro, L.A.; De Marco, P.; Juen, L. Sampling efficiency of a protocol to measure Odonata diversity in tropical streams. *PLoS ONE* **2021**, *16*, e0248216. [CrossRef]
41. Garrison, R.W.; von Ellenrieder, N.; Louton, J.A. Dragonfly genera of the New World: An illustrated and annotated key to the Anisoptera. In *Encyclopedia of South American Aquatic Insects: Odonata—Anisoptera*; Heckman, C.W., Ed.; The Johns Hopkins University Press: Baltimore, MD, USA, 2006.
42. Garrison, R.W.; Von Ellenrieder, N.; Louton, J.A. *Damselfly Genera of the New World: An Illustrated and Annotated Key to the Zygoptera*; The Johns Hopkins University Press: Baltimore, MD, USA, 2010.
43. Lencioni, F.A.A. *The Damselflies of Brazil: An illustrated Guide—The Non Coenagrionidae Families*; All Print Editora: São Paulo, Brazil, 2005.
44. Lencioni, F.A.A. *The Damselflies of Brazil: An Illustrated Guide—Coenagrionidae*; All Print Editora: São Paulo, Brazil, 2006.
45. Lencioni, F.A.A. *Damselflies of Brazil*; An Illustrated Identification Guide—Southeast Region; Camara Brazileira do Livro: Jacareí, Brazil, 2017.
46. Gotelli, N.J.; Ellison, A.M. *Princípios de Estatística Em Ecologia*; Artmed Editora: Porto Alegre, Brazil, 2011; p. 528. ISBN 8536324325.
47. Anderson, M.J.; Walsh, D.C.I. Permanova, Anosim, and the Mantel test in the face of heterogeneous dispersions: What null hypothesis are you testing? *Ecol. Monogr.* **2013**, *83*, 557–574. [CrossRef]
48. De Cáceres, M.; Sol, D.; Lapiedra, O.; Legendre, P. A framework for estimating niche metrics using the resemblance between qualitative resources. *Oikos* **2011**, *120*, 1341–1350. [CrossRef]
49. Oksanen, J.F.; Blanchet, G.; Friendly, M.; Kindt, R.; Dan McGlinn, P.L.; Minchin, P.R.; OHara, R.B.; Simpson, G.L.; Solymos, P.; Stevens, M.H.H.; et al. Vegan: Community Ecology Package 2017. R Package Version 2.4-3. Available online: https://CRANR-project.org/package=vegan (accessed on 21 March 2022).
50. De Cáceres, M.; Legendre, P.; Moretti, M. Improving indicator species analysis by combining groups of sites. *Oikos* **2010**, *119*, 1674–1684. [CrossRef]
51. Loiola, G.R.; De Marco, P. Behavioral ecology of Heteragrion consors Hagen (Odonata, Megapodagrionidae): A shade-seek Atlantic forest damselfly. *Rev. Bras. Èntomol.* **2011**, *55*, 373–380. [CrossRef]
52. Calvão, L.B.; Nogueira, D.; Montag, L.; Lopes, M.A.; Juen, L. Are Odonata communities impacted by conventional or reduced impact logging? *For. Ecol. Manag.* **2016**, *382*, 143–150. [CrossRef]
53. Pinto, P.; Kompier, T. In honor of conservation of the Brazilian Atlantic Forest: Description of two new damselflies of the genus Forcepsioneura discovered in private protected areas (Odonata: Coenagrionidae). *Zoologia* **2018**, *35*, 1–19. [CrossRef]
54. Faria, D.; Laps, R.R.; Baumgarten, J.; Cetra, M. Bat and Bird Assemblages from Forests and Shade Cacao Plantations in Two Contrasting Landscapes in the Atlantic Forest of Southern Bahia, Brazil. *Biodivers. Conserv.* **2006**, *15*, 587–612. [CrossRef]
55. Ribeiro, C.; Firme, B.; Araujo, A.S.; Sá, A.; Zander, F.; Teixeira, K.; Santos, L.R.; Rodrigues, M.E. Check-list of Odonata from the state of Bahia, Brazil: Ecological information, distribution, and new state records. *Odonatologica* **2021**, *50*, 161–186. [CrossRef]
56. Santos, C.; Santos, L.R.; Rodrigues, M.E. New records of the Critically Endangered Leptagrion acutum Santos, 1961 (Odonata, Coenagrionidae) from southern Bahia, Brazil. *Check List.* **2021**, *17*, 59–62. [CrossRef]
57. ICMBio; Brasília, D.F. *Livro Vermelho da Fauna Brasileira Ameaçada de Extinção*, 1st ed.; 2018. Available online: https://www.icmbio.gov.br/portal/images/stories/comunicacao/publicacoes/publicacoes-diversas/livro_vermelho_2018_vol7.pdf (accessed on 27 January 2022).
58. Dutra, S.; De Marco, P. Bionomic differences in odonates and their influence on the efficiency of indicator species of environmental quality. *Ecol. Indic.* **2015**, *49*, 132–142. [CrossRef]
59. Pires, M.M.; Müller, N.F.D.; Stenert, C.; Maltchik, L. Influence of different riparian vegetation widths and substrate types on the communities of larval Odonata (Insecta) in southern Brazilian streams. *Acta Limnol. Bras.* **2020**, *32*, e301. [CrossRef]

60. Oliveira-Junior, J.M.B.; Juen, L. The Zygoptera/Anisoptera Ratio (Insecta: Odonata): A New Tool for Habitat Alterations Assessment in Amazonian Streams. *Neotrop. Èntomol.* **2019**, *48*, 552–560. [CrossRef]
61. Ribeiro, C.; Juen, L.; Rodrigues, M.E. The Zygoptera/Anisoptera ratio as a tool to assess anthropogenic changes in Atlantic Forest streams. *Biodivers. Conserv.* **2021**, *30*, 1315–1329. [CrossRef]
62. Cassano, C.R.; Schroth, G.; Faria DDelabie, J.H.; Bede, L.; Oliveira, L.C.; Mariano-Neto, E. *Desafios e Recomendações Para a Conservação da Biodiversidade na Região Cacaueira do sul da Bahia*; Boletim Tecnico nº 205; CEPLAC/CEPEC: Ilhéus, Brazil, 2014.
63. De Almeida-Rocha, J.M.; Peres, C.A.; Monsalvo, J.A.B.; Oliveira, L.D.C. Habitat determinants of golden-headed lion tamarin (*Leontopithecus chrysomelas*) occupancy of cacao agroforests: Gloomy conservation prospects for management intensification. *Am. J. Primatol.* **2020**, *82*, e23179. [CrossRef]

diversity

MDPI

Article

Dragonflies (Odonata) in Cocoa Growing Areas in the Atlantic Forest: Taxonomic Diversity and Relationships with Environmental and Spatial Variables

Laís R. Santos [1,2,3] and Marciel E. Rodrigues [1,2,4,*]

1 Laboratory of Aquatic Organisms ("LOA"), Departament of Biological Sciences, Santa Cruz State University (UESC), Ilhéus 45662-900, Bahia, Brazil
2 Postgraduate Program in Tropical Aquatic Systems ("PPGSAT"), Department of Biological Sciences, Santa Cruz State University (UESC), Ilhéus 45662-900, Bahia, Brazil
3 Postgraduate Program in Ecology and Biodiversity Conservation ("PPGECB"), Department of Biological Sciences, Santa Cruz State University (UESC), Ilhéus 45662-900, Bahia, Brazil
4 Department of Exact and Technological Sciences, State University of Southwest Bahia (UESB), Vitória da Conquista 45083-900, Bahia, Brazil
* Correspondence: merodrigues@uesc.br

Abstract: In the south of Bahia state, a large part of the native Atlantic Forest areas has been modified for the cultivation of cocoa (*Theobroma cacao*). These crops are cultivated under the shade of the canopy of native trees, a system locally known as the "cabruca" agroforestry system. This study aimed to evaluate the relationship of Odonata assemblages (adults and larvae) in cocoa farming areas and to identify the relationships of these species with local and spatial environmental variables of the monitored sites. Altogether, adult and larvae were sampled at 22 sites. Physical and physicochemical water variables were recorded for each site. A total of 1336 dragonflies were collected, of which 20 were Zygoptera species and 30 were adult Anisoptera representatives. The different life stages were related to environmental variables such as conductivity, watercourse channel width, and dissolved oxygen. The space predictors were also associated with the assemblages, mainly for adults. The present study identified that cabruca areas maintain a great diversity of dragonflies, including species that are considered to be forest specialists and more sensitive to landscape changes. The characteristics of this cropping system are considered to be favorable for the conservation of the biodiversity of the Atlantic Forest.

Keywords: dragonflies; land uses; Atlantic Forest; *Theobroma cacao*; cabruca

Citation: Santos, L.R.; Rodrigues, M.E. Dragonflies (Odonata) in Cocoa Growing Areas in the Atlantic Forest: Taxonomic Diversity and Relationships with Environmental and Spatial Variables. *Diversity* **2022**, *14*, 919. https://doi.org/10.3390/d14110919

Academic Editors: M. Olalla Lorenzo-Carballa, Ricardo Koroiva and Luc Legal

Received: 7 September 2022
Accepted: 18 October 2022
Published: 28 October 2022

1. Introduction

Different land uses modify natural environments and negatively impact terrestrial and aquatic ecosystems, along with the associated biodiversity [1–3]. Among the different land uses, the conversion of forests into farmland and pastures has increased the degradation of aquatic ecosystems and water quality due to the removal of vegetation from the surrounding water bodies, silting of channels, and water pollution [3–6]. These impacts modify environments and can affect insect assemblages, thus, causing highly negative effects on aquatic ecosystems and their biodiversity [3,6,7].

Agriculture is one of the leading economic activities in Brazil. Additionally, this activity causes the highest rate of conversion of natural landscapes [3]. According to MapBiomas [8], 30.97% of Brazilian territory is used for agriculture, especially in the Pampas, Cerrado, and Atlantic Forest domains, where large areas have been converted to grow crops and graze livestock. In Atlantic Forest fragments, present in 17 Brazilian states, intensive exploitation has led to drastic reductions and modifications of native areas [3,9], and therefore, these Atlantic Forest fragments have been identified among the priority regions for conservation. The Atlantic Forest houses vast biological diversity and is highly susceptible to human intervention [10,11].

In this regard, the southern region of the state of Bahia can be highlighted, due to the cultivation of cocoa (*Theobroma cacao* L. 1753) since the 18th century. This type of cultivation has had an important economic impact on the region. Studies in cocoa growing areas have demonstrated that these areas have been able to maintain part of the region's flora and fauna, helping to conserve biodiversity [12–14]. In the agroforestry system used to cultivate cocoa, plants grow under the shade of the native trees in the forest, locally known as cabruca or cacao-cabruca areas [13–15]. This agroforestry system is considered to be favorable for the conservation of natural resources and local biodiversity, since it maintains part of the native forest structure and, consequently, protects terrestrial ecosystems, sustains some ecological services, and maintains the species diversity of fauna and flora [12–15].

Environmental changes caused by different land uses are directly related to the structuring and maintenance of biodiversity associated with aquatic ecosystems. Since any modification to the physical conditions of surrounding water bodies alters these environments and the physicochemical parameters of the water, local assemblages are also drastically modified [16–19]. These changes in the abiotic parameters of the water, caused by the reduced riparian vegetation, increase the entry of allochthonous material and sunlight [20–22]. Therefore, it is critical to understand the relationships of environmental and spatial variables among cacao-cabruca areas and the diversity of associated species, and how this type of cultivation affects biodiversity. This understanding can help to identify and to quantify the impacts (negative and positive) of the management and use of land, thus, supporting more effective decision making to protect species and their habitats [19,21].

Among aquatic invertebrate species, dragonflies have been widely used in studies to assess the effect of changes in environmental variables on their biodiversity in areas with different land uses [20,23–26]. This is due to the fact that there are species that can, more or less, tolerate changes in the natural environment, and therefore, are able to reflect the local conditions, and therefore, have been widely used as bioindicators of water quality and the impacts caused to the surroundings of aquatic ecosystems [27–30]. Dragonflies are organisms that are extremely dependent on aquatic ecosystems, using these environments for oviposition and development of larvae, in addition to depending on the surrounding terrestrial ecosystems, since adults use these environments to feed, to defend their territories for reproduction, and to perform important physiological functions such as thermoregulation [30–32].

These ecological and behavioral characteristics of the species allow them to reflect the characteristics and integrity of the ecosystems they live in [27,30–32], which makes the different species ideal as forest and open-area specialists and habitat generalists [33]. In this regard, the relationships among the different local and spatial environmental variables and the different Odonata species in cacao-cabruca areas should be evaluated to understand the effects of this type of cultivation on Odonata biodiversity. Moreover, the group can be used as a "surrogate" for other groups of aquatic insects [34].

Accordingly, in the present study, we evaluated the diversity of Odonata in cocoa cultivation areas and the relationships with local and spatial environmental variables between adults and larvae at the sampling sites. The following two specific objectives were proposed:

(I). To assess the diversity (richness and abundance) of adult and larvae dragonflies in cocoa farming areas among sampling sites. Our prediction is that cocoa farming areas, which are considered to be agroforestry systems, can maintain a wide range of Odonata species by also maintaining groups of species considered to be forest specialists, habitat generalists, and open-area specialists [14].

(II). To assess the influence of environmental physical variables, the physicochemical variables of water, and the spatial relationships in the structuring of dragonfly assemblages (adult and larvae individuals) in cocoa farms. In the present study, the local and physical variables such as canopy cover, channel width, bank structure, and amount of riparian vegetation sites, as well as abiotic water variables (dissolved oxygen, pH, and conductivity) and the distances between sampling sites were found to be important factors in the structuring of Odonata assemblages.

2. Materials and Methods

2.1. Study Area

The southern coastal area of the state of Bahia is inserted in the central corridor of the Atlantic Forest domain, between the south of Bahia and the north of Espírito Santo, which comprises coastal tablelands or tableland forests in a subunit of dense ombrophilous forest [35]. This region has high species diversity and some degree of endemism [35]. Agriculture is the main economic activity in the region [36] including cocoa cultivation, which is one of the five most relevant permanent crops. According to Köppen-Geiger, the climate in the region falls in the tropical rainforest Af classification (tropical super humid), with rainfall evenly distributed throughout the year.

This study was conducted in six municipalities, namely Una, Buerarema, São José da Vitória, Ilhéus, Uruçuca, and Itacaré, which make up the southern region of the state of Bahia. The adult specimens were collected in 22 streams, 16 of which were also used to collect the larval specimens (Figure 1) (Appendix A, Table A1). The samples were collected in first- to third-order streams from September to November 2019 and in July and August 2020. The selected properties for sampling belonged to organic cocoa producers of the Cooperativa Cabruca (an agricultural cooperative). In these areas, the cocoa plants are grown under the shade of native trees and in consortium with others crops, such as banana, açai berry, cupuaçu, vanilla, and palm oil. The properties are considered to be small with an organic cultivation system and without the use of pesticides. The channels of the lotic environments sampled were mostly located in the middle of the cocoa growing areas.

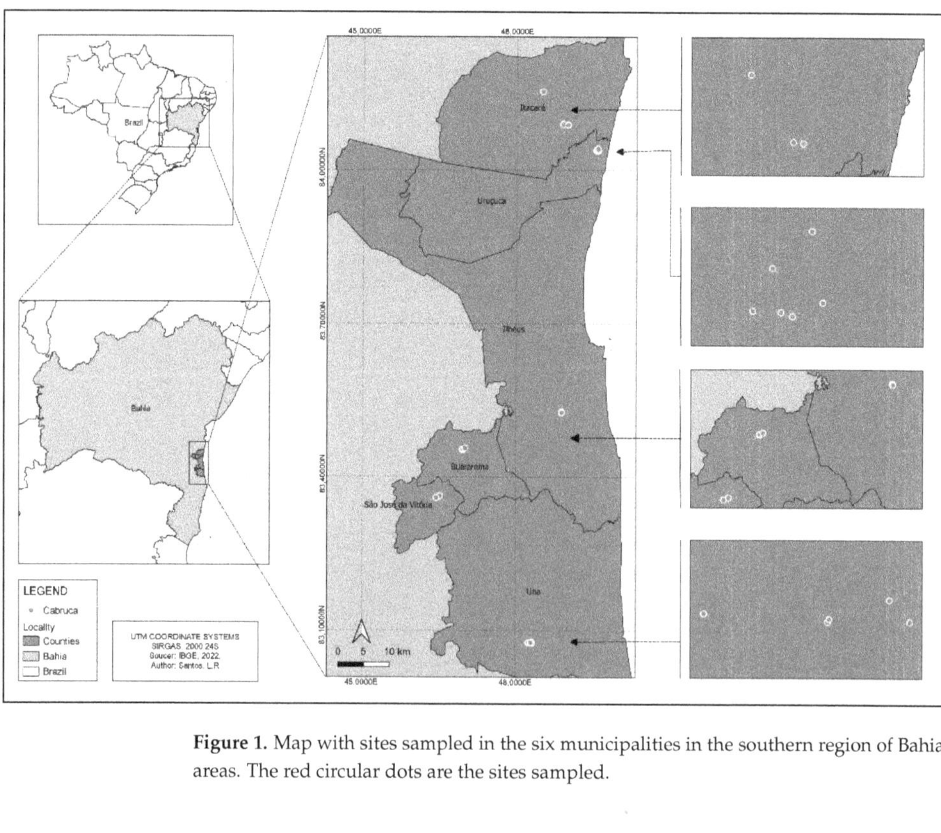

Figure 1. Map with sites sampled in the six municipalities in the southern region of Bahia in cabruca areas. The red circular dots are the sites sampled.

2.2. Sample Collection

The adult specimens were collected with an entomological net, in segments of 100 m on both banks of the streams using an entomological net and a sampling effort of 01:30 h for each sampled stream between 9:00 am and 4:00 pm. Two sampling campaigns were carried out at each stream to increase the representativeness of the assemblages at each sampling site.

The larvae were collected in areas of rapids and backwater in different habitats using a D-frame net and sieve, in all compartments of the environment. A 30 m segment on the shore was selected along each stream and 20 stretches of 1 m were sampled in proportion to the number of registered habitats in each stream (sand, leaves, particulate organic matter, gravel, consolidated bedrock), according to the method of Barbour [37]. Active searches were performed using the sieve, for 15 min, in the different habitats within the 100-m stretch in areas that had not been sampled with the D-frame net.

Subsequently, the specimens were sent to the Laboratory of Aquatic Organisms ("LOA") of the Santa Cruz State University (UESC) for curation and identification based on available keys [23,38–42]. The collected material is currently deposited in the Aquatic Insect Collection of UESC.

2.3. Environmental and Spatial Variables

2.3.1. Physical Variables of the Channels and Surroundings and Physicochemical Variables of the Water

The environmental variables were measured simultaneously with sample collection at each sampling point, where the physical structure of the habitats was evaluated using the habitat integrity index (HII), as proposed by Nessimian [43]. The HII is based on 12 habitat characteristics that evaluate the stream structure in relation to the characteristics of the forest areas, the land use pattern, retention mechanisms, aquatic vegetation, substrate, and debris. Finally, the obtained values of each metric were transformed using the method of Oliveira-Junior and Juen [27]. The sum of the 12 qualitative habitat characteristics varies from 0.1 to 1, where the lowest values are characteristic of locations with lower environmental integrity and values close to 1 refer to more intact or preserved locations. The HII has been widely used in ecological studies to evaluate the integrity of aquatic ecosystems and it is considered to be a good tool in studies that evaluate these changes in aquatic ecosystems [17].

In order to evaluate the amount of canopy coverage, three photographic images were obtained in the center of the stream every 10 m within the 30 m segment. In each of these segments, a photograph of the canopy was taken with a mobile phone camera positioned 30 cm over the water level, for a total of three images for every sampled stream. The images were uploaded for treatment with the Image J software to calculate the light and dark areas of each image. The software produces a black and white binary image of the photographs that differentiates the regions with light input from the canopy cover. The generated image was used to calculate the white area, which corresponds to the area of light input. After performing this process on the three images, the average light input between the images was calculated for each stream.

To obtain local riparian vegetation integrity data on both banks of the streams, three quadrants of 100 m^2 (10 m × 10 m) were selected along the 30 m stretch on each bank. Of these six quadrants, three quadrants were randomly selected to record the total number of native trees with more than 15 cm of circumference at breast height (CBH), followed by a calculation of the diameter at breast height (DBH). The number of cocoa individuals within each of the quadrants was also counted. In the field, we observed that this information could provide valuable insight into the relationships of Odonata assemblages at the sampling points.

The amount of forest cover was calculated from the information on the coordinates taken from each of the points sampled with the aid of GPS (Etrex 10). This information was loaded into the Qgis desktop software 3.16.9 Geographic Information System.

The images used to extract forest cover data were obtained from the MapBiomas land cover and use database, with a spatial resolution of 30 m (30 × 30). After the delimitation of the cultivation areas, buffers with a radius of 1000 m were created around the sampled points and then, the amount of coverage within each buffer was calculated (Appendix A, Table A1).

The width, depth, and current speed of the stream were measured along the 30 m stretch, totaling five measurements per point. A multiparameter probe (YSI Professional) was used to collect the physicochemical parameters of the water (temperature, pH, conductivity, dissolved oxygen (OD), and salinity).

All the values of the collected variables used in the analyses represent the means between the two sampling campaigns and the number of replicas at each point (Appendix A, Table A2). These variables were used as environmental predictors given their critical importance in the structuring of aquatic insect assemblages [20,26,34,44].

2.3.2. Spatial Variables

Spatial data were initially collected by calculating spatial filters from the geographical coordinates of each sampled stream. The coordinates were captured using a Garmin eTrex GPS. These coordinates were used to construct a Euclidean distance matrix using the "vegdist" function of the "vegan" package [45]. The spatial filters were calculated by analyzing the principal coordinates of neighbor matrices (PCNM), which, in turn, were also calculated using the "vegan" package with the "pcnm" function. This method can determine whether spatial predictors are structuring the distribution of assemblages [19]. The significant axes were selected using the "adespatial" package [45] in the R software. The PCNM vectors were submitted to the forward selection model [46]. The selected vectors were subsequently used in the redundancy analysis (RDA). All the methods mentioned above were used to test the influence of spatial variables on adult and larvae specimens.

2.3.3. Data Analysis

The data matrix containing the abundance of adult and larvae specimens was transformed, separately, into a relative abundance matrix using Hellinger's method to reduce the effect of large abundances [46]. Each stream was considered to be a sample unit, totaling 22 for adults and 16 for larvae. The relationship of group diversity with the with amount of forest cover was assessed by sorting the relative abundance of the adult species and the genera of the larvae with the amount of forest cover of the sampled areas using the "generic" function of R.

The local physical and physicochemical variables of the water were subjected to Spearman correlation analysis to exclude correlated variables (>70%). None of these variables presented this level of correlation, so none were excluded. Subsequently, these variables were used to perform a principal component analysis (PCA) using Euclidean distance to evaluate their relationship with the sampling sites. The axes of the PCA were tested with the broken stick method [47,48] to determine which axes would be used as environmental variables, and later in the analysis of the adult and larvae of the species collected.

The combinations between spatial and environmental predictors and the composition of adult and larvae assemblages were analyzed separately. The selected spatial and environmental vectors were used in the redundancy analysis (RDA) for each of the different life stages (adult and larvae). The RDA significance was tested by analysis of variance (ANOVA). All analyses were performed using the R software, with the "vegan" package [49] and "adespatial" package [45].

3. Results

A total of 689 adult Odonata individuals were captured, representing seven families and 50 species, 520 individuals distributed in 20 species of Zygoptera and 169 individuals and 30 species of Anisoptera (Appendix A, Table A2. Among the Zygoptera, the most representative species were *Argia chapadae* Calvert, 1909; *Hetaerina rosea* Selys, 1853; and *Heteragrion lencionii* Vilela, Farias & Santos, 2021, with, respectively, 154, 87, and 87 individuals. Among the Anisoptera, *Erythrodiplax fusca* Rambur, 1842, and *Perithemis thais* Kirby, 1889 totaled 62 and 18 individuals, respectively.

In relation to larvae, 647 specimens were collected, comprising five families and 11 genera of Zygoptera and three families and 23 genera of Anisoptera. The most representative genera were *Homeoura* and *Heteragrion* with 43 and 76 individuals, respectively, (Zygoptera), along with *Epigomphus* with 72 individuals (Anisoptera). Regarding the life stages (adult and larvae), 14 genera were collected, four genera of the suborder Zygoptera and 10 genera of the suborder Anisoptera. Sixteen genera were only found as adults and 20 genera were recorded only for the larvae (Appendix A, Tables A3 and A4).

The relationship between the relative abundance with the amount of forest cover in the cocoa areas indicated alteration of some species as a function of the amount of forest cover for adult and larvae in the evaluated areas (Figures 2 and 3). Some species and genera were related to areas with low amounts of forest cover in the sampled areas. The most open-area specialist genera and species targeted sites with a lower amount of forest cover, as in the case of the species *Telebasis willinki* Fraser, 1948; *Acanthagrion aepiolum* Tennessen, 2004; *Erythrodiplax famula* Erichson in Schomburgk, 1848; *Elasmothemis alcebiadesi* Santos, 1945; and *Diastatops nigra* Montgomery, 1940; and the genera *Castoraeschna*, *Roppaneura*, *Erythrodiplax*, *Enallagma*, *Lestes*, and *Tramea*, located on the left side of both graphs (Figures 2 and 3).

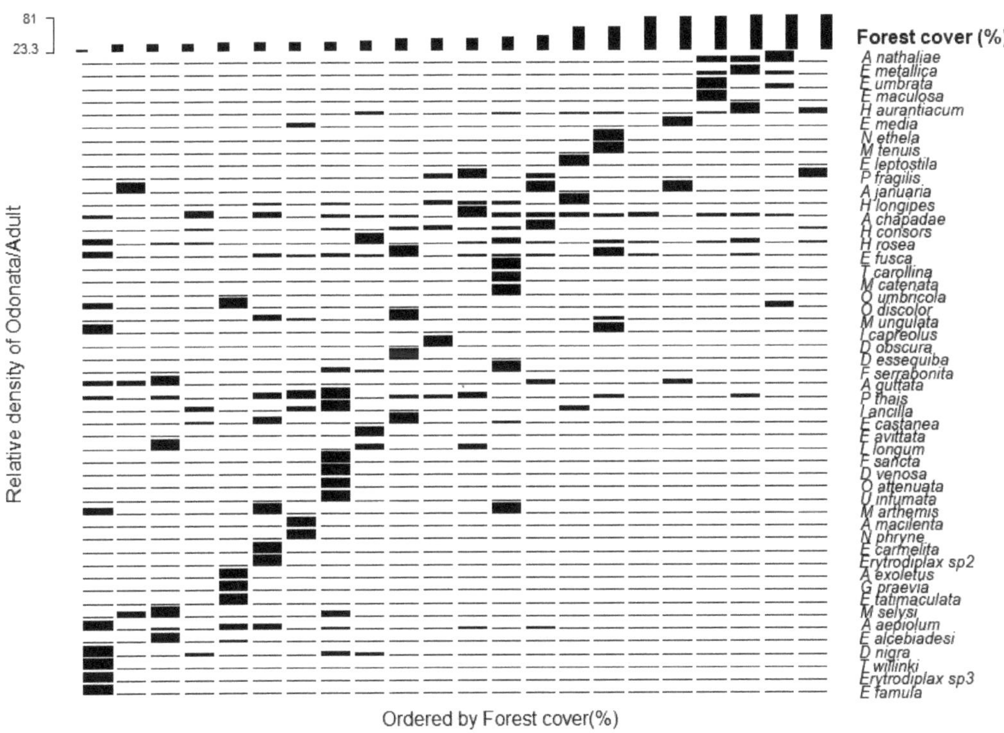

Figure 2. Relative abundance of adult individuals of Odonata with the amount of forest cover (%) for the streams sampled in cabruca areas.

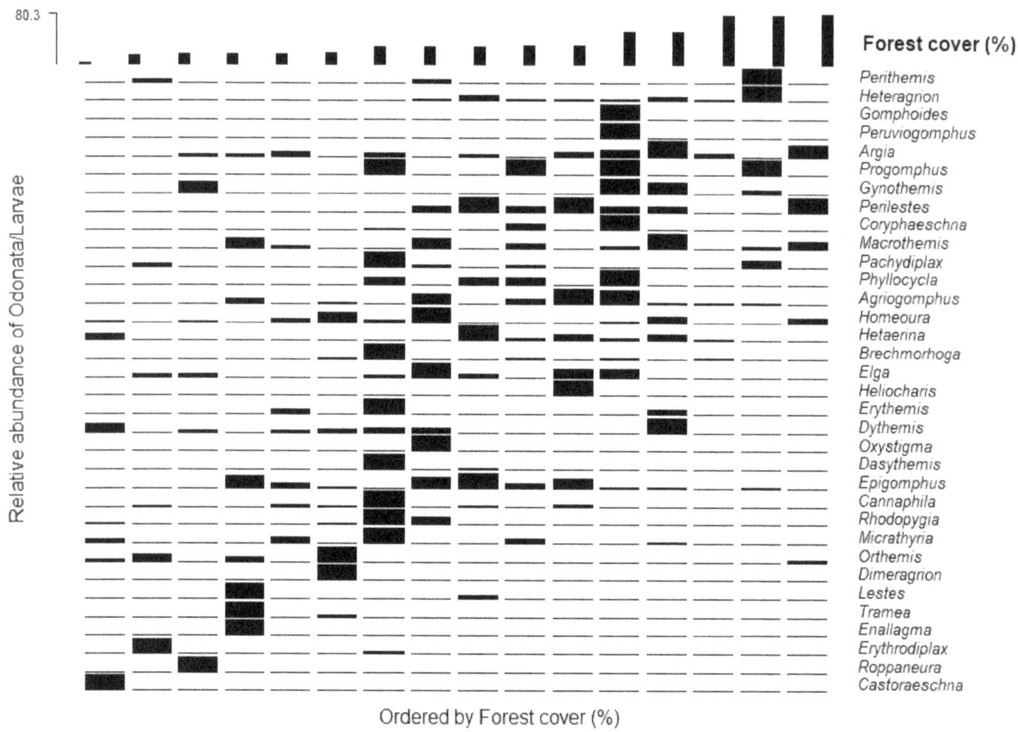

Figure 3. Relative abundance of larvae of Odonata with the amount of forest cover (%) for the streams sampled in cabruca areas.

The species of adults and genera of larvae considered to be habitat generalists are represented in the central strips of the charts, which show that these groups may have greater tolerance for the loss of forest cover surrounding streams in the cocoa growing areas (Figures 3 and 4). Finally, species and genera that were associated with cocoa growing areas with greater amounts of forest cover were *Aceratobasis nathaliae* Lencioni, 2004; *Epipleoneura metallica* Rácenis, 1955; *Erythrodiplax maculosa* Hagen, 1861; and *Heteragrion aurantiacum* Selys, 1862. The genera *Peruviogomphus*, *Gomphoide*, and *Heteragrion* were more abundant in the cacao-cabruca areas with a higher level of amount forest cover (Figure 4).

The environmental variables in the PCA analysis for adults presented the first four axes and the observed values were greater than those estimated by the broken stick method. The four axes of the PCA accounted for 58.48% of the cumulative proportion. The PC1 axis explained 17.47%, PC2 15.11%, PC3 14.53%, and PC4 11.37% (the eigenvalues for the axes were PC1 = 4.541, PC2 = 3.927, PC3 = 3.777, and PC4 = 2.957). The variables that contributed the most on each axis were greater than 0.6. For the PC1 axis, the variables were conductivity (−0.684), salinity (−0.758), and margin structure (0.630). For the PC2 axis, it was stream width (0.665). In the formation of the PC3 axis, the variables were depth (−0.679), undercut margin (0.630), aquatic vegetation (0.614), and HII (0.727). For the PC4 axis, the variables were dissolved oxygen (0.785) and luminosity (0.686).

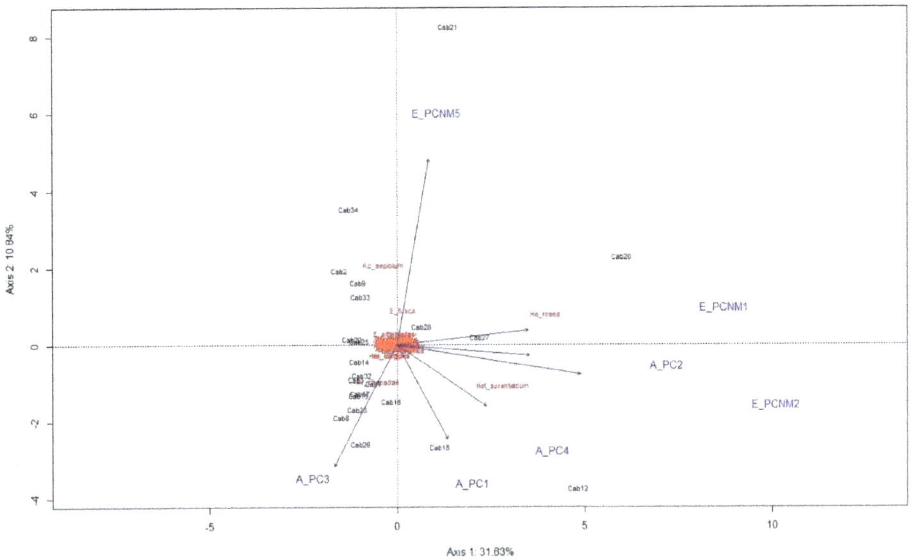

Figure 4. Result of the redundancy analysis showing the relationships among spatial variables and environmental variables with the sample sites and adult individuals of Odonata. A _= Environmental variable axes and E_ = Spatial variable axes.

For the larvae, the first three axes showed the best response in the PCA analysis according to the broken stick method. These accounted for 50.64% of the cumulative PCA proportion. The PCA1 axis explained 20.32%, PCA2 15.67%, and PC3 14.65% (respectively, the eigenvalues for the axes were 5.283; 4.073, and 3.807). The environmental variables that most influenced the formation of axes with values greater than 0.6 included conductivity (−0.639), dissolved oxygen (−0.747), salinity (−0.678), and margin structure (0.669) for PCA1, width (−0.787), channel/sediment (0.603), and current and backwater, or meanders (−0.646) for PCA2, and riparian forest width (0.621) and habitat integrity index (0.608) for PCA3.

For the spatial variables, the PCNM axes 1, 2, and 5 were selected by the forward selection method, corresponding to the axes with better correlation with the data when related to adult individuals. The obtained results were PCNM2 with correlation factor $R^2 = 0.199$ ($F = 4.97$, $p = 0.006$); PCNM1 with $R^2 = 0.116$ ($F = 3.24$, $p = 0.014$), and PCNM5 with $R^2 = 0.084$ ($F= 0.08$, $p = 0.018$). For the case of larvae, the selected axes were PCNM2 and -1. The obtained results were PCNM2 with $R = 0.200$ ($F = 3.493$, $p = 0.006$) and PCNM1 with $R^2 = 0.116$ ($F= 2.205$, $p= 0.030$) (Table 1).

Table 1. PCNM axes of spatial predictors selected by the forward selection model for adult and larvae among the streams in cocoa areas in the southern region of the state of Bahia.

		Axes	Order	R^2	R^2Cum	AdjR^2Cum	F	*p* Value
Adult	1	PCNM2	2	0.199	0.199	0.159	4.979	0.006
	2	PCNM1	1	0.116	0.316	0.244	3.245	0.014
	3	PCNM5	5	0.084	0.400	0.300	2.536	0.018
Larvae	1	PCNM4	4	0.235	0.235	0.195	5.867	0.038
	2	PCNM1	1	0.114	0.360	0.278	3.177	0.026

The result of the redundancy analysis (RDA) indicated that the combination of environmental and spatial variables accounted for 42.74% of the assemblage composition of adult individuals of Odonata in the cabruca areas. Axis 1 explained 31.63% and Axis 2 explained 10.84% of the data variation (Figure 4). The ANOVA showed that the ordering analysis generated by the RDA was statistically significant (F = 2.67, p = 0.001). The points most influenced by the evaluated axes were Sites 18 by Axis 1 of the PCA, Sites 8, 14, 15, 16, 17, 23, 26, and 32 by Axis 3 of the PCA, Sites 20, 27, and 28 by Axis 1 of the spatial analysis, and Sites 2, 9, 21, 33, and 34 by Axis 5 of the spatial analysis. The species associated with environmental and spatial variables were *Acanthagrion aepiolum* and *Erythrodiplax fusca* for the spatial axis (PCNM5) and *Hetearina rosea* for the spatial axis (PCNM1). The species *Heteragrion aurantiacum* was associated with the PC4 axis of the environmental variables and the species *Argia chapadae* with the PC3 axis. The partition analysis of the environmental and spatial variables indicated that the environmental variables accounted for 32.75%, the spatial variable for 40.06%, and the two together in the model explained 57.25% of the data variation (Table 2).

Table 2. Partition of the RDA for environmental and spatial variables related to adult individuals.

Variables	Df	R^2	Adj.R.Squared
Environment	4	0.32753	0.16930
Spatial	3	0.40061	0.30071
Environment + spatial	7	0.57255	0.35883
Residue			0.64117

The RDA performed with the assemblages of larvae of Odonata and the environmental and spatial variables resulted in 31.58% of the explanation obtained in the first four axes that contributed the most (Axis 1, 12.40% and Axis 2, 7.83%). The ranking generated by ANOVA did not show significant values (F = 1.269, p = 0.059) (Figure 5). However, from the graph, it was observed that the sites associated with environmental and spatial variables were Sites 7, 17 and 28 for the PC1 axis, Site 2 for the PC2 axis, Sites 27 and 34 for the axis PC3, and Site 16 for the PCNM2 axis. The partition analysis indicated that the environmental variable was responsible for 27%, the spatial variable for 19%, and both explained 36%; the variables are presented in Table 3.

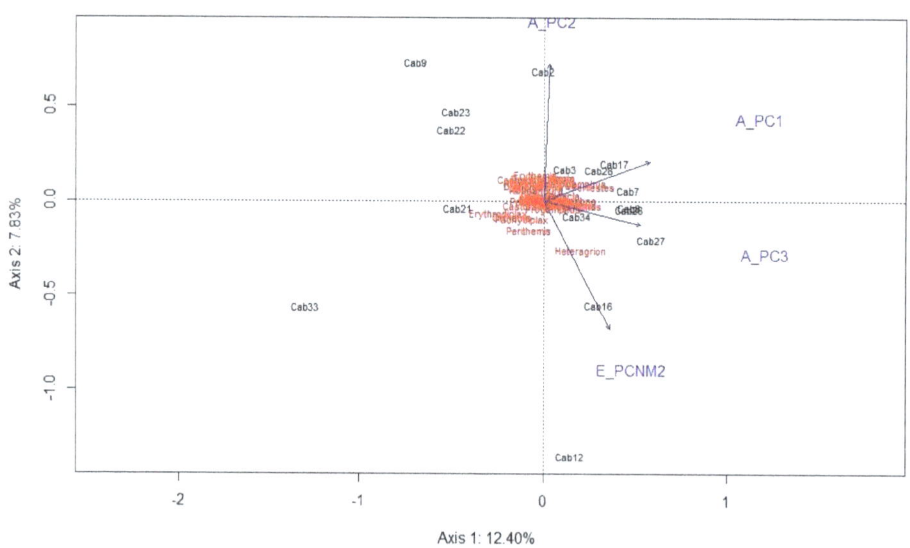

Figure 5. Results of the redundancy analysis showing the relationships among spatial variables, environmental variables, and larvae of Odonata. A_ = Environmental variable axes and E_ = Spatial variable axes.

Table 3. Partition of the RDA for environmental and spatial variables related to larvae.

Variables	Df	R^2	Adj.R.Squared
Environmental	3	0.27670	0.09588
Spatial	1	0.19581	0.13836
Environmental + Spatial	4	0.36794	0.13810
Residue			0.86190

4. Discussion

The results indicated that the cabruca areas maintain a vast diversity and richness of Odonata species (adults and larvae). Studies with the group in the Atlantic Forest that have compared different land uses have highlighted the importance of cocoa cultivation areas for the diversity of species in the region [14]. Notably, these areas contain species that are considered to be specialists of forests areas, such as *Epipleoneura metallica* Rácenis, 1955; *Aceratobasis nathaliae* Lencioni, 2004; *Heteragrion aurantiacum* Selys, 1862; *Forcepsioneura serrabonita* Pinto & Kompier, 2018; and *Perilestes fragilis* Hagen in Selys, 1862. According to the data, the distribution of these species in cocoa farms was related to greater environmental integrity of the sampled areas, which directly reflected the local environmental variables. In other words, the environmental and spatial variables in the cabruca areas played important roles in structuring these Odonata assemblages. Most of the environmental variables were common to the two evaluated life stages.

Regarding the relative abundance of species with the amount of forest cover in cocoa areas, a similar pattern of occurrence was observed between adult species and larval genera. The cabruca areas with the lowest amount of forest cover values showed greater relative abundance of adult species and larval genera, formed by species considered to be specialists of open areas, associated with areas with greater canopy opening, greater solar incidence, more lentic environments, and the presence of macrophytes [50,51]. Some examples are *Erythrodiplax latimaculata* Ris, 1911; *Telebasis willinki* Fraser, 1948; *Diastatops nigra* Montgomery, 1940; and *Elasmothemis alcebiadesi* Santos, 1945 for the adults, and the genera *Erythrodiplax*, *Enallagma*, and *Castoraeschna* for the larvae.

A broad range of species had its distributions along the entire gradient of the forest cover, and therefore, can be considered to be habitat generalist species [50,52], as in the cases of *Hetaerina rosea* Selys, 1853; *Argia chapadae* Calvert, 1909; and *Micrathyria ungulata* Förster, 1907; and the genera *Progomphus*; *Argia*, 1842; and *Hetaerina*. The relative abundance of some adult species and genera of larvae was associated with areas with a higher amount of forest cover, as in the cases of the species *Aceratobasis nathaliae* Lencioni, 2004; *Epipleoneura metallica* Rácenis, 1955; and *Heteragrion aurantiacum* Selys, 1862, usually considered to be forest specialist species, and the genera *Heteragrion*, *Gomphoide*; and *Peruviogomphus*, composed of more sensitive individuals that need greater environmental integrity and conditions more comparable to forest areas [53,54]. However, it is worth mentioning that some species considered to be specialists in forests that have already been recorded in the region in areas of native forest in the Atlantic Forest [14] were not recorded in the cocoa cultivation areas sampled. For example, *Aceratobasis cornicauda* Calvert, 1909; *Heliocharis amazona* Selys, 1853; *Kiautagrion acutum* Santos, 1961; and *Leptagrion macrurum* Burmeister, 1839, the last two species being phytotelmata [14]. Emphasizing that even crops that maintain a part of the native vegetation such as cabruca areas may not maintain some groups of species considered to be more sensitive or with more specialized habitats.

The environmental variables that most contributed to the formation of the PCA axes, for adults, included conductivity, salinity, dissolved oxygen, margin structure, stream width, channel undercut margin, depth, aquatic vegetation, and luminosity. For the larvae, the variables that most contributed to the formation of the PCA axes were almost the same as for the adults, except aquatic vegetation and undercut margin. These results emphasize the importance of the same set of environmental variables for dragonfly species (adults and larvae) in cocoa growing areas, and that the changes in these variables can modify the structuring of local assemblages in cabruca areas. These environmental variables have already been highlighted in other studies that have evaluated the structuring of Odonata assemblages in other types of land use or with different environmental modifications [20,24,26,29].

In cabruca areas, changes in the amount forest cover, riparian vegetation, canopy openings, erosion of stream banks, increase the number of vascular plants in the channels, and sediments in the stream, as well as the air and water temperature. These modifications may favor the abundance and permanence of species considered to be open-area specialists and habitat generalists [33], as in the cases of *Perithemis thais* Kirby, 1889; *Argia chapadae* Calvert, 1909; *Telebasis willinki* Fraser, 1948; and *Erythrodiplax latimaculata* Ris, 1911, which have also been recorded in other studies of altered areas [25,26,32,55,56].

Physicochemical variables of water also play important roles in structuring the assemblages of aquatic insects. Changes in water parameters and habitat affect most aquatic organisms, including Odonata [26]. The conductivity, salinity, and dissolved oxygen variables stand out in the structuring of Odonata assemblages (adults and larvae) and can serve as parameters to define the specific oviposition sites selected by adults and are closely associated with development from larvae stages of many Odonata species [42,57]. This is especially true for more sensitive species that depend on more pristine environments and more particular habitat conditions for their development. The water in more preserved environments has higher dissolved oxygen concentrations [58], which can help ensure the permanence of more sensitive species such as forest specialists. These forest specialists include *Heteragrion aurantiacum* Selys, 1862; *Forcepsioneura serrabonita* Pinto & Kompier, 2018; and *Perilestes fragilis* Hagen in Selys, 1862; as well as species of the genera *Heteragrion, Peruviogomphus,* and *Agriogomphus* [26,56,59–62].

Assemblage structuring between the sampling sites was also associated with the axes of spatial analysis. These results suggest that the spatial distance between the sampling points plays a critical role in the structuring of the assemblages in the cocoa cultivation sites. Areas of the same region have more homogeneous assemblages owing to the dispersal capacity of most of the sampled Odonata adults. The species *Acanthagrion aepiolum* Tennessen, 2004; *Erythrodiplax fusca* Rambur, 1842; and *Hetaerina rosea* Selys, 1853, which are open-area specialists (the first two) and habitat generalists (the third species), have also been associated with spatial filters. More generalist or open-area specialist species or species with greater dispersal capabilities are usually associated with spatial filters [26,33,51,63–65].

In general, cabruca areas are favorable environments for the conservation of habitat structure and for the colonization of more sensitive species. Moreover, such areas are considered to have sustainable agricultural cultivation, as revealed in several studies with other groups such as mammals, birds, and insects (dragonflies) [12,14,66]. However, few studies have related these areas to invertebrates in general and especially to aquatic invertebrates. In short, the results of this study emphasize that: (1) The cabruca areas manage to maintain a high diversity of Odonata and several species generally associated with more pristine areas. (2) The environmental and spatial variables are determinants in the structuring of Odonata assemblages in cabruca areas. More preserved areas or areas with higher amounts of forest cover maintain assemblages of species that are more sensitive to human impacts, whereas areas with lower amounts of forest cover have Odonata assemblages with more habitat generalist or open-area specialist species. (3) The relationship of the species with the amount of forest cover in the cabruca areas is reflected in the relative abundance of the species collected in the different sampled sites. From this perspective, the present study highlights that this cultivation system may be supporting the conservation of Odonata biodiversity in Atlantic Forest areas. Therefore, it is critical to ensure the integrity of these areas to maintain groups of more sensitive species. Notably, in Brazil, the recently approved Law 14,119 of 13 January 2021 establishes the National Policy on Payment for Environmental Services, which encourages recovery and recomposition through the planting of native species or by agroforestry systems. This law can further strengthen the relevance of cabruca areas as sites able to maintain large species diversity and as areas of economic, environmental, and social importance.

5. Conclusions

The present study identified that cabruca areas maintain a great diversity of dragonflies, including species that are considered forest specialists and more sensitive to landscape changes. Moreover, the local and spatial environmental characteristics proved to be important factors in the structuring of these assemblages in the cabruca areas. The characteristics of this cropping system are considered to be favorable for the conservation of the biodiversity of the Atlantic Forest. In this regard, conserving part or some of the characteristics of native habitats has contributed to maintain local species that are more sensitive to changes in natural landscapes, which confirms the importance of this agroforestry system for the conservation of dragonfly species in the Atlantic Forest areas.

Author Contributions: Conceptualization, M.E.R. and L.R.S.; methodology, M.E.R. and L.R.S.; formal analysis, M.E.R. and L.R.S.; investigation, M.E.R. and L.R.S.; resources, M.E.R.; data curation, M.E.R. and L.R.S.; writing—original draft preparation, M.E.R. and L.R.S.; writing—review and editing, M.E.R. and L.R.S.; supervision, M.E.R.; project administration, M.E.R.; funding acquisition, M.E.R. All authors have read and agreed to the published version of the manuscript.

Funding: This research was funded by Fundaçao de Amparo à Pesquisa do Estado da Bahia (FAPESB) for fellowships for the first author L.R.S. (process number 073.6787.2020.0007415-43). The Santa Cruz State University (UESC) and to the National Council for Scientific and Technological Development (CNPQ), for funding the research project (registered number UESC/PROPP 0220.1100.1693 and registered number CNPQ 423737/2018-0).

Institutional Review Board Statement: Not applicable.

Data Availability Statement: Not applicable.

Acknowledgments: The authors thank Francisco Valente Neto and Fernando Carvalho for their suggestions and criticisms of this manuscript, and the Cooperativa Cabruca for allowing research and support in field activities. Thanks are also due to the friends of the Odonatólogos da Bahia group for their support during collections, sorting, material identification, and friendship and to the post-graduate program in Tropical Aquatic Systems–PPGSAT/UESC. Thanks also to the Dean of Research and Graduate Studies (PROPP/UESC) for payment of the translation of the manuscript, and to Wagner Magalhães for reading, revising, and correcting the English in the text. To Saulo Araujo for helping with the figures.

Conflicts of Interest: The authors declare no conflict of interest. They also state that the funders had no role in the study design; in the collection, analysis or interpretation of data; in writing the manuscript; or in the decision to publish the results.

Appendix A

Table A1. Geographical coordinates of sites sampled in cocoa-growing areas in Bahia. Total value of the habitat integrity index (HII). Information on the sampling of adults and/or larvae at each of the sites sampled.

Points	Lat (utm)	Long (utm)	Total Value HII	Amount Forest Cover (%)	Adult Sampling	Larvae Sampling
Cab2	−14.43833	−39.04059	0.69	33.2	X	X
Cab3	−14.43848	−39.04015	0.67	41.9	X	X
Cab7	−14.43828	−39.04168	0.69	44.1	X	X
Cab8	−14.43667	−39.04091	0.7	41.8	X	X
Cab9	−14.43528	−39.0394	0.6	41.6	X	X
Cab12	−15.04876	−39.3252	0.73	79.4	X	X

Table A1. *Cont.*

Points	Lat (utm)	Long (utm)	Total Value HII	Amount Forest Cover (%)	Adult Sampling	Larvae Sampling
Cab14	−15.05157	−39.33198	0.6	78.6	X	-
Cab15	−14.89933	−39.10437	0.64	78.0	X	
Cab16	−14.90076	−39.10413	0.72	78.6	X	X
Cab17	−14.33459	−39.13932	0.75	80.3	X	X
Cab18	−14.39319	−39.09369	0.76	81.0	X	-
Cab20	−14.39199	−39.1025	0.66	37.0	X	-
Cab21	−14.39305	−39.09368	0.67	23.3	X	X
Cab22	−14.96656	−39.28364	0.68	35.1	X	X
Cab23	−14.96529	−39.28327	0.6	32.5	X	X
Cab25	−14.96356	−39.27904	0.81	33.6	X	-
Cab26	−14.43833	−39.04059	0.8	60.4	X	X
Cab27	−14.43848	−39.04015	0.8	43.7	X	X
Cab28	−14.43828	−39.04168	0.7	60.6	X	X
Cab32	−14.43667	−39.04091	0.78	31.0	X	-
Cab33	−14.43528	−39.0394	0.64	31.0	X	X
Cab34	−15.04876	−39.3252	0.7	33.0	X	X

Table A2. Measurements of local environmental variables (minimum and maximum values, average, and standard deviation) in the cacao-cabruca areas, in the streams sampled in the southern region of Bahia.

Variables	Number of Measurements per Sites	Minimum	Maximum	Average	Standard Deviation (SD)
Width (cm)	5	33.00	292.00	124.70	70.47
Depth (cm)	5	4.40	96.00	24.10	24.13
Velocity (m/s)	3	2.89	64.45	21.71	19.82
Temperature (°C)	2	20.00	24.70	22.22	1.01
Conductivity	2	29.10	90.80	60.45	21.04
Ph	2	5.31	8.91	6.64	0.94
Dissolved oxygen (mg/L)	2	0.53	65.44	11.24	15.30
Salinity	2	0.01	0.04	0.02	0.01
Total native trees	3	0	14.00	3.45	12.67
Native tress CAP * (cm)	3	0	291.00	116.34	73.44
Brightness	3	62,970	61,450,333	2,878,130	13,082,298

* CAP Circumference above the chest.

Table A3. Species recorded in the cacao areas, in the streams sampled in the southern region of Bahia.

SUBORDER	Family/Species	Abundance
ZYGOPTERA		
	CALOPTERYGIDAE	
	Hetaerina longipes Hagen in Selys, 1853	22
	Hetaerina rosea Selys, 1853	107
	COENAGRIONIDAE	
	Acanthagrion aepiolum Tennessen, 2004	86

Table A3. *Cont.*

SUBORDER	Family/Species	Abundance
ZYGOPTERA		
	Aceratobasis macilenta Rambur, 1842	1
	Aceratobasis nathaliae Lencioni, 2004	4
	Argia chapadae Calvert, 1909	144
	Epipleoneura metallica Rácenis, 1955	6
	Forcepsioneura sancta Hagen in Selys, 1860	1
	Forcepsioneura serrabonita Pinto & Kompier, 2018	8
	Idioneura ancilla Selys, 1860	6
	Ischnura capreolus Hagen, 1861	4
	Metaleptobasis selysi Santos, 1956	4
	Neoneura ethela Williamson, 1917	1
	Telagrion longum Selys, 1876	4
	Telebasis corallina Selys, 1876	2
	Telebasis willinki Fraser, 1948	1
	LESTIDAE	
	Archilestes exoletus Hagen in Selys, 1862	4
	HETERAGRIONIDAE	
	Heteragrion aurantiacum Selys, 1862	75
	Heteragrion consors Hagen in Selys, 1862	34
	PERILESTIDAE	
	Perilestes fragilis Hagen in Selys, 1862	6
ANISOPTERA	GOMPHIDAE	
	Gomphoides praevia St. Quentin, 1967	1
	LIBELLULIDAE	
	Anatya guttata Erichson in Schomburgk, 1848	6
	Anatya januaria Ris, 1911	3
	Dasythemis essequiba Ris, 1919	1
	Dasythemis venosa Burmeister, 1839	1
	Diastatops obscura Fabricius, 1775	2
	Diastatops nigra Montgomery, 1940	9
	Elasmothemis alcebiadesi Santos, 1945	6
	Elga leptostyla Ris, 1909	1
	Erythemis carmelita Williamson, 1923	1
	Erythrodiplax sp2 Brauer, 1868	1
	Erythrodiplax sp3 Brauer, 1868	1
	Erythrodiplax avittata Borror,1942	1
	Erythrodiplax castanea Burmeister, 1839	14
	Erythrodiplax famula Erichson in Schomburgk, 1848	1
	Erythrodiplax fusca Rambur, 1842	58
	Erythrodiplax latimaculata Ris, 1911	1
	Erythrodiplax maculosa Hagen, 1861	1
	Erythrodiplax media Borror, 1942	3
	Erythrodiplax umbrata Linnaeus, 1758	4
	Macrothemis tenuis Hagen, 1868	4
	Micrathyria artemis Ris, 1911	8
	Micrathyria catenata Calvert, 1909	1
	Micrathyria ungulata Förster, 1907	12
	Nephepeltia phryne Perty, 1833	1
	Oligoclada umbricola Borror, 1931	1
	Orthemis attenuata Erichson in Schomburgk, 1848	3
	Orthemis discolor Burmeister, 1839	4
	Perithemis thais Kirby, 1889	17
	Uracis infumata Rambur, 1842	2
	Total Abundance	689
	Zygoptera Abundance	520
	Anisoptera Abundance	169

Table A4. Genera recorded in the cacao-cabruca areas, in the streams sampled in the southern region of Bahia.

Subordem	Family/Genera	Abundance Larvae
Zygoptera		
	CALOPTERYGIDAE	
	Hetaerina	26
	COENAGRIONIDAE	
	Argia	33
	Enallagma	2
	Homeoura	43
	Heliocharis	1
	LESTIDAE	
	Lestes	5
	HETERAGRIONIDAE	
	Heteragrion	76
	Dimeragrion	1
	Oxystigma	6
	PERILESTIDAE	
	Perilestes	30
Anisoptera		
	AESHNIDAE	
	Roppaneura	4
	Castoraeschna	1
	Coryphaeschna	11
	GOMPHIDAE	
	Agriogomphus	36
	Gomphoides	2
	Epigomphus	72
	Peruviogomphus	1
	Phyllocycla	5
	Progomphus	4
	LIBELLULIDAE	
	Brechmorhoga	19
	Cannaphila	20
	Dasythemis	13
	Dythemis	24
	Elga	26
	Erythemis	19
	Erythrodiplax	6
	Gynothemis	11
	Macrothemis	22
	Micrathyria	19
	Orthemis	14
	Pachydiplax	16
	Perithemis	6
	Rhodopygia	16
	Tramea	7
	Total Abundance	647
	Zygoptera Abundance	229
	Anisoptera Abundance	418

References

1. WWF. *Living Forest Report*; WWF: Washington, DC, USA, 2011.
2. Mello, K.; Valente, R.A.; Randhir, T.O.; Santos, A.C.A.; Vettorazzi, C.A. Effects of land use and land cover on water quality of low-order streams in Southeastern Brazil: Watershed versus riparian zone. *Catena* **2018**, *167*, 130–138. [CrossRef]
3. Mello, K.; Taniwakir, H.; De Paula, F.R.; Valente, R.A.; Randhir, T.O.; Macedo, D.R.; Leal, C.G.; Rodrigues, B.R.; Hughes, R.M. Multiscale land use impacts on water quality: Assessment, planning, and future perspectives in Brazil. *J. Environ. Manag.* **2020**, *270*, 110879. [CrossRef] [PubMed]

4. Allan, J.D. Landscapes and riverscapes: The influence of land use on stream ecosystems. *Annu. Rev. Ecol. Evol. Syst.* **2004**, *35*, 257–284. [CrossRef]
5. Cruz, M.A.S.; Gonçalves, A.A.; Aragão, R.; Amorim, J.R.A.; Mota, P.V.M.; Srinivasan, V.S.; Garcia, C.A.B.; Figueiredo, E.E. Spatial and seasonal variability of the water quality characteristics of a river in Northeast Brazil. *Environ. Earth Sci.* **2019**, *78*, 1–11. [CrossRef]
6. Albert, J.S.; Destouni, G.; Duke-Sylvester, S.M.; Magurran, A.E.; Oberdorff, T.; Reis, R.E.; Winemiller, K.O.; Ripple, W.J. Scientists' warning to humanity on the freshwater biodiversity crisis. *Ambio* **2020**, *50*, 85–94. [CrossRef]
7. Tonello, G.; Decian, V.S.; Restello, R.M.; Hepp, L.U. The conversion of natural riparian forests into agricultural land affects ecological processes in Atlantic forest streams. *Limnologica* **2021**, *91*, 125927. [CrossRef]
8. MapBiomas. Available online: https://plataforma.brasil.mapbiomas.org/ (accessed on 7 September 2020).
9. Fundação SOS Mata Atlântica; INPE. *Atlas dos Remanescentes Florestais da Mata Atlântica: Período 2019/2020*; relatório técnico; Fundação SOS Mata Atlântica: São Paulo, Brazil, 2021.
10. Mittermeier, R.A.; Turner, W.R.; Larsen, F.W.; Brooks, T.M.; Gascon, C. Global Biodiversity Conservation: The Critical Role of Hotspots. In *Biodiversity Hotspots*; Zachos, F., Habel, J., Eds.; Springer: Berlin/Heidelberg, Germany, 2011.
11. Castuera-Oliveira, L.; Oliveira-Filho, A.T.; Eisenlohr, P.V. Emerging hotspots of tree richness in Brazil. *Acta Bot. Bras.* **2020**, *34*, 117–134. [CrossRef]
12. Cabral, J.P.; Faria, D.; Morante-Filho, J.C. Landscape composition is more important than local vegetation structure for understory birds in cocoa agroforestry systems. *Ecol. Manag.* **2021**, *481*, 118704. [CrossRef]
13. Cassano, C.R.; Schroth, G.; Faria, D.; Delabie, J.H.; Bede, L. Landscape and farm scale management to enhance biodiversity conservation in the cocoa producing region of southern Bahia, Brazil. *Biodivers. Conserv.* **2009**, *18*, 577–603. [CrossRef]
14. Santos, L.R.; Rodrigues, M.E. Land Uses for Pasture and Cacao Cultivation Modify the Odonata Assemblages in Atlantic Forest Areas. *Diversity* **2022**, *14*, 672. [CrossRef]
15. Cassano, C.R.; Schroth, G.; Faria, D.; Delabie, J.H.; Bede, L.; Oliveira, L.C.; Mariano-Neto, E. *Desafios e Recomendações para a Conservação da Biodiversidade na Região Cacaueira do Sul da Bahia*; Boletim tecnico nº 205; CEPLAC/CEPEC: Ilhéus, Brazil, 2014.
16. Malmqvist, B.; Rundle, S. Threats to the running water ecosystems of the world. *Environ. Conserv.* **2002**, *29*, 134–153. [CrossRef]
17. Brasil, L.S.; Lima, E.L.; Spigoloni, Z.A.; Ribeiro-Brasil, D.R.G.; Juen, L. The habitat integrity index and aquatic insect communities in tropical streams: A meta-analysis. *Ecol. Indic.* **2020**, *116*, 106495. [CrossRef]
18. Dias-Silva, K.D.; Brasil, L.S.; Juen, L.; Cabette, H.S.R.; Costa, C.C.; Freitas, P.V.; De Marco, P. Influence of local variables and landscape metrics on Gerromorpha (Insecta: Heteroptera) assemblages in savanna streams, Brazil. *Neotrop. Entomol.* **2020**, *49*, 191–202. [CrossRef] [PubMed]
19. Cunha, E.J.; Juen, L. Environmental drivers of the metacommunity structure of insects on the surface of tropical streams of the Amazon. *Austral Ecol.* **2020**, *45*, 586–595. [CrossRef]
20. Valente-Neto, F.; de Oliveira Roque, F.; Rodrigues, M.E.; Juen, L.; Swan, C.M. Toward a practical use of Neotropical odonates as bioindicators: Testing congruence across taxonomic resolution and life stages. *Ecol. Indic.* **2016**, *61*, 952–959. [CrossRef]
21. Brito, J.G.; Roque, F.O.; Martins, R.T.; Hamada, N.; Nessimian, J.L.; Oliveira, V.C.; Hughes, R.M.; de Paula, F.R.; Ferraz, S.F.B. Small forest losses degrade aquatic macroinvertebrate assemblages in the eastern Brazilian Amazon. *Biol. Conserv.* **2020**, *241*, 108263. [CrossRef]
22. Castro, D.M.P.; Dole´dec, S.; Callisto, M. Land cover disturbance homogenizes aquatic insect functional structure in neotropical savanna streams. *Ecol. Indic.* **2018**, *84*, 573–582. [CrossRef]
23. Lencioni, F.A.A. *Damselflies of Brazil: An Illustrated Identification Guide—1—Non Coenagrionidae Families*; All Print Editora: São Paulo, Brazil, 2005.
24. Silva, L.F.R.; Castro, D.M.P.; Juen, L.; Callisto, M.; Hughes, R.M.; Hermes, M.G. Functional responses of Odonata larvae to human disturbances in neotropical savanna headwater streams. *Ecol. Indic.* **2021**, *133*, 108367. [CrossRef]
25. Calvão, L.B.; Nogueira, D.S.; Montag, L.F.A.; Lopes, M.A.; Juen, L. Are Odonata communities impacted by conventional or reduced impact logging? *Ecol. Manag.* **2016**, *382*, 143–150. [CrossRef]
26. Mendes, T.P.; Fogac, L.; Alvarado, S.T.; Juen, L. Assessing habitat quality on alpha and beta diversityof Odonata larvae (Insect) in logging areas in Amazon forest. *Hydrobiologia* **2021**, *848*, 1147–1161. [CrossRef]
27. Oliveira-Junior, J.M.B.; Juen, L. The Zygoptera/Anisoptera Ratio (Insecta: Odonata): A New Tool for Habitat Alterations Assessment in Amazonian Streams. *Neotrop. Entomol.* **2019**, *48*, 552–560. [CrossRef] [PubMed]
28. Vorster, C.; Samways, M.J.; Simaika, J.P.; Kipping, J.; Clausnitzer, V.; Suhling, F.; Dijkstra, K.D.B. Development of a new continental-scale index for freshwater assessment based on dragonfly assemblages. *Ecol. Indic.* **2020**, *109*, 105819. [CrossRef]
29. Gómez-Tolosa, M.; Rivera-Velázquez, G.; Rioja-Paradela, T.M.; Mendoza-Cuenca, L.F.; Tejeda-Cruz, C.; López, S. The use of Odonata species for environmental assessment: A meta-analysis for the Neotropical region. *Environ. Sci. Pollut. Res.* **2021**, *28*, 1381–1396. [CrossRef] [PubMed]
30. Ribeiro, C.; Juen, L.; Rodrigues, M.E. The Zygoptera/Anisoptera ratio as a tool to assess anthropogenic changes in Atlantic Forest streams. *Biodivers. Conserv.* **2021**, *30*, 1315–1329. [CrossRef]
31. De Marco, J.P.; Batista, J.D.; Cabette, H.S.R. Community Assembly of Adult Odonates in Tropical Streams: An Ecophysiological Hypothesis. *PLoS ONE* **2015**, *10*, e0123023. [CrossRef] [PubMed]

32. Oliveira-Junior, J.M.B.; Shimano, Y.; Gardner, T.A.; Hughes, R.M.; De Marco, P., Jr.; Juen, L. Neotropical dragonflies (Insecta: Odonata) as indicators of ecological condition of small streams in the eastern Amazon. *Austral Ecol.* **2015**, *40*, 733–744. [CrossRef]
33. Carvalho, F.G.; Oliveira, R.F.; Barbosa, L.; Assis Montag, L.F.; Juen, L. Oil palm platation is not a suitable environment for most forest specialist species of Odonata in Amazonia. *Anim. Conserv.* **2018**, *21*, 526–533. [CrossRef]
34. Valente-Neto, F.; Rodrigues, M.E.; Roque, F.O. Selecting indicators based on biodiversity surrogacy and environmental response in a riverine network: Bringing operationality to biomonitoring. *Ecol. Indic.* **2018**, *94*, 198–206. [CrossRef]
35. Magalhães, J.H.R. Estrutura da comunidade arbórea de um remanescente de floresta madura nos tabuleiros costeiros do extremo sul da Bahia, Brasil. *Bol. Mus. Biol. Mello Leitão* **2018**, *40*, 93–122.
36. Superintendência de Estudos Econômicos da Bahia. Indicadores Territoriais: Território Identidade Litoral Sul da Bahia 2019. Available online: https://www.sei.ba.gov.br/images/informacoes_por/territorio/indicadores/pdf/litoralsul.pdf (accessed on 12 January 2021).
37. Barbour, M.T.; Gerritsen, J.; Griffith, G.E.; Frydenborg, R.; McCarron, E.; White, J.S.; Bastian, M.L. A framework for biological criteria for Florida streams using benthic macroinvertebrates. *J. N. Am. Benthol. Soc.* **1996**, *15*, 185–211. [CrossRef]
38. Garisson, R.W.; Ellenrieder, N.V.; Louton, J.A. *Dragonfly Genera of the New World: An Illustrated and Annotated Key to the Anisoptera*; The Johns Hopkins University Press: Baltimore, MD, USA, 2006.
39. Garisson, R.W.; Ellenrieder, N.V.; Louton, J.A. *Dragonfly Genera of the New World: An Illustrated and Annotated Key to the Zygoptera*; The Johns Hopkins University Press: Baltimore, MD, USA, 2010.
40. Lencioni, F.A.A. *Damselflies of Brazil: An Illustrated Identification Guide—2—Coenagrionidae Families*; All Print Editora: São Paulo, Brazil, 2006.
41. Lencioni, F.A.A. *Damselflies of Brazil—An Illustrated Identification Guide—Southeast Region*; Editora do Autor: Jacareí, Brazil, 2017.
42. Hamada, N.; Nessimian, J.L.; Querino, R.B. *Insetos Aquáticos na Amazônia Brasileira: Taxonomia, Biologia e Ecologia*; Editora INPA: Manaus, Brazil, 2019.
43. Nessimian, J.L.; Venticinque, E.; Zuanon, J.; De Marco, P., Jr.; Gordo, M.; Fidelis, L.; Batista, J.D.; Juen, L. Land use, habitat integrity, and aquatic insect assemblages in Central Amazonian streams. *Hydrobiologia* **2008**, *614*, 117–131. [CrossRef]
44. Cunha, E.J.; Assis, L.M.F.; Juen, L. Oil palm crops effects on environmental integrity of Amazonian streams and Heteroptera (Hemiptera) species diversity. *Ecol. Indic.* **2015**, *52*, 422–429. [CrossRef]
45. Dray, S.; Legendre, P.; Blanchet, G. Packfor: Forward Selection with Permutation (Canoco p. 46). R-Forge, The R Project for Statistical Computing. 2016. Available online: https://rdrr.io/rforge/packfor/ (accessed on 20 August 2022).
46. Blanchet, G.F.; Legendre, P.; Borcard, D. Forward Selection of Explanatory Variables. *Ecol. Soc. Am.* **2008**, *89*, 2623–2632. [CrossRef] [PubMed]
47. Legendre, P.; Legendre, L. *Numerical Ecology*; Elsevier: Amsterdam, The Netherlands, 2012; p. 1006.
48. Jackson, D.A. Stopping rules in principal components analysis, a comparison of heuristical and statistical approaches. *Ecology* **1993**, *74*, 2204–2214. [CrossRef]
49. Oksanen, J.; Blanchet, F.G.; Friendly, M.; Kindt, R.; Legendre, P.; McGlinn, D.; Minchin, P.R.; O'Hara, B.; Simpson, G.L.; Solymos, P.; et al. Vegan: Community Ecology Package. R Package Version 2.4-3. 2017. Available online: https://CRAN.R-project.org/package=vegan (accessed on 26 August 2021).
50. Carvalho, F.G.; Pinto, N.S.; Oliveira-Junior, J.M.B.; Juen, L. Efeitos da remoção da vegetação marginal nas comunidades de Odonata. *Acta Limnol. Bras.* **2013**, *25*, 10–18. [CrossRef]
51. Calvão, L.B.; Juen, L.; de Oliveira Junior, J.M.B.; Batista, J.D.; De Marco Júnior, P. Land use modifies Odonata diversity in streams of the Brazilian Cerrado. *J. Insect Conserv.* **2018**, *22*, 675–685. [CrossRef]
52. Guillermo-Ferreira, R.; Del-Claro, K. Resource defense polygyny by Hetaerina rosea Selys (Odonata: Calopterygidae): Influence of age and wing pigmentation. *Neotrop. Entomol.* **2011**, *40*, 78–84. [CrossRef]
53. Cabette, H.S.R.; Souza, J.R.; Shimano, Y.; Juen, L. Effects of changes in the riparian forest on the butterfly community (Insecta: Lepidoptera) in Cerrado areas. *Rev. Bras. Entomol.* **2017**, *61*, 43–50. [CrossRef]
54. Resende, D.C.; Marco, P. First description of reproductive behavior of the Amazonian damselfly Chalcopteryx rutilans (Rambur) (Odonata, Polythoridae). *Rev. Bras. Entomol.* **2010**, *54*, 436–440. [CrossRef]
55. Rodrigues, M.E.; Roque, F.O.; Ferreira, R.G.N.; Saito, V.S.; Samways, M.J. Egg-laying traits reflect shifts in dragonfly assemblages in response to different amount of tropical forest cover. *Insect Conserv. Divers.* **2018**, *11*, 231–240. [CrossRef]
56. Mendes, T.P.; Benone, N.L.; Juen, L. To what extent can oil palm plantations in the Amazon support assemblages of Odonata larvae? *Insect Conserv. Divers.* **2019**, *12*, 448–458. [CrossRef]
57. Corbet, P.S.; May, M.L. Fliers and perchers among Odonata: Dichotomy or multidimensional continuum? A provisional reappraisal the flier/percher template. *Int. J. Odonatol.* **2008**, *11*, 155–171. [CrossRef]
58. Fonseca, B.M.; de Mendonça-Galvão, L.; Padovesi-Fonseca, C.; de Abreu, L.M.; Fernandes, A.C.M. Nutrient baselines of Cerrado low-order streams: Comparing natural and impacted sites in Central Brazil. *Environ. Monit. Assess* **2014**, *186*, 19–33. [CrossRef]
59. Consatti, G.; Santos, D.M.; Renner, S.; Périco, E. Presença de Odonata em Áreas Preservadas e Não Preservadas nas Matas Ciliares do Rio Taquari, Rs. *Revista Iniciação Científica ULBRA* **2014**, *1*, 57–65.
60. Brasil, L.S.; Vieira, T.B.; De Oliveira-Junior, J.M.B.; Dias-Silva, K.; Juen, L. Elements of metacommunity structure in Amazonian Zygoptera among streams under different spatial scales and environmental conditions. *Ecol. Evol.* **2017**, *7*, 3190–3200. [CrossRef] [PubMed]

61. Luke, S.H.; Dow, R.A.; Butler, S.; Vun Khen, C.; Aldridge, D.C.; Foster, W.A.; Turner, E.C. The impacts of habitat disturbance on adult and larval dragonflies (Odonata) in rainforest streams in Sabah, Malaysian Borneo. *Freshw. Biol.* **2017**, *62*, 491–506. [CrossRef]
62. Pinto, A.P.; Kompier, T. In honor of conservation of the Brazilian Atlantic Forest: Description of two new damselflies of the genus Forcepsioneura discovered in private protected areas (Odonata: Coenagrionidae). *Zoologia* **2018**, *35*, e21351. [CrossRef]
63. Peruquetti, P.S.F.; De Marco Junior, P. Efeito da alteração ambiental sobre a comunidade de Odonata em riachos de Mata Atlântica de Minas Gerais, Brasil. *Rev. Bras. Zool.* **2002**, *19*, 317–327. [CrossRef]
64. Miguel, T.B.; Oliveira-Júnior, J.M.B.; Ligeiro, R.; Juen, L. Odonata (Insecta) as a tool for the biomonitoring of environmental quality. *Ecol. Indic.* **2017**, *81*, 555–566. [CrossRef]
65. Faria, D.; Laps, R.R.; Baumgarten, J.; Cetra, M. Bat and Bird Assemblages from Forests and Shade Cacao Plantations in Two Contrasting Landscapes in the Atlantic Forest of Southern Bahia, Brazil. *Biodivers. Conserv.* **2006**, *15*, 587–612. [CrossRef]
66. Ferreira, A.S.; Peres, C.A.; Dodonov, P.; Cassano, C.R. Multi-scale mammal responses to agroforestry landscapes in the Brazilian Atlantic Forest: The conservation value of forest and traditional shade plantations. *Agrofor. Syst.* **2020**, *94*, 2331–2341. [CrossRef]

diversity

MDPI

Article

Odonata from Iberá Wetland System (Corrientes, Argentina) Are Regional Biogeographic Schemes Useful to Assess Odonata Biodiversity and Its Conservation?

Alejandro del Palacio, Federico Lozano, Lia S. Ramos, María de las Mercedes Navarro and Javier Muzón *

Laboratorio de Biodiversidad y Genética Ambiental—BioGeA, Universidad Nacional de Avellaneda, Avellaneda B1870, Argentina
* Correspondence: jmuzon@undav.edu.ar

Abstract: Regionalization schemes reflect different macroscale distribution patterns and show large areas characterized by a common natural history, resulting in similar associations of biotic and abiotic features. Freshwater biota and terrestrial biota do not respond in the same way to environmental variables. The Iberá Depression, one of the largest wetlands in South America, is recognized in many schemes either as a functional unit or as an area with an ecotonal character. We used the distributional data of 128 species of Odonata, from a total of 103 collection sites from Corrientes and Misiones provinces, to test if Iberá functions as an ecological and functional unit, based on the Odonata distribution patterns. In addition, we tested if their distribution patterns fit into the most widespread regionalization schemes (hydrological basins, biogeographical provinces and ecoregions) used in Argentina. The Iberá Depression was not recovered as a functional unit; its sub-basins are more related to external basins than to each other. Neither the ecoregion nor the biogeographical schemes are suitable to explain the distribution patterns of the Odonata. The Odonata seem to respond to the availability of particular wetlands (e.g., ponds, streams, rivers, swamps, etc.), or to specific physical characteristics, such as the type of sediment, the availability of oxygen, etc., instead of to biogeographical or ecoregional schemes.

Keywords: Odonata; hydrological basin; biogeography; ecoregion system; Iberá; Argentina

Citation: del Palacio, A.; Lozano, F.; Ramos, L.S.; Navarro, M.d.l.M.; Muzón, J. Odonata from Iberá Wetland System (Corrientes, Argentina) Are Regional Biogeographic Schemes Useful to Assess Odonata Biodiversity and Its Conservation? *Diversity* **2022**, *14*, 842. https://doi.org/10.3390/d14100842

Academic Editors: M. Olalla Lorenzo-Carballa and Ricardo Koroiva

Received: 20 July 2022
Accepted: 1 October 2022
Published: 6 October 2022

1. Introduction

Biological inventories are the most direct way of knowing the diversity of a region [1]. They provide valuable information about the current status of the diversity, as they provide data on species richness, the presence of native, endemic, or threatened species, distribution patterns, etc. In addition, inventories are essential tools for monitoring diversity, which is essential to understand the influence of different impacts, natural or anthropogenic, or the effectiveness of different types of management actions, among others. It can be assumed that areas can be valued from the point of view of the singularity of their main biotic components and, then, priorities can be established in the development of conservation policies. On the other hand, when biological inventories are not complete or unavailable for a certain area, regionalization schemes based on biotic and/or abiotic components (e.g., phytogeographic, zoogeographic, biogeographic, ecoregional, hydrological basins) are common tools in order to assess both their biodiversity and conservation [2,3].

The rationale behind most of these regionalization schemes is to show different macroscale distribution patterns (i.e., plants, animals, biota and ecosystems), and to show large areas characterized by a common natural history, resulting in the similar associations of biotic and abiotic features. For almost all schemes, the main data used for their elaboration come from the distribution of terrestrial plants; this is even the case for the delimitation of ecoregions (i.e., areas based on macroscale patterns of ecosystems determined by climate) where vegetation is used as a surrogate for climate because it is considered a tangible expression of it [4]. Nevertheless, there are several problems with establishing these macroscale patterns. Most of them are related to the stability and the uniformity of the factors considered at different scales (e.g., the changing nature of the climate, the different stages of vegetation succession, potential vs. actual vegetation, anthropogenic modifications), heterogeneity between climate zones given by geomorphology, boundaries between units, available data, etc. In addition to this, it must be considered that not all taxa respond in the same way to the environmental variables and, therefore, their distribution patterns will not coincide. This is more evident or critical when considering the aquatic biota. The discrepancy between freshwater and terrestrial components takes another level of significance when it is noted that very few phyla have evolved to efficiently become terrestrial, becoming completely independent of the aquatic environment (e.g., some groups of Arthropoda and Chordata). This kind of discrepancy could lead to wrong conclusions or decisions regarding freshwater taxa. In addition, there are some issues when applying these schemes to study terrestrial faunal components: in general, it can be assumed that the distribution patterns of groups that are directly dependent on vegetation (for example, herbivorous animals, or animals that require a special plant physiognomy to nest or reproduce) will coincide more precisely with the current schemes. However, in those cases with less dependence on the vegetation, for example generalist predators or even those that have amphibian life cycles, such as Odonata, the non-coincidence with the schemes may be greater. Obviously, taxa such as Odonata establish multiple relationships with plants, not only aquatic, but also terrestrial, such as a preference for shady or sunny places, substrates for oviposition, etc.; but these relationships are not generally established at a specific level, rather with ecological types (i.e., trees, shrubs, grasses).

The order Odonata in Argentina is represented by 282 species, which can be divided into two main faunistic components [5]. One of them, with less richness, but with a higher level of endemism, is the Subantarctic component, distributed in Patagonia and characterized by presenting biogeographical relationships with Australia, New Zealand, and Tasmania [6]. The other one is the Neotropical component, in the north and center of the country between 34° and 36° S, where the southernmost limits of many widespread American genera are found (e.g., *Acanthagrion*, *Argia*, *Hetaerina*, *Erythemis*, *Miathyria*, *Micrathyria*, *Perithemis*, *Tauriphila*, and *Tramea*). One of the most diverse areas within the Neotropical component is the Iberá wetland system [7].

The Iberá wetland system (Figure 1a) represents one of the most biodiverse wetlands in southern South America, due to its extension, environmental singularity (i.e., *esteros* wetlands) and probably because of its ecotonal nature between the Paranaense forests and the Pampaean grasslands [7,8]. Several faunistic inventories have been published that account for this diversity [7–14]. It is a complex of marshes, ponds and swamps mixed with lotic environments connected by wide interface areas, with a changing physiognomy due to the changes in water levels. Iberá singularity is due also to the development of the *esteros*, a particular type of shallow wetland with a marked thermocline that supports large patches of floating vegetation called *embalsados*. The *embalsados*, or floating islands, are a common vegetal formation in the Parana basin, formed from a root cluster of a few dominant hydrophytes with high amounts of organic matter [15]. It is worth mentioning that Iberá has suffered during the last two years from strong pressure caused by natural and intentional fires. In 2022 alone 900 thousand hectares, which represents more than 10% of the surface of the province [16], were affected.

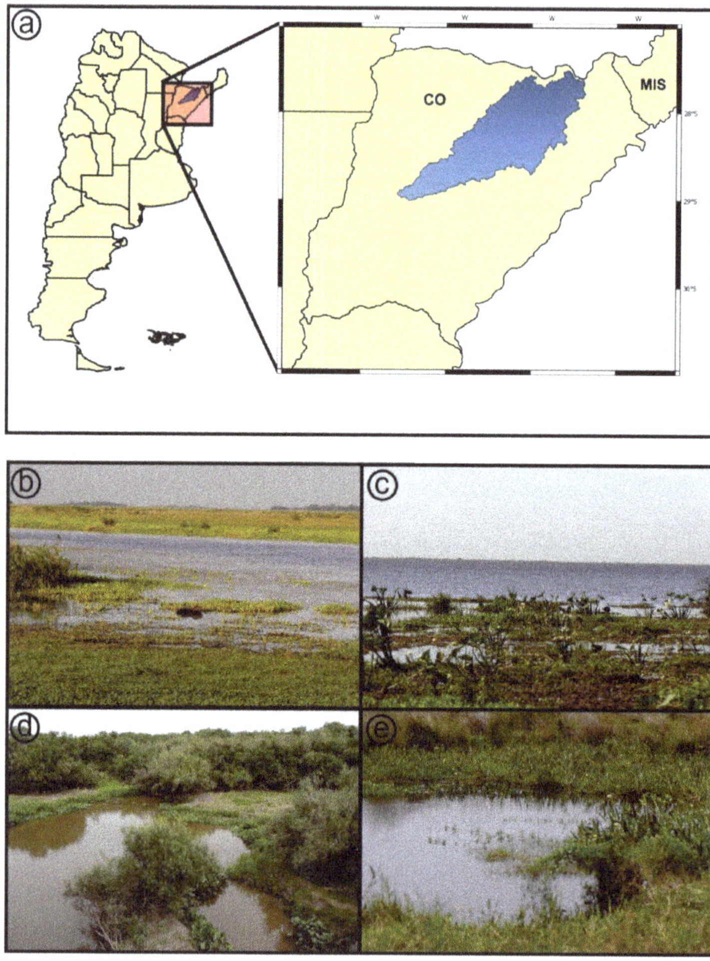

Figure 1. (**a**) Map of Argentina showing the Iberá Depression in the Corrientes province. (**b**) Photo of a typical wetland of the Iberá Depression (San Nicolás, Corrientes), (**c**) Photo of *Laguna Iberá* (Corrientes), (**d**) Photo of a typical wetland of Pay Ubre (Corrientes), (**e**) Photo of a typical wetland within Mburucuyá (Corrientes). CO: Corrientes province, MIS: Misiones province.

In addition, within the Iberá wetland system different regionalization schemes, based mainly on phytogeographical aspects, have given disparate results: following the classic biogeographic scheme [17] three provinces converge (Espinal, Chaco and Paranaense provinces); and following the ecoregional proposal [18], two ecoregions converge (*Campos y Malezales* and *Esteros del Iberá*). This complexity gives the area an ecotonal character [7]. These very different results allow us to evaluate their representativeness when analyzing a component of the aquatic biota. On a hydrological scale, the basins (especially the Iberá Depression) are delimited by physical barriers that hinder the distribution of most aquatic species. Due to this, a higher biotic similarity between sites within a basin is expected than between different basins.

The main objective of this work is to assess if Iberá, one of the most speciose areas in Argentina for odonates, and behaves as a functional and ecological unit based on the Odonata diversity, as well as to test how their distribution patterns fit into previous regionalization schemes (hydrological basins, biogeographical provinces and ecoregions). In addition, an updated inventory from Iberá is provided.

2. Materials and Methods

2.1. Study Area

The province of Corrientes is characterized by its remarkable and diverse wetlands (Figure 1b–e). Physiographically, the central sector of Corrientes constitutes a low-lying area that receives the name of *Región Deprimida* (Depressed Region) [19], which is different from the adjacent areas in terms of lithological and geomorphological features (Figure 2a). This area includes the Iberá Depression, a hydrological basin, and the marshy environments associated with the fluvial valleys of the Corriente, Batel Batelito and Santa Lucía rivers (Figure 2b) [20].

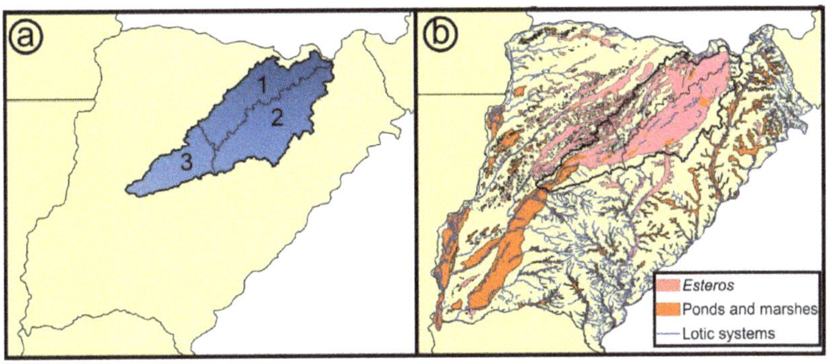

Figure 2. (**a**) Iberá Depression showing its sub-basins: 1: *Carambolas*, 2: *Iberá*, and 3: *Naciente del Río Corriente*, (**b**) Different types of wetlands within Corrientes province.

The Iberá Depression contains a complex of wetlands of about 14,000 km^2 with permanent or semi-permanent water, with different types of aquatic vegetation, all of which make the region one of the largest macro-wetlands in the world. It is located in the upper basin of the Corriente River and drains into the Paraná River. The Iberá Depression can be divided into three sub-basins: *Iberá*, *Carambolas* and *Naciente del Río Corriente* [21], each of them with particular dynamics (Figure 2a).

The term Iberá has also been used in different regionalization schemes due to its environmental complexity. Therefore, the geographical extent of Iberá strictly depends on how it is defined. Morrone et al. [22] have referred to the *Esteros del Iberá* province (Figure 3a) which is similar to the Riverine district, as defined by Apodaca et al. [23], they have referred to Iberá as a vast area covering the flood valleys of the Paraguay–Parana fluvial axis, from north-eastern Argentina and southern Paraguay to the Parana Delta, and the Uruguay River from southern Brazil to the Rio de la Plata.

Brown and Pacheco [18] have recognized the *Esteros del Iberá* Ecoregion (Figure 3b), a more restricted area that occupies 12,300 km^2 in the northwest of Corrientes province in Argentina. They have subdivided this area into two different zones, taking into account different vegetation structures: *Esteros del Iberá* and *Parque Chaqueño Correntino*.

There is a preliminary Odonata inventory from the Iberá Depression [7], in which 38 species within 21 genera were recorded from surveys done between 1999 and 2005.

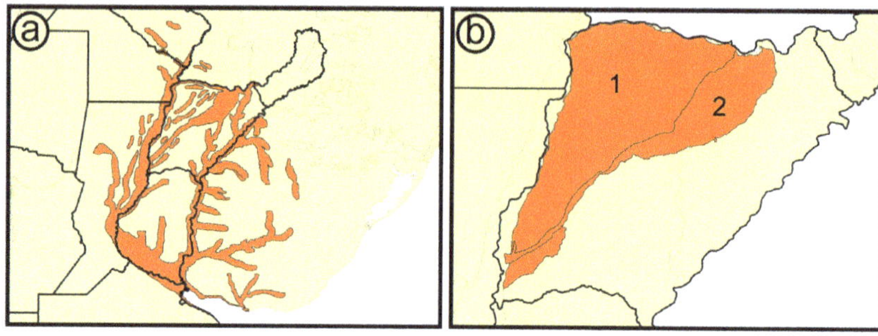

Figure 3. (**a**) *Esteros del Iberá* province *sensu* Morrone et al. [22], (**b**) *Esteros del Iberá* ecoregion *sensu* Brown and Pacheco [18]: 1: *Parque Chaqueño Correntino;* 2: *Iberá.*

2.2. Sampling Procedures

Odonata were collected with aerial nets, fixed by injection with 96% alcohol and then dehydrated with silica gel; once dry they were stored in plastic envelopes in the *Laboratorio de Biodiversidad y Genética Ambiental, Universidad Nacional de Avellaneda* (BioGeA) collection. Collections were made from September 2009 to April 2014, preferably between 10:00 and 14:00 on sunny days, when the adults were more active.

2.3. Regionalization Schemes

For the analyses, three regionalization schemes were used. These were selected based on the possibility of testing the internal relationships of the Iberá Depression. Therefore, Morrone et al. [22] and Apodaca et al. [23] were not used, since the Iberá Depression is contained within a much larger area, and no subdivisions have been provided by any of these authors:

1. Biogeographic provinces *sensu* Cabrera and Willink [17]: This scheme proposes a hierarchical partition, which is a divisive, non-agglomerative classification of the regions into domains, provinces, and districts. In this categorization, the successive levels mainly based on the phytogeographic hierarchy, from domain to district, are carried out by the presence of endemisms and predominance of families and species [18,24]. The vegetation of the Neotropical and Subantarctic sectors of Argentina is then classified into 12 provinces [24], three of them are represented in the area: *Espinal, Chaqueña* and *Paranaense* (Figure 4a).
2. Ecoregions *sensu* Brown and Pacheco [18]: This ecoregional scheme was developed by a panel of experts in flora and fauna. It recognizes 18 ecoregions for the country, three of them are represented in the study area: *Chaco Húmedo, Esteros del Iberá* and *Campos y Malezales* (Figure 3b).
3. Hydrological basins from Corrientes gathered from the *Subsecretaría de Recursos Hídricos* of Corrientes [21]: According to this scheme, the Iberá Depression is divided into three different sub-basins (*Iberá, Carambolas* and *Naciente del Río Corriente*). Surrounding these, there are eight basins that were used to test if the Iberá Depression functions as a unit (*Aguapey, Cuenca de arroyos del Paraná, Cuenca del Río Paraná hasta confluencia con Río Paraguay, Cuencas menores de Corrientes afluentes del Río Uruguay, Estero Batel, Estero Santa Lucia, Miriñay,* and *Río Corriente*) (Figure 4b).

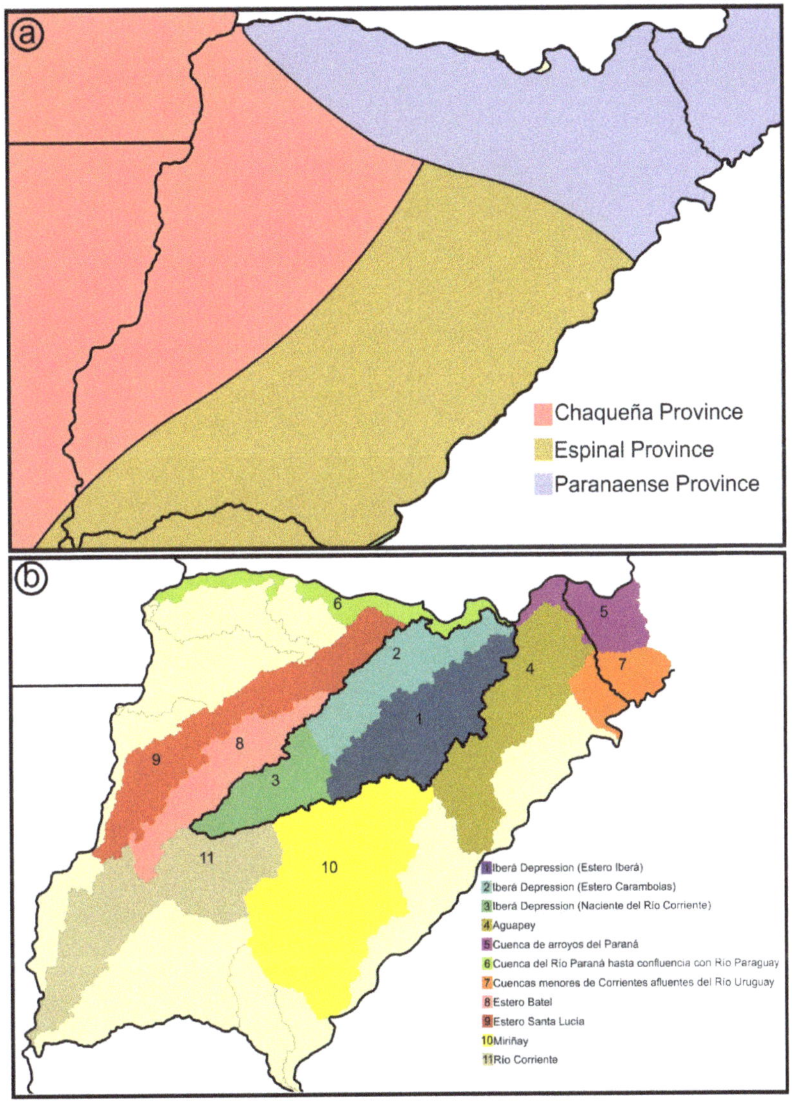

Figure 4. (**a**) Biogeographic provinces *sensu* Cabrera and Willink [17], (**b**) Map showing the Iberá Depression and surrounding hydrological basins, the selected basins in this study are numbered and highlighted in color.

2.4. Data Analysis

The distributional data of 128 species of Odonata (Supplementary Material Data S1) were gathered from the literature and the BioGeA Odonata collection (40% of all species recorded for Argentina [5]), which includes records from a total of 103 collection sites from Corrientes and Misiones provinces, visited at least once by the authors between 1999 and 2014 (Supplementary Data S2) (Figure 5).

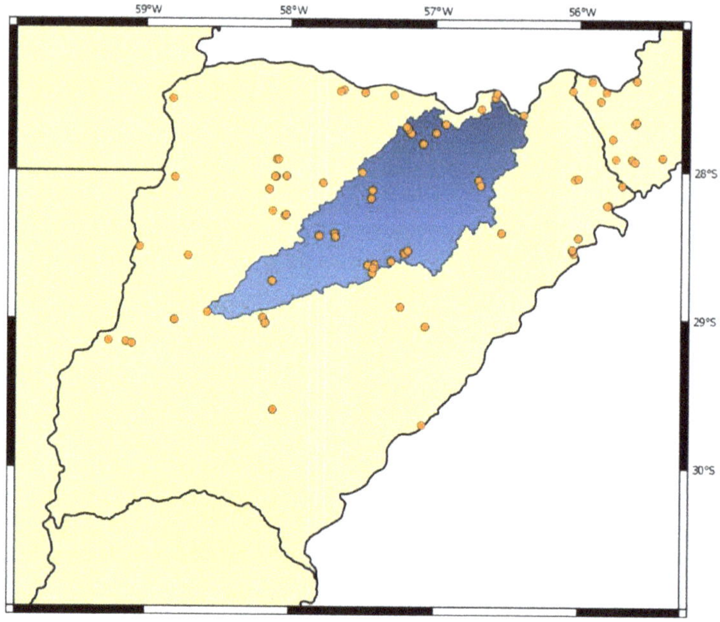

Figure 5. Localities sampled (see Supplementary Data S2 for georeferences).

In order to test whether the regionalization schemes explain the Odonata distribution patterns, the basins of the Iberá Depression were superimposed with the biogeographic and the ecoregional schemes in order to subdivide the basins into smaller units. Similarity analyses were performed using these fragments to see which groups were recovered. If the tested scheme explained the distribution of the Odonata, then the basin fragments should be grouped following the regionalization system. The intersection between the biogeographic provinces and the basins resulted in seven subunits (Figure 6a), one of which had to be discarded due to a lack of records. The intersection between the ecoregions and the basins resulted in eight subunits (Figure 7a), two of which had to be discarded due to a lack of records.

A data matrix of presence/absence for each of the 11 basins was constructed for the 128 species. We used the PAST (Paleontological Statistics) Version 4.10 [25] software for the UPGMA cluster analysis because it was the most suitable clustering method, since it had the highest cophenetic correlation (CC). The similarity Jaccard index was selected for the clustering, as it is preferred in cases when the differences in species richness between samples (or communities) need to be reflected in the measurement of β diversity [26].

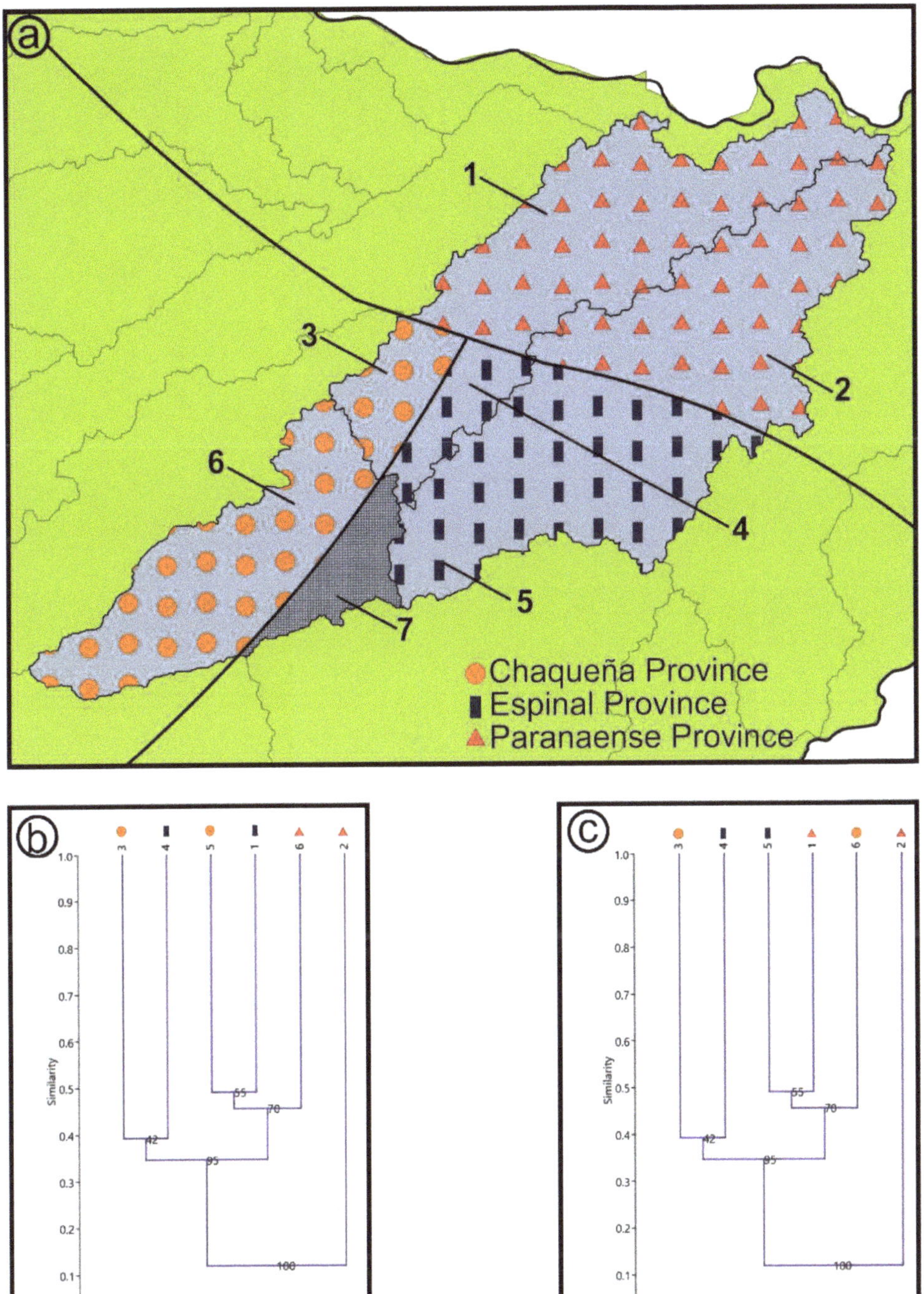

Figure 6. (**a**) Iberá Depression divided by the biogeographic provinces, (**b**) Dendrogram showing similarity between areas (Sorensen–Dice), (**c**) Dendrogram showing similarity between areas (Jaccard). The dendrograms showed a cophenetic correlation (CC) > 0.96.

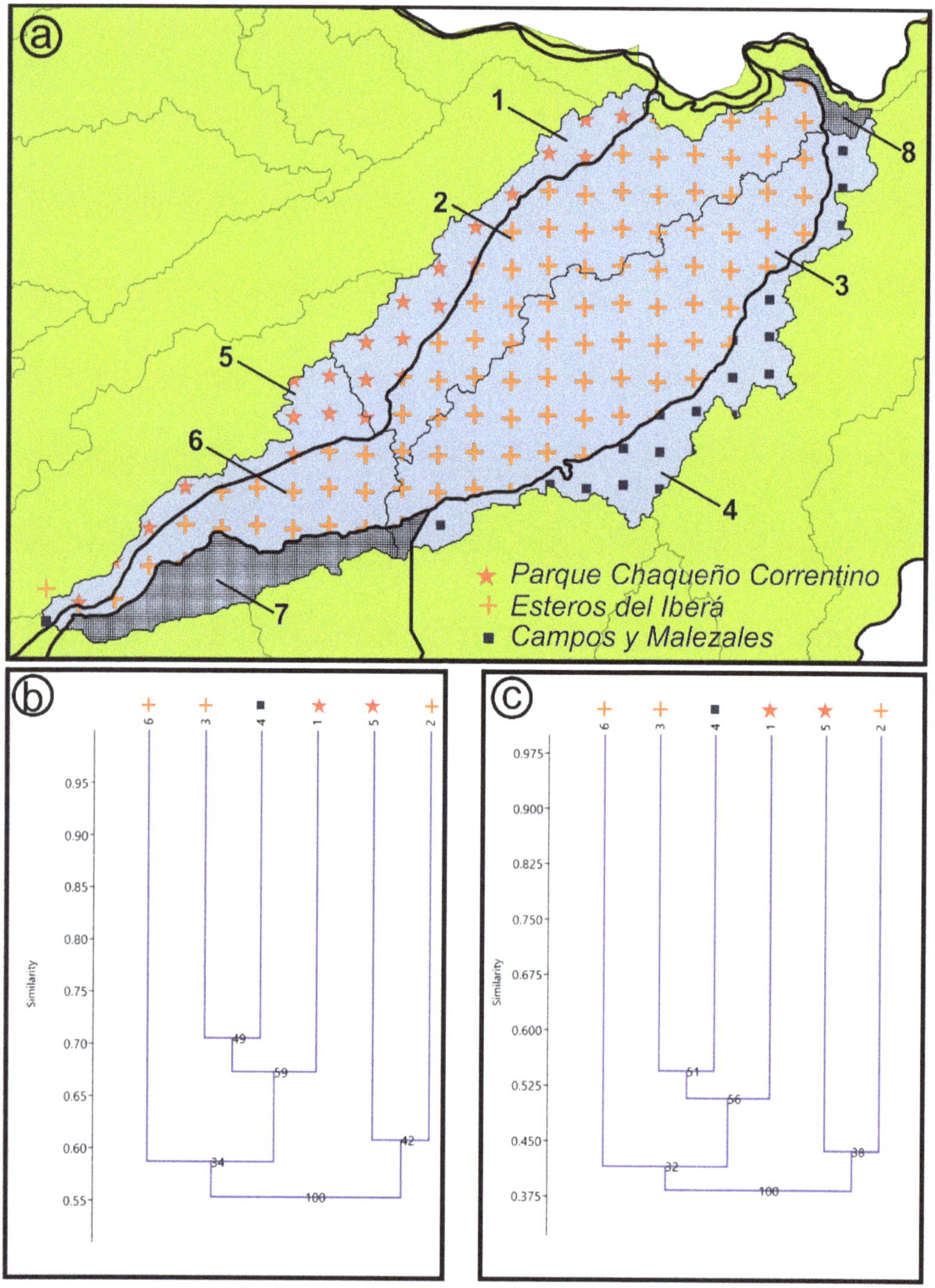

Figure 7. (**a**) Iberá Depression divided by the ecoregions, (**b**) Dendrogram showing similarity between areas (Sorensen–Dice), (**c**) Dendrogram showing similarity between areas (Jaccard). The dendrograms showed a CC > 0.96.

On the other hand, to avoid the "double-zero problem" (species absent from two sites), we selected the asymmetrical binary coefficient Sorensen–Dice, because the comparison excludes double zeros, which makes it preferable for ecological studies [27]. The Sorensen–Dice coefficient gives double weight to double presences, as absences may be due to various factors and do not necessarily reflect differences in the environment; double presence, on the contrary, is a strong indication of resemblance [28].

Maps were prepared with the free software Quantum Gis 3.24 [29] and the shapefiles were downloaded from the IGN (*Instituto Geográfico Nacional*, https://www.ign.gob.ar, accessed on 1 August 2022), whereas the basins were downloaded from HydroSHEDS website (https://www.hydrosheds.org, accessed on 1 April 2022) and modified according to the limits proposed in the *Atlas de Cuencas y Regiones Hídricas Superficiales de la República Argentina* [21].

3. Results

3.1. Checklist

A total of 61 species were recorded for the Iberá Depression (approximately one fourth of the richness of the country) in 30 genera. This represents an increase of 23 species and 9 genera since the inventory of 2008 (Supplementary Data S2), which recorded 38 species within 21 genera. Libellulidae is the most speciose family with 36 species recorded, followed by Coenagrionidae with 15.

3.2. Biogeographical Provinces

The areas defined by this scheme were not recovered in our cluster analysis, with none of the areas rearranging themselves according to the expected groups (Figure 6b,c).

3.3. Ecoregions

The areas defined by this scheme were also not recovered in our cluster analysis, with none of the areas rearranging themselves according to the expected groups (Figure 7b,c).

3.4. Basins

The analysis between the Iberá Depression and the external basins shows (according to the Jaccard and Sorencen–Dice indexes) two main groups. The Iberá Depression sub-basins show less similarity among them than with the external basins (Figure 8a,b). In this regard, two main clusters were found: Iberá and Carambolas grouped together with Santa Lucía ((9-1)2) and the group conformed by the Batel and Naciente del Río Corriente group (8-3) with Miriñay (10) as a "sister" group of both of them.

The β diversity analysis shows the highest similarity for the cluster Iberá-Santa Lucia ($J > 0.5$, $D > 0.7$) and Batel-Naciente del Río Corriente ($J > 0.5$, $D > 0.65$). On the other hand, the lowest similarity is shown to be the Aguapey ($J < 0.15$, $D < 0.2$), with the rest of the basins.

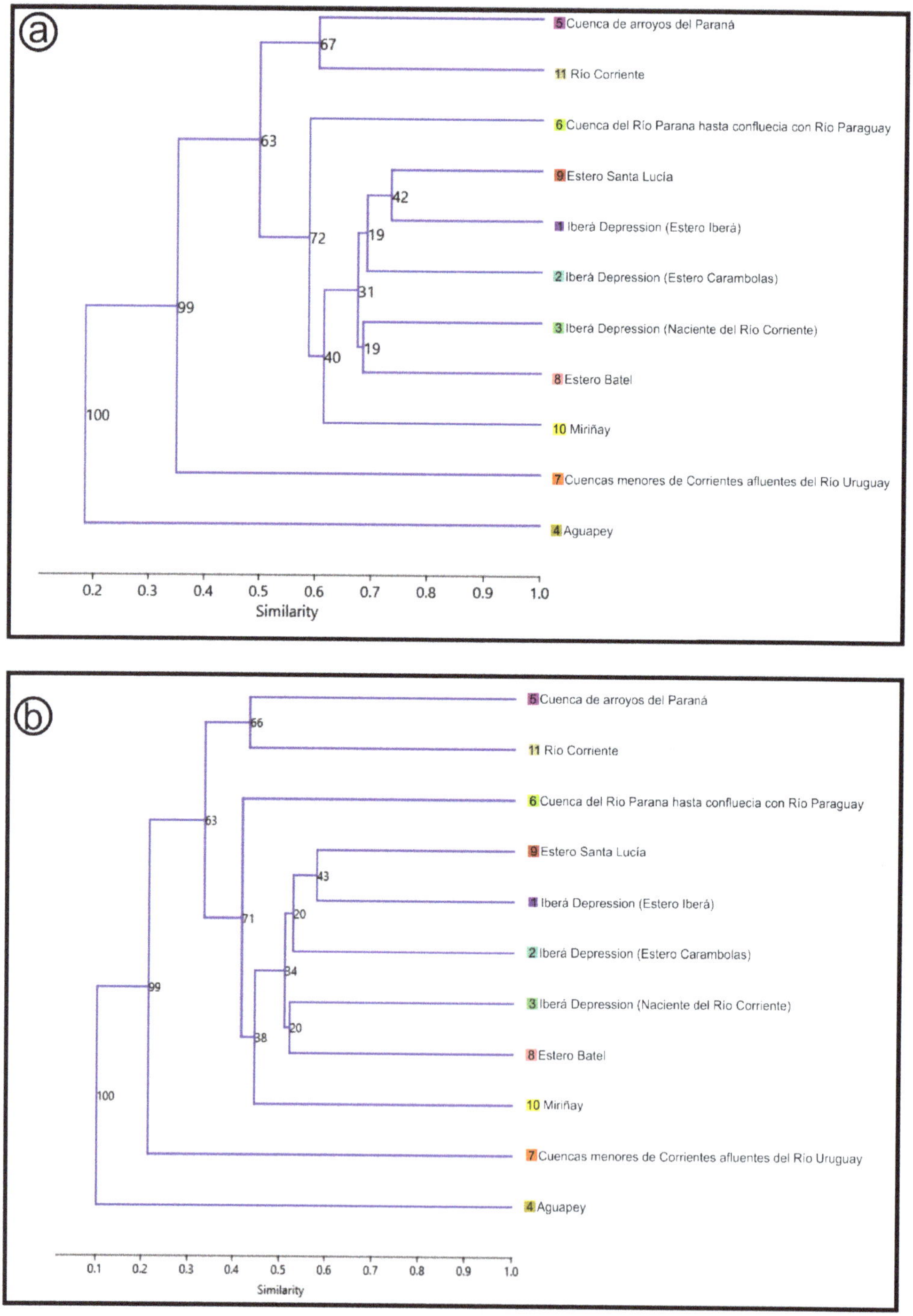

Figure 8. (**a**) Dendrogram showing similarity between hydrological basins (Sorensen–Dice), CC > 0.96, (**b**) Dendrogram showing similarity between hydrological basins (Jaccard), CC > 0.96.

4. Discussion

The affinities of the biota of Iberá have been discussed several times, based mainly on plant and animal components. Focusing on different arthropods, such as spiders and odonates, previous studies have attempted to validate, without much success, that this wetland system belongs to one of the biogeographical units postulated in previous schemes [7,14]. It has even recently been questioned regarding its nature as an ecoregion unit, postulating an ecotonal character based on a low degree of endemism and the multiple biogeographical units that converge in this region [14].

Since none of the areas established in the biogeographical and ecoregional proposals were recovered by the present results, we confirm that the odonates of Iberá do not follow any of the available regional biota-based schemes. The same can be said about the hydrological scheme, since our results show that the three internal basins of the Iberá Depression show higher similarities with the external basins than with each other.

The odonates, like other continental aquatic taxa, such as fish or rotifers [30–32], seem to be refractory to the widespread regionalization schemes, at least in South America. This is probably due to their amphibiotic cycle with a terrestrial phase with great flight capacity (including their ability to perform large migrations), and their role as generalist predators. From a more general perspective, inland aquatic ecosystems seem to support much coarser regionalization schemes than those based on terrestrial biota [22,23,31,32]; perhaps due to the buffering nature of the water and the possibility of physical connectivity between basins due to sporadic flooding.

For odonates, a possible interpretation of the distribution patterns observed in this study, and their mismatch with previous schemes, can be interpreted by looking at habitat types (Figure 2b). The odonate distribution patterns appear to reflect the type of aquatic habitat (lentic or lotic with similar environmental conditions), regardless of the specific composition of the surrounding terrestrial vegetation. In this sense, the two main groupings of areas (9-1-2 and 8-3) may be the result of the habitat typology present within these basins. In the case of the 9-1-2 group, the most predominant type of water mass is wetlands, with few and very small permanent streams. The second group (8-3) presents a mixture of rivers and numerous interconnected streams with numerous wetlands throughout its basin. In the peculiar case of Miriñay (10), its river originates in the Iberá Depression, the basin is characterized by a floodplain with impermeable soil and very little slope that favors the retention of surface water, forming an area of semi-permanent swamps (Figure 9a,b) [33]. According to these results, the main environmental attributes that seem to determine odonate distribution patterns are related to the availability of particular wetlands (e.g., lagoons, streams, rivers, swamps, etc.), or special physical characteristics, such as type of sediment, oxygen availability, etc.

Finally, it is worth mentioning that for any attempt to use widespread distribution patterns to propose or promote actions and/or policies for the conservation of aquatic biota, beyond the broad framework of faunal affinity (e.g., Neotropical or Austral), ecological data should be prioritized over geographic or climatic data, to achieve better results than those schemes based on terrestrial taxa. Those assessments whose objective is the protection or conservation of freshwater biota, such as environmental impact and monitoring, the conservation status of species, etc., must consider their ecological limitations with much more emphasis than their regional biogeographical context.

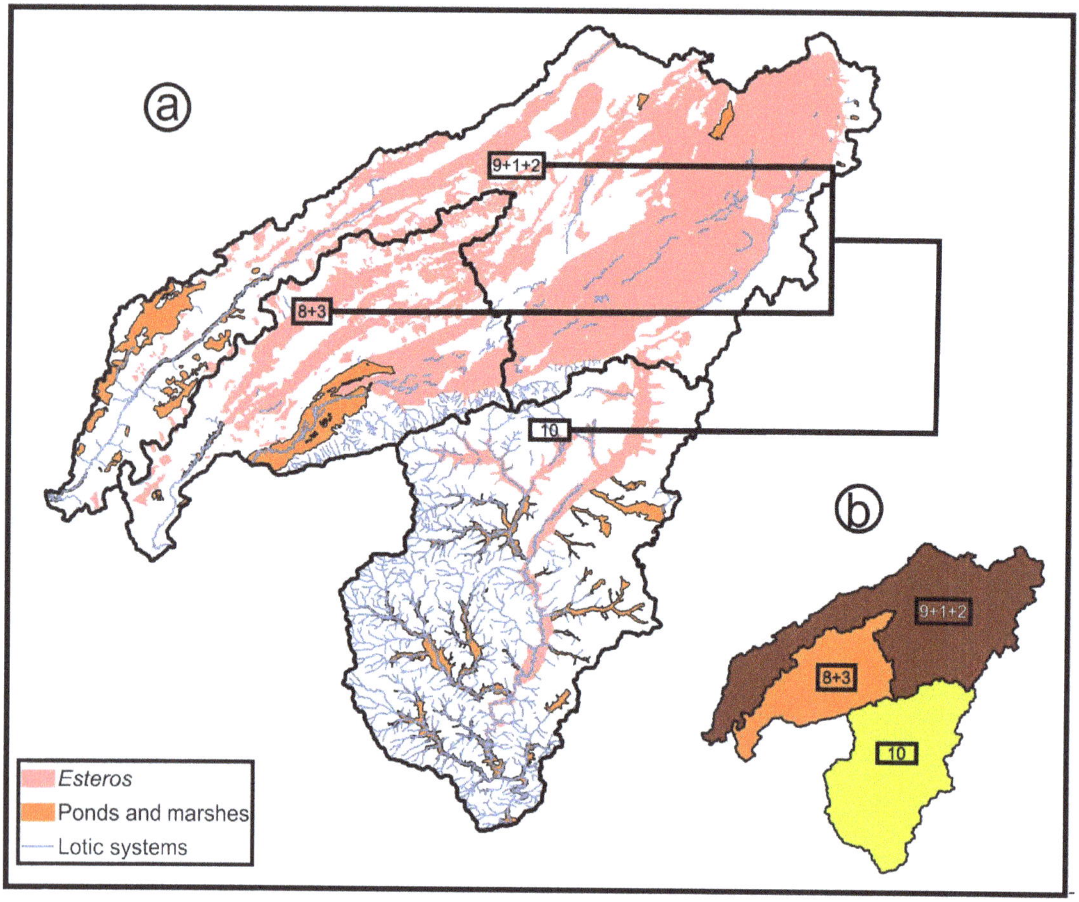

Figure 9. (**a**) Map of Iberá depression and external basins groups with lotic and lentic systems, (**b**) Groups of basins according to the results.

Supplementary Materials: The following supporting information can be downloaded at: https://www.mdpi.com/article/10.3390/d14100842/s1, Supplementary Data S1: Checklist of Odonata per basin used in this paper. *: new records for the Iberá De-pression compared with Muzón et al. [7]; Supplementary Data S2: List of localities sampled with georeferences.

Author Contributions: Conceptualization, J.M.; methodology, A.d.P., L.S.R. and J.M.; formal analysis, A.d.P.; data curation, A.d.P., F.L. and M.d.l.M.N.; writing—original draft preparation, A.d.P., F.L., L.S.R. and J.M.; writing—review and editing, A.d.P., F.L., L.S.R., J.M. and M.d.l.M.N. All authors have read and agreed to the published version of the manuscript.

Funding: This study was partially funded by CONICET (Consejo Nacional de Investigaciones Científicas y Técnicas, Argentina) and by the National Geographic Society (Grant # 7104-01).

Institutional Review Board Statement: Not applicable.

Data Availability Statement: Data are contained within the article; specimens analyzed in this study are deposited in the Entomological Collection of the Laboratorio de Biología y Genética Ambiental (BioGeA, Universidad Nacional de Avellaneda) and are available on request to the collection managers.

Conflicts of Interest: The authors declare no conflict of interest.

References

1. Noss, R.F. Indicators for Monitoring Biodiversity: A Hierarchical Approach. *Conserv. Biol.* **1990**, *4*, 355–364. [CrossRef]
2. Lehner, B.; Grill, G. Global River Hydrography and Network Routing: Baseline Data and New Approaches to Study the World's Large River Systems. *Hydrol. Process.* **2013**, *27*, 2171–2186. [CrossRef]
3. He, F.; Bremerich, V.; Zarfl, C.; Geldmann, J.; Langhans, S.D.; David, J.N.W.; Darwall, W.; Tockner, K.; Jähnig, S.C. Freshwater Megafauna Diversity: Patterns, Status and Threats. *Divers. Distrib.* **2018**, *24*, 1395–1404. [CrossRef]
4. Bailey, R.G. Identifying Ecoregion Boundaries. *Environ. Manage.* **2004**, *34*, 14–26. [CrossRef] [PubMed]
5. Lozano, F.; del Palacio, A.; Ramos, L.; Muzón, J. The Odonata of Argentina: State of knowledge and updated checklist. *Int. J. Odonatol.* **2020**, *23*, 113–153. [CrossRef]
6. Muzón, J.; Pessacq, P.; Lozano, F. The Odonata (Insecta) of Patagonia: A Synopsis of their current status with illustrated keys for their identification. *Zootaxa* **2014**, *3784*, 346–388. [CrossRef]
7. Muzón, J.; Von Ellenrieder, N.; Pessacq, P.; Lozano, F.; Garré, A.; Lambruschini, J.; Ramos, L.; Weigel Muñoz, S. Odonata from Iberá Wetlands (Corrientes, Argentina): Preliminary Inventory and Biodiversity. *Rev. Soc. Entomol. Argent.* **2008**, *67*, 59–67.
8. Rubio, G.D.; Moreno, C.E. Orb-Weaving Spider Diversity in the Iberá Marshlands, Argentina. *Neotrop. Entomol.* **2010**, *39*, 496–505. [CrossRef] [PubMed]
9. Casciotta, J.R.; Almirón, A.E.; Bechara, J. *Peces Del Iberá*; Grafikar: La Plata, Argentina, 2005.
10. Calcaterra, L.A.; Cuezzo, F.; Cabrera, S.M.; Briano, J.A. Ground Ant Diversity (Hymenoptera: Formicidae) in the Ibera Nature Reserve, the Largest Wetland of Argentina. *Ann. Entomol. Soc. Am.* **2010**, *103*, 71–83. [CrossRef]
11. Coscarón, M.C.; Melo, M.C.; Coddington, J.; Corronca, J. Estimating Biodiversity: A Case Study on True Bugs in Argentinian Wetlands. *Biodivers. Conserv.* **2009**, *18*, 1491–1507. [CrossRef]
12. Ingaramo, M.D.R.; Etchepare, E.G.; Álvarez, B.B.; Porcel, E. Riqueza y Composición de la Fauna de Anuros en la Región Oriental de la Reserva Natural Provincial Esteros Del Iberá, Corrientes, Argentina. *Rev. Biol. Trop.* **2012**, *60*, 759–769. [CrossRef]
13. Dufek, M.I.; Oscherov, E.B.; Damborsky, M.P. Actualización de La Distribución de Tábanos (Diptera: Tabanidae) En La Reserva Natural Esteros Del Iberá, Argentina. *Entomotropica* **2016**, *31*, 36–39.
14. Rubio, G.D.; Nadal, M.F.; Munévar, A.C.; Avalos, G.; Perger, R. Iberá Wetlands: Diversity Hotspot, Valid Ecoregion or Transitional Area? Perspective from a Faunistic Jumping Spiders Revision (Araneae: Salticidae). *Species* **2018**, *19*, 117–131.
15. Lozano, F.; del Palacio, A.; Ramos, L.; Granato, L.; Drozd, A.; Muzón, J. Recovery of local dragonfly diversity following restoration of an artificial lake in an urban area near Buenos Aires. *Basic Appl. Ecol.* **2022**, *58*, 88–97. [CrossRef]
16. Comunicación del CCT CONICET Incendios En Corrientes: Evaluarán Los Efectos a Mediano Plazo Sobre La Biodiversidad. Available online: https://nordeste.conicet.gov.ar/incendios-efectos-sobre-la-biodiversidad/ (accessed on 20 August 2022).
17. Cabrera, A.L.; Willink, A. *Biogeografía de América Latina*; Programa Regional de Desarrollo Científico y Tecnológico, Organización de los Estados Americanos: Washington, DC, USA, 1980.
18. Brown, A.; Pacheco, S. *Propuesta de Actualización del Mapa Ecorregional de la Argentina*; La situación: Ambient, Argentina, 2005; pp. 28–31.
19. Herbst, R.; Jorge, N.; Cruz, S. Mapa Litoestratigráfico de la Provincia de Corrientes. *Inst. Geol. Paleontol.* **1985**, *2*, 1–50.
20. Orfeo, O.; Neiff, J.J. Esteros del Iberá. Un Enorme Laboratorio a Cielo Abierto. *Sitios Interés Geológico La República Argent*: Argentina. 2008. Available online: http://repositorio.segemar.gov.ar/308849217/1343 (accessed on 1 October 2022).
21. Secretaria de Recursos Hídricos. Atlas de Cuencas y Regiones Hídricas Superficiales de la República Argentina; 2010. Available online: https://www.argentina.gob.ar/obras-publicas/hidricas/cartografia-hidrica-provincial (accessed on 1 October 2022).
22. Morrone, J.J.; Escalante, T.; Rodríguez-Tapia, G.; Carmona, A.; Arana, M.; Mercado-Gómez, J.D. Biogeographic regionalization of theNeotropical region: New map and shapefile. *Ann. Acad. Bras. Cienc.* **2022**, *94*, e20211167. [CrossRef]
23. Apodaca, M.J.; Katinas, L.; Guerrero, E.L. Hidden areas of endemism: Small units in the South-eastern Neotropics. *Syst. Biodivers.* **2019**, *17*, 425–438. [CrossRef]
24. Ribichich, A.M. El modelo clásico de la Fitogeografía de Argentina: UN Análisis Crítico. *Interciencia* **2002**, *27*, 669–675.
25. Hammer, Ø.; Harper, D.A.T.; Ryan, P.D. PAST: Paleontological Statistics Software Package for Education and Data Analysis. *Palaeontol. Electron.* **2001**, *4*, 1–9.
26. Cardoso, P.; Borges, P.A.; Veech, J.A. Testing the performance of beta diversity measures based on incidence data: The robustness to undersampling. *Divers. Distrib.* **2009**, *15*, 1081–1090. [CrossRef]
27. Singh, W. Robustness of three hierarchical agglomerative clustering techniques for Ecological Data. Master's Thesis, University of Iceland, Reykjavík, Iceland, 2008; p. 100.
28. Legendre, P.; Legendre, L. *Numerical Ecology*; Elsevier: Amsterdam, The Netherlands, 1998; Volume 40, ISBN 0444892508.
29. QGIS Development Team. QGIS Geographic Information System. *Open Source Geospat. Found. Proj.* 2022. Available online: http://qgis.osgeo.org (accessed on 1 August 2022).
30. López, H.L.; Menni, R.C.; Donato, M.; Miquelarena, A.M. Biogeographical Revision of Argentina (Andean and Neotropical Regions): An analysis using freshwater fishes. *J. Biogeogr.* **2008**, *35*, 1564–1579. [CrossRef]
31. Ringuelet, R.A. Zoogeografía y Ecología de los Peces de Aguas Continentales de la Argentina y consideraciones sobre las áreas ictiológicas de América del Sur. *Ecosur* **1975**, *2*, 1–122.

32. José de Paggi, S. Ecological and biogeographical remarks on the Rotifer Fauna of Argentina. *Inst. Nac. Limnol.* **1990**, *23*, 297–311.
33. Martínez, S.J. Cuenca del Río Miriñay. Available online: https://www.argentina.gob.ar/sites/default/files/43.pdf (accessed on 15 August 2022).

diversity

MDPI

Article

Genetic Diversity and Structure of *Anax imperator* Leach, 1815 Populations (Odonata: Aeshnidae) in Ponds at Regional and European Scales

Marceau Minot * and Aurélie Husté

ECODIV-INRAE, Department of Biology, UFR Sciences et Techniques University of Rouen-Normandy, 76000 Rouen, France; aurelie.huste@univ-rouen.fr
* Correspondence: m.minot@hotmail.fr

Abstract: Anthropogenic activities cause loss and fragmentation of natural habitats and have strong effects on population maintenance by increasing their isolation. Pond ecosystems are scattered waterbodies that can interact as a network connected by dispersal events of freshwater organisms. Identifying local genetic differentiations and understanding how gene flow occurs across these networks is essential to prevent risks associated with environmental perturbations. This study aimed to investigate genetic diversity and structure of *Anax imperator* Leach, 1815 populations at both regional and European scales using seven microsatellites markers. Seven populations of *A. imperator* were sampled in northwestern France and four populations were sampled in Italy (Sicily), Czech Republic, Switzerland and United Kingdom (U.K.). French populations presented a low genetic differentiation indicating a high gene flow and confirming dispersal events of this species between ponds at regional scale. No pattern of isolation by distance was found at the European scale. The populations presented a low genetic differentiation and no pattern of isolation by distance, suggesting historical or current movements of individuals. Only the U.K. population presented a significant genetic differentiation from other European populations, suggesting that the English Channel might act as a barrier to gene flow for *A. imperator*. However, Bayesian analysis showed that some dispersal events could occur between the U.K. and France (Normandy), probably facilitated by prevailing winds.

Keywords: dispersal barriers; dragonflies; genetic differentiation; pond networks; population structure

Citation: Minot, M.; Husté, A. Genetic Diversity and Structure of *Anax imperator* Leach, 1815 Populations (Odonata: Aeshnidae) in Ponds at Regional and European Scales. *Diversity* **2022**, *14*, 68. https://doi.org/10.3390/d14020068

Academic Editors: Michael Wink, M. Olalla Lorenzo-Carballa and Ricardo Koroiva

Received: 16 December 2021
Accepted: 16 January 2022
Published: 20 January 2022

1. Introduction

The spatial structure of populations is generally conditioned by intrinsic life traits (e.g., dispersal capacities), distances between sites and environmental factors such as physical barriers or climate gradients [1–3]. Anthropogenic activities modify landscape characteristics leading to loss and fragmentation of natural habitats [4]. Consequently, many wildlife populations live in isolated habitat patches and often suffer a loss of genetic diversity due to inbreeding [5,6]. Genetic studies are crucial to drive species conservation measures because a low genetic diversity also increases the risk of population extinction due to environmental perturbations and demographic stochasticity [7]. The local genetic diversity can vary independently from the geographical distribution of the considered species. Some wide-ranging species with high dispersal capacities can have different genetic diversities at smaller scales [8], while other ones with low dispersal ability can have low genetic differentiation at larger spatial scales [9,10]. In rare species, the genetic diversity can also be constraint by specific habitat requirements that often induce isolation of populations [11,12]. Overall, the genetic diversity depends on gene flow that is conditioned by the frequency of dispersal events between populations [13].

Most dispersal events cover only short distances (i.e., short distance dispersal; SDD) and take place within the boundaries of defined geographic or population limits [14].

However, these SDD events may also allow distant populations to connect by a 'step by step' dispersal process [15,16]. On the contrary, long distance dispersal (LDD) movements are generally rare and difficult to detect [17,18], except in migratory species [19,20]. The LDD often involve physical forces like wind and marine currents [21,22] or rely on organisms with higher dispersal abilities [16,23]. Dispersal events have major consequences on population dynamics. They enable species to colonize new patches and expand the occupancy on their territory. They also allow the maintenance of population dynamics in patches where species are already present and genetic mixing with other populations [14]. Both SDD and LDD are crucial for population maintenance because they limit the risk of persistent extinction on one site by spreading local temporal dynamics of extinction and colonization on multiple sites. Moreover, they preserve genetic diversity [13] which increases the resilience of populations to environmental changes (e.g., climate change, invasive species, habitat fragmentation) [24,25]. Estimating dispersal is often difficult in the field because these events are rare at the individual scale. However, genetic techniques can provide accurate estimates of gene flow between populations and therefore indirect measurements of dispersal [26,27].

Ponds are small waterbodies with high conservation interest because of their high biodiversity of aquatic plants, macroinvertebrates and amphibians [28,29]. Although ponds are often scattered elements in the landscape, they are regularly considered as working in networks. Interestingly, stepping-stone models (i.e., models in which individuals can move among an infinite array of populations) [30] can be applied to these ecosystems to investigate the genetic structure of pond populations [31,32]. The persistence of populations is constrained by the availability of suitable breeding ponds, the distance between them, as well as the nature of the habitats crossed, and the dispersal capacities of the considered species [33]. Finally, pond populations can be threatened by perturbations like water pollution or summer droughts [28,34,35]. Determining how gene flow is distributed across pond networks and identifying potential local genetic differentiations is essential to assess population decline and extinction risks associated with environmental perturbations and habitat fragmentation.

Odonates are insects with aquatic nymphal development and terrestrial (aerial) adults [36]. While some rare dispersal events have been reported at the nymphal stage, the vast majority of the movements are performed by flying adults [37]. Dispersal distances are difficult to quantify and can vary a lot depending on the dispersal ability of the species. For instance, most zygopteran species do not move over more than one kilometer during their lifetime, whereas some large Anisoptera can fly over several kilometers within minutes [38,39]. In many species, most individuals stay on the same pond during their whole lifetime [40–42], whereas other species like *Pantala flavescens* Fabricius, 1798 can undertake recurrent migration flights across the oceans [43]. Therefore, the degree of genetic differentiation between odonate populations depends on dispersal traits like body size or wing morphology, but also on their behaviour and ecological niche. For zygopteran species with similar body size, a specialist species like *Coenagrion mercuriale* Charpentier, 1840 [44] shows considerable genetic differentiation between populations at a local scale (i.e., within a distance of 24 km), whereas a very weak genetic structure was found at European and North-American scales for the generalist species *Ischnura elegans* Vander Linden, 1820 [45] and *I. hastata* Say, 1839 [46], respectively. Several landscape features might also act as a physical barrier to dispersal, limiting the gene flow between populations. For instance, dispersal movements of *C. mercuriale* are hindered by small hills, or patches of trees and shrubs [47,48].

Anax imperator Leach, 1815 is a large dragonfly species (i.e., body size between 7 and 8 cm). Its distribution ranges from South Africa to Sweden [49] and seems to expand very quickly (ca. 88 km per year) in response to the global climate change [50]. No migration on long distances is reported for this species [38] contrary to other Aeshnidae like *A. junius* Drury, 1773 which is known to migrate along the Eastern coast of the USA [51]. Nymphs and exuviae measure up to 5 cm [52] and can be found sometimes in high densities in large sun-exposed ponds with well-developed aquatic vegetation [53]. Mature adults present a territorial behaviour but are also very mobile around their mating sites. Movements of individuals have been recorded over few kilometers only, whereas this species is expected to undertake flights on much longer distances [54]. However, lack of information on long-distance dispersal events of *Anax imperator* can be explained by the difficulty to track insects during long periods with available capture-mark-recapture techniques.

The present study focused on eleven populations of *Anax imperator* from ponds in two Western European countries (i.e., France (Normandy) and United Kingdom) and three Central and Southern European countries (i.e., Switzerland, Czech Republic and Italy (Sicily)). Thus, genetic diversity and gene flow were investigated at both the regional (i.e., populations from Normandy) and the European scales. Samples consisted either of adult legs, nymphal legs, fresh exuviae (i.e., collected within the 24 h after emergence) or old exuviae (i.e., collected at an unknown date after ecdysis). Therefore, the DNA exploitability between fresh and old exuviae was also compared. Since *A. imperator* is a large dragonfly with high dispersal abilities between its breeding sites, only a weak genetic differentiation was expected between populations at a regional scale. Nevertheless, higher genetic differentiation was expected at the European scale. We also hypothesized that geographical barriers, and especially the English Channel or mountain chains such as the Alps, might limit gene flow between populations at the European scale.

2. Materials and Methods

2.1. Study Populations and Sample Collection

Samples were collected from seven localities in France (Normandy) and a single locality in the other countries (i.e., United Kingdom, Switzerland, Czech Republic and Italy (Sicily); Table 1. A total of 251 individuals (i.e., 6 to 39 per locality) was collected at or near ponds for DNA analysis. Depending on localities, different types of samples were collected: a hind-leg tarsus of adults, a hind-leg tibia of nymphs, fresh or old exuviae (Table 1). Some exuviae used for the analyses were obtained from individuals that were reared at the University of Rouen (Normandy, France). They were collected within the 24 h following emergence and were therefore qualified as "fresh". Other exuviae collected in the field were qualified as "old". Indeed, the date of emergence was unknown and since this exoskeleton can persist over several weeks in the vegetation [55,56], we had no idea of the delay between emergence and collection. After collection, all samples were stored in 99.5% ethanol until DNA extraction.

Table 1. Details on the 11 localities where populations of *Anax imperator* that were studied in Europe and the number of collected samples according to their source of DNA. NAs indicate no information on sex available.

Country	Pop	Site Name	Coordinates (WGS84)	Sex		Source of DNA			
				Males	Females	Adult Legs	Nymphal Legs	Fresh Exuviae	Old Exuviae
Italy	1	Sicily	37.086 N, 15.286 E	NA	NA	6			
Switzerland	2	Neuchâtel	47.002 N, 6.741 E	12	4	16			
Czech Republic	3	Kyjov	49.010 N, 17.128 E	7	21		16		12
France	4	Beaussault	49.682 N, 1.555 E	NA	NA		33		
France	5	Cerisy	49.199 N, −0.912 E	NA	NA		19		
France	6	Heudreville	49.133 N, 1.198 E	18	22			36	4
France	7	Bois-Guillaume	49.480 N, 1.102 E	25	14			39	
France	8	Marchésieux	49.178 N, −1.324 E	9	5		14		
France	9	Paluel	49.835 N, 0.624 E	8	7			15	
France	10	Bresle	49.914 N, 1.679 E	7	3				10
United Kingdom	11	York	53.964 N, −1.086 E	15	16				31
				Total (n = 251)		22	82	90	57

2.2. DNA Extraction and Microsatellite Genotyping

Collected legs were cut to smaller fragments using scissors. Then, DNA extraction was performed using QIAamp Micro kits (QIAGEN, Courtaboeuf, France), following the protocol provided with the kits. Collected exuviae were dried on a glass surface during the night before extraction to allow alcohol evaporation. The exuviae were cut with scissors to keep only the thorax, legs and the white tracheal lining from the abdomen. The material was placed in a 5 mL Eppendorf tube with three steel beads and homogeneously grinded with a MM400 mixer mill (Retsch, Éragny, FRance) for three minutes according to the protocol proposed by [57]. Finally, DNA extraction was performed using DNeasy Blood & Tissue Kits (QIAGEN, France), following the protocol provided with the kits except for the following changes: quantity of proteinase K was 25 µL, quantities of buffer AL and ethanol were 250 µL after incubation and as suggested by [57] and the elution step was performed twice with 50 µL AE buffer. Individuals were genotyped using 12 microsatellite loci previously developed for *A. imperator* [58]. Among these loci, two were derived from the sister species *A. parthenope* and showed successful amplification with *A. imperator*. Primers 5′-labelled with 3 fluorescent dyes (i.e., FAM, VIC and PET; Life Technologies SAS, Villebon-sur-Yvette, France) were used for amplification reaction in four separate PCR multiplexes in a thermocycler (Mastercycler nexus gradient Eppendorf, Montesson, France). Polymerase Chain Reactions (PCRs) were performed in total volumes of 12.5 µL containing 6.25 µL Qiagen Multiplex PCR Master Mix (QIAGEN, France), 4 µL RNAse-free water, 1.25 µL of one of the four multiplexed primer combinations (concentration of each primer: 2 µmol/µL) and 1 µL of DNA (10 ng/µL). PCR were performed using the following thermocycler program (QIAGEN): first an initial denaturation step at 95 °C for 15 min, then 35 cycles of denaturation consisting in 30 s at 93 °C, 90 s at an annealing temperature of 52 °C or 57 °C depending on the multiplexes used and an elongation at 72 °C for 60 s, and finally an extension at 65 °C for 30 min. Finally, 1 µL of each PCR product was added to a solution of 8.8 µL of formamide (Applied Biosystems, Villebon-sur-Yvette, France) and 0.2 µL of GeneScan 600 LIZ size standard (Applied Biosystems). Fragments were analysed by capillary electrophoresis using an ABI Prism 3500 Genetic Analyzer (Applied Biosystems). Then, results were analysed with the GeneMapper 4.1 software (Applied Biosystems).

2.3. Genetic Diversity

All analyses were performed using the R software [59]. Means are given ± SD.

The presence of null alleles was checked using the R package "PopGenReport" [60]. Deviation from expected Hardy–Weinberg Equilibrium (HWE) conditions for each locus and each population was tested using the R package "pegas" [61] and an exact test based on 10,000 Monte Carlo permutations of alleles. Linkage Disequilibrium (LD) between all locus-pair combinations was tested using the R package "genepop" (version 1.1.7) [62,63]. Markov chain parameters were 1000 dememorization, 100 batches and 1000 iterations per batch for each test. p-values in the detection of HWE and LD were corrected with a False Discovery Rate (FDR) procedure using Benjamini-Hochberg-Yekutieli method [64,65].

Observed heterozygosity (Ho), expected heterozygosity (He), numbers of alleles (Na), allelic richness (AR), and inbreeding coefficients (FIS) for each population were calculated using the R package "diveRsity" (version 1.9.90) [66]. The AR was calculated using the rarefaction method to correct for variation in sample size [67] and to avoid having to exclude a population from analyses. The 95% confidence intervals (CI) for FIS estimates were calculated using 10,000 bootstrap iterations.

The global measures of FIS and FST, as well as pairwise FST-values between all populations, were calculated using the diveRsity package. FST is considered as an effective measure for population genetic differentiation when using relatively small data sets with fewer than 20 loci [68,69]. All these F-statistics used the bias-corrected formulation of Weir and Cockerham [70]. Estimate 95% confidence intervals for all measures of differentiation were calculated using 10,000 bootstrap iterations.

2.4. Population Genetic Analyses and Geographic Structure

We used the pegas package [61] to perform an analysis of molecular variance (AMOVA) [71] based on Euclidian distances among individuals for all microsatellite loci. The AMOVA was conducted to partition total genetic variation across three hierarchical levels: among countries (i.e., U.K., France, Switzerland, Czech Republic and Sicily), among populations within countries and within populations. The statistical significance of the fixation indexes Φ was calculated using 10,000 permutations of data.

Genetic isolation-by-distance (IBD) is defined as a decrease in a genetic similarity among populations as the geographical distance between them increases. It was investigated considering a two-dimensional stepping-stone model and by studying the correlation between FST/(1-FST) and the natural-log-transformed (ln) geographic distance [72]. A Mantel test between a matrix of genetic differentiation between *A. imperator* populations (i.e., using FST/(1-FST)) and a matrix of Euclidean distances between these populations was performed using the package "ade4" with 10,000 permutations.

To investigate the genetic structure of the 11 populations of *A. imperator* sampled, a model-based clustering was performed using the STRUCTURE 2.3.4 program [73]. It uses a Bayesian Markov chain Monte Carlo (MCMC) method to identify genetic clusters (K) and assign individuals to these clusters. Each cluster is characterised by a set of allele frequencies at each locus. Individuals are assigned to these clusters based on the likelihood of their multilocus genotypes to belong to these genetic clusters by minimising deviations from Hardy–Weinberg equilibrium (HWE) and linkage disequilibrium (LD) [73]. We performed runs for a number of clusters (K) ranging from two to eight and with a number of 20 independent runs for each K. *Anax imperator* was expected to have high dispersal abilities leading to frequent exchanges of individuals between populations. Therefore, an admixture model with correlated allele frequencies was considered. The LOCPRIOR parameter was not considered, i.e., the geographic location of the individuals was not considered as an additional information. For each model, a burn-in period of 100,000 followed by 1,000,000 iterations was used to ensure convergence of the MCMC. The optimum number of clusters was identified using both the log-likelihood (lnP (K)) and the estimated ΔK for each K following [74]. The CLUster Matching and Permutation Program (CLUMPP) [75] was used to aggregate all STRUCTURE runs for the optimum identified value of K. STRUCTURE models, identification of K following the Evanno's method, CLUMPP analyses and visualisation of the individual Bayesian assignment probability for the optimum value of K were performed using the R package STRATAG [76].

Since French populations sampled were geographically close, we also investigated genetic structure in individuals using a Discriminant Analysis of Principal Components (DAPC) performed with the R package ADEGENET 2.1.3 [77].

Spatial genetic structure was also investigated using a spatial model in the R package GENELAND 4.9.2 [78]. Like STRUCTURE, GENELAND provides tools to identify clusters of individuals using Bayesian MCMC inferences with genetic data by maximizing Hardy–Weinberg equilibrium and minimizing linkage disequilibrium, but geographical coordinates of individuals are also considered to inform prior distribution. This spatial clustering method allows inference of the borders between inferred clusters and is a powerful method for detecting linear barriers to gene flow between populations [79]. The GENELAND analysis was performed using four independent runs and for each run, a number of clusters K ranging from Kmin = 1 to Kmax = 8 with 1,000,000 MCMC iterations, a burn-in period of 1000 and a thinning value of 100. We used a correlated allele frequency model that considered account the potential presence of null alleles. The best run was selected according to the highest average posterior probability given by GENELAND.

2.5. Migration Rates between Studied Populations

To estimate recent migration rates between populations (i.e., over the last several generations), analyses using MCMC were conducted in BAYESASS 3.0.4 software [80]. The model was first run considering default values of the mixing parameters for migration rates (i.e., 0.1), allele frequencies (i.e., 0.1) and inbreeding coefficients (i.e., 0.1). The acceptance rates given by BAYESASS of each of these three mixing parameters must be comprised optimally between 20% and 60% [81]. Since the acceptance rates were first higher than 60%, the model used ran with higher values of the mixing parameters (i.e., 0.7 for migration rate, 0.65 for allele frequencies and 0.8 for inbreeding coefficients) to ensure that all these acceptance rates fall in the acceptable range, and a burn-in of 1 × 106 for 1 × 107 iterations.

3. Results

3.1. Genetic Diversity

Two of the 12 microsatellites considered by [58], i.e., AiK04 and AiG03, were not retained because the first was monomorphic and the second was not successfully amplified. Loci AiJ04, AiL04 and AiM04 had high frequencies of null alleles (0.25, 0.25 and 0.23, respectively) and were also not retained. Then, individuals with more than three loci with missing values were removed from the data. Ten samples from fresh exuviae were excluded (i.e., 11.1% of the total number of fresh exuviae) and 18 samples from old exuviae were excluded (i.e., 31.6% of the total number of old exuviae). No sample from adult or nymphal legs had more than three missing loci. Finally, 223 individuals were considered for further analyses on seven markers (Table 2) and the total remaining missing values represented 6.6% of the loci.

Table 2. Genetic diversity measures (mean ± SE) in the 11 sampled populations of *Anax imperator* from Europe. Legend: n = number of sampled individuals, H_o = observed heterozygosity, H_e = expected heterozygosity, N_a = number of alleles, A_R = allelic richness, F_{IS} = inbreeding coefficient. Bolded H_o indicate populations presenting a significant departure from HWE condition. Bolded F_{IS} indicate a bootstrapped 95% confidence interval that does not overlap zero.

Country/Site	Pop	n	H_o	H_e	N_a	A_R	F_{IS}
Italy (Sicily)	1	6	0.71 ± 0.07	0.62 ± 0.05	29	4.14 ± 0.46	−0.17
Switzerland	2	16	0.65 ± 0.06	0.61 ± 0.06	42	4.02 ± 0.44	−0.06
Czech Republic	3	21	0.61 ± 0.07	0.65 ± 0.06	47	4.26 ± 0.57	0.06
France (Beau.)	4	33	0.61 ± 0.07	0.65 ± 0.06	56	4.39 ± 0.52	0.06
France (Cerisy)	5	19	0.59 ± 0.08	0.68 ± 0.05	46	4.40 ± 0.54	0.15
France (Heud.)	6	32	**0.47 ± 0.06**	0.67 ± 0.06	60	4.75 ± 0.54	**0.30**
France (Bois-G.)	7	34	0.59 ± 0.06	0.69 ± 0.05	54	4.63 ± 0.45	**0.16**
France (Marc.)	8	14	0.75 ± 0.07	0.72 ± 0.04	46	4.76 ± 0.47	−0.03
France (Paluel)	9	14	0.57 ± 0.06	0.66 ± 0.05	45	4.54 ± 0.38	0.15
France (Bresle)	10	9	0.43 ± 0.08	0.61 ±0.06	31	4.05 ± 0.53	0.25
United Kingdom	11	25	**0.33 ± 0.11**	0.63 ± 0.05	37	3.96 ± 0.35	**0.51**

Globally, populations showed substantial genetic variations. Estimates of observed and expected heterozygosity were close and ranged from 0.33 to 0.75 and from 0.61 to 0.72, respectively (Table 2). Only the population from U. K. presented an observed heterozygosity (0.33) significantly smaller than the expected heterozygosity (0.63). The total number of alleles over all loci ranged from 29 alleles in the population from Italy (Sicily), where the sample size was also reduced compared to the other locations, to 60 alleles in the Cerisy population (France; Table 2). A significant positive correlation was observed between the number of sampled individuals and the total number of alleles over all loci (Pearson correlation test: r = 0.84, p = 0.0013). Estimates of allelic richness per locus were similar between populations (Table 2).

Most markers met HWE conditions in each population except in the two populations showing departure from HWE conditions (i.e., U.K. and Heudreville, $p < 0.05$; Table 2). Because these departures of markers from HWE were not systematic in all populations, all markers were retained for further analyses. All other populations met HWE conditions ($p > 0.05$). FIS values showed significant deviation from zero in the two populations that did not meet HWE, indicating homozygosity excess in these two populations, and especially in the U.K. population (Table 2). Linkage disequilibrium (LD) tests for each pair of loci over all populations indicated no evidence for significant disequilibrium (all $p > 0.05$).

3.2. Population Genetic Differentiation

Global F_{IS} (0.1615, 95% CI = 0.1233–0.2000) and F_{ST} (0.0224, 95% CI = 0.0101–0.0366) were greater than zero. Pairwise F_{ST}-values ranged between −0.0151 and 0.1297.

A moderate genetic differentiation (i.e., 95% bootstrapped CI) was particularly found between U.K. and all other populations (all $F_{ST} > 0.081$). A moderate differentiation was also found between the Swiss and one of the French populations (i.e., Marchésieux, $F_{ST} = 0.046$; Table 3).

Table 3. F_{ST}-values between all populations. Bolded values indicate a bootstrapped 95% confidence interval that does not overlap zero. The mean F_{ST}-value is 0.0306 ± 0.0051.

	Italy (Sicily)	**Switz**	**Cz. Rep.**	**France (Beaus.)**	**France (Cerisy)**	**France (Heud.)**	**France (Bois-G.)**	**France (March.)**	**France (Paluel)**	**France (Bresle)**	**U.K.**
Italy (Sicily)	-	0.0293	0.0307	−0.0079	0.0163	0.0047	0.0104	0.0097	−0.0132	0.0254	**0.1055**
Switzerland		-	0.0128	0.0167	0.0111	0.0234	0.0192	**0.0446**	0.0126	0.0481	**0.1220**
Czech Republic			-	0.0110	0.0232	0.0272	0.0206	0.0291	0.0164	0.0509	**0.1148**
France (Beau.)				-	0.0037	0.0115	0.0143	0.0036	−0.0021	0.0191	**0.1297**
France (Cerisy)					-	−0.0059	0.0045	0.0049	−0.0063	−0.0055	**0.0903**
France (Heud.)						-	−0.0007	0.0019	−0.0031	−0.0151	**0.0843**
France (Bois-G.)							-	0.0183	0.0009	0.0193	**0.0841**
France (Marc.)								-	0.0052	0.0119	**0.0920**
France (Paluel)									-	0.0043	**0.1014**
France (Bresle)										-	0.0813
United Kingdom											-

The AMOVA analysis showed that most molecular genetic variation resulted from individual genetic variation within populations (92.73%; Table 4), the remainder (6.37%) resulting from genetic variation among countries ($p = 0.02$). Variation among populations within countries related only to French populations, since only one population was analyzed in the other countries. No significant genetic variation was found among populations sampled in France (0.90%, $p = 0.09$).

Table 4. Results of the AMOVA performed for the 11 population of *Anax imperator* sampled in the five European countries studied.

Source of Variation	**df**	**Sum of Squares**	**Variance Components**	**Percentage of Variance**	**Φ-Statistics**	***p*-Value**
Among countries	4	138.79	0.78	6.37	$\Phi_{CT} = 0.06$	0.02
Among populations within countries	6	82.21	0.11	0.90	$\Phi_{SC} = 0.01$	0.09
Within populations	212	2405.45	11.35	92.73	-	-
Total	222	2626.45	12.24	100.00	-	-

No evidence for isolation by distance was found among populations at the European scale, since the correlation between genetic and geographic distance matrices was not significant (Mantel test: $r = 0.33$, $p = 0.17$; Figure 1). However, three ellipses could be visually drawn to delimitate three point clouds: the U.K. population versus all others (1), French (Normandy) populations within each other (2) and populations from Central and Southern European countries (i.e., Switzerland, Czech Republic and Italy (Sicily)) versus French (Normandy) populations (3). Although no significant differentiation was found, these point clouds show a higher genetic differentiation between U.K. populations and all other populations (i.e., ellipse 1) than between French (Normandy) populations themselves (i.e., ellipse 2) and between French (Normandy) and the three populations from Central and Southern European countries (i.e., ellipse 3).

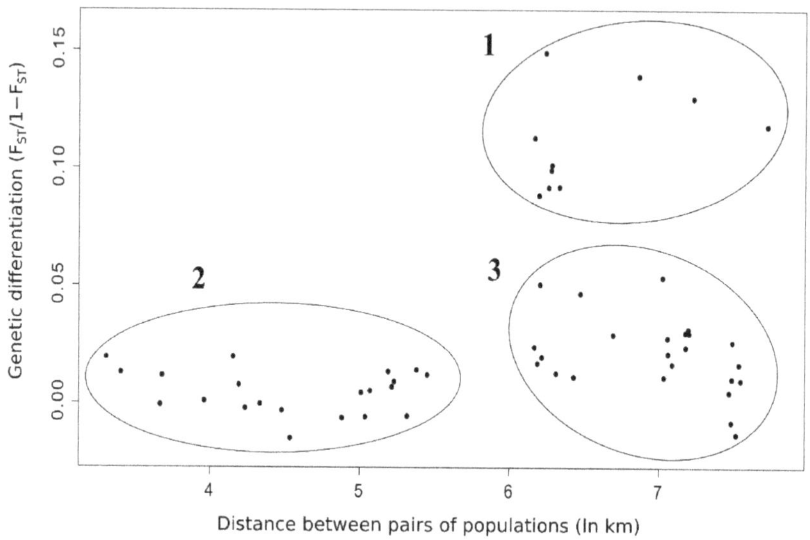

Figure 1. Relationship between pairwise population differentiation ($F_{ST}/1 \div F_{ST}$) and the geographic distance (ln km) separating populations. Ellipse 1 represents pairwise differentiations between U.K. and all other populations. Ellipse 2 represents pairwise differentiations between the French (Normandy) populations. Ellipse 3 represents pairwise differentiations between the French (Normandy) populations and the three populations from Central and Southern European countries (i.e., Switzerland, Czech Republic and Italy (Sicily) and among these three populations.

3.3. Spatial Genetic Structure

STRUCTURE analyses identified three genetic clusters since values of lnP (K) and ΔK showed a peak at K = 3 (Figure S1a,b). A first group contained populations from Switzerland, Czech Republic and Italy (Sicily) (Figure 2a,b). A second group contained the U.K. population (Figure 2a,b). A third group contained all French populations (Figure 2a,c).

DAPC on French populations only was performed retaining 70 Principal Components (PC) and two discriminant functions. It suggested three subclusters: one with the population of Bresle, a second with the population of Marchésieux and a third with the five other Normandy populations in which the population of Bois-Guillaume was slightly detached from the four other populations (Figure S2).

All independent runs performed in the spatial model given by GENELAND corroborated STRUCTURE results and identified three genetic clusters. The run with the highest average log posterior probability was retained. The MCMC converged within the 100,000 iterations. The U.K. and Bresle populations were assigned to one cluster with a probability of at least 0.7 (Figure S3a). Most of other Normandy populations except Beaussault were assigned to a second cluster with a probability of at least 0.7 (Figure S3b). Normandy population from Beaussault and populations from Switzerland, Czech Republic and Italy (Sicily) were assigned to a third cluster with a probability of at least 0.7 (Figure S3c).

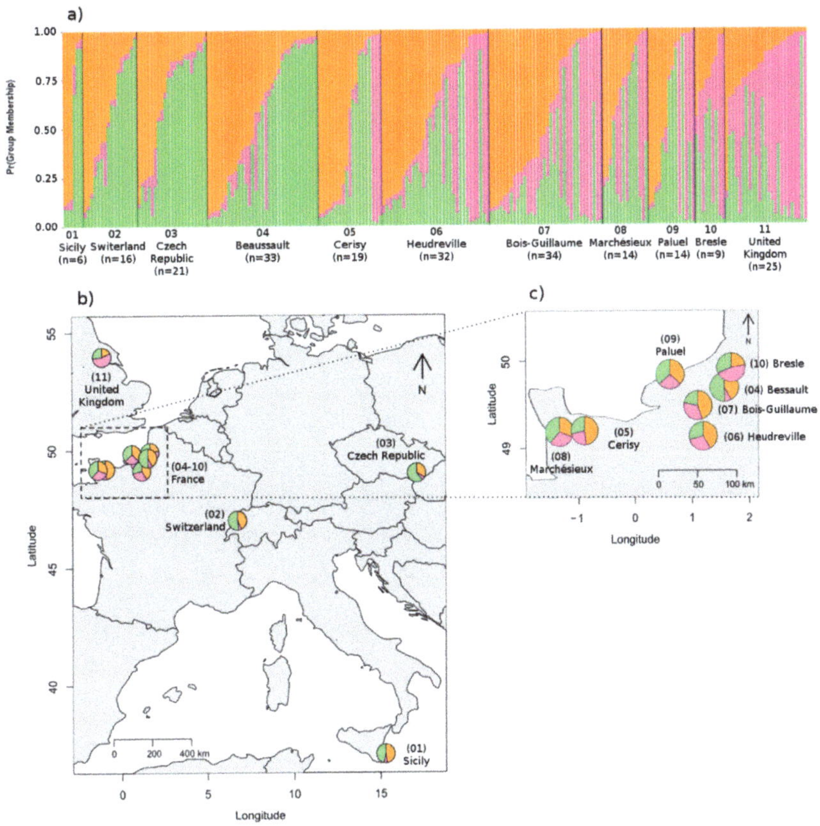

Figure 2. Results of individual assignments to each cluster by STRUCTURE. (**a**) Probabilities of individual membership in the 11 sampled populations, with bars representing individuals and colours representing the probability of belonging to the three genetic clusters identified with STRUCTURE. (**b**) Mean membership in the 11 sampled populations in Europe to each of the three clusters. French populations are delimited in a dotted rectangle. (**c**) Enlarged view of the French populations.

3.4. Recent Migration Rates among Populations

Bayesian analyses clearly showed several directional gene flows between the studied populations. The U.K. population was rather isolated from the other populations but it was a donor site only for the Bresle population from France (Table 5). All the three populations located in Central and Southern European countries (i.e., Switzerland, Czech Republic and Italy (Sicily)) showed no migratory exchange between them and were not sources for French populations. The French population from Beaussault, and especially the French population from Cerisy, were sources for Swiss, Sicilian and Czech populations (Table 5). In France especially, the population from Cerisy was identified as likely one of the main donors for the other four French populations. The populations from Heudreville and Beaussault seemed particularly implicated in source-sink processes, since they acted both as donor and target populations. The French populations from Paluel and Bois-Guillaume were identified as receivers only. The French population from Marchésieux seemed isolated from all other populations since no exchange was identified (Table 5; Figure 3).

Table 5. Bayesian modelling of potential bias in direction of dispersal (gene flow) among the 11 populations of *Anax imperator* in Europe. Numbers represent the proportion that disperses between sites (bold indicates self-recruitment). Values <0.05 (5%) are in grey. Italic indicates pairs of sites with ≥10% exchange. These values represent historical gene flow, and do not provide any information about contemporary levels of dispersal among sites.

Target	Potential Donor Site										
	Italy (Sicily)	Switz.	Cz. Rep.	France (Beaus.)	France (Cerisy)	France (Heud.)	France (Bois-G.)	France (March.)	France (Paluel)	France (Bresle)	U.K.
Italy (Sicily)	**0.68**	0.02	0.02	0.05	*0.10*	0.02	0.02	0.02	0.02	0.02	0.02
Switzerland	0.01	**0.68**	0.02	0.07	*0.12*	0.01	0.03	0.01	0.01	0.01	0.01
Czech Republic	0.01	0.02	**0.69**	0.04	*0.17*	0.01	0.02	0.01	0.01	0.01	0.01
France (Beau.)	<0.01	0.02	0.03	**0.69**	*0.18*	0.01	0.01	0.01	<0.01	<0.01	<0.01
France (Cerisy)	0.01	0.02	0.03	0.03	**0.78**	0.02	0.02	0.01	0.01	0.01	0.04
France (Heud.)	<0.01	0.01	0.01	0.03	*0.13*	**0.72**	0.04	<0.01	<0.01	<0.01	0.02
France (Bois-G.)	<0.01	0.01	0.02	0.02	*0.12*	0.06	**0.73**	<0.01	<0.01	<0.01	0.01
France (Marc.)	0.02	0.02	0.02	0.04	*0.10*	0.04	0.02	**0.69**	0.01	0.01	0.03
France (Paluel)	0.01	0.02	0.02	0.05	0.08	0.05	0.02	0.02	**0.68**	0.01	0.02
France (Bresle)	0.02	0.02	0.02	0.04	0.04	0.03	0.02	0.02	0.02	**0.68**	*0.11*
United Kingdom	<0.01	0.01	0.01	0.01	0.02	0.02	0.02	0.01	<0.01	0.01	**0.87**

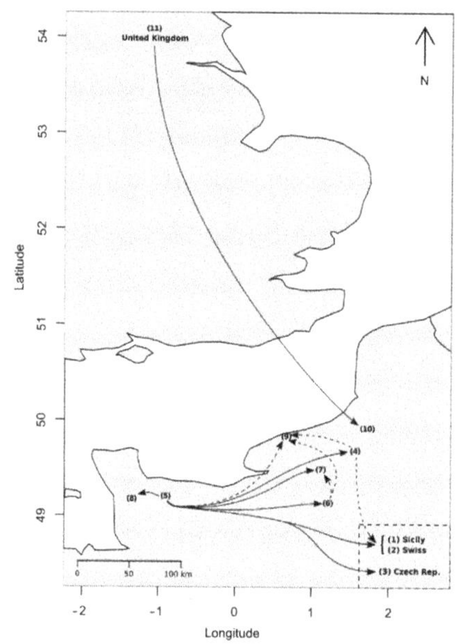

Figure 3. Map of the gene flow between the 11 populations of *Anax imperator* in Europe. Arrows indicate gene flow between pairs of sites with ≥5% exchange (dashed lines) and ≥10% exchange (solid lines). Numbers in parentheses are numbers assigned to populations in Table 1.

4. Discussion

Anax imperator populations sampled in the Normandy region in France had a low level of genetic differentiation (i.e., F_{ST}-values all below 0.03) compared to previous studies on European odonates at a local or regional scale (i.e., F_{ST} up to 0.08, 0.10, 0.24 and 0.28 for *Leucorrhinia dubia* Vander Linden 1825, *Coenagrion scitulum* Rambur 1842, *C. mercuriale,* respectively) [45,82–84]. Such results confirm the high mobility of *A. imperator* [85] and its efficient dispersal between ponds at the regional scale [54]. However, at the European scale, a moderate level of genetic differentiation was found between populations. In particular, the population sampled in the United Kingdom presented the highest genetic differentiation from the other populations sampled (i.e., F_{ST} up to 0.13). Moreover, movements of individuals seem not to occur between all populations at local and European scales, meaning that all ponds may not play the same role in maintaining exchanges and a posteriori population viability.

All populations presented an observed heterozygosity close to expected levels, except the population sampled in the U.K. In this population, observed heterozygosity was lower than expected heterozygosity and allelic richness was lower compared to other populations. This lower genetic diversity was also associated with a significant degree of inbreeding. Low genetic diversity and inbreeding suggest a possible isolation of this population with only few exchanges of individuals with other populations from the same region. They can also indicate a previous population bottleneck or a founder effect following a recent colonization [7]. Indeed, the U.K. population of *A. imperator* was sampled near the city of York, at the Northern margin of its distribution range. In the current context of climate change [86], many species are shifting their distributions to higher altitudes or toward the poles [87]. A northward shift of range margins was already reported for the distribution of many odonates in England, including *A. imperator* that moved 85 km to the North between 1960–1970 and 1985–1995 periods [88]. Therefore, the observed low genetic diversity might be a consequence of a recent colonization of that sampling site. The two French populations from Bois-Guillaume and Heudreville also presented a significant degree of inbreeding. This was quite unexpected because these populations were among the largest that were sampled in the Normandy region. The pond in Bois-Guillaume is located in a suburban landscape suggesting a possible negative effect of the surrounding urbanization on exchanges between this population and the other ponds. Possible negative effects of human activities movements were already reported for other insect species [89,90]. The population from Heudreville was sampled in a large well-vegetated pond and with only few surrounding waterbodies [91] which may have prevented exchanges with other populations.

In pond networks, exchanges of individuals are a determining factor for the persistence of local populations. These networks are often referred to as metapopulations [92] in which some ponds act as sources or other as sinks [93]. In this study, the genetic variation was much higher within populations than between them, but a significant variation was found among countries. Among French (Normandy) populations, the genetic diversity was very similar. One of them (i.e., Bresle population) showed a genetic diversity very close to the U.K population, and another (i.e., Beaussault population) was closely related to the populations of Central and Southern European countries, especially the Swiss one. Within the French (Normandy) populations, one population (i.e., Cerisy) was identified as the main donor for all other populations, except the Bresle one, suggesting higher exchanges of this population with the U.K. population than with other Normandy populations. The Cerisy population was identified as a source of genetic variability for the rest of Normandy, whereas three other populations were only receivers of gene flow. The populations studied are probably supported by a network of many ponds. Therefore, we can hypothesize that the density of ponds around the Cerisy forest or the large size of the pond in Heudreville allows maintenance of large populations. From these source populations, some individuals may leave to smaller sink populations in areas where pond density is lower. However, this hypothesis requires further population demographic studies to be explored [94,95]. Overall, the results indicate a high gene flow in *A. imperator* between all sampled populations from continental Europe. These populations may be connected by long distance movements [96], but also by high rates of short distance movements between ponds, leading to a stepping-stone dispersal [97].

Increasing genetic isolation with distance is a common relationship that often shapes the genetic structure of populations [2]. At the European scale, this pattern was already reported for odonates (e.g., *Ischnura elegans*) [45], but also other flying insects (*Operophtera brumata* Linnaeus, 1958) [27] or flying mammals (*Myotis daubentonii* Kuhl, 1817) [98]. In the present study, this pattern was not found among the populations of *A. imperator* at the European scale. For instance, even if the U.K. population presented a genetic differentiation from the French (Normandy) populations located ca. 500 km away and populations situated further east, no differentiation was found between the French populations and the populations of Central and Southern European countries. However, this result suggests that the English Channel could act as a physical barrier to the gene flow of *A. imperator* [99]. No genetic difference due to the English Channel or the Baltic Sea was found in smaller odonate species such as *I. elegans* [45]. This difference in results may be due to the fact that the populations of *I. elegans* can reach very high densities at some sites, which increases the observed intra-population variation and limits genetic drifts. For instance, in a survey of 20 ponds in Normandy, the larval density of *I. elegans*, was about 6 times higher than that of *A. imperator* [100]. The small populations of *A. imperator* may, therefore, be more prone to genetic differentiation in case of reduced gene flows [7].

Gene flows occurred from the U.K. population to one French population (i.e., Bresle) and from several French (Normandy) populations to all populations from Central and Southern European countries. These movements follow the same direction as the westerly winds brought by the gulf stream across the English Channel and prevailing in a major part of France. Whether by supporting active migrations of large species e.g., [101,102] or blowing small species through long distances [96], the wind probably plays a major role in the dispersal of odonates [38]. *Anax imperator* is a large and mobile dragonfly that was never reported as a migratory species, i.e., as a species flying on long-distances between emergence places and new habitats where reproduction take place [38]. However, we can hypothesize that some individuals might occasionally be able to undertake long-distance flights helped by wind currents, similarly to the regular migration movements described for the sister species *Anax junius* in the U.S.A [51]. A few rare long-distance movements supported by prevailing winds could cause some U.K. ponds to become sources of migrants for some French (Normand) ones, and other Normand ponds to become migration sources for further populations in Central and Southern European countries.

Genetic studies on odonates mostly use fresh material, especially legs of adults [43,44] or sometimes heads of adults [45,102]. However, this collection method is relatively invasive. It was recommended to avoid it for species with high conservation value [103] and negative effects on survival were observed on small odonate species on which several legs were lost [104]. Alternative methods based on DNA extraction from exuviae are, therefore, increasingly used [82,84]. This non-invasive method has the advantage of ensuring that individuals have grown in the studied site, while the origin cannot always be assessed for adults [105]. For larger species, collecting exuviae is also easier than catching flying adults that fly very fast over ponds during the reproductive period. Nevertheless, the persistence of DNA in these exoskeletons is poorly known. Especially, prolonged exposure to sunlight or enzymatic action on hydrated exuviae after rain may lead to a significant reduction of DNA yields [106]. In this study, we were able to compare the DNA yields of fresh exuviae reared at the laboratory with that of other 'old' exuviae sampled in situ. Although the total amount of DNA was much lower in 'old' exuviae, most samples (i.e., 31.6% of 'old' exuviae excluded versus 11.1% from 'fresh' exuviae excluded) could be used for this microsatellite study. We therefore recommend this method for further studies, at least on large species that are likely to contain more genetic material.

Legs of nymphs provide also a reliable source of DNA, but are still seldom used in population genetic studies on dragonflies but see [12,107]. Contrary to adults or exuviae, nymphs can be sampled during all weather conditions and all seasons of the year, a feature that can simplify the schedule of field sessions. Moreover, in many species, high nymphal densities ease the collection of a large number of samples, whereas flying adults may be hard to catch, especially those of Aeshnidae. Although identification of some species may be difficult in the field, we suggest that nymphal DNA sampling should be considered more in further studies, especially those on large dragonfly species.

Overall, this study provides insights into gene flow of *A. imperator* populations at both regional and European scales. Our results highlight the role of the English Channel as a potential barrier to dispersal, especially for movements from France (Normandy) to the U.K. They also suggest a probable role of the wind for long-distance movements of odonates (e.g., Gulf Stream). Only a small fraction of ponds harboring *A. imperator* in Europe were sampled and more investigations would be useful to confirm the relationship between individuals and major wind currents. However, a high gene flow was found between continental populations, which may indicate that the distance between ponds at the European scale do not prevent dispersal movements of this large dragonfly. Dispersal probably occurs on a large spatial scale via successional movements from pond to pond at local scales. Nevertheless, the current pattern of genetic diversity may also mirror historical exchanges between populations rather than a contemporary gene flow [108]. Since the number of European ponds underwent a dramatic decline during the last century [34], the gene flow described in the present study may no longer be relevant today. Further studies comparing genetic markers with different mutation rates would be needed to address this question and disentangle historical and contemporary connectivity between European ponds [109].

Supplementary Materials: The following are available online at https://www.mdpi.com/article/10.3390/d14020068/s1, Figure S1: Estimation of number of clusters (K) with STRUCTURE using (a) mean of estimated Ln probabilities of data (±SD) for each K-value and (b) Delta K for each K-value (Evanno's method), Figure S2: Discriminant Analysis of Principal Components results of Anax imperator individuals of sampled French populations, Figure S3: Results of spatial Geneland analysis on the 11 populations of Anax imperator. Each figure corresponds to a cluster identified by Geneland. Black dots indicate the position of the populations (see Figure 1). Black lines indicated the posterior probabilities of membership in the three clusters, with darker colours (red) indicating highest posterior probabilities of belonging to the cluster.

Author Contributions: Conceptualization, M.M. and A.H.; methodology, M.M. and A.H.; field work and preparation of the samples: M.M.; laboratory work, M.M. and A.H.; software, A.H.; validation, M.M. and A.H.; formal analysis, A.H.; investigation, M.M. and A.H.; resources, A.H.; data curation, A.H.; writing—original draft preparation, M.M. and A.H.; writing—review and editing: M.M. and A.H.; supervision, A.H.; project administration, A.H.; funding acquisition, A.H. All authors have read and agreed to the published version of the manuscript.

Funding: This study was funded by the Seine-Normandy Water Agency via the research project CROSS. M. Minot was supported by a grant provided by the French Ministry of Higher Education and Research through the Normandy's Doctoral School of Integrative Biology, Health and Environment (EdN BISE 497, Rouen University).

Institutional Review Board Statement: Not applicable.

Data Availability Statement: The data presented in this study will be publicly available from the corresponding author upon publication.

Acknowledgments: M. Aubert and A. Husté (ECODIV, EA 1293 INRAe) were the co-supervisors. We want to thank the contributors that accepted to send us samples for this study. Thanks are especially due to Andrea Corso from Sicily (Italy), Sebastien Tschanz from Neuchâtel (Switzerland) and Martin Roberts from York (United Kingdom). We also thank L. Mundell for reviewing the English language.

Conflicts of Interest: The authors declare no conflict of interest.

References

1. Thomas, C.D.; Kunin, W.E. The Spatial Structure of Populations. *J. Anim. Ecol.* **1999**, *68*, 647–657. [CrossRef]
2. Sexton, J.P.; Hangartner, S.B.; Hoffmann, A.A. Genetic Isolation by Environment or Distance: Which Pattern of Gene Flow Is Most Common? *Evolution* **2013**, *68*, 1–15. [CrossRef]
3. Norman, A.J.; Stronen, A.V.; Fuglstad, G.-A.; Ruiz-Gonzalez, A.; Kindberg, J.; Street, N.R.; Spong, G. Landscape Relatedness: Detecting Contemporary Fine-Scale Spatial Structure in Wild Populations. *Landsc. Ecoogyl.* **2017**, *32*, 181–194. [CrossRef]
4. Maxwell, S.L.; Fuller, R.A.; Brooks, T.M.; Watson, J.E. Biodiversity: The Ravages of Guns, Nets and Bulldozers. *Nat. News* **2016**, *536*, 143–145. [CrossRef] [PubMed]
5. Frankham, R.; Ballou, J.D.; Briscoe, D.A. *Introduction to Conservation Genetics*; Cambridge University Press: Cambridge, UK, 2010.
6. Dennis, R.L.H.; Dapporto, L.; Dover, J.W.; Shreeve, T.G. Corridors and Barriers in Biodiversity Conservation: A Novel Resource-Based Habitat Perspective for Butterflies. *Biodivers. Conserv.* **2013**, *22*, 2709–2734. [CrossRef]

7. Allendorf, F.W.; Luikart, G.H.; Aitken, S.N. *Conservation and the Genetics of Populations*, 2nd ed.; Wiley-Blackwell: Hoboken, NJ, USA, 2012.
8. Bergamaschi, A.C.B.; Lama, M.A.D.; Bergamaschi, A.C.B.; Lama, M.A.D. Gene Variation and Genetic Differentiation among Populations of the Solitary Mud Dauber Wasp *Trypoxylon* (*Trypargilum*) *albitarse* Fabricius 1804 (Hymenoptera, Crabronidae). *Genet. Mol. Biol.* **2015**, *38*, 519–526. [CrossRef]
9. De Meester, L.; Gómez, A.; Okamura, B.; Schwenk, K. The Monopolization Hypothesis and the Dispersal–Gene Flow Paradox in Aquatic Organisms. *Acta Oecologica* **2002**, *23*, 121–135. [CrossRef]
10. Habel, J.C.; Junker, M.; Schmitt, T. High Dispersal Ability and Low Genetic Differentiation in the Widespread Butterfly Species *Melanargia galathea*. *J. Insect Conserv.* **2010**, *14*, 467–478. [CrossRef]
11. Heinken, T.; Weber, E. Consequences of Habitat Fragmentation for Plant Species: Do We Know Enough? Perspectives in Plant Ecology. *Evol. Syst.* **2013**, *15*, 205–216. [CrossRef]
12. Kohli, M.K.; Sahlén, G.; Kuhn, W.R.; Ware, J.L. Extremely Low Genetic Diversity in a Circumpolar Dragonfly Species, *Somatochlora sahlbergi* (Insecta: Odonata: Anisoptera). *Sci. Rep.* **2018**, *8*, 15114. [CrossRef]
13. Baguette, M.; Blanchet, S.; Legrand, D.; Stevens, V.M.; Turlure, C. Individual Dispersal, Landscape Connectivity and Ecological Networks. *Biol. Rev.* **2013**, *88*, 310–326. [CrossRef]
14. Jordano, P. What Is Long-Distance Dispersal? And a Taxonomy of Dispersal Events. *J. Ecol.* **2017**, *105*, 75–84. [CrossRef]
15. Le Corre, V.; Kremer, A. Cumulative Effects of Founding Events during Colonisation on Genetic Diversity and Differentiation in an Island and Stepping-Stone Model. *J. Evol. Biol.* **1998**, *11*, 495–512. [CrossRef]
16. Coughlan, N.E.; Kelly, T.C.; Jansen, M.A.K. "Step by Step": High Frequency Short-Distance Epizoochorous Dispersal of Aquatic Macrophytes. *Biol. Invasions* **2017**, *19*, 625–634. [CrossRef]
17. Nathan, R. Long-Distance Dispersal of Plants. *Science* **2006**, *313*, 786–788. [CrossRef]
18. Da Fonte, L.F.M.; Mayer, M.; Lötters, S. Long-Distance Dispersal in Amphibians. *Front. Biogeogr.* **2019**, *11*, 1–14. [CrossRef]
19. Newton, I. *The Migration Ecology of Birds*; Elsevier/Acad. Press: Amsterdam, The Netherlands, 2008.
20. Brink, K.; Gough, P.; Royte, J.; Schollema, P.P.; Wanningen, H. *From Sea to Source 2.0. Protection and Restoration of Fish Migration in Rivers Worldwide*; World Fish Migration Foundation: Groningen, The Netherlands, 2018.
21. Bowen, B.W.; Karl, S.A. Population Genetics and Phylogeography of Sea Turtles. *Mol. Ecol.* **2007**, *16*, 4886–4907. [CrossRef]
22. Haest, B.; Hüppop, O.; van de Pol, M.; Bairlein, F. Autumn Bird Migration Phenology: A Potpourri of Wind, Precipitation and Temperature Effects. *Glob. Chang. Biol.* **2019**, *25*, 4064–4080. [CrossRef]
23. Green, A.J.; Figuerola, J. Recent Advances in the Study of Long-Distance Dispersal of Aquatic Invertebrates via Birds. *Divers. Distrib.* **2005**, *11*, 149–156. [CrossRef]
24. Stockwell, C.A.; Hendry, A.P.; Kinnison, M.T. Contemporary Evolution Meets Conservation Biology. *Trends Ecol. Evol.* **2003**, *18*, 94–101. [CrossRef]
25. Kelly, E.; Phillips, B.L. Targeted gene flow for conservation. *Conserv. Biol.* **2016**, *30*, 259–267. [CrossRef] [PubMed]
26. Ross, K.G. How to Measure Dispersal: The Genetic Approach. The Example of Fire Ants. In *Dispersal*; Oxford University Press: Oxford, NY, USA, 2001; pp. 29–42.
27. Leggett, H.C.; Jones, E.O.; Burke, T.; Hails, R.S.; Sait, S.M.; Boots, M. Population Genetic Structure of the Winter Moth, *Operophtera brumata* Linnaeus, in the Orkney Isles Suggests Long-Distance Dispersal. *Ecol. Entomol.* **2011**, *36*, 318–325. [CrossRef]
28. Biggs, J.; von Fumetti, S.; Kelly-Quinn, M. The Importance of Small Waterbodies for Biodiversity and Ecosystem Services: Implications for Policy Makers. *Hydrobiologia* **2017**, *793*, 3–39. [CrossRef]
29. Hill, M.J.; Biggs, J.; Thornhill, I.; Briers, R.A.; Gledhill, D.G.; White, J.C.; Wood, P.J.; Hassall, C. Urban Ponds as an Aquatic Biodiversity Resource in Modified Landscapes. *Glob. Chang. Biol.* **2017**, *23*, 986–999. [CrossRef]
30. Kimura, M.; Weiss, G.H. The Stepping Stone Model of Population Structure and the Decrease of Genetic Correlation with Distance. *Genetics* **1964**, *49*, 561–576. [CrossRef]
31. Watts, A.G.; Schlichting, P.; Billerman, S.; Jesmer, B.; Micheletti, S.; Fortin, M.-J.; Funk, C.; Hapeman, P.; Muths, E.L.; Murphy, M.A. How Spatio-Temporal Habitat Connectivity Affects Amphibian Genetic Structure. *Front. Genet.* **2015**, *6*, 1–13. [CrossRef] [PubMed]
32. Youngquist, M.B.; Inoue, K.; Berg, D.J.; Boone, M.D. Effects of Land Use on Population Presence and Genetic Structure of an Amphibian in an Agricultural Landscape. *Landsc. Ecol.* **2017**, *32*, 147–162. [CrossRef]
33. Thornhill, I.; Batty, L.; Hewitt, M.; Friberg, N.R.; Ledger, M.E. The Application of Graph Theory and Percolation Analysis for Assessing Change in the Spatial Configuration of Pond Networks. *Urban Ecosyst.* **2018**, *21*, 213–225. [CrossRef]
34. Oertli, B.; Biggs, J.; Céréghino, R.; Grillas, P.; Joly, P.; Lachavanne, J.-B. Conservation and Monitoring of Pond Biodiversity: Introduction. *Aquat. Conserv. Mar. Freshw. Ecosyst.* **2005**, *15*, 535–540. [CrossRef]
35. Anderson, T.L.; Ousterhout, B.H.; Peterman, W.E.; Drake, D.L.; Semlitsch, R.D. Life History Differences Influence the Impacts of Drought on Two Pond-Breeding Salamanders. *Ecol. Appl.* **2015**, *25*, 1896–1910. [CrossRef]
36. Suhling, F.; Sahlén, G.; Gorb, S.; Kalkman, V.J.; Dijkstra, K.-D.B.; van Tol, J. Order Odonata. In *Thorp and Covich's Freshwater Invertebrates*; Elsevier: Amsterdam, The Netherlands, 2015; pp. 893–932. [CrossRef]
37. Bilton, D.T.; Freeland, J.R.; Okamura, B. Dispersal in Freshwater Invertebrates. *Annu. Rev. Ecol. Syst.* **2001**, *32*, 159–181. [CrossRef]
38. Corbet, P.S. *Dragonflies: Behaviour and Ecology of Odonata*; Harley Books: Colchester, UK, 1999.

39. Suhling, F.; Martens, A.; Suhling, I. Long-Distance Dispersal in Odonata: Examples from Arid Namibia. *Austral Ecol.* **2017**, *42*, 544–552. [CrossRef]
40. Conrad, K.F.; Willson, K.H.; Harvey, I.F.; Thomas, C.J.; Sherratt, T.N. Dispersal Characteristics of Seven Odonate Species in an Agricultural Landscape. *Ecography* **1999**, *22*, 524–531. [CrossRef]
41. Rouquette, J.R.; Thompson, D.J. Patterns of Movement and Dispersal in an Endangered Damselfly and the Consequences for Its Management. *J. Appl. Ecol.* **2007**, *44*, 692–701. [CrossRef]
42. Le Gall, M.; Chaput-Bardy, A.; Husté, A. Context-Dependent Local Movements of the Blue-Tailed Damselfly, *Ischnura elegans*: Effects of Pond Characteristics and the Landscape Matrix. *J. Insect Conserv.* **2017**, *21*, 243–256. [CrossRef]
43. Troast, D.; Suhling, F.; Jinguji, H.; Sahlén, G.; Ware, J. A Global Population Genetic Study of *Pantala flavescens*. *PLoS ONE* **2016**, *11*, e0148949. [CrossRef] [PubMed]
44. Lorenzo-Carballa, M.O.; Ferreira, S.; Sims, A.M.; Thompson, D.J.; Watts, P.C.; Cher, Y.; Damoy, V.; Evrard, A.; Gelez, W.; Vanappelghem, C. Impact of Landscape on Spatial Genetic Structure and Diversity of *Coenagrion mercuriale* (Zygoptera:Coenagrionidae) in Northern France. *Freshw. Sci.* **2015**, *34*, 1065–1078. [CrossRef]
45. Wellenreuther, M.; Sánchez-Guillén, R.A.; Cordero-Rivera, A.; Svensson, E.I.; Hansson, B. Environmental and Climatic Determinants of Molecular Diversity and Genetic Population Structure in a Coenagrionid Damselfly. *PLoS ONE* **2011**, *6*, e20440. [CrossRef]
46. Lorenzo-Carballa, M.O.; Hadrys, H.; Cordero-Rivera, A.; Andrés, J.A. Population Genetic Structure of Sexual and Parthenogenetic Damselflies Inferred from Mitochondrial and Nuclear Markers. *Heredity* **2012**, *108*, 386–395. [CrossRef] [PubMed]
47. Purse, B.V.; Hopkins, G.W.; Day, K.J.; Thompson, D.J. Dispersal Characteristics and Management of a Rare Damselfly. *J. Appl. Ecol.* **2003**, *40*, 716–728. [CrossRef]
48. Watts, P.C.; Saccheri, I.J.; Kemp, S.J.; Thompson, D.J. Population Structure and the Impact of Regional and Local Habitat Isolation upon Levels of Genetic Diversity of the Endangered Damselfly *Coenagrion mercuriale* (Odonata: Zygoptera). *Freshw. Biol.* **2006**, *51*, 193–205. [CrossRef]
49. Boudot, J.P.; Kalkman, V.J. *Atlas of the European Dragonflies and Damselflies*; KNNV Publishing: Zeist, The Netherlands, 2015.
50. Flenner, I.; Sahlén, G. Dragonfly community re-organisation in boreal forest lakes: Rapid species turnover driven by climate change? *Insect Conserv. Divers.* **2008**, *1*, 169–179. [CrossRef]
51. Knight, S.M.; Pitman, G.M.; Flockhart, D.T.T.; Norris, D.R. Radio-Tracking Reveals How Wind and Temperature Influence the Pace of Daytime Insect Migration. *Biol. Lett.* **2019**, *15*, 20190327. [CrossRef]
52. Minot, M.; Le Gall, M.; Husté, A. Biometry of the Large Dragonfly *Anax imperator* (Odonata: Aeshnidae): A Study of Traits from Larval Development to Adults. *Eur. J. Entomol.* **2019**, *116*, 269–280. [CrossRef]
53. Corbet, P.S. The Life-History of the Emperor Dragonfly *Anax imperator* Leach (Odonata: Aeshnidae). *J. Anim. Ecol.* **1957**, *26*, 1–69. [CrossRef]
54. Minot, M.; Besnard, A.; Husté, A. Habitat Use and Movements of a Large Dragonfly (Odonata: *Anax Imperator*) in a Pond Network. *Freshw. Biol.* **2021**, *66*, 241–255. [CrossRef]
55. Lubertazzi, M.A.A.; Ginsberg, H.S. Persistence of Dragonfly Exuviae on Vegetation and Rock Substrates. *Northeast. Nat.* **2009**, *16*, 141–147. [CrossRef]
56. Roland, H.-J. Haltbarkeit von Anax-Exuvien Am Ort Der Emergenz. *Libellula* **2010**, 3, 231–240.
57. Keller, D.; Brodbeck, S.; Holderegger, R. Characterization of Microsatellite Loci in *Leucorrhinia caudalis*, a Rare Dragonfly Endangered throughout Europe. *Conserv. Genet. Resour.* **2009**, *1*, 179. [CrossRef]
58. Hadrys, H.; Timm, J.; Streit, B.; Giere, S. A Panel of Microsatellite Markers to Study Sperm Precedence Patterns in the Emperor Dragonfly *Anax imperator* (Odonata: Anisoptera). *Mol. Ecol. Notes* **2007**, *7*, 296–298. [CrossRef]
59. R Core Team. *R: A Language and Environment for Statistical Computing*; R Foundation for Statistical Computing: Vienna, Austria, 2019.
60. Adamack, A.T.; Gruber, B. Landgenreport: A new R function to simplify landscape genetic analysis using resistance surface layers. *Mol. Ecol. Resour.* **2015**, *15*, 1172–1178.
61. Paradis, E. Pegas: An R Package for Population Genetics with an Integrated–Modular Approach. *Bioinformatics* **2010**, *26*, 419–420. [CrossRef] [PubMed]
62. Raymond, M.; Rousset, F. An Exact Test for Population Differentiation. *Evolution* **1995**, *49*, 1280–1283. [CrossRef]
63. Rousset, F. Genepop'007: A Complete Re-Implementation of the Genepop Software for Windows and Linux. *Mol. Ecol. Resour.* **2008**, *8*, 103–106. [CrossRef]
64. Benjamini, Y.; Hochberg, Y. Controlling the False Discovery Rate: A Practical and Powerful Approach to Multiple Testing. *J. R. Stat. Soc. Ser. B (Methodol.)* **1995**, *57*, 289–300. [CrossRef]
65. Benjamini, Y.; Yekutieli, D. The Control of the False Discovery Rate in Multiple Testing under Dependency. *Ann. Stat.* **2001**, *29*, 1165–1188. [CrossRef]
66. Keenan, K.; McGinnity, P.; Cross, T.F.; Crozier, W.W.; Prodöhl, P.A. DiveRsity: An R Package for the Estimation of Population Genetics Parameters and Their Associated Errors. *Methods Ecol. Evol.* **2013**, *4*, 782–788. [CrossRef]
67. Kalinowski, S.T. Counting Alleles with Rarefaction: Private Alleles and Hierarchical Sampling Designs. *Conserv. Genet.* **2004**, *5*, 539–543. [CrossRef]

68. Gaggiotti, O.E.; Lange, O.; Rassmann, K.; Gliddon, C. A Comparison of Two Indirect Methods for Estimating Average Levels of Gene Flow Using Microsatellite Data. *Mol. Ecol.* **1999**, *8*, 1513–1520. [CrossRef] [PubMed]
69. Whitlock, M.C.; Mccauley, D.E. Indirect Measures of Gene Flow and Migration: $F_{ST} \neq 1/(4Nm + 1)$. *Heredity* **1999**, *82*, 117–125. [CrossRef]
70. Weir, B.S.; Cockerham, C.C. Estimating F-Statistics for the Analysis of Population Structure. *Evolution* **1984**, *38*, 1358–1370.
71. Excoffier, L.; Smouse, P.E.; Quattro, J.M. Analysis of Molecular Variance Inferred from Metric Distances among DNA Haplotypes: Application to Human Mitochondrial DNA Restriction Data. *Genetics* **1992**, *131*, 479–491. [CrossRef] [PubMed]
72. Rousset, F. Genetic Differentiation and Estimation of Gene Flow from F-Statistics under Isolation by Distance. *Genetics* **1997**, *145*, 1219–1228. [CrossRef]
73. Pritchard, J.K.; Stephens, M.; Donnelly, P. Inference of Population Structure Using Multilocus Genotype Data. *Genetics* **2000**, *155*, 945–959. [CrossRef] [PubMed]
74. Evanno, G.; Regnaut, S.; Goudet, J. Detecting the Number of Clusters of Individuals Using the Software Structure: A Simulation Study. *Mol. Ecol.* **2005**, *14*, 2611–2620. [CrossRef] [PubMed]
75. Jakobsson, M.; Rosenberg, N.A. CLUMPP: A Cluster Matching and Permutation Program for Dealing with Label Switching and Multimodality in Analysis of Population Structure. *Bioinformatics* **2007**, *23*, 1801–1806. [CrossRef]
76. Archer, F.I.; Adams, P.E.; Schneiders, B.B. Stratag: An r Package for Manipulating, Summarizing and Analysing Population Genetic Data. *Mol. Ecol. Resour.* **2017**, *17*, 5–11. [CrossRef]
77. Jombart, T. Adegenet: A R Package for the Multivariate Analysis of Genetic Markers. *Bioinformatics* **2008**, *24*, 1403–1405. [CrossRef]
78. Guillot, G.; Mortier, F.; Estoup, A. Geneland: A Computer Package for Landscape Genetics. *Mol. Ecol. Notes* **2005**, *5*, 712–715. [CrossRef]
79. Blair, C.; Weigel, D.E.; Balazik, M.; Keeley, A.T.; Walker, F.M.; Landguth, E.; Cushman, S.A.M.; Murphy, M.; Waits, L.; Balkenhol, N. A Simulation-Based Evaluation of Methods for Inferring Linear Barriers to Gene Flow. *Mol. Ecol. Resour.* **2012**, *12*, 822–833. [CrossRef]
80. Wilson, G.A.; Rannala, B. Bayesian Inference of Recent Migration Rates Using Multilocus Genotypes. *Genetics* **2003**, *163*, 1177–1191. [CrossRef]
81. Gamerman, D. *Markov Chain Monte Carlo: Stochastic Simulation for Bayesian Inference*; Chapman & Hall: New York, NY, USA, 1997.
82. Keller, D.; Brodbeck, S.; Flöss, I.; Vonwil, G.; Holderegger, R. Ecological and Genetic Measurements of Dispersal in a Threatened Dragonfly. *Biol. Conserv.* **2010**, *143*, 2658–2663. [CrossRef]
83. Dolný, A.; Šigutová, H.; Ožana, S.; Choleva, L. How Difficult Is It to Reintroduce a Dragonfly? Fifteen Years Monitoring *Leucorrhinia dubia* at the Receiving Site. *Biol. Conserv.* **2018**, *218*, 110–117. [CrossRef]
84. Swaegers, J.; Janssens, S.B.; Ferreira, S.; Watts, P.C.; Mergeay, J.; McPeek, M.A.; Stoks, R. Ecological and Evolutionary Drivers of Range Size in *Coenagrion* Damselflies. *J. Evol. Biol.* **2014**, *27*, 2386–2395. [CrossRef]
85. Levett, S.; Walls, S. Tracking the Elusive Life of the Emperor Dragonfly *Anax imperator* Leach. *J. Br. Dragonfly Soc.* **2011**, *27*, 59–68.
86. IPCC Climate Change 2013: The Physical Science Basis. *Contribution of Working Group I to the Fifth Assessment Report of the Intergovernmental Panel on Climate Change*; Cambridge University Press: Cambridge, UK; New York, NY, USA, 2013.
87. Hickling, R.; Roy, D.B.; Hill, J.K.; Fox, R.; Thomas, C.D. The Distributions of a Wide Range of Taxonomic Groups Are Expanding Polewards. *Glob. Chang. Biol.* **2006**, *12*, 450–455. [CrossRef]
88. Hickling, R.; Roy, D.B.; Hill, J.K.; Thomas, C.D. A Northward Shift of Range Margins in British Odonata. *Glob. Chang. Biol.* **2005**, *11*, 502–506. [CrossRef]
89. Jha, S. Contemporary Human-Altered Landscapes and Oceanic Barriers Reduce Bumble Bee Gene Flow. *Mol. Ecol.* **2015**, *24*, 993–1006. [CrossRef]
90. Vanden Broeck, A.; Maes, D.; Kelager, A.; Wynhoff, I.; WallisDeVries, M.F.; Nash, D.R.; Oostermeijer, J.G.B.; Van Dyck, H.; Mergeay, J. Gene Flow and Effective Population Sizes of the Butterfly *Maculinea alcon* in a Highly Fragmented, Anthropogenic Landscape. *Biol. Conserv.* **2017**, *209*, 89–97. [CrossRef]
91. PRAM Normandie. Available online: http://recensementmare.pramnormandie.com/API/ (accessed on 11 January 2022).
92. Gilpin, E.M.; Hanski, I. Metapopulation Dynamics: Empirical and Theoretical Investigations. *Biol. J. Linn. Soc.* **1991**, *42*, 3–16.
93. Pulliam, H.R. Sources, Sinks, and Population Regulation. *Am. Nat.* **1988**, *132*, 652–661. [CrossRef]
94. Peery, M.Z.; Beissinger, S.R.; House, R.F.; Bérubé, M.; Hall, L.A.; Sellas, A.; Palsbøll, P.J. Characterizing Source–Sink Dynamics with Genetic Parentage Assignments. *Ecology* **2008**, *89*, 2746–2759. [CrossRef] [PubMed]
95. Thomaz, A.T.; He, Q. When Are Populations Not Connected like a Circuit? Identifying Biases in Gene Flow from Coalescent Times. *Mol. Ecol. Resour.* **2019**, *19*, 1381–1384. [CrossRef]
96. Suhling, F.; Martens, A.; Marais, E. How to Enter a Desert—Patterns of Odonata Colonisation of Arid Namibia. *Int. J. Odonatol.* **2009**, *12*, 287–308. [CrossRef]
97. Saura, S.; Bodin, Ö.; Fortin, M.-J. EDITOR'S CHOICE: Stepping Stones Are Crucial for Species' Long-Distance Dispersal and Range Expansion through Habitat Networks. *J. Appl. Ecol.* **2014**, *51*, 171–182. [CrossRef]
98. Atterby, H.; Aegerter, J.N.; Smith, G.C.; Conyers, C.M.; Allnutt, T.R.; Ruedi, M.; MacNicoll, A.D. Population Genetic Structure of the Daubenton's Bat (*Myotis daubentonii*) in Western Europe and the Associated Occurrence of Rabies. *Eur. J. Wildl. Resour.* **2010**, *56*, 67–81. [CrossRef]

99. Razgour, O.; Rebelo, H.; Puechmaille, S.J.; Juste, J.; Ibáñez, C.; Kiefer, A.; Burke, T.; Dawson, D.A.; Jones, G. Scale-Dependent Effects of Landscape Variables on Gene Flow and Population Structure in Bats. *Divers. Distrib.* **2014**, *20*, 1173–1185. [CrossRef]
100. Minot, M.; Aubert, M.; Husté, A. Pond Creation and Restoration: Patterns of Odonate Colonization and Community Dynamics. *Biodivers. Conserv.* **2021**, *30*, 4379–4399. [CrossRef]
101. May, M.L. A Critical Overview of Progress in Studies of Migration of Dragonflies (Odonata: Anisoptera), with Emphasis on North America. *J. Insect Conserv.* **2013**, *17*, 1–15. [CrossRef]
102. Chaput-Bardy, A.; Lemaire, C.; Picard, D.; Secondi, J. In-Stream and Overland Dispersal across a River Network Influences Gene Flow in a Freshwater Insect, *Calopteryx splendens*. *Mol. Ecol.* **2008**, *17*, 3496–3505. [CrossRef] [PubMed]
103. Monroe, E.M.; Lynch, C.; Soluk, D.A.; Britten, H.B. Nonlethal Tissue Sampling Techniques and Microsatellite Markers Used for First Report of Genetic Diversity in Two Populations of the Endangered *Somatochlora hineana* (Odonata: Corduliidae). *Ann. Entomol. Soc. Am.* **2010**, *103*, 1012–1017. [CrossRef]
104. Cordero-Rivera, A.; Perez, F.E.; Andres, J.A. The Effect of Handling Damage, Mobility, Body Size, and Fluctuating Asymmetry on Lifetime Mating Success of *Ischnura graellsii* (Rambur) (Zygoptera: Coenagrionidae). *Odonatologica* **2002**, *31*, 117–128.
105. Raebel, E.M.; Merckx, T.; Riordan, P.; Macdonald, D.W.; Thompson, D.J. The Dragonfly Delusion: Why It Is Essential to Sample Exuviae to Avoid Biased Surveys. *J. Insect Conserv.* **2010**, *14*, 523–533. [CrossRef]
106. Watts, P.C.; Thompson, D.J.; Daguet, C.; Kemp, S.J. Exuviae as a Reliable Source of DNA for Population-Genetic Analysis of Odonates. *Odonatologica* **2005**, *34*, 183–187.
107. Monroe, E.M.; Britten, H.B. Conservation in Hine's Sight: The Conservation Genetics of the Federally Endangered Hine's Emerald Dragonfly, *Somatochlora hineana*. *J. Insect Conserv.* **2014**, *18*, 353–363. [CrossRef]
108. Incagnone, G.; Marrone, F.; Barone, R.; Robba, L.; Naselli-Flores, L. How Do Freshwater Organisms Cross the "Dry Ocean"? A Review on Passive Dispersal and Colonization Processes with a Special Focus on Temporary Ponds. *Hydrobiologia* **2015**, *750*, 103–123. [CrossRef]
109. Epps, C.W.; Keyghobadi, N. Landscape Genetics in a Changing World: Disentangling Historical and Contemporary Influences and Inferring Change. *Mol. Ecol.* **2015**, *24*, 6021–6040. [CrossRef] [PubMed]

diversity

MDPI

Article

Evolution and Biogeographic History of Rubyspot Damselflies (Hetaerininae: Calopterygidae: Odonata)

Samantha Standring [1,*], Melissa Sánchez-Herrera [2,3], Rhainer Guillermo-Ferreira [4], Jessica L. Ware [3], Yesenia Margarita Vega-Sánchez [5], Rebecca Clement [6], Jonathan P. Drury [7], Gregory F. Grether [8], Antonio González-Rodríguez [5], Luis Mendoza-Cuenca [9], Cornelio A. Bota-Sierra [10,11] and Seth Bybee [12]

1 Entomology Department, University of California Riverside, Riverside, CA 92521, USA
2 Faculty of Natural Sciences, Universidad del Rosario, Bogotá 111221, Colombia
3 American Museum of Natural History, New York, NY 10024, USA
4 Departamento de Ciências Biológicas, Universidade Federal do Triângulo Mineiro, Uberaba 38025-180, Brazil
5 Instituto de Investigaciones en Ecosistemas y Sustentabilidad, Universidad Nacional Autónoma de México, Morelia 04510, Mexico
6 Department of Biological Sciences, The George Washington University, Washington, DC 20052, USA
7 Department of Biosciences, Durham University, Durham DH1 3LE, UK
8 Department of Ecology and Evolutionary Biology, University of California Los Angeles, Los Angeles, CA 90095, USA
9 Facultad de Biología, Universidad Michoacana de San Nicolás de Hidalgo, Morelia 58030, Mexico
10 Grupo de Entomología Universidad de Antioquia (GEUA), Universidad de Antioquia, Medellín 050010, Colombia
11 Alabama Museum of Natural History & UA Museums Department of Research and Collections, The University of Alabama, Tuscaloosa, AL 35487, USA
12 Department of Biology and Monte L. Bean Museum, Brigham Young University, Provo, UT 84602, USA
* Correspondence: ssmit038@ucr.edu

Citation: Standring, S.; Sánchez-Herrera, M.; Guillermo-Ferreira, R.; Ware, J.L.; Vega-Sánchez, Y.M.; Clement, R.; Drury, J.P.; Grether, G.F.; González-Rodríguez, A.; Mendoza-Cuenca, L.; et al. Evolution and Biogeographic History of Rubyspot Damselflies (Hetaerininae: Calopterygidae: Odonata). *Diversity* **2022**, *14*, 757. https://doi.org/10.3390/d14090757

Academic Editors: M. Olalla Lorenzo-Carballa and Ricardo Koroiva

Received: 19 July 2022
Accepted: 7 September 2022
Published: 14 September 2022

Abstract: The damselflies Hetaerininae, a subfamily of Calopterygidae, comprise four genera distributed from North to South America: *Hetaerina*, *Mnesarete*, *Ormenophlebia* and *Bryoplathanon*. While several studies have focused on the intriguing behavioral and morphological modifications within *Hetaerina*, little of the evolutionary history of the group is well understood. Understanding the biogeographical history of Hetaerininae is further complicated by uncertainty in important geological events, such as the closure of the Central American Seaway (CAS). We generated a phylogenetic hypothesis to test the relationships and divergence times within Hetaerininae using IQtree and BEAST2 and found that *Mnesarete* and *Ormenophlebia* render *Hetaerina* paraphyletic. Reclassification of the genera within Hetaerininae is necessary based on our results. We also tested the fit to our dataset of two different hypotheses for the closure of CAS. Our results supported a gradual closure, starting in the Oligocene and ending in the Pliocene. Using Ancestral Character State Reconstruction, we found that the rubyspot, which is associated with higher fecundity in several species, was ancestral for Hetaerininae and subsequently lost four times. Estimates of diversification in association with the rubyspot are needed to understand the plasticity of this important character. Forest habitat was the ancestral state for Hetaerininae, with transitions to generalist species of *Hetaerina* found primarily in the Mesoamerican region. These results add to our understanding of the relationship between morphology, biogeography and habitat in a charismatic group of damselflies.

Keywords: biogeography; Zygoptera; wing coloration; mating behavior

1. Introduction

Extant Odonata (damselflies and dragonflies) represent some of the earliest branching lineages of winged insects [1]. While some species are long distance migrants, others do not stray far from their natal nymphal water source; indeed, dispersal capabilities are heterogeneous among this diverse clade of ~6300 species [2,3]. The Zygoptera (damselflies) are an extant suborder of Odonata and comprise over 3000 species distributed globally [4], with a hotspot in tropical Central and South America. Within Zygoptera, the Calopterygidae comprises over 150 species, including many species with metallic bodies and conspicuous wing coloration. The family is divided among three subfamilies: the clearwings or Caliphaeinae Tillaryd and Fraser, 1939, the demoiselles or Calopteryginae Selys, 1859 and the Hetaerininae Tillyard and Fraser, 1939.

Hetaerininae damselflies comprise four genera: *Hetaerina* Hagen, 1853 (rubyspot damselflies), *Mnesarete* Cowley, 1934, *Ormenophlebia* Garrison, 2006 and *Bryoplathanon* Garrison, 2006 (see Figure 1). Hetaerinines are easy to recognize due to their combination of dense wing venation, metallic green or reddish body coloration and unique male caudal appendage morphology [5]. We will use Hetaerininae and the common name rubyspot interchangeably, as the greatest diversity within hetaerinines is found within *Hetaerina*, the genus traditionally referred to as rubyspot damselflies.

Figure 1. Habitus images showing diversity between the four genera of Hetaerininae. (**A**) *Ormenophlebia imperatrix*, photo by Jim Johnson; (**B**) *Hetaerina amazonica*, photo by R. Guillermo-Ferreira; (**C**) *Mnesarete guttifera*, photo by R. Guillermo-Ferreira; (**D**) *Bryoplathanon globifer*, photo by Tom Kompier.

The nearly 70 species of Hetaerininae are distributed from North to South America, with some species occupying only forested habitats and others occupying both forest and grassland. Throughout their range, habitat and climate vary from arid desert to tropical rainforest. Variations in habitat over geological time may have caused barriers following dispersal, leading to speciation. Another likely barrier preventing dispersal was the Central American Seaway (CAS) [6], a body of water that separated North and South America until approximately 20–4 mya. As damselflies, in general, are relatively weak fliers compared to dragonflies and migratory insects [3,7,8], the likelihood that Hetaerininae could travel between Central and South America over open ocean, before the CAS was closed, seems low. However, a phylogenetic hypothesis does not exist for this subfamily, limiting our ability to test how these barriers may have influenced Hetaerininae speciation. If the Caribbean Sea was a barrier to lineage dispersal, then we expect that dispersal events would take place only after the closure of the CAS by continental landmasses or stepping stone islands.

Timing for the closure of the CAS may also have greatly influenced Hetaerininae speciation. Two main hypotheses for the timing of the closure of the CAS exist: (1) a gradual closure starting during the Oligocene and ending in the Pliocene (5.3–2.6 mya), with islands connecting landmasses [9]; (2) an abrupt closure during the Middle Miocene (16.0–11.6 mya) [10].

While studies have tested divergence times for Hetaerininae, they were limited in taxa inclusion and data, making them challenging to use in biogeographic studies. The earliest example is Dumont [11], who estimated the absolute rates of divergence times using the program r8s and suggested the age of Hetaerininae to be approximately 60 mya. The nodes were calibrated with enforcement of a minimum and maximum age constraint based on the fossil evidence. However, r8s assumes autocorrelated rates based on a user-inputted phylogeny, and some note that r8s-estimated node ages may be older than the estimates generated with Bayesian methods and may be less consistent among analyses (e.g., [12]), thus, the age of Hetaerininae may be far younger. Suvorov et al. [13] and Kohli et al. [14] found the age of Hetaerininae to be 43 million and 40 million years in age, respectively, using transcriptomes.

A younger age, as found by Suvorov et al. [13] and Kohli et al. [14], would mean an increased likelihood that dispersal between Central America and South America occurred early on in Hetaerininae evolution; while, with an older age of 60 million years, well before the CAS was closed, dispersal between these two landmasses would have been less likely early on.

For several species of rubyspot damselflies (e.g., *Hetaerina americana*, *H. titia* and *H. cruentata*), reproductive behavior, territoriality and overall fitness displays are well documented [15–21]. Males are territorial, and experimental manipulation of the size of their rubyspot has been shown to affect their fitness by increasing male fecundity [7,20,22] and reducing male survival and successful prey capture [23]. The rubyspot has been used to distinguish between two highly similar genera within Hetaerininae: *Hetaerina* and *Mnesarete* [5,24]. This taxonomic use of the rubyspot to separate the two groups assumes that the fitness benefits of red wing spots are present in all environments of Hetaerininae. That would suggest that once wing spots appear they are unlikely to disappear. However, populations of *H. aurora* have been found both with and without the rubyspot, and *H. titia* varies seasonally in wing color and size, supporting an alternative hypothesis that they might be advantageous as signals in some environments, but disadvantageous in other environments. In this scenario, we might expect to see repeated gains and/or losses of red wing pigmentation, making use of the basal rubyspot as an identifying character misleading. Additionally, we would expect to see an association between the basal rubyspot and habitat type. However, despite the behavioral and taxonomic importance of the rubyspot, evolution and habitat association of both the basal and apical rubyspots are not well understood and have never been studied across the entire subfamily.

Herein, we generate a phylogenetic hypothesis to: (1) test whether the CAS created a barrier to Hetaerininae dispersal using two different closure hypotheses; (2) explore the evolution of basal and apical wing coloration and preferred habitat (forest vs. generalist species) of Hetaerininae; (3) test the monophyly of the species groups, genera (*Hetaerina*, *Mnesarete* and *Ormenophlebia*) and the relationships between the genera.

2. Materials and Methods

2.1. Taxon Sampling

We sampled extensively across *Hetaerina* and *Mnesarete*, including 214 ingroup and outgroup taxa. The taxon sampling consisted of 13 of the 24 species of *Mnesarete* (40 specimens), 31 of the 39 species of *Hetaerina* (152 specimens) and two of the four species of the rare *Ormenophlebia* (three specimens), for a total of 46 ingroup species (192 specimens). We were not able to obtain *Bryoplathanon* spp. for the analysis. Multiple individuals from each species at different geographic locations were included to evaluate the monophyly for the proposed taxonomic genera and species within the family. Geographic origin, collector and Genbank Accession Numbers are summarized in Table S2. We identified specimens using the taxonomic keys provided in the revisions of this group [5,24]. Rosser Garrison confirmed the IDs.

2.2. DNA Extraction, Amplification and Sequencing

DNA extraction was performed primarily on the legs of dried specimens provided from both museums and personal collections. These included the Florida State Collection of Arthropods, as well as co-authors and collaborators at Rutgers University, Newark, USA, UFSCAR and GCEUMSNH (Universidad Michoacana de San Nicolás de Hidalgo, Morelia, Mexico). For some species, a small portion of flight muscle tissue was also included. Additional species were collected in the field. All field-collected specimens were preserved in 95% ETOH following collection and deposited at the Monte Bean Museum Insect Cryo Collection.

For the majority of the taxa, DNA extraction, amplification and sequencing was performed at Brigham Young University (Provo, UT, USA). Three mitochondrial loci (COI 5′ end, COI 3′ end, 12S) and two nuclear loci (ITS1/ITS2, ef1a) were targeted for amplification using primers specifically designed for *Hetaerina* (Table S3; [25,26]). Gene fragments were amplified using standard polymerase chain reaction (PCR) techniques. Yield and potential contamination were monitored by gel electrophoresis. Sequencing was performed at the BYU DNA sequencing center. Collaborators also provided sequence data and additional sequence data was gathered from GenBank for 21 specimens (Table S2).

2.3. Phylogenetic Analyses

Sequence data was uploaded to Geneious 11.1.5 (https://www.geneious.com) (accessed on 12 October 2020), where it was assembled, edited, aligned and concatenated. All genes were aligned using MAFFT [27]. Phylogenetic relationships were reconstructed using maximum likelihood partitioned analyses in IQ-tree v1.6.12 [28]. Best-fit partitioning schemes were estimated using ModelFinder [29] in IQ-tree, allowing partitions to be merged to reduce over-parameterization and increase the model fit. One thousand replicates of Ultrafast bootstrap (BS) and SH-like approximate likelihood ratio test (SH-aLRT) were performed to estimate the node support [30,31]. The ggplot2, ggtree and ape packages in R v4.0.5 were used to view and analyze the phylogeny, and devtools, harrietr and phytools were used to summarize support values [32–35].

2.4. Divergence Dating Analysis

We ran a relaxed log-normal clock divergence dating analysis in BEAST v2.6.2 [36] using the birth death tree model to account for the possibility that rates changed continuously along the branches. We reduced our taxon sampling in the molecular matrix to include one sample per species, choosing the sample with the largest number of loci to minimize missing data among taxa, including a total of 62 individuals in the analysis. We used a fixed starting tree, generated in IQ-tree and made ultrametric in R using chronos in ape [33] to reduce computation time. We used ModelFinder [29] to estimate site models and merge similar partitions. Based on our results, COI 5′ end and COI 3′ end were merged for the BEAST2 analysis. As the default uniform clock prior is improper (i.e., the substitution rate cannot be less than 0), we used a lognormal prior for ucldMean and increased the speed of convergence by providing a mean value of 0.000001 [37]. Analysis of the log file with Tracer v1.7.1 [38] confirmed the rejection of the strict clock model with the observation that the 95% credible interval of ucld.stdev excluded zero.

Chosen fossils, their accession numbers, publication date and justifications for node calibrations are shown in Table 1 below. For fossil priors, we chose the oldest fossil crown member of a clade when multiple were available; all fossils were chosen with Parham et al. [39] best practices in mind. A lognormal distribution was used for three fossil priors; for this distribution, we used the minimum possible age in the fossils range as the zero offset and chose parameters such that the median was the maximum age range for the fossil. We used *Sinocalopteryx shangyongensis*, the oldest known Calopterygidae fossil, to set the maximum age for a uniform prior of our ingroup [40].

Table 1. Age of selected nodes and fossils included in analysis. See Figure S6 for numbered nodes.

Fossil	Accession Number	Publication	Fossil Placement	Age and Justification	Prior	Shape	Mean Age	CI
Calopteryx andancensis	PaleoDB collection 113893	Nel and Brisac [41]	Node 68	9.0–5.3 mya; Steininger et al. [42]	Lognormal	Offset = 5.3 Mean = 13.4	25.4	19.1, 32.0
Sapho legrandi	PaleoDB collection 194946	Nel and Petrulevičius [43]	Node 73	27.82–24.8 mya; Steininger et al. [42]	Lognormal	Offset = 24.8 Mean = 31.4	25.7	24.8, 27.9
Chlorocypha cordasevae	PaleoDB collection 105962	Nel et al. [44]	Node 119	11.1–9.4 mya; Steininger et al. [42]	Lognormal	Offset = 9.4 Mean = 13.2	42.5	31.2, 53.3
Sinocalopteryx shangyongensis	PaleoDB collection 194570	Lin et al. [40]	Node 65	56–5.3 mya	Uniform	Min = 5.3 Max = 56	53.5	47.6, 56.0
MRCA of Hetaerininae			Node 74				36.2	30.1, 42.1
North & Central American clade			Node 76				29.3	24.1, 34.8
South American clade			Node 87				28.4	23.1, 33.6
Ormenophlebia split			Node 82				23.2	18.5, 28.1
Ormenophlebia diverged			Node 86				4.7	2.6, 7.1

Three identical but separate BEAST2 analyses were run for 100,000,000 generations to ensure the mixing of the data. These results were combined, and the log files were evaluated in Tracer v1.7.1 to confirm the mixing of the data and to check ESS values before combining the tree file results. Ten percent burnin was an appropriate cut off for this dataset after observing the distribution, and so we removed 10% burnin from the tree files following the evaluation of log files in Tracer.

2.5. Ancestral Character State Reconstruction

Wing characters were coded based on voucher specimens. Maximum likelihood Ancestral Character State Reconstruction (ACSR) was run using the equal rates model in ace:ape [33]. Characters were mapped onto the ultrametric dated phylogeny generated in BEAST2 (see *Divergence dating analysis*).

Habitat data were divided into two categories: generalist (species is present in non-forested areas) and forest specialist (species only found in forested areas). Habitat type was determined based on a literature review [5,24,45–48] and personal observation of adult reproductive behavior.

Characters were as follows:

1. Male, Hindwing, color: hyaline wings (0), color only basally on wing (1), entire wing colored (2)
2. Male, Hindwing, apical color: absent (0), apical melanization only (1), multiple cells colored black or red (2)
3. Male, Forewing, apical color: absent (0), apical melanization only (1), multiple cells colored black or red (2)
4. Male, apical color: absent from forewings and hindwings (0), present in hindwing (1), present in hindwings and forewings (2). Presence was indicated by any color present, melanization or multiple cells.
5. Habitat: generalist (0), forest specialist (1)

2.6. Biogeography Analysis

We tested the fit of two different closure hypotheses of the CAS on the evolution of Hetaerininae using time stratified analyses in BioGeoBEARSv1.1.2 [49]. The analysis was limited to the ingroup only and run using standard options with a maximum range size of six. Areas were delimited based on Morrone et al. [50] and determined for each species based on Garrison [5,24] and collection records. We tested three different scenarios: two-time stratified analyses for the CAS closure, and a control with no distance constraints. The two-time stratified analyses tested the following hypotheses for the CAS closure timing: (1) a gradual closure starting during the Oligocene and ending in the Pliocene 4 mya, with islands connecting land masses [9]; (2) an abrupt closure during the Middle Miocene 14 mya [10]. Minimum geographic distance between landmasses was determined using paleogeography literature [9,10,51]. Time multiplication matrices for all the scenarios were coded as follows: separation between landmasses greater than 200 km was coded as 0, no separation was coded as 1 and variation between the two was coded between 0 and 1, depending on the distance between landmasses. A distance of 200 km and greater was treated as a barrier to dispersal, as that distance prevents dispersal of Hetaerininae to Cuba in the present day.

We also tested the fit of three different models in BioGeoBEARS; likelihood-based Dispersal-Extinction Cladogenesis (DEC), likelihood version of the Dispersal-Vicariance Analysis (DIVALIKE) and a likelihood range evolution model BAYAREALIKE [49] (Table S5). Each of the three models were tested with and without founder-event speciation (+J). We picked the best fitting biogeographical model for within and among scenarios using the lowest AICc values. Recent work has shown that +J models can be included in AICc comparisons [52]. The best selected reconstructed areas models for each scenario were mapped over the best time calibrated phylogeny. See File S1 for BioGeoBEARS scripts and results.

3. Results

3.1. Phylogenetic Analysis

ModelFinder merged the models for four partitions, resulting in seven partitions being included in the analysis, with protein coding loci coded for each position (see Supplementary Table S1). The total alignment length was 4979 sites, with 36.8% missing data.

The monophyly of both Calopterygidae and Hetaerininae was recovered with high support (91.9% SH-aLRT and 99% BS and 100% SH-aLRT and BS, respectively). The *Hetaerina pilula* clade was a sister to all other Hetaerininae. *Hetaerina* was not recovered as monophyletic, with both *Mnesarete* and *Ormenophlebia* nested within the genus. *Mnesarete* was also recovered as polyphyletic. Within the Hetaerininae clade, our results suggest that most of the species are highly supported monophyletic groups (Figure 2). There are two major clades within the phylogeny, one largely South American and one largely Mesoamerican (Figure 3).

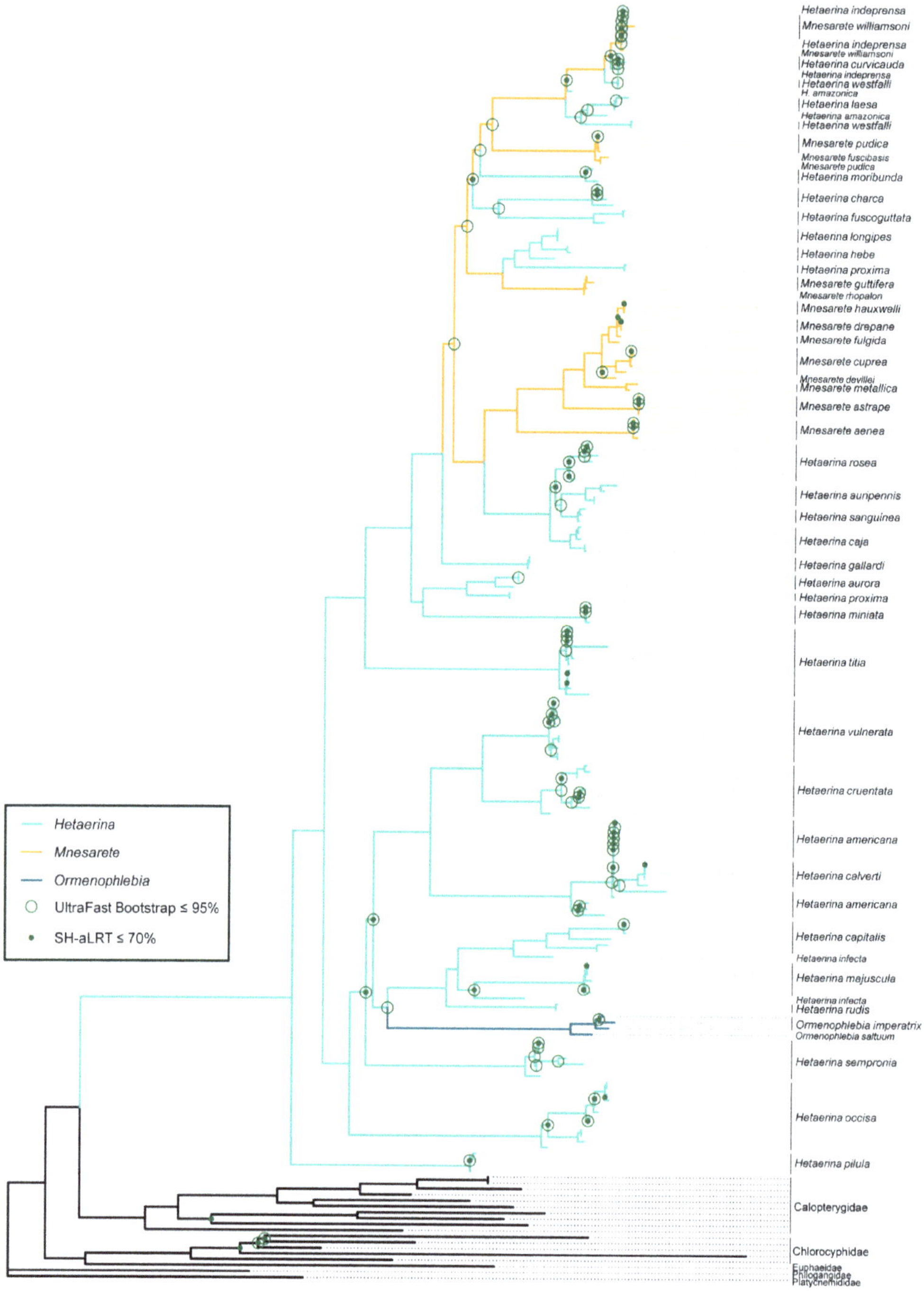

Figure 2. Maximum likelihood phylogenetic reconstruction of Hetaerininae estimated in IQtree. Relationships within and between species shown. Light blue branches: *Hetaerina* spp. Yellow branches: *Mnesarete* spp. Dark blue branches: *Ormenophlebia* spp. Large empty green circle: UltraFast bootstrap support less than or equal to 95%. Small filled green circle: SH-aLRT support less than or equal to 70%.

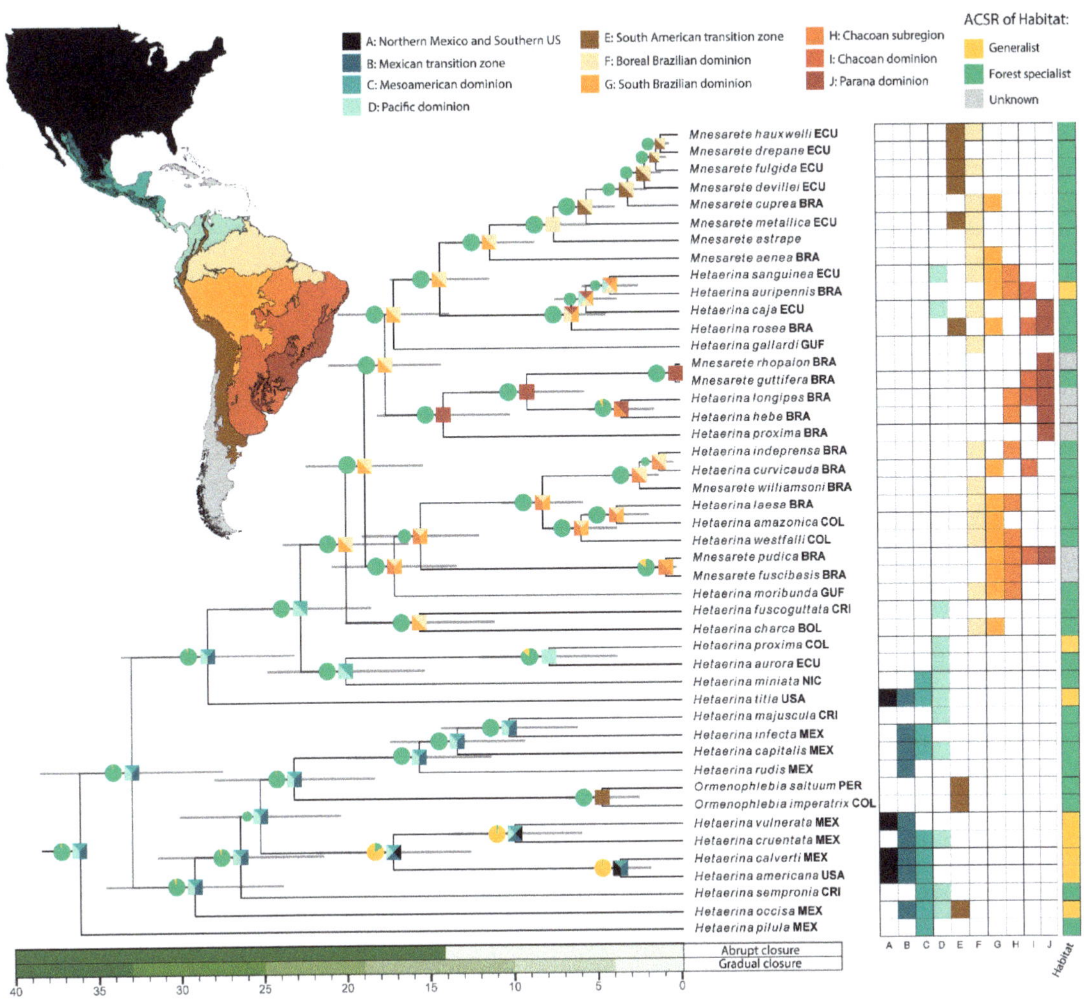

Figure 3. Hetaerininae biogeography analysis on time-calibrated phylogenetic tree. Biogeographic analysis conducted in BioGeoBEARS v1.1.2 with a maximum range size of six. Bayesian time-calibrated tree estimated in BEAST2 v2.6.6. Color-coded matrix corresponds with current species distribution. Squares at nodes represent the ancestral range with the highest probability from the Pliocene BAYAREALIKE +J analysis. Likelihood of dispersal between Central and South America for two hypotheses (abrupt Middle Miocene closure and gradual closure) shown. Dark green represents low likelihood of dispersal, light green easy dispersal. Map based on Morrone et al., 2022 area map [50]. Pie charts represent ACSR of habitat, generalist or forest specialist. Far right column on matrix corresponds to character states coded for ACSR of habitat.

3.2. Ancestral Character State Reconstruction: Wing Coloration

Male coloration, or the "rubyspot", at the base of the hindwing was ancestral to Hetaerininae (Figure 4). The rubyspot was lost at least four times within Hetaerininae. Independent losses were reconstructed in *Ormenophlebia* (6 mya), *Mnesarete williamsoni* (1 mya), the clade of *M. rhopalon* and *M. gutiffera* (1 mya) and the clade with *M. drepane*, *M. cuprea*, *M. metallica*, *M. astrape*, *M. aenea*, *M. fulgida*, *M. devillei* and *M. hauxwelli* (12 mya).

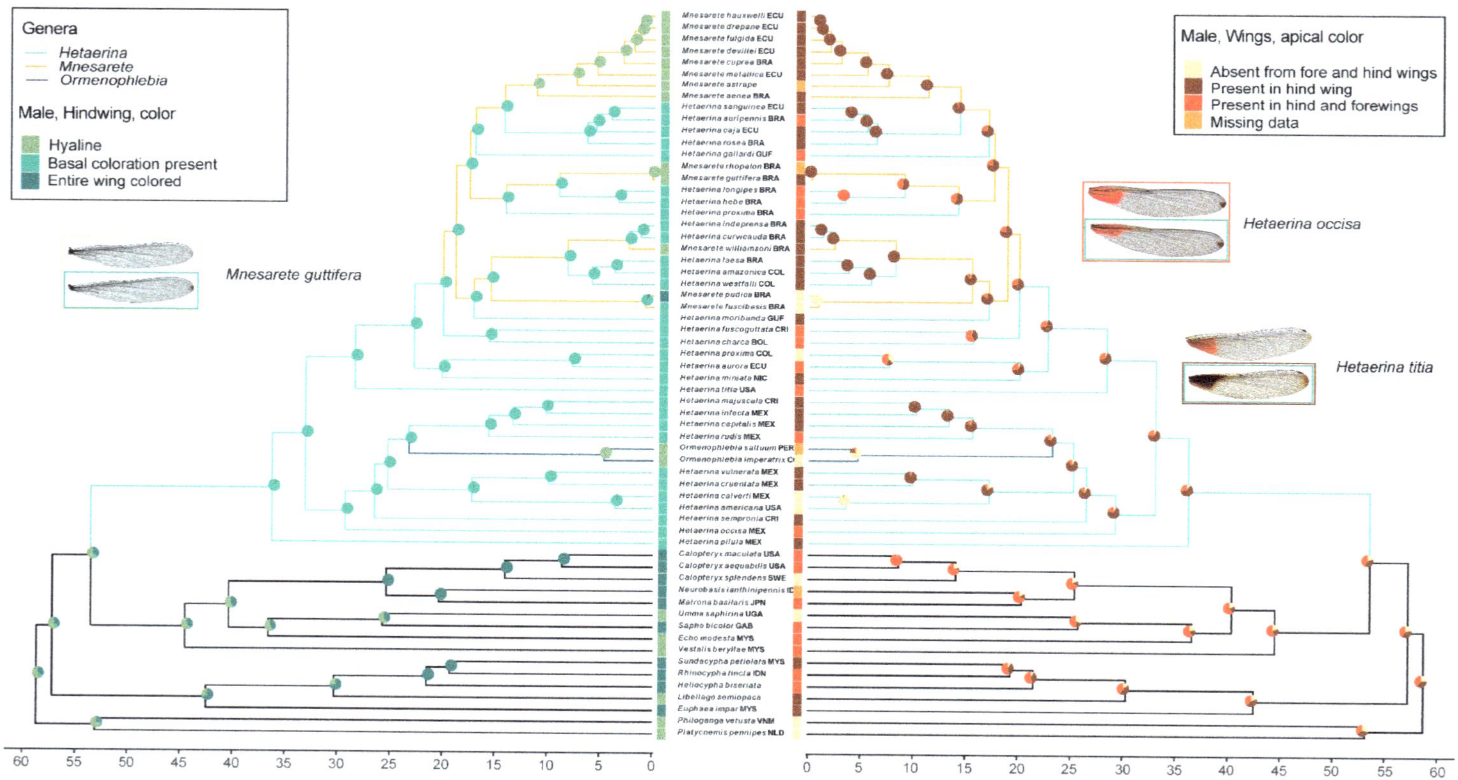

Figure 4. Likelihood ancestral character state reconstruction of coloration in male wings. Characters mapped onto divergence dated phylogeny generated in BEAST2. Pie charts show probability of presence or absence of state at each node. **Left:** ACSR of male hindwing color mapped onto phylogeny. Green: Hyaline wings. Teal: Basal coloration present. Dark blue: Entire wing colored. Basal coloration has the highest likelihood for the ancestral node of Hetaerininae. **Right:** ACSR of apical color in male wings mapped onto phylogeny. Tan: Apical color absent from forewings and hindwings. Dark red: Apical color present in hindwing. Red: Apical color present in hindwings and forewings. Orange: Missing data. Wings for three species shown, forewing on top, hindwing on bottom: *Mnesarete guttifera*, *Hetaerina occisa* and *Hetaerina titia*. Colored boxes show character state.

Male coloration apically in the hindwing was ancestral to Hetaerininae, and completely lost three times (in the past 2–6 my), with transitions to melanization only occurring five times (Figure S3). In contrast, male coloration apically in the forewing was probably absent ancestrally and gained five times (1–16 mya) (Figure S4). Apical coloration in both the hindwings and forewings has been gained eleven times, and the complete loss of coloration apically in both the forewings and hindwings occurred three times (2–6 mya).

3.3. Ancestral Character State Reconstruction: Habitat

The ancestral habitat for Hetaerininae was as a forest specialist, with five transitions to generalist habitat behavior, occupying either grassland or grassland and forest (Figure 3). One of the transitions occurred ~17 mya in the basal clade and included *H. vulnerata*, *H. cruentata*, *H. calverti* and *H. americana*. Within this clade, species were widespread, occurring from the Nearctic to the Neotropics. Other transitions between habitat preference were at the species level.

3.4. Divergence Dating and Biogeography

Four partitions, with a total of 3488 sites, were included in the analysis, with 20.5% missing data. Log files showed ESS values above 6000 for all parameters and mixing across all analyses. The most recent common ancestor (MRCA) for all the Calopterygidae originated 53.5 mya [CI 47.6, 56.0] during the early Eocene epoch (Figure 5 and Table 1). The MRCA of *H. pilula* and all other members of Hetaerininae originated 36.2 mya [CI 30.1, 42.1] (Figure 5) during the late Eocene. The two well-supported geographical clades, the primarily North and Central American clade (10 of 12 spp) and the primarily South American clade (31 of 33 spp), originated concurrently during the Oligocene epoch 29.3 mya [CI 24.1, 34.8] and 28.4 mya [CI 23.1, 33.6], respectively (Figure 5). Both of these clades diversified mostly during the Miocene; however, the South American clade also showed diversification during the Pliocene and Pleistocene epochs (Figure 5). Within the North/Central American clade, *Ormenophlebia* split off from the *Hetaerina* relatives 23.2 mya [CI 18.5, 28.1] and diverged approximately 4.7 mya [CI 2.6, 7.1]. On the other hand, within the South American clade, despite the polyphyletic recovery of *Mnesarete* species, the MRCA of all these lineages originated ~21 mya. All the lineages containing *Mnesarete* diversified in the Miocene, Pliocene, Pleistocene and Holocene epochs (Figure 5).

For each of the two tested scenarios for the CAS closure (i.e., an abrupt Middle Miocene closure and a gradual closure) and the control, the BAYAREALIKE +J model had the best fit AICc values (Table S5). In the time stratified analyses, closure starting during the Oligocene and ending in the Pliocene had the lowest AICc value. The ancestral areas reconstructed were consistent for all of the CAS closure scenarios. The most likely areas for the MRCA for all Hetaerininae and the other two geographical clades are the Mexican transition zone, Mesoamerican dominion and the Pacific dominion (Figure 3).

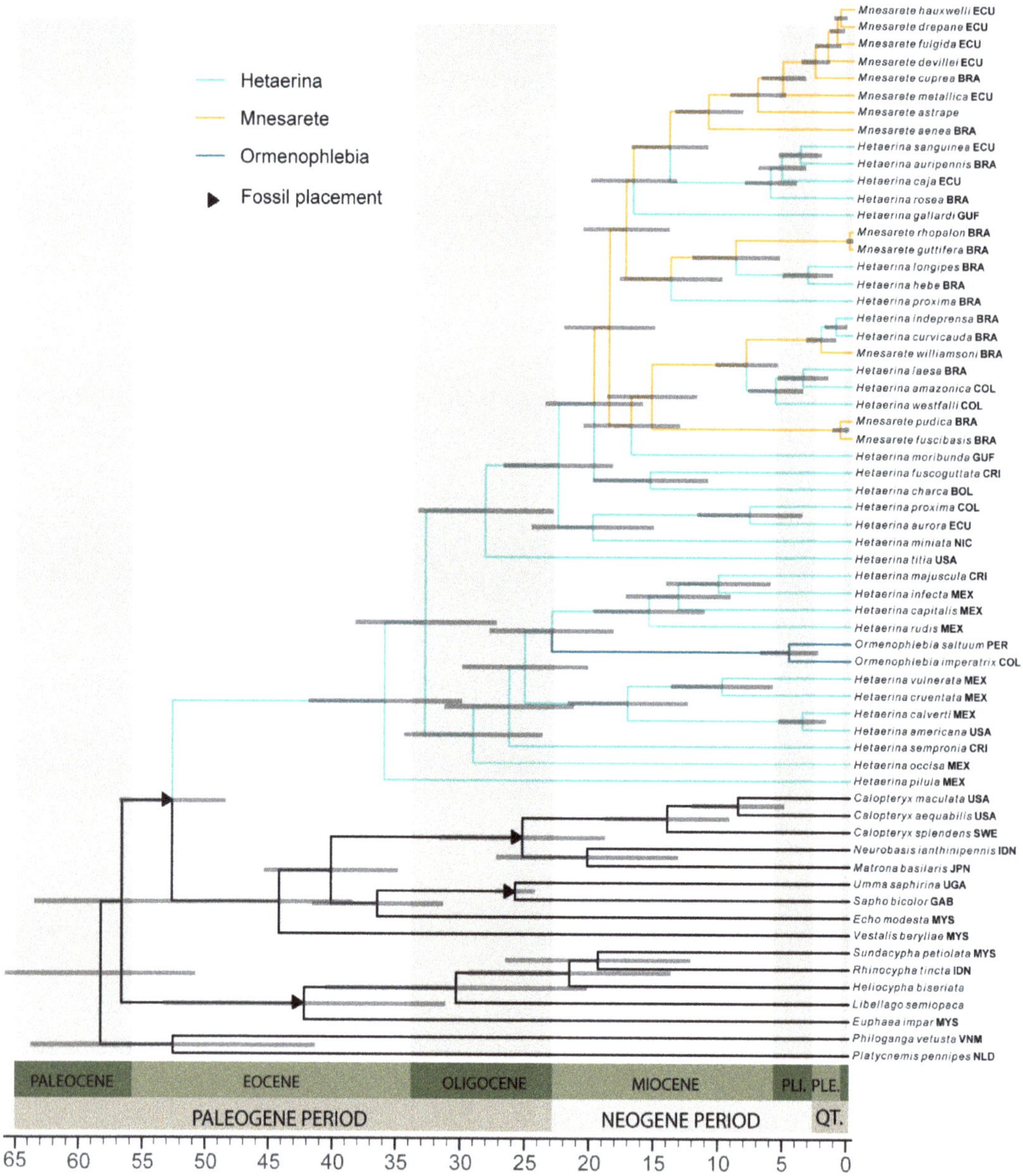

Figure 5. Phylogenetic relationships and time calibration of Hetaerininae. Bayesian time-calibrated tree estimated in BEAST2 v2.6.6 with median node ages, 95% high probability density (HPD) intervals for each node displayed in gray bars. Fossil-calibrated nodes represented with black triangles.

4. Discussion

Biogeography: Hetaerininae split off from the rest of Calopterygidae ~36 mya. Our reconstructions do not definitely place the pre-CAS ancestor in North or South America, with Hetaerininae having an ancestral area that comprised the Mexican transition zone, Mesoamerican dominion and the Pacific region. However, they are consistent in having a later dispersal into southern South America and northern North America and a gradual connection between the regions. Support for the gradual closure model suggests that

Hetaerininae dispersal was not restricted exclusively to overland dispersal and must have included some transoceanic dispersal. Surprisingly, this indicates that the Caribbean was not a complete barrier to dispersal.

The MRCA of *Ormenophlebia* dispersed to the southern range of the Pacific dominion ~5 mya. Starting during the Miocene, the Pebas System encompassed numerous wetlands and lakes covering >1 million km^2 [53] and may have facilitated further dispersal south of *Ormenophlebia* to the South American transition zone. As the Pebas wetland transitioned to a fluvial-dominated system ~10 mya, this would have promoted further dispersal. As dispersal was possible but unlikely ~20 mya, two separate clades formed within Hetaerininae: one primarily northern and one primarily southern (Figure 3). Shortly after dispersal to South America, the Andean uplift would have been another barrier to dispersal back to North and Central America. The Pebas System likely influenced dispersal in the South American clade as well, facilitating dispersal throughout the Boreal Brazilian dominion and the South Brazilian dominion [54].

Habitat and distribution: Interestingly, the generalist species of *Hetaerina* appear to be found primarily in the Mesoamerican region and are more widespread than most of the forest specialists. Towards the end of the Miocene, the temperature began dropping in North America, coinciding with a decrease in swamp forest and an increase in evergreen mixed forest [55]. This change in climate may have selected for species in North America with more diverse habitat preferences, while still retaining the ability to occupy tropical forests. This increased flexibility would then have allowed them to disperse more easily and occupy a wider range. In contrast, the increasingly tropical climate in South America may have supported the high diversification seen following dispersal to South America (Figure 3).

Phylogeny and taxonomy: Overall, the phylogenetic reconstruction shows unique patterns for the neotropical subfamily Hetaerininae. Our results suggest that *Mneserate* and *Ormenophlebia* are nested within *Hetaerina* (Figure 2); *Mnesarete* is paraphyletic and closely related to the mostly South American *Hetaerina* clade, while *Ormenophlebia* is monophyletic but nested within the mostly Central American clade (Figure 2). The general morphology of both *Mnesarete* and *Hetaerina* is similar, and they both possess cross-veins in the mid-basal space of their wings [5,24]; however, most of what is currently known as *Hetaerina* have a distinctive red basal spot on their wings [5]. Our results support Garrison's [5,24] view that the definition of *Mnesarete* is problematic and that this genus is paraphyletic. Further, while the majority of species form monophyletic clades, a few do not. The newly described *Hetaerina calverti* [56] is nested within *H. americana* in our analyses. *Mnesarete fulgida, M. drepane* and *M. hauxwelli* form a clade, with *M. fulgida* nested within *M. drepane. Mnesarete fuscibasis* is nested within *M. pudica* and renders it paraphyletic. *Hetaerina indeprensa, M. williamsoni* and *H. curvicauda* form a clade, with *H. indeprensa* nested within *M. williamsoni* and *H. curvicauda*. These results may be due to missing data. This is especially an issue when all members of a species do not have overlapping loci sequenced (e.g., *H. indeprensa*).

Geographic isolation has clearly played a large role in speciation in this cryptic group. For example, there are two separate clades of *M. westfalli*, one from Brazil and the other from Colombia. Similarly, *H. cruentata* is monophyletic but has two distinct clades, one from Central America (Costa Rica and Panama) and the other from Mexico. *Hetaerina occisa* is monophyletic but also forms two clades, one from Mexico and the other from Ecuador, Colombia and Peru. *H. infecta* is recovered as polyphyletic, and *H. hebe* is recovered as paraphyletic, but again this may be due to missing data as the taxa only had one locus for phylogenetic reconstruction.

What is clear is that further molecular and morphological studies that investigate the taxonomy of the group are needed. Our data calls for a new revision for all the neotropical calopterygid damselflies to establish valid monophyletic groupings. The task is not small and if our data are supported with further study, will likely require doing away with several genera, as *Hetaerina* Hagen in Selys 1853 has priority over *Mnesarete* Cowley 1934, *Ormenophlebia* Garrison 2006 and *Bryoplathanon* Garrison 2006.

Wing "ruby" spot evolution: Our results suggest that wing spot gain and loss is not correlated with habitat preference or phylogenetic position. In several species of *Hetaerina*, male red spots are sexually selected traits that mediate territorial interactions between males [16,57–62]. Wing spot traits (e.g., color and size) also play a role in mediating aggressive interactions between members of different species in the genus, and, in some instances, have been targeted by selection acting to reduce energetically costly interspecific fighting [7,61,62], but see [63,64] for instances where interspecific territoriality is maintained by selection. It is intriguing that one of the few *Mnesarete* species with extensively colored male wings also exhibits male-male territoriality and elaborate courtship displays [65]. Our phylogenetic analyses demonstrate that wing spots have been lost four times in this clade (Figure 4). Given the importance of wing spots in territorial interactions, further research into the reproductive strategies of species without wing spots is imperative to understanding the selective pressures that led to the loss of wing spots.

5. Conclusions

A revised taxonomical classification is needed, and color is not a diagnostic character for classification purposes in Hetaerininae. Morphological traits, such as male cerci and female intersternites, may provide morphological evidence for a stronger classification scheme. Further research on the behavior and ecology of species with hyaline wings is needed to better understand why wing coloration was lost in some lineages. Our results support a gradual dispersal of Hetaerininae from North America to South America that began in the Oligocene and that the Caribbean was not a barrier to dispersal. We suggest that extension of the Isthmus of Panama during the Oligocene, shortening the distance between Central America and South America, assisted in their dispersal. However, the lack of Hetaerininae on Caribbean islands suggests transoceanic dispersal to be challenging for this group. Further work is needed to fully understand their complicated dispersal history. An interesting avenue for research that our study only superficially examined is the evolution of Hetaerininae coloration (both body and wing) and habitat, specifically the light environment of forest vs. grassland habitat. This combined with estimates of diversification should allow for greater insight of the ecological shifts in their evolutionary history.

Supplementary Materials: The following supporting information can be downloaded at: https://www.mdpi.com/article/10.3390/d14090757/s1, Figure S1: Maximum likelihood IQtree results with full node support; Figure S2: Male, Hindwing color, Ancestral Character State Reconstruction; Figure S3: Male, Hindwing, apical color, Ancestral Character State Reconstruction; Figure S4: Male, Forewing, apical color, Ancestral Character State Reconstruction; Figure S5: Male, apical color in fore and hind wings, Ancestral Character State Reconstruction; Figure S6: Dated phylogenetic results with numbered nodes; Table S1: Partitions for IQTree analysis; Table S2: Taxa included in analysis; Table S3: Primers used in analysis; Table S4: Taxa included in BEAST2 and BioGeoBEARS analyses; Table S5: BioGeoBEARS model testing results; Table S6: Ancestral Character State Reconstruction matrix. Sequence data submitted to GenBank. File S1: BioGeoBEARS scripts and results.

Author Contributions: Conceptualization, M.S.-H., R.G.-F., J.L.W. and S.B.; Data curation, S.S. and Y.M.V.-S.; Formal analysis, S.S. and C.A.B.-S.; Investigation, S.S. and R.C.; Methodology, S.S. and M.S.-H.; Project administration, J.L.W., R.C. and S.B.; Resources, M.S.-H., R.G.-F., A.G.-R., Y.M.V.-S., J.P.D., G.F.G. and S.B.; Supervision, M.S.-H. and S.B.; Validation, Y.M.V.-S., C.A.B.-S. and S.B.; Visualization, S.S.; Writing—original draft, S.S.; Writing—review & editing, S.S., M.S.-H., J.L.W., Y.M.V.-S., A.G.-R., R.C., J.P.D., G.F.G., L.M.-C. and S.B. All authors have read and agreed to the published version of the manuscript.

Funding: This research was funded by the National Science Foundation (DEB-1265714; DBI-1564386).

Data Availability Statement: The data presented in this study are openly available in GenBank, see Table S2 for accession numbers.

Acknowledgments: We would like to acknowledge Rosser Garrison for all of the work he has done within Hetaerininae, and for his help in confirming identifications. R.G.-F. thanks CNPq (Proc. 307836/2019-3) for a productivity grant.

Conflicts of Interest: The authors declare no conflict of interest.

References

1. Misof, B.; Liu, S.; Meusemann, K.; Peters, R.S.; Donath, A.; Mayer, C.; Frandsen, P.B.; Ware, J.; Flouri, T.; Beutel, R.G.; et al. Phylogenomics Resolves the Timing and Pattern of Insect Evolution. *Science* **2014**, *346*, 763–767. [CrossRef] [PubMed]
2. Suárez-Tovar, C.M.; Sarmiento, C.E. Beyond the Wing Planform: Morphological Differentiation between Migratory and Nonmigratory Dragonfly Species. *J. Evol. Biol.* **2016**, *29*, 690–703. [CrossRef] [PubMed]
3. Corbet, P.S. *Dragonflies: Behaviour and Ecology of Odonata*; Harley Books: Colchester, UK, 1999.
4. Bybee, S.M.; Kalkman, V.J.; Erickson, R.J.; Frandsen, P.B.; Breinholt, J.W.; Suvorov, A.; Dijkstra, K.D.B.; Cordero-Rivera, A.; Skevington, J.H.; Abbott, J.C.; et al. Phylogeny and Classification of Odonata Using Targeted Genomics. *Mol. Phylogenet. Evol.* **2021**, *160*, 107115. [CrossRef] [PubMed]
5. Garrison, R.W. A Synopsis of the Genus Hetaerina with Descriptions of Four New Species. *Trans. Am. Entomol. Soc.* **1990**, *116*, 175–259.
6. McGirr, R.; Seton, M.; Williams, S. Kinematic and Geodynamic Evolution of the Isthmus of Panama Region: Implications for Central American Seaway Closure. *Bull. Geol. Soc. Am.* **2021**, *133*, 867–884. [CrossRef]
7. Grether, G.F. Sexual Selection and Survival Selection on Wing Coloration and Body Size in the Rubyspot Damselfly Hetaerina Americana. *Evolution* **1996**, *50*, 1939–1948. [CrossRef]
8. Córdoba-Aguilar, A. Adult Survival and Movement in Males of the Damselfly Hetaerina Cruentata (Odonata: Calopterygidae). *Fla. Entomol.* **1994**, *77*, 256–264. [CrossRef]
9. Jaramillo, C.A. Evolution of the Isthmus of Panama: Biological, Palaeoceanographic and Palaeoclimatological Implications. In *Mountains, Climate and Biodiversity*; Hoorn, C., Perrigo, A., Antonelli, A., Eds.; John Wiley & Sons Ltd.: Hoboken, NJ, USA, 2018; pp. 323–338.
10. Montes, C.; Cardona, A.; Jaramillo, C.; Pardo, A.; Silva, J.C.; Valencia, V.; Ayala, C.; Pérez-Angel, L.C.; Rodriguez-Parra, L.A.; Ramirez, V.; et al. Middle Miocene Closure of the Central American Seaway. *Science* **2015**, *348*, 226–229. [CrossRef]
11. Dumont, H.J.; Vanfleteren, J.R.; de Jonckheere, J.F.; Weekers, P.H.H. Phylogenetic Relationships, Divergence Time Estimation, and Global Biogeographic Patterns of Calopterygoid Damselflies (Odonata, Zygoptera) Inferred from Ribosomal DNA Sequences. *Syst. Biol.* **2005**, *54*, 347–362. [CrossRef]
12. Mulcahy, D.G.; Noonan, B.P.; Moss, T.; Townsend, T.M.; Reeder, T.W.; Sites, J.W.; Wiens, J.J. Estimating Divergence Dates and Evaluating Dating Methods Using Phylogenomic and Mitochondrial Data in Squamate Reptiles. *Mol. Phylogenet. Evol.* **2012**, *65*, 974–991. [CrossRef]
13. Suvorov, A.; Scornavacca, C.; Fujimoto, M.S.; Bodily, P.; Clement, M.; Crandall, K.A.; Whiting, M.F.; Schrider, D.R.; Bybee, S.M. Deep Ancestral Introgression Shapes Evolutionary History of Dragonflies and Damselflies. *Syst. Biol.* **2022**, *71*, 526–546. [CrossRef] [PubMed]
14. Kohli, M.; Letsch, H.; Greve, C.; Béthoux, O.; Deregnaucourt, I.; Liu, S.; Zhou, X.; Donath, A.; Mayer, C.; Podsiadlowski, L.; et al. Evolutionary History and Divergence Times of Odonata (Dragonflies and Damselflies) Revealed through Transcriptomics. *iScience* **2021**, *24*, 103324. [CrossRef] [PubMed]
15. Alcock, J. Post-Copulatory Mate Guarding by Males of the Demselfly Hetaerina Vulnerata Selys (Odonata: Calopterygidae). *Anim. Behav.* **1982**, *30*, 99–107. [CrossRef]
16. Grether, G.F.; Grey, R.M. Novel Cost of a Sexually Selected Trait in the Rubyspot Damselfly Hetaerina Americana: Conspicuousness to Prey. *Behav. Ecol.* **1996**, *7*, 465–473. [CrossRef]
17. Córdoba-Aguilar, A.; Serrano-Meneses, M.A.; Contreras-Garduño, J.; Raihani, G. The Lek Mating System of Hetaerina Damselflies (Insecta: Calopterygidae). *Behaviour* **2009**, *146*, 189–207. [CrossRef]
18. Guillermo-Ferreira, R.; Del-Claro, K. Resource Defense Polygyny by Hetaerina Rosea Selys (Odonata: Calopterygidae): Influence of Age and Wing Pigmentation. *Neotrop. Entomol.* **2011**, *40*, 78–84. [CrossRef] [PubMed]
19. Anderson, C.N.; Grether, G.F. Multiple Routes to Reduced Interspecific Territorial Fighting in Hetaerina Damselflies. *Behav. Ecol.* **2011**, *22*, 527–534. [CrossRef]
20. Contreras-Garduño, J.; Villanueva, G.; Alonso-Salgado, A. Phenoloxidase Production: The Importance of Time after Juvenile Hormone Analogue Administration in Hetaerina Americana (Fabricius) (Zygoptera: Calopterygidae). *Odonatologica* **2012**, *41*, 1–88.
21. Córdoba-Aguilar, A.; González-Tokman, D.M. The Behavioral and Physiological Ecology of Adult Rubyspot Damselflies (Hetaerina, Calopterygidae, Odonata). *Adv. Study Behav.* **2014**, *46*, 311–341. [CrossRef]
22. Grether, G.F. Intrasexual Competition Alone Favors a Sexually Dimorphic Ornament in the Rubyspot Damselfly Hetaerina Americana. *Evolution* **1996**, *50*, 1949–1957. [CrossRef]
23. Grether, G.F. Survival Cost of an Intrasexually Selected Ornament in a Damselfly. *Proc. R. Soc. Lond.* **1997**, *264*, 207–210. [CrossRef]
24. Garrison, R.W. A Synopsis of the Genera Mnesarete Cowley, Bryoplathanon Gen. Nov., and Ormenophlebia Gen. Nov. (Odonata: Calopterygidae). *Contrib. Sci.* **2006**, *50*, 61–84.
25. Dijkstra, K.D.B.; Kalkman, V.J.; Dow, R.A.; Stokvis, F.R.; van Tol, J. Redefining the Damselfly Families: A Comprehensive Molecular Phylogeny of Zygoptera (Odonata). *Syst. Entomol.* **2014**, *39*, 68–96. [CrossRef]
26. Weekers, P.H.H.; de Jonckheere, J.F.; Dumont, H.J. Phylogenetic Relationships Inferred from Ribosomal ITS Sequences and Biogeographic Patterns in Representatives of the Genus Calopteryx (Insecta: Odonata) of the West Mediterranean and Adjacent West European Zone. *Mol. Phylogenet. Evol.* **2001**, *20*, 89–99. [CrossRef] [PubMed]

27. Katoh, K.; Misawa, K.; Kuma, K.I.; Miyata, T. MAFFT: A Novel Method for Rapid Multiple Sequence Alignment Based on Fast Fourier Transform. *Nucleic Acids Res.* **2002**, *30*, 3059–3066. [CrossRef] [PubMed]
28. Nguyen, L.-T.; Schmidt, H.A.; von Haeseler, A.; Minh, B.Q. IQ-TREE: A Fast and Effective Stochastic Algorithm for Estimating Maximum-Likelihood Phylogenies. *Mol. Biol. Evol.* **2015**, *32*, 268–274. [CrossRef]
29. Kalyaanamoorthy, S.; Minh, B.Q.; Wong, T.K.F.; von Haeseler, A.; Jermiin, L.S. ModelFinder: Fast Model Selection for Accurate Phylogenetic Estimates. *Nat. Methods* **2017**, *14*, 587–589. [CrossRef]
30. Hoang, D.T.; Chernomor, O.; von Haeseler, A.; Minh, B.Q.; Vinh, L.S. UFBoot2: Improving the Ultrafast Bootstrap Approximation. *Mol. Biol. Evol.* **2018**, *35*, 518–522. [CrossRef]
31. Guindon, S.; Dufayard, J.F.; Lefort, V.; Anisimova, M.; Hordijk, W.; Gascuel, O. New Algorithms and Methods to Estimate Maximum-Likelihood Phylogenies: Assessing the Performance of PhyML 3.0. *Syst. Biol.* **2010**, *59*, 307–321. [CrossRef]
32. Yu, G. Using Ggtree to Visualize Data on Tree-Like Structures. *Curr. Protoc. Bioinform.* **2020**, *69*, e96. [CrossRef]
33. Paradis, E.; Schliep, K. Ape 5.0: An Environment for Modern Phylogenetics and Evolutionary Analyses in R. *Bioinformatics* **2019**, *35*, 526–528. [CrossRef] [PubMed]
34. Wickham, H.; Hester, J.; Chang, W. Devtools: Tools to Make Developing R Packags Easier. R Package Version 1.1.6. 2018. Available online: https://CRAN.R-project.org/package=devtools (accessed on 12 April 2022).
35. Revell, L.J. Phytools: An R Package for Phylogenetic Comparative Biology (and Other Things). *Methods Ecol. Evol.* **2012**, *3*, 217–223. [CrossRef]
36. Bouckaert, R.; Vaughan, T.G.; Barido-Sottani, J.; Duchêne, S.; Fourment, M.; Gavryushkina, A.; Heled, J.; Jones, G.; Kühnert, D.; de Maio, N.; et al. BEAST 2.5: An Advanced Software Platform for Bayesian Evolutionary Analysis. *PLoS Comput. Biol.* **2019**, *15*, e1006650. [CrossRef] [PubMed]
37. Ho, S.Y.W.; Lo, N. The Insect Molecular Clock. *Aust. J. Entomol.* **2013**, *52*, 101–105. [CrossRef]
38. Rambaut, A.; Drummond, A.J.; Xie, D.; Baele, G.; Suchard, M.A. Posterior Summarization in Bayesian Phylogenetics Using Tracer 1.7. *Syst. Biol.* **2018**, *67*, 901–904. [CrossRef]
39. Parham, J.F.; Donoghue, P.C.J.; Bell, C.J.; Calway, T.D.; Head, J.J.; Holroyd, P.A.; Inoue, J.G.; Irmis, R.B.; Joyce, W.G.; Ksepka, D.T.; et al. Best Practices for Justifying Fossil Calibrations. *Syst. Biol.* **2012**, *61*, 346–359. [CrossRef]
40. Lin, Q.B.; Petrulevičius, J.F.; Huang, D.Y.; Nel, A.; Engel, M.S. First Fossil Calopterygoidea (Odonata: Zygoptera) from Southeastern Asia: A New Genus and Species from the Paleogene of China. *Geobios* **2010**, *43*, 349–353. [CrossRef]
41. NEL, A.; BRISAC, P. Découverte Du Premier Calopteryx Fossile Du Miocène de l'Ardèche (Odonota, Calopterygidae). *Bull. Soc. Entomol. Fr.* **1994**, *99*, 27–30. [CrossRef]
42. Steininger, F.F.; Berggren, W.A.; Kent, D.V.; Bernor, R.L.; Sen, S.; Agusti, J. Circum-Mediterranean Neogene (Miocene and Pliocene) Marine-Continental Chronologic Correlations of European Mammal Units. In *The Evolution of Western Eurasian Neogene Mammal Faunas*; Bernor, R.L., Fahlbusch, V., Mitmann, H.-W., Eds.; Columbia University Press: New York, NY, USA, 1996; pp. 7–46.
43. Nel, A.; Petrulevičius, J.F. Afrotropical and Nearctic Genera of Odonata in the French Oligocene: Biogeographic and Paleoclimatic Implications (Insecta: Calopterygidae, Aeshnidae). *Ann. Soc. Entomol. Fr.* **2010**, *46*, 228–236. [CrossRef]
44. Nel, A.; Gross, M.; Engel, M.S. First Fossil Occurrence of the Jewel Damselflies (Odonata: Chlorocyphidae): A New Species from the Late Miocene of Styria, Austria. *Ann. Soc. Entomol. Fr.* **2017**, *53*, 280–285. [CrossRef]
45. Renner, S.; Périco, E.; Ely, G.J.; Sahlén, G. Preliminary Dragonfly (Odonata) Species List from the Pampa Biome in Rio Grande Do Sul, Brazil, with Ecological Notes for 19 New Records for the State. *Biota Neotrop.* **2017**, *17*, e20170374. [CrossRef]
46. Vilela, D.S.; Koroiva, R.; Tosta, T.H.A.; Novaes, M.C.; Guillermo-Ferreira, R. Dragonflies and Damselflies from the West of Minas Gerais, Brazil: Checklist and New Records. *Biota Neotrop.* **2020**, *20*, e20190851. [CrossRef]
47. Bota-Sierra, C.A.; Flórez-V, C.; Escobar, F.; Sandoval-H, J.; Novelo-Gutiérrez, R.; Londoño, G.A.; Cordero-Rivera, A. The Importance of Tropical Mountain Forests for the Conservation of Dragonfly Biodiversity: A Case from the Colombian Western Andes. *Int. J. Odonatol.* **2021**, *24*, 233–247. [CrossRef]
48. Abbot, J.C.; Bota-Sierra, C.A.; Guralnick, R.; Kalkman, V.; González-Soriano, E.; Novelo-Gutiérrez, R.; Bybee, S.M.; Ware, J.L.; Belitz, M.W. Diversity of Nearctic Dragonflies and Damselflies (Odonata). *Diversity* **2022**, *14*, 575. [CrossRef]
49. Matzke, N.J. *Probabilistic Historical Biogeography: New Models for Founder-Event Speciation, Imperfect Detection, and Fossils Allow Improved Accuracy and Model-Testing*; University of California: Berkeley, CA, USA, 2013.
50. Morrone, J.J.; Escalante, T.; Rodriguez-Tapia, G.; Carmona, A.; Arana, M.; Mercado-Gomez, J.D. Biogeographic Regionalization of the Neotropical Region: New Map and Shapefile. *An. Acad. Bras. Cienc.* **2022**, *94*, e20211167. [CrossRef] [PubMed]
51. Iturralde-Vinent, M.A.; MacPhee, R.D.E. Paleogeography of the Caribbean Region: Implications for Cenozoic Biogeography. *Bull. Am. Mus. Nat. Hist.* **1999**, *238*, 1–72.
52. Matzke, N.J. Statistical Comparison of DEC and DEC+J Is Identical to Comparison of Two ClaSSE Submodels, and Is Therefore Valid. *J. Biogeogr.* **2021**, 1–26. [CrossRef]
53. Wesselingh, F.P.; Salo, J.A. A Miocene Perspective on the Evolution of the Amazonian Biota. *Scr. Geol.* **2006**, *133*, 439–458.
54. Hoorn, C.; Wesselingh, F.P.; ter Steege, H.; Bermudez, M.A.; Mora, A.; Sevink, J.; Sanmartín, I.; Sanchez-Meseguer, A.; Anderson, C.L.; Figueiredo, J.P.; et al. Amazonia through Time: Andean Uplift, Climate Change, Landscape Evolution, and Biodiversity. *Science* **2010**, *330*, 927–931. [CrossRef]

55. Kotthoff, U.; Greenwood, D.R.; McCarthy, F.M.G.; Müller-Navarra, K.; Prader, S.; Hesselbo, S.P. Late Eocene to Middle Miocene (33 to 13 Million Years Ago) Vegetation and Climate Development on the North American Atlantic Coastal Plain (IODP Expedition 313, Site M0027). *Clim. Past* **2014**, *10*, 1523–1539. [CrossRef]
56. Vega-Sánchez, Y.M.; Mendoza-Cuenca, L.F.; Gonzálezrodríguez, A. Hetaerina Calverti (Odonata: Zygoptera: Calopterygidae) Sp. Nov., a New Cryptic Species of the American Rubyspot Complex. *Zootaxa* **2020**, *4766*, 485–497. [CrossRef] [PubMed]
57. Anderson, C.N.; Grether, G.F. Character Displacement in the Fighting Colours of Hetaerina Damselflies. *Proc. R. Soc. B Biol. Sci.* **2010**, *277*, 3669–3675. [CrossRef] [PubMed]
58. Anderson, C.N.; Grether, G.F. Interspecific Aggression and Character Displacement of Competitor Recognition in Hetaerina Damselflies. *Proc. Biol. Sci. R. Soc.* **2010**, *277*, 549–555. [CrossRef] [PubMed]
59. González-Tokman, D.; Córdoba-Aguilar, A.; González-Santoyo, I.; Lanz-Mendoza, H. Infection Effects on Feeding and Territorial Behaviour in a Predatory Insect in the Wild. *Anim. Behav.* **2011**, *81*, 1185–1194. [CrossRef]
60. Guillermo-Ferreira, R.; Bispo, P.C. Male and Female Interactions during Courtship of the Neotropical Damselfly Mnesarete Pudica (Odonata: Calopterygidae). *Acta Ethol.* **2012**, *15*, 173–178. [CrossRef]
61. Drury, J.P.; Grether, G.F. Interspecific Aggression, Not Interspecific Mating, Drives Character Displacement in the Wing Coloration of Male Rubyspot Damselflies (Hetaerina). *Proc. R. Soc. B Biol. Sci.* **2014**, *281*, 20141737. [CrossRef]
62. Grether, G.F.; Drury, J.P.; Berlin, E.; Anderson, C.N. The Role of Wing Coloration in Sex Recognition and Competitor Recognition in Rubyspot Damselflies (*Hetaerina* Spp.). *Ethology* **2015**, *121*, 674–685. [CrossRef]
63. Drury, J.P.; Anderson, C.N.; Castillo, M.B.C.; Fisher, J.; McEachin, S.; Grether, G.F. A General Explanation for the Persistence of Reproductive Interference. *Am. Nat.* **2019**, *194*, 268–275. [CrossRef]
64. Grether, G.F.; Drury, J.P.; Okamoto, K.W.; McEachin, S.; Anderson, C.N. Predicting Evolutionary Responses to Interspecific Interference in the Wild. *Ecol. Lett.* **2020**, *23*, 221–230. [CrossRef]
65. Pena-Firme, P.; Guillermo-Ferreira, R. Females of the Red Damselfly Mnesarete Pudica Are Attracted to More Ornamented Males and Attract Rival Males. *Sci. Rep.* **2020**, *10*, 14320. [CrossRef]

diversity

MDPI

Article

Testing the Effect of Sampling Effort on Inferring Phylogeographic History in *Psolodesmus mandarinus* (Calopterygidae, Odonata)

Liang-Jong Wang [1,*], Yen-Wei Chou [1] and Jen-Pan Huang [2,*]

1 Division of Forest Protection, Taiwan Forestry Research Institute, Taipei 10066, Taiwan
2 Biodiversity Research Center, Academia Sinica, Taipei 11529, Taiwan
* Correspondence: ljwang23@ms17.hinet.net (L.-J.W.); airbugshuang@gmail.com (J.-P.H.)

Abstract: Phylogeographic studies have revealed spatial genetic structure and inferred geographical processes that may have generated genetic diversity and divergence. These study results have implications not only on the processes that generate intraspecific and interspecific diversity but also on the essential integrals for defining evolutionary entities (e.g., species). However, the resulting phylogeographic inferences might be impacted by the sampling design, i.e., the number of individuals per population and the number of geographic populations studied. The effect of sampling bias on phylogeographic inferences remains poorly explored. With a comprehensive sampling design (including 186 samples from 56 localities), we studied the phylogeographic history of a Taiwanese endemic damselfly, *Psolodesmus mandarinus*, with a specific focus on testing the impact of the sampling design on phylogeographic inference. We found a significant difference in the genetic structure of eastern and western populations separated by the Central Mountain Range (CMR) of Taiwan. However, isolation by the CMR did not lead to reciprocally monophyletic geographic populations. We further showed that, when only a subset of individuals was randomly included in the study, monophyletic geographic populations were obtained. Furthermore, historical demographic expansion could become undetectable when only a subset of samples was used in the analyses. Our results demonstrate the impact of sampling design on phylogeographic inferences. Future studies need to be cautious when inferring the effect of isolation by a physical barrier.

Keywords: Zygoptera; molecular phylogeny; network; population genetics; sampling effect

Citation: Wang, L.-J.; Chou, Y.-W.; Huang, J.-P. Testing the Effect of Sampling Effort on Inferring Phylogeographic History in *Psolodesmus mandarinus* (Calopterygidae, Odonata). *Diversity* **2022**, *14*, 809. https://doi.org/10.3390/d14100809

Academic Editors: M. Olalla Lorenzo-Carballa and Ricardo Koroiva

Received: 15 August 2022
Accepted: 27 September 2022
Published: 28 September 2022

Publisher's Note: MDPI stays neutral with regard to jurisdictional claims in published maps and institutional affiliations.

1. Introduction

Different sampling efforts may impact phylogeographic inferences [1–4]. For example, coalescent simulations based on multilocus data have been shown to more accurately reconstruct population history than those resulting from the use of single locus datasets [5]. Additionally, sampling an insufficient number of individuals from each population/locality may also impact the estimated population genetic parameters (e.g., the genetic diversity parameter θ), which may lead to biased parameter values for simulation-based studies (e.g., approximate Bayesian computation, ABC, methods [1]) and therefore support erroneous phylogeographic histories. Because most of the conventional phylogeographic studies rely heavily on the inferred gene tree, particularly the mitochondrial gene tree [5], the sensitivity of phylogenetic reconstruction due to taxa sampling could have profound effects on the reconstructed phylogeographic history [6]. The importance of inferences from phylogegraphic studies extend beyond semantic issues. For example, identifying areas of high genetic diversity, e.g., historical climatic refugia, and distinct genetic entities, e.g., cryptic species, can significantly influence conservation strategies [7]. However, rather than testing for insufficient sampling, many phylogeographic studies intrinsically assume that their sampling design represents the true distribution of genetic diversity across the geographic distribution of the studied organism.

The subtropical island of Taiwan accommodates high biodiversity (both species and genetic diversity [8–10]) and has been the focus of extensive phylogeographic studies in the past two decades because of its recent, yet drastic tectonic history, which may have generated high levels of intraspecific genetic diversity (e.g., [8–10]). One of the main topics has been the effect of the Central Mountain Range (CMR) on driving population subdivision. Specifically, three phylogeographic patterns have been identified across multiple different evolutionary lineages: (1) different geographic populations separated by mountain ranges form monophyletic lineages, indicating that the CMR (or mountain ranges in general) can effectively promote allopatric divergence; (2) significant genetic structure is found between geographic populations, which implies reduced gene flow because of the CMR; and (3) no significant geographic genetic structure (see Table 1 for a non-comprehensive summary). Although biological and ecological differences between organismal groups have often been argued to be responsible for the different phylogeographic patterns (e.g., freshwater associated species are often attributed to phylogeographic pattern 1 [11]; see Table 1), such differences in phylogeographic patterns could also result from differences in sampling effort. For example, when multiple molecular markers have been included in a study, different phylogeographic patterns have often been inferred (e.g., [9,11–14]). Additionally, studies that reveal insignificant geographic genetic structure (pattern 3) between eastern and western populations tend to include a smaller number of individuals (Table 1).

Table 1. Examples of Taiwanese phylogeographic studies and their evolutionary inferences.

Organism	Sample Size	# Localities	Inferred Pattern *	Reference
Bamboo viper	201	40	2	[12]
Bat1	108	50	2	[11]
Bat2	146	50	1	[11]
Bat3	234	50	2	[11]
Bat4	164	50	2	[11]
Toad	279	27	2	[15]
Damselfly1	159	32	2	[14]
Damselfly2 $	60	20	1	[16]
Flying squirrel1	40	20	3	[17]
Flying Squirrel2	35	18	3	[17]
Freshwater Crab	88	18	1	[18]
Freshwater Prawn	195	20	1	[19]
Frog	198	31	1	[20]
Spider	189	18	3	[21]
Small mammal	71	29	1	[22]
Stag beetle	52	25	1	[9]
Freshwater fish	71	16	1	[23]
Tree frog	564	33	1	[18]

* Isolation by the CMR leads to reciprocal monophyly (1), significant genetic structure (2), or no genetic differentiation (3) between eastern and western populations. $ The same species, *P. mandarinus*, utilized in this study.

In this study, we aimed to test the effect of sampling effort, specifically focusing on the sample size of individuals from eastern and western populations separated by the CMR, on the resulting mitochondrial phylogeography. Note that we understand that mitochondrial phylogeography can be erroneous because of, for example, the existence of nuclear copies of mitochondrial DNA (NUMTs), which has been recently identified in Odonata [24]. However, and while we fully acknowledge the limit of mitochondrial phylogeography [25,26] and the benefit of multilocus data and coalescent-based analyses for statistic phylogeography [2,5], mitochondrial gene genealogy is still, if not predominant, included in the majority of phylogeographic studies in animals. Specifically, mitochondrial phylogeography is often the first dataset that can be obtained to form testable hypothesis and can be readily compared across multiple co-distributed taxa given the cornucopia of published data [3]. Furthermore, molecular-based species delimitation and the identification of cryptic genetic groups/species both rely heavily on mitochondrial datasets [27]. By assessing the effect of sampling effort on the resulting inferences based on the pattern of mitochondrial gene topology and population structure, our results will have broader impacts on not only phylogeographic studies *per se*, but also on how consistent the different types of biological entities that are identified as distinct genetic clusters are in molecular systematics that involve different sample sizes.

The endemic damselfly *Psolodesmus mandarinus* of Taiwan is a common and large-sized odonate that can be found close to creeks and small streams from low to mid-elevations in the mountain regions. There are three subspecies in Taiwan and the nearby Yaeyama islands, identified based on wing color patterns [28,29]. The Yaeyama subspecies, *P. m. kuroiwae*, is genetically distinct and divergent from the other two Taiwanese subspecies and has been elevated to full species status [16,30]. The two Taiwanese morphological subspecies are geographically structured to the northern, southern and eastern parts of Taiwan (Figure 1). However, intermediate forms can often be found. Unsurprisingly, the two Taiwanese subspecies did not form monophyletic mitochondrial groups in a previous study [16]; instead, the mitochondrial gene tree revealed two geographic lineages separated by the CMR [16]. However, one population from the east, Tongmen, has individuals from both the eastern and western lineages. The sampling from eastern Taiwan was limited in the previous study, and thus the extent of the geographic distribution of the two genetic lineages and the phylogeographic history of the species may not be correctly inferred. In this study, we expanded the geographic taxon sampling (a total of 124 localities; Figure 2 and Table 2) and increased the length of the sequenced mitochondrial region (a total of three mitochondrial loci; 1959 bp long) to study the mitochondrial genetic diversity and the geographic distribution of the genetic diversity. Specifically, we tested (1) whether the observed pattern of geographic lineages can be an artifact of limited sampling, (2) the effect of limited sampling on demographic inferences, and (3) based on our new data, we discuss the phylogeographic history of *P. mandarinus*.

Figure 1. Populations of *Psolodesmus mandarinus* in Taiwan. (**A–C**) males. (**D–F**) females. (**A**) North Taiwan. Pamierh Park, Shihlin District, Taipei city. (**B**) South Taiwan. Shanping, Liukuei District, Kaohsiung city. (**C**) East Taiwan. Hsiama, Haituan Township, Taitung County. (**D**) North Taiwan. Wulai, Wulai District, New Taipei city. (**E**) South Taiwan. Neiwen, Neiwen Township, Pintung County. (**F**) East Taiwan. Fenglin, Fenglin Township, Hualien County.

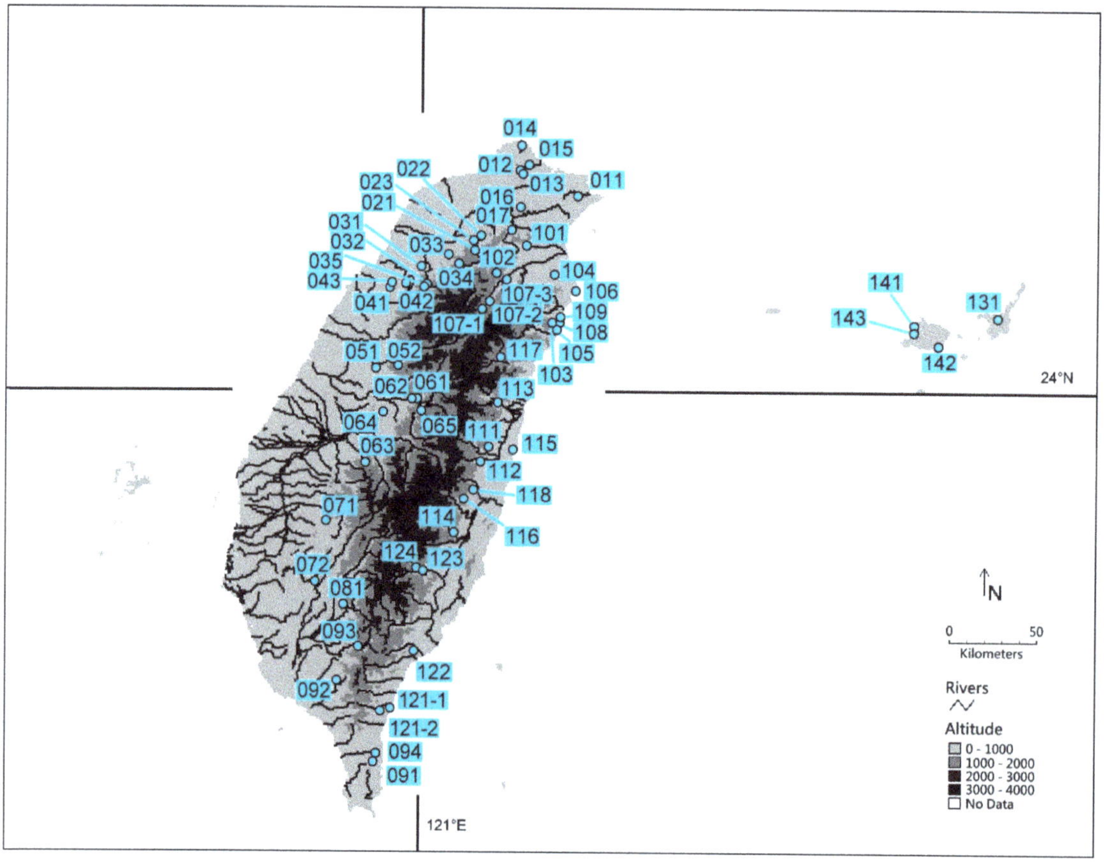

Figure 2. Distribution of sampling locations of *P. madarinus* in Taiwan and *P. kuroiwae* in Japan. Sampling locations are marked with open circles. The numbers of sampling locations are the same as in Table 2. The map was made using DIVA-GIS (https://www.diva-gis.org/; accessed on 1 June 2016).

Table 2. Sampling localities and their haplotype information.

No.	Acronym	Locality	GPS Coordinates	Altitude	Haplotype •	Accession Numbers
Taiwan						
11	TI	Tinglanku, Shuanghsi District, New Taipei city	25°01′04.4″ N 121°52′32.0″ E	42 m	H01 (2)	KM360534
12	PI	Pingtenli, Shihlin District, Taipei city	25°08′24.6″ N 121°34′43.1″ E	500 m	H02 (1)	KM360535
13	PA	Pamierh Park, Shihlin District, Taipei city	25°07′20.6″ N 121°35′35.5″ E	330 m	H01 (2)	KM360534
14	AL	Alipang, Shihmen District, New Taipei city	25°15′50.5″ N 121°35′05.2″ E	140 m	H01 (1), H03 (2)	KM360534, KM360536
15	LU	Lukuping, Wanli District, New Taipei city	25°10′07.9″ N 121°37′18.3″ E	419 m	H01 (4)	KM360534
16	YI	Yinhotung, Hsintien District, New Taipei city	24°57′30.5″ N 121°34′55.9″ E	212 m	H01 (1), H04 (1)	KM360534, KM360537

Table 2. *Cont.*

No.	Acronym	Locality	GPS Coordinates	Altitude	Haplotype •	Accession Numbers
17	WU	Wulai, Wulai District, New Taipei city	24°50′20.6″ N 121°32′08.4″ E	219 m	H01 (4)	KM360534
21	JU	Junghua, Fuhsing Township, Taoyuan County	24°44′05.5″ N 121°21′02.1″ E	505 m	H01 (1), H05 (1)	KM360534, KM360538
22	LI	Liuhsia, Fuhsing Township, Taoyuan County	24°48′34.3″ N 121°22′25.5″ E	364 m	H01 (2), H06 (1)	KM360534, KM360539
23	FU	Fuhsing, Fuhsing Township, Taoyuan County	24°47′20.9″ N 121°20′22.8″ E	369 m	H07 (3)	KM360540
31	PE	Peipu, Peipu Township, Hsinchu County	24°39′27.8″ N 121°04′45.5″ E	264 m	H07 (4)	KM360540
32	SH	Shihlu, Chienshih Township, Hsinchu County	24°33′58.6″ N 121°06′23.7″ E	1110 m	H07 (2)	KM360540
33	CS	Chienshihhsienho, Chienshih Township, Hsinchu County	24°42′48.7″ N 121°12′32.4″ E	300 m	H07 (1), H08 (1), H09 (1)	KM360540-KM360542
34	CH	Chienshih, Chienshih Township, Hsinchu County	24°40′11.1″ N 121°15′57.7″ E	851 m	H01 (1), H07 (1), H10 (1)	KM360534, KM360540, KM360543
35	KU	Kuanwu, Wufeng Township, Hsinchu County	24°33′48.3″ N 121°05′35.8″ E	812 m	H07 (2), H09 (1)	KM360540, KM360542
41	ST	Shihtanpeitawo, Shihtan Township, Miaoli County	24°33′09.5″ N 120°54′58.5″ E	272 m	H07 (1)	KM360540
42	NA	Nanchuang, Nanchuang Township, Miaoli County	24°34′16.9″ N 121°00′00.1″ E	332 m	H07 (4)	KM360540
43	TO	Touwu, Touwu Township, Miaoli County	24°34′40.8″ N 120°55′34.3″ E	179 m	H07 (1), H08 (1), H11 (1)	KM360540, KM360541, KM360544
51	HS	Hsinshe, Hsinshe District, Taichung city	24°08′54.9″ N 120°50′38.7″ E	605 m	H07 (3), H12 (1)	KM360540, KM360545
52	KK	Kukuan, Hoping District, Taichung city	24°09′28.3″ N 120°57′38.9″ E	687 m	H07 (2), H13 (1)	KM360540, KM360546
61	PP	Penpusi, Puli Township, Nantou County	23°59′41.4″ N 121°03′41.1″ E	735 m	H14 (2)	KM360547
62	KY	Kuanyinpupu, Puli Township, Nantou County	23°59′32.9″ N 121°02′06.0″ E	646 m	H07 (1), H17 (1)	KM360540, KM360550
63	HT	Hsitou, Luku Township, Nantou County	23°40′27.8″ N 120°47′26.9″ E	1082 m	H07 (7), H11 (1), H16 (1)	KM360540, KM360544, KM360549
64	LH	Lienhuachih, Yuchih Township, Nantou County	23°55′26.1″ N 120°53′03.5″ E	735 m	H07 (2), H15 (1)	KM360540, KM360548
65	JE	Jenai, Jenai Township, Nantou County	23°55′42.4″ N 121°04′56.1″ E	1120 m	H07 (2)	KM360540
71	CP	Chungpu, Chungpu Township, Chiayi County	23°23′13.2″ N 120°35′34.1″ E	816 m	H07 (3)	KM360540
72	NH	Nanhua Dam, Nanhua District, Tainan city	23°04′38.3″ N 120°32′03.5″ E	198 m	H07 (3)	KM360540
81	SP	Shanping, Liukuei District, Kaohsiung city	22°58′00.1″ N 120°41′02.5″ E	660 m	H07 (4)	KM360540
91	MU	Mutan, Mutan Township, Pintung County	22°10′45.5″ N 120°50′26.5″ E	280 m	H18 (3)	KM360551

Table 2. *Cont.*

No.	Acronym	Locality	GPS Coordinates	Altitude	Haplotype •	Accession Numbers
92	TA	Taiwu, Taiwu Township, Pintung County	22°35′11.8″ N 120°38′55.3″ E	395 m	H19 (3), H20 (2), H21 (1)	KM360552-KM360554
93	WT	Wutai, Wutai Township, Pintung County	22°45′22.8″ N 120°45′34.2″ E	438 m	H07 (2)	KM360540
94	NE	Neiwen, Neiwen Township, Pintung County	22°13′24.4″ N 120°51′22.1″ E	321 m	H18 (3)	KM360551
101	TP	Tsaopi, Yuanshan Township, Yilan County	24°45′41.4″ N 121°36′42.6″ E	603 m	H01 (1)	KM360534
102	MI	Mingchih, Tatung Township, Yilan County	24°37′54.6″ N 121°27′11.7″ E	1047 m	H01 (2)	KM360534
103	SM	Shenmihu, Nanao Township, Yilan County	24°22′41.3″ N 121°44′48.8″ E	1100 m	H01 (1)	KM360534
104	TU	Sanfu, Tungshan Township, Yilan County	24°37′03.1″ N 121°45′23.9″ E	140 m	H01 (2)	KM360534
105	KF	Kufeng, Nanao Township, Yilan County	24°20′41.0″ N 121°46′15.7″ E	18 m	H01 (3), H22 (1)	KM360534, KM360555
106	SU	Suao, Nanao Township, Yilan County	24°32′18.6″ N 121°51′55.4″ E	314 m	H01 (3)	KM360534
107-1	SE	Province Highway 7A, Nanao Township, Yilan County	24°26′41.3″ N 121°23′02.5″ E	1088 m	H26 (1)	KM360559
107-2	SE	Province Highway 7A, Nanao Township, Yilan County	24°29′09.7″ N 121°25′30.5″ E	781 m	H01 (1), H26 (1)	KM360534, KM360559
107-3	SE	Province Highway 7A, Nanao Township, Yilan County	24°35′37.7″ N 121°30′32.5″ E	355 m	H01 (1)	KM360534
108	NN2	Nanao II, Nanao Township, Yilan County	24°22′57.3″ N 121°47′02.0″ E	220 m	H01 (2), H22 (1), H23 (1), H24 (1), H25 (1)	KM360534, KM360555-KM360558
109	NN1	Nanao I, Nanao Township, Yilan County	24°24′03.2″ N 121°47′09.7″ E	190 m	H01 (1), H23 (1), H27 (1)	KM360534, KM360556, KM360560
111	FE	Fenglin, Fenglin Township, Hualien County	23°45′29.9″ N 121°25′23.6″ E	249 m	H32 (3), H37 (1), H38 (1), H40 (1), H41 (1), H47 (1), H50 (1), H51 (1), H55 (1)	KM360565, KM360570, KM360571, KM360573, KM360574, KM360580, KM360583, KM360584, KM360588
112	KL	Kuangfulintao, Wanjung Township, Hualien County	23°40′57.0″ N 121°22′58.1″ E	229 m	H39 (1), H45 (1), H53 (1), H54 (1), H56 (1)	KM360572, KM360578, KM360586, KM360587, KM360589
113	TM	Tungmen, Hsiulin Township, Hualien County	23°58′39.4″ N 121°28′22.0″ E	198 m	H28 (1), H29 (1)	KM360561, KM360562
114	NNN	Nanan, Chohsi Township, Hualien County	23°19′35.7″ N 121°14′26.3″ E	445 m	H30 (3), H31 (1), H32 (1)	KM360563-KM360565
115	CY	Chienying, Fenglin Township, Hualien County	23°44′52.6″ N 121°32′53.9″ E	160 m	H43 (1), H44(1)	KM360576, KM360577
116	JS	Juisui, JuiSui Township, Hualien County	23°29′45.0″ N 121°17′43.8″ E	1141 m	H41 (2), H42 (1), H43 (3), H46 (1), H48 (1)	KM360574-KM360576, KM360579, KM360581

Table 2. *Cont.*

No.	Acronym	Locality	GPS Coordinates	Altitude	Haplotype •	Accession Numbers
117	HP	Hsipao, Hsiulin Township, Hualien County	24°12′26.3″ N 121°28′54.6″ E	939 m	H33 (1), H34 (1), H35 (1), H36 (1)	KM360566-KM360569
118	FY	Fuyuan, JuiSui Township, Hualien County	23°32′40.7″ N 121°20′37.1″ E	898 m	H43 (1), H45 (1), H49 (1), H52 (1)	KM360576, KM360578, KM360582, KM360585
121-1	TT	Tachu Main Stream, Tawu Township, Taitung County	22°25′59.2″ N 120°52′45.4″ E	288 m	H18 (3), H67 (1)	KM360551, KM360600
121-2	TTB	Tachu Tributary, Tawu Township, Taitung County	22°26′59.0″ N 120°55′46.1″ E	123 m	H18 (3)	KM360551
122	CI	Chihpen, Peinan Township, Taitung County	22°44′09.2″ N 121°03′00.8″ E	137 m	H68 (1)	KM360601
123	TY	Tsiayunchiao, Haituan Township, Taitung County	23°08′21.1″ N 121°05′59.8″ E	475 m	H69 (1), H70 (1)	KM360602, KM360603
124	HM	Hsiama, Haituan Township, Taitung County	23°09′09.3″ N 121°03′53.7″ E	680 m	H57 (1), H58 (1), H59 (1), H60 (1), H61 (1), H62 (1), H63 (1), H64 (1), H65 (1), H66 (1)	KM360590-KM360599
Japan						
131	OM	Ishigaki, Mt.Omoto	24°25′15″ N 124°11′02″ E	300 m	H72 (3), H73 (1), H74 (1), H75 (1), H76 (1), H77 (1)	H75: KM360604
141	SO	Iriomote, Sonai	24°23′17″ N 123°44′59″ E	50 m	H78 (1), H82 (1), H85 (1)	H82: KM360606
142	OH	Iriomotea, Otomi	24°17′09″ N 123°52′55″ E	80 m	H79 (1), H81 (2), H83 (1), H84 (1)	H79: KM360605, H84: KM360607
143	SR	Iriomotea, Sirahama	24°21′35″ N 123°45′06″ E	60 m	H80 (1), H86 (1), H87 (1), H88 (1), H89 (1)	

• Haplotypes were identified based on the concatenated sequences.

2. Materials and Methods

2.1. Sampling, DNA Extraction, Sequencing, and Alignment

Specimens of *Psolodesmus mandarinus* (186 specimens) were collected from 124 localities in Taiwan (Table 2, Figure 2), and *Psolodesmus kuroiwae* (21 specimens) were collected from four localities on Ishigaki and Iriomote Islands between 1999 and 2008. Most specimens were dried-preserved, and 2–5 legs for each specimen were removed and preserved in 95% ethanol for molecular studies. The voucher specimens were deposited in the Laboratory of Systematic Entomology and Forest Biodiversity, Taiwan Forestry Research Institute, Taipei, Taiwan. Total genomic DNA was extracted from one or two legs of each specimen using DNeasy Blood & Tissue Kits (Qiagen, Hilden, Germany) or the QuickExtract™ DNA Extraction Solution Kit (Epicentre, Madison, WI, USA) following the manufacturer's protocol. DNA samples were stored at −20 °C. We sequenced three mitochondrial loci (COI, tRNA-Leu, COII) with three sets of primers designed in this study (Table 3). Polymerace chain reactions(PCR) were carried out in a total volume of 25 μL, containing 10× reaction buffer, 0.25 mM dNTPs, 2.0 mM MgCl2, 0.4 μM of each primer, 0.2 μL of Super-Therm polymerase (Hoffman-La-Roche, USA), 12.8 μL ddH$_2$O, and 3 μL of DNA template in an GeneAmp PCR System 9700 (Applied Biosystems, Foster City, CA, USA). The PCR profile was as follows: denaturing at 94 °C for 5 min, 35 cycles of amplification at 94 °C for 50 s followed by 50 °C for 50 s and 72 °C for 50 s, and a final extension at 72 °C for 7 min. PCR products were stained with ethidium bromide and visualized under UV light using

1.0% agarose gel after electrophoresis. PCR products were purified using a Gel/PCR DNA Fragments Extraction Kit (Geneaid, Taipei, Taiwan) and sequenced in both directions on an ABI PRISM™ 3730 automatic sequencer (Perkin Elmer, USA) at the Genomics BioSci & Tech, Taiwan. The three overlapping segments of amplified DNA were manually concatenated into a single sequence. Concatenated sequences were aligned without gaps using Clustal W [31]. We checked for pre-matured stop codons in the COI and COII sequences in order to identify possible NUMTs using Mega 6.0 [32]. Sequences generated in this study have been deposited in GenBank (KM360534-KM360607).

Table 3. Primers used in this study.

Set Name	Primer Name	Primer Sequence (5′-3′)	Direction	Length	Amplification Region (Mt Gene)
Pmk-005	Pmk-F001 (Pmk-COI-1684F) Pmk-R005 (Pmk-COI-2346R)	CCCACGACTAAACAACATAAG GGAACAGCAATTACTATTGTGG	forward reverse	663 bp	COI
Pmk-006	Pmk-F006 (Pmk-COI-2178F) Pmk-R006 (Pmk-COI-2917R)	CCCAAGAAAGAGGAAAGAAG GAATCTATGTTCTGTTGGTGG	forward reverse	740 bp	COI
Pmk-007	Pmk-F007 (Pmk-COI2895F) Pmk-R007 (Pmk-COI3708)	CACCACCAACAGAACATAG GTCATCTAGTGAGGCTTCAC	forward reverse	814 bp	COI-tRNA-Leu-COII

2.2. Phylogenetic and Network Analyses

The phylogenetic analyses of samples from *P. mandarinus* and the congener *P. kuroiwae* were performed using Maximum Likelihood (ML) and Bayesian Inference (BI). The best-fitting model was chosen as the model of molecular evolution in both the ML and the BI by the hierarchical likelihood ratio tests in jModeltest 2.1 [33]. ML phylogenetic reconstruction was conducted using Mega 6.0 [32], and ML branch supports were calculated with 1000 bootstrap replicates [34]. BI phylogeny was performed using MrBayes v.3.2 [35]. MCMC runs for 10 million generations were repeated twice, with trees sampled every 100 generations. The first 25,000 trees in each run were discarded as burn-in, and the remaining trees were used to construct Bayesian consensus trees. A statistical parsimony haplotype network was constructed using TCS v. 1.21 [36]. The maximum mutational step was set at 130 for connections between haplotypes. An additional run with a parsimony probability set at 0.95 was performed to test the statistical supports of connections. Gaps were treated as the fifth character state regardless of the length.

2.3. Population Genetic Analyses

Each individual sample was assigned to either the eastern or the western population according to hypothesized phylogeographic breaks (Figure 2). The southern regions of Taiwan may have phylogeographic affinities to either eastern or western Taiwan, which varies across studied organisms. Here, we assigned individuals from southern Taiwan to the western region in that studies using aquatic or riverine organisms often reveal a southwestern genetic clade (e.g., [18,37]). The summary statistic of population size (θ) estimated based on the number of segregating sites per site was calculated for the eastern and western populations separately using the theta.s function implemented in the R package pegas [38]. A neutrality test of the aligned sequences in different populations was performed using the Tajima's D index via the tajima.test function in pegas. Furthermore, genetic differentiation between populations was calculated using the diff_stats function in the mmod package [39]. Specifically, Nei's GST, Jost's D, and ΦST were calculated from our mitochondrial dataset.

2.4. Testing the Effect of Sampling Effort on Inferring Population Structure Phylogenetic Reconstruction

In order to test the effect of sampling effort on phylogeographic studies, we randomly sampled our sequences with different numbers of individuals to represent the eastern and western populations by a customized R script using R (https://www.r-project.org/;

accessed on 1 June 2016). Specifically, we randomly subsampled the eastern and western populations of our DNA sequences (a total of 300 subsampled sequence alignments) 100 hundred times for 10, 20, and 50 individuals. A phylogenetic tree was reconstructed for each alignment using the neighbor-joining method with a TN93 model with a gamma variable for rate correction via the dist.dna and nj functions rooted using the midpoint method implemented in the R package ape [40]. The reciprocal monophyly of eastern and western populations in each reconstructed tree was then assessed using the is.monophyletic function in ape. We then reported how often reciprocal monophyly was observed with different levels of sampling effort. Furthermore, the population size (θ) of each population and genetic differentiation between populations (ΦST) were also calculated for each subsampled alignment. Whether different sampling efforts can result in significantly different values of population genetic parameters was tested using the 100 replicates of each dataset.

3. Results

3.1. Sequence Alignment, Phylogenetic, and Network Analyses

A total of 186 individuals were successfully sequenced; 86 of them from the western population and the remaining 100 from the eastern population. The sequence alignment was 1959 bp in length with 43 and 73 parsimony uninformative and informative sites, respectively. A total of 70 unique haplotypes were found in *P. mandarinus,* and 19 were identified in *P. kuroiwae*. Four haplotypes of *P. kuroiwae* were selected as outgroups for the phylogenetic analyses. The monophyly of *P. mandarinus* was supported in the phylogenetic analyses (Figure 3). The phylogeny reconstructed based on mitochondrial loci (COI-tRNA-Leu-COII) revealed that two distinct haplotype clades (widespread and East) exist in *P. mandarinus* in Taiwan, which were separated by a deep phylogenetic split. The widespread clade included the haplotypes distributed throughout Taiwan. The geographical distributions of two high-frequency haplotypes H01 (38 individuals) and H07 (48 individuals) ranged from North–East (Nanao, Yilan County) to North (Chienshih, Hsinchu County) and from North (Chienshih, Hsinchu County) to South (Shanping, Kaohsiung city) (Figure 4 & Table 2), respectively. The eastern clade that was restricted to eastern Taiwan ranged from southern Yilan County (No. 109, Nanao I) to northern Taitung County (No. 124, Hsiama). The eastern clade contained two subclades, one smaller subclade distributed in Nanao, southern Yilan County (three haplotypes: H22, H23, H25) and one larger subclade distributed in Hualien County to Hsiama, northern Taitung County. The site Hsiama (No. 124 in Haituan Township, Taitung County, Table 1) had 10 haplotypes, which were geographically widely distributed haplotypes that belong to most of the subclades within the two main clades.

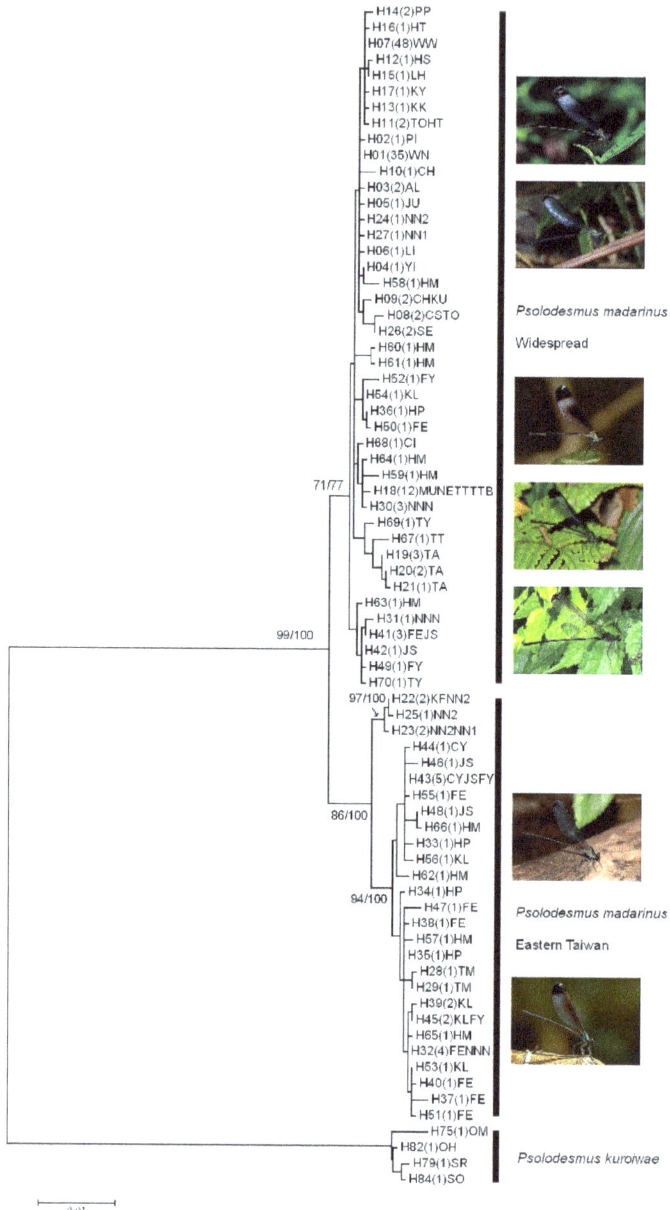

Figure 3. The phylogenetic tree of *Psolodesmus madarinus* haplotypes from the maximum likelihood analysis based on the HKY + G model. Bootstrap values from the maximum likelihood (ML) analyses together with the posterior probabilities from the Bayesian analysis (BI) are indicated (ML/BI) near the branches. Groups of haplotypes are labelled according to whether they are found widespread in Taiwan or endemic to eastern Taiwan.

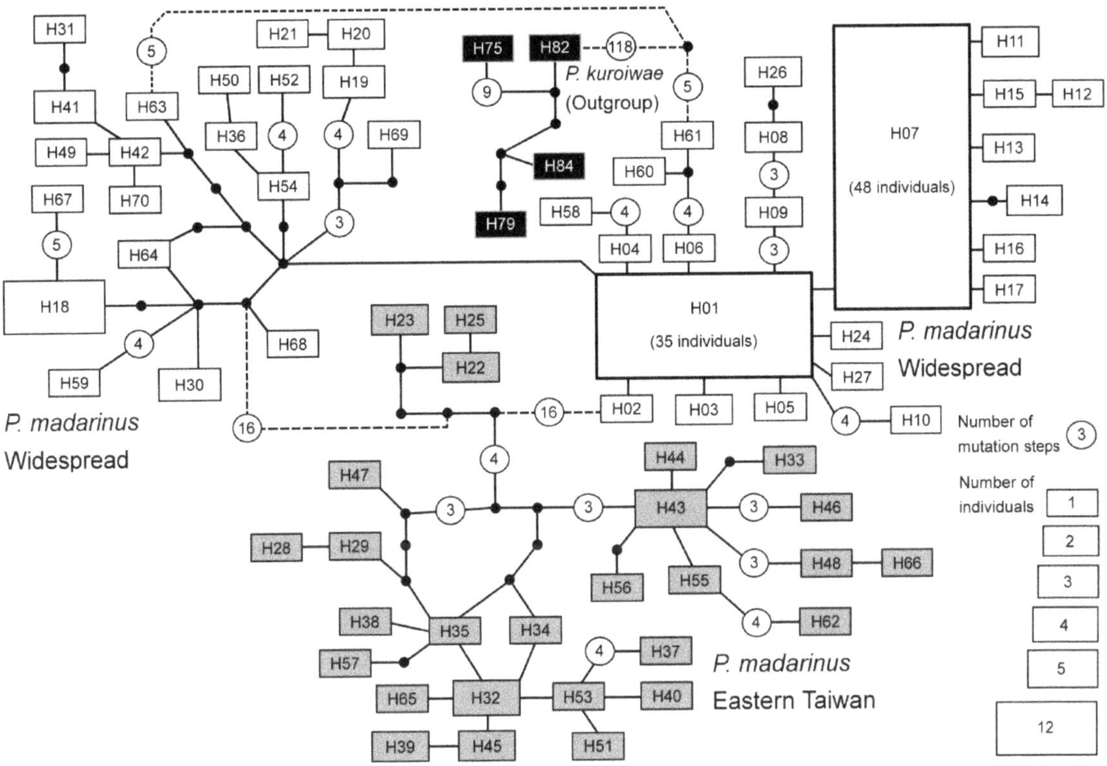

Figure 4. Statistical parsimony network of *P. madarinus* haplotypes. Haplotype groups that are widespread in Taiwan, endemic to eastern Taiwan or belong to *P. kuroiwae* (outgroup) are marked with white, grey and black rectangles, respectively. The sizes of the rectangles for each haplotype are proportional to the number of individuals found carrying each haplotype. Black dots represent hypothetical and unobserved intermediate haplotypes. Solid lines between haplotypes represent one mutational step. For haplotypes connected with more than three mutational steps, open circles with numbers are applied, indicating the number of mutational steps between haplotypes. Dashed lines represent connections between haplotypes with statistical probability < 95%.

3.2. Population Genetic Analyses

The estimated θs for the western and eastern populations were 4.775418 and 19.50795, respectively. The estimated Tajima's Ds were −2.15586 ($p = 0.01$) and −0.4803459 ($p = 0.67$), respectively; the p value assumes that D follows a beta distribution after rescaling on [0,1]. Therefore, the eastern population may have a larger estimated effective population size than the western population because of the higher estimated genetic diversity; on the other hand, only the western population may have experienced a recent population expansion because of a significantly negative Tajima's D value. The calculated Nei's GST, Jost's D, and ΦST from the dataset were 0.02, 0.81, and 0.85, respectively and the significance of population subdivision was estimated as 0.000999 based on 1000 permutations. That is, a statistically significant population structure between the western and the eastern populations was identified.

3.3. *The Effect of Sampling Effort on Phylogenetic Reconstruction and Inferring Population Structure*

Phylogenetic relationships reconstructed using the datasets that randomly selected 10, 20, and 50 individuals from western and eastern populations did not result in reciprocal monophyletic geographic groups (0 out of the 300 datasets). Furthermore, none of the datasets supported a monophyletic eastern lineage. Nevertheless, 8 out of the 100 datasets that randomly drew 10 individuals per population resulted in a monophyletic western lineage, while results from the datasets that randomly chose 20 and 50 individuals per population did not support a monophyletic western lineage (0 out of 200 datasets). Therefore, a monophyletic geographic lineage of *P. mandarinus* can result when a small sample size is used.

Most of the subsampled alignments led to smaller θ values than the original alignment for both western and eastern populations (99, 100, and 94 times for the western population and 99, 99, and 94 times for the eastern population from datasets using only 10, 20, and 50 individuals per population, respectively). On the other hand, the calculated Tajima's D values were very often higher for the subsampled alignments than for the original dataset (100, 100, and 82 times for the western population and 80, 92, and 93 times for the eastern population from datasets using only 10, 20, and 50 individuals per population, respectively). Three out of the 100 subsampled alignments containing 10 individuals per population resulted in significant ($p < 0.05$) negative values of Tajima's D for the western population, while only one showed significant negative values for the eastern population. For the 20 individuals per population dataset, 29 of the 100 datasets indicated significant negative Tajima's D values, while none resulted in significant negative D values for the western and eastern populations, respectively. Finally, 89 of the 100 datasets that subsampled 50 individuals per population resulted in significant negative Tajima's Ds for the western population, while none of them was significant for the eastern population. Hence, small sampling sizes in a phylogeographic study result in a smaller inferred effective population size and biased pattern of demographic expansion in *P. mandarinus* (Figure 5).

The estimated ΦST values between western and eastern populations were very often higher from the subsampled alignments than the original dataset (Figure 5). However, this pattern was more significant in datasets that contained more individuals per population. Specifically, the 10 individuals per population dataset resulted in 76 ΦST values (out of 100) larger than the estimate from the original dataset; the 20 individuals per population dataset resulted in 95 values that were larger than the original; finally, the ΦST values estimated from 50 individual datasets were all larger than that estimated from the original dataset. The datasets with a sample size of 10 individuals had the highest standard deviations (SDs = 0.096, 0.074, and 0.027 for 10, 20, and 50 individuals per population dataset, respectively), showing a wider range of estimate values, while similar mean values of ΦST were found among datasets (0.197, 0.214, and 0.214).

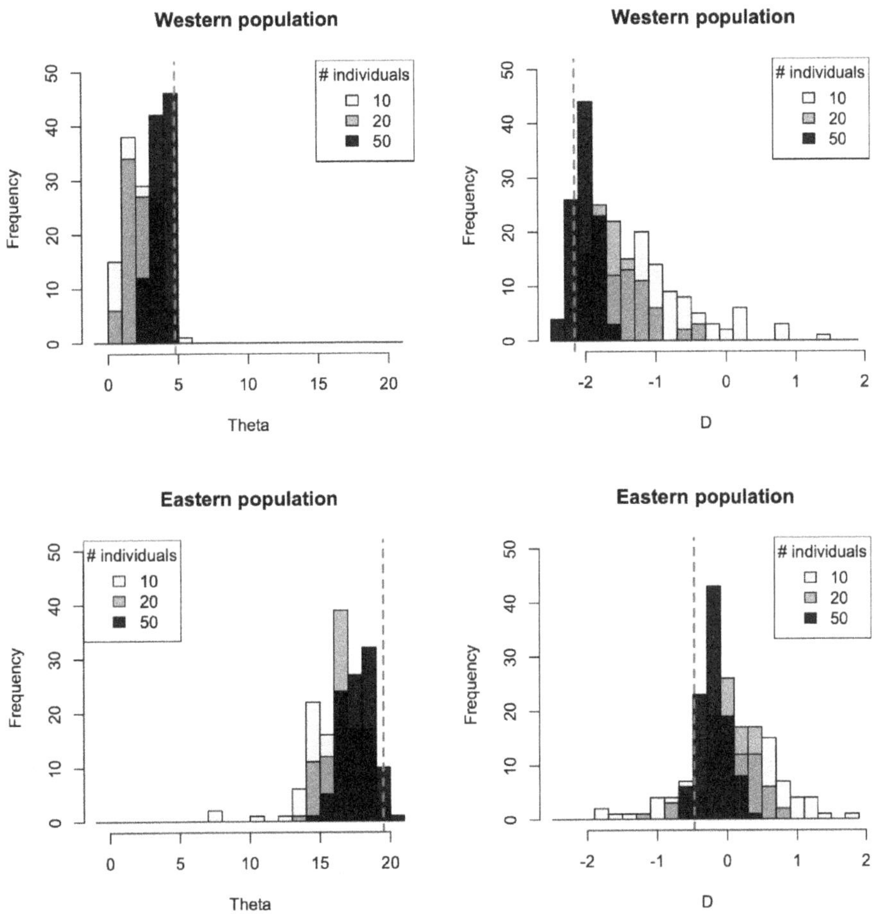

Figure 5. The values of θ and Tajima's D calculated for western and eastern populations of *Psolodesmus mandarinus*. The red dashed lines indicate the values estimated from the original dataset (86 individuals from western and 100 individuals from eastern populations).

4. Discussion

The sampling effect on the resulting evolutionary inferences has long been a point of argument in phylogeographic studies. However, few studies have explicitly tested the effect of different sampling intensities on the possible inferences using empirical data sets. We studied the phylogeographic history of *Psolodesmus mandarinus* using 186 samples from the entire geographic distribution with a mitochondrial locus containing 1959 sites. We found two distinct lineages from the phylogenetic and network analyses (Figures 3 and 4; c.f., [16]), where one of the lineages was geographically widespread while the other was restricted to the eastern part of the CMR. Although the eastern and western populations subdivision by the CMR does not lead to reciprocal monophyly in the *P. mandarinus* system, geographic genetic structure is apparent, indicating that the CMR does have a strong effect on isolating geographic populations from either side of the mountain range. We further showed that although reciprocal monophyly cannot be found from our complete dataset, by subsampling a subset of individuals from eastern and western populations, a monophyletic western lineage could sometimes be found. The estimated population genetic parameters could be biased when a subset of individuals was used in the analyses; for example, a significant demographic expansion inferred for the western population became less apparent when only a subset of samples was included. Our results imply that different sampling efforts (specifically, the number of individuals per population) may in part explain different phylogeographic histories inferred among different empirical studies in Taiwan (Table 1). We discuss the ramifications of our findings on phylogeographic inferences and provide a revised phylogeographic history of *P. mandarinus* in the following sections.

4.1. The Effect of Sampling Effort on the Reconstructed Phylogeographic History

Our results clearly indicate that by subsampling our mitochondrial dataset, not only can the reconstructed phylogenetic inferences (i.e., whether a monophyletic geographic group can be observed) be impacted, but also the estimated population genetic parameters (Figures 5 and 6). Since the number of individuals sampled per geographic population/locality varies across phylogeographic studies (c.f., this study and [16]; see also Table 1), our results imply that different phylogeographic inferences made across different empirical studies may result from different sampling efforts. Many studies have attributed different phylogeographic patterns observed among systems to differences in species-specific ecological and biological properties, while an alternative hypothesis that such differences can result simply because of unequal sampling efforts has rarely been tested [1,2]. We have shown that different sampling efforts can indeed result in very different phylogeographic inferences—specifically, (1) whether a geographic population may appear to be monophyletic, (2) whether a population expansion event can be identified, and (3) whether a significant geographic genetic structure can be detected.

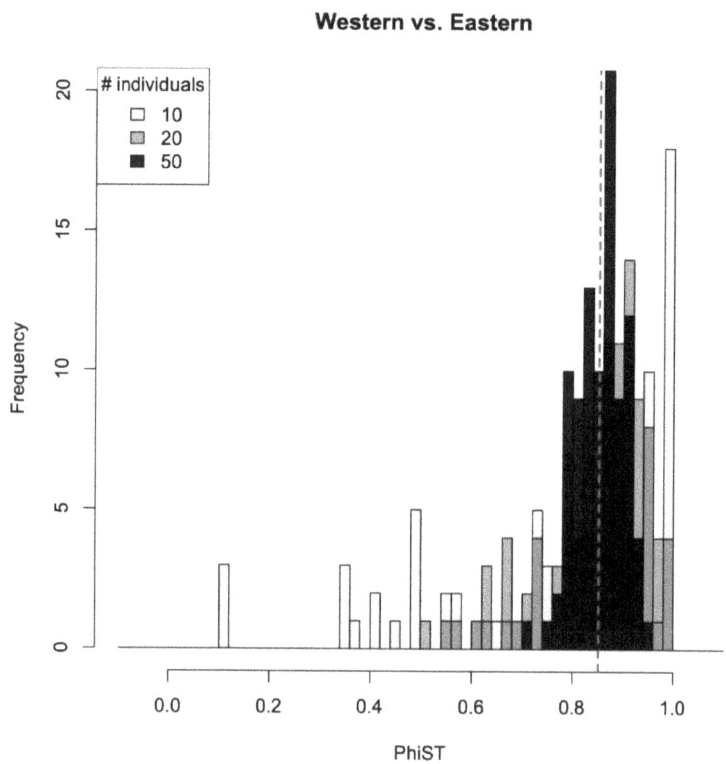

Figure 6. The fixation indices (ΦST) calculated between western and eastern populations. The red dashed line indicates the original estimated value (86 individuals from western and 100 individuals from eastern populations).

By reviewing selected phylogeographic studies in Taiwan (Table 1), we also found that studies that inferred the phylogeographic patterns of (1: reciprocal monophyly) and (3: no genetic divergence between population) often had a smaller sampling size (in terms of the number of individuals or the number of localities) than studies that resulted in pattern (2: apparent genetic divergence between geographic populations) (Figure 7). That is, without considering the biological and ecological differences among the selected empirical phylogeographic systems (Table 1), sampling size difference alone may explain at least in part the different phylogeographic patterns found in different empirical studies. Specifically, our various subsampling designs revealed that a small sampling size can lead to a higher probability of observing monophyletic geographic groups (phylogeographic pattern 1) and a less apparent geographic genetic structure (phylogeographic pattern 3). Note that we are not discrediting the importance of biological and ecological differences among evolutionary lineages that can lead to different phylogeographic patterns. It has been shown that with more than 500 sampled individuals and 33 localities, there was no significant geographic genetic structure in a treefrog species [37]. Our study instead goes against making direct links between the biological and ecological properties of the study system with the inferred phylogeographic pattern because the inferred phylogeographic pattern may be affected by many other stochastic factors [2], and sampling effort could be one such factor.

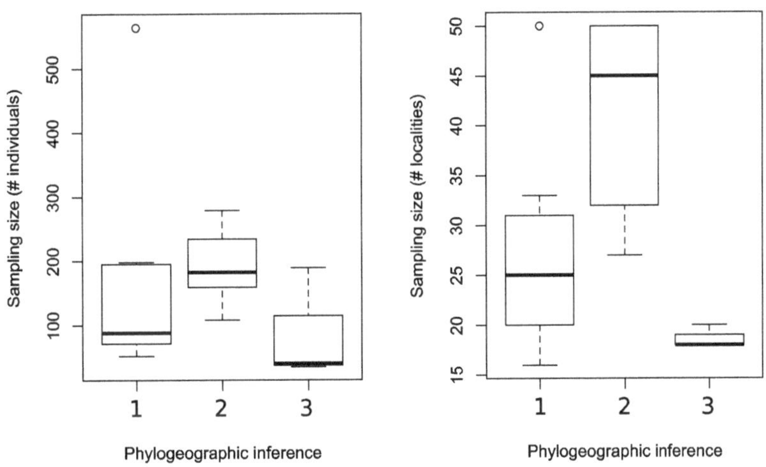

Figure 7. Summary of sampling sizes, i.e., number of individuals and number of localities, from selected empirical phylogeographic studies that resulted in different evolutionary inferences. 1. Studies that showed monophyletic geographic populations. 2. Studies that revealed significant genetic differentiation between geographic populations. 3. Studies that inferred panmictic geographic populations.

4.2. The Phylogeographic History of *Psolodesmus mandarinus*

Our results agree with the inferences made by a previous phylogeographic study on *P. mandarinus*: (1) the Yaeyama taxon forms a distinct evolutionary lineage, (2) there is an eastern Taiwan restricted lineage and a widespread lineage of the Taiwanese samples, and (3) the two Taiwanese lineages do not correspond to the subspecies assignment based on wing morphology [16]. However, by sequencing additional fragments from the mitochondrial genome and sampling more individuals and localities, we unraveled additional genetic diversity represented by the significant increase in the number of haplotypes (Figures 3 and 4 and Table 2). While only seven haplotypes were identified in [16], we have identified a total of 71 haplotypes. We further showed that there is a higher genetic diversity found in eastern Taiwan in addition to the fact that the eastern population harbors individuals from the two main evolutionary lineages. Our results therefore imply that eastern Taiwan is likely the geographic origin of the Taiwanese *P. mandarinus*. The phylogeographic pattern and inference are in direct contrast to what has been hypothesized for another widespread damselfly species, *Euphaea formosa*, in Taiwan [14]. The genus *Psolodesmus* belongs to the family Calopterygidae, which is most abundant in temperate regions; on the other hand, the diversity center of the family Euphaeidae, to which the genus *Euphaea* belongs, is in the tropics [41]. Taiwan is a subtropical island that harbors species of both temperate and tropical origins that may have immigrated into Taiwan via different historical routes [42,43]. *P. mandarinus* and *E. formosa* might have colonized Taiwan via different historical routes because of their differences in ecological preferences and geographic origins.

While the eastern population may have a stable effective population size through recent history, a population size expansion was inferred for the western CMR population of *P. mandarinus*. The inferred contrasting demographic histories also imply that the western population was founded by the eastern population, where the recently founded population went through size expansion event after colonizing previously unoccupied habitats. On the other hand, the low to mid- elevation riverine habitats of western Taiwan cover a larger geographic area than those of eastern Taiwan (Figure 1). Furthermore, the river systems of western Taiwan might have been connected to form a much larger river system during periods of lowered sea level [14]. An increase in the population size of the western population may thus also be impacted by geo-historical events. While we could not effectively test which factor may have played a major role in the demographic history of *P. mandarinus*, it is likely that both evolutionary history and geo-historical events were involved in shaping the population genetic diversity and divergence as shown in other Taiwanese taxa (e.g., [8–10,14]).

5. Conclusions

Phylogeographic studies depend on the sampling design, e.g., the number of geographic populations and the number of individuals per population, to understand the spatial variation of intraspecific genetic diversity and to make inferences regarding the origin and maintenance of biodiversity. We showed that different sampling designs can impact the pattern revealed by the genetic data and thus lead to different inferences regarding the effect of a geographic barrier. We further demonstrated that different phylogeographic patterns observed among biological systems in Taiwan, although often being attributed to their biological differences, may simply be the result of different geographic sampling strategies. We argue that future phylogeographic studies require not only the careful design of spatial sampling strategies but also the testing of sampling effects on the resulting inferences as we have shown in our study.

Author Contributions: Conceptualization, L.-J.W. and J.-P.H.; methodology, L.-J.W., J.-P.H. and Y.-W.C.; data analysis, L.-J.W., J.-P.H. and Y.-W.C.; investigation, L.-J.W.; resources, L.-J.W.; data curation, L.-J.W., Y.-W.C. and J.-P.H.; writing—original draft preparation, L.-J.W. and J.-P.H.; writing—review and editing, L.-J.W., J.-P.H. and Y.-W.C. All authors have read and agreed to the published version of the manuscript.

Funding: This study was partially supported by the granted project No. 103AS-13.3.4.-FI-G2 from the Taiwan Forestry Research Institute.

Institutional Review Board Statement: Not applicable.

Informed Consent Statement: Not applicable.

Data Availability Statement: The sequence data presented in this study are openly available in GenBank of NCBI (National Center for Biotechnology Information) at [https://www.ncbi.nlm.nih.gov (accessed on 1 June 2016)]. See Table 2 for accession numbers.

Acknowledgments: We are grateful to Shih-Chieh Huang, Jyh-Jong Cherng, Hsien-Chih Wu, Ye-Zhi Ai, Chen-Hsiang Chen and Yung-Jen Chang for their help in the fieldwork. Two reviewers provided helpful suggestions that improved the clarity of the manuscript.

Conflicts of Interest: The authors declare no conflict of interest.

References

1. Templeton, A.R. Statistical phylogeography: Methods of evaluating and minimizing inference errors. *Mol. Ecol.* **2004**, *13*, 789–809. [CrossRef] [PubMed]
2. Knowles, L.L. Statistical phylogeography. *Annu. Rev. Ecol. Evol. Syst.* **2009**, *40*, 593–612. [CrossRef]
3. Gutiérrez-García, T.; Vázquez-Domínguez, E. Comparative phylogeography: Designing studies while surviving the process. *BioScience* **2011**, *61*, 857–868. [CrossRef]
4. Meirmans, P.G. Seven common mistakes in population genetics and how to avoid them. *Mol. Ecol.* **2015**, *24*, 3223–3231. [CrossRef]
5. Nielsen, R.; Beaumont, M.A. Statistical inferences in phylogeography. *Mol. Ecol.* **2009**, *18*, 1034–1047. [CrossRef]
6. Heath, T.A.; Hedtke, S.M.; Hillis, D.M. Taxon sampling and the accuracy of phylogenetic analyses. *J. Syst. Evol.* **2008**, *46*, 239–257.
7. Malaney, J.L.; Cook, J.A. Using biogeographical history to inform conservation: The case of Preble's meadow jumping mouse. *Mol. Ecol.* **2013**, *22*, 6000–6017. [CrossRef]
8. Yang, Y.J.; Lin, Y.S.; Wu, J.L.; Hui, C.F. Variation in mitochondrial DNA and population structure of the Taipei treefrog *Rhacophorus taipeianus* in Taiwan. *Mol. Ecol.* **1994**, *3*, 219–228. [CrossRef]
9. Huang, J.P.; Lin, C.P. Diversification in subtropical mountains: Phylogeography, Pleistocene demographic expansion, and evolution of polyphonic mandibles in Taiwanese stag beetle, *Lucanus formosanus*. *Mol. Phylogenet. Evol.* **2010**, *57*, 1149–1161. [CrossRef]
10. Wang, Y.H.; Hsiao, Y.W.; Lee, K.H.; Tseng, H.Y.; Lin, Y.P.; Komaki, S.; Lin, S.M. Acoustic differentiation and behavioral response reveals cryptic species within *Buergeria* treefrogs (Anura, Rhacophoridae) from Taiwan. *PLoS ONE* **2017**, *12*, e0184005. [CrossRef]
11. Kuo, H.C.; Chen, S.F.; Fang, Y.P.; Flanders, J.; Rossiter, S.J. Comparative rangewide phylogeography of four endemic Taiwanese bat species. *Mol. Ecol.* **2014**, *23*, 3566–3586. [CrossRef] [PubMed]
12. Creer, S.; Malhotra, A.; Thorpe, R.A.; Chou, W.H. Multiple causation of phylogeographical pattern as revealed by nested clade analysis of the bamboo viper (*Trimeresurus stejnegeri*) within Taiwan. *Mol. Ecol.* **2001**, *10*, 1967–1981. [CrossRef] [PubMed]
13. Creer, S.; Thorpe, R.A.; Malhotra, A.; Chou, W.H.; Stenson, A.G. The utility of AFLP for supporting mitochondrial DNA phylogeographical analyses in the Taiwanese bamboo viper, *Trimeresurus stejnegeri*. *J. Evol. Biol.* **2004**, *17*, 100–107. [CrossRef] [PubMed]
14. Huang, J.P.; Lin, C.P. Lineage-specific late Pleistocene expansion of an endemic subtropical gossamer-wing damselfly, *Euphaea formosa*, in Taiwan. *BMC Evol. Biol.* **2011**, *11*, 94. [CrossRef] [PubMed]
15. Yu, T.L.; Lin, H.D.; Weng, C.F. A New Phylogeographic Pattern of Endemic *Bufo bankorensis* in Taiwan Island Is Attributed to the Genetic Variation of Populations. *PLoS ONE* **2014**, *9*, e98029. [CrossRef]
16. Lin, S.C.; Chen, Y.F.; Shieh, S.H.; Yang, P.S. A revision of the status of *Psolodesmus mandarinus* based on molecular and morphological evidence (Odonata: Calopterygidae). *Odonatologica* **2014**, *43*, 51066.
17. Oshida, T.; Lin, L.K.; Chang, S.W.; Chen, Y.J.; Lin, J.K. Phylogeography of two sympatric giant flying squirrel subspecies, *Petaurista alborufus lena* and *P. philippensis grandis* (Rodentia: Sciuridae), in Taiwan. *Biol. J. Linn. Soc.* **2011**, *102*, 404–419. [CrossRef]
18. Shih, H.T.; Hung, H.C.; Schubart, C.D.; Chen, C.A.; Chang, H.W. Intraspecific genetic diversity of the endemic freshwater crab *Candidopotamon rathbunae* (Decapoda, Brachyura, Potamidae) reflects five million years of the geological history of Taiwan. *J. Biogeogr.* **2006**, *33*, 980–989. [CrossRef]
19. Liu, M.Y.; Tzeng, C.S.; Lin, H.D. Phylogeography and the genetic structure of the land-locked freshwater prawn *Macrobrachium asperulum* (Crustacea: Decapoda: Palaemonidae) in Taiwan. *Hydrobiologia* **2011**, *671*, 1–12. [CrossRef]
20. Jang-Liaw, N.H.; Chou, W.H. Phylogeography of the fanged dicroglossine frog, *Limnonectes fujianensis* (Anura, Ranidae), in Taiwan. *Zoolog. Sci.* **2011**, *28*, 254–263. [CrossRef]
21. Lee, J.W.; Jiang, L.; Su, Y.C.; Tso, I.M. Is central mountain range a geographic barrier to the giant wood spider Nephila pilipes (Araneae: Tetragnathidae) in Taiwan? A population genetic approach. *Zool. Stud.* **2004**, *43*, 112–122.
22. Oshida, T.; Lee, J.K.; Lin, L.K.; Chen, Y.J. Phylogeography of Pallas's squirrel in Taiwan: Geographical isolation in an arboreal small mammal. *J. Mammal.* **2006**, *87*, 247–254. [CrossRef]
23. Wang, T.Y.; Liao, T.Y.; Tzeng, C.S. Phylogeography of the Taiwanese Endemic Hillstream Loaches, *Hemimyzon formosanus* and *H. taitungensis* (Cypriniformes: Balitoridae). *Zool. Stud.* **2007**, *46*, 547–560.
24. Ožana, S.; Dolný, A.; Pánek, T. Nuclear copies of mitochondrial DNA as a potential problem for phylogenetic and population genetic studies of Odonata. *Syst. Enotmol.* **2022**, *47*, 591–602. [CrossRef]
25. Ballard, J.M.O.; Whitloch, M.C. The incomplete natural history of mitochondria. *Mol. Ecol.* **2004**, *13*, 729–744. [CrossRef]
26. Zink, R.M.; Barrowclough, G.F. Mitochondrial DNA under siege in avian phylogeography. *Mol. Ecol.* **2008**, *17*, 2107–2121. [CrossRef]
27. Fujisawa, T.; Barraclough, T.G. Delimiting species using single-locus data and the general mixed Yule coalescent approach: A revised method and evaluation on simulated data sets. *Syst. Biol.* **2013**, *62*, 707–724. [CrossRef]
28. Lieftinck, M.A.; Lien, J.C.; Maa, T.C. *Catalogue of Taiwanese Dragonflies (Insecta: Odonata)*; Asian Ecological Society: Taichung, China, 1984.
29. Wang, L.J. *Dragonflies of Taiwan*; JemJen Publishing: New Taipei, China, 2000.
30. Karjalainen, S.; Hämäläinen, M. *Demoiselle Damselflies. Winged Jewels of Silvery Streams*; Caloptera Publishing: Helsinki, Finland, 2013.

31. Thompson, J.D.; Higgins, D.G.; Gibson, T.J. CLUSTAL W: Improving the sensitivity of progressive multiple sequence alignment through sequence weighting, position specific gap penalties and weight matrix choice. *Nucleic Acids Res.* **1994**, *22*, 4673–4680. [CrossRef]
32. Tamura, K.G.; Stecher, G.; Peterson, D.; Filipski, A.; Kumar, S. MEGA6: Molecular Evolutionary Genetics Analysis Version 6.0. *Mol. Biol. Evol.* **2013**, *30*, 2725–2729. [CrossRef]
33. Darriba, D.; Taboada, G.L.; Doallo, R.; Posada, D. jModelTest 2: More models, new heuristics and parallel computing. *Nat. Methods* **2012**, *9*, 772. [CrossRef]
34. Felsenstein, J. Confidence limits on phylogenies: An approach using the bootstrap. *Evolution* **1985**, *39*, 783–791. [CrossRef] [PubMed]
35. Ronquist, F.; Huelsenbeck, J.P. MrBayes 3: Bayesian phylogenetic inference under mixed models. *Bioinformatics* **2003**, *19*, 1572–1574. [CrossRef] [PubMed]
36. Clement, M.; Posada, D.; Crandall, K.A. TCS: A computer program to estimate gene genealogies. *Mol. Ecol.* **2000**, *9*, 1657–1660. [CrossRef] [PubMed]
37. Lin, H.D.; Chen, Y.R.; Lin, S.M. Strict consistency between genetic and topographic landscapes of the brown tree frog (*Buergeria robusta*) in Taiwan. *Mol. Phylogenet. Evol.* **2012**, *62*, 251–262. [CrossRef]
38. Paradis, E. Pegas: An R package for population genetics with an integrated-modular approach. *Bioinformatics* **2010**, *26*, 419–420. [CrossRef]
39. Winter, D.J. MMOD: An R library for the calculation of population differentiation statistics. *Mol. Ecol. Resour.* **2012**, *12*, 1158–1160. [CrossRef]
40. Paradis, E. APE: Analyses of phylogenetics and evolution in R language. *Bioinformatics* **2004**, *20*, 289–290. [CrossRef]
41. Dijkstra, K.D.B.; Kalkman, V.J.; Dow, R.A.; Stokvis, F.R.; Van, T.J. Redefining the damselfly families: A comprehensive molecular phylogeny of Zygoptera (Odonata). *Syst. Entomol.* **2014**, *39*, 68–96. [CrossRef]
42. Liao, C.C.; Chen, C.H. Investigation of floristic similarities between Taiwan and terrestrial ecoregions in Asia using GBIF data. *Bot. Stud.* **2017**, *58*, 15. [CrossRef]
43. He, J.; Gao, Z.; Su, Y.; Lin, S.; Jiang, H. Geographical and temporal origins of terrestrial vertebrates endemic to Taiwan. *J. Biogeogr.* **2018**, *45*, 2458–2470. [CrossRef]

diversity

MDPI

Article

The Quality of Sequence Data Affects Biodiversity and Conservation Perspectives in the Neotropical Damselfly *Megaloprepus caerulatus*

Wiebke Feindt [1,*] and Heike Hadrys [1,2]

1 Division of Ecology and Evolution, University of Veterinary Medicine Hannover, Bünteweg 17d, 30559 Hannover, Germany
2 Division of Invertebrate Zoology, American Museum of Natural History, Central Park West at 79th Street, New York, NY 10024, USA
* Correspondence: wiebke.feindt@tiho-hannover.de or wiebke.feindt@ecolevol.de; Tel.: +49-511-953-8882

Abstract: Ideally, the footprint of the evolutionary history of a species is drawn from integrative studies including quantitative and qualitative taxonomy, biogeography, ecology, and molecular genetics. In today's research, species delimitations and identification of conservation units is often accompanied by a set of—at minimum—two sequence markers appropriate for the systematic level under investigation. Two such studies re-evaluated the species status in the world's largest Odonata, the Neotropical damselfly *Megaloprepus caerulatus*. The species status of the genus *Megaloprepus* has long been debated. Despite applying a highly similar set of sequence markers, the two studies reached different conclusions concerning species status and population genetic relationships. In this study, we took the unique opportunity to compare the two datasets and analyzed the reasons for those incongruences. The two DNA sequence markers used (16S rDNA and CO1) were re-aligned using a strict conservative approach and the analyses used in both studies were repeated. Going step by step back to the first line of data handling, we show that a high number of unresolved characters in the sequence alignments as well as internal gaps are responsible for the different outcomes in terms of species delimitations and population genetic relationships. Overall, this study shows that high quality raw sequence data are an indispensable requirement, not only in odonate research.

Keywords: molecular data handling; sequence alignments; species delimitation; conservation units

Citation: Feindt, W.; Hadrys, H. The Quality of Sequence Data Affects Biodiversity and Conservation Perspectives in the Neotropical Damselfly *Megaloprepus caerulatus*. *Diversity* **2022**, *14*, 1056. https://doi.org/10.3390/d14121056

Academic Editors: M. Olalla Lorenzo-Carballa and Ricardo Koroiva

Received: 28 September 2022
Accepted: 22 November 2022
Published: 1 December 2022

1. Introduction

Molecular genetic studies using mitochondrial and nuclear DNA markers strongly rely on high quality sequence data to reconstruct evolutionary patterns such as phylogenetic relationships, population structures, modes of adaptation, and species diversity in taxonomic groups. Specifically, when evaluating a hypothesis of species delimitation, the use of a set of DNA sequence markers has become standard. The advantages of Sanger sequencing include direct comparisons of genes for estimating genetic diversities and reconstructing phylogenies [1–5].

The repeatability of results under the most conserved scheme is a *conditio sine qua non* in science. In molecular genetics, one should expect, when using the same set of DNA markers and DNA samples coming from similar tissues/individuals, similar or the same results as an outcome. However, the quality of raw sequence data, interpretation of ambiguous data and alignment problems often affect the outcomes of reconstructions [6–8]. A recent example is two studies on the neotropical damselfly genus *Megaloprepus* (Zygoptera: Odonata), which apply the same set of DNA markers but reach contradictory results. In this genus, species status has long been questioned, not only because of differences in morphological patterns and a broad distribution across tropical Latin America, despite a highly conserved ecological niche, but also because of molecular genetic data [9]. In the

19th century, the first three species were described in this genus: *M. caerulatus*, *M. latipennis* and *M. brevistigma* [10]. Later, the species status of *M. latipennis* and *M. brevistigma* was refused [11] but the monotypic status of the genus was still under debate [9,12–14].

In a first attempt to evaluate the species status of *Megaloprepus* using genetic data, Feindt et al. [12] analyzed two mitochondrial sequence markers (NADH-dehydrogenase subunit 1 (*nad1*) and 16S ribosomal DNA (16S rDNA)) from more than 100 tissue samples of four populations in Mesoamerica: Los Tuxtlas Biosphere Reserve (Mexico), Corcovado National Park and Biological Research Station La Selva (Costa Rica), and Barro Colorado Island (Panama). The results showed a strong genetic isolation of populations with genetic distances comparable to species level differences in other odonate species (6.8% and 7.5% in ND1 and between 4% and 4.9% in 16S rDNA) [12]. The authors conclude that the genus *Megaloprepus* consists of at least three distinct species [12].

The second genetic dataset in a follow-up study by Fincke and colleagues [15] included tissue samples from 56 to 68 specimens of eight populations throughout Mesoamerica: Los Tuxtlas Biosphere Reserve (México), Cusoco National Park (Honduras), El Jaguar reserve and Bartola Reserve (Nicaragua), Corcovado National Park and Biological Research Station La Selva (Costa Rica), Barro Colorado Island (Panama), and Rio Canande Reserve (Ecuador). Here, the resulting phylogeny showed three clades similar to the previous study [12], except for the samples from Barro Colorado Island, which appeared as three paraphyletic clades [15]. However, in contrast to the phylogenetic clades, the genealogical haplotype networks revealed a high number of shared haplotypes between populations, despite high geographic distances. In addition, Fincke et al. [15] found a higher variability in the slower evolving 16S rDNA gene compared to the Cytochrome c oxidase subunit 1 (CO1) gene, a phenomenon not observed in odonates or other groups before, [5,16–21]. Furthermore, Finke et al. combined their 16S rDNA data with the Feindt et al. 16S rDNA data for their phylogenetic reconstruction. Finke et al. [15] described the population from the Corcovado National Park (Costa Rica), collected by them, as genetically distinct from the population at the same locality collected during a similar time period by Feindt et al. [12]—a result difficult to explain in biological terms. In short, the two studies draw remarkably different conclusions about the delimitation of species within the genus *Megaloprepus*, which in turn leads to a significant different perspective for conservation and biodiversity (3 versus 1 conservation unit).

To identify potential causes of these different results, we took the unique opportunity in this study to look deeper into the data handling and analyses of the two datasets. We focused on two main questions: (i) does adding (or not removing) unresolved characters and gaps from sequence alignments increase haplotype diversity, and "dilute" topologies in phylogenetic tree and genealogical network reconstructions? (ii) Does the reported higher variability in 16S rDNA hold up to the reanalysis? Consequently, the causal mechanisms that may have led to the above results and their different interpretation of phylogenetic trees and haplotype networks are explored—not only for those two datasets—but also as an example for a general awareness concerning a sensible use of genetic data.

2. Materials and Methods

2.1. Data Mining

The comparison of genetic data handling is based on the DNA marker genes 16S rDNA and CO1. For this purpose, we included the dataset from Fincke et al. [15]—hereafter Study-A, and the data of Feindt et al. [12] plus additionally sequenced CO1—hereafter Study-B (see Table S1 for species identifications and origin, accession numbers, and additional information).

For Study-A, the 16S rDNA dataset was provided on request as an alignment directly from Fincke et al., since sequences were not published on GenBank (NCBI, [22]). This dataset appeared as an alignment fraction prepared for a tree analysis, which therefore may include many question marks at the same positions inside the alignment. Usually, those questions marks in a bigger alignment stand for missing information sites, especially when several distinct species and genes are combined and/or if sequences of different lengths are used. The CO1 sequences were downloaded from GenBank [22]. Here it is important to mention that these CO1 sequences are labeled as unverified with the comment: "similar to cytochrome oxidase subunit I". Such labels signify that the open reading frame is interrupted, potentially indicating sequencing errors. All sequences were verified as *Megaloprepus* via BLAST searches [23]. According to Fincke et al., primers are mentioned in [24]. Unfortunately, this reference only includes two primer sets for 16S rDNA and none for CO1. This fact makes objective comparisons and repetitions impossible.

For Study-B, 16S rDNA sequences were downloaded from GenBank [22]. Consequently, Study-A and Study-B, as well as the present study, use the same 16S dataset containing 106 individual sequences. In addition, because CO1 was not originally included in Study-B, CO1 marker genes were newly Sanger sequenced, following the steps described by Bergmann et al. [25], except that sequencing was performed at the Yale University (DNA Analysis Facility on Science Hill), and uploaded to GenBank [22]. In this study, each specimen that is included in the 16S rDNA Study-B dataset was also sequenced to obtain the CO1 dataset, and consequently, 106 CO1 sequences are analyzed here (see Table S1 for all accession numbers). For the 16S rDNA, gene fragment standard primers (P784 and P785) first described in Simon et al. [26] were used, and for CO1, the barcoding region using the primer set LCO 1490 and HCO 2198 [27] was sequenced.

2.2. Sequence Editing and Alignments

To identify the source of incongruence between the two study results concerning population genetic and phylogenetic relationships, three different alignments for each sequence marker were generated (total six alignments):

(i) Alignments 1 include the original Study-A sequences (for CO1 and 16S rDNA separately),
(ii) Alignments 2 consist of the edited Alignments 1 (again, separately for CO1 and 16S rDNA), and
(iii) Alignments 3 include the combined sequence data from both studies (Alignments 2 from Study-A plus Study-B alignments for either CO1 or 16S rDNA).

All alignments were performed using MUSCLE [28] under standard conditions (gap open -400, gap extend 0) in the most conservative way.

For the Alignments 1, the sequence data from Study-A were used without any editing except for length cuts (16S rDNA Alignment 1 and CO1 Alignment 1, see Supplementary Material: Files S1 and S2). Because the original sequences were of different length, our cutting was intended to obtain the same length for all sequences and the longest possible overlap. We cut sequences in conserved regions within the alignment.

For generating the Alignments 2, the original sequence data from Alignment 1 was manually edited. The reasons, therefore, are the inaccuracies observed within individual sequences such as large gaps and ambiguous bases. For the 16S rDNA Alignment 2, this editing included a verification of gaps and/or missing data. First, all question marks were removed, and the dataset was realigned. This alignment partly removed internal gaps. However, in the next step we deleted individual sequences, which still had internal gaps, probably due to incomplete sequencing or non-readable chromatograms, and repeated the alignment. Finally, sequences were cut to the same length. In addition, we intended to align the 16S rDNA dataset using the secondary structure as reference, which, however, did not result in a better alignment.

For the CO1 Alignment 2, the approach was slightly different. Because of the high content of N bases throughout all individual sequences, the data handling and editing was as follows: We first deleted all sequences with N bases on different positions in the alignment. In addition, sequences containing equivocal bases such as R, S or M were strictly removed as those carry the risk of identifying false haplotypes. Because mitochondria are maternally inherited, heterozygous positions are not expected to lead to all mitochondrial genes of one individual being identical. In a next step, a verification of the N bases, which were at the same positions between sequences, was needed to assure a correct alignment. Therefore, the published mitochondrial genome of *M. caerulatus* (GenBank: KU958377 [29]) was used as a seed sequence. Reverse complement on the Study-A sequences was performed to ensure the correct reading frame and a strict removal of those N's allowed a solid alignment. Finally, a length cut was performed (File S4: CO1 Alignment 2).

In a second attempt, we intended to align Study-A raw-sequences by 'simply' removing N's and gaps and translating the dataset into amino acid sequences. This attempt left us with no alignment because the translation to amino acid sequences was interrupted. However, the final Alignments 2 and 3 translate correctly into amino acid sequences.

For Alignments 3, the edited Study-A 16S rDNA and CO1 Alignments 2 (Files S3 and S4) were combined with the corresponding sequence data of Study-B. Datasets were aligned and cut to the same length (Files S5 and S6: 16S rDNA and CO1 Alignment 3).

For all alignments, we report parsimony-informative sites performed in PAUP* vers. 4.0b8 [30] and the number of singleton sites determined in MEGA 11 [31], as the latter may indicate sequencing errors. MEGA defines a site as a singleton site "if at least three sequences contain unambiguous nucleotides or amino acids" [31].

2.3. Population Structure, Genealogical Network, and Phylogenetic Analyses

Genealogical relationships between individuals and populations were reconstructed for each of the 6 alignments, repeating the original procedures from studies A and B. Briefly, genetic diversity was estimated using DnaSP [32] and sequence divergence between and within groups was computed in MEGA 7 [33] using the Kimura 2-parameter (K2P) model [34]. Minimum spanning haplotype networks were calculated in POPART v. 1.7 [35] to visually display relationships and haplotype distribution. In addition, haplotype networks analyses were performed using the statistical parsimony software TCS by applying a parsimony connection limit of 95% for the haplotype distribution (TCS vers. 1.13, [36]). Finally, the genetic hierarchical population structure was studied via F_{ST}-values [37] and an analysis of molecular variance (AMOVA, [38]) using 10,000 permutations for statistical significance was performed in Arlequin 3.5 [39]. In the AMOVA, we decided to test for two hierarchic levels: (i) no grouping among populations and (ii) three groups. These assumptions are based on the results from the Study-A and on the hypothesis of three distinct groups, as suggested by Study-B.

Since the CO1 Alignment 1 included ambiguities, some analyses could not be performed because programs rejected the dataset. The following analyses are therefore not presented for the CO1 Alignment 1: the summary statistics via DnaSP [32], the AMOVA using Arlequin 3.5 [39], and the TCS analysis (TCS vers. 1.13 [36]).

Finally, phylogenies were reconstructed for each sequence marker separately, using the Alignments 3. In each tree search, three closely related sister species to *Megaloprepus*; *Coryphagrion grandis*, *Mecistogaster linearis* and *Mecistogaster lucretia* [40,41] were used as outgroups (Table S1). Phylogenetic tree reconstruction was performed via maximum likelihood inference using IQ-TREE [42] while allowing ModelFinder [43] to estimate the best substitution model. Branch confidence was estimated with 1000 bootstrap replicates [44]. Furthermore, we carried out phylogenetic reconstruction using Maximum Parsimony (MP) and Bayesian inference (BI) following the steps described in Study-B [12]. Briefly, MP was performed in PAUP* vers. 4.0b8 [30] using a full heuristic search (50% majority-rule, 1000 bootstrap replicates and reconnection branch swapping option (TBR)). MrBayes vers. 3.7 [45] was used to perform Bayesian analysis, where the most likely model of nucleotide substitution was tested separately for each locus in ModelTest vers. 3.7 [46] using the Akaike Information Criterion (AICc model selection) [47]. Finally, we used the following settings: two independent runs were performed under the best fit-model (JC for 16S rDNA and HKY+G for CO1) for 20×10^6 generations and each four Markov chains; trees were sampled every 1000 generations—but the first 20,000 trees were discarded as 'burn-in'. Posterior probabilities and consensus topology is based on the remaining trees.

3. Results

3.1. Alignments

Study-A datasets originally included 56 individual sequences for 16S rDNA and 68 sequences for CO1 from eight populations from Mexico to Ecuador [15], whereas Study-B included 106 individual sequences from four populations (Table S1). The following four populations were also sampled in both studies in a similar time period: the Biosphere Reserve Los Tuxtlas in Mexico, the Biological Research Station La Selva in Costa Rica, Corcovado National Park, also in Costa Rica, and Barro Colorado Island in Panama, and should by theory give very similar genetic results.

3.1.1. 16S

The 16S rDNA Alignment 1 contained variable nucleotide positions that were partly masked—as variable nucleotide positions in one population were equipped with question marks in the other population and vice versa. In addition, since we were interested in verifying population structures and species splits, we had to delete six individual sequences because of unclear origin. These individual sequences were not mentioned in the Study-A species list (RU_621, RU_828, or MC_530 (Table S1)). One sequence was deleted because it was downloaded from GenBank [22]. The 16S rDNA Alignment 1 therefore included only 49 sequences from seven populations with a total length of 577 bp. In total, 70 parsimony-informative characters out of 101 variable characters and 30 singleton sites were detected (see Supplementary File S1).

For the 16S rDNA Alignment 2, during editing, ten additional sequences had to be deleted from the Alignment 1 due to large internal gaps (such as RU631_HN, RU632_BCI, RU803_LT). Consequently, the final 16S rDNA Alignment 2 comprised 39 sequences. The alignment length was reduced to 325 bp because sequences from Nicaragua were significantly shorter than all others. An analysis without these sequences would have allowed an alignment length of 456 bp (see Supplementary Files S1 and S3) but would not have allowed for a comparison of genetic diversity among those populations. The resulting alignment contains 27 variable positions, whereof 26 are parsimony-informative sites, and there is only one singleton site.

Finally, the 16S rDNA Alignment 3—a combined alignment of both studies, contained 150 sequences from seven populations, with a total alignment length of 321 bp and 28 variable characters (26 parsimony-informative sites) (File S5). Furthermore, two singleton sites were found. See Table 1 for the final number of sequences per population in the Alignments 3. Screenshots from the 16S rDNA Alignment 1 and Alignment 3 show the alignment status (Supplementary Material Figures S1 and S2, File S7).

Table 1. Final number of sequences included in the Alignments 3 per study (A and B), sampling location, and genetic marker.

Sampling Location	CO1		16S rDNA	
	N Study-A	N Study-B	N Study-A	N Study-B
Biosphere Reserve Los Tuxtlas, Mexico (RBLT)	1	18	4	18
Cusuco National Park, Honduras (HN)	/	/	/	/
Nicaragua * (NI)	11	/	13	/
Nicaragua * (NI-b)	1	/	2	/
Biological Research Station La Selva, Costa Rica (LS)	6	30	15	30
Corcovado National Park 'Field Station Sirena', Costa Rica (CNP)	6	29	/	29
Barro Colorado Island, Panama (BCI)	3	29	4	29
Rio Canande Reserve, Ecuador (CAN)	1	/	1	/
N total	29	106	39	106

* In Nicaragua there are two sampling points: Natural Reserve El Jaguar and Reserve Bartola (belonging to the Biological Reserve Indio Maíz). However, the information about which sequences originate from which sampling locality is missing in Study-A. Here, the populations were labeled NI and NI-b. According to our in-house barcodes, NI-b belongs to Reserve Bartola—Biological Reserve Indio Maíz.

3.1.2. CO1

For the CO1 Alignment 1—contrary to the study design, in which we were only aiming to perform length cuts—seven sequences were removed to be able to reach a total alignment length of 419 bp. This removal was because of general sequence length where some of those sequences were as short as 117 bp, as well as the general overlapping area with sequences not starting with the same codon. Therefore, the final alignment included 61 individual sequences (File S2). The alignment contained 58 parsimony-informative characters while 26 variable positions are not parsimony-informative and 26 singleton sites were found.

Because of the strict reviewing and editing, Alignment 2 included only 29 individual sequences (42.6% of the original dataset) from seven populations (Tables 1 and S1). The total alignment length resulted in 395 bp and included 54 variable characters, of which 42 are parsimony-informative sites. Twelve singleton sizes were found (File S4).

The final Alignment 3 included 135 individuals from seven populations, where 106 individual sequences originate from the newly sequenced 106 individuals from Study-B. It had a length of 395 bp, including 55 parsimony informative sites of a total of 59 variable characters, as well as four singleton sites (File S6).

3.2. Population Structure, Genealogical Network, and Phylogenetic Analyses

The application of a strict and conservative alignment approach to both datasets of Study-A demonstrated that the results of Study-A could not be replicated for any of the alignments. All haplotype networks and the phylogenetic tree topologies mirror the network and tree topologies of Study-B—most obvious, however, in Alignment 2 and Alignment 3. This effect appeared even though sequences in the Alignments 1 were not edited and the fact that the bioinformatic algorithms are the same.

The number of polymorphic sites, singleton sites, nucleotide diversity, and haplotype diversity per population site was, as expected, higher in Alignments 1 than in Alignments 2 or Alignments 3 for both sequence markers (Tables 2 and 3). In particular, the analyses of the 16S rDNA Alignment 1 (BCI and LS populations) show a high number of haplotype diversity most certainly due to the ambiguities within the sequences (Table 2). In the Alignments 2 and 3, haplotype diversity was reduced. This effect was most obvious in the BCI, La Selva and Nicaragua (NI) populations (Table 2). Furthermore, the genetic variability within the populations was high (genetic differentiation within groups via the Kimura 2-parameter (K2P) model) in the two Alignments 1 (CO1 and 16S rDNA, Table S2). Up to 1.69% was found in BCI in 16S rDNA and 0.75% in NI-b in CO1, and in this alignment, the unusual effect of a higher 16S genetic variability was repeatable. (Table S2). In contrast, in the Alignments 2 and 3 for CO1 and 16S rDNA, this high variability within populations was not observed. The highest variability within populations with 0.18% was found in the BCI (Panama) population in the 16S rDNA Alignment 3, 0.28% in the Los Tuxtlas population (Mexico) and 0.20% in BCI in the CO1 Alignment 3 (Table S2).

The genetic distance values between groups using the Kimura 2-parameter (K2P) model are in accordance with geography and sampling schemes, cf. [12]. Moreover, the values indicate geographic isolation among populations, which appear arranged in three clusters (with high genetic distances between clusters and low genetic distances within clusters). This pattern is visible in the analyses of all alignments (Alignments 1 to 3) for CO1 and 16S rDNA, although it is most obvious in the Alignments 3. The identified clusters are as follows: (i) RBLT, (ii) NI and CNP, and (iii) NI-b, LS, BCI and CAN (Table S2). Although the length of each marker gene studied in the Alignments 3 is due to the previous editing of the Alignment 2 relatively short, the observed genetic distances between clusters within a range of 4.26% to 5.06% in 16S rDNA and from 7.61% to 9.62% in CO1 are comprehensible for a species level (Table S2). The F_{ST}-values support these genetic distances. A significant isolation among all clusters and even populations was observed in all Alignments with F_{ST}-values of a minimum of F_{ST} = 0.84 (in 16S rDNA Alignment 1 between the Nicaragua populations NI and NI-b) (Table S2). Within clusters, the F_{ST}-values are inconsistent. On the other hand, between La Selva in Costa Rica and CAN Ecuador—geographically the most distant populations (app. 3300 km), the F_{ST}-value is relatively low at F_{ST} = 0.40, and the F_{ST}-value is between BCI and La Selva (app. 540 km) F_{ST} = 0.85.

Table 2. Summary statistics for the 16S rDNA gene fragment of *Megaloprepus*. Shown are N: number of individuals, S: number of polymorphic sites (parsimony informative), H: number of haplotypes, Hd (±SD): haplotype diversity (standard deviation) and π (±SD): nucleotide diversity (standard deviation).

	16S rDNA					
	Population *	**N**	**S**	**H**	**Hd (±SD)**	**π (±SD)**
Alignment 1 577 bp	BCI	7	15	4	0.714 (0.181)	0.015 (0.007)
	RBLT	5	4	2	0.400 (0.237)	0.006 (0.003)
	LS	18	11	4	0.399 (0.138)	0.003 (0.001)
	CAN	2	0	2	0.00	0.00
	HN	1	0	1	0.00	0.00
	NI	14	3	3	0.385 (0.149)	0.002 (0.003)
	NI-b	2	3	2	1.000 (0.500)	0.008 (0.006)
	Total	49	29	10	0.759 (0.047)	0.021 (0.003)
Alignment 2 325 bp	BCI	4	2	3	0.833 (0.222)	0.003 (0.001)
	RBLT	4	0	1	0.00	0.00
	LS	15	1	2	0.248 (0.131)	0.001 (0.000)
	CAN	1	0	1	0.00	0.00
	NI	13	3	2	1.00 (0.250)	0.008 (0.000)
	NI-b	2	0	1	0.00	0.00
	Total	39	26	7	0.722 (0.047)	0.029 (0.002)
Alignment 3 321 bp	BCI	33	2	3	0.544 (0.036)	0.002 (0.000)
	CNP	29	1	2	0.069 (0.063)	0.000 (0.000)
	RBLT	22	1	2	0.312 (0.106)	0.001 (0.000)
	LS	45	1	2	0.202 (0.073)	0.001 (0.000)
	CAN	1	0	1	0.00	0.00
	NI	13	0	1	0.00	0.00
	NI-b	2	0	1	0.00	0.00
	Total	145	27	10	0.831 (0.015)	0.029 (0.001)

* Sampling localities from north to south are: RBLT (Biosphere Reserve Los Tuxtlas, Veracruz, Mexico), HN (Cusuco National Park, Honduras), NI (El Jaguar reserve, Nicaragua), NI-b Biological Reserve Indio Maíz "Bartola Reserve", Nicaragua, LS (Biological Research Station La Selva, Costa Rica), CNP (Corcovado National Park "Sirena Field Station", Costa Rica), BCI (Barro Colorado Island, Panama) and CAN (Rio Canande Reserve, Esmeraldas Province, Ecuador).

Table 3. Summary statistics for the CO1 gene fragment of *Megaloprepus*. Shown are N: number of individuals, S: number of polymorphic sites (parsimony informative), H: number of haplotypes, Hd (±SD): haplotype diversity (standard deviation) and π (±SD): nucleotide diversity (standard deviation).

	CO1					
	Population *	**N**	**S**	**H**	**Hd (±SD)**	**π (±SD)**
Alignment 1 ** 419 bp	BCI	7	/	/	/	/
	CNP	11	/	/	/	/
	RBLT	5	/	/	/	/
	LS	21	/	/	/	/
	CAN	2	/	/	/	/
	NI	13	/	/	/	/
	NI-b	2	/	/	/	/
	Total	61	/	/	/	/
Alignment 2 395 bp	BCI	3	1	2	0.667 (0.314)	0.002 (0.001)
	CNP	6	0	1	0.00	0.00
	RBLT	1	0	1	0.00	0.00
	LS	6	3	2	0.333 (0.215)	0.003 (0.002)
	CAN	1	0	1	0.00	0.00
	NI	11	0	1	0.00	0.00
	NI-b	1	0	1	0.00	0.00
	Total	29	54	9	0.786 (0.049)	0.052 (0.004)
Alignment 3 395 bp	BCI	32	6	5	0.502 (0.093)	0.002 (0.006)
	CNP	35	0	1	0.00	0.00
	RBLT	19	3	4	0.713 (0.058)	0.003 (0.000)
	LS	36	3	2	0.056 (0.052)	0.000 (0.000)
	CAN	1	0	1	0.00	0.00
	NI	11	0	1	0.00	0.00
	NI-b	1	0	1	0.00	0.00
	Total	135	59	15	0.826 (0.017)	0.055 (0.002)

* Sampling localities from north to south are: RBLT (Biosphere Reserve Los Tuxtlas, Veracruz, Mexico), HN (Cusuco National Park, Honduras), NI (El Jaguar reserve, Nicaragua), NI-b Biological Reserve Indio Maíz "Bartola Reserve", Nicaragua, LS (Biological Research Station La Selva, Costa Rica), CNP (Corcovado National Park "Sirena Field Station", Costa Rica), BCI (Barro Colorado Island, Panama) and CAN (Rio Canande Reserve, Esmeraldas Province, Ecuador). ** DnaSp could not analyze CO1 Alignment 1 because of the unidentifiable base content.

3.3. Genealogical Network Analyses

The minimum spanning haplotype networks performed in PopArt [35] show for all alignments three distinct clusters—although less explicit in the 16S and CO1 Alignments 1. The networks reflect the genetic distances precisely (Figures 1 and 2). As the most northern population, samples from Mexico (RBLT) comprise the first group. The second cluster again contains the CNP population (Costa Rica) and NI population (Nicaragua). The third cluster includes the populations from Nicaragua (NI-b), Costa Rica (LS), Panama (BCI) and Ecuador (CAN). Unfortunately, PopArt [35] does not include a "cut-off connection limit", as in the statistical parsimony approach (TCS analysis). The latter defines, depending on previously set similarity levels, either connected or separated haplotypes. TCS analyses [36] using a 95% connection limit also showed that haplotypes split into three distinct networks. Unfortunately, the TCS analysis was not possible for the CO1 Alignment 1. However, a split into three distinct clusters was obtained in the Alignments 2 (results shown for Alignments 3 in Figure 3, and results for the 16S rDNA Alignments 1 and 2, as well as CO1 Alignment 2 are shown in Figure S3, File S7). It supports the genetic sub-structuring demonstrated in Study-B.

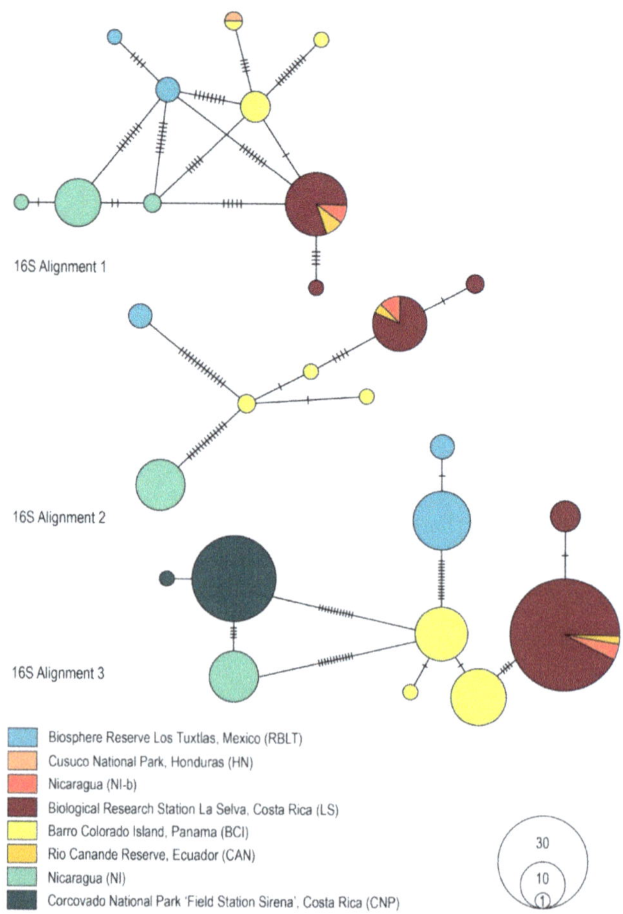

Figure 1. Minimum spanning haplotype networks showing the genealogical relationships within the genus *Megaloprepus* for the three different 16S rDNA alignments. POPART vers. 1.7 [35] was used to build the networks. Populations are color-coded in accordance with their origin.

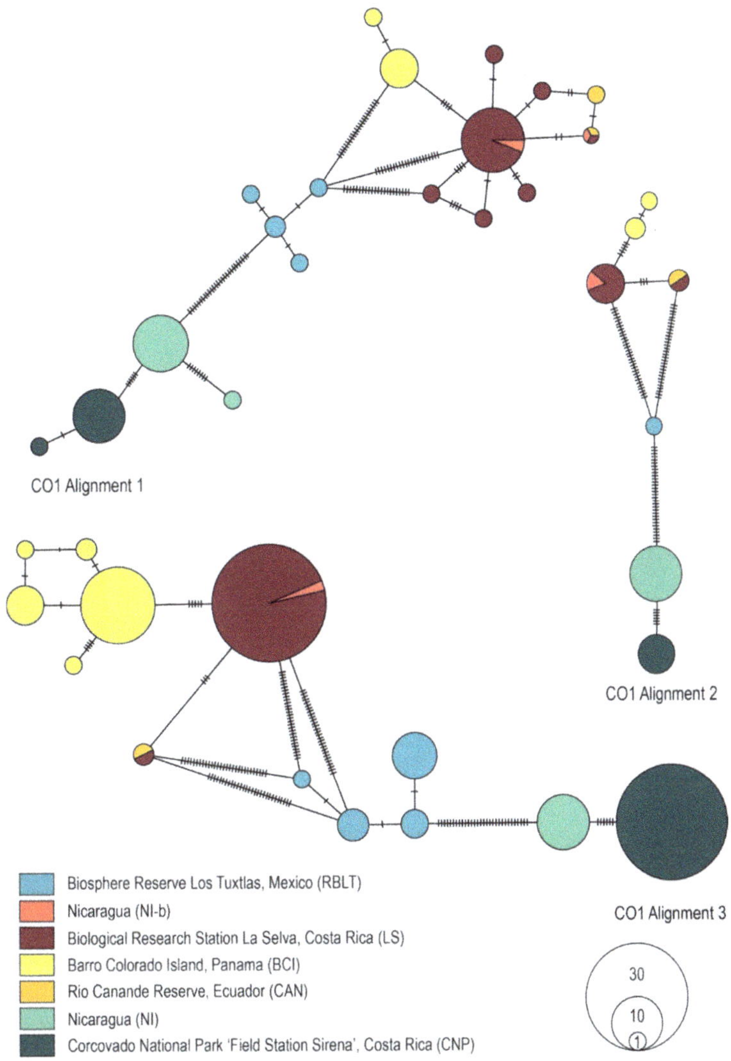

Figure 2. Minimum spanning haplotype networks showing the genealogical relationships within the genus *Megaloprepus* for the three different CO1 alignments. POPART vers. 1.7 [35] was used to build the networks. Populations are color-coded in accordance with their origin.

Comparing haplotype networks for the Alignments 1 shown in the present study and the original networks from Study-A reveals a somewhat similar genealogical structure for 16S rDNA (cf. [15] and Figure 1). However, the close relatedness between CNP and NI-b from Study-A could not be confirmed here, as NI-b appears closely related to LS. In contrast, a different pattern was observed for CO1. While Study-A obtained only five haplotypes, we found 18 haplotypes in the present network (cf. [15] and Figure 2), a surprising result that cannot be explained by editing the alignments but is now in accordance with the more conserved 16S rDNA marker gene.

Finally, the AMOVA analyses of the different Alignments are in accordance with the previous results, although for the Alignment 1 the AMOVA could be performed only for the 16S rDNA dataset. For 16S rDNA in the no-grouping setting, the percentage of variation among groups was higher in Alignment 2 and 3 than in Alignment 1, whereas a high variation within groups was found in Alignment 1. However, significant differentiation between populations was observed in all three alignments, with Φ_{ST} values close to one. Taking the three groups into account, the highest level of variation was explained among groups while the variation among populations within groups was substantially lower, and significant differentiation among regions and among populations within regions was observed (Table S3a,b). For CO1, the results are very similar for Alignments 2 and 3: a significant differentiation among populations (Φ_{ST} values close to 1) was obtained in the no-grouping setting, and the highest variation is found among groups.

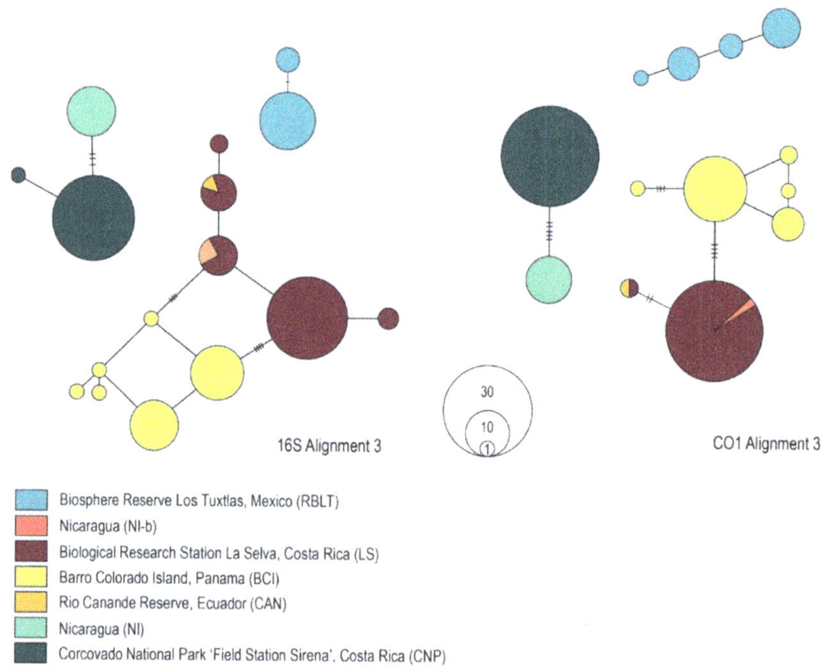

Figure 3. Genealogical relationships among *Megaloprepus* populations for the 16S rDNA Alignment 3 (**left**) and the CO1 Alignment 3 (**right**) based on statistical parsimony (95% connection limit) in TCS vers. 1.2.1 [36]. Populations are color-coded according to their origin.

3.4. Phylogenetic Tree Reconstruction

For the phylogenetic tree reconstruction with IQ-TREE [42], ModelFinder identified TIM3+F+I for 16S rDNA and TN+F+G4 for CO1 as the best-fit substitution models. The resulting phylogenetic trees from all three different tree calculations show similar results for both sequence markers with strong support for most nodes (Figures 4 and 5, MP and BI trees are shown in Figures S4–S7, File S7). The topologies show three main clades corresponding to the haplotype networks with RBLT as one clade, LS, BCI, NI and CAN as the second clade, and CNP and NI-b as the third clade. Here, it becomes obvious that individuals from one population are clustering together regardless of the origin of genetic samples—either Study A or B. This result shows a discrepancy with the phylogeny obtained in Study-A. Here, the two CNP datasets (one from Study-A and a second from Study-B) appeared as sister clades. Furthermore, the Study-B samples from CNP, LS and RBLT did not align well into the tree topology, appearing as separate sub-clades.

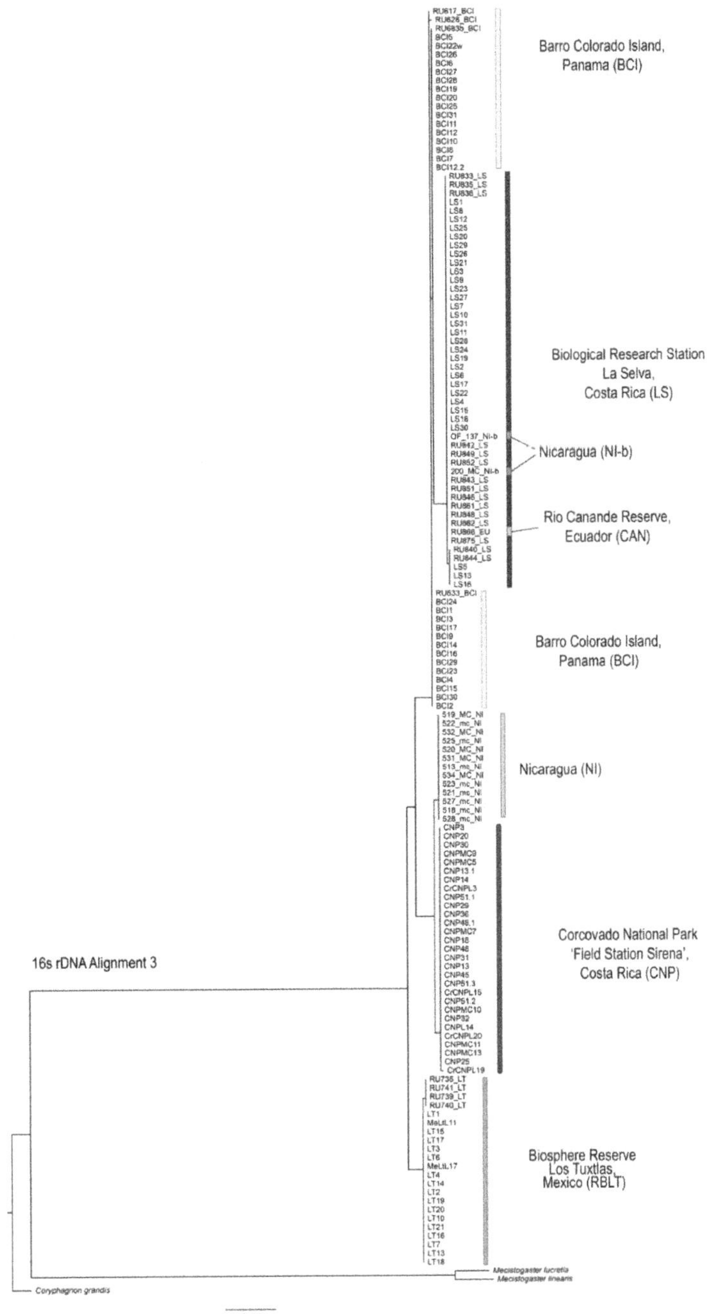

Figure 4. Phylogeny using the 16S rDNA sequence marker showing the relationships of the different *Megaloprepus* samples from Study-A and Study-B (Alignment 3) performed via maximum likelihood inference using IQ-TREE [42] with 1000 bootstrap replicates. *Coryphagrion grandis, Mecistogaster linearis* and *Mecistogaster lucretia* are outgroups. For the species IDs please compare Table S1. Color codes match with Figures 1–3 and are in accordance with the sample origin.

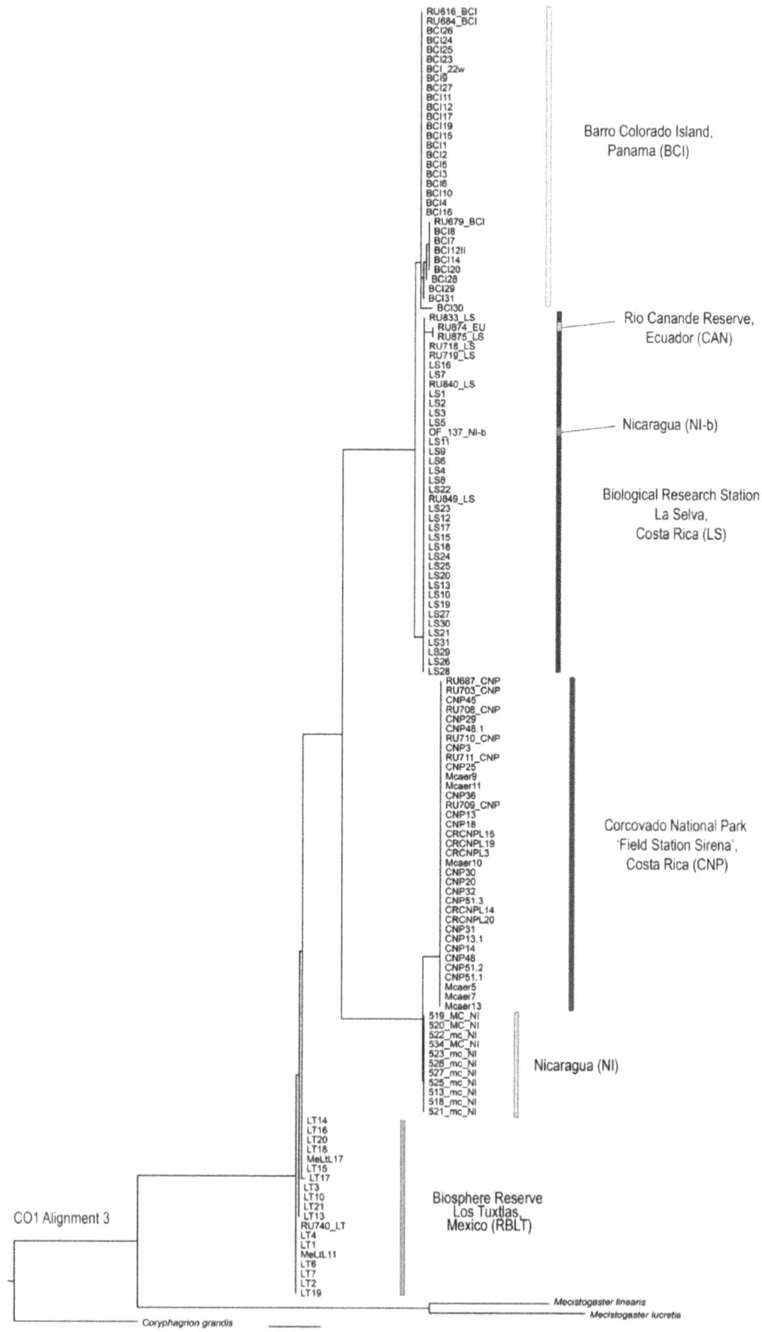

Figure 5. Phylogeny using the CO1 sequence marker showing the relationships of the different *Megaloprepus* samples from Study-A and Study-B (Alignment 3) performed via maximum likelihood inference using IQ-TREE [42] with 1000 bootstrap replicates. *Coryphagrion grandis*, *Mecistogaster linearis* and *Mecistogaster lucretia* are outgroups. For the species IDs please compare Table S1. Color codes match with Figures 1–3 and are in accordance with the sample origin.

4. Discussion

Contemporary biodiversity research covers an enormous area, from cellular processes to species level and global ecology research. Evolutionary genetics can principally bridge the different levels of observation but unfortunately we have very few model systems available for doing so [48]. Odonata (dragonflies and damselflies) have been long established and prominent model systems in ecological and evolutionary research and are promising candidates for bridging some gaps. Odonates have revolutionized studies of sexual selection and fitness correlates by using genetic markers [49–51]. Despite this head start, the multitude of new comparative NGS genomic approaches for "next generation" ecological and biodiversity studies has not yet entered odonate research and most genetic studies in odonates still rely on single DNA marker genes mostly generated by Sanger sequencing.

Single DNA sequence markers are still highly valuable for the fast evaluation of biodiversity patterns, and for species delimitation and species identification studies, via character-based barcoding approaches [4,5,52,53]. Here, the quality of sequence data is crucial. Consequently, when applying Sanger sequencing, comprehensive and strict sequence evaluation methods and alignment approaches are a *conditio sine qua non* for reliable data analyses.

The quality of sequence data depends on many factors and mistakes can occur in any step from specimen collection to final data interpretation. As an example, for resolving taxonomic questions in Odonata by means of DNA barcoding, two main pitfalls have been described: (i) inaccuracies in the identification of voucher specimens and (ii) sequencing errors [54,55]. A recent publication showed that sequences deposited at GenBank described as Odonata specimens were actually dipteran DNA [55]. In another example, Lorenzo-Carballa et al. identified in a molecular revision of *Ischnura aurora* questionable DNA sequences [56]. Odonates are a good example for false molecular data interpretation in non-model organisms but clearly not the only one [57,58].

There is a multitude of reasons for sequencing errors, also for Sanger sequencing, which is generally characterized by a high correctness [59]. Probable causes include remaining primer dimers in the cycle-sequencing product, editing errors (including miss-called nucleotides), (cross-) contamination with DNA of other animals [60], and the amplification of pseudogenes (through poor primer design/low primer specificity) [61].

The present study took advantage of two similar datasets to address the historic question of the species status within the damselfly genus *Megaloprepus*. The two mitochondrial gene datasets revealed contrary genealogical relationships among populations and biogeographical clusters, with obvious consequences for species delimitation and the determination of conservation management units. *Megaloprepus caerulatus* has long been considered as a single species genus. Two studies have changed this *status quo*: while Feindt et al. [12] "simply" suggested that the genus *Megaloprepus* consists of more than one species, Fincke et al. [15] assigned subspecies status to three historically described species: *M. caerulatus*, *M. c. brevistigma*, *M. c. latipennis* [10,11]. A new recent study, however, defined four species within the genus *Megaloprepus* applying quantitative and qualitative morphology and molecular genetics [62]. Here, *M. caerulatus*, *M. brevistigma*, and *M. latipennis* received species status and a new species *M. diaboli* was described [62]. Such different results have significant consequences for the future conservation of the species, since the number of conservation units is different (one against four). Although *Megaloprepus* is not listed in the IUCN Red List of Threatened Species, an evaluation of conservation status is much needed. This is because the four *Megaloprepus* species occur in old grown tropical rainforests, population sizes appear to be small, they breed in tree holes of big old trees, they are sensitive towards heat, and do not cross big light gaps or open areas [9,63]. Neotropical rainforests are under great threat with high forest loss inside and outside protected areas [64], most likely driven by selective logging and therefore increasing edge effects. Those facts in combination clearly underline the need for conservation efforts. If the subspecies status had validity, the distributional range of *Megaloprepus caerulatus* would be from southern Mexico to Peru. This range, however, is now shared by four species but with local endemics and most likely only few overlapping zones [62]. There is a need for studies focusing on the anthropogenic impact on those species in more detail; however, the negative effects of small, isolated forest habitats with increasing temperatures and selective logging of big trees are apparently severe.

In this study, raw sequences were re-analyzed by applying a conservative alignment approach to highlight the consequences of not eliminating sequence ambiguities. Our strict editing of the original Study-A data reflects in both Alignments 1 (16S rDNA and CO1) the results of Study-B instead of Study-A, showing clear population sub-structuring, similar to the analyses of Alignment 2 and 3. Manual sequence editing demanded a deletion of nearly 30% of the sequences in 16S and more than 50% in CO1 from the original datasets. Sequences containing ambiguous bases must clearly be removed from a dataset to identify haplotypes, address diversity indices, or identify species units with certainty, as in mitochondrial DNA heterozygous loci are not expected. This has been underlined firstly by DnaSP [32], Arlequin 3.5 [39], and TCS vers. 1.13 [36], rejecting the CO1 Alignment 1, and secondly by the our genetic diversity estimates (when sequences are removed from the dataset diversity indices should decrease). Direct comparisons are possible between the 16S Alignment 1 and Alignment 2 in populations with more than one individual, such as BCI and RBLT. Here the number of polymorphic sites (parsimony informative), the number of singleton sites, and the number of haplotypes decreased significantly. Adding the Study-B data did not increase the diversity indices exponentially. This was observed, for example, in the RBLT population in 16S rDNA. In Alignment 2, only one haplotype from four individual sequences was identified, but in Alignment 3, two haplotypes from twenty-two individual sequences were obtained. Similarly, in CO1 from Alignment 2 with six individual sequences for CNP and from Alignment 3 with 35 individual sequences, only one haplotype was detected. The genetic distances within populations behaved similarly when sequences with ambiguities were removed. Leaving in ambiguous characters may lead to a higher haplotype diversity and diluted tree topologies, while strict editing (removal of sequences with unresolved characters and alignment corrections) may have the opposite effect by not detecting some haplotypes. We favor the conservative sequence validation approach, which reduces sample size and sequence length but increases reliability.

The haplotype networks presented in Study-A [15] contrast with all networks obtained in the present study. First, the suggested higher variability of the 16S rDNA gene compared to the CO1 gene fragment could not be confirmed; second, the relatedness of populations contrasts with the phylogeny presented (no clear sub-structuring among populations); and third, the observed mutational steps between populations are smaller than in Study-B. One explanation could be the algorithm used in Study-A, by PopArt [35]. The program masks any columns in the alignment with gaps or ambiguous characters. This way, sequences that are not truly identical become identical after these columns are removed (pers. com. J. Leigh). This effect could have occurred in the 16S rDNA marker because sequence variances between populations were masked through the inserted Ns which are found in the raw dataset.

The fact that in Study-A [15] the CO1 marker appears to evolve more slowly than the more conserved 16S rDNA gene fragment contradicts current knowledge in Odonata [1,18,65]. The protein coding CO1 gene accumulates mutations over time faster than the more conserved ribosomal 16S rDNA [1]. The high number of singleton sites observed within the 16S rDNA Alignment 1 in comparison to the CO1 Alignment 1 and Alignments 2 and 3 could be a reason for the higher variability of the 16S rDNA gene in Study-A. However, different mitochondrial genes perform differently and a given sequence marker is chosen depending on the research question, mutation rates and availability of primers or lab resources [1]. In odonates, for example, CO1 and 16S rDNA are—besides numerous other mitochondrial and nuclear genes—well-established marker genes for population genetic and phylogenetic studies [1,4,5,41,56–59,66–69]. The 16S rDNA gene is composed of highly conserved as well as variable regions (helices) [70], but overall shows less variability among groups than CO1. Recently also the *nad 2* and the *A+T-rich* mitochondrial control region have been successfully used for species and population level research in Odonates [1,71] and should be included in future multi-gene approaches.

The haplotype networks and the phylogeny presented here show a clear grouping of all *Megaloprepus* individuals into three distinct clades with genetic distances at the species level, which supports and confirms the previous results of Feindt et al. [12,62]. The phylogeny of Fincke et al. [15] showed a similar topology, although the authors described a strict separation of the Study-A and Study-B CNP samples in their analyses. Unfortunately, a phylogenetic reconstruction for a concatenated 16S rDNA and CO1 dataset could not be estimated, because rigorous editing of the Study-A datasets did not leave enough specimens with sequences for both genes present. This could have allowed for a better insight into the discrepancies of the CNP population and the different grouping. When building the 16S rDNA Alignment 3 with the combined datasets, a reverse complement had to performed on one fraction of the sequence data to be able to align the two datasets. If comparing different strains of a gene (5′ to 3′ direction vs. 3′ to 5′ direction), gaps in the alignment could cause the observed pattern in Study-A. Alignments can strongly influence tree topologies and reporting alignment building methods in detail makes a study stronger and more comprehensible [65]. These results underline the great need for high-quality data and high scientific standards in data handling.

In the present study, the CO1 sequences from Study-A [15] were flagged as "CO1-like". GenBank [22] assigns sequences with this label when the open reading frame of a sequence is somehow incomplete and automatic translation is not possible. As mentioned above, simple sequencing errors, (cross-) contamination but also pseudogenes may be responsible for "CO1-like" sequences. A (cross-) contamination could be excluded because during BLAST searches [23] most sections of the sequence except the N bases aligned 95–96% to existing CO1 sequences of the same species and up to 90% to other Odonata species. Pseudogenes are nuclear copies of mitochondrial derived genes (numts [72]) that accumulate mutations over time, disrupting the open reading frames. As genes that moved to the nuclear genome and lost function (non-coding), numts are interrupting phylogenetic analyses simply due to the violation of comparing homologous DNA [61]. Furthermore, when numts are analyzed and compared in population genetic studies, or with the aim of proving cryptic species, sequence comparisons are inconclusive such that faulty high genetic divergences can be obtained [61,73,74]. Only recently has the existence of numts been described in *Leucorrhinia* species (Odonata) [74]. In the present case, we did not observe stop codons inside the "CO1-like" sequences; rather, N bases were found at different positions inside the sequence. Unfortunately, the chromatograms are not available for exact troubleshooting but—from our perspective—those N bases appear to be sequencing errors from the sequencer itself, or are a result of manual editing (as they were of different lengths: 1–3 N bases in a row). Therefore, we did not further test for pseudogenes. Generally, in the case that editing may become necessary, chromatograms shall be reviewed manually to distinguish sequence quality (e.g., precise single peaks versus double peaks, miss-calls, or non-identifiable peaks) and low-quality chromatograms with unclear base calling should be redone. Data handling and analytical training in molecular evolution using nucleotide sequences are crucial for avoiding low sequence quality being used.

In addition, the availability of high-quality data is critical. Today, GenBank [22] and the Barcode of Life Data Systems (BOLD, [75]) are the main public resources for sequence data. Making sequencing data available belongs to good scientific standards and it should be mandatory for any publishing process, but this does not necessarily tell us about the quality of the data. There should be a greater awareness of this simple fact. Providing the original unedited chromatogram files along with the final data could help researchers to evaluate the data quality.

5. Conclusions

This study highlights the importance of high-quality raw sequence data as a backbone for phylogenetic tree and genealogical network analyses, as well as the use of strict hypothesis-related bioinformatic algorithms. This is especially true when it comes to species delimitation and discovery, which may have a significant impact on conservation management efforts.

Today, four species within the genus *Megaloprepus* are valid—all with sensitive habitat requirements. Biodiversity measures within the genus are based on the sequence data. However, future research is needed to identify the anthropogenic impact on these species and establish conservation measures.

Supplementary Materials: The following supporting information can be downloaded at: https://www.mdpi.com/article/10.3390/d14121056/s1, File S1: 16S_Alignment1, File S2: CO1_Alignment1, File S3: 16S_Alignment2, File S4: CO1_Alignment2, File S5: 16S_Alignment3, File S6: CO1_Alignment3, File S7: Figure S1. Screenshots of the 16S rDNA Alignment 1 and 16S rDNA Alignment 3. Figure S2. Screenshots of the CO1 Alignment 1 and CO1 Alignment 3. Figure S3. Genealogical relationships among Megaloprepus populations for the 16S rDNA Alignment 1 and 2, and the CO1 Alignment 2 based on statistical parsimony (95% connection limit) in TCS vers. 1.2.1. Figure S4. Phylogeny based on Bayesian inference using MrBayes vers. 3.7 for 16S rDNA Alignment3 the with posterior probabilities shown on the corresponding nodes. Figure S5. Phylogeny based on Bayesian inference using MrBayes vers. 3.7 for CO1 Alignment3 the with posterior probabilities shown on the corresponding nodes. Figure S6. Maximum parsimony tree obtained with PAUP* vers. 4.0b8 for the 16S rDNA Alignment3 with 1000 bootstrap replicates. Figure S7. Maximum parsimony tree obtained with PAUP* vers. 4.0b8 for the CO1 Alignment3 with 1000 bootstrap replicates. Table S1: Overview of species ID's, sampling localities, and NCBI Accession numbers used for the present study. Table S2: Genetic distances (Kimura 2-parameter (K2P) model) between and within populations of Megaloprepus and as a measure of gene flow between populations the FST-values are shown. Table S3a: AMOVA design and results: for the 16S rDNA sequence marker under different settings (no groupings vs. 3 groups). Table S3b: AMOVA design and results: for the CO1 sequence marker under different settings (no groupings vs. 3 groups1).

Author Contributions: Conceptualization, W.F. and H.H.; formal analysis, data curation, and visualization, W.F.; writing, reviewing, and editing, W.F. and H.H. All authors have read and agreed to the published version of the manuscript.

Funding: This Open Access publication was founded by the Deutsche Forschungsgemeinschaft (DFG, German Research Foundation)—491,094,227 "Open Access Publication Funding" and the University of Veterinary Medicine Hannover, Foundation.

Institutional Review Board Statement: Not applicable.

Data Availability Statement: DNA sequences generated in this study have all been submitted to GenBank (https://www.ncbi.nlm.nih.gov/genbank/). Additional data are available in the supplementary material.

Acknowledgments: We are grateful to Bernd Schierwater for providing helpful comments and four anonymous reviewers for their excellent reviews.

Conflicts of Interest: The authors declare no conflict of interest.

References

1. Cheng, Y.C.; Chen, M.Y.; Wang, J.F.; Liang, A.P.; Lin, C.P. Some mitochondrial genes perform better for damselfly phylogenetics: Species- and population-level analyses of four complete mitogenomes of Euphaea sibling species. *Syst. Entomol.* **2018**, *43*, 702–715. [CrossRef]
2. Damm, S.; Schierwater, B.; Hadrys, H. An integrative approach to species discovery in odonates: From character-based DNA barcoding to ecology. *Mol. Ecol.* **2010**, *19*, 3881–3893. [CrossRef] [PubMed]
3. Graham, C.H.; Ron, S.R.; Santos, J.C.; Schneider, C.J.; Moritz, C. Integrating phylogenetics and environmental niche models to explore speciation mechanisms in dendrobatid frogs. *Evolution* **2004**, *58*, 1781–1793. [CrossRef]
4. Paknia, O.; Bergmann, T.; Hadrys, H. Some 'ant'swers: Application of a layered barcode approach to problems in ant taxonomy. *Mol. Ecol. Resour.* **2015**, *15*, 1262–1274. [CrossRef]
5. Rach, J.; Bergmann, T.; Paknia, O.; DeSalle, R.; Schierwater, B.; Hadrys, H. The marker choice: Unexpected resolving power of an unexplored CO1 region for layered DNA barcoding approaches. *PLoS ONE* **2017**, *12*, e0174842. [CrossRef]

6. Lemmon, A.R.; Brown, J.M.; Stanger-Hall, K.; Lemmon, E.M. The effect of ambiguous data on phylogenetic estimates obtained by maximum likelihood and Bayesian inference. *Syst. Biol.* **2009**, *58*, 130–145. [CrossRef] [PubMed]
7. Ogden, T.H.; Rosenberg, M.S. Multiple sequence alignment accuracy and phylogenetic inference. *Syst. Biol.* **2006**, *55*, 314–328. [CrossRef] [PubMed]
8. Simon, S.; Blanke, A.; Meusemann, K. Reanalyzing the Palaeoptera problem—The origin of insect flight remains obscure. *Arthropod Struct. Dev.* **2018**, *47*, 328–338. [CrossRef] [PubMed]
9. Fincke, O.M. Use of forest and tree species, and dispersal by giant damselflies (Pseudostigmatidae): Their prospects in fragmented forests. In *Forest and Dragonflies, 4th WDA International Symposium of Odonatology*; Pensoft: Sofia, Bulgaria, 2006; pp. 103–125.
10. Selys, L.E.d. Synopsis des Agrionines. Première Légion—Pseudostigma. *Bull. L'académie R. Sci. Lett. Beaux-Arts Belg.* **1860**, *2*, 9–27.
11. Selys, L.E.d. Révision du synopsis des Agrionines, premiére partie comprenant des légions Psuedostigma—Podagrion—Platycnemis et Protoneura. *Mémoire Cour. Académie R. Belg.* **1886**, *38*, 233.
12. Feindt, W.; Fincke, O.; Hadrys, H. Still a one species genus? Strong genetic diversification in the world's largest living odonate, the Neotropical damselfly *Megaloprepus caerulatus*. *Conserv. Genet.* **2014**, *15*, 469–481. [CrossRef]
13. Ris, F. Libellen (Odonata) aus der Region der amerikanischen Kordilleren von Costarica bis Catamarca. *Arch. Nat.* **1916**, *82A*, 1–197.
14. Schmidt, E. Odonata nebst Bemerkungen über die *Anomisma* und *Chalcopteryx* des Amazonas-Gebiets. In *1941–1942 Beiträge zur Fauna Perus Nach der Ausbeute der Hamburger Südperu Expedition 1936*; G. Fischer: Hamburg, German, 1942; Volume 2, pp. 225–276.
15. Fincke, O.M.; Xu, M.; Khazan, E.S.; Wilson, M.; Ware, J.L. Tests of hypotheses for morphological and genetic divergence in *Megaloprepus* damselflies across Neotropical forests. *Biol. J. Linn. Soc.* **2018**, *125*, 844–861. [CrossRef]
16. Damm, S.; Hadrys, H. *Trithemis morrisoni* sp. nov. and *T. palustris* sp. nov. from the Okavango and Upper Zambezi Floodplains previously hidden under *T. stictica* (Odonata: Libellulidae). *Int. J. Odonatol.* **2009**, *12*, 131–145. [CrossRef]
17. De Mandal, S.; Chhakchhuak, L.; Gurusubramanian, G.; Kumar, N.S. Mitochondrial markers for identification and phylogenetic studies in insects—A Review. *DNA Barcodes* **2014**, *2*, 1–9. [CrossRef]
18. Papadopoulou, A.; Anastasiou, I.; Vogler, A.P. Revisiting the insect mitochondrial molecular clock: The mid-Aegean trench calibration. *Mol. Biol. Evol.* **2010**, *27*, 1659–1672. [CrossRef]
19. Nicolas, V.; Schaeffer, B.; Missoup, A.D.; Kennis, J.; Colyn, M.; Denys, C.; Tatard, C.; Cruaud, C.; Laredo, C. Assessment of three mitochondrial genes (16S, Cytb, CO1) for identifying species in the Praomyini tribe (Rodentia: Muridae). *PLoS ONE* **2012**, *7*, e36586. [CrossRef] [PubMed]
20. Fujisawa, T.; Vogler, A.P.; Barraclough, T.G. Ecology has contrasting effects on genetic variation within species versus rates of molecular evolution across species in water beetles. *Proc. R. Soc. B Biol. Sci.* **2015**, *282*, 20142476. [CrossRef]
21. Shearer, T.; Van Oppen, M.; Romano, S.; Wörheide, G. Slow mitochondrial DNA sequence evolution in the Anthozoa (Cnidaria). *Mol. Ecol.* **2002**, *11*, 2475–2487. [CrossRef]
22. Clark, K.; Karsch-Mizrachi, I.; Lipman, D.J.; Ostell, J.; Sayers, E.W. GenBank. *Nucleic Acids Res.* **2016**, *44*, D67–D72. [CrossRef] [PubMed]
23. Johnson, M.; Zaretskaya, I.; Raytselis, Y.; Merezhuk, Y.; McGinnis, S.; Madden, T.L. NCBI BLAST: A better web interface. *Nucleic Acids Res.* **2008**, *36*, W5–W9. [CrossRef]
24. Ware, J.; May, M.; Kjer, K. Phylogeny of the higher Libelluloidea (Anisoptera: Odonata): An exploration of the most speciose superfamily of dragonflies. *Mol. Phylogenetics Evol.* **2007**, *45*, 289–310. [CrossRef] [PubMed]
25. Bergmann, T.; Rach, J.; Damm, S.; DeSalle, R.; Schierwater, B.; Hadrys, H. The potential of distance-based thresholds and character-based DNA barcoding for defining problematic taxonomic entities by CO1 and ND1. *Mol. Ecol. Resour.* **2013**, *13*, 1069–1081. [CrossRef] [PubMed]
26. Simon, C.; Frati, F.; Beckenbach, A.; Crespi, B.; Liu, H.; Flook, P. Evolution, weighting, and phylogenetic utility of mitochondrial gene sequences and a compilation of conserved polymerase chain reaction primers. *Ann. Entomol. Soc. Am.* **1994**, *87*, 651–701. [CrossRef]
27. Folmer, O.; Black, M.; Hoeh, W.; Lutz, R.; Vrijenhoek, R. DNA primers for amplification of mitochondrial cytochrome c oxidase subunit I from diverse metazoan invertebrates. *Mol. Mar. Biol. Biotechnol.* **1994**, 3, 294–299. [PubMed]
28. Edgar, R.C. MUSCLE: Multiple sequence alignment with high accuracy and high throughput. *Nucleic Acids Res.* **2004**, *32*, 1792–1797. [CrossRef] [PubMed]
29. Feindt, W.; Osigus, H.-J.; Herzog, R.; Mason, C.E.; Hadrys, H. The complete mitochondrial genome of the neotropical helicopter damselfly *Megaloprepus caerulatus* (Odonata: Zygoptera) assembled from next generation sequencing data. *Mitochondrial DNA Part B* **2016**, *1*, 497–499. [CrossRef] [PubMed]
30. Swofford, D.L. PAUP * Phylogenetic Analysis Using Parsimony (* and Other Methods). 2002. Available online: http://www2.ib.unicamp.br/profs/sfreis/SistematicaMolecular/Aula09MetodoParcimonia/Leituras/ThePhylogeneticHandbookParcimonia.pdf (accessed on 27 September 2022).
31. Tamura, K.; Stecher, G.; Kumar, S. MEGA11: Molecular evolutionary genetics analysis version 11. *Mol. Biol. Evol.* **2021**, *38*, 3022–3027. [CrossRef]
32. Librado, P.; Rozas, J. DnaSP v5: A software for comprehensive analysis of DNA polymorphism data. *Bioinformatics* **2009**, *25*, 1451–1452. [CrossRef]

33. Kumar, S.; Stecher, G.; Tamura, K. MEGA7: Molecular evolutionary genetics analysis version 7.0 for bigger datasets. *Mol. Biol. Evol.* **2016**, *33*, 1870–1874. [CrossRef]
34. Kimura, M. A simple method for estimating evolutionary rates of base substitutions through comparative studies of nucleotide sequences. *J. Mol. Evol.* **1980**, *16*, 111–120. [CrossRef] [PubMed]
35. Leigh, J.W.; Bryant, D. popart: Full-feature software for haplotype network construction. *Methods Ecol. Evol.* **2015**, *6*, 1110–1116. [CrossRef]
36. Clement, M.; Posada, D.; Crandall, K. TCS: A computer program to estimate gene genealogies. *Mol. Ecol.* **2000**, *9*, 1657–1660. [CrossRef] [PubMed]
37. Wright, S. *Evolution and the Genetics of Populations, Volume 2, The Theory of Gene Frequencies*; University of Chicago Press: Chicago, IL, USA, 1969.
38. Excoffier, L.; Smouse, P.E.; Quattro, J.M. Analysis of molecular variance inferred from metric distances among DNA haplotypes: Application to human mitochondrial DNA restriction data. *Genetics* **1992**, *131*, 479–491. [CrossRef] [PubMed]
39. Excoffier, L.; Lischer, H.E. Arlequin suite ver 3.5: A new series of programs to perform population genetics analyses under Linux and Windows. *Mol. Ecol. Resour.* **2010**, *10*, 564–567. [CrossRef] [PubMed]
40. Groeneveld, L.F.; Clausnitzer, V.; Hadrys, H. Convergent evolution of gigantism in damselflies of Africa and South America? Evidence from nuclear and mitochondrial sequence data. *Mol. Phylogenetics Evol.* **2007**, *42*, 339–346. [CrossRef] [PubMed]
41. Toussaint, E.F.; Bybee, S.M.; Erickson, R.J.; Condamine, F.L. Forest giants on different evolutionary branches: Ecomorphological convergence in helicopter damselflies. *Evolution* **2019**, *73*, 1045–1054. [CrossRef]
42. Nguyen, L.-T.; Schmidt, H.A.; von Haeseler, A.; Minh, B.Q. IQ-TREE: A fast and effective stochastic algorithm for estimating maximum-likelihood phylogenies. *Mol. Biol. Evol.* **2014**, *32*, 268–274. [CrossRef]
43. Kalyaanamoorthy, S.; Minh, B.Q.; Wong, T.K.; von Haeseler, A.; Jermiin, L.S. ModelFinder: Fast model selection for accurate phylogenetic estimates. *Nat. Methods* **2017**, *14*, 587. [CrossRef]
44. Hoang, D.T.; Chernomor, O.; Von Haeseler, A.; Minh, B.Q.; Vinh, L.S. UFBoot2: Improving the ultrafast bootstrap approximation. *Mol. Biol. Evol.* **2017**, *35*, 518–522. [CrossRef] [PubMed]
45. Ronquist, F.; Huelsenbeck, J.P. MrBayes 3: Bayesian phylogenetic inference under mixed models. *Bioinformatics* **2003**, *19*, 1572–1574. [CrossRef] [PubMed]
46. Posada, D. jModelTest: Phylogenetic model averaging. *Mol. Biol. Evol.* **2008**, *25*, 1253–1256. [CrossRef] [PubMed]
47. Akaike, H. Information theory and an extension of the maximum likelihood principle. In *Selected Papers of Hirotugu Akaike*; Springer: Berlin/Heidelberg, Germany, 1998; pp. 199–213.
48. Tautz, D.; Arctander, P.; Minelli, A.; Thomas, R.H.; Vogler, A.P. A plea for DNA taxonomy. *Trends Ecol. Evol.* **2003**, *18*, 70–74. [CrossRef]
49. Fincke, O.M.; Hadrys, H. Unpredictable offspring survivorship in the damselfly, *Megaloprepus coerulatus*, shapes parental behavior, constrains sexual selection, and challenges traditional fitness estimates. *Evolution* **2001**, *55*, 762–772. [CrossRef] [PubMed]
50. Hadrys, H.; Wargel, A.; Giere, S.; Kraus, B.; Streit, B. A panel of microsatellite markers to detect and monitor demographic bottlenecks in the riverine dragonfly *Orthetrum coerulescens* F. *Mol. Ecol. Notes* **2007**, *7*, 287–289. [CrossRef]
51. Hadrys, H.; Timm, J.; Streit, B.; Giere, S. A panel of microsatellite markers to study sperm precedence patterns in the emperor dragonfly *Anax imperator* (Odonata: Anisoptera). *Mol. Ecol. Notes* **2007**, *7*, 296–298. [CrossRef]
52. DeSalle, R.; Schierwater, B.; Hadrys, H. MtDNA: The small workhorse of evolutionary studies. *Front. Biosci.-Landmark* **2017**, *22*, 873–887. [CrossRef] [PubMed]
53. Koroiva, R.; Pepinelli, M.; Rodrigues, M.E.; de Oliveira Roque, F.; Lorenz-Lemke, A.P.; Kvist, S. DNA barcoding of odonates from the Upper Plata basin: Database creation and genetic diversity estimation. *PLoS ONE* **2017**, *12*, e0182283. [CrossRef]
54. Koroiva, R.; Kvist, S. Estimating the barcoding gap in a global dataset of cox1 sequences for Odonata: Close, but no cigar. *Mitochondrial DNA Part A* **2018**, *29*, 765–771. [CrossRef]
55. Vega-Sánchez, Y.M.; Lorenzo-Carballa, M.O.; Vilela, D.S.; Guillermo-Ferreira, R.; Koroiva, R. Comment on "Molecular identification of seven new Zygopteran genera from South China through partial cytochrome oxidase subunit I (COI) gene". *Meta Gene* **2020**, *25*, 100759. [CrossRef]
56. Lorenzo-Carballa, M.O.; Sanmartín-Villar, I.; Cordero-Rivera, A. Molecular and Morphological Analyses Support Different Taxonomic Units for Asian and Australo-Pacific Forms of *Ischnura aurora* (Odonata, Coenagrionidae). *Diversity* **2022**, *14*, 606. [CrossRef]
57. Meiklejohn, K.A.; Damaso, N.; Robertson, J.M. Assessment of BOLD and GenBank—Their accuracy and reliability for the identification of biological materials. *PLoS ONE* **2019**, *14*, e0217084. [CrossRef]
58. Arabi, J.; Cruaud, C.; Couloux, A.; Hassanin, A. Studying sources of incongruence in arthropod molecular phylogenies: Sea spiders (Pycnogonida) as a case study. *Comptes Rendus Biol.* **2010**, *333*, 438–453. [CrossRef] [PubMed]
59. Goodwin, S.; McPherson, J.D.; McCombie, W.R. Coming of age: Ten years of next-generation sequencing technologies. *Nat. Rev. Genet.* **2016**, *17*, 333. [CrossRef]
60. Wilson, C.G.; Nowell, R.W.; Barraclough, T.G. Cross-contamination explains "inter and intraspecific horizontal genetic transfers" between asexual bdelloid rotifers. *Curr. Biol.* **2018**, *28*, 2436–2444.e2414. [CrossRef]
61. Buhay, J.E. "COI-like" sequences are becoming problematic in molecular systematic and DNA barcoding studies. *J. Crustac. Biol.* **2009**, *29*, 96–110. [CrossRef]

62. Feindt, W.; Hadrys, H. The damselfly genus Megaloprepus (Odonata: Pseudostigmatidae): Revalidation and delimitation of species-level taxa including the description of one new species. *Zootaxa* **2022**, *5115*, 487–510. [CrossRef] [PubMed]
63. Khazan, E.S. Tests of biological corridor efficacy for conservation of a Neotropical giant damselfly. *Biol. Conserv.* **2014**, *177*, 117–125. [CrossRef]
64. Hansen, M.C.; Wang, L.; Song, X.-P.; Tyukavina, A.; Turubanova, S.; Potapov, P.V.; Stehman, S.V. The fate of tropical forest fragments. *Sci. Adv.* **2020**, *6*, eaax8574. [CrossRef] [PubMed]
65. Brower, A.V.; Desalle, R. Practical and theoretical considerations for choice of a DNA sequence region in insect molecular systematics, with a short review of published studies using nuclear gene regions. *Ann. Entomol. Soc. Am.* **1994**, *87*, 702–716. [CrossRef]
66. Damm, S.; Dijkstra, K.-D.B.; Hadrys, H. Red drifters and dark residents: The phylogeny and ecology of a Plio-Pleistocene dragonfly radiation reflects Africa's changing environment (Odonata, Libellulidae, Trithemis). *Mol. Phylogenetics Evol.* **2010**, *54*, 870–882. [CrossRef] [PubMed]
67. Damm, S.; Hadrys, H. A dragonfly in the desert: Genetic pathways of the widespread *Trithemis arteriosa* (Odonata: Libellulidae) suggest male-biased dispersal. *Org. Divers. Evol.* **2012**, *12*, 267–279. [CrossRef]
68. Dijkstra, K.-D.B.; Kalkman, V.J.; Dow, R.A.; Stokvis, F.R.; Van Tol, J. Redefining the damselfly families: A comprehensive molecular phylogeny of Zygoptera (Odonata). *Syst. Entomol.* **2014**, *39*, 68–96. [CrossRef]
69. Vega-Sánchez, Y.M.; Mendoza-Cuenca, L.F.; González-Rodríguez, A. Complex evolutionary history of the American Rubyspot damselfly, Hetaerina americana (Odonata): Evidence of cryptic speciation. *Mol. Phylogenetics Evol.* **2019**, *139*, 106536. [CrossRef] [PubMed]
70. Misof, B.; Anderson, C.; Buckley, T.; Erpenbeck, D.; Rickert, A.; Misof, K. An empirical analysis of mt 16S rRNA covarion-like evolution in insects: Site-specific rate variation is clustered and frequently detected. *J. Mol. Evol.* **2002**, *55*, 460–469. [CrossRef]
71. Bronstein, O.; Kroh, A.; Haring, E. Mind the gap! The mitochondrial control region and its power as a phylogenetic marker in echinoids. *BMC Evol. Biol.* **2018**, *18*, 80. [CrossRef] [PubMed]
72. Lopez, J.V.; Yuhki, N.; Masuda, R.; Modi, W.; O'Brien, S.J. Numt, a recent transfer and tandem amplification of mitochondrial DNA to the nuclear genome of the domestic cat. *J. Mol. Evol.* **1994**, *39*, 174–190. [CrossRef]
73. Song, H.; Buhay, J.E.; Whiting, M.F.; Crandall, K.A. Many species in one: DNA barcoding overestimates the number of species when nuclear mitochondrial pseudogenes are coamplified. *Proc. Natl. Acad. Sci. USA* **2008**, *105*, 13486–13491. [CrossRef] [PubMed]
74. Ožana, S.; Dolný, A.; Pánek, T. Nuclear copies of mitochondrial DNA as a potential problem for phylogenetic and population genetic studies of Odonata. *Syst. Entomol.* **2022**, *47*, 591–602. [CrossRef]
75. Ratnasingham, S.; Hebert, P.D. BOLD: The Barcode of Life Data System (http://www.barcodinglife.org/). *Mol. Ecol. Notes* **2007**, *7*, 355–364. [CrossRef]

diversity

MDPI

Article

Meiotic Analysis of Gomphidae Species Sheds Light on the Large X Chromosome of the Family (Anisoptera, Odonata)

Liliana M. Mola [1,*], Pablo J. Rebagliati [2], María F. Fourastié [3] and Silvia S. Agopian [4]

[1] Laboratorio de Citogenética y Evolución, Departamento de Ecología, Genética y Evolución, Instituto de Ecología, Genética y Evolución (CONICET-UBA), Facultad de Ciencias Exactas y Naturales, Universidad de Buenos Aires, Intendente Güiraldes 2160, Ciudad Autónoma de Buenos Aires C1428EGA, Argentina
[2] Facultad de Bromatología, Universidad Nacional de Entre Ríos, Sede Centro. 25 de mayo 709, Gualeguaychú E2822ABA, Entre Ríos, Argentina
[3] Instituto de Ecología, Genética y Evolución (CONICET-UBA), Facultad de Ciencias Exactas y Naturales, Universidad de Buenos Aires, Intendente Güiraldes 2160, Ciudad Autónoma de Buenos Aires C1428EGA, Argentina
[4] Independent Researcher, Ciudad Autónoma de Buenos Aires C1176ABM, Argentina
* Correspondence: lilimola@yahoo.com.ar

Abstract: In most Anisoptera families, the modal diploid number is 25 in males (24 autosomes + X), and the X chromosome is one of the smallest elements of the complement. The family Gomphidae is an exception, as it has a modal diploid number of 23 (22 + X), and the X chromosome is the largest of the complement and of medium-to-large size in many species. We studied the meiosis of three gomphid species from Argentina: *Aphylla* cf. *distinguenda* (Campion, 1920), *Phyllocycla propinqua* Belle, 1972 and *Phyllocycla* sp. Chromosome number is 2n = 23, n = 11 + X, except for *Phyllocycla propinqua*, showing n = 10 + X. The X chromosome of these species is medium-sized and presents heteropyknotic blocks of different sizes. Despite the small number of gomphid species analysed, there is a clear trend of increasing size of the X chromosome with the increasing amount of heterochromatin. Our results, together with those from the literature, suggest that its large size might have been due to a progressive accumulation of repetitive DNA and heterochromatinisation and not to fusion, as previously suggested. This led us to propose that the ancestral number coincided with the modal number of Gomphidae. A revision of the derived sex-determining systems in Odonata is also provided.

Keywords: holokinetic chromosomes; gomphids X chromosome evolution; sex-determination systems; *Aphylla*; *Phyllocycla*

Citation: Mola, L.M.; Rebagliati, P.J.; Fourastié, M.F.; Agopian, S.S. Meiotic Analysis of Gomphidae Species Sheds Light on the Large X Chromosome of the Family (Anisoptera, Odonata). *Diversity* **2022**, *14*, 874. https://doi.org/10.3390/d14100874

Academic Editors: M. Olalla Lorenzo-Carballa, Ricardo Koroiva and Michael Wink

Received: 3 August 2022
Accepted: 6 October 2022
Published: 17 October 2022

1. Introduction

Gomphidae is the second largest family of the suborder Anisoptera, with approximately 1000 species. Among the most species-rich genera, *Aphylla, Phyllocycla, Phyllogomphoides* and *Progomphus* are mainly distributed in the Neotropics but are also found in the Nearctic region [1].

Odonata exhibits some particular cytogenetic features, such as holokinetic chromosomes (i.e., without primary constriction or centromere) and equatorial division of the X chromosome, but the type of meiosis of autosomes is controversial. Some authors establish that meiosis is post-reductional (i.e., sister chromatids separate in the first division and homologues in the second one). They are based on the orientation of the bivalents with subterminal ciasmata on the equatorial plane, the equational division of the heteromorphic autosomal and neo-sex bivalents, and autosomal trivalents at metaphase I. These lead to the presence of heteromorphic chromatids or three chromatids, respectively, at all metaphases II [2]. Instead, Nokkala and collaborators [3] consider that the meiotic division of the autosomes is pre-reductional (canonical) based on the study of one species that presents only interstitial chiasmata and their interpretation of the migration of the homologous telomeres at anaphase I. Odonata is also characterised by having a single chiasma in all the bivalents and a noticeably small autosome pair (m-chromosomes) of about half the size of the following pair, which shows a regular meiotic behaviour [2].

The chromosome number has been determined in more than 600 species of the order belonging to 23 families [4]. In the Neotropical region, cytogenetic data have been reported for about 235 species from eight countries (Argentina, Bolivia, Brazil, Chile, Perú, Suriname, Uruguay and Venezuela) [4].

Although the haploid number is relatively constant (12, 13 or 14) in nearly 93% of the species, the chromosome number ranges from n = 3 in *Macrothemis hemichlora* (Burmeister, 1839) to n = 21 in *Orthemis nodiplaga* Karsch, 1891 (Libellulidae). In the families of the suborder Anisoptera, the modal haploid number is 12 in Gomphidae, 13 in Cordulegasteridae, Corduliidae, Libellulidae and Macrodiplacidae, and 14 in Aeshnidae [2,4–6].

About 95% of the species possess an XX/X0 sex-determination chromosome system with male heterogamety. The X chromosome is generally the smallest of the complement or the second smallest element after the m chromosomes. In contrast, it is the largest of the complement in some Gomphidae species and different theories have been postulated to explain its unusual size [7–17]. The sex chromosome systems may have originated from fusions or insertions, such as the neo-XY/neo-XX system and the multiple $X_1X_2Y/X_1X_1X_2X_2$ system in *Micrathyria ungulata* [2,4,6,11,18].

C-banding revealed that most autosomes have heterochromatic blocks in both telomeric regions, which are either small or large and symmetric or asymmetric [19]. The X chromosome of males is entirely C-positive, and exhibits intermediate staining or possesses C-positive bands localised in the terminal or interstitial regions [6,15,19–24].

In the present study, we analysed the meiotic behaviour and the characteristics of the X chromosome in three species of Gomphidae: *Aphylla* cf. *distinguenda* (Campion, 1920), *Phyllocycla propinqua* Belle, 1972 and *Phyllocycla* sp., and discuss the origin of the large X chromosomes and the diploid number of the common ancestor of Gomphidae. Moreover, we provide a review of the derived sex-determining systems in the order.

2. Material and Methods

The present study was conducted on three adult males of *Aphylla* cf. *distinguenda* and one adult male of *Phyllocycla* sp. from Tigre in the Lower Delta of the Paraná River (30°28′00″ S 62°49′59″ W) (Buenos Aires Province), and six adult males of *Phyllocycla propinqua* from Arroyo León, Department of Eldorado (26°24′04″ S 54°37′07″ W) (Misiones Province), Argentina. Administration of National Parks of Argentina issued the permit for the collection and transport of material in the protected area.

Within the cosmopolitan family Gomphidae, the genus *Phyllocycla* is distributed from southern Mexico to Uruguay and northern Argentina, and the genus *Aphylla* from southeast United States to Uruguay and northern Argentina [25].

The specimens were etherised in the field, their abdomen was longitudinally incised on the dorsal side and they were whole fixed in 3:1 (absolute ethanol: glacial acetic acid). Later, the gonads were dissected out and immersed in fresh fixative for 24 h before storage in 70% ethanol at 4 °C. For meiotic studies, a piece of gonad was placed in 45% acetic acid for 2 to 3 min to facilitate cell spreading and slides were made by the squash technique in iron propionic haematoxylin.

The chromosome number of *Phyllocycla* sp. and *Aphylla* cf. *distinguenda* was previously communicated in [2], as well as a preliminary study of the meiosis of *Phyllocycla propinqua* was described in [26].

3. Results

Aphylla cf. *distinguenda* (2n = 23, n = 11 + X). At spermatogonial metaphase there are 23 chromosomes; the X chromosome is the largest of the complement and lies at the centre of an autosomal ring, where the distinguishable pairs of homologues appear to be close to each other (Figure 1A). During prophase I until diplotene, the large X chromosome shows a large positively heteropyknotic telomeric region, a slim subterminal isopycnotic region and a medium positively heteropyknotic telomeric region (Figure 1B,C). From diplotene onwards, bivalents have one chiasma in submedial or (less frequently) medial position and decrease gradually in size, except for the *m* bivalent, which is half the size of the lower bivalent; the size of the X chromosome is similar to that of the medium bivalents (Figure 1D). At prophase II, the autosomes adopt the typical epsilon ξ-like shape and the X chromosome is composed of a single chromatid (Figure 1E). At metaphase II, the X chromosome lies on the equatorial plate together with the autosomes (Figure 1F).

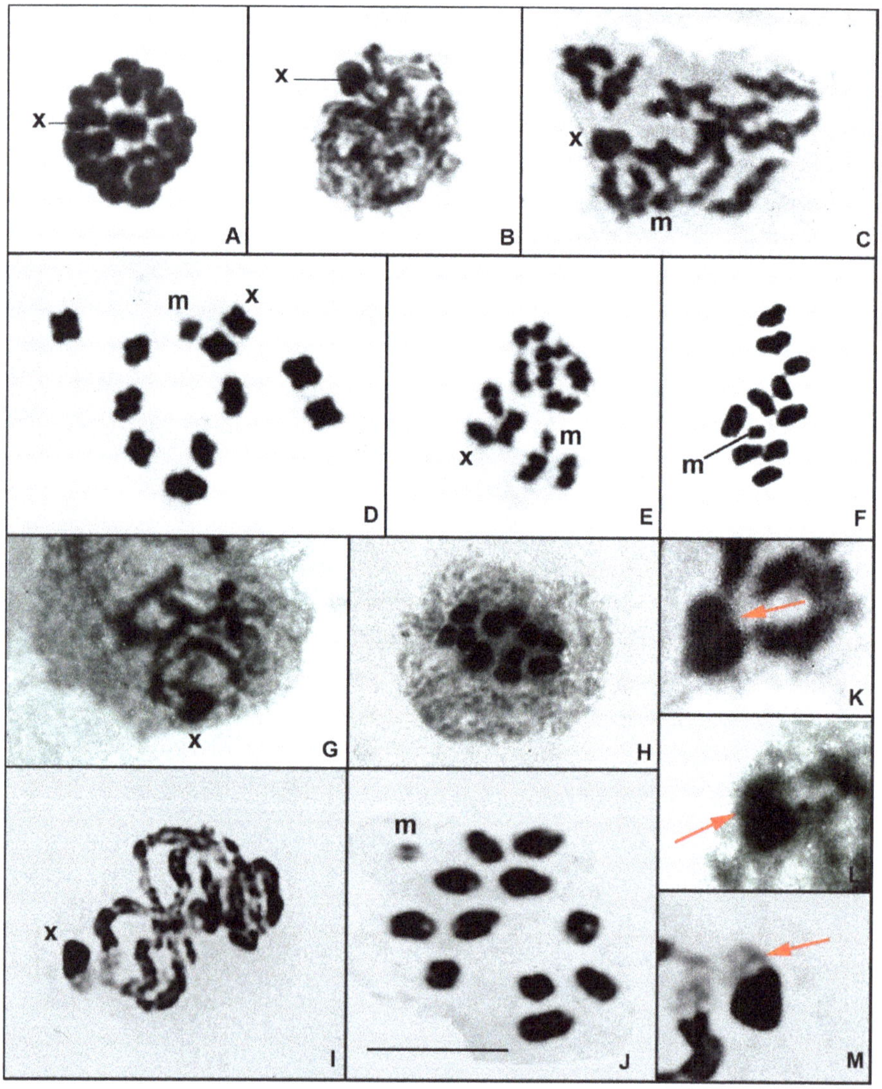

Figure 1. *Aphylla* cf. *distinguenda* (2n = 23, n = 11+ X) (**A–F,K**), *Phyllocycla propinqua* (n = 10 + X) (**G,H,L**) and *Phyllocycla* sp. (n = 11 + X) (**I,J,M**). A—Spermatogonial prometaphase, **B**—Pachytene, **C**—Diplotene, **D**—Diakinesis, **E**—Prophase II, **F**—Metaphase II with n = 11, **G**—Pachytene, **H**—Metaphase I, **I**—Late pachytene, **J**—Prometaphase I, **K–M**—Magnifications of X chromosomes from figures **C**, **G**, **I, respectively**. Arrows point isopycnotic regions. Bar: **A–J** = 10 um, **K–M** = 5 um.

One of the individuals studied showed a variation in the chromosome number at the first and second meiotic divisions. Of 28 diplotenes and diakineses analysed, 23 cells presented 11 + X and 5 10 + X, whereas of 27 prometaphases and metaphases II, 2 presented 11 + X and 25 10 + X (Figure 1E,F).

Phyllocycla propinqua (n = 10 + X). At pachytene, the X chromosome is large and shows a large positively heteropyknotic telomeric region, a slim subterminal isopycnotic region and a small positively heteropyknotic telomeric region similar to *Aphylla distinguenda* (Figure 1G). At diakinesis and subsequent meiotic stages, the sex chromosome becomes isopycnotic with the bivalents and its size is similar to that of the median bivalents (Figure 1H). Bivalents show a terminal chiasma, and a slightly larger bivalent can be recognized, whereas the remaining ones decrease gradually and the m chromosomes are absent (Figure 1H). All prometaphases show 11 chromosomes, at metaphase II, no chromosome is observed out of the equatorial plate, and at anaphase II, all chromosomes migrate synchronously.

Phyllocycla sp. (n = 11 + X). From the early prophase II onwards, the X chromosome is large and positively heteropycnotic (Figure 1I). At pachytene the X chromosome has a large positively heteropycnotic region and a small terminal isopycnotic region (Figure 1I). Bivalents possess a single terminal chiasma and decrease gradually in size, except for the bivalent formed by the small m chromosomes, which are negatively heteropycnotic (Figure 1). At diakinesis, the X chromosome becomes isopycnotic and its size is similar to that of the median bivalents; it is hardly recognisable at prometaphase and metaphase I.

4. Discussion

Cytogenetic studies conducted on 76 species of Odonata revealed a modal number of n = 12 (11 + X) (Table 1). A reduction in modal number through fusions has been reported in five species, resulting in haploid numbers of 10 + X and 9 + X (Table 1). Interpopulation variation in chromosome number has been recorded only in two species (*Asiagomphus melaenops* and *Trigomphus melampus*), probably due to species misidentification.

Table 1. Chromosomal data by Gomphidae species.

	Species	n (Male)	X Size in Mitosis	X Size in Meiosis	Locality	References
1	*Anisogomphus bivittatus* (Selys, 1854)	11 + X 11 + X 11 + X	- LL -	S S S	India India Nepal	[27] [28] [29] as *Temnogomphus bivittatus* (Selys, 1854)
2	*A. occipitalis* (Selys, 1854)	11 + X	-	S -	Nepal India	[29] [30]
3	*Aphylla* cf. *distinguenda* (Campion, 1920)	11 + X	LL	M	Argentina	[2] this work
4	*A. edentata* Selys, 1869	11 + X	-	-	Bolivia	[5]
5	*A. producta* Selys, 1854	11 + X	-	-	Bolivia	[5]
6	*A. theodorina* (Navas, 1933)	11 + X	LL	LL	Brazil	[13]
7	*A. williamsoni* (Gloyd, 1936)	11 + X	M	M	USA	[31]
8	*Arigomphus lentulus* (Needham, 1902)	11 + X	-	-	USA	[32] as *Gomphus lentulus* Needham, 1902
9	*A. pallidus* (Rambur, 1842)	11 + X	-	-	USA	[5] as *Gomphus pallidus* Rambur, 1842

Table 1. *Cont.*

	Species	n (Male)	X Size in Mitosis	X Size in Meiosis	Locality	References
10	*A. submedianus* (Williamson, 1914)	11 + X	-	-	USA	[32] as Gomphus submedianus Williamson, 1914
11	*Asiagomphus melaenops* (Selys, 1854)	9 + X 9 + X 11 + X 11 + X	- - LL LL	- M L L	Japan Japan Japan Japan	[33] [34] [35] [14] all as *Gomphus melaenops* Selys, 1854
12	*Burmagomphus* cf. *arboreus* Lieftinck, 1940	11 + X	-	-	India	[30]
13	*B. divaricatus* Lieftinck, 1964	11 + X	-	M	India	[36]
14	*B. pyramidalis* Laidlaw, 1922	11 + X 11 + X	-	S-M M	India India	[30,37] [36]
15	*B. sivalikensis* Laidlaw, 1922	11 + X	-	M	India	[36]
16	*B. williamsoni* Förster, 1914	11 + X	-	M	India	[36]
17	*Davidius fujiama* Fraser, 1936	11 + X 2n = 24 F	H	LL	Japan	[15]
18	*D. moiwanus* (Okumura, 1935)	11+X	H	M	Japan	[15] as *D. m. moiwanus* (Okumura)
19	*D. nanus* (Selys, 1869)	11 + X 11 + X 11 + X	- - M	S M M	Japan Japan Japan	[38] as *Gomphus hakiensis* Oguma, 1926 [14] [16]
20	*Dromogomphus spinosus* (Selys, 1854)	11 + X	-	-	USA	[32]
21	*D. spoliatus* (Hagen, 1857)	11 + X	-	-	USA	[32]
22	*Epigomphus llama* Calvert, 1903	9 + X	-	-	Bolivia	[5]
23	*Erpetogomphus designatus* Hagen, 1857	11 + X	-	-	USA	[5]
24	*E. diadophis* Calvert, 1905	11 + X	-	-	USA	[5]
25	*E. ophibolus* Calvert, 1905	11 + X	-	M	Mexico	[39]
26	*Gomphoides* sp.	11 + X	-	-	Bolivia	[5]
27	*Gomphus confraternus* Selys, 1873	11 + X	-	-	USA	[32]
28	*G. exilis* Selys, 1854	11 + X 2n = 24F	-	- -	USA Canada	[32] [11]
29	*G. graslini* Rambur, 1842	11 + X	LL	-	France	[10] [11]
30	*G. pulchellus* Selys, 1840	11 + X	-	M-L	France	[40]
31	*G. vulgatissimus* (Linnaeus, 1758)	11 + X	-	s	Russia	[20]
32	*Ictinogomphus decoratus* (Selys, 1854)	11 + X	LL	M-L	Singapur	[41] as *I. decoratus melaenops*

Table 1. *Cont.*

	Species	n (Male)	X Size in Mitosis	X Size in Meiosis	Locality	References
33	*I. pertinax* (Hagen in Selys, 1854)	11 + X	-	M	Japan	[14]
34	*I. rapax* (Rambur, 1942)	11 + X 11 + X 11 + X	H - LL	LL - LL	India India India	[7] as *Ictinus rapax* [42–45] [9] as *Ictinus rapax* Omura 1949
35	*Nepogomphus modestus* (Selys, 1878)	11 + X 11 + X	- M	M M	India India	[46] [22]
36	*Nihonogomphus ruptus* (Selys, 1858)	11 + X	-	S	Russia	[20]
37	*N. viridis* Oguma, 1926	11 + X 11 + X	- H	L L	Japan Japan	[35] [16]
38	*Octogomphus specularis* (Hagen, 1859)	11 + X	-	-	USA	[32]
39	*Onychogomphus forcipatus* (Linnaeus, 1758)	11 + neo-XY			Austria	[11]
40	*O. saundersii* Selys, 1854	11 + neo-XY			India	[12,30,37,47] as *O. s. duaricus* Fraser, 1924
41	*O schmidti* Fraser, 1937	11 + neo-XY			India	[12,30,47,48]
42	*Ophiogomphus bison* Selys, 1873	11 + X/12 + X	-	-	USA	[32]
43	*O. cecilia* (Fourcroy, 1785)	12 F 11 + X 11 + X	- - H	LL LL LL	Finland Russia Russia	[8] as *O. serpentinus* Charp. Syn *Aeschna serpentina* Charpentier, 1825 [19] [17] as *O. c. cecilia* (Four.)
44	*O. colubrinus* Selys, 1854	11 + X	-	-	USA	[32]
45	*O. obscurus* Bartenev, 1909	11 + X	-	-	Russia	[49]
46	*O. occidentalis* Hagen, 1882	11 + X	-	-	USA	[32]
47	*O. rupinsulensis* (Walsh, 1862)	11 + X	-	S-M	USA	[32]
48	*O spinicornis* Selys, 1878	11 + X	-	LL	China	[50] as *O. spinicorne*
49	*Paragomphus capricornis* (Förster, 1914)	11 + X	-	L	Thailand	[51]
50	*P. lineatus* (Selys, 1850)	11 + X 11 + X 11 + X	- - L	M - M	Nepal India India	[29] [30] [22]
51	*Phanogomphus lividus* (Selys, 1854)	11 + X	-	-	USA	[32] as *Gomphus lividus* Selys, 1854
52	*Ph. militaris* (Hagen, 1858)	11 + X	-	-	USA	[32] as *Gomphus militaris* Hagen, 1858

Table 1. *Cont.*

	Species	n (Male)	X Size in Mitosis	X Size in Meiosis	Locality	References
53	*Ph. spicatus* (Selys, 1854)	11 + X	-	-	USA	[32] as *Gomphus spicatus* Selys, 1854
54	*Phyllocycla propinqua* Belle, 1972	10 + X	-	M	Argentina	[51] this work
55	*Phyllocycla* sp.	11 + X	-	-	Bolivia	[5]
56	*Phyllocycla* sp.	11	-	-	Brazil	[52]
57	*Phyllocycla* sp.	11 + X	-	M	Argentina	[2] this work
58	*Phyllogomphoides undulatus* (Needham, 1944)	11 + X	-	S	Surinam	[53]
59	*Progomphus borealis* McLachlan, 1873	11 + X	-	-	USA	[32]
60	*P. intricatus* (Hagen, 1857)	11 + X 11 + neo-XY	- -	- -	Bolivia Brazil	[5] [52]
61	*P. obscurus* (Rambur, 1842)	11 + X	-	-	USA	[32]
62	*P. phyllochromus* Ris, 1918	11 + X	-	-	Bolivia	[5]
63	*Scalmogomphus bistrigatus* (Hagen, 1854)	11 + X 11 + X	- -	LL LL	Nepal India	[29] [30,54] both as *Onychogomphus bistrigatus* (Hagen, 1854)
64	*Shaogomphus postocularis* (Selys, 1869)	11 + X 11 + X	- -	- S	Japan Russia	[16,35] both as *Gomphus postocularis* Selys,1869 [20] as *Gomphus epophtalmus* Selys, 1872
65	*Sieboldius albardae* Selys, 1886	11 + X 11 + X 11 + X	- H H	LL LL LL	Japan Japan Japan	[35] [14] [16]
66	*Sinictinogomphus clavatus* (Fabricius, 1775)	11 + X	LL	M	Japan	[14] as *Ictinogomphus clavatus* (Fabricius, 1775)
67	*Stylogomphus suzukii* (Oguma, 1926)	11 + X	- -	S -	Japan Japan	[55] [42] both as *Gomphus suzukii* Oguma, 1926
68	*Stylurus flavipes* (Charpentier, 1825)	11 + X	-	-	Russia	[48]
69	*S. plagiatus* (Selys, 1854)	11 + X	-	-	USA	[32] as *Gomphus plagiatus* Selys, 1854
70	*S. scudderi* (Selys, 1873)	11 + X	-	-	USA	[32] as *Gomphus scudderi* Selys, 1873

Table 1. *Cont.*

	Species	n (Male)	X Size in Mitosis	X Size in Meiosis	Locality	References
71	*S. townesi* Gloyd, 1936	11 + neo-XY			USA	[31] as *Gomphus townesi* Gloyd, 1936
72	*Trigomphus citimus* (Needham, 1931)	10 + X	-	-	Japan	[33]
		10 + X	-	S	Japan	[34]
		10 + X	-	L	Japan	[14] all as *Gomphus citimus tabei* Asahina, 1949
73	*T. interruptus* (Selys, 1854)	9 + X	-	L	Japan	[14]
74	*T. melampus* (Selys, 1869)	9 + X	-	M	Japan	[55] as *Gomphus melampus* Selys, 1869
		10 + X	-	L	Japan	[55] as *Gomphus unifasciatus* Oguma, 1926
		9 + X	-	M-L	Japan	[35] as *Gomphus melampus bifasciatus* Asahina
		11 + X	-	S	Japan	[34] as *Gomphus m. bifasciatus* Asahina
		10 + X	LL	M-S	Japan	[16]
75	*T. ogumai* Asahina, 1949	10 + X	-	s	Japan	[14]
76	*Zonophora callipus* Selys, 1869	11 + X	LL	M	Surinam	[53]

Notes: **s**—the smallest chromosome, **S**—among small chromosomes/bivalents, **M**—among medium chromosomes/bivalents, **L**—among large chromosomes/bivalents, **LL**—the largest chromosome/bivalent, **H**—huge chromosome. This table is adapted from Table 1, Gomphidae of [4], with information regarding the chromosome X length and with the addition of data of new bibliography.

Five of the 24 species of *Aphylla* so far described have been studied cytogenetically, showing n = 11 + X (Table 1) [1]. In *A. theodorina*, *A.* cf. *distinguenda* (analysed here), *Ictinogomphus decoratus*, *I. rapax* and *Zonophora callipus*, the median to large X chromosome is located in the centre of the mitotic metaphase plate [7,13,41,53]. This is an unusual feature as in most species with small X chromosomes, it does not adopt any particular arrangement at mitosis.

An intraindividual variation in chromosome number was detected in a specimen of *Aphylla* cf. *distinguenda*. This may be related to the fact that the testes of odonates consist of globular cysts arranged around a central duct running the length of each gonad [56,57]. A spermatogenic wave begins at a given point and then proceeds slowly towards the rest of the gonad, so that most cells in each cyst are at the same developmental stage. On this basis, the difference in the chromosome number found in the individual of *Aphylla* cf. *distinguenda* was probably due to the abnormal segregation of a bivalent at meiosis I, leading to its loss in the cell or cells that gave rise to the cysts with 10 + X analysed. Such reduction would be represented at the second meiotic division. An increase in the proportion of cells with a missing chromosome at meiosis II may be the result of a sampling error, but this is unlikely because we analysed the same number of cells at diakinesis. Another possible explanation is that individual cysts were differently affected by different environmental conditions, and that the most represented cysts were those missing one chromosome at meiosis II.

In *Phyllocycla*, cytogenetic studies have been performed in South American specimens from five of the 31 species so far described (Table 1) [1]. The modal number of the family is present in *Phyllocycla* sp. studied here and in *Phyllocycla* sp. studied by [5]. Souza Bueno [52] suggested a sex-determining mechanism other than X for a specimen of *Phyllocycla* sp. (n = 11), but it cannot be identified because the diploid number was not reported. On the other hand, *Phyllocycla propinqua* (n = 10 + X) shows a reduction in the number of autosomes.

This was probably due to autosomal fusion in homozygous condition, as suggested by the presence of a larger bivalent not observed in other species.

Another distinct feature of the species analysed cytogenetically is the spatial arrangement of the large X chromosome at meiosis II, as it is aligned on the equatorial plate at metaphase II and migrates synchronously with the autosomes at anaphase II. In contrast, in species with small X chromosome it is usually outside the metaphase plate at metaphase II and migrates asynchronously with the autosomes (generally lying ahead) at anaphase II.

4.1. Ancestral Chromosome Number of Gomphidae

Kiauta [11] assumed that the ancestral number of the family was 2n = 25 and that the process of reduction in chromosome number occurred in three successive steps, all of which involved the sex chromosome. First, the X chromosome might have fused to an autosome (A), giving rise to the neo-XY system and reducing the chromosome number to 24. In the second step, the neo-Y (A′) chromosome probably fused to another autosome (B); this resulted in a neo-neo-Y chromosome and a neo-X/neo-neo-Y system, with the consequent reduction in the chromosome number to 23. In the third step, the autosomal portion (A) of the neo-X might have been translocated to the autosomal homologue (B′) (originating a chromosome homologous to the neo-neo-Y), thereby restoring the X0 system without change in the chromosome number (2n = 23). It is worth mentioning that the X chromosome most likely retained its original small size after these arrangements.

According to [11], the first two steps were supported by cytogenetic evidence. With regard to the third step for which no evidence was available, the author argued that it was necessary for originating the chromosome complement found in most species of Gomphidae with a detached X chromosome. Kiauta [11] cited *Onychogomphus forcipatus* as the primary example to support the hypothesis of chromosome number reduction from 25 to 24. The author assumed that the variation in chromosome number (12–13 elements) observed among meiotic cells of a same specimen was due to a "reversible fusion" of the X chromosome, that is to say that it was fused to an autosome in some cells and unfused in others. On the other hand, Mola [58] performed the cytological analysis of *Rhionaeschna bonariensis* (12 + neo-XY) and confirmed that such karyotypic variation was due to the presence of univalents derived from desynapsis of the sex bivalent. This explanation could also account for the karyotypic variation in *Onychogomphus forcipatus*. To support the complement reduction from 24 to 23 chromosomes, Kiauta [11] hypothesised that the X chromosome of *Gomphus graslini*, *Ophiogomphus cecilia* and *Ictinogomphus rapax* is large because it was formed by fusion, thereby being a neo-X. Taking into account the rearrangements proposed by this author, at meiosis are expected to observe different configurations. The first configuration is a trivalent formed by the pairing of the neo-neo-Y with the autosome B′ and the neo X. If chiasma formation fails, there are two different configurations. One of them is a heteromorphic bivalent formed by the pairing of the neo-neo-Y with the autosome B′ and a large univalent corresponding to the neo-X. The other is a heteromorphic bivalent formed by the pairing of the neo-neo-Y with the neo-X and a univalent corresponding to the autosome B′. However, *Ictinogomphus rapax* and *Ophiogomphus cecilia* show 11 homomorphic bivalents and the large X chromosome at meiosis [7,9,17,19]. Only the characteristics of the 23 mitotic chromosomes are available for *Gomphus graslini* [11].

Tyagi [12,47] also conducted evolutionary studies of the karyotype of Gomphidae based on species of *Onychogomphus*. This author included the neo-neo-X/neo-neo-Y system to the scheme proposed by Kiauta [11] and suggested that it resulted from the neo-X/neo-neo-Y through the fusion of an autosome with the neo-X, giving rise to the neo-neo-X. The existence of this neo-system was strongly suggested by the finding of a reduced chromosome number in some spermatogonial cells, and the variation in the chromosome number of different meiotic cells was assumed to derive from "unstable fusions" (i.e., "reversible fusions").

Kiauta [11] and Tyagi [47] supported the hypothesis that the ancestral chromosome number of Gomphidae was 25 mainly due to the presence of some species with a very large X chromosome and other species with a complement of 2n = 24 and a neo-XY sex-determination system. However, the theoretical steps are not reflected in the course of meiosis and some of their observations can be interpreted from a different perspective (as mentioned above). In this context, the origins of the large X chromosome and the neo-XY sex-determination system are discussed below.

Later, Perepelov and Bugrov [17] proposed that the large X chromosome of *Ophiogomphus cecilia* originated from the fusion of the original X chromosome with two chromosomes of the same autosomal pair. This hypothesis seems unlikely as it would result in a genetic imbalance in both sexes: in females by the duplication of one chromosome pair and in males by the duplication of one chromosome.

Considering the discussion presented above, together with the fact that the modal chromosome number of the family is 23 (n = 11 + X) in about 86% species, we propose that the ancestral chromosome number of Gomphidae coincides with the modal chromosome number of the family.

4.2. *Characterisation of the Large X Chromosome of Gomphidae*

The size of the sex chromosome has been described or illustrated at the mitosis or meiosis of about 60% of Gomphidae species (Table 1). In most odonates, the X chromosome is the smallest or the second smallest of the complement if the m chromosomes are present. In Gomphidae, the X chromosome is the smallest of the complement in only two species (*Gomphus vulgatissimus* and *Trigomphus ogumai*) [14,20]. It may be the size of the smaller bivalents [20,27–29,34,38,53,55] or the largest of the complement [7–9,13–17,19,29,30,35,50,54], but in most cases its size is similar to that of the medium and large bivalents (Table 1). In mitosis, it may even be considerably larger (huge chromosome) than the other chromosomes of the complement (Table 1).

The distribution of the constitutive heterochromatin in the autosomes of gomphids matches with that of the other families analysed cytogenetically. The heterochromatin is found in the telomeric region of all the chromosomes (except for *Nepogomphus modestus*, with a bivalent lacking C-bands), and the number of repeats may vary among species, chromosomes or different telomeric regions of a same chromosome [6,17,20,22,28,36].

The X chromosome in species of other families with C-banding may be entirely C-positive, may show small C-terminal bands on one or both telomeric regions or may be isopycnotic [6]. Although the X chromosome is entirely C-positive in about half of the species analysed, this is possibly due to a higher degree of contraction (facultative heterochromatinisation in males) rather than to the presence of constitutive heterochromatin.

In Gomphidae, the amount and distribution of constitutive heterochromatin on the X chromosome have provided evidence to explain the origin of its large size. The X chromosome of *Gomphus vulgatissimus* is the smallest of the complement and has a low amount of heterochromatin in both telomeric regions [20]. In *Shaogomphus postocularis* the X chromosome presents a large terminal heterochromatic region covering about half of the chromosome and in *Nihonogomphus ruptus* it has large terminal heterochromatic blocks in both telomeric regions. In these two species, the size of the X chromosome is similar to that of the smaller bivalents [20]. In *Paragomphus lineatus*, the X chromosome is almost entirely heterochromatic and shows an euchromatic submedial region, and in *Nepogomphus modestus* it contains a large heterochromatic region and a small euchromatic segment in terminal position. In both species the X chromosome is of similar size to that of medium bivalents [22]. In *Ophiogomphus cecilia* the sex chromosome contains a large inhomogeneous heterochromatic portion and an euchromatic one with three intercalar heterochromatic segments and in *Davidius fujiama* it presents three large heterochromatic blocks, two of which are located in terminal position. In these two species the X chromosome is the largest of the complement. Although the other two species of *Davidius* were not analysed by C-banding, in meiosis their X chromosome exhibits three darker areas similar to those present in *D. fujiama* [15–17]. Intensely stained blocks can be frequently distinguished in preparations with no banding patterns, which may correspond to C bands [6].

Therefore, the heteropycnotic regions observed at prophase I in the X chromosome of *Aphylla* cf. *distinguenda*, *P. propinqua* and *Phyllocycla* sp. analysed here could be considered as regions of constitutive heterochromatin.

Despite the fact that the heterochromatin of the X chromosome has been studied in a small number of gomphid species, there is a clear trend towards an increase in heterochromatin amount with increasing X chromosome size. This allows us to propose that its large size would have originated by progressive accumulation of repetitive DNAs and heterochromatinisation rather than to fusions, as previously suggested [11,12,17,47].

4.3. Derived Sex-Determining Systems

In Odonata, no reports have been published on sex-determining systems originated by fragmentation of the X chromosome, as documented for other insects with holokinetic chromosomes such as Heteroptera and Lepidoptera [59].

In Odonata families with a small X chromosome (except Gomphidae), the identification of a heteromorphic sex bivalent at the different meiotic stages may be a difficult task, depending on both the size of the autosome with which it might have fused and on the degree of contraction of the bivalents. In about half of the species with the neo- XY system, the sex bivalent in males is homomorphic throughout meiosis and its presence can be mainly inferred from the even number of chromosomes in spermatogonial cells (Table 2). On the contrary, in the other species the sex bivalent is heteromorphic in meiosis I and II, or it is recognised only at diplotene and diakinesis (Table 2).

Table 2. Chromosomal data of species, subspecies, populations or individuals with derived sex determination systems.

Family Species	Suborder	2n Male	n Male	H	SBS	N	Locality	References
	Anisoptera							
Aeshnidae								
Aeshna caerulea (Ström, 1783)			11 + neo-XY	Y	SS	2	Finland	[60]
A. grandis (Linnaeus, 1758)		27	13 + X			-	USSR	[61]
		25	12 + X			-	USSR	[62]
		25	X			-	Finland	[20]
		26	12 + neo-XY	Y	LL	23	Finland	[8,60]
		26F	12 + neo-XX			-		
			12 + neo-XY	Y	LL	8	Netherlands	[10,11,63]
		26	12 + neo-XY	Y	LL	-	Russia	[21]
A. juncea (Linnaeus, 1758)			12 + X			-	USSR	[62]
		26	12 + neo-XY	N	-	14	Finland	[60]
		26F		-	-	3		
			12 + neo-XY	D	L	6	Italy	[57]
		26	12 + neo-XY	Y ‡	L	4	Russia	[21]
A.serrata Hagen, 1856			12 + neo-XY	N	-	1	Finland	[60] as *A.osiliensis fennica*

Table 2. *Cont.*

Family Species	Suborder	2n Male	n Male	H	SBS	N	Locality	References
A. viridis Eversmann, 1836		- 26F 26	12 + neo-XY 12 + neo-XY	N - D ‡	- - LL	3 2 2	Finland Russia	[60] [19]
Anax ephippiger (Burmeister, 1839)		14 14F	6 + neo-XY	D	M	-	India	[64] as *Hemianax ephippiger*
Caliaeschna microstigmata (Schneider, 1845)			6 + neo-XY	N	-	1	Greece	[65]
Gynacanta interioris Williamson, 1923		26	12 + neo-XY	D	M	2	Brazil	[13]
Rhionaeschna bonariensis (Rambur, 1842)		26	12 + neo-XY	Y Y	LL LL	5 2	Argentina Uruguay	[58,66] both as *Aeschna bonariensis*
R. planaltica (Calvert, 1845)		16	7 + neo-XY	Y	SS	2	Argentina	[58,66] both as *Aeschna cornigera planaltica*
Gomphidae								
Onychogomphus forcipatus (Linnaeus, 1758)		24	11 + neo-XY	Y	LL	-	Austria	[11]
† *O. saundersii* Selys, 1854		22? 23?	11 + neo-XY	Y	LL	10	India	[37] as *O. saundersi duaricus*
O. schmidti Fraser, 1937		22	11 + neo-XY	-	LL	-	India	[47,48]
Progomphus intricatus (Hagen in Selys, 1858)		23 24	11 + X 11 + neo-XY	- N	- -	- 3	Bolivia Brazil	[5] [52]
Stylurus townesi Gloyd, 1936		23?	11 + neo-XY	Y	LL	1	USA	[31] as *Gomphus townesi*
Libellulidae								
Crocothemis servilia (Hagen, 1857) *servilia*		25	12 + X				India, Nepal, China, Philippines, Japan Singapore, Korea, Thailand	[7,29,30,42,43,67–75]
C. servilia mariannae ssp. n.		24	11 + neo-XY 11 + neo-XY 11 + neo-XY 11 + neo-XY	N D Y Y	- S SS SS	3 5 - 25	Japan Japan Japan Japan	[76] [69] [74] [75]
Elasmothemis williamsoni (Ris, 1919)		22	11 + neo-XY	N	-	2	Surinam	[53] as *Dythemis williamsoni*
Erythrodiplax media Borror, 1942		 22F 22	10 + X 11 10 + neo-XY	 D	 L	- 1 8	Bolivia, Brazil Brazil Argentina	[5,13,52] [77] [78]
Macrothemis hemichlora (Burmeister,1839)		6	2 + neo-XY	N	-	-	Bolivia	[5]

Table 2. *Cont.*

Family Species	Suborder	2n Male	n Male	H	SBS	N	Locality	References
Micrathyria longifasciata Calvert, 1909		24	11 + neo-XY	Y	LL	8	Argentina	[79]
M. ungulata Foerster, 1907		23	**10 + X_1X_2Y**	Y	M	2	Argentina	[18]
Neurothemis tulia (Drury, 1773)		28	13 + neo-XY 12 + X	Y	U	- 4	India Thailand	[54,67] [73] all *as N.t.tulia*
Orthemis aequilibris Calvert, 1909		12	5 + neo-XY	N	-	1	Surinam	[53]
O.ambinigra Calvert, 1909		12	5 + neo-XY	N	-	19	Argentina	[2,6,80]
O. discolor (Burmeister, 1839)		23 25 23 24 F	11 + X 11 + neo-XY 10 + neo-XY 11 + X			4	Surinam Perú, Surinam, Brazil, Argentina Argentina	[53] as *O. ferruginea* [2,13,52,53,77] all as *O. ferruginea* [6]
O. levis Calvert, 1906		7	**2II + III**	N	-	2	Bolivia	[5]
Orthemis sp.		10	4 + neo-XY	N	-	4	Bolivia	[5] as *O. ferrugínea*
Pseudothemis zonata (Burmeister, 1839)		-	11 + neo-XY	Y	LL	6	Japan	[76]
Trithemis aurora (Burmeister, 1839)		25	12 + X 9 + neo-XY	— -	— -	— 4	India Nepal	[30,81] [29]
Corduliidae								
Somatochlora borisi Marinov, 2001			10 + neo-XY	N	-	7	Bulgaria	[82]
	Zygoptera							
Coenagrionidae								
Ischnura lobata Needham, 1930			13 + neo-XY	Y	LL	5	China	[50]
Leptagrion macrurum (Burmeister, 1839)		30	neo-XY	-	-	2	Brazil	[83]
Mecistogaster sp.2		12	5 + neo-XY	N	-	-	Bolivia	[5]
Lestidae								
Lestes vigilax Selys, 1862			**9II + III**	**-**	-	1	USA	[31]
Megapodagrionidae								
Heteragrion sp. b		26	12 + neo-XY	D	M	2	Brazil	[52]

Notes: **H:** heteromorphism of the sex bivalent: Y yes, N no, D until Diakinesis. ‡: With C Banding. **SBS**: Sex bivalent size: LL the largest of the complement, L among large bivalents, M among medium bivalents, S among small bivalents, SS the smallest of the complement, U sex chromosomes as univalents in all meiotic stages. **N**: number of individuals analysed. **F**: female. **II**: bivalent. **III**: trivalent. †: *O. saundersii* Selys, 1854 or *Nychogomphus duaricus* (Fraser, 1924). Both names were assigned by Kuznetsova and Golub [4] for the species studied by Tyagi [37] as *O. saundersi duaricus*.

Neo-systems are rare, as they have been recorded in 35 species, subspecies, populations or some individuals from a population of a total of more than 600 cytogenetically analysed species of damselflies and dragonflies [4] (Table 2). They have a heterogeneous distribution and are found in only 5 species of three families of Zygoptera and in 30 species of four families of Anisoptera (Table 2). Some genera stand out by the presence of neo-systems in over 25% of their analysed species, such as *Aeshna, Rhionaeschna, Onychogomphus* and *Orthemis* [4] (Table 2).

In Gomphidae, the presence of the neo XY system is always associated with an increase in the modal diploid number of males. The increase in the diploid number may be due to an autosomal fragmentation that in homozygous condition would increase the diploid number to 25. The origin of the neo XY system is due to the fusion of the X chromosome with an autosome that would reduce the diploid number to 24. These two rearrangements may be independent, which is to say that the X chromosome does not necessarily have to fuse with one of the fragmented autosomes. The sex bivalent is the largest of the complement in the three species of *Onychogomphus* and in *Stylurus townesi*; suggesting that the fusion involved the X chromosome and the largest autosome [11,31,37,47,48]. The fact that the three species of *Onychogomphus* studied exhibit the same autosomal pair involved in the fusion may indicate that this rearrangement occurred in a common ancestor. In *Stylurus townesi* the fusion most likely had an independent origin since other three species of the genus have the modal haploid number (11 + X) of the family (Table 1). The presence of a neo-XY system in individuals of *Progomphus intricatus* from Brazil was postulated on the basis of the diploid number; however, its presence could not be detected at meiosis [52]. The occurrence of a chromosome number higher than the modal number of the family together with a neo-XY system has been reported not only for Gomphidae but also for *Neurothemis tulia* (Libellulidae); *Ischnura lobata; Leptagrion macrurum* (Coenagrionidae) and *Heteragrion* sp. b (Megapodagrionidae) [50,52,54,67,83].

In seven other species, the presence of the neo-XY system is associated with a remarkable reduction in the entire chromosome complement (diploid numbers between 6 and 16) (Table 2). However, the existence of some species (e.g., *Perithemis lais* (Perty, 1834), *Anax guttatus* (Burmeister, 1839) and *Rhionaeschna intricata* (Martin, 1908)) with reduced complements not involving the sex chromosome may indicate that it is not predisposed towards fusion [5,13,68].

Summarizing, the neo-XX/neo-XY sex-determining system has been found in 24 species (or 26, considering *Orthemis levis* and *Lestes vigilax*, see below), in one subspecies proposed by Kiauta [69] (*Crocothemis servilia mariannae* ssp. n.), in one or more populations of six species and in one or some individuals from a population of *Orthemis discolor* [6,53] (Table 2).

The multiple $X_1X_1X_2X_2/X_1X_2Y$ system is another derived sex-determining system present in *Micrathyria ungulata* [18] and probably in *Orthemis levis* and *Lestes vigilax* based on the presence of a trivalent in the individuals studied [5,31]. An alternative possibility is that the latter species have a neo-XY system and an autosomal trivalent, though no heterozygosity has ever been reported for an autosomal fusion or fragmentation.

5. Conclusions

The study of the meiosis and the characteristics of the heterochromatin in the X chromosome of three species of Gomphidae, together with data from the bibliography led us to propose that the ancestral diploid number of the family was 23 and that the unusually large size of the sex chromosome was due to an increase in heterochromatin rather than to structural rearrangements, as previously claimed. We also propose that the increase in diploid number in species with neo-XY systems in gomphids originated by mechanisms of autosomal fragmentation and X-autosome fusion. Moreover, the analysis of the neo-sex determining systems allowed us to pose that two other species of odonates that present trivalents could have a multiple X1X1X2X2/X1X2Y sex-determining system.

Author Contributions: Conceptualization, L.M.M. and S.S.A.; methodology, S.S.A. and P.J.R.; Investigation, S.S.A., P.J.R., M.F.F. and L.M.M.; writing—original draft preparation, L.M.M.; writing—review and editing, S.S.A., P.J.R., M.F.F. and L.M.M.; funding acquisition, L.M.M. All authors have read and agreed to the published version of the manuscript.

Funding: This research was supported by grants from the National Council of Scientific and Technological Research (CONICET), Multi-year Investigation Project (PIP) N° 11220200102115CO.

Institutional Review Board Statement: Not applicable.

Data Availability Statement: The data presented in this study are available on request from the corresponding author.

Acknowledgments: We would like to thank to J. Muzón for the identification of the specimens of *Phyllocycla propinqua* and to A. Rodrigues Capítulo for the identification of the specimens of *Phyllocycla* sp.

Conflicts of Interest: The authors declare no conflict of interest.

References

1. Novelo-Gutiérrez, R.; Ramírez, A.; González-Soriano, E. Chapter 14.2—Superfamily Gomphoidea. In *Thorp and Covich's Freshwater Invertebrates, 4th ed*; Hamada, N., Thorp, J.H., Rogers, D.C., Eds.; Academic Press: Cambridge, MA, USA, 2018; Volume 3, pp. 377–397.
2. Mola, L.M. *Cytogenetics of American Odonata. In Odonata: Biology of Dragonflies*, 1st ed.; Tyagi BK. Scientific Publishers: Jodhpur, India, 2007; pp. 153–173.
3. Nokkala, S.; Laukkanen, A.; Nokkala, C. Mitotic and meiotic chromosomes in *Somatochlora metallica* (Corduliidae, Odonata). The absence of localized centromeres and inverted meiosis. *Hereditas* **2002**, *136*, 7–12. [CrossRef] [PubMed]
4. Kuznetsova, V.G.; Golub, N.V. A checklist of chromosome numbers and a review of karyotype variation in Odonata of the world. *Comp. Cytogenet.* **2020**, *14*, 501–540. [CrossRef] [PubMed]
5. Cumming, R.B. Cytogenetic Studies in the Order Odonata. Ph.D. Thesis, University of Texas, Austin, TX, USA, 1964.
6. Mola, L.M.; Fourastié, M.F.; Agopian, S.S. High karyotypic variation in *Orthemis* Hagen, 1861 species, with insights about the neo-XY in *Orthemis ambinigra* Calvert, 1909 (Libellulidae, Odonata). *Comp. Cytogenet.* **2021**, *15*, 355–374. [CrossRef] [PubMed]
7. Asana, J.J.; Makino, S. A comparative study of the chromosomes in the Indian dragonflies. *J. Fac. Sci. Hokkaido Univ. Zool.* **1935**, *4*, 67–86.
8. Oksala, T. Zytologische Studien an Odonaten. III. Die Ovogenese. *Ann. Acad. Sci. Fenn.* **1945**, *9*, 1–132.
9. Omura, T. The spermatogenesis of an Indian dragonfly, *Ictinus rapax* (Rambur) with special reference to the behaviour of the spermatozoa in the cyst. *Biol. J. Okayama Univ.* **1952**, *1*, 103–146.
10. Kiauta, B. Morphology and kinetic behaviour of the odonate sex chromosomes, with a review of the distribution of sex determining mechanisms in the order. *Genen. Phaenen.* **1968**, *12*, 21–24.
11. Kiauta, B. Sex chromosomes and sex determining mechanisms in Odonata, with a review of the cytological conditions in the family Gomphidae, and reference to the karyotypic evolution in the order. *Genetica* **1969**, *40*, 127–157. [CrossRef]
12. Tyagi, B.K. Cytogenetics, karyosystematics and cytophylogeny of the Indian Odonata. *Indian Rev. Life Sci.* **1986**, *6*, 215–229.
13. Ferreira, A.; Kiauta, B.; Zaha, A. Male germ cell chromosomes of thirty-two Brazilian dragonflies. *Odonatologica* **1979**, *8*, 5–22.
14. Kiauta, B.; Kiauta, M. List of species, with chromosome number and preliminary notes on the karyotypes of the Odonata, collected in May, 1979 and August, 1980 by the members of the Kansai Research Group of Odonatology. Personal Communication, Utrecht, Holland, 14 July 1982.
15. Suzuki, K.J.; Saitoh, K. Germ-line chromosomes of two species of *Dadivíus* with special reference to the sex chromosomes (Anisoptera, Gomphidae). *Odonatologica* **1988**, *17*, 275–280.
16. Suzuki, K.J.; Saitoh, K. A revised chromosome study of Japanese Odonates (I). Chromosomes of 14 species belonging to nine families. *Sci. Rep. Hirosaki Univ.* **1990**, *37*, 38–49.
17. Perepelov, E.; Bugrov, A.G. C-heterochromatin in chromosomes of *Ophiogomphus cecilia cecilia* (Four.) (Anisoptera: Gomphidae) with notes on the sex chromosome origin in the species. *Caryologia* **2001**, *54*, 169–172. [CrossRef]
18. Mola, L.M.; Papeschi, A.G.; Carrillo, E.T. Cytogenetics of seven species of dragonflies. *Hereditas* **1999**, *131*, 147–153. [CrossRef]
19. Perepelov, E.A.; Bugrov, A.G.; Warchałowska-Śliwa, E. C banded karyotypes of some dragonfly species from Russia. *Folia Biol.* **1998**, *46*, 137–142.
20. Perepelov, E.A.; Bugrov, A.G.; Warchalowska-Sliwa, E. C-banded karyotypes of some dragonfly species from Russia. II. The families Cordulegasteridae, Corduliidae and Gomphidae. *Folia Biol.* **2001**, *49*, 175–178.
21. Perepelov, E.; Bugrov, A.G. Constitutive heterochromatin in chromosomes of some Aeshnidae, with notes on the formation of the neo-XY/neo-XX mode of sex determination in *Aeshna* (Anisoptera). *Odonatologica* **2002**, *31*, 77–83.
22. Walia, G.K.; Chahal, S.S. Distribution of constitutive heterochromatin and nucleolar organizer regions in two species of family Gomphidae (Odonata: Anisoptera). *Nucleus* **2014**, *57*, 223–227. [CrossRef]

23. Kuznetsova, V.G.; Maryańska-Nadachowska, A.; Shapoval, N.A.; Anokhin, B.A.; Shapoval, A.P. Cytogenetic characterization of eight Odonata species originating from the Curonian Spit (the Baltic Sea, Russia) using C-banding and FISH with 18S rDNA and telomeric (TTAGG)n probes. *Cytogenet. Genome Res.* **2018**, *153*, 147–157. [CrossRef]
24. Walia, G.K.; Devi, M. Distribution of constitutive heterochromatin in four species of genus *Copera* of family Platycnemididae (Odonata: Zygoptera) from India. *Int. J. Life Sci.* **2018**, *6*, 457–461.
25. Garrison, R.W.; von Ellenrieder, N.; Louton, J.A. *Dragonfly Genera of the New World. An illustrated and Annotated Key to the Anisoptera*, 1st ed.; The Johns Hopkins University Press: Baltimore, MD, USA, 2006; 368p.
26. De Gennaro, D. Análisis Meiótico y Caracterización de la Heterocromatina en Especies Argentinas de Anizoptera (Odonata). Licenciatura´s Thesis, Universidad de Buenos Aires, Buenos Aires, Argentina, 2004.
27. Das, C. Studies on the association between non-homologous chromosomes during meiosis in four species of the Indian dragonflies (Odonata). *J. Zool. Soc. India* **1956**, *8*, 119–132.
28. Walia, G.K.; Chahal, S.S. Linear differentiation of chromosomes of *Anisogomphus bivittatus* Selys, 1854 from India (Odonata: Anisoptera: Gomphidae). *Int. J. Entomol.* **2020**, *5*, 120–122.
29. Kiauta, B. *Cytotaxonomy of Dragonflies, with Special Reference to the Nepalese Fauna*; Lectures Delivered at the Tribhuvan University, Kathmandu, II; Nepal Research Center: Kathmandu, Nepal, 1975; 78p.
30. Tyagi, B.K. Studies on the Chromosomes of Odonata of Dun Valley (Dehradun, India). Ph.D. Thesis, University of Garhwal, Srinagar, 1978. In Tyagi 1982.
31. Kiauta, B.; van Brink, J.M. Male chromosome complements of some Florida dragonflies, United States. *Odonatologica* **1978**, *7*, 15–25.
32. Cruden, R.W. Chromosome numbers of some North American dragonflies (Odonata). *Can. J. Genet. Cytol.* **1968**, *10*, 200–214. [CrossRef] [PubMed]
33. Toyoshima, H.; Hirai, H. Studies on chromosomes of four dragonflies from Kagawa Prefecture. *Kagawa Biol.* **1953**, *1*, 17–19.
34. Hirai, H. Chromosomes of six species of dragonflies. *Zool. Mag.* **1956**, *65*, 198–202.
35. Omura, T. A comparative study of the spermatogenesis in the Japanese dragonfly II: Family Aeschnidae, Gomphidae and Calopterygidae. *Biol. J. Okayama Univ.* **1957**, *3*, 1–86.
36. Walia, G.K.; Chahal, S.S.; Singh, N. Cytogenetic studies on three species of genus *Burmagomphus* of family Gomphidae (Odonata: Anisoptera) from India. *Intern. J. Zool. Investig.* **2021**, *7*, 272–277.
37. Tyagi, B.K. A note on the karyotypes of *Burmagomphus pytamidalis* Laidlow and *Onychogomphus saundersi duaricus* Faser (Anizoptera; Gomphidae). *Odonatologica* **1977**, *6*, 277–282.
38. Kichijo, H. Chromosomes of *Tachopteryx pryeri* and *Gomphus hakiensis* (Odonata, Aeshnidae). *Jpn. J. Genet.* **1939**, *15*, 287–289. [CrossRef]
39. Kiauta, B. The chromosomes of four Neotropical dragonflies from Mexico. *CIS* **1970**, *11*, 8–9.
40. Kiauta, B. Notes on new or little known dragonfly karyotypes IV. Spermatocyte chromosomes of *Calopterix splendens splendens* (Harris) (Zygoptera. Calopterygidae), *Gomphus pulchellus* Selys, and *Libellula depressa* Linnaeus (Anisoptera: Gomphidae, Libellulidae) from Northern France. *Genen. Phaenen.* **1973**, *16*, 55–60.
41. Kiauta, B.; Kiauta, M. The chromosome numbers of eleven dragonfly species from Singapore. *Not Odonatol.* **1982**, *1*, 164–165.
42. Makino, S.A. comparative study of the chromosomes in the Indian dragonflies. *Jpn. J. Genet.* **1935**, *11*, 234–235. [CrossRef]
43. Kichijo, H. Insect chromosomes. IV. Order of dragonflies, Pt. 2. *Nagasaki Med. J.* **1942**, *10*, 1639–1648, In Kuznetsova & Golub 2020.
44. Dasgupta, J. Cytological studies of some Indian dragonflies. II: A study of the chromosomes during meiosis in thirty species of Indian Odonata (Insecta). *Proc. Zool. Soc.* **1957**, *10*, 1–65, In Kuznetsova & Golub 2020.
45. Bagga, S. Cytology and Cytochemistry of Gametogenesis of India Dragonflies. Ph.D. Thesis, Delhi University, Delhi, India, 1961. In Tyagi 1982.
46. Walia, G.K.; Sandhu, R.; Goyal, S. Cytogenetical analysis of *Nepogomphus modestus* from Palampur area of Himachal Pradesh, India (Gomphidae: Anisoptera). *Chromosome Sci.* **2006**, *9*, 99–100.
47. Tyagi, B.K. Cytotaxonomy of the genus Onychogomphus Selys (Odonata: Anisoptera, Gomphidae), with a Special Reference to the Evolution of the Sex-Determining Mechanism and Reduced Chromosome Number in the family Gomphidae. In *Proceedings of the First Indian Symposium of Odonatology*; Mathavan, S., Ed.; Madurai Kamaraj University: India, 1985; pp. 217–228.
48. Tyagi, B.K. The chromosome numbers and sex-determining mechanisms newly recorded in thirteen Indian dragonflies (Odonata). *CIS* **1978**, *25*, 5–7.
49. Perepelov, E.; Bugrov, A.G. The constituent geterochromatin in karyotypes of dragonflies. *Belyshevia* **2001**, *1*, 10–13, In Kuznetsova & Golub 2020.
50. Zhu, H.; Wu, J. Notes on the male germ cell karyotypes of some Odonata from the Shanxi Province, China. *Not. Odonatol.* **1986**, *2*, 118–120.
51. Kiauta, B.; Kiauta, M.A.J.E. The chromosome numbers of some Odonata from Thailand. *Not. Odonatol.* **1983**, *2*, 27–28.
52. Souza Bueno, A.M. Estudios Vromosomicos na Ordem Odonata. Master's Thesis, Universidad Estatal Paulista, São Paulo, Brazil, 1982.
53. Kiauta, B. The karyotypes of some Anisoptera from Surinam. *Odonatologica* **1979**, *2*, 267–283.
54. Tyagi, B.K. Cytotaxonomy of Indian dragonflies. *Indian Rev. Life Sci.* **1982**, *2*, 149–161.

55. Oguma, K. A comparative study of the spermatocyte chromosome in allied species of the dragonfly. *J. Fac. Sci. Hokkaido Univ. Zool.* **1930**, *6*, 1–32.
56. Srivastava, B.K. Male reproductive organs of *Crocothemis servilia servilia* Drury (Libellulidae: Odonata). *Zool. Anz.* **1963**, *170*, 386–392.
57. Kiauta, B. Studies on the germ cell chromosome cytology of some cytotaxonomically interesting or hitherto not studied Odonata from the autonomous region Friuli-Venezia Giulia (northern Italy). *Atti. Mus. Civ. Stor. Nat. Trieste* **1971**, *27*, 65–127.
58. Mola, L.M. Post-reductional meiosis in *Aeshna* (Aeshnidae, Odonata). *Hereditas* **1995**, *122*, 47–55. [CrossRef]
59. Mola, L.M.; Papeschi, A.G. Holokinetic chromosomes at a glance. *J. Basic Appl. Genet.* **2006**, *17*, 17–33.
60. Oksala, T. *Zytologische Studien an Odonaten: 1. Chromosomenverhältnisse bei der Gattung Aeschna, mit Besonderer Berücksichtigung der Postreduktionellen Teilung der Bivalente*, 1st ed.; Suomalainen Tiedeakatemia: Helsinki, Finland, 1943; Volume 4, p. 64.
61. Fuchsówna, J.; Sawczyńska, J. Observations on the heterochromosomes during the spermatogenesis in dragonflies (Odonata). Pt. I. *Aeschna grandis* L. and *Libellula quadrimaculata* L. *Arch. Sci. Soc.* **1928**, *9*, 177–197.
62. Makalowskaja, W.N. Comparative karyological studies of dragonflies (Odonata). *Arch. Russes. Anat. Histol. Embryol.* **1940**, *25*, 24–39.
63. Kiauta, B. Considerations on the evolution of the chromosome complement in Odonata. *Genetica* **1967**, *38*, 430–446. [CrossRef] [PubMed]
64. Seshachar, B.R.; Bagga, S. Chromosome number and sex-determining mechanism in dragonfly *Hemianax ephippiger* (Burmeister). *Cytologia* **1962**, *27*, 443–449. [CrossRef]
65. Kiauta, B. Notes on new or little known dragonfly karyotypes, 2. Male germ cell chromosomes of four East Mediterranean species: *Lestes barbarus* (Fabricius), *Calopteryx splendens amasina* Bartenev (Zygoptera: Lestidae, Calopterygidae), *Caliaeschna microstigma* (Schneider) and *Orthetrum taeniolatum* (Schneider) (Anisoptera: Aeshnidae, Libellulidae). *Genen. Phaenen.* **1972**, *15*, 95–98.
66. Mola, L.M.; Papeschi, A.G. Karyotype evolution in *Aeshna* (Aeshnidae: Odonata). *Hereditas* **1994**, *121*, 185–189. [CrossRef]
67. Ray Chaudhuri, S.P.; Dasgupta, J. Cytological studies on the Indian dragonflies I. Structure and behaviour of chromosomes in six species of dragonflies (Odonata). *Proc. Zool. Soc. Bengal.* **1949**, *2*, 81–93.
68. Kiauta, B.; Kiauta, M.A.J.E. The chromosome numbers of sixteen dragonfly species from the Arun Valley, Eastern Nepal. *Not. Odonatol.* **1982**, *9*, 143–146.
69. Kiauta, B. The status of the Japanese *Crocothemis servilia* (Drury) as revealed by karyotypic morphology (Anisoptera: Libellulidae). *Odonatologica* **1983**, *12*, 381–388.
70. Yadav, J.S. A note on the karyotypic variability in *Crocothemis erythraea* (Brullé) and *C. servilia* (Drury) (Anisoptera, Libellulidae). *Not. Odonatol.* **1979**, *1*, 77–79.
71. Kiauta, B.; Kiauta, M.A.J.E. On a small collection of dragonfly karyotypes from the Philippines. *Odonatologica* **1980**, *9*, 237–245.
72. Kiauta, B.; Kiauta, M. Further notes on Philippine odonate karyotypes. *Not. Odonatol.* **1983**, *2*, 14–15.
73. Katatani, N. On the chromosomes of dragonflies, 1. Synopsis on the studies in some Japanese dragonflies. *Aeschna* **1987**, *20*, 21–31.
74. Higashi, K.; Kayano, H. The distribution of distinct karyomorphs of *Crocothemis servilia* Drury (Anisoptera, Libellulidae) in Kyushu and the south-western islands of Japan. *Jpn. J. Entomol.* **1993**, *61*, 1–10.
75. Higashi, K.; Lee, C.E.; Kayano, H.; Kayano, A. Korea strait delimiting distribution of distinct karyomorphs of *Crocothemis servilia* (Drury) (Anisoptera: Libellulidae). *Odonatologica* **2001**, *30*, 265–270.
76. Omura, T. A comparative study of the spermatogenesis in the Japanese dragonflies. I. Family Libellulidae. *Biol. J. Okayama Univ.* **1955**, *2*, 95–135.
77. Kiauta, B.; Boyes, J.W. Cytology of ten South American Libellulidae, with cytophylogenetic consideration of the genera *Orthemis* Hagen and *Erythrodipax* Brauer (Odonata, Anisoptera). *Genetica* **1972**, *43*, 407–421. [CrossRef]
78. Mola, L.M. Meiotic studies in nine species of *Erythrodiplax* (Libellulidae, Odonata). Neo-XY sex chromosome system in *Erythrodiplax media*. *Cytologia* **1996**, *61*, 349–357. [CrossRef]
79. Agopian, S.S.; Mola, L.M. Intra and interspecific karyotype variability in five species of Libellulidae (Odonata, Anisoptera). *Caryologia* **1988**, *41*, 69–78. [CrossRef]
80. Agopian, S.S.; Mola, L.M. Bandeo C en Odonata. In Proceedings of the XV Congreso de la Sociedad Argentina de Genética, Corrientes, Argentina, 16–21 September 1984; p. 17.
81. Oguma, K.; Asana, J.J. Additional data to our knowledge on the dragonfly chromosome with a note on the occurrence of X-Y chromosome in the ant-lion (Neuroptera). *J. Fac. Sci. Hokkaido Univ. Zool.* **1932**, *1*, 133–142.
82. Grozeva, S.M.; Marinov, M.G. Cytogenetic study of *Somatochlora borisi* Marinov, 2001 (Odonata: Corduliidae), and three relative species. *Acta Zool. Bulg.* **2007**, *59*, 53–58.
83. Kiauta, B. The karyotype of the damselfly, *Leptagrion macrurum* (Burmeister, 1839), and its possible origin, with a note on the cytotaxonomic affinities of the genus (Zygoptera: Coenagrionidae). *Odonatologica* **1972**, *1*, 31–35.

diversity

MDPI

Review

A Bibliometric Analysis of the Global Research in Odonata: Trends and Gaps

José Max Barbosa Oliveira-Junior [1,2,*], Tainã Silva Rocha [2], Suellen Furtado Vinagre [3,4], Jair Costa Miranda-Filho [3,4], Cristian Camilo Mendoza-Penagos [4,5], Karina Dias-Silva [2,6], Leandro Juen [2,4] and Lenize Batista Calvão [2,4,7]

1 Programa de Pós-Graduação em Sociedade, Natureza e Desenvolvimento (PPGSND), Programa de Pós-Graduação em Sociedade, Ambiente e Qualidade de Vida (PPGSAQ), Programa de Pós-Graduação em Biodiversidade (PPGBEES), Instituto de Ciências e Tecnologia das Águas (ICTA), Universidade Federal do Oeste do Pará (UFOPA), Rua Vera Paz, s/n (Unidade Tapajós) Bairro Salé, Santarém 68040-255, Pará, Brazil
2 Programa de Pós-Graduação em Ecologia (PPGECO), Instituto de Ciências Biológicas (ICB), Universidade Federal do Pará (UFPA), Rua Augusto Correia, N.1, Bairro Guamá, Belém 66075-110, Pará, Brazil
3 Programa de Pós-Graduação em Ecologia Aquática e Pesca (PPGEAP), Instituto de Ciências Biológicas (ICB), Universidade Federal do Pará (UFPA), Rua Augusto Correia, N.1, Bairro Guamá, Belém 66075-110, Pará, Brazil
4 Laboratório de Ecologia e Conservação (LABECO), Instituto de Ciências Biológicas (ICB), Universidade Federal do Pará (UFPA), Rua Augusto Correia, N.1, Bairro Guamá, Belém 66075-110, Pará, Brazil
5 Programa de Pós-Graduação em Zoologia (PPGZOOL), Universidade Federal do Pará (UFPA), Rua Augusto Correia, N.1, Bairro Guamá, Belém 66075-110, Pará, Brazil
6 Programa de Pós-Graduação em Biodiversidade e Conservação (PPGBC), Universidade Federal do Pará (UFPA), Rua Coronel José Porfírio, N.2515, Bairro São Sebastião, Altamira 68372-040, Pará, Brazil
7 Programa de Pós-Graduação em Ciências Ambientais (PPGCA), Universidade Federal do Amapá (UNIFAP), Rodovia Juscelino Kubitscheck, km 02, Jardim Marco Zero, Macapá 68903-419, Amapá, Brazil
* Correspondence: jose.mbo@ufopa.edu.br

Citation: Oliveira-Junior, J.M.B.; Rocha, T.S.; Vinagre, S.F.; Miranda-Filho, J.C.; Mendoza-Penagos, C.C.; Dias-Silva, K.; Juen, L.; Calvão, L.B. A Bibliometric Analysis of the Global Research in Odonata: Trends and Gaps. *Diversity* **2022**, *14*, 1074. https://doi.org/10.3390/d14121074

Academic Editors: M. Olalla Lorenzo-Carballa, Ricardo Koroiva and Alan N. Andersen

Received: 20 July 2022
Accepted: 2 December 2022
Published: 5 December 2022

Abstract: Insects of the order Odonata have been used as indicators of environmental quality in different aquatic systems around the world. In this context, we conducted a bibliometric analysis to understand the general patterns of research on Odonata published in the past decade (2012–2021). We extracted literature from the Web of Science (WoS) in the advanced search option and used search terms related to Odonata plus search strings for each term. A total of 2764 Odonata publications were identified. The journals with the most published articles on Odonata were *Zootaxa*, *International Journal of Odonatology* and *Odonatologica*. The countries with the most Odonata publications were the USA, Brazil and China. Most studies were conducted on streams, ponds and rivers. Ecology, taxonomy and behavior were the main study topics. Of the total articles on Odonata, 982 involved Zygoptera and 946 Anisoptera. Another 756 studies were focused on both suborders. The increase in ecological and taxonomic studies of Odonata reflects the dynamic characteristics of this order, and its relatively well-defined systematics, especially in the case of adults. Despite the recent increase in the number of publications, there are still many gaps related to topics such as biogeography, parasitism, competition within and between species, evolutionary and phylogenetic relationships, as well as studies of the eggs (e.g., their development) and larval exuviae (e.g., their morphological features).

Keywords: Anisoptera; Zygoptera; dragonflies; damselflies; tendencies and shortfalls; global research; scientific production

1. Introduction

Aquatic insects have been employed as indicators of environmental quality in various types of freshwater systems worldwide [1]. Among the aquatic insect orders, Odonata (dragonflies and damselflies—see Supplementary Material Figure S1) have stood out because of their high habitat specificity and well-resolved taxonomy [2–4]. Furthermore, compared to aquatic macroinvertebrates, the use of Odonata adults for biomonitoring has several advantages. For instance, most species can be recognized quickly and captured in the field; they are distributed in a wide range of habitats, are sensitive to changes in water quality and ecological conditions of the surrounding environment, and the species assembly is typically large enough for assessments, especially in the tropics [5,6]. In addition, there are some Odonata species with antagonistic interactions, allowing the development of environmental quality indices [7,8].

Equally important, dragonflies arouse both cultural and academic interest [9]. The charisma of these animals, their grand flight maneuvers and vibrant colors attract the attention of many people, which explains the increasing number of partnership networks between researchers and dragonfly lovers [10–12], as well as citizen science programs [13–15]. Furthermore, because of their rich evolutionary history [16,17] and their ecological [18] and taxonomic particularities [3,4], dragonflies have been the focus of numerous investigations.

Currently, around 6376 species of odonates are known worldwide, with estimations suggesting that between 1000 and 1500 species have yet to be discovered [19,20]. Except for Antarctica, dragonflies are distributed on all continents, with the greatest diversity found in the Tropics (Paleotropics and Neotropics) [19]. Throughout their distribution range, it is possible to find them associated with different lentic (ponds, swamps, marshes, pools, wells) or lotic (rivers, streams, waterfalls, springs) water bodies, where they perform their development, hunting and breeding activities [2].

Odonates have an amphibious life cycle, meaning that part of their life is spent in the water as larvae and the other part is spent in the environment adjacent to the water bodies, as flying adults [2]. Both life stages have unique characteristics, causing them to respond differently to environmental changes. For example, Odonata larvae are more sensitive to changes in water's physical and chemical characteristics [21], whereas adults are more sensitive to changes in riparian vegetation [22]. Therefore, Odonata are widely used as indicators of environmental quality. They respond very well to changes in ecosystems, which can be evaluated using different methods such as: surrogates; taxa/species richness; species composition and the ratio between the suborders based on adult studies conducted at certain localities [23–26]; the developmental stage [21,27]; multimeric indices [28,29]; fluctuating asymmetry [30]; behavioral diversity [31]; ethodiversity [32]; phylogenetic diversity [33,34]; morphology [25,34] or the taxonomic level used (for establishing cost-benefit monitoring programs), see [27,35].

There has been an increase in odonate research globally since the beginning of the century, with an increase of 76.27% in publications between 2000 and 2013 [36]. The main Odonata research focus has been ecology, followed by taxonomy, morphology, phylogeny, and biomonitoring [36]. Despite the increase in published studies, most research is published in peer-reviewed journals with restricted and difficult access to their content (i.e., without open access) [37,38].

Therefore, we used a bibliometric analysis to understand the general patterns of Odonata research published between 2012 and 2021. Bibliometric analysis is a popular and rigorous method for exploring and analyzing large volumes of scientific data [39]. Specifically, we assessed nine questions: (i) what was the year and (ii) which were the journals were the articles were published; (iii) in which countries did the studies occur; (iv) in which habitats where the studies conducted; (v) what was the research focus; (vi) which suborders were studied; (vii) which life stage was studied; (viii) what was the taxonomic resolution, and (ix) which were the most commonly used keywords? This type of analysis provides information that enables clarification of the context of Odonata research and serves as a basis for directing future research efforts towards areas that need it most.

2. Materials and Methods

2.1. Database Search Using Keywords

We used a systematic method to identify, analyze, and summarize studies published on Odonata from 2012 to 2021 in the Web of Science (WoS) database (main collection—

https://www.webofscience.com/ accessed on 3 March 2022). WoS is one of the world's leading citation databases, and an increasing number of articles have used (or at least mentioned) this database for academic research [40].

We extracted the literature by using the WoS advanced search option and Odonata related keywords with synonyms or terms with related meaning, and search strings for each term. The search strings for the terms were connected by the boolean OR operator, using the search type "Topical" type in the WoS. The studies were compiled using the following keywords: odonat* OR dragonfl* OR damselfl* OR anisopter* OR zygopter* OR anisozygopter* (Figure 1). The search was carried out on 3 March 2022.

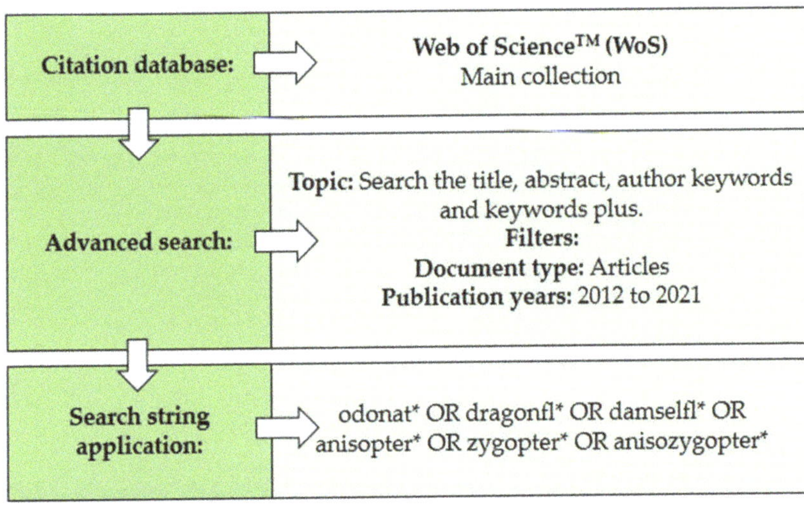

Figure 1. Methodology overview, with search terms and search strings (boolean operator) used to obtain global research on Odonata, available from the Web of Science database (2012–2021).

2.2. Criteria to Include or Exclude Studies

We created a spreadsheet that included the results from the WoS database. We then downloaded all the studies in the spreadsheet with institutional support from the Federal University of Western Pará (Ufopa) via the Federated Academic Community (CAFe). We thoroughly analyzed the title, abstract, keywords, materials and methods, and results of selected studies, looking for approaches that included Odonata. Only documents containing the following three scopes were considered: (1) studies focused on Odonata, regardless of the life stage or study subject; (2) articles (excluding literature reviews, books or book chapters); (3) papers published between 2012 and 2021 (both years included). The study selection was carried out and revised by all authors to ensure the correct exclusion and inclusion.

Subsequently, from each document we extracted the following information: (i) year of publication; (ii) publishing journal; (iii) continent and country where the study was conducted; (iv) habitat where the study was conducted; (v) research focus; (vi) the target Odonata suborder/s; (vii) life stage studied; (viii) taxonomic resolution; and (ix) article keywords.

Information on taxonomic resolution, level of organization, type of study and analyzed life stages were classified as follows [36]:

Taxonomic resolution: (a) species; (b) genus; (c) family; or (d) order.

Type of study: (a) ecological—studies involving theoretical approaches, modeling, macroecology or intra/interspecific relationships; (b) taxonomic—studies that described or redescribed species, identification keys or inventories; (c) phylogenetic—studies emphasizing relationships among taxa; (d) morphometric—studies that emphasized the description of bodily structures in larvae or adults; (e) teaching—studies focused on teaching activities (e.g., games). Any paper incorporating two or more of these approaches was counted separately for each type of study.

Life stage: (a) egg; (b) larva; (c) exuvia; or (d) adult.

2.3. Countries Where the Studies Were Conducted

For experiments carried out in laboratories, we defined the studies host country as the country where the laboratory was located. For field research in the natural environment, the place where the study was conducted was identified as the host country. To assess this, we checked the "Methods" section of each article for information on where the work was developed.

2.4. Data Analysis

To assess our nine (i–ix) objective questions, we checked the information extracted from the articles in the database. We express the following data through histograms: (i) temporal trend (year); (ii) journal of publications; (iii) number of publications by country; (iv) habitat type studied; (v) research focus; (vi) suborder; (vii) life stage; and (viii) taxonomic resolution. Finally, we performed a keyword cloud analysis of the articles (ix), using the online program ShapeWordle (https://www.shapewordle.com/ accessed on 2 July 2022) [41]. We created a cloud, where each word is sized according to its number of occurrences, in which the program assigns a weight from 0 to 1 (from lowest to highest occurrence). A maximum word limit of 50 was set. For each continent, we selected the five countries with the highest number of articles, and then plotted in a bar graph the three research focuses with the highest number of articles.

3. Results and Discussion

We found a total of 4121 published studies in the Web of Science database (step 1). We removed 1357 studies that did not comply with any of the categories we defined: not dealing with Odonata; and not being an article (e.g., books, book chapter) (step 2). The 2764 remaining papers were used for further analyses (step 3) (Figure 2).

3.1. Temporal Trend (Year), Journal Publications and Keywords Cloud

The lowest number of published articles occurred in 2014 (n = 244; 8.83%); however, after 2014 there was an increase in the number of publications, with 2019 being the year with the highest number (n = 302; 10.93%) (Figure 3). The journal with the highest number of published articles on Odonata was *Zootaxa*, with a total of 284 publications (10.27%), followed by the *International Journal of Odonatology* (226 publications; 8.18%), and *Odonatologica* (202 publications; 7.31%). Other important journals (an additional 497 journals) included fewer than 56 articles each (Figure 4, Table S1). The five most prominent words highlighted in the keywords cloud were 'Odonata' (weight = 1.00), 'dragonfly' (weight = 0.95), 'damselfly' (weight = 0.56), 'species' (weight = 0.53) and 'Zygoptera' (weight = 0.38) (Figure 5, Table S2).

There was an increase in the number of publications and the frequencies and trends changed over the time analyzed in this study. Although there are years with low production during the period analyzed, the trend towards an increase in the number of studies on dragonflies is maintained when compared with previous analyses [36] but reaches a plateau after 2017. This growth is about 36% in relation to the immediately preceding decade, when a 76% growth was recorded [36]. This interpretation must be taken with caution because during the period analyzed in this study, ~1000 more articles were produced than in the previous decade.

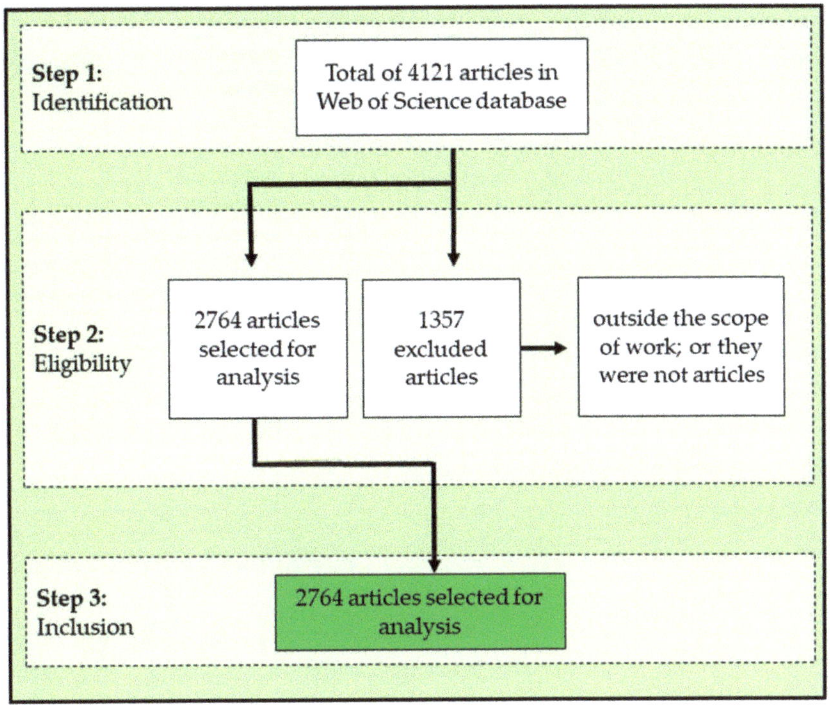

Figure 2. Flowchart of the steps used to extract the final number of articles about Odonata from the Web of Science database, from 2012 to 2021.

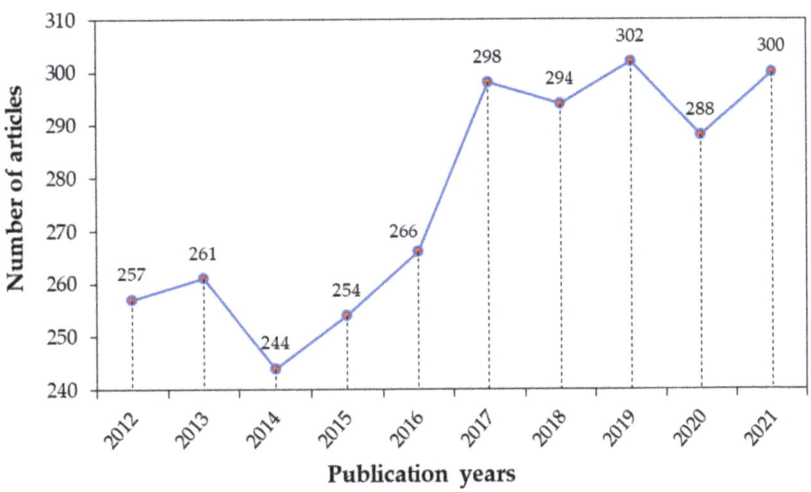

Figure 3. Scientific articles on Odonata available from the Web of Science database, per year of publication (2012–2021).

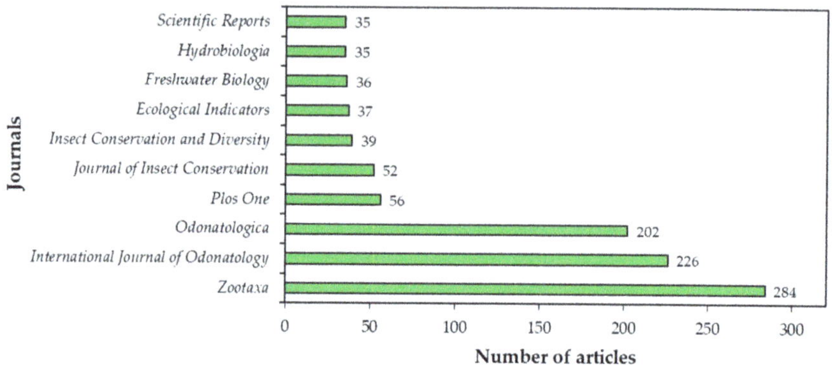

Figure 4. Scientific articles on Odonata available from the Web of Science database, per journal of publication. Plotted are the ten journals with the highest number of publications between 2012 and 2021.

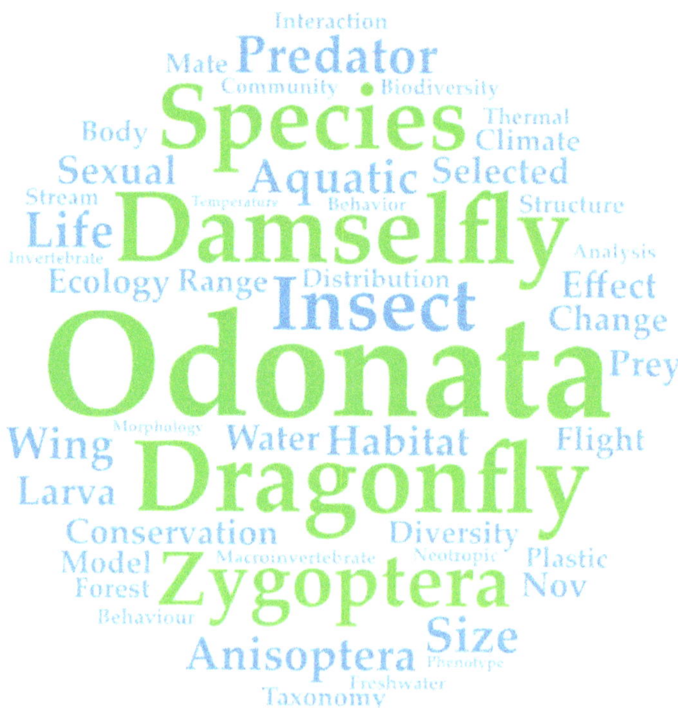

Figure 5. Keywords Cloud of scientific articles on Odonata, available from the Web of Science database (2012–2021). The five most frequent keywords are highlighted in green.

Zootaxa, *International Journal of Odonatology* and *Odonatologica*, are the journals with the highest number of papers published on Odonata. All are international, specialized, and peer-reviewed journals, but there are differences between them, which can be used to better interpret the results. *Zootaxa* is a broad scope journal, with a preference for papers in zoology (any animal taxa), mainly focused on the systematic review of groups and/or description of new taxa, in a fast, high-quality format. Its wide scope, frequency of publication, and review system have allowed this journal to have a positive impact on the advancement of knowledge of many groups, including Odonata [42]. On the other hand, it was expected that the *International Journal of Odonatology* and *Odonatologica*, specialized journals in Odonatology, would also have many publications. Both accept any type of research related to dragonflies, so their spectrum includes broader fields of research such as ecology, conservation, ethology, and reproduction.

However, other journals publishing a broader range of papers also showed increases, suggesting diverse and continued growth in Odonata research. Moreover, these journals cover a broad spectrum of research topics ranging from experimental to applied sciences. This is partly due to the numerous advances in knowledge about this insect group, indicating that dragonflies are an important model organism for research in ecology and evolution [6,18,23,43,44]. Equally important is the organization of researchers in professional societies, such as the Dragonfly Society of the Americas (https://www.dragonflysocietyamericas.org/ accessed on 4 July 2022), Sociedad de Odonatologia Latinoamericana (https://www.odonatasol.org/ accessed on 4 July 2022) and the Worldwide Dragonfly Association (https://worlddragonfly.org/ accessed on 4 July 2022), which facilitate greater interaction and academic partnerships, especially amongst young odonatologists.

The most prominent keywords reflect the context in which the research was conducted [45]. Thus, it makes sense that our findings of the most frequent keywords in the articles were Odonata, dragonfly, damselfly, species, and Zygoptera. We expected that the most frequent word would be Odonata. However, dragonfly is the common and most widespread name for odonates globally [9]. Finally, the keywords indicate that the species level was the most commonly used taxonomic resolution in the articles, and that Zygoptera was the most studied suborder, presumably because of their generally greater sensitivity to disturbance and subsequent use in ecological impact studies.

3.2. Spatial Trend of Publications (Across Countries)

The ten countries with the most publications were the USA (n = 346), Brazil (n = 272), China (n = 204), Germany (n = 135), Japan (n = 132), Canada (n = 122), Sweden (n = 121), France (n = 110), Mexico (n = 103) and India (n = 83) (Figure 6). Several other countries had >10 publications (Figure 6; Table S3), indicating the global interest on the Odonata as a target organism for research.

Regarding the countries with the largest number of publications, for the American continents, the USA (first place) and Brazil (second place) remain the countries with the largest number of publications. This reaffirms the importance of the work developed by these two countries, which have a broad tradition of research contributing considerably to the knowledge of Neotropical dragonflies [36]. The case is similar in the African context, where the countries that maintain the highest number of publications are South Africa and Algeria, places where leading odonatologists have been established for many years. Likewise in Europe and Asia, the greater number of publications are in countries with a long history and tradition of Odonata research such as Germany in Europe, or China and Japan in Asia [36].

3.3. Research Focus

Most studies focused on Odonata ecology (n = 717), followed by studies on taxonomy (n = 584), behavior (n = 576), morphology (n = 343) and ecological monitoring (n = 207) (Figure 7; Table S4).

The studies show a great diversity of research areas in which dragonflies have been used as research targets. However, it was expected that the largest number of publications would be focused on ecology and biodiversity studies, because many journals publish articles on Odonata that are not specific to taxonomy or phylogeny. Miguel et al. [36] found the same trend worldwide. Moreover, areas such as ecology and behavior are constantly growing, so periodically new metrics, approaches and methodologies are published, generating interest to replicate them in different parts of the world [46,47]. Ecology is the main research focus on the American and African continents. In Asia and Oceania, taxonomy is the main type of study, while in Europe, behavior studies stand out (Figure 8; Table S5).

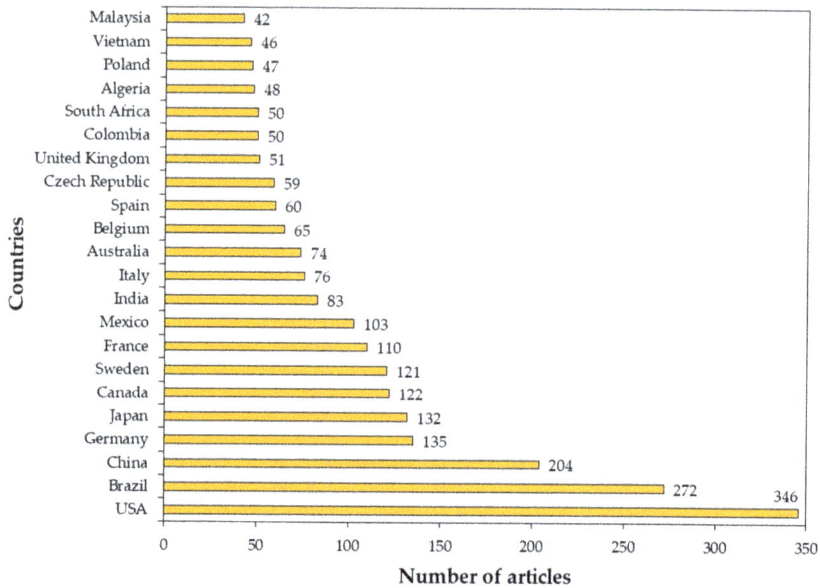

Figure 6. Global production of scientific articles on Odonata, available from the Web of Science database (2012 to 2021). The figure shows only those countries with >40 publications each.

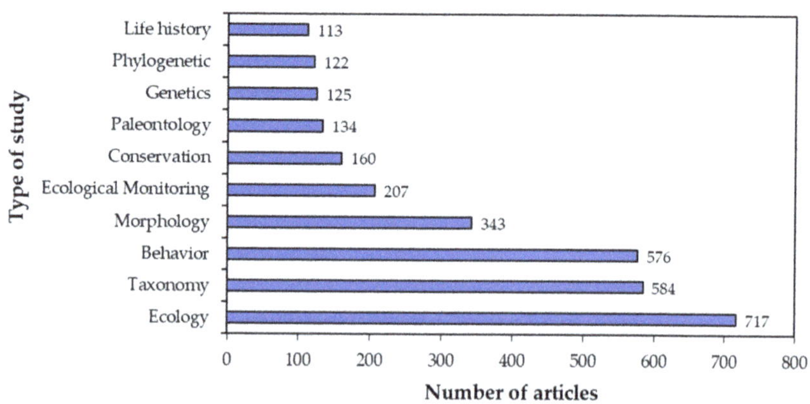

Figure 7. Research focus (top 10) to the scientific articles published on Odonata, available from the Web of Science database (2012–2021).

Although studies in genetics have increased, there are disadvantages in their replicability, especially in developing countries, because of both financial and equipment limitations [36]. Finally, topics of decreasing scientific interest and funding are associated with fewer publications (e.g., natural history and basic species biology).

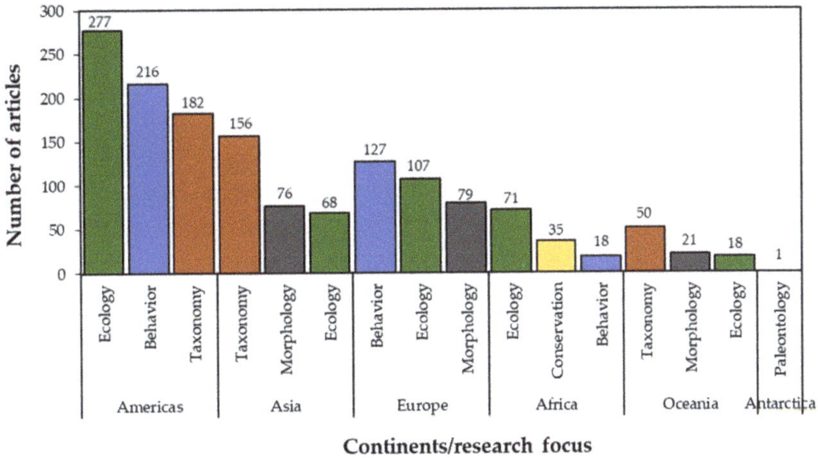

Figure 8. The top three study topics, by continent, in the scientific articles published on Odonata, available from the Web of Science database (2012–2021).

The focus of research on each continent is varied and likely related to who lives where. For example, the fact that the Americas contain the largest number of researchers in ecology, behavior and taxonomy is directly related to the establishment and growth of different research groups, mainly in Argentina, Brazil, Colombia, the United States, and Mexico (e.g., Dragonfly Society of the Americas and Sociedad de Odonatologia Latinoamericana) [36,48–51]. Asia and Oceania each show the same research patterns, mainly taxonomy, morphology, and ecology. In all three continents, there is a history of taxonomic studies which is maintained today. In Europe, the main research focus is behavior, ecology, and morphology. European odonate biodiversity has been well known for a century, and only after more than 100 years was a new species described there [52]. Africa is the only continent where the main lines of research are also focused on conservation. This is certainly related to the studies developed by several research groups that are focused on conservation, especially in South Africa, such as the research group led by Michael J. Samways and John P. Simaika. M.J. Samways has published extensively on various aspects of Odonata ecology and conservation [53], especially regarding landscape ecology and insect conservation in general [54]. J.P. Simaika has more than a decade of experience working in rivers, lakes, wetlands and artificial ponds in Africa. Their research has led of public conservation policies at the international level [54].

3.4. Study Habitat Types

The highest number of published articles on Odonata were conducted in the field (Figure 9), i.e., in streams (n = 668), ponds (n = 437), rivers (n = 318), and lakes (n = 278), but many were also part of laboratory experiments (n = 364) or involved fossilized material (n = 125) (Figure 9). Markedly fewer studies have been conducted in pools (n = 42), reservoirs (n = 20), mesocosms (n = 10), and plants (n = 3) (Figure 9; Table S6).

Because of the strong relationship between dragonflies and aquatic environments, it is reasonable that research on Odonata focused on some of these environments. Most of the research has been conducted in lotic environments, such as rivers and streams, systems that are under constant anthropogenic threats [55]. The type of impact, as well as the degree of intensity, affects the complex dynamics of functioning and interconnection in this habitat, generating serious effects on their health as well as on the biodiversity that inhabits them [56]. Because of this, numerous environmental laws across the world stress the evaluation and monitoring of lotic bodies as a priority [1,57].

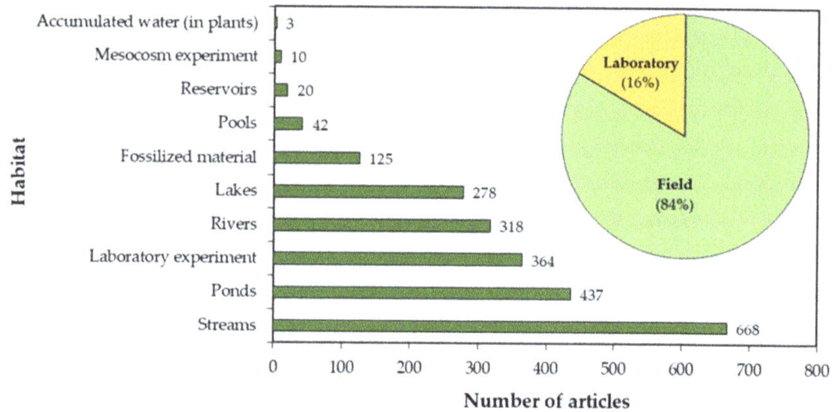

Figure 9. Top 10 habitat types for scientific articles on Odonata, available from the Web of Science database (2012–2021).

3.5. Suborder, Life Stage and Taxonomic Resolution Most Used in Studies

Of the total articles on Odonata, 982 involved the Zygoptera, 946 Anisoptera and 15 the Anisozygoptera. Another 756 studies were focused on both main suborders, i.e., Anisoptera and Zygoptera (Figure 10a). Most studies (n = 1662) were focused only on adults, 714 only on the nymphs, 160 on adults and nymphs, and 40 on nymphs and eggs. It is noteworthy that several articles focused on more than one life stage (Figure 10b). Most articles (n = 2381) used species-level taxonomy (Figure 11).

Most studies focused on Odonata, including the suborders Anisoptera and Zygoptera in their analyses. Numerous studies included both suborders to compare the responses among their species [23,44]. When a single suborder was analyzed, the Zygoptera and Anisoptera were used with similar frequencies. Thus, there was no clear pattern or preference for a specific suborder. One reason for this tendency is the well resolved taxonomy in both suborders [43], because a resolved taxonomy is the basis for asking different questions in areas such as evolution, systematics, or ecology.

Adults or larvae were the stages most commonly used in research. Presumably, this is because adult Odonata are visible in the field, their collection requires only an insect net, excellent taxonomic keys for most species are available and easily accessible, and the adults can be identified to the species level [3,4,9]. This is partially true for larvae, the second stage with the most publications, particularly in regions with a long tradition in larval dragonfly research [58]. On the other hand, in regions where this tradition is more recent, there are fewer larval studies [36]. Furthermore, only the larvae of 1/3 of the Odonata species are described [59] and larvae detection and transportation from the field to the laboratory is more complex [36]. Research studies on both adults and larvae are numerous, which is especially useful when one may want to compare responses between the two phases to provide a more comprehensive analysis [27]. There has also been research on three or more phases (Figure 10), presumably when one wants to evaluate ontogenetic development, for which the collection of eggs and the development of larvae are needed [60]. This also applies in the case of larvae and exuviae and larvae and adults, where the association of the different stages is important for taxonomic description [61].

Despite not being an official life stage, research on exuviae shows great potential for tracing the route of environmental contaminants [62] or for adults that are difficult to find or capture [63]. Exuviae also provide proof of life cycle completion at particular habitats [62]. However, exuvia research requires investing much field time, considerable care in transportation and storage of such fragile specimens, and identification problems similar to those with larvae. Incidentally, the results reinforce the evidence that shows how different life stages in Odonata are useful for evaluating different research questions. Finally, it is not surprising that fossil studies are numerous in Odonata because there are many odonate fossil records, which provide the basis for studies of evolutionary relationships within the order [64].

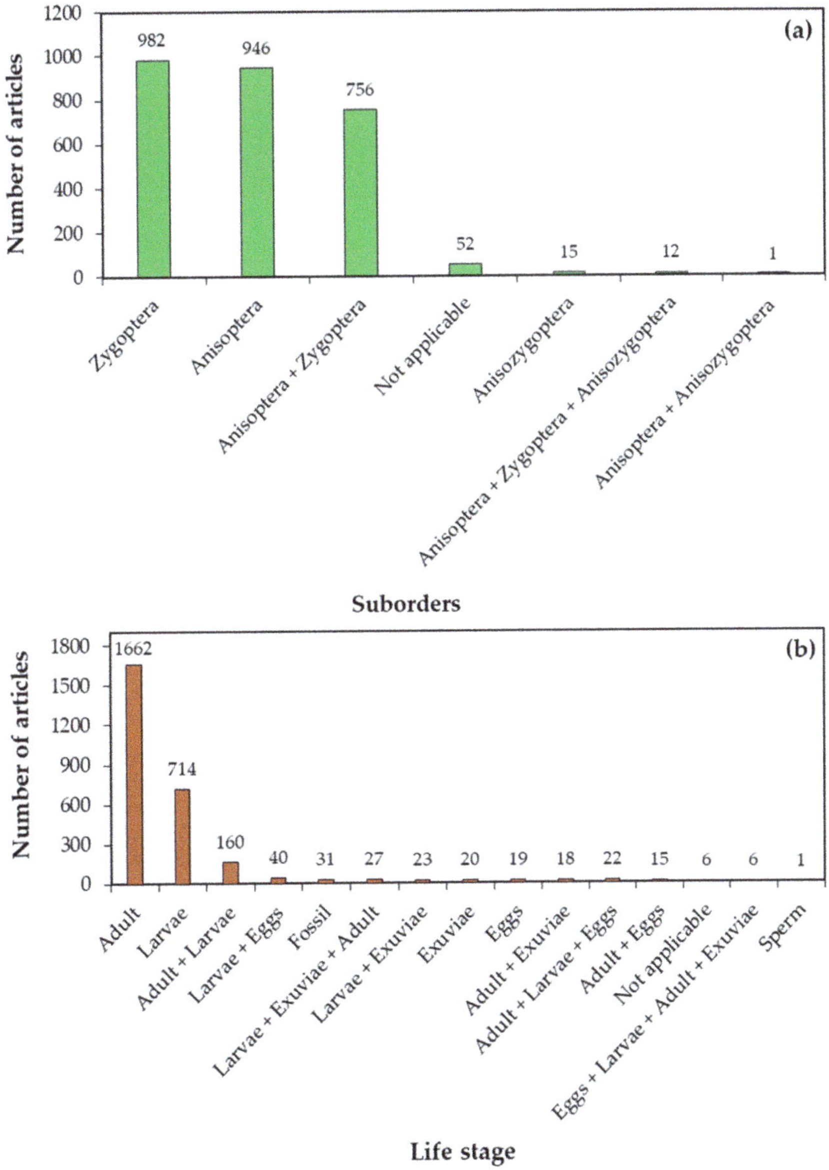

Figure 10. Number of scientific articles on Odonata, available from the Web of Science database (2012–2021), classified by: (**a**) suborders; and (**b**) life stage and/or analyzed material. Not applicable refers to studies where the life stage is not indicated.

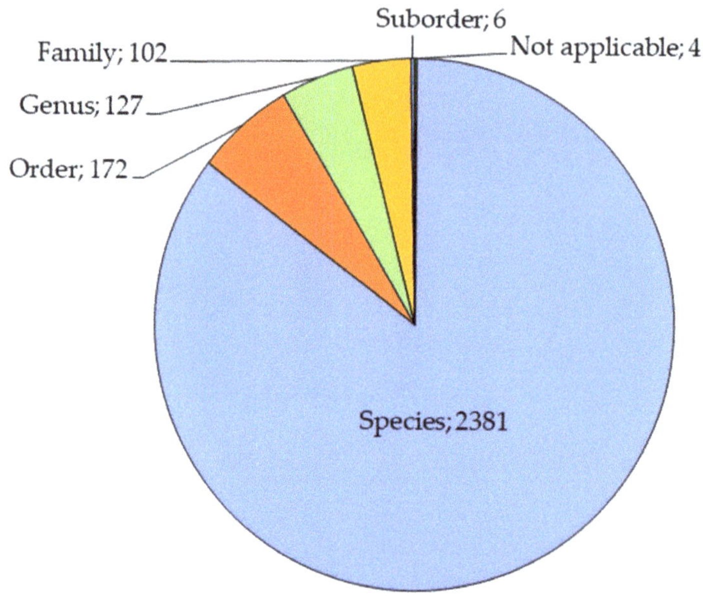

Figure 11. Number of scientific articles on Odonata, available from the Web of Science database (2012–2021), classified by taxonomic resolution. Not applicable refers to studies that do not indicate the taxonomic resolution.

Species is the taxonomic level most often used in most studies, but genus, family, and order are used as well. Identification to species is important because it facilitates more accurate information about a particular taxon. However, this requires more time, taxonomic expertise, and financial resources, which is a disadvantage for projects that need an immediate response or are underfunded [65]. Therefore, there is much discussion in the literature regarding the preferable taxonomic level for studies. For example, Jansen et al. [66] point out that it is challenging to classify organisms into higher-level taxonomic groups, such as families, because many family-specific features may not be distinguishable.

Although most Odonata papers in the WoS database use the species level of taxonomic resolution, Mendoza-Penagos et al. [35] demonstrated that the family level provides an effective tool odonate for biomonitoring of tropical streams by providing ecologically meaningful information. Future studies should evaluate the costs and benefits of diagnosing impacts by comparing multiple taxonomic levels against a common disturbance gradient [67,68].

4. Conclusions

Our results indicate an increase in published research on Odonata available on the WoS, and on a range of topics as diverse as ecology, biomonitoring, genetics, and environmental education. The increase in ecological studies on Odonata may reflect the dynamic characteristics of this order, and its relatively well-defined systematics, especially in the case of adults. Despite the increased number of publications in the WoS database, there are still many spatial gaps (e.g., poorly studied regions/countries), and gaps in study focus, such as basic biology (e.g., life cycle, anatomy, physiology, habitat), biogeography, parasitism, competition within and between species, evolutionary and phylogenetic relationships, and Odonata eggs. This demonstrates that some areas are seriously neglected. However, such studies are essential for a better understanding of how species may respond to different factors (historical, biogeographic, ecological) and to increase the background information necessary in other types of studies. It is especially necessary to increase the number of researchers and research on the larval stage of most Odonata species, as well as the potential effects of climate change on larval and adult stages. Although most Odonata diversity is found in the tropics, historically, countries with greater purchasing or economic power have a larger number of publications [69]. Thus, money and the lack of professional training are important gaps to overcome. One way to alleviate this is to initiate temperate–tropical partnerships for training new people, strengthening tropical research institutions, and conducting more joint research in the tropics.

Supplementary Materials: The following are available online at https://www.mdpi.com/article/10.3390/d14121074/s1, Figure S1: Adult specimens of Odonata (Insecta): (a) *Perithemis thais* (Anisoptera: Libellulidae); and (b) *Hetaerina moribunda* (Zygoptera: Calopterygidae). Source: Cristian C. Mendoza-Penagos. Table S1: Scientific articles on Odonata available from the Web of Science database (between 2012 and 2021), per journal of publication. Table S2: List and weight of the 50 most common keywords in scientific articles on Odonata (Insecta), available from the Web of Science (WoS) (2012–2021). For elaboration, the online program Shape Wordle was used. Table S3: Global production of scientific articles on the Odonata (Insecta), available from the Web of Science database (2012 to 2021), by country in which the research was carried out. Table S4: Contribution of the different types of study (research focus) to the scientific articles published on Odonata, available from the Web of Science database (2012–2021). Table S5: Contribution of the different types of study (research focus) to the scientific articles published on Odonata, available from the Web of Science database (2012–2021), by continent in which the research was carried out. Table S6: Type of habitat for scientific articles on Odonata, available from the Web of Science database (2012–2021).

Author Contributions: Conceptualization, J.M.B.O.-J., C.C.M.-P., K.D.-S., L.J. and L.B.C.; methodology, J.M.B.O.-J., T.S.R., S.F.V., J.C.M.-F., C.C.M.-P., K.D.-S., L.J. and L.B.C.; formal analysis, J.M.B.O.-J., T.S.R., S.F.V., J.C.M.-F., C.C.M.-P., K.D.-S., L.J. and L.B.C.; investigation, J.M.B.O.-J., T.S.R., S.F.V., J.C.M.-F., C.C.M.-P., K.D.-S., L.J. and L.B.C.; resources, J.M.B.O.-J., L.J. and K.D.-S.; data curation, J.M.B.O.-J., T.S.R., S.F.V., J.C.M.-F., C.C.M.-P., K.D.-S., L.J. and L.B.C.; writing—original draft preparation, J.M.B.O.-J., T.S.R., S.F.V., J.C.M.-F., C.C.M.-P., K.D.-S., L.J. and L.B.C.; writing—review and editing, J.M.B.O.-J., T.S.R., S.F.V., J.C.M.-F., C.C.M.-P., K.D.-S., L.J. and L.B.C.; visualization, J.M.B.O.-J., T.S.R., S.F.V., J.C.M.-F., C.C.M.-P., K.D.-S., L.J. and L.B.C.; supervision, J.M.B.O.-J.; project administration, J.M.B.O.-J.; funding acquisition, J.M.B.O.-J. All authors have read and agreed to the published version of the manuscript.

Funding: This research received no external funding.

Institutional Review Board Statement: Not applicable.

Data Availability Statement: Not applicable.

Acknowledgments: Robert M. Hughes reviewed the English and content. J.M.B.O.-J thanks the Programa de Pós-Graduação em Sociedade, Natureza e Desenvolvimento (PPGSND/UFOPA) and Edital N° 13/2020 CAPES (PDPG Amazônia Legal) for the academic support. We thank the anonymous reviewers and academic editors for their valuable contributions. L.J. is thankful to Conselho Nacional de Desenvolvimento Científico e Tecnológico (CNPq) for granting research support through productivity grants (Process No. 304710/2019-9).

Conflicts of Interest: The authors declare no conflict of interest. The funders had no role in the design of the study; in the collection, analyses, or interpretation of data; in the writing of the manuscript, or in the decision to publish the results.

References

1. Bonada, N.; Prat, N.; Resh, V.H.; Statzner, B. Developments in aquatic insect biomonitoring: A comparative analysis of recent approaches. *Annu. Rev. Entomol.* **2006**, *51*, 495–523. [CrossRef] [PubMed]
2. Corbet, P.S. *Dragonflies: Behaviour and Ecology of Odonata*, 1st ed.; Harley Books: Colchester, UK, 1999; 850p.
3. Garrison, R.W.; Ellenrieder, N.V.; Louton, J.A. *Dragonfly genera of the New World: An Illustrated and Annotated Key to the Anisoptera*; The Johns Hopkins University Press: Baltimore, MA, USA, 2006; 490p.
4. Garrison, R.W.; Von Ellenrieder, N.; Louton, J.A. *Damselfly genera of the New World: An Illustrated and Annotated Key to the Zygoptera*; Johns Hopkins University Press: Baltimore, MA, USA, 2010; 490p.
5. Samways, M.J.; Steytler, N.S. Dragonfly (Odonata) distribution patterns in urban and forest landscapes, and recommendations for riparian management. *Biol. Conserv.* **1996**, *78*, 279–288. [CrossRef]

6. Chovanec, A.; Waringer, J. Ecological integrity of river–floodplain systems—Assessment by dragonfly surveys (Insecta: Odonata). *Regul. Rivers Res. Manag.* **2001**, *17*, 493–507. [CrossRef]
7. Oliveira-Junior, J.M.B.; Juen, L. The Zygoptera/Anisoptera Ratio (Insecta: Odonata): A New Tool for Habitat Alterations Assessment in Amazonian Streams. *Neotropic. Entomol.* **2019**, *48*, 552–560. [CrossRef] [PubMed]
8. Batista, J.D.; Ferreira, V.R.S.; Cabette, H.S.R.; De Castro, L.A.; De Marco, P.; Juen, L. Sampling efficiency of a protocol to measure Odonata diversity in tropical streams. *PLoS ONE* **2021**, *16*, e0248216. [CrossRef] [PubMed]
9. May, M.L. Odonata: Who they are and what they have done for us lately: Classification and ecosystem services of dragonflies. *Insects* **2019**, *10*, 62. [CrossRef]
10. Sociedad de Odonatología Latinoamericana. Available online: https://www.odonatasol.org/ (accessed on 8 February 2022).
11. Dragonfly Society of the Americas. Available online: https://www.dragonflysocietyamericas.org/ (accessed on 8 February 2022).
12. Worldwide Dragonfly Association. Available online: https://worlddragonfly.org/ (accessed on 8 February 2022).
13. Ožana, S.; Burda, M.; Hykel, M.; Malina, M.; Prášek, M.; Bárta, D.; Dolný, A. Dragonfly Hunter CZ: Mobile application for biological species recognition in citizen science. *PLoS ONE* **2019**, *14*, e0210370. [CrossRef]
14. Patten, M.A.; Hjalmarson, E.A.; Smith-Patten, B.D.; Bried, J.T. Breeding thresholds in opportunistic Odonata records. *Ecol. Indic.* **2019**, *106*, 105460. [CrossRef]
15. Bried, J.; Ries, L.; Smith, B.; Patten, M.; Abbott, J.; Ball-Damerow, J.; Cannings, R.; Cordero-Rivera, A.; Córdoba-Aguilar, A.; De Marco, P.; et al. Towards global volunteer monitoring of odonate abun-dance. *BioScience* **2020**, *70*, 914–923. [CrossRef]
16. Carle, F.L.; Kjer, K.M.; May, M.L. Evolution of Odonata, with special reference to Coenagrionoidea (Zygoptera). *Arthropod Syst. Phylogeny* **2008**, *66*, 37–44.
17. Felker, A.S. New Damselflies of the Family Kennedyidae (Odonata: Protozygoptera) from the Upper Permian of the Vologda Region. *Paleontol. J.* **2021**, *55*, 396–404. [CrossRef]
18. Córdoba-Aguilar, A. *Dragonflies and Damselflies, Model Organisms for Ecological and Evolutionary Research*, 1st ed.; Oxford University Press: Oxford, UK, 2008; 290p.
19. Kalkman, V.J.; Clausnitzer, V.; Dijkstra, K.-D.B.; Orr, A.G.; Paulson, D.R.; van Tol, J. Global diversity of dragonflies (Odonata) in freshwater. *Hydrobiologia* **2008**, *595*, 351–363. [CrossRef]
20. Suhling, F.; Sahlén, G.; Gorb, S.; Kalkman, V.J.; Dijkstra, K.D.B.; van Tol, J. Order Odonata. In *Thorp and Covich's Freshwater Invertebrates: Ecology and General Biology*, 5th ed.; Thorp, J.H., Covich, A.P., Eds.; Academic Press: Cambridge, MA, USA, 2015; pp. 893–932.
21. Mendes, T.P.; Luiza-Andrade, A.; Cabette, H.S.R.; Juen, L. How does environmental variation affect the distribution of dragonfly larvae (Odonata) in the Amazon-Cerrado transition zone in Central Brazil? *Neotropic. Entomol.* **2018**, *47*, 37–45. [CrossRef] [PubMed]
22. Carvalho, F.G.; Pinto, N.S.; Oliveira Junior, J.M.B.; Juen, L. Effects of marginal vegetation removal on Odonata communities. *Acta Limnol. Bras.* **2013**, *25*, 10–18. [CrossRef]
23. Oliveira-Junior, J.M.B.; Dias-Silva, K.; Teodósio, M.A.; Juen, L. The Response of Neotropical Dragonflies (Insecta: Odonata) to Local and Regional Abiotic Factors in Small Streams of the Amazon. *Insects* **2019**, *10*, 446. [CrossRef]
24. Oliveira-Junior, J.M.B.; Juen, L. Structuring of dragonfly communities (Insecta: Odonata) in eastern Amazon: Effects of environmental and spatial factors in preserved and altered streams. *Insects* **2019**, *10*, 322. [CrossRef]
25. Oliveira-Junior, J.M.B.; Teodósio, M.A.; Juen, L. Patterns of co-occurrence and body size in dragonflies and damselflies (Insecta: Odonata) in preserved and altered Amazonian streams. *Austral Entomol.* **2021**, *60*, 436–450. [CrossRef]
26. Oliveira-Junior, J.M.B.; De Marco, P.; Dias-Silva, K.; Leitão, R.P.; Leal, C.G.; Pompeu, P.S.; Gardner, T.A.; Hughes, T.A.; Juen, L. Effects of human disturbance and riparian conditions on Odonata (Insecta) assemblages in eastern Amazon basin streams. *Limnologica* **2017**, *66*, 31–39. [CrossRef]
27. Valente-Neto, F.; Roque, F.O.; Rodrigues, M.E.; Juen, L.; Swan, C. Toward a practical use of Neotropical odonates as bioindicators: Testing congruence across taxonomic resolution and life stages. *Ecol. Indic.* **2016**, *61*, 952–959. [CrossRef]
28. Miguel, T.B.; Oliveira-Junior, J.M.B.; Ligeiro, R.; Juen, L. Odonata (Insecta) as a tool for the biomonitoring of environmental quality. *Ecol. Indic.* **2017**, *81*, 555–566. [CrossRef]
29. Valente-Neto, F.; Rodrigues, M.E.; Roque, F.O. Selecting indicators based on biodiversity surrogacy and environmental response in a riverine network: Bringing operationality to biomonitoring. *Ecol. Indic.* **2018**, *94*, 198–206. [CrossRef]
30. Pinto, N.S.; Juen, L.; Cabette, H.S.R.; De Marco, P. Fluctuating Asymmetry and Wing Size of Argia tinctipennis Selys (Zygoptera: Coenagrionidae) in Relation to Riparian Forest Preservation Status. *Neotropic. Entomol.* **2012**, *41*, 178–185. [CrossRef] [PubMed]
31. Resende, B.O.; Ferreira, V.R.S.; Brasil, L.S.; Calvão, L.B.; Mendes, T.P.; Carvalho, F.G.; Mendoza-Penagos, C.C.; Bastos, R.C.; Brito, J.S.; Oliveira-Junior, J.M.B.; et al. Impact of environmental changes on the behavioral diversity of the Odonata (Insecta) in the Amazon. *Sci. Rep.* **2021**, *11*, 9742. [CrossRef]
32. Guillermo-Ferreira, R.; Juen, L. Odonate ethodiversity as a bioindicator of anthropogenic impact. *Int. J. Odonatol.* **2021**, *24*, 149–157. [CrossRef] [PubMed]
33. Geraldo de Carvalho, F.; Duarte, L.; Nakamura, G.; Dubal dos Santos Seger, G.; Juen, L. Changes of Phylogenetic and Taxonomic Diversity of Odonata (Insecta) in Response to Land Use in Amazonia. *Forests* **2021**, *12*, 1061. [CrossRef]

34. Bastos, R.C.; Brasil, L.S.; Oliveira-Junior, J.M.B.; Geraldo Carvalho, F.; Lennox, G.D.; Barlow, J.; Juen, L. Morphological and phylogenetic factors structure the distribution of damselfly and dragonfly species (Odonata) along an environmental gradient in Amazonian streams. *Ecol. Indic.* **2021**, *122*, 107257. [CrossRef]
35. Mendoza-Penagos, C.C.; Calvão, L.B.; Juen, L. A new biomonitoring method using taxonomic families as substitutes for the suborders of the Odonata (Insecta) in Amazonian streams. *Ecol. Indic.* **2021**, *124*, 107388. [CrossRef]
36. Miguel, T.B.; Calvão, L.B.; Vital, M.V.C.; Juen, L. A scientometric study of the order Odonata with special attention to Brazil. *Int. J. Odonatol.* **2017**, *20*, 27–42. [CrossRef]
37. Tennant, J.P.; Waldner, F.; Jacques, D.C.; Masuzzo, P.; Collister, L.B.; Hartgerink, C.H. The academic, economic and societal impacts of Open Access: An evidence-based review. *F1000 Res.* **2016**, *5*, 632. [CrossRef]
38. Piwowar, H.; Priem, J.; Larivière, V.; Alperin, J.P.; Matthias, L.; Norlander, B.; Farley, A.; West, J.; Haustein, S. The state of OA: A large-scale analysis of the prevalence and impact of Open Access articles. *PeerJ* **2018**, *6*, e4375. [CrossRef]
39. Donthu, N.; Kumar, S.; Mukherjee, D.; Pandey, N.; Lim, W.M. How to conduct a bibliometric analysis: An overview and guidelines. *J. Bus. Res.* **2021**, *133*, 285–296. [CrossRef]
40. Zhu, J.; Liu, W. A tale of two databases: The use of Web of Science and Scopus in academic papers. *Scientometrics* **2020**, *123*, 321–335. [CrossRef]
41. Wang, Y.; Chu, X.; Zhang, K.; Bao, C.; Li, X.; Zhang, J.; Fu, C.-W.; Hurter, C.; Deussen, O.; Lee, B. ShapeWordle: Tailoring Wordles using Shape-aware Archimedean Spirals. *IEEE Trans. Vis. Comput. Graph.* **2019**, *26*, 991–1000. [CrossRef]
42. Zhang, Z.Q. Contributions of Zootaxa to biodiversity discovery: An overview of the first twenty years. *Zootaxa* **2021**, *4979*, 6–16. [CrossRef] [PubMed]
43. Bybee, S.M.; Kalkman, V.J.; Erickson, R.J.; Frandsen, P.B.; Breinholt, J.W.; Suvorov, A.; Dijkstra, K.D.B.; Cordero-Rivera, A.; Skevington, J.H.; Abbott, J.C.; et al. Phylogeny and classification of Odonata using targeted genomics. *Mol. Phylogenet. Evol.* **2021**, *160*, 107115. [CrossRef]
44. Oliveira-Junior, J.M.B.; Shimano, Y.; Gardner, T.A.; Hughes, R.M.; De Marco, P.; Juen, L. Neotropical dragonflies (Insecta: Odonata) as indicators of ecological condition of small streams in the eastern Amazon. *Austral Ecol.* **2015**, *40*, 733–744. [CrossRef]
45. Verbeek, A.; Debackere, K.; Luwel, M.; Zimmermann, E. Measuring progress and evolution in science and technology–I: The multiple uses of bibliometric indicators. *Int. J. Manag. Rev.* **2002**, *4*, 179–211. [CrossRef]
46. Bybee, S.; Córdoba-Aguilar, A.; Duryea, M.C.; Futahashi, R.; Hansson, B.; Lorenzo-Carballa, M.O.; Schilder, R.; Stoks, R.; Suvorov, A.; Svensson, E.I.; et al. Odonata (dragonflies and damsel-flies) as a bridge between ecology and evolutionary genomics. *Front. Zool.* **2016**, *13*, 46. [CrossRef]
47. Cordero-Rivera, A. Behavioral diversity (ethodiversity): A neglected level in the study of biodiversity. *Front. Ecol. Evol.* **2017**, *5*, 7. [CrossRef]
48. Palacino-Rodríguez, F. Two decades of progress in over one hundred years of study: Present status of Odonata research in Colombia. *Odonatologica* **2016**, *45*, 327–334. [CrossRef]
49. Lozano, F.; del Palacio, A.; Ramos, L.; Muzón, J. The Odonata of Argentina: State of knowledge and updated checklist. *Int. J. Odonatol.* **2020**, *23*, 113–153. [CrossRef]
50. González-Soriano, E.; Novelo-Gutiérrez, R. Biodiversidad de Odonata en MéxicoBiodiversity of Odonata in Mexico. *Rev. Mex. Biodivers.* **2014**, *85*, 243–251. [CrossRef]
51. White, E.L.; Hunt, P.D.; Schlesinger, M.D.; Corser, J.D.; deMaynadier, P.G. *A Conservation Status Assessment of Odonata for the Northeastern United States*; New York Natural Heritage Program: Albany, NY, USA, 2014. Available online: https://rcngrants.org/sites/default/files/final_reports/RCN%202011-3%20final%20report.pdf (accessed on 11 November 2022).
52. López-Estrada, E.K.; Barona Fernández, J.; Cardo- Maeso, N.; Teruel Montejano, S.; Díaz-Martínez, C. *Onychogomphus cazuma* sp. nov. from Spain: Molecular and morphological evidence supports the discovery of a new European dragonfly species (Odonata: Gomphidae). *Odonatatologica* **2021**, *49*, 125–154. [CrossRef]
53. Bried, J.T.; Samways, M.J. A review of odonatology in freshwater applied ecology and conservation science. *Freshw. Sci.* **2015**, *34*, 1023–1031. [CrossRef]
54. Samways, M.J.; Barton, P.S.; Birkhofer, K.; Chichorro, F.; Deacon, C.; Fartmann, T.; Fukushima, C.S.; Gaigher, R.; Habel, J.C.; Hallmann, C.A.; et al. Solutions for humanity on how to conserve insects. *Biol. Conserv.* **2020**, *242*, 108427. [CrossRef]
55. Leal, C.G.; Lennox, G.D.; Ferraz, S.B.; Ferreira, J.; Gardner, T.A.; Thomson, J.R.; Berenguer, E.; Lees, A.C.; Hughes, R.M.; Nally, R.M.; et al. Integrated terrestrial-freshwater planning doubles conservation of tropical aquatic species. *Science* **2020**, *370*, 6512. [CrossRef] [PubMed]
56. Calvão, L.B.; Nogueira, D.S.; Montag, L.F.A.; Lopes, M.A.; Juen, L. Are Odonata communities impacted by conventional or reduced impact logging? *For. Ecol. Manag.* **2016**, *382*, 143–150. [CrossRef]
57. Buss, D.F.; Carlisle, D.; Chon, T.S.; Culp, J.; Harding, J.S.; Keizer-Vlek, H.E.; Robinson, W.A.; Strachan, S.; Thirion, C.; Hughes, R.M. Stream biomonitoring using macroinvertebrates around the globe: A comparison of large-scale programs. *Environ. Monit. Assess.* **2015**, *187*, 4132. [CrossRef]
58. Tennessen, K.J. *Dragonfly Nymphs of North America: An Identification Guide*, 1st ed.; Springer: New York, NY, USA, 2019; 620p.
59. Hamada, N.; Nessimian, J.L.; Querino, R.B. *Insetos Aquáticos na Amazônia Brasileira: Taxonomia, Biologia e Ecologia*, 1st ed.; Editora do INPA: Manaus, AM, Brasil, 2014; 724p.

60. Schenk, K.; Söndgerath, D. Influence of egg size differences within egg clutches on larval parameters in nine libellulid species (Odonata). *Ecol. Entomol.* **2005**, *30*, 456–463. [CrossRef]
61. Paul, S.; Kakkassery, F.K. Taxonomic and diversity studies on odonate nymphs by using their exuviae. *J. Entomol. Zool. Stud.* **2013**, *1*, 47–53.
62. Simon, E.; Tóthmérész, B.; Kis, O.; Jakab, T.; Éva Szalay, P.; Vincze, A.; Baranyai, E.; Harangi, S.; Miskolczi, M.; Dévai, G. Environmental-Friendly Contamination Assessment of Habitats Based on the Trace Element Content of Dragonfly Exuviae. *Water* **2019**, *11*, 2200. [CrossRef]
63. Silva-Méndez, G.; Riso, S.; Lorenzo-Carballa, M.O.; Cordero-Rivera, A. Sampling larvae, exuviae or adults of Odonata for ecological studies: A test of methods in permanent rivers in the Iberian Peninsula. *Odonatologica* **2022**, *51*, 63–81. [CrossRef]
64. Rehn, A.C. Phylogenetic analysis of higher-level relationships of Odonata. *Syst. Entomol.* **2003**, *28*, 181–240. [CrossRef]
65. Vianna, D.M.; De Marco, P. Higher-taxon and cross-taxon surrogates for odonate biodiversity in Brazil. *Nat. ConServação* **2012**, *10*, 34–39. [CrossRef]
66. Jansen, J.; Hill, N.A.; Dunstan, P.K.; Eléaume, M.P.; Johnson, C.R. Taxonomic Resolution, Functional Traits, and the Influence of Species Groupings on Mapping Antarctic Seafloor Biodiversity. *Front. Ecol. Evol.* **2018**, *6*, 81. [CrossRef]
67. Whitter, T.R.; Van Sickle, J. Macroinvertebrate tolerance values and an assemblage tolerance index (ATI) for western USA streams and rivers. *J. N. Am. Benthol. Soc.* **2010**, *29*, 852–866. [CrossRef]
68. Valente-Neto, F.; Martinez, B.T.; Ferreira, A.; Neto, F.S.; de Souza, F.L.; Souza, R.P.; Escarpinati, S.C.; Hughes, R.M.; Roque, F.O. Incorporating costs, thresholds and spatial extents for selecting stream bioindicators in an ecotone between two Brazilian biodiversity hotspots. *Ecol. Indic.* **2021**, *127*, 107761. [CrossRef]
69. Rodríguez-Navarro, A.; Brito, R. The link between countries' economic and scientific wealth has a complex dependence on technological activity and research policy. *Scientometrics* **2022**, *127*, 2871–2896. [CrossRef]

MDPI
St. Alban-Anlage 66
4052 Basel
Switzerland
www.mdpi.com

Diversity Editorial Office
E-mail: diversity@mdpi.com
www.mdpi.com/journal/diversity

www.ingramcontent.com/pod-product-compliance
Lightning Source LLC
Chambersburg PA
CBHW061547220926
51850CB00005B/278
* 9 7 8 3 7 2 5 8 0 7 0 9 3 *